# Handbook on Advancements in Smart Antenna Technologies for Wireless Networks


Chen Sun
*ATR Wave Engineering Laboratories, Japan*

Jun Cheng
*Doshisha University, Japan*

Takashi Ohira
*Toyohashi University of Technology, Japan*



Information Science REFERENCE

**INFORMATION SCIENCE REFERENCE**

Hershey · New York

Director of Editorial Content: Kristin Klinger
Senior Managing Editor: Jennifer Neidig
Managing Editor: Jamie Snavely
Assistant Managing Editor: Carole Coulson
Typesetter: Jeff Ash and Larissa Vinci
Cover Design: Lisa Tosheff
Printed at: Yurchak Printing Inc.

Published in the United States of America by
Information Science Reference (an imprint of IGI Global)
701 E. Chocolate Avenue, Suite 200
Hershey PA 17033
Tel: 717-533-8845
Fax: 717-533-8661
E-mail: cust@igi-global.com
Web site: http://www.igi-global.com

and in the United Kingdom by
Information Science Reference (an imprint of IGI Global)
3 Henrietta Street
Covent Garden
London WC2E 8LU
Tel: 44 20 7240 0856
Fax: 44 20 7379 0609
Web site: http://www.eurospanbookstore.com



Library of Congress Cataloging-in-Publication Data

Handbook on advancements in smart antenna technologies for wireless networks / Chen Sun, Jun Cheng, and Takashi Ohira, editors.

p. cm.

Summary: "This book is a comprehensive reference source on smart antenna technologies featuring contributions with in-depth descriptions of terminologies, concepts, methods, and applications related to smart antennas in various wireless systems"--Provided by publisher.

Includes bibliographical references and index.

ISBN 978-1-59904-988-5 (hardcover) -- ISBN 978-1-59904-989-2 (ebook)

1. Adaptive antennas--Handbooks, manuals, etc. 2. Wireless communication systems--Equipment and supplies--Handbooks, manuals, etc. 3. Microwave antennas--Handbooks, manuals, etc. 4. Radio--Transmitters and transmission--Handbooks, manuals, etc. I. Sun, Chen, 1977- II. Cheng, Jun, 1964- III. Ohira, Takashi, 1955-

TK7871.67.A33.H36 2008

621.382'4--dc22

2008008473

British Cataloguing in Publication Data
A Cataloguing in Publication record for this book is available from the British Library.

All work contributed to this book set is original material. The views expressed in this book are those of the authors, but not necessarily of the publisher.

*If a library purchased a print copy of this publication, please go to http://www.igi-global.com/agreement for information on activating the library's complimentary electronic access to this publication.*

# Table of Contents

**Foreword** ..... xv

**Preface** ..... xvi

**Acknowledgment** ..... xx

**Section I**
**Algorithms**

**Chapter I**
Eigencombining: A Unified Approach to Antenna Array Signal Processing ..... 1
*Constantin Siriteanu, Seoul National University, Korea*
*Steven D. Blostein, Queen's University, Canada*

**Chapter II**
Robust Adaptive Beamforming ..... 33
*Zhu Liang Yu, Nanyang Technological University, Singapore*
*Meng Hwa Er, Nanyang Technological University, Singapore*
*Wee Ser, Nanyang Technological University, Singapore*
*Huawei Chen, Nanyang Technological University, Singapore*

**Chapter III**
Adaptive Beamforming Assisted Receiver ..... 60
*Sheng Chen, University of Southampton, UK*

**Chapter IV**
On the Employment of SMI Beamforming for Cochannel Interference Mitigation in Digital Radio ..... 82
*Thomas Hunziker, University of Kassel, Germany*

**Chapter V**
Random Array Theory and Collaborative Beamforming ..... 94
*Hideki Ochiai, Yokohama National University, Japan*
*Patrick Mitran, University of Waterloo, Canada*
*H. Vincent Poor, Princeton University, USA*
*Vahid Tarokh, Harvard University, USA*

**Chapter VI**
Advanced Space-Time Block Codes and Low Complexity Near Optimal Detection for Future Wireless Networks ..... 107
*W. H. Chin, Institute for Infocomm Research, Singapore*
*C. Yuen, Institute for Infocomm Research, Singapore*

# Detailed Table of Contents



This chapter unifies the principles and analyses of conventional signal processing algorithms for receive-side smart antennas, and compares their performance and numerical complexity. The chapter starts with a brief look at the traditional single-antenna optimum symbol-detector, continues with analyses of conventional smart antenna algorithms, i.e., statistical beamforming (BF) and maximal-ratio combining (MRC), and culminates with an assessment of their recently-proposed superset known as eigencombining or eigenbeamforming. BF or MRC performance fluctuates with changing propagation conditions, although their numerical complexity remains constant. Maximal-ratio eigencombining (MREC) has been devised to achieve best (i.e., near-MRC) performance for complexity that matches the actual channel conditions. The authors derive MREC outage probability and average error probability expressions applicable for any correlation. Particular cases apply to BF and MRC. These tools and numerical complexity assessments help demonstrate the advantages of MREC versus BF or MRC in realistic scenarios.



# Chapter II

In this chapter, we first review the background, basic principle and structure of adaptive beamformers. Since there are many robust adaptive beamforming methods proposed in literature, for easy understanding, we organize them into two categories from the mathematical point of view: one is based on quadratic optimization with linear and nonlinear constraints; the another one is max-min optimization with linear and nonlinear constraints. With the max-min optimization technique, the state-of-the-art robust adaptive beamformers are derived. Theoretical analysis and numerical results are presented to show their superior performance.

Adaptive beamforming is capable of separating user signals transmitted on the same carrier frequency, and thus provides a practical means of supporting multiusers in a space-division multiple-access scenario. Moreover, for the sake of further improving the achievable bandwidth efficiency, high-throughput quadrature amplitude modulation (QAM) schemes have become popular in numerous wireless network standards, notably, in the recent WiMax standard. This contribution focuses on the design of adaptive beamforming assisted detection for the employment in multiple-antenna aided multiuser systems that employ the high-order QAM signalling. Traditionally, the minimum mean square error (MMSE) design is regarded as the state-of-the-art for adaptive beamforming assisted receiver. However, the recent work (Chen et al., 2006) proposed a novel minimum symbol error rate (MSER) design for the beamforming assisted receiver, and it was demonstrated that this MSER design provides significant performance enhancement, in terms of achievable symbol error rate, over the standard MMSE design. This MSER beamforming design is developed fully in this contribution. In particular, an adaptive implementation of the MSER beamforming solution, referred to as the least symbol error rate algorithm, is investigated extensively. The proposed adaptive MSER beamforming scheme is evaluated in simulation, in comparison with the adaptive MMSE beamforming benchmark.



# Chapter IV

Many common adaptive beamforming methods are based on a sample matrix inversion (SMI). The schemes can be applied in two ways. The sample covariance matrices are either computed over preambles, or the sample basis for the SMI and the target of the beamforming are identical. A vector space representation provides insight into the classic SMI-based beamforming variants, and enables elegant derivations of the well-known second-order statistical properties of the output signals. Moreover, the vector space representation is helpful in the definition of appropriate interfaces between beamfoming and soft-decision signal decoding in receivers aiming at adaptive cochannel interference mitigation. It turns out that the performance of standard receivers incorporating SMI-based beamforming on short signal intervals and decoding of BICM (bit-interleaved coded modulation) signals can be significantly improved by proper interface design.



# Chapter V

In wireless sensor networks, the sensor nodes are often randomly situated, and each node is likely to be equipped with a single antenna. If these sensor nodes are able to synchronize, it is possible to beamform by considering sensor nodes as a random array of antennas. Using probabilistic arguments, it can be shown that random arrays formed by dispersive sensors can form nice beampatterns with a sharp main lobe and low sidelobe levels. This chapter reviews the probabilistic analysis of linear random arrays, which dates back to the early work of Y. T. Lo (1964), and then discusses recent work on the statistical analysis of two-dimensional random arrays originally derived in the framework of wireless sensor networks.



Space-time block coding is a way of introducing multiplexing and diversity gain in wireless systems equipped with multiple antennas. There are several classes of codes tailored for different channel conditions. However, in almost all the cases, maximum likelihood detection is required to fully realize the diversity introduced. In this chapter, we present the fundamentals of space-time block coding, as well as introduce new codes with better performance. Additionally, we introduce the basic detection algorithms which can be used for detecting space-time block codes. Several low complexity pseudo-maximum likelihood algorithms will also be introduced and discussed.



# Chapter VII

Modulated codes (MC) are error correction codes (ECC) defined on the complex field and therefore can be naturally combined with an intersymbol interference (ISI) channel. It has been previously proved that for any finite tap ISI channel there exist MC with coding gain comparing to the uncoded AWGN channel. In this chapter, we first consider space-time MC for memory channels, such as multiple transmit and receive antenna systems with ISI. Similar to MC for single antenna systems, the space-time MC can be also naturally combined with a multiple antenna system with ISI, which provides the convenience of the study. Some lower bounds on the capacities C and the information rates of the MC coded systems are presented. We also introduce an MC coded zero-forcing decision feedback equalizer (ZF-DFE) where the channel is assumed known at both the transmitter and the receiver. The optimal MC design based on the ZF-DFE are presented.



# Chapter VIII

This chapter analyzes the problem of blind channel estimation under Space-Time Block Coded transmissions. In particular, a new blind channel estimation technique for a general class of space-time block codes is proposed. The method is solely based on the second-order statistics of the observations, and its computational complexity reduces to the extraction of the main eigenvector of a generalized eigenvalue problem. Additionally, the identifiability conditions associated to the blind channel estimation problem are analyzed, which is exploited to propose a new transmission technique based on the idea of code diversity or combination of different codes. This technique resolves the ambiguities in most of the practical cases, and it can be reduced to a non-redundant precoding consisting in a single set of rotations or permutations of the transmit antennas. Finally, the performance of the proposed techniques is evaluated by means of several simulation examples.



In this chapter, we describe a compact array antenna. Beamforming is achieved by tuning the load reactances at parasitic elements surrounding the active central element. The existing beam forming algorithms for this reactively controlled parasitic array antennas require long training time. In comparison with these algorithms, a faster beamforming algorithm, based on simultaneous perturbation stochastic approximation (SPSA) theory with a maximum cross-correlation coefficient (MCCC) criterion, is proposed in this chapter. The simulation results validate the algorithm. In an environment where the signal-to-interference ratio (SIR) is 0 dB, the algorithm converges within 50 iterations and achieves an output SINR of 10 dB. With the fast beamforming ability and its low power consumption attribute, the antenna makes the mass deployment of smart antenna technologies practical. To give a comparison of the beamforming algorithm with one of the standard beamforming algorithms for a digital beamforming (DBF) antenna array, we compare the proposed algorithm with the least mean square (LMS) beamforming algorithm. Since the parasitic array antenna is in nature an analog antenna, it cannot suppress correlated interference. Here, we assume that the interferences are uncorrelated.

This chapter introduces joint beamforming (or precoding) and space-time coding for multiple input multiple output (MIMO) channels. First, we explain key ideas of beamforming and space- time coding and then we discuss the joint design of beamformer and space-time codes and its benefits. Beamforming techniques play a key role in smart antenna systems as they can provide various features, including spatially selective transmissions to increase the capacity and coverage increase. STC techniques can offer both coding gain and diversity gain over MIMO channels. Thus, joint beamforming and STC is a more practical approach to exploit both spatial correlation and diversity gain of MIMO channels. We believe that joint design will be actively employed in future standards for wireless communications.



# Chapter XIV

This chapter introduces the adaptive modulation and coding (AMC) as a practical means of approaching the high spectral efficiency theoretically promised by multiple-input multiple-output (MIMO) systems. It investigates the AMC MIMO systems in a generic framework and gives a quantitative analysis of the multiplexing gain of these systems. The effects of imperfect channel state information (CSI) on the AMC MIMO systems are pointed out. In the context of imperfect CSI, a design of robust near-capacity AMC MIMO system is proposed and its good performance is verified by simulation results. The proposed adaptive system is compared with the non-adaptive MIMO system, which shows the adaptive system approaches the channel capacity closer.



# Chapter XV

This chapter takes a closer look at a class of MIMO detention methods, collectively referred to as relaxation detectors. These detectors provide computationally advantageous alternatives to the optimal maximum likelihood detector. Previous analysis of relaxation detectors have mainly focused on the implementation aspects, while resorting to Monte Carlo simulations when it comes to investigating their performance in terms of error probability. The objective of this chapter is to illustrate how the performance of any detector in this class can be readily quantified thought its diversity gain when applied to an i.i.d. Rayleigh fading channel, and to show that the diversity gain is often surprisingly simple to derive based on the geometrical properties of the detector.



# Chapter XVI

This chapter discusses four different optimization problems of practical importance for transmission in point to multipoint networks with a multiple input transmitter and multiple output receivers. Existing solutions to each of the problems are adapted to a multi-carrier transmission scheme by considering the special structure of the resulting space-frequency channels. Furthermore, for each of the problems, suboptimum approaches are presented that almost achieve optimum performance and, at the same time, do not have the iterative character of optimum algorithms, i.e., they deliver a solution in a fixed number of steps. The purpose of this chapter is to give an overview on optimum design of point to multipoint networks from an information theoretic perspective and to introduce non-iterative algorithms that are a good practical alternative to the sometimes costly iterative algorithms that achieve optimality.

## Section III
## Applications of Smart Antennas

# Foreword

Smart antennas have undergone dramatic changes since they were first considered for wireless systems. At first they were extremely expensive, using discrete components with analog combining, and suitable only for high-cost military systems. In the 1980's and 1990's, however, came the development of smart antenna concepts for commercial systems, and increases in digital signal processing complexity that pushed single-antenna wireless systems close to their theoretical (Shannon) capacity. Smart antennas are now seen as the key concept to further, many-fold, increases in both the link capacity, through spatial multiplexing, and system capacity, through interference suppression and beamforming, as well as increasing coverage and robustness. Furthermore, decreases in integrated circuit cost and antenna advancements have made smart antennas attractive in terms of both cost and implementation even on small devices. Thus, the last few years have seen an exponential growth in smart antenna research, the inclusion of smart antenna technology into wireless standards, and the beginnings of widespread deployment of smart antennas in wireless networks. This explosive growth has created a strong need for a handbook summarizing the most recent advancements to keep wireless engineers up-to-date on current developments.

This book presents 25 chapters covering key recent advancements in smart antenna technology. It consists of four sections: algorithms, performance issues, applications, and experiments and implementation. The first section provides an overview of existing smart antenna combining algorithms, followed by recent advancements including those on eigenbeamforming, robust adaptive beamforming using the min-max criterion, minimum symbol-error-rate beamforming, and sample matrix inversion, each of which can provide substantially improved performance. For sensor networks, the concept of collaborative beamforming is introduced, which can beamform using arbitrarily located sensors. Next, space-time coding is discussed and recent advancements including space-time coding with low complexity, memory, and blind channel estimation are presented. Finally, techniques for adaptive beamforming and direction finding in compact arrays using a single-port electronically steerable parasitic array radiator are presented.

The next section discusses performance issues with smart antennas. The first chapter describes why electromagnetic effects need to be included in the analysis of smart antenna systems. Then the key issues for transmit diversity: feedback, sounding, and calibration, are discussed and new approaches for better performance are presented. Next joint beamforming and space-time coding, and then adaptive coding and modulation with smart antennas, are analyzed. Finally, two new approaches of relaxation detectors and non-iterative multiple-user spatial multiplexing techniques are presented.

Applications of smart antennas are presented in the third section. The first system application described is CDMA, where more effective modeling and simulation techniques are presented, and then the improvements with cross-layer optimization with scheduling is discussed. Next, the application of smart antennas in mobile ad hoc networks is analyzed. The next chapter shows how smart antennas can be used to achieve unconditional security in generating secret keys for encryption of communications. Finally, the use of smart antennas in radio frequency identification readers is shown to have the potential to improve dramatically reader performance.

The last section describes the implementation of smart antenna testbeds and new experimentally-implemented techniques. The first chapter provides an overview of existing testbeds and then describes the implementation of a testbed using commercially available components, including a field programmable gate array. A testbed for ad hoc networks with smart antennas is also presented. A new technique for a wideband spatial beamformer that does not use frequency filters or tapped delay lines is then described and demonstrated. Finally, the implementation of a compact array as a single-port electronically steerable parasitic array radiator is described.

*Jack Winters, Jack Winters Communications, LLC, USA*

# Preface

The dramatic growth of the wireless communication industry is creating a huge market opportunity. Wireless operators are currently searching for new technologies that can be implemented in the existing wireless communication infrastructure to provide broader bandwidth per user channel, better quality, and new value-added services. Employing smart antennas presents an elegant and relatively economical way to improve the performance of wireless transmission (Winters, 1998; Soni, 2002; Bellofiore, 2002a; Bellofiore, 2002b; Diggavi, 2004).

Deployed at base stations in the existing wireless infrastructure, smart antennas bring outstanding improvement in capacity to radio communication systems, which have severely limited frequency resources, by employing an efficient beamforming scheme (Tsoulos, 1997). New value-added services, such as position location (PL) services for emergency calls, fraud detection, and intelligent transportation systems are also being implemented in real-world applications thanks to the direction-finding ability of smart antennas (Tsoulos, 1999).

Smart antennas can also be efficiently used at mobile terminals. Employed at mobile terminals (e.g., notebook PCs, PDAs) in *ad hoc* networks or wireless local-area networks (WLAN), the direction-finding ability permits the design of packet routing protocols that can determine the optimal manner of packet relaying (Nasipuri, 2000). The beamforming or interference-suppression ability makes it possible to increase throughput at network nodes where it is limited by interference from neighboring nodes (Winters, 2006).

A multiple-input multiple-output (MIMO) wireless communication channel can by built by installing antenna arrays that provide uncorrelated signal outputs at both transmitters and receivers. A MIMO system's capacity for channel information increases with the number of arrayed antenna elements (Telatar, 1999). Transmitting a space-time block coded waveform over a MIMO system dramatically increases the data rate over wireless channels (Naguib, 2000).

To take advantage of smart antennas' potential, recent designs of high-data-rate wireless transmission, distributed sensor networks, wireless network protocols, wireless security, software-defined radio, cognitive radios, and radio frequency identification (RFID) systems have pursued the integration of smart antennas as one of the key technologies.

Researchers in both academia and industry are actively studying smart antenna architectures, algorithms and practical implementations. This handbook aims to provide the readers with a single comprehensive guide to the issues of smart antennas in wireless communication scenarios, covering the wide spectrum of topics related to state-of-the-art smart antenna technologies in wireless systems/networks.

To serve this purpose, this handbook features 25 chapters authored by leading experts in both academia and industry, offering in-depth descriptions of terminologies and concepts relevant to smart antennas in a variety of wireless systems. Furthermore, the handbook explores the challenges facing smart antenna technologies in various wireless propagation environments and application scenarios, including system modeling, algorithms, performance evaluation, practical implementation issues and applications, future research and development trends, and market potential.

The handbook's chapters are organized into four interrelated sections: Algorithms, Performance Issues, Applications of Smart Antennas in Wireless Networks and Systems, and Experiments and Implementations. The following gives an overview of each chapter's contents. The handbook's chapters are organized into four interrelated sections: Algorithms, Performance Issues, Applications of Smart Antennas, and Experiments and Implementations. The following gives an overview of each chapter's contents."

## Section I: Algorithms

**Chapter I** gives a unified analysis of receiver-side beamforming and maximum ratio combining (MRC) algorithms from the viewpoint of eigenbeamforming. Results suggest that the performance of beamforming and MRC fluctuate with the variation in wireless propagation environments. The proposed eigenbeamforming method provides a unified approach to designing smart antenna algorithms.

**Chapter II** studies the performance of robust Capon beamformers using the max-min optimization method. Results show that the robust Capon beamformers are robust against array-steering vector errors and provide a relatively high output signal-to-interference plus noise ratio (SINR). However, the results also indicate that the robust Capon beamformers still cannot achieve the optimal output SINR.

**Chapter III** presents detection techniques applying adaptive beamforming for use in multiple-antenna/multi-user systems that employ high-order QAM signaling. A novel minimum symbol error rate (MSER) design for a beamforming-assisted receiver is proposed. Furthermore, an adaptive implementation of the MSER beamforming is examined. Results show that the MSER beamforming design offers a higher user capacity and is more robust in a near-far scenario than the conventional MMSE beamforming design. Moreover, it is shown that the adaptive implementation of MSER beamforming operates successfully in fast fading conditions and consistently outperforms the adaptive LMS beamforming benchmarker.

**Chapter IV** investigates a sample covariance matrix (SCM)-based beamforming, i.e., sample matrix inversion (SMI) beamforming, for receiver-side interference mitigation in wireless networks where co-channel interference (CCI) is encountered. With the help of a vector space representation, enhanced interfaces between beamforming and signal decoding have been devised for scenarios with block-wise stationary CCI and transmission signals both with and without preambles. Results show that the error rate performance at the decoder output can be significantly improved by employing the SMI-based beamforming on short signal intervals and decoding BICM (bit-interleaved coded modulation) signals.

**Chapter V** considers the beamforming issue in *ad hoc* networks with arbitrarily located sensors. Under ideal assumptions such as the absence of mutual coupling and perfect synchronization, results show that such random arrays can form good beam patterns with sharp main lobes and low sidelobe peaks. The probabilistic performance of planar random arrays (or collaborative beamforming) with a view toward application to wireless ad hoc sensor networks is also analyzed.

**Chapter VI** presents the fundamentals of space-time block coding and, moreover, introduces new codes with better performance. The basic detection algorithms that can be used to detect space-time block codes are discussed. Furthermore, several low complexity pseudo-maximum likelihood algorithms are proposed and discussed. The study proves that these proposed schemes are able to closely match the performance of maximum likelihood detection while only requiring a small fraction of the computational cost.

**Chapter VII** considers space-time modulated codes (MC) for memory channels, such as those used for multiple-transmit and -receive antenna systems with intersymbol interference (ISI). A joint decoding method for space-time MC encoded channels, i.e., the joint zero-forcing decision feedback equalizer (ZF-DFE), is presented. Analytical and numerical results show that the reliable information rates that can be achieved by the MC coded channels based on standard random coding techniques are larger than those of the channels themselves when the channel SNR is relatively low.

**Chapter VIII** analyzes the problem of blind channel estimation under space-time block coded transmissions. A new blind channel estimation criterion is proposed. Analysis shows that this technique reduces the problem of extracting the main eigenvector of a generalized eigenvalue (GEV) and does not introduce additional ambiguities. Numerical evaluation shows that the performance of the proposed blind approaches is close to that of a coherent receiver.

**Chapter IX** studies the adaptive beamforming of a compact array antenna, the electronically parasitic array radiator (Espar) antenna. This antenna has one active element connected to the radio-frequency (RF) port and multiple surrounding parasitic elements loaded with tunable reactance. Beamforming is achieved by tuning the load reactances. A faster beamforming algorithm, based on simultaneous perturbation stochastic approximation (SPSA) theory with a maximum cross-correlation coefficient (MCCC) criterion, is proposed here. Results show that the proposed algorithm achieves sufficient interference suppression.

**Chapter X** presents Direction of arrival (DoA) estimation with the compact array antenna, the Espar antenna, using methods based on reactance switching. DoA estimation methods by an ESPAR antenna are proposed based on three types of algorithms: power pattern cross correlation (PPCC), reactance-domain (RD) multiple signal classification (MUSIC), and RD estimation of signal parameters via rotational invariance techniques (ESPRIT). These three methods exploit the reactance diversity provided by an Espar antenna to correlate different antenna output signals measured at different times and for different reactance values.

## Section II: Performance Issues

**Chapter XI** takes a look at multi-antenna communication systems from the electromagnetic point of view, ranging from adaptive array antennas to MIMO systems. It shows that when introducing multiple antennas into a system, the electromagnetic effect needs to be considered. Analysis shows that even though the mutual coupling degrades the performance of an adaptive system by destroying the wavefront of the signals, it improves performance by increasing the order of singular values in the channel decomposition for a MIMO system, thus yielding a more reliable multiplexing gain.

**Chapter XII** describes the underlying principle, evolving techniques, and corresponding industrial applications of transmit beamforming of MIMO systems, which exploits channel state information (CSI) at the transmitter. In particular, it discusses the codebook-based feedback techniques with various quantization complexities and feedback overheads. Application examples of these techniques in 3GPP, IEEE 802.11n, and 80216d/e are studied. Results show that MIMO beamforming delivers more than 2 dB gain for most practical antenna configurations.

**Chapter XIII** discusses the joint beamforming and space-time coding techniques used to exploit the spatial correlation and diversity gain of MIMO channels, respectively. The beamforming system directly increases the link's signal-to-noise ratio (SNR) while the space-time coding provides the coding gain and diversity gain to improve link performance. The practical implementation issues such as imperfect CSI for these joint beamforming and space-time coding techniques are also discussed.

**Chapter XIV** analyzes adaptive modulation and coding (AMC) as a practical means of approaching the high spectral efficiency theoretically promised by MIMO systems. Using a generic framework, the study gives a quantitative analysis of the system's multiplexing. In the context of imperfect CSI, an adaptive turbo coded MIMO system is proposed and its performance is evaluated. It is shown that this system achieves a near-capacity performance and is robust against the CSI imperfection.

**Chapter XV** investigates a class of relaxation detectors that are approximations of the optimal maximum-likelihood detector. The study illustrates how the performance of any detector in this class can be readily quantified through its diversity gain when applied to an independent and identically-distributed (i.i.d.) Rayleigh fading channel. It is shown that the diversity gain is easy to derive based on the geometrical properties of the detector.

**Chapter XVI** discusses different optimization problems of practical importance for transmission in point-to-multipoint networks with a multiple-input transmitter and multiple output receivers. Optimum transmission parameters of these schemes are computed by iterative algorithms involving a complexity that strongly depends on the a priori unknown number of iterations required to reach convergence. To closely approximate the performance of optimum approaches, suboptimum allocation algorithms are presented. Results show that computation of the optimum transmission parameters requires a complexity similar to that of only one iteration of the optimum approaches, and thus users are assigned decoupled spatial dimensions, which makes it possible to reduce the required signaling overhead.

## Section III: Applications of Smart Antennas

**Chapter XVII** studies the performance of smart antennas in a code division multiple access (CDMA) cellular network. An effective analytical model and simulation techniques that provide rapid and accurate assessment of the performance of CDMA systems employing a smart antenna are presented. The close match of the results from the analytical model and from simulation verifies the usefulness of the analytical model. Furthermore, results show that smart antennas play a significant role in improving the performance of cellular CDMA systems.

**Chapter XVIII** considers a cellular downlink packet data system employing the space-time block coded MIMO scheme. The cross-layer performance of typical scheduling algorithms and a point-to-point power control scheme over a time division multiplexing (TDM)-based shared MIMO channel are evaluated for a CDMA high data rate (HDR) system. Analysis shows that the multi-user diversity gain increases the aggregate throughput and reduces the transmission error rate. It is also shown that space-time block coding/MIMO and one-bit and multi-bit power control improve the physical and media access protocol (MAC) layer performance but may limit the multiuser diversity gain or the potential throughput of schedulers for delay-tolerant bursty data services.

**Chapter XIX** presents the implementations of multi-beam antenna (MBA) techniques for wireless *ad hoc* network applications. Both multiple fixed-beam antennas (MFBAs) and multi-channel smart antennas (MCSAs) are discussed. The performance in terms of node throughput and the probability of concurrent communications are examined while incorporating two random-access scheduling (RAS) schemes in the contention resolution process for the node priority issue and throughput maximization, respectively.

**Chapter XX** proposes an application of smart antennas to generating secret keys for encryption of communications over wireless networks. The scheme uses a smart antenna, such as the Espar antenna, at the access point (AP). Intentionally generating random directional beam patterns creates channel fluctuation, which is transformed into a random key for encryption of communications. Experimental results show that the system has the potential to achieve "unconditional security."

**Chapter XXI** presents the applications of smart antenna technologies to radio frequency identification (RFID) systems. A 3×2-element planar phased array antenna has been designed in a compact package for RFID readers. The antenna covers the 860–960 GHz frequency band with more than 10 dB input return loss, 12 dBi broadside gain, and up to 40° elevation beam scanning with a 4-bit reflection-type phase shifter array.

## Section IV: Experiments and Implementations

**Chapter XXII** gives a comprehensive review of different types of testbeds for MIMO wireless communication systems. Furthermore, the design and development of a 2x2 MIMO testbed that uses antennas built in-house, commercially available RF chips for an RF front end, and a Field Programmable Gate Array (FPGA) for based signal processing are described. The developed testbed is verified and tested with Alamouti quadrature phase shift keying (QPSK) signaling.

**Chapter XXIII** presents a testbed for implementing directional MAC protocols with smart antennas in wireless networks. The testbed makes it possible to investigate performance of MAC protocols in the real environment. It incorporates a compact array antenna, the Espar antenna, as a practical smart antenna, IEEE802.15.4/ZigBee, global positioning system (GPS) and gyro modules to allow easy installment of different MAC protocols.

**Chapter XXIV** introduces a wideband spatial beamformer as an alternative approach for a wideband smart antenna without tapped-delay lines and frequency filters. A prototype antenna is developed. Furthermore, an experiment is carried out to verify the concept of the proposed wideband spatial beamformer. The experiment's results show that the wideband spatial beamformer has sufficient beam steering capability with a relatively simple technique and without using filters for the delay lines.

**Chapter XXV** studies three working modes, omni-, sector and adaptive modes, for a compact array antenna. The Espar antenna is implemented as a representative compact array antenna. Experiments are carried out to verify the omni-, sector and adaptive beam patterns of the Espar antennas.

In these 25 chapters, this timely publication provides an indispensable reference for people interested in smart antennas at all levels as well as for those working within the fields of wireless communications. In short, the handbook was prepared to help readers understand smart antennas as a key technology in modern wireless communication systems. It is our hope that this handbook will not only serve as a valuable reference for students, educators, faculty members, researchers, engineers and research strategists in the field but also guide them toward envisioning the future research and development of smart antenna technologies.

## REFERENCES


Bellofiore, S., Balanis, C. A., Foutz, J., & Spanias, A. S. (2002a), Smart-antenna systems for mobile communication networks, part 1. overview and antenna design, *IEEE Antennas and Propagation Magazine*, 44(3), 145-154.

Bellofiore, S., Foutz, J., Balanis, C. A., & Spanias, A. S. (2002b), Smart-antenna systems for mobile communication networks. part 2. beamforming and network throughput, *IEEE Antennas Propagation Magazine*, 44(4), 106-114.

Diggavi, S. N., Al-Dhahir, N., Stamoulis, A., & Calderbank, A. R. (2004), Great expectations: The value of spatial diversity in wireless networks, *Proceedings of the IEEE*, 99(2), 219-270.

Naguib, A. F., Seshadri, N., & Calderbank, A. R. (2000), Increasing data rate over wireless channels, *IEEE Signal Processing Magazine*, 17(3), 76–92.

Nasipuri, A., Ye, S., You, J., & Hiromoto, R. E. (2000, September), *A MAC protocol for mobile ad hoc networks using directional antennas*," Paper presented at the IEEE Wireless Communications and Networking Conference.

Soni, R. A., Buehrer, R. M., & Benning, R. D. (2002), Intelligent antenna system for cdma2000, *IEEE Signal Processing Magazine*, 19(4), 54-67.

Telatar, E. (1999), Capacity of multi-antenna Gaussian channels, European Transactions on Telecommunications, 10(6), 589–595.

Tsoulos, G. V., Beach, M., & McGeehan, J. (1997), Wireless personal communications for the 21st century: European technological advances in adaptive antennas, *IEEE Communications Magazine*, 35(9), 102-109.

Tsoulos, G. V. (1999), Smart antennas for mobile communication systems: benefits and challenges, *Electronics & Communication Engineering Journal*, 11(2), 84-94.

Winters, J. H. (1998), Smart antennas for wireless systems, *IEEE Personal Communications Magazine*, 5(1), 23-27.

Winters, J. H. (2006), Smart antenna techniques and their application to wirless ad hoc networks, *IEEE Wireless Communications*, 13(4), 77-83.

# Acknowledgment

The editors would like to acknowledge the help of all involved in the collation and review process of the book, without whose support the project could not have been satisfactorily completed.

Most of the authors of chapters included in this book also served as referees for chapters written by other authors. Thanks go to all those who provided constructive and comprehensive reviews.

Special thanks also go to the publishing team at IGI Global, whose contributions throughout the whole process from inception of the initial idea to final publication have been invaluable. In particular to Ross Miller, Jessica Thompson and Rebecca Beistline, who continuously prodded via e-mail for keeping the project on schedule.

In closing, we wish to thank all of the authors for their insights and excellent contributions to this handbook.

*Chen Sun, PhD*
*ATR Wave Engineering Laboratories, Japan*

*Jun Cheng, PhD*
*Doshisha University, Japan*

*Takashi Ohira, PhD*
*Toyohashi University of Technologies, Japan*

*May 2008*

# Section I
# Algorithms

# Chapter I
# Eigencombining:
## A Unified Approach to Antenna Array Signal Processing

**Constantin Siriteanu**
*Seoul National University, Korea*

**Steven D. Blostein**
*Queen's University, Canada*

## ABSTRACT

*This chapter unifies the principles and analyses of conventional signal processing algorithms for receive-side smart antennas, and compares their performance and numerical complexity. The chapter starts with a brief look at the traditional single-antenna optimum symbol-detector, continues with analyses of conventional smart antenna algorithms, i.e., statistical beamforming (BF) and maximal-ratio combining (MRC), and culminates with an assessment of their recently-proposed superset known as eigencombining or eigenbeamforming. BF or MRC performance fluctuates with changing propagation conditions, although their numerical complexity remains constant. Maximal-ratio eigencombining (MREC) has been devised to achieve best (i.e., near-MRC) performance for complexity that matches the actual channel conditions. The authors derive MREC outage probability and average error probability expressions applicable for any correlation. Particular cases apply to BF and MRC. These tools and numerical complexity assessments help demonstrate the advantages of MREC versus BF or MRC in realistic scenarios.*

## INTRODUCTION

***General perspective.*** Andrew Viterbi is credited with famously stating that "spatial processing remains as the most promising, if not the last frontier, in the evolution of multiple access systems" (Roy, 1998, p. 339). Multiple-antenna-transceiver communications systems, also known as single-input multiple-output (SIMO), multiple-input single-output (MISO), or multiple-input multiple-output (MIMO) systems, which exploit the spatial dimension of the radio channel, promise tremendous benefits over the traditional single-input single-output (SISO) transceiver concept, in terms of data rate, subscriber capacity, cell coverage, link quality, transmit power, etc. Such benefits can be achieved with **smart antennas**, i.e., SIMO, MISO, and MIMO systems that combine baseband signals for optimum performance (Paulraj, Nabar, & Gore, 2005).

Herein, we consider receive smart antennas (i.e., the SIMO case) deployed in noise-limited scenarios with frequency-flat multipath fading (El Zooghby, 2005, Section 3.3) (Jakes, 1974) (Vaughan & Andersen, 2003, Chapter 3), for which the following signal combining techniques have conventionally been proposed:

- ***Statistical beamforming (BF)***, i.e., digitally steering a radio beam along the dominant eigenvector of the correlation matrix of the channel fading gain vector (S. Choi, Choi, Im, & Choi, 2002) (El Zooghby, 2005, Eqn. (5.23), p. 126, Eqns. (5.78–80), p. 148) (Vaughan & Andersen, 2003, Section 9.2.2). BF enhances vs. SISO the *average*, over the fading and noise, signal-to-noise ratio (SNR) by an *array gain* factor that is ultimately proportional to the antenna correlation and is no greater than the number of antenna elements. Since BF requires the estimation of only the projection of the channel gain vector onto the eigenvector mentioned above, it has low numerical complexity. However, BF is effective only for highly-correlated channel gains, i.e., when the intended signal arrives with narrow azimuth angle spread (AS).
- ***Maximal-ratio combining (MRC)***, i.e., maximizing the output SNR *conditioned* on the fading gains (Brennan, 2003; Simon & Alouini, 2000). This SNR is computed by averaging over the noise only, i.e., conditioning on the channel gains. When the intended-signal AS is large enough to significantly reduce antenna correlation, MRC can greatly outperform BF as a result of *diversity gain* and array gain, at the cost of much higher numerical complexity incurred due to channel estimation for each antenna element.

Note that, for fully correlated (i.e., coherent) channel gains, both BF and MRC reduce to the classical notion of "beamforming" whereby a beam is formed towards the intended signal arriving from a discrete direction (Monzingo & Miller, 1980; Trees, 2002; Godara, 2004).

Statistical beamforming and diversity combining principles have traditionally been classified, studied, and applied separately, leading to disparate and limited performance analyses of BF and MRC. Furthermore, since BF and MRC optimize the average SNR and the conditioned SNR, respectively, they have opposing performance-maximizing spatial correlation requirements, as well as significantly different, correlation-independent, numerical complexities (Siriteanu, Blostein, & Millar, 2006; Siriteanu, 2006; Siriteanu & Blostein, 2007). Because correlation varies in practice due to variable AS (Algans, Pedersen, & Mogensen, 2002), BF or MRC performance fluctuates, whereas numerical complexity remains constant. Therefore, MRC can actually waste processing resources and power, whereas BF can often perform poorly (Siriteanu *et al.*, 2006; Siriteanu, 2006; Siriteanu & Blostein, 2007).

Limitations of stand-alone BF or MRC deployments can be overcome by jointly exploiting their principles, under the unifying framework of *eigencombining*. ***Maximal-ratio eigencombining (MREC)*** first applies the Karhunen-Loeve Transform (KLT) with several dominant eigenvectors of the channel correlation matrix to recast the received signal vector in a reduced-dimension space, and then optimally combines the new, uncorrelated, signals (Alouini, Scaglione, & Giannakis, 2000; Brunner, Utschick, & Nossek, 2001; F. A. Dietrich & Utschick, 2003; Jelitto & Fettweis, 2002; Siriteanu & Blostein, 2007). The number of eigenvectors used for the KLT is referred to as the MREC order. Minimum and maximum orders render MREC equivalent with BF and MRC, respectively (Alouini *et al.*, 2000; Dong & Beaulieu, 2002; Siriteanu & Blostein, 2007). The KLT decorrelating effect simplifies the performance analysis for MREC, i.e., also for BF and MRC, over the entire correlation range (Alouini *et al.*, 2000; Dong & Beaulieu, 2002; Siriteanu & Blostein, 2007). Eigengain decorrelation also simplifies fading factor estimation and combining implementation over MRC, thus reducing the numerical complexity (Alouini *et al.*, 2000; Siriteanu & Blostein, 2007). For the medium-to-high correlation values (i.e., 0.5 – 0.9) often incurred at base-stations in typical urban scenarios (Siriteanu & Blostein, 2007), MREC can reduce problem dimension vs. MRC, further reducing numerical complexity, while offering near-optimum performance, and thus outperforming BF. Consequently, MREC of order selected to suit the channel and noise statistics or the system load can improve signal processing efficiency over BF and MRC (Siriteanu *et al.*, 2006; Siriteanu, 2006; Siriteanu & Blostein, 2007).

***Chapter outline and objectives.*** The next subsection provides more background information on BF, MRC, and MREC. Then, a signal model is described that incorporates additive noise as well as spatial fading caused by signal arrival with AS, for a base station in typical urban scenarios. The traditional SISO approach is then described, and expressions for symbol-detection performance measures such as the outage probability (OP) and average error probability (AEP) are derived. The conventional antenna array signal processing concepts of BF and MRC are studied afterward, for ideal and adverse fading correlation conditions, and their numerical complexities are compared for actual implementations, which require channel estimation. Next, the BF and MRC principles are unified under the framework of MREC, which is shown to simplify the MRC analysis for channel correlation conditions that render difficult direct MRC study. AEP and OP expressions that are derived for MREC but also cover SISO, BF, and MRC, as well as numerical complexity evaluations, serve to demonstrate the benefits of adaptive-eigencombining-based smarter antennas for realistic scenarios with random AS.

## FURTHER BACKGROUND, MOTIVATION, AND LITERATURE REVIEW

SIMO, MISO, and MIMO smart antennas deploying diversity combining, statistical beamforming, space-time coding, and spatial multiplexing can provide tremendous performance and capacity improvements over SISO (S. Choi *et al.*, 2002; El Zooghby, 2005; Goldberg & Fonollosa, 1998; Paulraj *et al.*, 2005; Rooyen, Lotter, & Wyk, 2000; Simon & Alouini, 2000; Stridh, Bengtsson, & Ottersten, 2006; Tse & Viswanath, 2005). However, these multi-antenna algorithms require powerful and, thus, power-hungry baseband processing, and their performance is highly dependent on spatial correlation, which is affected by radio propagation conditions (Salz & Winters, 1994). Nonetheless, latest drafts of standards for wireless communications systems specify multi-antenna transceivers for cellular systems, e.g., 3GPP and 3GPP2, and for area networks, e.g., IEEE802.11n and IEEE802.16e (Hottinen, Kuusela, Hugl, Zhang, & Raghothaman, 2006). In this chapter, we concentrate on improving the performance and efficiency of baseband signal processing for receive smart antennas by jointly exploiting the principles of statistical beamforming (S. Choi *et al.*, 2002; El Zooghby, 2005; Goldberg & Fonollosa, 1998; Rooyen *et al.*, 2000; Stridh, Bengtsson, & Ottersten, 2006) and diversity combining (Brennan, 2003; Simon & Alouini, 2000).

The concept of beamforming originates in the radar literature (Applebaum, 1976), where the intended signal was assumed to arrive from a unique direction, i.e., coherently, without spatial fading (Salz & Winters, 1994). Signals picked up by the receiving antenna array can then be processed optimally with a combiner obtained from the deterministic *array steering vector* (El Zooghby, 2005, Section 5.1.4) (Godara, 2004, Section 2.1.1) (Goldberg & Fonollosa, Section 4.1), to form antenna pattern beams that effectively enhance the intended signal, and thus yield array gain (Godara, 2004, Section 2.2.4). Nevertheless, in practice, signals arrive at the base station with nonzero AS (Algans *et al.*, 2002; 3GPP, 2003; Pedersen, Mogensen, & Fleury, 2000; Vaughan & Andersen, 2003), which produces spatial fading, i.e., loss of coherence between the channel gains at the various antenna elements (Salz & Winters, 1994). Spatial fading can yield diversity gain through maximal-ratio combining (MRC), which maximizes the SNR conditioned on the fading, by projecting the received signal vector onto an estimate of the channel gain vector (Brennan, 2003; Jakes, 1974; Lee, 1982; Simon & Alouini, 2000). However, fading estimation based on pilot-symbol-aided modulation (PSAM) at the transmitter and interpolation at the receiver (Siriteanu & Blostein, 2004; Siriteanu *et al.*, 2006; Siriteanu, 2006) can demand significant processing resources in the case of MRC (Siriteanu *et al.*, 2006; Siriteanu, 2006; Siriteanu & Blostein, 2007). Though less complex than MRC, statistical beamforming (BF), which projects the received signal vector onto the dominant eigenvector of the spatial fading correlation matrix, only maximizes the *average* SNR and therefore is effective only in high-correlation environments (S. Choi *et al.*, 2002, Section III.A) (El Zooghby, 2005, Section 5.3.3) (Goldberg & Fonollosa, 1998, Section 5) (Rooyen *et al.*, 2000, Chapters 5,6) (Stridh, Bengtsson, & Ottersten, 2006, Section III.A) (Vaughan & Andersen, 2003, Section 9.2.2).

Since statistical beamforming and diversity combining have traditionally been addressed separately, joint BF and MRC studies and performance comparisons are few and incomplete (El Zooghby, 2005, Sections 7.6–7) (Hottinen, Tirkkonen, & Wichman, 2003, Section 2.2) (Rooyen *et al.*, 2000, Section 6.4) (Vaughan & Andersen, 2003, Section 9.2.2,9.3.4). Furthermore, existing studies of BF provide incomplete evaluations of the effect on performance of noncoherent channel gains (S. Choi *et al.*, 2002; J. Choi & Choi, 2003) (El Zooghby, 2005, Sections 6.3, 7.6–7) (Rooyen *et al.*, 2000, Sec- tion 5.1.4) (Vaughan & Andersen, 2003, Section 9.2.2). For MRC, on the other hand, performance studies are available even for correlated channel gains, but they do not cover the entire correlation range continuously (Brennan, 2003, Section 8) (F. A. Dietrich & Utschick, 2003) (Jakes, 1974) (Lee, 1982, Section 10.6) (Simon & Alouini, 2000, Section 9.6).

The low complexity of BF and its ability to produce significant array gain for narrow AS have made this algorithm the preferred choice for high spatial correlation scenarios (El Zooghby, 2005, Sections 5.1.4, 5.3.3). Otherwise, the much more complex MRC has been deployed, to yield array and diversity gains. Thus, unfavorable actual correlation results in poor BF and MRC performance (Brennan, 2003, Section VIII) (El Zooghby, 2005, Sections 5.1.1, 9.2) (Rooyen *et al.*, 2000, Sections 6.1.2, 6.2.1, 6.4) (Simon & Alouini, 2000, Section 9.6) (Vaughan & Andersen, 2003, Sections 9.2.2.1–2). The correlation between channel gains at different antenna elements is affected by propagation conditions, i.e., power azimuth spectrum (p.a.s.) type and AS, as well as by antenna geometry (Algans *et al.*, 2002; El Zooghby, 2005; Salz & Winters, 1994; Vaughan & Andersen, 2003). Therefore, unfavorable AS or inadequate interelement distance can drastically reduce or completely eliminate the performance gains achievable in theory with BF and MRC over SISO. As already mentioned, actual deployment of MRC consumes significant processing resources on estimating the individual channel gain factors (Siriteanu *et al.*, 2006; Siriteanu, 2006; Siriteanu & Blostein, 2007) (Vaughan & Andersen, 2003, Section 9.2.1.3), whereas BF requires the estimation of a single fading coefficient (S. Choi *et al.*, 2002, Section III) (Goldberg & Fonollosa,1998, Section 5) (Stridh, Bengtsson, & Ottersten, 2006, Section 2) (Siriteanu *et al.*, 2006; Siriteanu & Blostein, 2007) (Vaughan & Andersen, 2003, Section 9.2.1.3). Furthermore, since in practice the AS fluctuates — slowly, compared to the channel fading (Algans *et al.*, 2002) (Brunner *et al.*, 2001, Section I) (Alouini *et al.*, 2000, Section 3.3) (Siriteanu & Blostein, 2007, Section II) — the performance of BF or MRC varies, while numerical computational complexity remains constant (Siriteanu *et al.*, 2006; Siriteanu, 2006; Siriteanu & Blostein, 2007).

These disadvantages of conventional receive smart antennas have enticed researchers to devise the more cost-effective approach herein entitled "eigencombining", though also known in the literature as "principal components combining" (Alouini *et al.*, 2000) or as "eigenbeamforming" (Brunner *et al.*, 2001). Unlike in MRC, where the antenna signals are directly combined, eigencombining processes signals obtained by projecting the received signals onto dominant eigenvectors of the channel gain correlation matrix.

Eigencombining has recently been proposed for antenna array receivers as a more versatile technique whose performance and computational requirements can follow the channel statistics (Brunner *et al.*, 2001; J. Choi & Choi, 2003; Jelitto & Fettweis, 2002; F. A. Dietrich & Utschick, 2003; Siriteanu *et al.*, 2006; Siriteanu, 2006; Siriteanu & Blostein, 2007). The origins of eigencombining can be traced to beamspace (data-independent) beamforming (Blogh & Hanzo, 2002, Section 3.2.8) (Godara, 2004, Section 2.6) (Trees, 2002, Sections 3.10, 6.9, 7.10), and particularly to principal-component or eigenspace (data-dependent, adaptive) beamforming (Trees, 2002, Sections 6.8, 7.9), which were proposed for antenna array signal-processing dimension reduction. Eigencombining has been promoted for SIMO transceivers as an enhancement to BF for scenarios with non-zero AS, as well as a lower-complexity alternative to MRC for scenarios with non-rich scattering (Alouini *et al.*, 2000; Brunner *et al.*, 2001; J. Choi & Choi, 2003; Jelitto & Fettweis, 2002; F. A. Dietrich & Utschick, 2003; Siriteanu *et al.*, 2006; Siriteanu, 2006; Siriteanu & Blostein, 2007). The statistics of the channel fading vary slowly compared to the Doppler-induced fading (Sampath, Erceg, & Paulraj, 2005, Section 5.B) (Siriteanu & Blostein, 2007, Section II). Therefore, eigendecomposition-updating (Alouini *et al.*, 2000, Section 3.3) (Goldberg & Fonollosa, 1998, Section 7.2) computations inherent to eigencombining can be distributed over long intervals (Brunner *et al.*, 2001, Section I) and do not add significantly to the per-symbol complexity (Siriteanu & Blostein, 2007, Table II), as shown in (Siemens, 2000).

Transmit-side statistical eigenprocessing is better known in MISO systems as eigenbeamforming (Brunner *et al.*, 2001; Rensburg & Friedlander, 2004) and in MIMO systems as either precoding (Bahrami & Le-Ngoc, 2006; Sampath *et al.*, 2005; Vu & Paulraj, 2006) or eigenbeamforming (Hottinen *et al.*, 2003; Zhou & Giannakis, 2003). Eigenbeamforming based on knowledge of the actual channel vector or matrix has also been proposed, in (Hottinen *et al.*, 2003, Section 2.3.2) (Hottinen *et al.*, 2006) (Paulraj *et al.*, 2005, Section 5.4.4) (Tse & Viswanath, 2005, Section 7.1.1). Accompanied by space-time coding, MISO or MIMO *statistical* precoding promises high performance for low complexity and very low receiver-to-transmitter feedback rate (Brunner *et al.*, 2001, Section 4) (Hottinen *et al.*, 2003, Sections 10.2–3) (Rensburg & Friedlander, 2004; Sampath *et al.*, 2005; Vu & Paulraj, 2006; Zhou & Giannakis, 2003). A range of design criteria — e.g., capacity maximization, error probability minimization — have yielded closely-resembling precoding approaches whereby data is sent over dominant eigenvectors of the transmit-side correlation matrix using water-filling techniques (Bahrami & Le-Ngoc, 2006; Sampath *et al.*, 2005; Vu & Paulraj, 2006; Zhou & Giannakis, 2003). Transmit-side statistical eigenprocessing has shown promise for MISO WCDMA systems (Hottinen et al., 2003, Section 10.2-3), and has already been specified for MIMO in the IEEE 802.16e standard (Hottinen *et al.*, 2006).

From a receive-side perspective, this chapter reviews conventional signal combining approaches and places their priciples and analyses under the unifying framework of eigencombining. Expressions applicable for any correlation conditions are then derived for the outage probability and the average error probability of MREC, MRC, and BF, allowing for realistic performance comparisons. The numerical complexities of actual implementations of BF, MRC, and MREC are also compared. It emerges that, besides unifying and simplifying the analyses of BF and MRC, MREC can be adapted to channel and noise statistics for more efficient antenna array signal processing than with stand-alone BF or MRC.

## SIGNAL AND CHANNEL MODELS

### Received Signal Model

This chapter focuses on receiver-side multibranch signal combining (i.e., SIMO). Although we will present numerical results only for base-station antenna arrays, the subsequent analysis applies for any other multibranch receivers, e.g., subscriber-station antenna arrays or CDMA Rake receivers (El Zooghby, 2005, Section 2.3.3.3) (Patenaude, Lodge, & Chouinard, 1999) (Simon & Alouini, 2000, Section 7.1, p. 160).

Hereafter we assume that $L$ replicas of the transmitted signal are available at the receiver, affected by multipath fading (El Zooghby, 2005; Jakes, 1974; Lee, 1982; Ertel, Cardieri, Sowersby, Rappaport, & Reed, 1998; J. C. Liberti

& Rappaport, 1999; Paulraj *et al.*, 2005; Tse & Viswanath, 2005) and additive noise. These replicas are further referred to as *branches*. After demodulation, matched-filtering, and symbol-rate sampling, the baseband complex-valued received signal vector can be written based on (Proakis, 2001, Eqn. 14.4–1, p. 822) as

$$\tilde{\mathbf{y}} = \sqrt{E_s}\, b\tilde{\mathbf{h}} + \tilde{\mathbf{n}}, \tag{1}$$

where $b$ is the M-ary phase-shift-keying (M–PSK) (Simon & Alouini, 2000, Section 3.1.3, p. 35) random, equiprobable, zero-mean, unit-variance transmitted symbol, and $E_s$ is the average energy transmitted per symbol, while $\tilde{\mathbf{h}}$ and $\tilde{\mathbf{n}}$ are the complex-valued, mutually uncorrelated **channel gain** and receiver interference-plus-noise vectors, respectively. The $L\times 1$-dimensional complex-valued vectors from (1) are detailed below:

$$\tilde{\mathbf{y}} = [\tilde{y}_1\ \tilde{y}_2\ \ldots\ \tilde{y}_L]^T,\quad \tilde{\mathbf{h}} = [\tilde{h}_1\ \tilde{h}_2\ \ldots\ \tilde{h}_L]^T,\quad \tilde{\mathbf{n}} = [\tilde{n}_1\ \tilde{n}_2\ \ldots\ \tilde{n}_L]^T. \tag{2}$$

The components of the received signal vector, i.e., the branches, can be written as

$$\tilde{y}_i = \sqrt{E_s}\, b\tilde{h}_i + \tilde{n}_i,\quad i = 1:L \overset{\Delta}{=} 1,\ldots,L. \tag{3}$$

The interference-plus-noise vector is assumed to be a zero-mean circularly-symmetric complex-valued Gaussian (ZMC-SCG) (Paulraj *et al.*, 2005, p. 39) spatially- and temporally-white random vector, with variance $N_0$ per component. For simplicity, we will refer to $\tilde{\mathbf{n}}$ simply as the noise vector. Its distribution is described using the following notation:

$$\tilde{\mathbf{n}} \sim \mathcal{N}_c(\mathbf{0}, N_0\mathbf{I}). \tag{4}$$

## Fading Channel Model

Since we later deal with the impact of spatial fading correlation on multi-antenna array receiver performance, we propose here a simple, relevant, and realistic channel model, based on (3GPP, 2003; Algans *et al.*, 2002; Vaughan & Andersen, 2003). Assume that the transmitted signal reaches the receiver after propagation over Q paths possibly arriving with distinct azimuth angles. Even when there is no azimuth dispersion, several paths could still exist due to scattering over a small area around the subscriber station (Vaughan & Andersen, 2003, Section 9.2.1.2). Assuming that all these paths arrive within the same symbol interval (i.e., flat fading), the channel gain vector can then be written as (El Zooghby, 2005, Eqn. (5.9), p. 123) (Ertel *et al.*, 1998, p. 16)

$$\tilde{\mathbf{h}} = \frac{1}{\sqrt{Q}}\sum_{q=1}^{Q}\alpha_q\,\mathbf{a}(\theta_q), \tag{5}$$

where $\alpha_q$ represents the complex-valued attenuation of the path with angle of arrival (AoA) $\theta_q$ with respect to the antenna array broadside (the line perpendicular on the line connecting the array elements), and $\mathbf{a}(\theta_q)$ is the corresponding so-called *array steering vector* (El Zooghby, 2005, Section 5.1.4) (Godara, 2004, Section 2.1.1) (Goldberg & Fonollosa, Section 4.1). For a uniform linear array — ULA (Trees, 2002, Section 2.3) — with normalized interelement distance, i.e., ratio between the physical interelement distance and half of the transmitted-signal carrier wavelength, denoted hereafter as $d_n$, the array steering vector can be written as (Brunner *et al.*, 2001, p. 4) (Goldberg & Fonollosa, Eqn. (4))

$$\mathbf{a}(\theta_q) = \left[1\quad e^{-j\pi d_n \sin\theta_q}\quad \ldots\quad e^{-j\pi(L-1)d_n \sin\theta_q}\right] \tag{6}$$

We follow the Gaussian wide-sense stationary uncorrelated scattering model (Ertel *et al.*, 1998, p. 16). Then, for $Q >$ 10, the channel gain vector from (5) is approximately a ZMCSCG vector, i.e.,

$$\tilde{\mathbf{h}} \sim \mathcal{N}_c(\mathbf{0}, \mathbf{R}_{\tilde{\mathbf{h}}}), \tag{7}$$

where

$$\mathbf{R}_{\tilde{\mathbf{h}}} \overset{\Delta}{=} E\{\tilde{\mathbf{h}}\tilde{\mathbf{h}}^H\} \tag{8}$$

is the channel gain vector correlation (in this case, also covariance) matrix (Proakis, 2001, p. 33), which is Hermitian. Refer to (Trees, 2002, Appendix A) for the matrix properties invoked herein. Note that ZMCSCG channel gains imply Rayleigh fading (Simon & Alouini, 2000, Section 2.2.1.1).

The numerical results shown throughout this work assume a Toeplitz structure for $\mathbf{R}_{\tilde{\mathbf{h}}}$, i.e., the elements on each left-right diagonal are equal. Then, the first line of $\mathbf{R}_{\tilde{\mathbf{h}}}$ provides the information on all its elements. This applies when the signals are received with a ULA.

The elements on the main diagonal of this matrix, i.e.,

$$\left(\mathbf{R}_{\tilde{\mathbf{h}}}\right)_{i,i} = E\{|\tilde{h}_i|^2\} \overset{\Delta}{=} \sigma^2_{\tilde{h}_i}, \quad i = 1 : L, \tag{9}$$

are the autocorrelations and also the variances (Proakis, 2001, p. 32) of the individual channel gains. They are assumed equal for the numerical results shown in this work. Such assumption is valid for antenna arrays, but not for CDMA Rake receiver taps (Alouini *et al.*, 2000) (Patenaude, Lodge, & Chouinard, 1999). Nonetheless, the subsequent analysis applies for any multibranch receiver.

Using (3), (4), and (9), the signal-to-noise ratio (SNR) for the $i$ th branch given the corresponding channel gain, i.e., the **conditioned SNR** or the symbol-detection SNR, can be written as

$$\tilde{\gamma}_i \overset{\Delta}{=} \frac{E_s}{N_0} |\tilde{h}_i|^2, \tag{10}$$

with average over the fading distribution, i.e., **average SNR**, given by

$$\tilde{\Gamma}_i \overset{\Delta}{=} \frac{E_s}{N_0} \sigma^2_{\tilde{h}_i} = \log_2 M \, \gamma_b \, \sigma^2_{\tilde{h}_i}, \tag{11}$$

where $\gamma_b \overset{\Delta}{=} \frac{1}{\log_2 M} \frac{E_s}{N_0}$ represents the *average bit-SNR.*

As shown later, propagation conditions (wave scattering) determine antenna correlation, which impacts the performance of antenna array signal processing algorithms. The eigendecomposition of the channel correlation matrix illustrates the distribution of intended-signal power along the eigenvectors of $\mathbf{R}_{\tilde{\mathbf{h}}}$.

Let us discuss a few features of the correlation matrix and its eigendecomposition (Trees, 2002, Appendix A). The Hermitian matrix $\mathbf{R}_{\tilde{\mathbf{h}}}$ has real-valued non-negative eigenvalues, which we consider ordered as

$$\lambda_1 \geq \lambda_2 \geq \ldots \geq \lambda_L \geq 0. \tag{12}$$

The corresponding eigenvectors of $\mathbf{R}_{\tilde{\mathbf{h}}}$, denoted as $\mathbf{u}_i$, $i$ = 1:$L$, form an orthonormal basis in $\mathbb{C}^L$, i.e., the space of complex-valued $L$-dimensional vectors. The *eigendecomposition* of $\mathbf{R}_{\tilde{\mathbf{h}}}$ is described by

$$\mathbf{R}_{\tilde{\mathbf{h}}} = \mathbf{U}_L \mathbf{\Lambda}_L \mathbf{U}_L^H = \sum_{i=1}^{L} \lambda_i \mathbf{u}_i \mathbf{u}_i^H, \tag{13}$$

where $\mathbf{\Lambda}_L$ is a diagonal matrix having the eigenvalues of $\mathbf{R}_{\tilde{\mathbf{h}}}$ on the main diagonal, and $\mathbf{U}_L \overset{\Delta}{=} [\mathbf{u}_1 \quad \mathbf{u}_2 \quad \ldots \quad \mathbf{u}_L]$ is a unitary matrix. The trace of the channel vector correlation matrix, i.e.,

$$tr(\mathbf{R}_{\tilde{\mathbf{h}}}) \overset{\Delta}{=} \sum_{i=1}^{L} (\mathbf{R}_{\tilde{\mathbf{h}}})_{i,i} = \sum_{i=1}^{L} \sigma_{\tilde{h}_i}^2 = \lambda_1 + \lambda_2 + \ldots + \lambda_L,$$

is a measure of the received intended-signal power. Eigenvectors corresponding to larger eigenvalues are denoted as *dominant*.

The following propositions have simple proofs (not shown here).

**Proposition 1** *The elements of the* ***channel gain*** *vector* $\tilde{\mathbf{h}}$ *are coherent – i.e., we can write* $\tilde{\mathbf{h}} = h_1 \mathbf{u}_1$, *where* $h_1 \overset{\Delta}{=} \mathbf{u}_1^H \tilde{\mathbf{h}}$ *— if and only if (iff)* $\lambda_1 = tr(\mathbf{R}_{\tilde{\mathbf{h}}})$. In general, two random variables are referred to as coherent or fully correlated if the absolute value of their correlation coefficient is 1.

**Proposition 2** *The elements of* $\tilde{\mathbf{h}}$ *are uncorrelated, i.e.,* $(\mathbf{R}_{\tilde{\mathbf{h}}})_{i,j} \overset{\Delta}{=} E\{\tilde{h}_i \tilde{h}_j^*\} = \lambda \delta_{i,j}$, *iff the eigenvalues of* $\mathbf{R}_{\tilde{\mathbf{h}}}$ *are all equal, i.e.,*

$$\lambda_i = \lambda = \frac{1}{L} tr(\mathbf{R}_{\tilde{\mathbf{h}}}),\ i = 1:L. \ (\delta_{i,j} = 1 \text{ for } i = j, \text{ and } \delta_{i,j} = 0 \text{ for } i \neq j.)$$

If the eigenvalues are all equal then $\mathbf{R}_{\tilde{\mathbf{h}}} = \lambda \mathbf{I}$, and we can consider $\mathbf{U}_L = \mathbf{I}_L$.

## Power Azimuth Spectrum and Azimuth Angle Spread Models

*Definitions.* In wireless communications, wave scattering (Ertel *et al.*, 1998; Jakes, 1974; Lee, 1982; Tse & Viswanath, 2005; Vaughan & Andersen, 2003) induces azimuthal angle dispersion, which has been thoroughly characterized in recognition of its important effect on spatial fading correlation and thus on antenna array performance (El Zooghby, 2005, Section 3.8) (Salz & Winters, 1994).

The channel spatial selectivity, i.e., the antenna decorrelation, is affected by the **power azimuth spectrum** (p.a.s.), which is the distribution of received intended-signal power vs. the azimuth angle (Algans *et al.*, 2002, Eqn. (2)) (El Zooghby, 2005, p. 69) (Pedersen, Mogensen, & Fleury, 2000, Eqn. (10)) (Vaughan & Andersen, 2003, Section 4.6), i.e.,

$$P(\theta) = \frac{1}{Q} \sum_{q=1}^{Q} E\{|\alpha_q|^2\} \delta(\theta - \theta_q), \tag{14}$$

for the channel gain vector described by (5). Averaging is over $\alpha_q$, and $\delta(\cdot)$ represents the Dirac delta function. The square-root second central moment of the p.a.s. is hereafter referred to as ***azimuth spread*** **(AS)** as in (3GPP, 2003, Section 4.6.2) (Algans *et al.*, 2002, p. 525) (El Zooghby, 2005, Eqn. (3.10), p. 71) (Pedersen, Mogensen, & Fleury, 2000, p. 438) (Vaughan & Andersen, 2003, Section 4.6.1).

***Power Azimuth Spectrum Model.*** The Laplacian p.a.s. accurately models actual radio channel measurements for rural, sub/urban, and indoor scenarios (Algans *et al.*, 2002) (Pedersen, Mogensen, & Fleury, 2000) (Vaughan & Andersen, 2003), and was therefore adopted for simulation purposes by wireless systems standardization organizations (3GPP, 2003, Section 4.5.4) (El Zooghby, 2005, p. 70). Based on (3GPP, 2003, Section 4.5.4) (El Zooghby, 2005, Eqn. (3.8), p. 70) (Pedersen, Mogensen, & Fleury, 2000, Eqn. (17)) (Vaughan & Andersen, 2003, Section 4.6.2), we can write a truncated Laplacian p.a.s. as

$$P(\theta)=\begin{cases}\dfrac{1}{1-\exp\left\{-\dfrac{\pi}{\sigma/\sqrt{2}}\right\}}\cdot\dfrac{1}{2\cdot\sigma/\sqrt{2}}\cdot\exp\left\{-\dfrac{|\theta-\theta_c|}{\sigma/\sqrt{2}}\right\} & \text{, for } \theta\in[\theta_c-\pi,\theta_c+\pi\,], \\ 0 & \text{, otherwise,}\end{cases} \tag{15}$$

where $\theta_c$ is the mean AoA, while $\sigma$ approximates the AS. The numerical results shown in this chapter rely on this p.a.s. model.

For this truncated Laplacian p.a.s., mathematical operations similar to those in (Salz & Winters, 1994, Appendix) produce the following expressions for the real and imaginary parts of the $(m, n)$th element of $\mathbf{R}_{\tilde{h}}$ for a ULA:

$$\begin{aligned}\Re\{(\mathbf{R}_{\tilde{h}})_{m,n}\} &= J_0(z_{m,n})+2\sum_{k=1}^{\infty}J_{2k}(z_{m,n})\frac{\cos(2k\theta_c)}{1+\left[2k\sigma/\sqrt{2}\right]^2} \\ \Im\{(\mathbf{R}_{\tilde{h}})_{m,n}\} &= \frac{1+\exp\left\{-\dfrac{\pi}{\sigma/\sqrt{2}}\right\}}{1-\exp\left\{-\dfrac{\pi}{\sigma/\sqrt{2}}\right\}}\cdot 2\sum_{k=0}^{\infty}J_{2k+1}(z_{m,n})\frac{\sin[(2k+1)\theta_c]}{1+\left[(2k+1)\sigma/\sqrt{2}\right]^2},\end{aligned} \tag{16}$$

where $J_i(\cdot)$ is the $i$th-order Bessel function of the first kind (Abramowitz & Stegun, 1995, §9.1.21, p. 360), and $z_{m,n}\overset{\Delta}{=}\pi(m-n)d_n$. Numerical results shown in this work assume $d_n = 1$ and $\theta_c = 0$; the latter yields real-valued interelement correlation.

***Azimuth Spread Distribution Model for Typical Urban Scenarios.*** Measurements have shown that the AS depends on the environment, antenna array location and height, and is time-varying (Algans *et al.*, 2002). Typical sub/urban base-station AS — measured in degrees — is well modeled by a random variable with lognormal distribution, i.e., (3GPP, 2003; Algans *et al.*, 2002) (El Zooghby, 2005, Eqn. (3.21), p. 75)

$$AS=10^{\varepsilon x+\mu};\quad x\sim\mathcal{N}(0,1). \tag{17}$$

For numerical results shown towards the end of the chapter we consider the typical urban scenario measured in (Algans *et al.*, 2002) and described in Table 1, for which $\varepsilon = 0.47$, and $\mu = 0.74$. There, the **azimuth spread** takes predominantly small-to-moderate values, e.g., $Pr(1^\circ < AS < 20^\circ)\approx 0.8$ (3GPP, 2003, Figure 5.6) (Algans *et al.*, 2002, Figure 1).

Besides Doppler-induced temporal channel fading, subscriber station motion also causes AS variations (Algans *et al.*, 2002, Section V) (Vaughan & Andersen, 2003, Section 4.6.1). The expression for the AS correlation over distance (or, equivalently, time) was empirically determined for typical urban scenarios in (Algans *et al.*, 2002) as

$$\rho_{AS}(d)=e^{-d/d_{AS}}, \tag{18}$$

where $d$ is the distance traveled by the subscriber station, and $d_{AS}$ is the AS decorrelation distance, i.e., the distance over which the AS correlation decreases by a factor of three. This distance is typically several orders of magnitude larger than the fading coherence distance (Sampath *et al.*, 2005, p. 407) (Siriteanu & Blostein, 2007, p. 918) (Vu & Paulraj, 2006, p. 2320).

*Table 1. Typical urban scenario features*

| | |
|---|---|
| Buildings height/distribution | 4-6 floors/uniform density |
| Street layout | irregular |
| Line-of-sight | not present |
| Subscriber station – base station distance | 0.2 to 1.1 km |
| Antenna location/height | above rooftop level/32 m-high |

*Figure 1. Top: Correlation between adjacent ULA elements, ρ, computed using (16), and the eigenvalues of $\mathbf{R}_{\tilde{h}}$, for $d_n$ = 1, $\theta_c = 0^o$, and Laplacian power azimuth spectrum. Bottom: Partial sums of eigenvalues, in percentage of $tr(\mathbf{R}_{\tilde{h}})$.*

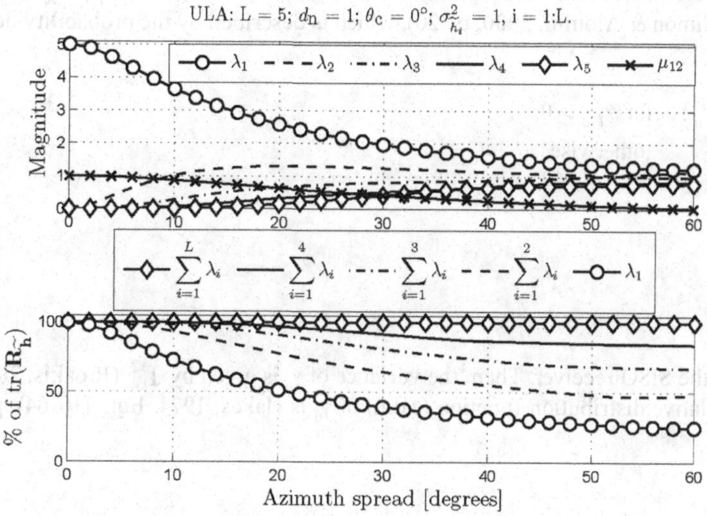


## Azimuth Spread Effect on Channel Gain Correlation

We conclude this section on signal and channel models with a numerical illustration. Consider a ULA with $L = 5$ and $d_n = 1$, and unit-variance channel gains. Then, for $\theta_c = 0^\circ$, the top subplot in Figure 1 shows the correlation, ρ, between any two adjacent antennas, computed with (16), and the eigenvalues, $\lambda_i, i = 1:L$, of the channel gain correlation matrix, $\mathbf{R}_{\tilde{h}}$. On the horizontal axis we represent the AS. The bottom subplot displays the partial eigenvalue sums,

$$\sum_{i=1}^{N} \lambda_i, N = 1:L.$$

For zero AS the channel gains are coherent and $\lambda_1 \neq 0, \lambda_{2:L} = 0$. For low AS, the channel gains are highly correlated and the received intended-signal power, proportional to $tr(\mathbf{R}_{\tilde{h}})$, is mostly concentrated along a few eigenvectors. Increasing AS yields decreasing antenna correlation, and the intended-signal power is distributed more uniformly along more eigenvectors.

# CONVENTIONAL RECEIVERS

In this section we introduce conventional receiver algorithms and study the impact of channel correlation on their performance. The numerical complexity associated with actual implementations of these algorithms is also evaluated.

## The Single-Input Single-Output (SISO) Concept

Hereafter we will use the index 0 to indicate SISO variables and performance measures. For SISO, the received signal model is similar to that shown in (3). Assuming perfectly known channel at the receiver, the combiner $w_0 = \tilde{h}_0^*$, i.e., the complex-conjugate of the actual channel gain, maximizes the conditioned SNR, i.e.,

$$\gamma_0 \overset{\Delta}{=} \frac{E_s}{N_0} |\tilde{h}_0|^2 . \tag{19}$$

Let us denote with $\sigma_0^2$ the variance of the SISO channel gain, i.e., $\sigma_0^2 = E\{|\tilde{h}_0|^2\}$.

For our assumptions of Rayleigh fading, $\gamma_0$ can be shown (Simon & Alouini, 2000, p. 18) (Proakis, 2001, p. 44) to have a central chi-square distribution (Proakis, 2001, p. 41) with 2 degrees of freedom, i.e., $\gamma_0 \sim \chi^2(2)$, or, equivalently, an exponential distribution (Simon & Alouini, 2000, p. 20), which is described by the probability density function (p.d.f.)

$$p(\gamma_0) = \begin{cases} 1/\Gamma_0 \exp(-\gamma_0/\Gamma_0) & ,\text{for } \gamma_0 \geq 0, \\ 0 & ,\text{otherwise}, \end{cases} \tag{20}$$

where

$$\Gamma_0 \overset{\Delta}{=} E\{\gamma_0\} = \frac{E_s}{N_0}\sigma_0^2$$

is the average SNR for the SISO receiver. Then, the variance of $\gamma_0$ is given by $\Gamma_0^2$ (Proakis, 2001, Eqn. (2.1-12), p. 42). Furthermore, the cumulative distribution function (c.d.f.) of $\gamma_0$ is (Jakes, 1974, Eqn. (10-64), p. 310) (Lee, 1982, Eqn. (5.2-15), p. 319)

$$P(\gamma_0) = \begin{cases} 1 - \exp(-\gamma_0/\Gamma_0) & ,\text{for } \gamma_0 \geq 0, \\ 0 & ,\text{otherwise}, \end{cases} \tag{21}$$

which measures the probability that the conditioned output SNR is smaller than a threshold, and thus represents an alternate definition of the ***outage probability*** (Simon & Alouini, 2000, p. 5).

Given the conditioned SNR at the output of an optimal combiner, $\gamma_0$, the conditioned symbol **error probability** for M–PSK can be written as (Simon & Alouini, 2000, Eqn. 8.22, p. 198)

$$P_e(\gamma_0) = \frac{1}{\pi}\int_0^{\frac{M-1}{M}\pi} \exp\left(-\gamma_0 \frac{g_{PSK}}{\sin^2\phi}\right) d\phi, \tag{22}$$

where

$$g_{PSK} \overset{\Delta}{=} \sin^2\frac{\pi}{M}.$$

Note that similar finite-limit integral expressions in exponential functions also describe other modulations (Simon & Alouini, 2000, Chapter 8).

In a fading channel, the conditioned error probability defined in (22) is a random variable, which suggests two different symbol-detection performance measures (Simon & Alouini, 2000, Section 1.1.2, p. 5): 1) the already mentioned

**outage probability (OP)**, actually defined as the probability that the conditioned error rate exceeds a given threshold; 2) the **average error probability (AEP)**, defined as the average over the fading of the conditioned error probability.

The following AEP derivation technique is taken from (Simon & Alouini, 2000, Chapter 9). By definition, the AEP is

$$P_e \stackrel{\Delta}{=} \int_0^{\infty} P_e(\gamma_0) p(\gamma_0)\, d\gamma_0 \tag{23}$$

Substituting $P_e(\gamma_0)$ from (22) in (23) yields the AEP expression as

$$P_e = \frac{1}{\pi}\int_0^{\frac{M-1}{M}\pi}\int_0^{\infty} \exp\left(-\gamma_0 \frac{g_{PSK}}{\sin^2\phi}\right) p(\gamma_0)\, d\gamma_0\, d\phi. \tag{24}$$

Since the moment generating function (m.g.f.) of $\gamma_0$ is by definition $M_{\gamma_0}(s) \stackrel{\Delta}{=} E\{e^{s\gamma_0}\}$ (Simon & Alouini, 2000, Eqn. 2.4, p. 18), the AEP can be written as

$$P_e = \frac{1}{\pi}\int_0^{\frac{M-1}{M}\pi} M_{\gamma_0}\left(-\frac{g_{PSK}}{\sin^2\phi}\right) d\phi. \tag{25}$$

For our Rayleigh fading assumptions the m.g.f. of $\gamma_0$ is given by (Simon & Alouini, 2000, Table 2.2, p. 19)

$$M_{\gamma_0}(s) = (1 - s\Gamma_0)^{-1}. \tag{26}$$

Then, Eqn. (25) yields the following simple finite-limit integral (thus, nonclosed-form) SISO AEP expression for M–PSK:

$$P_{e,SISO} = \frac{1}{\pi}\int_0^{\frac{M-1}{M}\pi}\left(1 + \Gamma_0 \frac{g_{PSK}}{\sin^2\phi}\right)^{-1} d\phi. \tag{27}$$

Similar results are possible for other modulations, as well as for Ricean and Nakagami-*m* fading (Alouini *et al.*, 2000) (Simon & Alouini, 2000, Table 9.1, p. 269).

Above we have looked at the SISO case. Hereafter, SIMO signal processing algorithms are described and analyzed.

## Maximum-Average-SNR (Statistical) Beamforming (BF) and Array Gain

***Statistical Beamforming Procedure.*** The objective of statistical beamforming (BF) is to find a combiner $\tilde{\mathbf{w}}$ for $\tilde{\mathbf{y}}$ in (1) that maximizes the SNR obtained by averaging over fading and noise. After the linear combining

$$\tilde{\mathbf{w}}^H \tilde{\mathbf{y}} = \sqrt{E_s}\, b \tilde{\mathbf{w}}^H \tilde{\mathbf{h}} + \tilde{\mathbf{w}}^H \tilde{\mathbf{n}} \tag{28}$$

the output power, averaging over noise and fading, is

$$E\{|\tilde{\mathbf{w}}^H \tilde{\mathbf{y}}|^2\} = E_s E\{|\tilde{\mathbf{w}}^H \tilde{\mathbf{h}}|^2\} + E\{|\tilde{\mathbf{w}}^H \tilde{\mathbf{n}}|^2\} = E_s \tilde{\mathbf{w}}^H \mathbf{R}_{\tilde{\mathbf{h}}} \tilde{\mathbf{w}} + N_0 \tilde{\mathbf{w}}^H \tilde{\mathbf{w}}, \tag{29}$$

which yields the average output SNR as

$$SNR_{\text{avg}}(\tilde{\mathbf{w}}) \overset{\Delta}{=} \frac{E_s}{N_0} \frac{\tilde{\mathbf{w}}^H \mathbf{R}_{\tilde{\mathbf{h}}} \tilde{\mathbf{w}}}{\tilde{\mathbf{w}}^H \tilde{\mathbf{w}}}. \quad (30)$$

The second ratio in (30) is a Rayleigh quotient (Golub & Loan, 2000), whose properties yield

$$\frac{E_s}{N_0}\lambda_L \leq SNR_{\text{avg}}(\tilde{\mathbf{w}}) \leq \frac{E_s}{N_0}\lambda_1. \quad (31)$$

The lower bound is achieved with a combiner proportional to $\mathbf{u}_L$, i.e., $\tilde{\mathbf{w}} \propto \mathbf{u}_L$. The upper bound is achieved with $\tilde{\mathbf{w}}_{BF} \propto \mathbf{u}_1$, which therefore represents maximum-average-SNR, or statistical, beamforming (BF). Coherent detection requires

$$\tilde{\mathbf{w}}_{BF} = h_1 \mathbf{u}_1, \quad (32)$$

where $h_1 \overset{\Delta}{=} \mathbf{u}_1^H \tilde{\mathbf{h}}$. Recovery of a BPSK symbol, for instance, is attempted as follows:

$$\hat{b}_{BF} = sign\left\{\Re\left[\tilde{\mathbf{w}}_{BF}^H \tilde{\mathbf{y}}\right]\right\} \quad (33)$$

*BF Performance Analysis. Array Gain.* After substituting (32) into (28) we can write the conditioned output SNR for BF as

$$\gamma_1 \overset{\Delta}{=} \frac{E_s}{N_0} |\mathbf{u}_1^H \tilde{\mathbf{h}}|^2 = \frac{E_s}{N_0} |h_1|^2. \quad (34)$$

It has been demonstrated in (Alouini *et al.*, 2000, Section 4.1) that $h_1 \sim \mathcal{N}_c(0,\lambda_1)$, so that $\gamma_1$ is exponentially distributed, with average

$$\Gamma_1 \overset{\Delta}{=} E\{\gamma_1\} = \frac{E_s}{N_0}\lambda_1. \quad (35)$$

Thus, the BF conditioned output SNR has the same distribution type as for each of the branches, but with a different average — see the SISO output SNR p.d.f. in (20). The average output SNR is an appropriate performance measure for SISO and BF because it completely describes the p.d.f. of the conditioned output SNR. Communications systems terminology typically denotes as *array gain* (Godara, 2004, Section 2.2.4) (Paulraj *et al.*, 2005, p. 91) (Vaughan & Andersen, 2003, Section 9.2.2) an increase in average SNR due to multibranch combining. The BF **array gain** is then (in decibels)

$$G_{A,BF,dB} = 10\log_{10}(\Gamma_1/\Gamma_0). \quad (36)$$

Propositions 1 and 2 indicate that

$$\Gamma_1 = \frac{E_s}{N_0}\lambda_1 \in \left[\frac{E_s}{N_0}\frac{1}{L}tr(\mathbf{R}_{\tilde{\mathbf{h}}}), \frac{E_s}{N_0}tr(\mathbf{R}_{\tilde{\mathbf{h}}})\right], \quad (37)$$

where the lower and upper bounds are achieved for uncorrelated and coherent branches, respectively. For unit-variance channel gains we have $tr(\mathbf{R}_{\tilde{\mathbf{h}}}) = L$, so that

$$G_{A,BF,dB} \in \left[0, 10\log_{10}L\right], \tag{38}$$

i.e., the BF combiner in (32) yields a maximum array gain of $G_{A,BF,dB} = 10\log_{10}L$ for coherent channel gains, and no array gain whatsoever for uncorrelated branches (Vaughan & Andersen, 2003, Sections 9.2.2.1–2, 9.3.4). Therefore, BF should only be employed for low AS.

Figure 2 shows the p.d.f.s of the conditioned SNR for SISO and for a ULA BF antenna with $L$ = 5 coherent and unit-variance channel gains. Note that the abscissa is in dB. Compared to SISO, BF yields higher probabilities for large conditioned-SNR values and lower probabilities for small conditioned-SNR values.

Now, the same approach that has lead to the SISO AEP expression from (27), produces for M–PSK and BF the AEP expression

$$P_{e,BF} = \frac{1}{\pi}\int_0^{\frac{M-1}{M}\pi}\left(1+\Gamma_1\frac{g_{PSK}}{\sin^2\phi}\right)^{-1} d\phi, \tag{39}$$

which applies for any interbranch correlation, i.e., for any AS. Clearly, larger $\Gamma_1$ improves BF symbol-detection performance. Propositions 1 and 2, along with (37), confirm that best BF performance is achieved when the channel gains are coherent, i.e., for zero AS. Note also that for $\Gamma_1 = \Gamma_0$, i.e., equal average transmitted power for BF and SISO, or for $L$ = 1, the BF AEP expression (39) reduces to the SISO AEP expression from (27).

Figure 3 plots the AEPs, computed with (39), vs. the average bit-SNR, for a BF ULA with $L$ = 5 and for SISO, for BPSK and Rayleigh fading, unit-variance, coherent channel gains ($AS$ = 0). The left-shift of the BF AEP plot by about 7 dB relative to the SISO AEP plot is due to BF array gain.

*Figure 2. The p.d.f. of the conditioned output SNR for SISO — i.e., $\gamma_0$ from (19) — and for BF of L = 5 coherent branches — i.e., $\gamma_1$ from (34) — obtained using (20) accordingly, for $\sigma_0^2 = \sigma_{\tilde{h}_i}^2 = 1$, $\forall i = 1:L$, and average bit-SNR $\gamma_b$ = 10dB.*

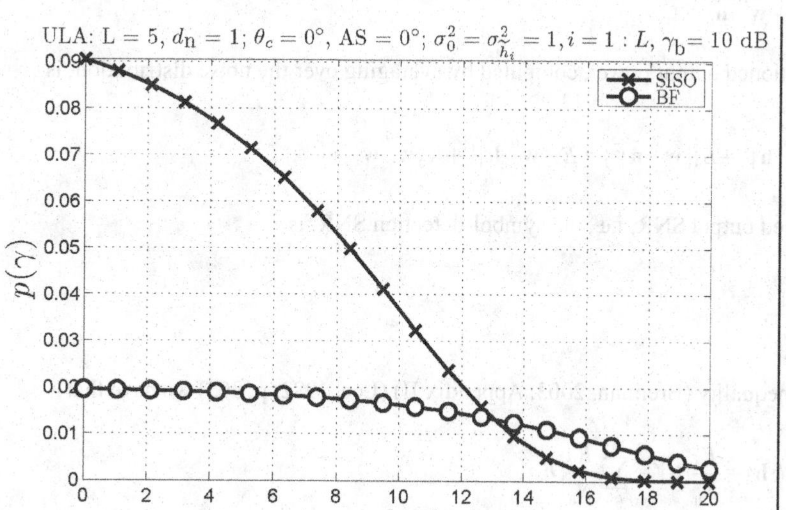

*Figure 3. AEP computed with (39) vs. average bit-SNR for BF ULA with L = 5 and for SISO, for BPSK and Rayleigh fading, coherent, unit-variance channel gains.*

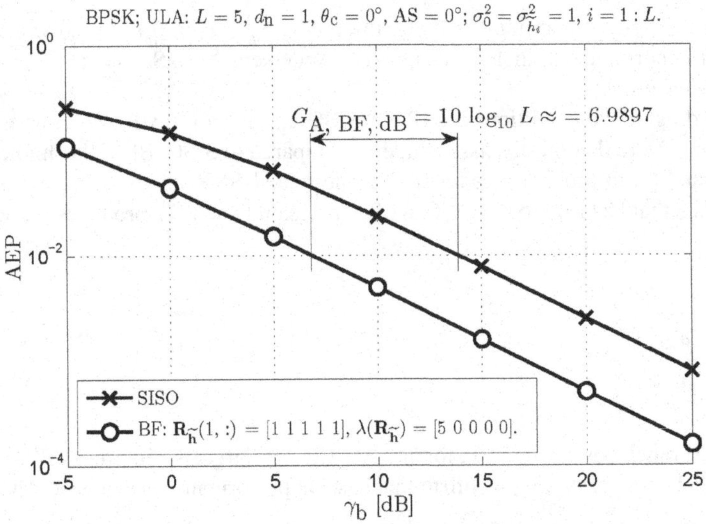


## Maximal-Ratio Combining (MRC), Amount of Fading, Diversity Gain, and Diversity Order

While BF originates in the demand to improve reception in line-of-sight links, diversity combining procedures, such as maximal-ratio combining (MRC), were devised precisely with the objective of dealing with the fading caused by rich scattering.

***MRC Procedure.*** Again, the linear combination of the received signal vector described by (1) with a weight vector $\tilde{\mathbf{w}}$ yields

$$\tilde{\mathbf{w}}^H \tilde{\mathbf{y}} = \sqrt{E_s}\, b\, \tilde{\mathbf{w}}^H \tilde{\mathbf{h}} + \tilde{\mathbf{w}}^H \tilde{\mathbf{n}}. \tag{40}$$

The combiner conditioned output power, computed by averaging over the noise distribution, is

$$E\{|\tilde{\mathbf{w}}^H \tilde{\mathbf{y}}|^2\} = E_s |\tilde{\mathbf{w}}^H \tilde{\mathbf{h}}|^2 + E\{|\tilde{\mathbf{w}}^H \tilde{\mathbf{n}}|^2\} = E_s |\tilde{\mathbf{w}}^H \tilde{\mathbf{h}}|^2 + N_0 \tilde{\mathbf{w}}^H \tilde{\mathbf{w}}, \tag{41}$$

so that the conditioned output SNR, i.e., the symbol-detection SNR, is

$$SNR(\tilde{\mathbf{w}}) \stackrel{\Delta}{=} \frac{E_s}{N_0} \frac{|\tilde{\mathbf{w}}^H \tilde{\mathbf{h}}|^2}{\tilde{\mathbf{w}}^H \tilde{\mathbf{w}}}. \tag{42}$$

From the Schwarz inequality (Brennan, 2003, Appendix II) (Lee, 1982, p. 305) we have that

$$\max_{\tilde{\mathbf{w}} \in \mathbb{C}^L} SNR(\tilde{\mathbf{w}}) = SNR(k\,\tilde{\mathbf{h}}) = \frac{E_s}{N_0} |\tilde{\mathbf{h}}|^2 = \sum_{i=1}^{L} \frac{E_s}{N_0} |\tilde{h}_i|^2 . \tag{43}$$

Since the proportionality factor $k$ from (43) does not affect the SNR, the weight vector

$$\tilde{\mathbf{w}}_{MRC} = \tilde{\mathbf{h}} \tag{44}$$

maximizes the combiner-output conditioned-SNR, which is then given by

$$\tilde{\gamma} = \sum_{i=1}^{L} \tilde{\gamma}_i, \tag{45}$$

i.e., the sum of the individual branch SNRs. This justifies the appellative of *maximal-ratio combining* (MRC) for this approach (Brennan, 2003). Maximum-likelihood recovery of a BPSK transmitted symbol, for instance, is attempted as follows:

$$\hat{b}_{MRC} = sign\left\{\Re\left[\tilde{\mathbf{w}}_{MRC}^{H}\tilde{\mathbf{y}}\right]\right\} = sign\left\{\Re\left[\tilde{\mathbf{h}}^{H}\tilde{\mathbf{y}}\right]\right\}. \tag{46}$$

Averaging in (45) over the fading then yields the **average SNR** for MRC as

$$\tilde{\Gamma} \overset{\Delta}{=} E\{\tilde{\gamma}\} = \sum_{i=1}^{L} E\{\tilde{\gamma}_i\} = \sum_{i=1}^{L} \tilde{\Gamma}_i, \tag{47}$$

which, for identically distributed channel gains, indicates that MRC achieves maximum, $L$-fold, **array gain** over SISO, i.e., $G_{A,MRC,dB} = 10\log_{10}L$ (Vaughan & Andersen, 2003, Section 9.2.2.2, Eqn. 9.2.17). Note that, unlike BF, MRC achieves this maximum array gain regardless of the channel gain correlations. This is because, unlike in BF, the MRC weights are (coherent with) the corresponding channel gains, irrespective of the correlation between channel gains. Nevertheless, the correlation between the channel gains does impact other symbol-detection performance measures for MRC, as will be shown further below.

***MRC Performance Analysis for Uncorrelated and Identically Distributed Channel Gains.*** For the sake of simplicity, many previous MRC analyses have assumed, as we do below, that the Gaussian channel gains are uncorrelated, i.e., independent. For now they are also assumed identically distributed, which implies that all branches have the same average SNR, i.e., $\tilde{\Gamma}_i = \tilde{\Gamma}_1, \forall i = 1:L$. Then, it can be shown (Proakis, 2001, Eqn. (2.1-110), p. 41) that the conditioned SNR at the MRC output, i.e., $\tilde{\gamma}$ from (45), has a chi-square distribution with $2L$ degrees of freedom, i.e., $\tilde{\gamma} \sim \chi^2(2L)$, or, equivalently, a Gamma distribution, described by the p.d.f. (Jakes, 1974, Eqn. (5.2-14), p. 319) (Lee, 1982, Eqn. (10-61), p. 310)

$$p(\tilde{\gamma}) = \begin{cases} \dfrac{1}{(L-1)!\tilde{\Gamma}_1}\left(\tilde{\gamma}/\tilde{\Gamma}_1\right)^{L-1}\exp\left(-\tilde{\gamma}/\tilde{\Gamma}_1\right) & ,\text{for } \tilde{\gamma} \geq 0, \\ 0 & ,\text{otherwise,} \end{cases} \tag{48}$$

or by the c.d.f. (Jakes, 1974, Eqn. (5.2-15), p. 319) (Lee, 1982, Eqn. (10-64), p. 310)

$$P(\tilde{\gamma}) = \begin{cases} 1 - \exp\left(-\tilde{\gamma}/\tilde{\Gamma}_1\right)\sum_{l=1}^{L}\dfrac{1}{(l-1)!}\left(\tilde{\gamma}/\tilde{\Gamma}_1\right)^{l-1} & ,\text{for } \tilde{\gamma} \geq 0, \\ 0 & ,\text{otherwise.} \end{cases} \tag{49}$$

***Evaluating Diversity Combining Performance: Deep-Fade Probability, Amount of Fading.*** Receiver performance is determined by the probability of a deep fade, i.e., the probability of subunitary conditioned output SNR value given a high $\tilde{\Gamma}_1$ value (Tse & Viswanath, 2005, p. 55). The above c.d.f. indicates that $Pr(\tilde{\gamma} \leq 0\, dB \,|\, \tilde{\Gamma}_1 = 30\, dB)$ is $10^{-3}$ for SISO (equivalent to MRC with $L = 1$), $10^{-6.3}$ for MRC of $L = 2$ branches, and only $10^{-9.8}$ for MRC of $L = 3$ branches. Although MRC yields full array gain, this performance improvement is largely due to fading severity reduction, which is explained next.

Fading severity is quantifiable analytically independently of $\tilde{\Gamma}_1$ and the threshold SNR by the ***amount of fading* (AF)**, which, for a generic combiner with conditioned output SNR denoted as $\tilde{\gamma}$, is given by (Simon & Alouini, 2000, Eqn. 2.5, p. 18)

$$AF \stackrel{\Delta}{=} \frac{\operatorname{var}(\tilde{\gamma})}{\left(E\{\tilde{\gamma}\}\right)^2} = \frac{E\{\tilde{\gamma}^2\} - \left(E\{\tilde{\gamma}\}\right)^2}{\left(E\{\tilde{\gamma}\}\right)^2}. \tag{50}$$

The Rayleigh SISO case yields $AF_{SISO} = 1$. On the other hand, using (47) and (Proakis, 2001, Eqn. (2.1-12), p. 42) it can be shown that, for MRC of $L$ branches with independent and identically distributed (i.i.d.) Rayleigh fading channel gains,

$$AF_{MRC} = \frac{1}{L} \tag{51}$$

i.e., MRC reduces fading severity $L$-fold.

When comparing combining methods employing a different number of branches, such as MRC and SISO, the variance of the output SNR will capture solely the fading-reducing effect of diversity combining if we require that the combining methods yield the same average output SNR (Proakis, 2001, Eqn. (14.4-34), p. 829). As mentioned earlier, for Rayleigh fading, given the SISO average SNR $\Gamma_0$, the variance of the SISO conditioned SNR is $\Gamma_0^2$. If for MRC with i.i.d. channel gains we set $\tilde{\Gamma}_i = \Gamma_0/L,\ \forall i = 1:L$ — so that $\Gamma_0 = \sum_{i=1}^{L} \tilde{\Gamma}_i$ — then the variance of the MRC conditioned output SNR can be shown to be $\Gamma_0^2 / L$, which indicates an $L$-fold fading severity reduction over SISO.

For this situation, Figure 4 shows the p.d.f. of the conditioned output SNR, computed with (48), for SISO and for MRC with $L$ = 5,9,13,17 i.i.d. branches, given the average output SNR value $\Gamma_0 = \sum_{i=1}^{L} \tilde{\Gamma}_i$dB. The peak of the p.d.f. increases with $L$, whereas the standard deviation decreases, i.e., fading severity diminishes. Figure 4 suggests that

$$\lim_{L\to\infty} p(\tilde{\gamma}) = \begin{cases} 1 & ,\text{for } \tilde{\gamma} = \Gamma_0, \\ 0 & ,\text{otherwise,} \end{cases} \tag{52}$$

which describes the symbol-detection SNR for a SISO nonfading channel.

***MRC Performance Analysis for Uncorrelated and Nonidentically Distributed Channel Gains.*** Let us now derive the MRC AEP for the more encompassing case of uncorrelated channel gains with possibly unequal variances. Using (45) along with the channel gain independence property, the m.g.f. of the conditioned output SNR for MRC can be written as (Simon & Alouini, 2000, Eqn. (9.11), p. 269)

$$M_{\tilde{\gamma}}(s) = \prod_{i=1}^{L} M_{\tilde{\gamma}_i}(s) = \prod_{i=1}^{L} \left(1 - s\tilde{\gamma}_i\right)^{-1}. \tag{53}$$

The AEP derivation procedure used earlier for SISO yields in this case the following AEP expression (Simon & Alouini, 2000, Section 9.2.3.2)

$$P_{\text{e,MRC,indep.}} = \frac{1}{\pi} \int_0^{\frac{M-1}{M}\pi} \prod_{i=1}^{L} \left(1 + \tilde{\Gamma}_i \frac{g_{PSK}}{\sin^2\phi}\right)^{-1} d\phi. \tag{54}$$

For MRC of i.i.d. channel gains, i.e., when $\sigma_{\tilde{h}_i}^2 = \sigma_{\tilde{h}_1}^2,\ \forall i = 1:L$, the above becomes

$$P_{\text{e,MRC,i.i.d.}} = \frac{1}{\pi} \int_0^{\frac{M-1}{M}\pi} \left(1 + \tilde{\Gamma}_1 \frac{g_{PSK}}{\sin^2\phi}\right)^{-L} d\phi. \tag{55}$$

*Figure 4. The p.d.f. of the conditioned output SNR for SISO and for MRC of $L = 5,9,13,17$ i.i.d branches, with equal output average SNR, $\Gamma_0 = \sum_{i=1}^{L} \widetilde{\Gamma}_i = 10$ dB.*

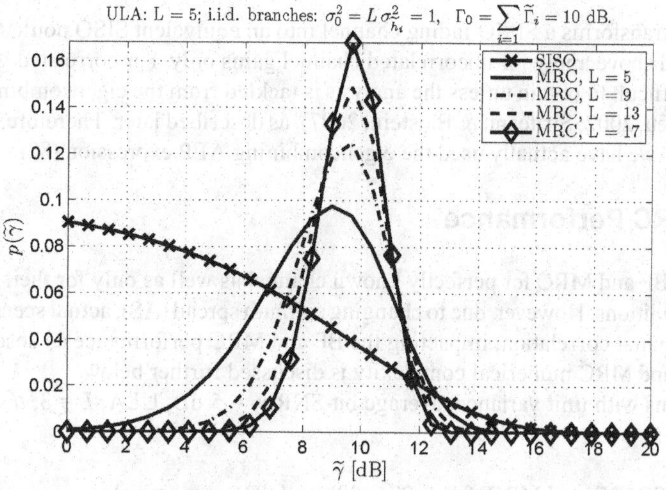


***Diversity gain.*** For BPSK transmitted signal and MRC of $L = 1{:}5$ i.i.d. branches with $\sigma_{h_i}^2 = 1$, $i = 1 : L$, Figure 5 displays the AEP computed using (55), vs. the average bit-SNR. For $AEP = 10^{-2}$, MRC with $L = 5$ outperforms SISO by about 15 dB. Of this, about 7 dB is array gain due to coherent combining, i.e., $G_{A,MRC,dB} = 10\log_{10} L$. The remaining 8 dB gain is due to fading severity reduction, and is called ***diversity gain*** (Paulraj *et al.*, 2005, Section 5.2, p. 86). Note that the diversity gain increases with lower AEP and with increasing $L$. Nevertheless, the diversity gain gradient diminishes as the number of branches increases. This phenomenon and the cost associated with a diversity branch will in practice limit the feasible number of antenna elements.

***Diversity order.*** For BPSK and independent unit-variance Rayleigh fading channel gains, the MRC AEP expression from (55) can be approximated at high average bit-SNR as (Siriteanu & Blostein, 2004, Eqn. (12))

$$P_{\text{e,MRC,i.i.d.}} \approx \frac{(2L)!}{2^{2L+1}(L!)^2}\left(\frac{E_s}{N_0}\right)^{-L}. \tag{56}$$

The SNR exponent in such asymptotic AEP expressions, i.e., the high-SNR AEP curve slope, is commonly referred to as *diversity order* (Paulraj *et al.*, 2005, Section 5.2) (Vaughan & Andersen, 2003, Section 9.3.4). Thus, (56) indicates that MRC has diversity order equal to the number of i.i.d. combined branches. Since

$$10\log_{10} P_{\text{e,MRC,i.i.d.}} \propto -L\left[\frac{E_s}{N_0}\right]_{\text{in dB}}, \tag{57}$$

a 10 dB SNR increase will decrease the AEP by a factor of $10^L$ at high SNR, which is confirmed by Figure 5.

On the other hand, for a given value of $tr(\mathbf{R}_{\widetilde{\mathbf{h}}})$, manipulation of (55) yields

$$\lim_{L\to\infty} P_{\text{e,MRC,i.i.d.}} = \frac{1}{\pi}\int_0^{\frac{M-1}{M}\pi} \exp\left\{-\frac{E_s}{N_0} tr(\mathbf{R}_{\widetilde{\mathbf{h}}})\frac{g_{PSK}}{\sin^2\phi}\right\} d\phi, \tag{58}$$

which, based on (22), is the error probability for a SISO nonfading channel with SNR equal to

$$\frac{E_s}{N_0} tr\left(\mathbf{R}_{\bar{\mathbf{h}}}\right).$$

Thus, diversity combining transforms a SIMO fading channel into an equivalent SISO nonfading channel.

The expressions derived above apply for uncorrelated channel gains only. For correlated gains, MRC performance measure expressions are difficult to obtain unless the analysis is tackled from the eigencombining perspective (Alouini *et al.*, 2000; Dong & Beaulieu, 2002; Siriteanu & Blostein, 2007), as described later. Therefore, for the numerical results depicted hereafter for MRC we have actually used the eigencombining AEP expression (82).

## Realistic BF and MRC Performance

Above, we have evaluated BF and MRC for perfectly known channel as well as only for their respective performance-maximizing correlation conditions. However, due to changing azimuth spread (AS), actual scenarios can feature variable and possibly unfavorable spatial correlation, impacting the BF and MRC performance as described next. The effect of channel estimation on BF and MRC numerical complexity is discussed further below.

For QPSK, channel gains with unit variance, average bit-SNR $\gamma_b$ = 5 dB, ULA, $L$ = 3, $d_n$ = 1, and $\theta_c$ = 0, Figure 6 shows vs. AS the following:

- **Top:** the AEPs for SISO, BF, and MRC, from (27), (39), and (82), respectively.
- **Bottom:** the correlations between adjacent and nonadjacent ULA elements, denoted as $\rho_{1,2}$ and $\rho_{1,3}$, respectively; the eigenvalues $\lambda_i, i = 1 : L$, of the channel correlation matrix $\mathbf{R}_{\bar{\mathbf{h}}}$.

Note that BF performance is degrading with increasing AS because this decreases the correlation between channel gains, which distributes intended-signal power along an increasing number of eigenvectors of $\mathbf{R}_{\bar{\mathbf{h}}}$. Thus, the correlation decreases between the applied BF weights from (32) and the channel gains from (3), which reduces the BF array gain. For adjacent channel gain correlation lower than 0.9, BF can be greatly outperformed by MRC. At very high AS, BF yields SISO-like performance. On the other hand, MRC performance improves with increasing AS because the diversity gain increases with decreasing correlation between the channel gains. MRC nears its best performance for adjacent element correlation lower than about 0.5, which is achieved for $AS > 25^\circ$ in this case of $\lambda_c / 2$-spaced antenna elements. Recall that in typical urban scenarios the AS is random with $Pr(1^\circ < AS < 20^\circ) \approx 0.8$. Thus, although MRC may significantly outperform BF, it only seldom achieves maximum diversity gain.

*Figure 5. SISO and MRC AEP for BPSK transmitted signal and Rayleigh i.i.d. unit-variance channel gains; the high-SNR slope reveals the diversity order*

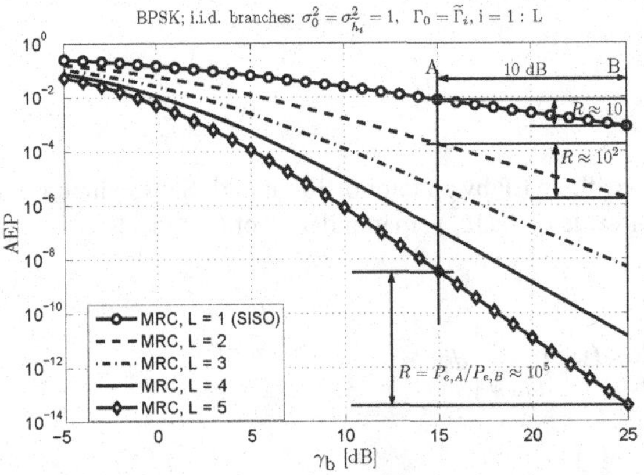

## Realistic BF and MRC Numerical Complexity

Let us now discuss our assumptions about the channel knowledge at the receiver, and the numerical complexity incurred to obtain this knowledge and to implement statistical beamforming (BF) and maximal-ratio combining (MRC). The factors affecting the spatial correlation typically change much more slowly than the Doppler-shift-induced temporal fading—see the earlier discussion on azimuth spread temporal correlation. Then, a sufficient number of uncorrelated (Goldberg & Fonollosa, 1998, Section 7.2) samples of the received signal vector are available for accurate channel eigendecomposition updating (Alouini, Scaglione, & Giannakis, 2000, Eqn. (15)) (Brunner *et al.*, 2001, Sections 4.2, 4.3) (Goldberg & Fonollosa, 1998, Eqn. (28)) (Vu & Paulraj, 2006). Therefore, throughout this chapter we assume perfectly known spatial fading eigendecomposition described by (13). Note further that the updating operations can be distributed over long intervals and thus do not significantly increase the per-symbol computational volume (Siemens, 2000).

Although the symbol-detection performance is analyzed for perfectly known channel gains in this chapter, similar procedures readily extend to optimum combining implementation given channel gain estimates and statistical knowledge about the fading and noise, as shown in (Siriteanu & Blostein, 2007, Section III.D). The effect of channel estimation on the performance of a suboptimal, but more practical, implementation, which disregards fading and noise statistics in combiner weight computation, is analyzed in (Siriteanu & Blostein, 2004) (Siriteanu & Blostein, 2007, Appendix). These references contain comprehensive information on the performance of BF and MRC for estimated channel and optimum and suboptimum implementation. Below we evaluate only the numerical complexity incurred by channel estimation and combining implementation in BF and MRC.

Actual implementation of MRC requires the estimation of $\tilde{h}_i$, $i = 1{:}L$, whereas BF only requires the estimation of $h_1$ if $\mathbf{u}_1$ is perfectly known. These time-varying factors are typically estimated using pilot-symbol-aided modulation (PSAM) at the transmitter and pilot-sample interpolation at the receiver. Pilot symbols are embedded in each slot of the transmitted stream. At the receiver, $T$ pilot samples closest to the symbol for which the fading gain is estimated are used for interpolation. Considered herein are the following typical interpolators: 1) a simple but suboptimal filter whose time-response approaches the sampling function, i.e., $sinc(x) = \sin(\pi x) / (\pi x)$; the estimation method employing this interpolation vector is hereafter denoted as SINC PSAM; 2) the more complex but optimal minimum mean-squared-error approach; the estimation method employing this interpolation method is hereafter denoted as MMSE PSAM. Detailed descriptions of SINC and MMSE PSAM appear in (Siriteanu & Blostein, 2004, Section III.B) (Siriteanu *et al.*, 2006, Section 2.6) (Siriteanu, 2006, Sections 2.5, 3.6).

*Figure 6. Top: AEP performance for QPSK, unit-variance channel gains, and average bit-SNR $\gamma_b = 5$ dB, for SISO and for BF and MRC ULA with $L = 3$; Bottom: Correlations between the channel gains corresponding to adjacent and extreme antenna elements, and the eigenvalues of $\mathbf{R}_{\tilde{\mathbf{h}}}$.*

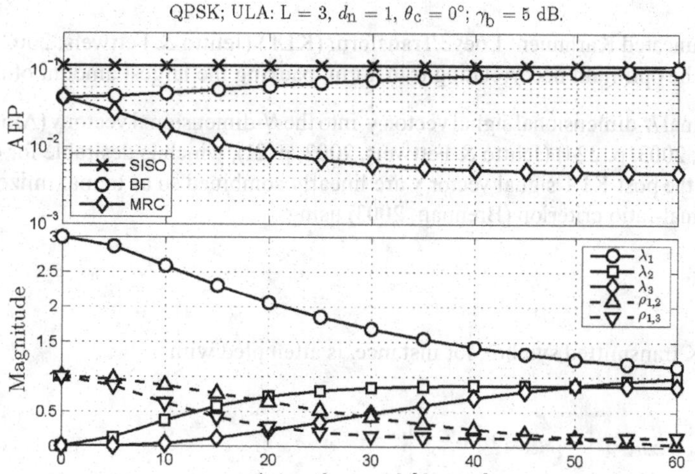

Table 2 reproduces from (Siriteanu & Blostein, 2007, Table II) the numerical complexities of BF and MRC in terms of the number of complex multiplications and additions required per detected symbol for SINC and MMSE PSAM channel estimation and optimum or suboptimum combining implementations. Unlike the performance, the **numerical complexity** of MRC and BF is independent of correlation conditions. The right-most column in Table 2, which displays the MRC-vs.-BF complexity ratio for $L = 4$ and $T = 11$, indicates that MRC can be much more complex than BF. For instance, for SINC PSAM and suboptimum combining, MRC is 3 times more complex than BF. For the same settings, comparisons shown in (Siriteanu *et al.*, 2006) revealed that a field-programmable gate array (FPGA) chip programmed with a fixed-point MRC design consumes about 3 times more dynamic power than its BF counterpart. On the other hand, MRC is more than 11 times more complex than BF for MMSE PSAM because this approach requires joint estimation of all the channel gains using matrix operations (Siriteanu, 2006, Section 3.6.2). For SINC PSAM the channel gains are estimated separately, using only vector inner products (Siriteanu & Blostein, 2004, Eqn. (42)) (Siriteanu, 2006, Eqn. (3.110), p. 83).

The above performance and complexity comparisons indicate that BF and MRC should only be deployed for high (> 0.9) and low (< 0.5) correlation values, respectively. However, as mentioned earlier, actual values predominantly fall between these extremes and are slowly time-varying. Required then is a more flexible approach, hereafter entitled *eigencombining,* that jointly exploits BF and MRC principles, as described next (Siemens, 2000; Alouini *et al.*, 2000; Brunner *et al.*, 2001; J. Choi & Choi, 2003; F. A. Dietrich & Utschick, 2003; Hottinen *et al.*, 2003; Jelitto & Fettweis, 2002; Siriteanu & Blostein, 2007).

## MAXIMAL-RATIO EIGENCOMBINING (MREC): SUPERSET OF BF AND MRC

**Maximal-ratio eigencombining (MREC)** of order $N$, $1 \le N \le L$, denoted hereafter with MREC$_N$, consists of two steps:

- The $L \times N$, full column rank, matrix $\mathbf{U}_N \stackrel{\Delta}{=} [\mathbf{u}_1 \quad \mathbf{u}_2 \quad \ldots \quad \mathbf{u}_N]$ comprising the $N$ dominant eigenvectors of $\mathbf{R}_{\tilde{\mathbf{h}}}$ transforms the received signal vector $\tilde{\mathbf{y}}$ from (1) into the $N$-dimensional vector

$$\mathbf{y} = \sqrt{E_s}\, b\mathbf{h} + \mathbf{n}, \tag{59}$$

  where

$$\mathbf{y} \stackrel{\Delta}{=} \mathbf{U}_N^H \tilde{\mathbf{y}}, \quad \mathbf{h} \stackrel{\Delta}{=} \mathbf{U}_N^H \tilde{\mathbf{h}}, \quad \mathbf{n} \stackrel{\Delta}{=} \mathbf{U}_N^H \tilde{\mathbf{n}}. \tag{60}$$

  This represents a truncated Karhunen–Loeve Transform (KLT) (Jelitto & Fettweis, 2002, p. 21) (Hottinen *et al.*, 2003, p. 224), i.e., the optimum decorrelating transform, packing the largest amount of average intended-signal power from the original $L$-dimensional signal vector $\tilde{\mathbf{y}}$ into the $N$-dimensional vector $\mathbf{y}$ (Alouini *et al.*, 2000, Section 3.1) (Hottinen *et al.*, 2003, p. 224) (Jelitto & Fettweis, 2002, p. 21), which is desirable for dimension reduction.
- The components of the post-KLT signal vector $\mathbf{y}$ are linearly combined so as to maximize the conditioned output SNR (i.e., the maximal-ratio criterion (Brennan, 2003) using

$$\mathbf{w}_{MREC} = \mathbf{h}. \tag{61}$$

  Recovery of a BPSK transmitted symbol, for instance, is attempted with

$$\hat{b}_{MREC} = sign\{\Re\left[\mathbf{w}_{MREC}^H \mathbf{y}\right]\} = sign\{\Re\left[\mathbf{h}^H \mathbf{y}\right]\}. \tag{62}$$

### *Eigencombining*

Since eigencombining unifies the core principles of statistical beamforming and diversity combining, MREC is referred to as a superset of BF and MRC. Actually, $\text{MREC}_{N=1}$ represents **statistical beamforming (BF)**, whereas **maximal-ratio combining (MRC)** is equivalent with *full MREC*, i.e., $\text{MREC}_{N=L}$ (Alouini *et al.*, 2000, Section 4.2) (Dong & Beaulieu, 2002, Section II.A) (Siriteanu, 2006, Section 3.9.3) (Siriteanu & Blostein, 2007, Section III.D.2). Thus, any performance measure expression derived for MREC can be specialized to BF and MRC.

The components of $\mathbf{y}$ from (59), denoted hereafter as *eigenbranches*, are given by

$$y_i \stackrel{\Delta}{=} \mathbf{u}_i^H \,\tilde{\mathbf{y}} = \sqrt{E_s}\; b\, h_i + n_i, \forall i = 1:N, \tag{63}$$

where

$$h_i \stackrel{\Delta}{=} \mathbf{u}_i^H \,\tilde{\mathbf{h}}, \quad n_i \stackrel{\Delta}{=} \mathbf{u}_i^H \,\tilde{\mathbf{n}}. \tag{64}$$

The components of $\mathbf{h}$, $h_i$, $i = 1 : N$, are hereafter referred to as channel *eigengains*. For our assumptions, the eigengains have zero mean. They are mutually uncorrelated because

$$E\{h_i h_j^*\} = \mathbf{u}_i^H\, E\{\tilde{\mathbf{h}}\tilde{\mathbf{h}}^H\}\,\mathbf{u}_j = \mathbf{u}_i^H\, \mathbf{R}_{\tilde{\mathbf{h}}}\, \mathbf{u}_j = \mathbf{u}_i^H \left(\sum_{l=1}^{L} \lambda_l \mathbf{u}_l \mathbf{u}_l^H\right) \mathbf{u}_j = \sum_{l=1}^{L} \lambda_l \mathbf{u}_i^H\, \mathbf{u}_l \mathbf{u}_l^H\, \mathbf{u}_j = 0, \quad \forall i \neq j, \tag{65}$$

and have autocorrelations (variances) given by

$$\sigma_{h_i}^2 \stackrel{\Delta}{=} E\{|h_i|^2\} = \mathbf{u}_i^H\, E\{\tilde{\mathbf{h}}\tilde{\mathbf{h}}^H\}\mathbf{u}_i = \mathbf{u}_i^H \left[\sum_{l=1}^{L} \lambda_l \mathbf{u}_l \mathbf{u}_l^H\right] \mathbf{u}_i = \sum_{l=1}^{L} \lambda_l \mathbf{u}_i^H\, \mathbf{u}_l \mathbf{u}_l^H \mathbf{u}_i = \lambda_i, \tag{66}$$

for any fading distribution (Alouini *et al.*, 2000). Thus,

$$\mathbf{R}_{\mathbf{h}} \stackrel{\Delta}{=} E\{\mathbf{h}\mathbf{h}^H\} = \mathbf{\Lambda}_N = diag\{\lambda_i\}_{i=1}^N. \tag{67}$$

Initial assumptions of Rayleigh fading and ZMCSCG noise yield

$$\mathbf{h} \sim \mathcal{N}_C(\mathbf{0}, \mathbf{\Lambda}_N), \tag{68}$$

so that the eigengains are independent, and

$$\mathbf{n} \sim \mathcal{N}_c(\mathbf{0}, N_0 \mathbf{I}_N), \tag{69}$$

so that the post-KLT noise is spatially (as well as temporally) white.

As mentioned earlier, $\mathbf{\Lambda}_L$ and $\mathbf{U}_L$ are hereafter assumed perfectly known. For the performance analysis shown below, the eigengains are also assumed perfectly known.

*Table 2. Per-symbol* **numerical complexity** *(no. of complex multiplications/additions)*

| Implementation | Interpolation | MRC | BF | Ratio, for $L = 4$, $T = 11$ |
|---|---|---|---|---|
| Suboptimum | SINC | $L(T+1)$ | $(L+T+1)$ | 3 |
| | MMSE | $L(LT+1)$ | $(L+T+1)$ | 11.25 |
| Optimum | SINC | $L(L+T+1)$ | $(L+T+2)$ | 3.75 |
| | MMSE | $L(LT+L+1)$ | $(L+T+2)$ | 11.5 |

## Conditioned Output SNR

The conditioned SNR of the $i$th eigenbranch described by (63) is given by

$$\gamma_i \stackrel{\Delta}{=} \frac{E_s}{N_0} |h_i|^2, \quad i = 1 : N. \tag{70}$$

Since $h_i \sim \mathcal{N}_c(0, \lambda_i)$, the distribution of $\gamma_i$ is $\chi^2(2)$ or exponential, with average

$$\Gamma_i \stackrel{\Delta}{=} E\{\gamma_i\} = \frac{E_s}{N_0}\lambda_i$$

and variance $\Gamma_i^2$. The p.d.f., c.d.f., and m.g.f. for the eigenbranch conditioned SNRs have expressions analogous to those for SISO shown in (20), (21), and (26), respectively. Importantly, $\gamma_i$, $i = 1 : N$, are mutually independent.

As done for MRC in the derivation of (45), it can be shown for MREC that $\mathbf{w}_{MREC}$ from (61) maximizes the **conditioned output SNR**, i.e.,

$$\gamma = \sum_{i=1}^{N} \gamma_i, \tag{71}$$

justifying the title of *maximal-ratio eigencombining.*

## Array Gain and Amount of Fading

The MREC **average output SNR** is then

$$\Gamma \stackrel{\Delta}{=} E\{\gamma\} = \sum_{i=1}^{N} \Gamma_i. \tag{72}$$

The **array gain**—defined in (36), for BF—can be written as follows for MREC and identically distributed channel gains, i.e, for $\sigma_{\tilde{h}_i}^2 = \sigma_0^2 = \frac{1}{L} tr\mathbf{R}_{\tilde{\mathbf{h}}}$, $i = 1 : L$:

$$G_{A,MREC,N,dB} \stackrel{\Delta}{=} 10\log_{10}\frac{\Gamma}{\Gamma_0} = 10\log_{10}\left[L\left(\frac{\sum_{i=1}^{N}\lambda_i}{\sum_{i=1}^{L}\lambda_i}\right)\right] \in [10\log_{10}N, 10\log_{10}L]. \tag{73}$$

The lower and upper bounds are attained for uncorrelated and coherent channel gains, respectively. For BF, i.e., for $MRC_{N=1}$, Eqn. (73) yields $G_{A,BF,dB} \in [0, 10\log_{10}L]$, i.e., (38). For MRC, i.e., for $MREC_{N=L}$, Eqn. (73) yields $G_{A,MRC,dB} = 10\log_{10}L$, irrespective of the channel gain correlations, as already discussed.

To examine the fading-reduction capabilities of MREC, the **amount of fading** (AF)—defined in (50)—is now computed for the MREC output. Using the statistical independence of the individual eigenbranch SNRs, the variance of the MREC conditioned output SNR from (71) can be found as:

$$\sum_{i=1}^{N} \Gamma_i^2.$$

Then, for the MREC AF we can write:

$$\frac{1}{N} \leq AF_{MREC,N} = \frac{\sum_{i=1}^{N} \Gamma_i^2}{\left(\sum_{i=1}^{N} \Gamma_i\right)^2} = \frac{\sum_{i=1}^{N} \lambda_i^2}{\left(\sum_{i=1}^{N} \lambda_i\right)^2} \leq 1. \tag{74}$$

According to Cauchy's inequality (Abramowitz & Stegun, 1995, §3.2.9, p. 11), the lower bound is achieved for identically distributed eigenbranches, i.e., when $\lambda_i = \lambda$, $\forall\, i = 1 : N$. For $N = L$, Proposition 2 indicates that the lower bound is achieved for i.i.d. channel gains, confirming that MRC of $L$ branches reduces the fading severity vs. SISO by a factor of $L$. The upper bound in (74) is achieved when $\lambda_1 \neq 0$ and $\lambda_i = 0$, $\forall\, i = 2 : L$, i.e., for coherent channel gains. For $N = 1$, Eqn. (74) confirms that BF does not reduce fading.

## Conditioned Output SNR Probability Density Function

Since the eigengains are independent, Eqns. (71) and (26) readily yield the m.g.f. of $\gamma$ as

$$M_\gamma(s) = \prod_{i=1}^{N} \frac{1}{1 - s\Gamma_i}. \tag{75}$$

To simplify subsequent mathematical manipulation, let us hereafter employ the Laplace transform (Abramowitz & Stegun, 1995, §29.1.1, p. 1020) of the p.d.f. of $\gamma$, i.e., $F_\gamma(s) \stackrel{\Delta}{=} M_\gamma(-s)$, which we refer to as the *reversed moment generating function* (r.m.g.f.) of $\gamma$.

Eigenvalues of $\mathbf{R}_{\tilde{\mathbf{h}}}$ can become (nearly) equal, e.g., for certain AS values, as shown in Figures 1 and 6. Therefore, let $\{\Xi_1, \Xi_2, \ldots, \Xi_{N_d}\}$ denote the distinct values in the set of eigenbranch average SNRs $\{\Gamma_1, \Gamma_2, \ldots, \Gamma_N\}$. Then, Eqn. (75) yields

$$F_\gamma(s) = \prod_{k=1}^{N_d} \frac{1}{(1 + s\Xi_k)^{r_k}}, \tag{76}$$

where $r_k$ represent the algebraic multiplicities of $\Xi_k$, $k = 1{:}N_d$, with

$$\sum_{k=1}^{N_d} r_k = N.$$

The r.m.g.f. from (76) can be expressed as (Siriteanu, 2006, Eqn. (3.165), p. 114)

$$F_\gamma(s) = \frac{1}{A} \sum_{k=1}^{N_d} \sum_{l=1}^{r_k} c_{k,l} \frac{1}{\left(s + \frac{1}{\Xi_k}\right)^l}, \tag{77}$$

where $A \stackrel{\Delta}{=} \prod_{k=1}^{N_d} \Xi_k^{r_k} = \prod_{i=1}^{N} \Gamma_i$, and the factor $c_{k,l}$ is given by (Siriteanu, 2006, Eqn. (3.166), p. 114)

$$c_{k,l} = (-1)^{r_k - l} \cdot \sum_{\Omega_l} \prod_{\substack{j=1 \\ j \neq k}}^{N_d} d_j \cdot \left(\frac{1}{\Xi_j} - \frac{1}{\Xi_k}\right)^{-(r_j + i_j)}, \tag{78}$$

for $k = 1: N_d$, $l = 1: r_k$, where $\Omega_i$ stands for the set of integers

$$\{i_j, j = 1: N_d, j \neq k \mid 0 \leq i_j \leq r_k - l, \sum_{\substack{j=1 \\ j\neq k}}^{N_d} i_j = r_k - l\},$$

and

$$d_j = \frac{(r_j - 1 + i_j)!}{(r_j - 1)! i_j!}$$

is the binomial coefficient.

The inverse Laplace transform of (77) yields for the p.d.f. of the MREC **conditioned output SNR** $\gamma$ the following closed-form (Siriteanu, 2006, Eqn. (3.169), p. 115)

$$p(\gamma) = \frac{1}{A} \sum_{k=1}^{N_d} \sum_{l=1}^{r_k} c_{k,l} \cdot \frac{\gamma^{l-1} e^{-\gamma/\Xi_k}}{(l-1)!}, \tag{79}$$

by using (Abramowitz & Stegun, 1995, §29.3.10, p. 1022). This expression applies for correlated branches even when some eigenvalues are equal. Furthermore, for $L = 1$ (79) describes SISO, for $N = 1$ it describes BF, and for $N = L$ it describes MRC. When all eigenvalues are distinct, this reduces to (Lee, 1982, Eqn. 10-60, p. 308).

Figure 7 shows conditioned-SNR p.d.f. plots — obtained using (79) — for SISO and a $\text{MREC}_{N=1:L}$ ULA with $L$= 5, for equal average output SNR of 10 dB, and Laplacian p.a.s. with $\theta_c = 0$ and $AS = 10^\circ$. The plots indicated with MREC, N = 1, and MREC, N = 5, represent BF and MRC, respectively. Since the AS is nonzero, $\text{MREC}_{N=2:L}$ benefits from diversity gain, which changes the p.d.f. plot shape compared to SISO and BF (i.e., the p.d.f. peaks at non-zero SNR). Note that the p.d.f.s for $\text{MREC}_{N=3:L}$ nearly overlap, suggesting that MRC-like performance is achievable with low-order MREC.

## Outage Probability

The c.d.f. of $\gamma$ can now be obtained from (79) as

$$P(\gamma) = \frac{1}{A} \sum_{k=1}^{N_d} \sum_{l=1}^{r_k} c_{k,l} \Xi_k^l \left[1 - e^{-\gamma/\Xi_k} \sum_{n=1}^{l} \frac{(\gamma/\Xi_k)^{n-1}}{(n-1)!}\right]. \tag{80}$$

A widely-applicable outage probability (OP) closed-form expression for MREC (BF, MRC), given the threshold SNR $\gamma_{th}$, results then from

$$P_o \stackrel{\Delta}{=} Pr(\gamma \leq \gamma_{th}) = P(\gamma_{th}). \tag{81}$$

## Average Error Probability

The procedure used to derive the MRC AEP expression from (54) for uncorrelated branches, along with (75) and (76), yields the following MREC AEP expression for M–PSK (Alouini *et al.*, 2000, Eqn. (21)) (Siriteanu, 2006, Eqn. (3.151), p. 109):

$$P_{\text{e,MREC,N}} = \frac{1}{\pi} \int_0^{\frac{M-1}{M}\pi} F_\gamma\left(\frac{g_{PSK}}{\sin^2\phi}\right) d\phi = \frac{1}{\pi} \int_0^{\frac{M-1}{M}\pi} \prod_{i=1}^{N} \left(1 + \Gamma_i \frac{g_{PSK}}{\sin^2\phi}\right)^{-1} d\phi \tag{82}$$

*Figure 7. The p.d.f. of the conditioned output SNR for SISO and for a* $MREC_{N=1:L}$ *ULA with* $L = 5$*, for* $AS = 10^\circ$ *and* $\Gamma_0 = \sum_{i=1}^{L} \tilde{\Gamma}_i = 10$ *dB.*

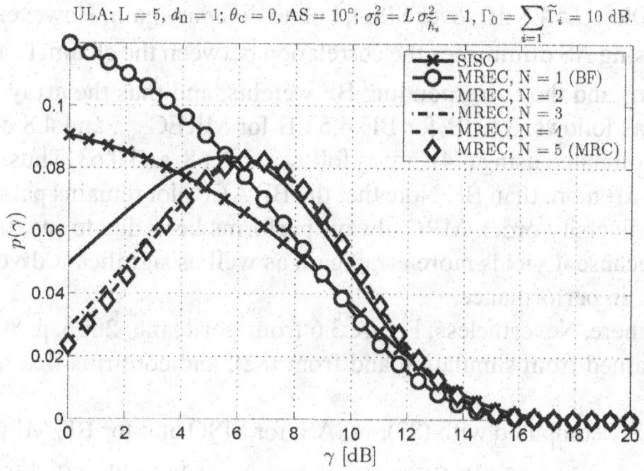


This nonclosed-form, finite-limit integral expression for the average **error probability** can be easily evaluated numerically. Now, using (77), the above can be recast as (Siriteanu, 2006, Eqn. (3.167), p. 114)

$$P_{\mathrm{e,MREC,N}} = \frac{1}{A}\sum_{k=1}^{N_d}\sum_{l=1}^{r_k} c_{k,l} \cdot \Xi_k^l \cdot I_l(\Xi_k), \tag{83}$$

where (Siriteanu, 2006, Eqn. (3.168), p. 114)

$$I_l(\Xi_k) \overset{\Delta}{=} \frac{1}{\pi}\int_0^{\frac{M-1}{M}\pi}\left[1+\Xi_k \frac{g_{PSK}}{\sin^2\phi}\right]^{-l} d\phi \tag{84}$$

is expressible in closed-form as in (Siriteanu, 2006, Section 3.10.2.1) based on (Simon & Alouini, 2000, Eqns. (5A.17–19), pp. 127–128). Note that (83) is not as straightforward to program on a computer as (82), because the factors $c_{k,l}$ given by (78) depend on the relative magnitudes of the eigenvalues of $\mathbf{R}_{\tilde{\mathbf{h}}}$.

Bear in mind that all the performance measure expressions derived above for MREC apply to SISO, BF, and MRC for $L = 1$, $N = 1$, and $N = L$, respectively. For correlated channel gains, the analysis of MRC based on its equivalence with full MREC is, due to the decorrelating effect of the KLT, much simpler than the analysis of MRC considered on its own (Alouini *et al.*, 2000; Dong & Beaulieu, 2002; Siriteanu & Blostein, 2007).

## Performance Comparison

For unit-variance channel gains consider SISO and a BF, MREC, and MRC ULA with $L = 3$ for the following two cases:

- $AS_1 = 0$, i.e., coherent channel gains; then, the first row and the eigenvalue matrix of (Toeplitz) $\mathbf{R}_{\tilde{\mathbf{h}}}$ are given by $(\mathbf{R}_{\tilde{\mathbf{h}}})_{1,j=1:L} = [1 \;\; 1 \;\; 1]$ and $\Lambda_L = \mathrm{diag}\{3, 0, 0\}$.

- $AS_2 = 15^\circ$; then, $(\mathbf{R}_{\tilde{\mathbf{h}}})_{1,j=1:L} = [1.\ 0.76\ 0.42]$ and $\mathbf{\Lambda}_L = \text{diag}\{2.30, 0.58, 0.12\}$.

Figure 8 displays the AEPs computed using (82). For coherent branches, BF, $\text{MREC}_{N=2}$, and MRC actually coincide. There is maximum array gain, i.e., $10\log_{10}L \approx 4.8$ dB, over SISO, but no diversity gain. However, BF provides maximum array gain only for $AS_1 = 0^\circ$. Increasing AS diminishes the correlation between the channel gains, which decreases the correlation between the channel gains and the corresponding BF weights, and thus the array gain. For $AS_2 = 15^\circ$ the array gains computed with (73) are as follows: 3.6 dB for BF, 4.6 dB for $\text{MREC}_{N=2}$, and 4.8 dB (maximum) for MRC. The respective **amounts of fading**, computed using (74), are as follows: 1, 0.68, and 0.63. Thus, $\text{MREC}_{N=2}$ yields most of the available array gain, and about 1 dB more than BF. Note that the BF AEP plot remains parallel to that of SISO, even for non-zero AS, confirming a unit diversity order. MRC always performs best, due to maximum array and diversity gains. $\text{MREC}_{N=2}$ outperforms BF, because it yields more array gain as well as significant **diversity gain**. Actually, for small $\gamma_b$, $\text{MREC}_{N=2}$ approaches MRC in performance.

Simulation results are not shown here. Nevertheless, Figure 3.6 from (Siriteanu, 2006, p. 80) shows close agreement between the average error rates obtained from simulation and from (82), and confirms that the performance of MRC coincides with that of full MREC.

Figure 9 depicts the AEP for QPSK computed with (82), vs. AS, for SISO and for BF, MRC, and MREC ULA with $L = 3$, for unit-variance channel gains and $\gamma_b = 5$ dB. BF performance degrades with AS increasing around $0$, unlike that of $\text{MREC}_{N>1}$. Furthermore, for $AS < 10^\circ$, near-MRC performance can be achieved with $\text{MREC}_{N=2}$. For estimated eigen/gains, analysis, simulation, and FPGA-based fixed-point-implementation results from (Siriteanu & Blostein, 2004; Siriteanu *et al.*, 2006; Siriteanu, 2006; Siriteanu & Blostein, 2007) suggest that MREC of carefully-selected small order can yield MRC-like performance, i.e., much better performance than BF.

## Numerical Complexity Comparison

Let us compare the BF, MRC, and MREC numerical complexities for SINC/MMSE PSAM eigen/gain estimation and sub/optimum combining implementation — see (Siriteanu & Blostein, 2004; Siriteanu, 2006; Siriteanu & Blostein, 2007) for more details on these estimation and implementation techniques.

Unlike the channel gains, $\tilde{h}_i$, $i = 1 : L$, required for actual MRC, the eigengains, $h_i$, $i = 1 : N$, required for BF or MREC, are always uncorrelated, and therefore they are estimated separately, at lower complexity. Then, the numerical complexity of $\text{MREC}_N$ (excluding the eigendecomposition, for the reasons explained earlier) is simply $N$ times that of BF shown in Table 2 — see also (Siriteanu & Blostein, 2007, Table II).

For the example case with $L = 5$, $T = 11$, Table 2 indicates that suboptimum MRC (Siriteanu, 2006, Section 3.7.2) (Siriteanu & Blostein, 2007, Section III.C.1) is about 4.7 times more complex with MMSE PSAM than with SINC PSAM. This is because correlated fading requires joint estimation of $\tilde{h}_i$, $i = 1 : L$, for MMSE PSAM, but not for SINC PSAM (Siriteanu, 2006, Section 3.6.2). For optimum implementation (Siriteanu, 2006, Appendix A), MRC is about 3.5 times more complex with MMSE PSAM than with SINC PSAM. On the other hand, SINC and MMSE PSAM-based MREC have the same complexity, for either optimum or suboptimum implementation. The KLT also makes similar the complexities of the MREC optimum (Siriteanu, 2006, Section 3.7.1) (Siriteanu & Blostein, 2007, Section III.D.1) and suboptimum implementations (Siriteanu, 2006, Section 3.7.2) (Siriteanu & Blostein, 2007, Section III.C.4).

Finally, for MMSE PSAM, $L = 5$, and $T = 11$, full MREC has about three times lower complexity than MRC, for either optimum or suboptimum implementation. Moreover, $\text{MREC}_{N=2}$ has eight times lower complexity than MRC. On the other hand, for SINC PSAM, although full MREC is slightly more complex than MRC, $\text{MREC}_{N=2}$ still offers about a two-fold complexity reduction vs. MRC.

# SMARTER ANTENNA ARRAYS WITH ADAPTIVE MREC

## MREC Order Selection

The benefits of eigencombining can only be obtained by appropriate order selection. The procedure is analogous to the water-filling-based selection of transmission eigendirections in transmit-side eigenbeamforming or precoding (Sampath *et al.*, 2005; Vu & Paulraj, 2006; Zhou & Giannakis, 2003). Signal power is transmitted along an eigenvector only if the

*Figure 8. AEP computed with (82) for SISO and for BF, MREC, and MRC ULA with L = 3, for* $AS_1 = 0^\circ$ *and* $AS_2 = 15^\circ$

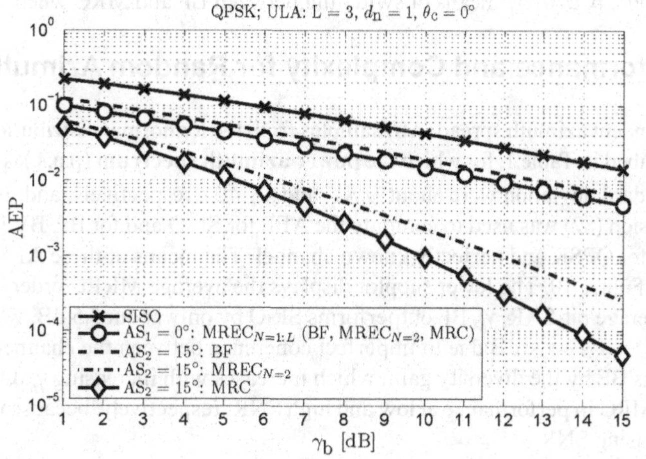


*Figure 9. Top: AEP performance for QPSK, unit-variance channel gains, and average bit-SNR* $\gamma_b$ *= 5dB, for SISO and for BF, MRC, and MREC ULA with L = 3; Bottom: Correlations between the channel gains corresponding to adjacent and extreme antenna elements, and the eigenvalues of* $\mathbf{R}_{\tilde{\mathbf{h}}}$.

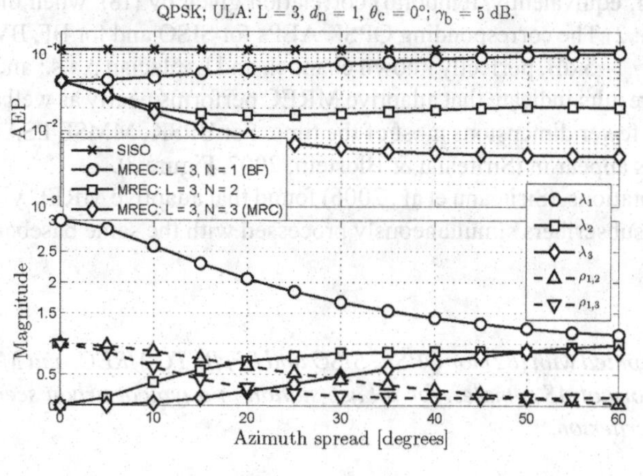


channel is "good" along it. However, the actual power-loading that occurs in transmit-side eigenbeamforming does not have a counterpart in receive-side eigencombining.

Consider the simple *bias–variance tradeoff criterion* (BVTC) (Jelitto & Fettweis, 2002, p. 22), wherein the MREC order is determined from

$$\min_{N=1:L}\left[E_s \cdot \sum_{i=N+1}^{L} \lambda_i + N_0 \cdot N\right]. \tag{85}$$

Essentially, the BVTC balances the loss incurred by removing the weakest $(L-N)$ intended-signal contributions (the first term) against the residual-noise contribution (the second term). The BVTC output depends on the relative magnitudes of the eigenvalues, i.e., on the spatial correlation, as well as on the symbol SNR, $E_s / N_0$. Order-selection criteria that also depend on the actual channel estimation method, symbol-detection performance (or quality-of-service), and system load appear in (Siriteanu *et al.*, 2006; Siriteanu & Blostein, 2007). Coarse adaptation to the propagation conditions was proposed in (El Zooghby, 2005, p. 270) by means of switching between BF and MRC when correlation reaches 0.5.

## Adaptive-MREC Performance and Complexity for Random Azimuth Spread

We generated 10,000 independent azimuth spread (AS) samples from the lognormal distribution described by (17) for the typical urban scenario described in Table 1, for which the **power azimuth spectrum** (p.a.s.) is Laplacian — see (15). The AS sequence mean and standard deviation were about $9.81^\circ$ and $13.40^\circ$, respectively, and $Pr(1^\circ \le AS \le 20^\circ) \approx 0.82$. For each AS sample, expression (82) was used to compute the AEP for SISO and for BF, BVTC-based adaptive MREC, and MRC ULA with $L = 5$, for QPSK and perfectly known channel. The means of these AEPs over the AS samples are shown in the top subplot of Figure 10. The lower subplot displays the average MREC order output by the BVTC. Note first that for high enough average bit-SNR, $\gamma_b$, BF outperforms SISO by only about 5.5 dB, which implies a loss of array gain for BF of about $10\log_{10}L - 5.5 \approx 1.5$ dB due to imperfect coherence between the channel gains and the applied BF combiner. MRC outperforms BF by the diversity gain, which increases with increasing $\gamma_b$. On the other hand, adaptive MREC approaches BF and MRC in performance at low and high SNR, respectively, because the BVTC outputs increasing MREC order with increasing SNR.

Similar results appear for BPSK, MMSE PSAM channel estimation, and optimum combining implementation in (Siriteanu & Blostein, 2007, Figure 2). There, for $\gamma_b = 4$ dB, the average BVTC output is $N = 2$, which makes MREC two times more complex than BF and eight times less complex than MRC, as discussed earlier. On the other hand, $\gamma_b = 4$ dB yields $AEP = 2 \cdot 10^{-2}$ for $\text{MREC}_{N=2}$. For $AEP \approx 2 \cdot 10^{-2}$, adaptive MREC outperforms BF by more than 3 dB, and is outperformed by MRC by less than 0.5 dB.

Figure 11 displays for a typical urban scenario the following numerical results: 1) Lognormal **AS** samples generated using (17), with spatial (or, equivalently, temporal) correlation given by (18), when the subscriber moves over the distance shown on the abscissa; 2) The corresponding QPSK AEPs for SISO and for BF, BVTC-based adaptive MREC, and MRC ULA with $L = 5$, for $\gamma_b = 5$ dB, perfectly known channel, and Laplacian p.a.s.; and 3) The MREC order output by the BVTC, vs. time. These results indicate that adaptive MREC performs nearly as well as MRC (and thus much better than BF), while processing fewer dimensions most of the time. For BPSK, MMSE PSAM, and optimum combining implementation, similar results appear in (Siriteanu & Blostein, 2007, Figure 3).

For FPGA-based implementations, (Siriteanu et al., 2006) found that adaptive MREC yields near-MRC performance while doubling the number of subscribers simultaneously processed with the same baseband resources at the base-sta-

*Figure 10. Top: Mean AEP computed with (82) for QPSK, SISO and BF, BVTC MREC, and MRC ULA with L = 5, obtained by averaging over 10000 lognormal AS samples, for a base-station in a typical urban scenario. Bottom: Mean MREC order selected with the BVTC criterion.*

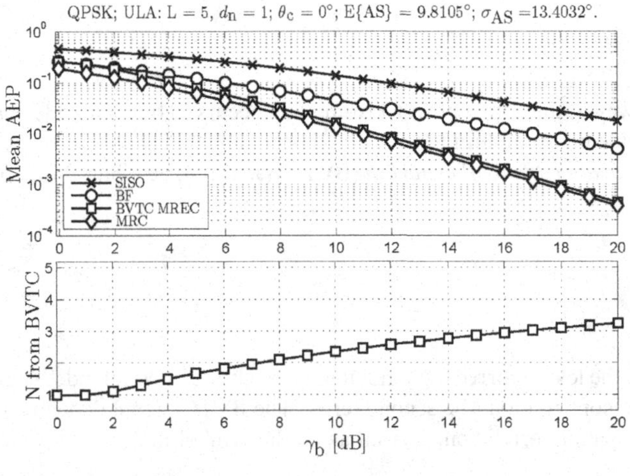

*Figure 11. Top: Base-station AS vs. distance traveled by the subscriber, for a typical urban scenario; Middle: AEP computed for QPSK with (82), for SISO and BF, BVTC MREC, and MRC ULA with L = 5, for $\gamma_b$ = 5 dB; Bottom: MREC order N selected with the BVTC, vs. time.*

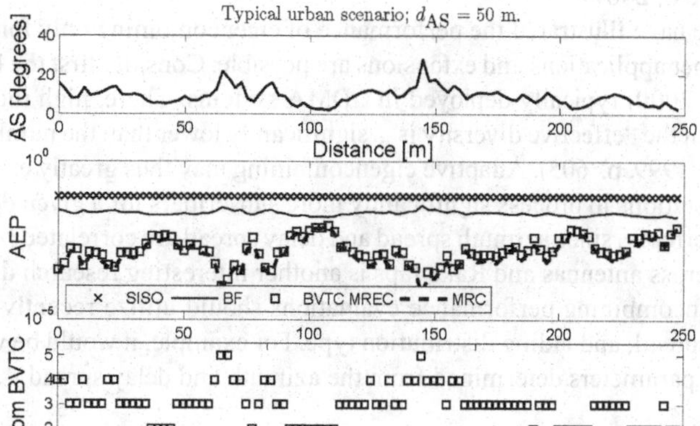


tion. Resources wasted with MRC on low-AS subscribers can with adaptive MREC benefit high-AS subscribers. Thus, adaptive MREC can efficiently benefit from *azimuth spread diversity*.

In conclusion, through its adaptability to channel propagation and noise conditions — as well as to subscriber quality-of-service and system processing power limitations (Siriteanu *et al.*, 2006; Siriteanu & Blostein, 2007) — MREC closes the performance and complexity gaps between BF and MRC, and eliminates the unwarranted risks and costs of their stand-alone deployments.

## RELATED APPLICATIONS AND FURTHER RESEARCH

For the numerical results shown in this chapter we have simulated AS values similar to those measured in actual urban scenarios at the base station (Algans *et al.*, 2002; Vaughan & Andersen, 2003). Nevertheless, multi-antenna transceivers are also envisioned at the subscriber stations (El Zooghby, 2005, Chapter 9), although there adoption pace is slow, due to high cost and power consumption (Hottinen *et al.*, 2006). Although the subscriber station typically experiences much higher AS than the base station (3GPP, 2003, Table 4.1) (El Zooghby, 2005, Sections 3.6.5, 9.2), the limited size of the former restricts antenna array interelement distance to values that may yield channel gain correlations (3GPP, 2003, Table 4.2) that are too large for MRC and too small for BF. Higher returns may then be possible with adaptive MREC that finely matches the actual channel and noise conditions, compared to switching between BF and MRC for a certain correlation threshold (El Zooghby, 2005, p. 270).

Throughout this chapter, we have analyzed the BF, MRC, and adaptive MREC performance for receiving antenna array receivers (i.e., SIMO systems) assumed to have perfect channel knowledge, and have compared their numerical complexities for estimated fading. The BF, MRC, and adaptive MREC performance is analyzed for imperfectly known fading as well as for optimum and suboptimum combining implementations in (Siriteanu & Blostein, 2004; Siriteanu, 2006; Siriteanu & Blostein, 2007). Performance and resource usage evaluations for a fixed-point FPGA-based implementation appear in (Siriteanu *et al.*, 2006). A description of optimum eigencombining implementation for interference-limited systems appears in (Siriteanu, 2006, Section 6.5.4). Simulation results for SIMO eigencombining for interference-limited scenarios and estimated channel eigenstructure are shown in (Brunner *et al.*, 2001).

An interesting issue that has been pointed out in (Siriteanu & Blostein, 2004) (Siriteanu & Blostein, 2007, Section III.C), and should be given further consideration, is that, for the practical case of suboptimum channel estimation and suboptimum combining implementation, MREC can actually outperform MRC. Thus, knowledge of channel eigendecomposition — which can be estimated more accurately than the actual channel gains, especially in high-mobility scenarios (Brunner *et al.*, 2001; Vu & Paulraj, 2006) — can in practice not only reduce complexity, but even enhance the performance. This effect is more pronounced at high spatial correlation and low SNR (F. A. Dietrich & Utschick, 2003, Figure 2) (Jelitto & Fettweis, 2002, Figure 6) (Siriteanu & Blostein, 2004, Figures 2, 3, 4, 7). Further research is also necessary on order selection criteria that have recently been proposed in (F. Dietrich & Utschick, 2002) (Siriteanu et al., 2006) (Siriteanu & Blostein, 2007).

Although in this chapter we have illustrated the performance of eigencombining only for spatial receivers and have not considered interference, other applications and extensions are possible. Consider first the Rake receivers (Patenaude *et al.*, 1999; Simon & Alouini, 2000) typically deployed in CDMA systems. There, high intertap correlation found by measurements has indicated that the "effective diversity is ... significantly lower than the number of maxima in the power delay profile" (Patenaude *et al.*, 1999, p. 605). Adaptive eigencombining may thus greatly extend subscriber station battery life, as well as allow base stations to process significantly more subscribers for a given deployment and operational cost, compared to MRC. Furthermore, since azimuth spread and delay spread are correlated (Algans *et al.*, 2002, Section VI.B), joint eigencombining across antennas and Rake taps is another interesting research direction.

Finally, more realistic eigencombining performance evaluations should utilize recently-documented relationships among azimuth spread, delay spread, and fading distribution type. For example, it would be worth investigating Rice or Nakagami fading models with parameters determined from the azimuth and delay spread.

## SUMMARY AND CONCLUSION

The chapter has provided an introduction to principal signal processing methods employed for receive-side smart antenna technology. We have compared statistical beamforming (BF) and maximal-ratio combining (MRC) with the conventional single-antenna transceiver, and have found that great performance benefits are achievable due to array and diversity gain, in favorable propagation conditions. For unfavorable conditions, BF and MRC performance can hardly justify their deployment and operational costs. Furthermore, although their numerical complexity remains fixed, BF and MRC performance fluctuates when propagation conditions vary as in actual scenarios.

Instead of BF and MRC, adaptive eigencombining should be considered in the development of smarter antenna technologies. Maximal-ratio eigencombining (MREC) unifies the signal processing principles and analyses of BF and MRC. Expressions for symbol-detection performance measures such as the outage probability and average error probability have been derived that apply throughout the fading correlation range. Numerical results presented for realistic scenarios demonstrate the performance-improving and complexity-reducing benefits of adaptive MREC over conventional BF or MRC.

## REFERENCES


3GPP. (2003). *Technical Specification Group Radio Access Network. Spatial Channel Model for Multiple Input Multiple Output (MIMO) Simulations, Release 6* (Tech. Rep. No. TS 25.996). 3rd Generation Partnership Project (3GPP).

Abramowitz, M., & Stegun, I. A. (Eds.). (1995). *Handbook of mathematical functions with formulas, graphs and mathematical tables.* New York, NY 10014: Dover Publications, Inc.

Algans, A., Pedersen, K. I., & Mogensen, P. E. (2002, April). Experimental analysis of the joint statistical properties of azimuth spread, delay spread, and shadow fading. *IEEE Journal on Selected Areas in Communications, 20*(3), 523-531.

Alouini, M.-S., Scaglione, A., & Giannakis, G. B. (2000, September). PCC: principal components combining for dense correlated multipath fading environments. In *Proc. IEEE vehicular technology conference, (VTC '00)* (Vol. 5, p. 2510-2517).

Applebaum, S. (1976, September). Adaptive arrays. *IEEE Transactions on Antennas and Propagation, 24*(5), 585 - 598.

Bahrami, H. R., & Le-Ngoc, T. (2006, December). Precoder design based on correlation matrices for MIMO systems. *IEEE Transactions on Wireless Communications, 5*(12), 3579 - 3587.

Blogh, J. S., & Hanzo, L. (2002). *Third-generation systems and intelligent wireless networking: Smart antennas and adaptive modulation.* Chichester, West Sussex, England: John Wiley and Sons.

Brennan, D. G. (2003, February). Linear diversity combining techniques. *Proceedings of the IEEE, 91*(2), 331 - 356.

### *Eigencombining*

Brunner, C., Utschick, W., & Nossek, J. A. (2001). Exploiting the short-term and long-term channel properties in space and time: eigenbeamforming concepts for the BS in WCDMA. *European Transactions on Telecommunications. Special Issue on Smart Antennas*, 12(5), 365-378.

Choi, J., & Choi, S. (2003, May). Diversity gain for CDMA systems equipped with antenna arrays. *IEEE Transactions on Vehicular Technology, 52*(3), 720-725.

Choi, S., Choi, J., Im, H.-J., & Choi, B. (2002, September). A novel adaptive beamforming algorithm for antenna array CDMA systems with strong interferers. *IEEE Transactions on Vehicular Technology, 51*(5), 808-816.

Dietrich, F., & Utschick, W. (2002). On the effective spatio-temporal rank of wireless communication channels. In *Proc. 13th IEEE international symposium on personal, indoor and mobile radio communications, (PIMRC '02)* (Vol. 5, p. 1982-1986).

Dietrich, F. A., & Utschick, W. (2003, September). Maximum ratio combining of correlated Rayleigh fading channels with imperfectly known channel. *IEEE Communications Letters, 7*(9), 419-421.

Dong, X., & Beaulieu, N. (2002, January). Optimal maximal ratio combining with correlated diversity branches. *IEEE Communications Letters, 6*(1), 22-24.

El Zooghby, A. (2005). *Smart antenna engineering.* Norwood, MA: Artech House.

Ertel, R. B., Cardieri, P., Sowersby, K. W., Rappaport, T. S., & Reed, J. H. (1998, February). Overview of spatial channel models for antenna array communication systems. *IEEE Personal Communications,5*(1), 10 - 22.

Godara, L. C. (2004). *Smart antennas.* Boca Raton, FL: CRC Press.

Goldberg, J. M., & Fonollosa, J. R. (1998). Downlink beamforming for spatially distributed sources in cellular mobile communications. *Signal Processing, 65*(2), 181-197.

Golub, G. H., & Loan, C. F. van. (2000). *Matrix computations.* New York, NY: The John Hopkins University Press.

Hottinen, A., Kuusela, M., Hugl, K., Zhang, J., & Raghothaman, B. (2006, August). Industrial embrace of smart antennas and MIMO. *IEEE Wireless Communications, 13*(4), 8-16.

Hottinen, A., Tirkkonen, O., & Wichman, R. (Eds.). (2003). *Multi-antenna tranceiver techniques for 3G and beyond.* Chichester, West Sussex, England: John Wiley and Sons.

Jakes, W. C. (Ed.). (1974). *Microwave mobile communications.* New York, NY: John Wiley and Sons.

J. C. Liberti, J., & Rappaport, T. S. (1999). *Smart antennas for wireless communications: IS-95 and third generation CDMA applications.* Upper Saddle River, NJ: Prentice Hall PTR.

Jelitto, J., & Fettweis, G. (2002, December). Reduced dimension space-time processing for multi-antenna wireless systems. *IEEE Wireless Communications, 9*(6), 18-25.

Lee, W. C. Y. (1982). *Mobile communications engineering.* Englewood Cliffs, NJ: McGraw-Hill.

Monzingo, R. A., & Miller, T. W. (1980). *Introduction to adaptive arrays.* New York: John Wiley and Sons.

Patenaude, F., Lodge, J., & Chouinard, J.-Y. (1999, March). Eigen analysis of wide-band fading channel impulse responses. *IEEE Transactions on Vehicular Technology, 48*(2), 593 - 606.

Paulraj, A., Nabar, R., & Gore, D. (2005). *Introduction to space-time wireless communications.* Cambridge, UK: Cambridge University Press.

Pedersen, K. I., Mogensen, P. E., & Fleury, B. H. (2000, March). A stochastic model of the temporal and azimuthal dispersion seen at the base station in outdoor propagation environments. *IEEE Transactions on Vehicular Technology, 49*(2), 437-447.

Proakis, J. G. (2001). *Digital communications* (Fourth ed.). New York, NY: McGraw-Hill, Inc.

Rensburg, C. van, & Friedlander, B. (2004, November). Transmit diversity for arrays in correlated Rayleigh fading. *IEEE Transactions on Vehicular Technology, 53*(6), 1726-173.

Rooyen, P. van, Lotter, M., & Wyk, D. van. (2000). *Space-time processing for CDMA mobile communications.* Norwell, MA: Kluwer Academic Publishers.

Roy, R. H. (1998, March). An overview of smart antenna technology: the next wave in wireless communications. In *Proceedings of the IEEE Aerospace Conference, 1998* (Vol. 3, p. 339-345).

Salz, J., & Winters, J. H. (1994, November). Effect of fading correlation on adaptive arrays in digital mobile radio. *IEEE Transactions on Vehicular Technology, 43*(4), 1049-1057.

Sampath, H., Erceg, V., & Paulraj, A. (2005, March). Performance analysis of linear precoding based on field trials results of MIMO-OFDM system. *IEEE Transactions on Wireless Communications, 4*(2), 404-409.

Siemens. (2000). *Text proposal for RAN WG1 report on Tx diversity solutions for multiple antennas* (Tech. Rep. No. TSGR1#15 R1-00-1126). 3rd Generation Partnership Project (3GPP).

Simon, M. K., & Alouini, M.-S. (2000). *Digital communication over fading channels. A unified approach to performance analysis.* Baltimore, Maryland: John Wiley and Sons.

Siriteanu, C. (2006). *Maximal-ratio eigen-combining for smarter antenna array wireless communication receivers.* Unpublished doctoral dissertation, Queen's University, Kingston, Canada.

Siriteanu, C., & Blostein, S. D. (2004, January–April). Maximal-ratio eigencombining: a performance analysis. *Canadian Journal of Electrical and Computer Engineering, 29*(1/2), 15-22.

Siriteanu, C., & Blostein, S. D. (2007, March). Maximal-ratio eigen-combining for smarter antenna arrays. *IEEE Transactions on Wireless Communications, 6*(3), 917 - 925.

Siriteanu, C., Blostein, S. D., & Millar, J. (2006). FPGA-based communications receivers for smart antenna array embedded systems. *EURASIP Journal on Embedded Systems. Special Issue on Field- Programmable Gate Arrays in Embedded Systems, 2006*, Article ID 81309, 13 pages.

Stridh, R., Bengtsson, M., & Ottersten, B. (2006, April). System evaluation of optimal downlink beamforming with congestion control in wireless communication. *IEEE Transactions on Wireless Communications, 5*(4), 743 - 751.

Trees, H. L. V. (2002). *Optimum array processing: Part IV of detection, estimation, and modulation theory.* New York, NY: John Wiley & Sons, Inc.

Tse, D., & Viswanath, P. (2005). *Fundamentals of wireless communication.* Cambridge, UK: Cambridge University Press.

Vaughan, R., & Andersen, J. B. (2003). *Channels, propagation and antennas for mobile communications.* London: The Institution of Electrical Engineers.

Vu, M., & Paulraj, A. (2006, June). Optimal linear precoders for MIMO wireless correlated channels with nonzero mean in space-time coded systems. *IEEE Transactions on Signal Processing, 54*(6), 2318-2332.

Zhou, S., & Giannakis, G. B. (2003, July). Optimal transmitter eigen-beamforming and space-time block coding based on channel correlations. *IEEE Transactions on Information Theory, 49*(7), 1673 - 1690.

# Chapter II
# Robust Adaptive Beamforming

**Zhu Liang Yu**
*Nanyang Technological University, Singapore*

**Meng Hwa Er**
*Nanyang Technological University, Singapore*

**Wee Ser**
*Nanyang Technological University, Singapore*

**Huawei Chen**
*Nanyang Technological University, Singapore*

## INTRODUCTION

**In this chapter, we first review the background, basic principle and structure of adaptive beamformers. Since there are many robust adaptive beamforming methods proposed in literature, for easy understanding, we organize them into two categories from the mathematical point of view: one is based on quadratic optimization with linear and nonlinear constraints; the another one is max-min optimization with linear and nonlinear constraints. With the max-min optimization technique, the state-of-the-art robust adaptive beamformers are derived. Theoretical analysis and numerical results are presented to show their superior performance.**

## BACKGROUND

The array signal processing has been studied for some decades as an attractive method for signal detection and estimation in hash environment. An array of sensors can be flexibly configured to exploit spatial and temporal characteristics of signal and noise and has many advantages over single sensor. It has many applications in radar, radio astronomy, sonar, wireless communication, seismology, speech acquisition, medical diagnosis and treatment (Tsoulos, 2001) (Krim & Viberg, 1996) (Van Veen & Buckley, 1988), etc.

There are two kinds of array beamformers: fixed beamformer and adaptive beamformer. The weight of fixed beamformer is pre-designed and it does not change in applications. The adaptive beamformer automatically adjusts its weight according to some criteria. It significantly outperforms the fixed beamformer in noise and interference suppression. A typical representative is the linearly constrained minimum variance (LCMV) beamformer (Compton, 1988) (Hudson, 1981) (Johnson & Dudgeon, 1993) (Monzingo & Miller, 1980) (Naidu, 2001). A famous representative of the LCMV is the Capon beamformer (Capon, 1969). In ideal cases, the Capon beamformer has high performance in interference and noise suppression provided that the array steering vector (ASV) is known. However, the ideal assumptions of adaptive beamformer may be violated in practical applications (Vural, 1979) (Jablon, 1986a) (Jablon, 1986b) (Cox, Zeskind, & Owen, 1988) (Chang & Yeh, 1993). The performance of the adaptive beamformers highly degrades when there are array imperfections such as steering direction error, time delay error, phase errors of the array sensors, multipath propagation effects, wavefront distortions. This is known as the target signal cancellation problem. Tremendous work has been done to improve the robustness of adaptive beamformer (Nunn, 1983) (Er & Cantoni, 1983) (Buckley & Griffiths, 1986) (Er & Cantoni, 1986c) (Er, 1988) (Thng, Cantoni, & Leung, 1993) (Thng, Cantoni, & Leung, 1995) (Zhang & Thng, 2002) (Er & Cantoni, 1986b) (Er & Cantoni, 1990) (Cox, Zeskind, & Owen, 1987) (Vorobyov, Gershman, & Luo, 2003) (Lorenz & Boyd, 2005) (Li, Stoica, & Wang, 2003) (Stoica, Wang, & Li, 2003) (Affes & Grenier, 1997) (Er & Ng, 1994).

To overcome the problem of target signal cancellation caused by the steering direction error, multiple-point constraints (Hudson, 1981)(Nunn, 1983) were introduced in adaptive array. The idea of this approach is intuitive. With multiple gain constraints at different directions in the vicinity of the assumed one, the array processor becomes robust in the region where constraints are imposed. However, the available number of constraints is limited because the constraints consume the degrees of freedom (DOFs) of array processor for interference suppression.

Another class of solution is to introduce the derivative constraints into the array processor (Hudson, 1981) (Er & Cantoni, 1983) (Buckley & Griffiths, 1986) (Er & Cantoni, 1986c) (Er, 1988) (Thng et al., 1993) (Thng et al., 1995) (Zhang & Thng, 2002). With the derivative constraints, the array response is almost flat in the vicinity of target direction. The beamformer has widened beamwidth in the target direction. With a small steering direction error, the beamformer does not cancel the target signal. However, the widened beamwidth is achieved at the cost of reduced capability in interference suppression because the additional derivative constraints consume the DOFs of beamformer. Derivative constraints can be used to obtain not only a flat response of array processor, but also a flat null in the assumed signal direction in blocking matrix design (Fudge, 1996).

A new set of constraints for robust array processor against the steering error was also proposed in (Er & Cantoni, 1986b)(Er & Cantoni, 1990). The idea is to minimize the weighted mean square deviation between the desired array response and the response of the processor over the variations in parameters, such as the steering error, the phase errors and the array geometry error, etc. Although the constraints derived by this approach are quadratic (Er, 1985), a set of linear constraints was derived approximately (Er & Cantoni, 1986b). In the approximation of quadratic constraints to linear constraints, a problem arises that how many constraints should be selected. A method to determine the number of necessary linear constraints and to select the constraints was proposed in (Er & Cantoni, 1986b).

Techniques restraining excess coefficients growth were also proposed in array processor to achieve robust performance. When array processor cancels target signal, the norm of the filter coefficients grows to a large value beyond the normal value for noise and interference suppression. In (Cox et al., 1987), an inequality constraint is imposed on the coefficients norm of adaptive beamformer to limit the growth of tap coefficients. The excess coefficients growth problem can also be solved by using noise injection method (Jablon, 1986a). Artificially generated noise is added to reference signals of adaptive filters. Although the artificial noise causes estimation errors in the beamformer coefficients, it prevents tap coefficients from growing excessively, resulting in robustness against array imperfections. A similar approach called the leaky least mean square (LMS) algorithm can also be used (Claesson & Nordholm, 1992) for this purpose.

Other robust beamforming methods include the calibration based approaches (Fudge & Linebarger, 1994). The calibration can generally eliminate the inherent error of the array processor, such as geometry error, sensor response error, etc. However, it cannot eliminate the dynamic errors, such as the steering error when the source is moving in a vicinity of the assumed direction. Target tracking methods (Affes & Grenier, 1997)(Er & Ng, 1994) were introduced in array processor so that the look direction is steered to the continuously estimated direction-of-arrival (DOA). One problem is that this method may mistrack to the interference in the absence of target signal unless some other methods are used to limit the tracking region.

The robust beamforming methods discussed above solve part of robust beamforming problems. More research works still need to be carried out, especially in real applications. In (Er & Ng, 1994), a new approach was proposed for robust beamforming in the presence of steering direction error. It iteratively searches for the optimal direction by maximizing the mean output power of the Capon beamformer using first-order Taylor series approximation in terms of steering direction error. This method does not suffer from performance loss in interference/noise suppression. However, its performance degrades when there exist multiple errors, such as the steering direction error, the array geometry error and the array sensor phase error, because the array steering vector in (Er & Ng, 1994) is assumed to be a vector function of steering direction only. When multiple imperfections exist, the assumed model of the ASV is violated. In (Yu & Er, 2006a) (Yu, 2006), a new model of the ASV is adopted. All of these array imperfections are modeled as general phase errors (GPEs).

The GPEs are broad-sense errors which simplify the mathematical expression of the ASV in terms of array imperfections. Therefore, the ASV can be described as a vector function of the GPEs. A new approach is derived based on this model to search for the optimal estimates of the GPEs by maximizing the output power of the Capon beamformer. The estimated GPEs can be used to compensate the errors in the actual ASV. Since there are no additional constraints introduced in adaptive beamforming, this new method does not suffer from performance loss in interference/noise suppression.

Recently, robust methods with clear theoretical background have been proposed (Vorobyov et al., 2003)(Lorenz & Boyd, 2005)(Li et al., 2003) (Stoica et al., 2003). An uncertainty set of the ASV is used. The true ASV is assumed to be in an ellipsoid centered at the nominal ASV. The designed beamformers are robust against arbitrary variation of the true ASV within an assumed uncertainty set. These beamformers (Vorobyov et al., 2003) (Lorenz & Boyd, 2005) (Li et al., 2003) (Stoica et al., 2003) are equivalent and belong to the diagonal loading approach (Li et al., 2003). The diagonal loading factor can be calculated from the constraint equation. In this chapter, we give an overview of these robust beamformers and will show that these beamformers can also be derived based on max-min optimization. Improvement of these beamformers is also presented by using steering vector projection technique. Some of the comments in this chapter can also be found in Yu (2006), Yu and Er (2006a; 2006b) etc.

## OVERVIEW OF ADAPTIVE BEAMFORMER

### Signal Model

An array comprising of M sensors is shown in Figure 1. The observed signal at the $m$th sensor, $x_m(t)$, can be expressed as the sum of all the plane wave signals and the background noise

$$x_m(t) = \sum_{k=0}^{K-1} a_k(t) e^{-j\omega_0 \tau_{mk}} + n_m(t) \tag{1}$$

where $a_k(t)$ is the $k$th signal and $n_m(t)$ is the complex random noise received by the $m$th sensor. $\omega_0$ is the carrier frequency and $\tau_{mk}$ is the time delay for the $k$th signal at the $m$th sensor.

Let $\mathbf{x}(t)$ be the observed signal vector

$$\mathbf{x}(t) = \begin{bmatrix} x_1(t) & x_2(t) & \cdots & x_M(t) \end{bmatrix}^T \tag{2}$$

where $(\bullet)^T$ denotes the vector transpose. Using matrix notation, we have

$$\mathbf{x}(t) = \sum_{k=0}^{K-1} a_k(t)\mathbf{s}(\theta_k) + \mathbf{n}(t) \tag{3}$$

where $\mathbf{s}(\theta_k)$ is the array steering vector given by

$$\mathbf{s}(\theta_k) = \begin{bmatrix} e^{-j\omega_0\tau_{1k}} & e^{-j\omega_0\tau_{2k}} & \cdots & e^{-j\omega_0\tau_{Mk}} \end{bmatrix}^T \tag{4}$$

and $\mathbf{n}(t)$ is the received noise vector given by

$$\mathbf{n}(t) = \begin{bmatrix} n_1(t) & n_2(t) & \cdots & n_M(t) \end{bmatrix}^T \tag{5}$$

The sensor-collected signals, $x_m(t), m = 1, \cdots, M$, are weighted by $w_m, m = 1, \cdots, M$ and summed up to form the output signal $y(t)$

$$y(t) = \sum_{m=1}^{M} w_m^* x_m(t) = \mathbf{w}^H \mathbf{x}(t) \tag{6}$$

where $\mathbf{w}$ denotes the complex weight vector, $(\bullet)^*$ denotes the conjugation and $(\bullet)^H$ denotes the Hermitian transpose,

$$\mathbf{w} = \begin{bmatrix} w_1 & w_2 & \cdots & w_M \end{bmatrix}^T \tag{7}$$

For the sources that can be modeled by stationary stochastic processes, the mean output power of the array system is given by

$$p(\mathbf{w}) = E\left\{|y(t)|^2\right\} = \mathbf{w}^H \mathbf{R}\mathbf{w} \tag{8}$$

where $E\{\bullet\}$ is the expectation operator, and $\mathbf{R}$ is the array covariance matrix defined by

$$\mathbf{R} = E\left\{\mathbf{x}(t)\mathbf{x}^H(t)\right\} \tag{9}$$

The covariance matrix $\mathbf{R}$ is the statistical second-order property of the impinging signals. In real applications, $\mathbf{R}$ is estimated using the received array snapshots. The estimated array covariance matrix, $\hat{\mathbf{R}}$, is given by

$$\hat{\mathbf{R}} = \frac{1}{N_s}\sum_{k=1}^{N_s} \mathbf{x}(k)\mathbf{x}^H(k) \tag{10}$$

where $N_s$ is the number of snapshots. When $N_s$ approaches infinity, the estimated $\mathbf{R}$ asymptotically approaches the true one.

If all the impinging sources and the noise are mutually uncorrelated, $\mathbf{R}$ is a non-negative matrix given by

$$\mathbf{R} = \sum_{k=0}^{K-1} \sigma_k^2 \mathbf{s}(\theta_k)\mathbf{s}^H(\theta_k) + \mathbf{R}_n = \mathbf{R}_s + \mathbf{R}_n \tag{11}$$

where $\sigma_k^2 = E\{|a_k(t)|^2\}$ is the power of the $k$th source, and $\mathbf{R}_n = \sigma_n^2\mathbf{I}$ is the covariance matrix of the noise, where $\sigma_n^2$ is the power of the noise and $\mathbf{I}$ is an identity matrix. $\mathbf{R}_s$ is the covariance matrix of the impinging directional sources. In the following context, we always assume that $\theta_0$ is the direction of the target source and $s(\theta_k)$ can be abbreviated as $\mathbf{s}_k$.

The eigen-decomposition (EVD) of the covariance matrix plays an important role in array processing and thus its structure will be explored further. The matrix $\mathbf{R}$ can be decomposed and expressed in the following form:

$$\mathbf{R} = \mathbf{U}\boldsymbol{\Sigma}\mathbf{U}^H = \sum_{m=1}^{M} \lambda_m \mathbf{u}_m \mathbf{u}_m^H \tag{12}$$

where $\boldsymbol{\Sigma} = diag\{\lambda_1, \lambda_2, \cdots, \lambda_M\}$ is the eigenvalue matrix and $\mathbf{U} = [\mathbf{u}_1 \;\; \mathbf{u}_2 \;\; \cdots \;\; \mathbf{u}_M]$ is the corresponding eigenvector matrix. An important property is that, the eigenvalue $\lambda_m$ of $\mathbf{R}$ is

$$\lambda_m = \begin{cases} \gamma_m + \sigma_n^2, 1 \le m \le K \\ \sigma_n^2, K+1 \le m \le M \end{cases} \tag{13}$$

where $\gamma_m$ is the $m$th eigenvalue of the matrix $\mathbf{R}_s$. Splitting the eigenvector matrix into

$$\mathbf{U} = [\mathbf{U}_s \;\; \mathbf{U}_n] \tag{14}$$

where

$$\begin{aligned} \mathbf{U}_s &= [\mathbf{u}_1 \;\; \mathbf{u}_2 \;\; \cdots \;\; \mathbf{u}_K] \\ \mathbf{U}_n &= [\mathbf{u}_{K+1} \;\; \mathbf{u}_{K+2} \;\; \cdots \;\; \mathbf{u}_M] \end{aligned} \tag{15}$$

*Figure 1. Array processor with M sensors*

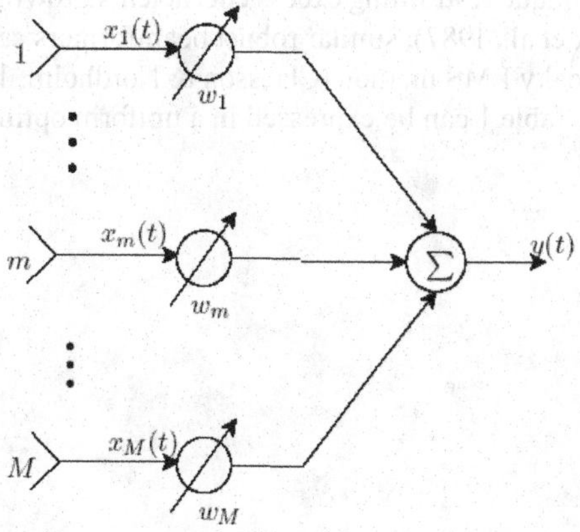


The matrix $\mathbf{U}_s$ is the eigen-basis of the so-called signal subspace, and the matrix $\mathbf{U}_n$ is the eigen-basis of the noise subspace. The ASV of each signal is a linear combination of the column vectors of $\mathbf{U}_s$, i.e.,

$$\mathbf{s}_k \in span(\mathbf{U}_s) \tag{16}$$

## Linearly Constrained Minimum Variance Array Beamformers

A famous representative of LCMV beamformers, the Capon beamformer(Capon, 1969), is designed to minimize its mean output power subject to a constraint on unity gain in the look direction as an indirect way of rejecting noise and interferences incident on the array. The optimal weight vector $\hat{\mathbf{w}}$ is the solution of the following constrained optimization problem

$$\begin{cases} \min_{\mathbf{w}} \mathbf{w}^H \mathbf{R} \mathbf{w} \\ s.t.\ \mathbf{s}_0^H \mathbf{w} = 1 \end{cases} \tag{17}$$

where $\mathbf{s}_0$ is the nominal ASV corresponding to the assumed target direction $\theta_0$. Using the Lagrange multipliers methodology (Bazaraa & Shetty, 1979), the optimal weight vector $\hat{\mathbf{w}}$ is given as

$$\hat{\mathbf{w}} = \frac{\mathbf{R}^{-1}\mathbf{s}_0}{\mathbf{s}_0^H \mathbf{R}^{-1} \mathbf{s}_0} \tag{18}$$

The optimal output power $\hat{p}$ of the array beamformer is

$$\hat{p} = \frac{1}{\mathbf{s}_0^H \mathbf{R}^{-1} \mathbf{s}_0} \tag{19}$$

## Conventional Robust Adaptive Array Beamformers

In order to overcome the target signal cancellation problem caused by the steering direction error, the multiple-point constraints (Takao, 1976)(Hudson, 1981)(Nunn, 1983), the derivative constraints (Hudson, 1981)(Er & Cantoni, 1983)(Buckley

& Griffiths, 1986)(Er & Cantoni, 1986c) (Er, 1988)(Thng et al., 1993)(Thng et al., 1995)(Zhang & Thng, 2002) and a new set of constraints (Er & Cantoni, 1986b) (Er & Cantoni, 1990) (Buckley, 1987) (Er, 1985) are introduced in robust adaptive beamforming. Moreover, the technique restraining excess coefficients growth was also proposed in array processor to achieve robust performance (Cox et al., 1987), similar robust beamformers can be obtained by using the noise injection method (Jablon, 1986a) and the leaky LMS method (Claesson & Nordholm, 1992).

Most of the above approaches listed in Table 1 can be expressed in a uniform optimization problem as

$$\begin{cases} \min\limits_{\mathbf{w}} \mathbf{w}^H \mathbf{R} \mathbf{w} \\ s.t.\ \mathbf{C}_0^H \mathbf{w} = \mathbf{f} \\ \quad f(\mathbf{w}) \le 0 \end{cases} \tag{20}$$

The linear constraints include the linear constraints derived for robust beamformer, e.g., the multiple-point constraints (Hudson, 1981) (Nunn, 1983) and the derivative constraints (Hudson, 1981) (Er & Cantoni, 1983) (Buckley & Griffiths, 1986) (Er & Cantoni, 1986c) (Er, 1988), etc. Another constraint set $f(\mathbf{w}) \le 0$ contains nonlinear inequality constraints for robust beamformer. Most of the nonlinear constraints derived so far for robust beamforming are of quadratic form (Er & Cantoni, 1986b) (Er & Cantoni, 1990) (Jablon, 1986a) (Cox et al., 1987).

The solution of the universal optimization problem (20) can be derived as follows. Ignoring the inequality constraint in (20), the solution $\hat{\mathbf{w}}$ of the optimization problem (20) is given by

*Table 1. Robust beamformers with different kinds of constraints*

| Constraints | Constraints Type | References | Comments |
|---|---|---|---|
| Multiple-point constraints | Linear constraints | (Takao, 1976)<br>(Hudson, 1981)<br>(Nunn, 1983) | |
| Derivative Constraints | Linear constraints | (Hudson, 1981)<br>(Er & Cantoni, 1983)<br>(Buckley & Griffiths, 1986)<br>(Er & Cantoni, 1986c)<br>(Er, 1988)<br>(Thng et al., 1993)<br>(Thng et al., 1995)<br>(Zhang & Thng, 2002) | In Thng's work, there are also non-linear constraints. |
| Regional Constraints | Quadratic constraints<br>Or linear constraints | (Er & Cantoni, 1986b)<br>(Er & Cantoni, 1990)<br>(Buckley, 1987)<br>(Er, 1985) | The quadratic constraints can be transformed to linear constraints |
| Restraining norm of coefficients | Quadratic constraints | (Cox et al., 1987) | |
| Noise rejection method | Quadratic constraints | (Jablon, 1986a) | |

$$\hat{\mathbf{w}} = \mathbf{R}^{-1}\mathbf{C}_0(\mathbf{C}_0^H\mathbf{R}^{-1}\mathbf{C}_0)^{-1}\mathbf{f} \tag{21}$$

If $f(\hat{\mathbf{w}}) \le 0$, the inequality constraint is inactive in optimization, i.e., $\hat{\mathbf{w}}$ is the solution of (20). Otherwise, the optimal solution of (20) is obtained on the boundaries of the constraints. The solution is obtained by solving the following optimization problem

$$\begin{cases} \min_{\mathbf{w}} \mathbf{w}^H\mathbf{R}\mathbf{w} \\ s.t.\ \mathbf{C}_0^H\mathbf{w} = \mathbf{f} \\ \quad f(\mathbf{w}) = 0 \end{cases} \tag{22}$$

This quadratic programming problem with equality constraints can be solved using the Lagrange multiplier methodology (Bazaraa & Shetty, 1979).

## ROBUST BEAMFORMING BASED ON MAX-MIN OPTIMIZATION

In this section, we discuss some contemporary robust adaptive beamformers (Vorobyov et al., 2003) (Lorenz & Boyd, 2005) (Li et al., 2003) (Li, Stoica, & Wang, 2004) (Stoica et al., 2003) (Yu & Er, 2006a) (Yu & Er, 2006b) (Yu, 2006). Most of these beamformers are equivalent to the following optimization problem

$$\begin{cases} \max_{\Psi} \min_{\mathbf{w}} \mathbf{w}^H\mathbf{R}\mathbf{w} \\ s.t\ \mathbf{C}_0^H(\Psi)\mathbf{w} = \mathbf{f} \\ \quad f(\mathbf{w}, \Psi) \le 0 \end{cases} \tag{23}$$

where $\Psi$ stands for the parameters of array imperfections.

These robust beamformers can be derived by maximizing the output power of the Capon beamformer subject to some constraints on the array imperfections. If the error in the ASV can be modeled as a vector function of some parameters, like the steering direction error (Er & Ng, 1994), the time-delay error (Zou, Yu, & Lin, 2004), and the general phase errors (Yu, Zou, & Er, 2004), robust beamformer can be constructed by maximizing the output power of the Capon beamformer to these parameters in their feasible ranges. Efficient gradient descent method (Bazaraa & Shetty, 1979) can be derived to search for the optimal parameters. With these estimated optimal parameters, the error in the ASV can be compensated. The signal cancellation effect on the array output is then reduced. We will not discuss these methods in detail in this chapter. Instead, we will discuss methods against arbitrary array steering vector error. By maximizing output power of the Capon beamformer to arbitrary steering vector error, the optimal solution has the similar mathematical form of diagonal loading method with the diagonal loading factor can be calculated from constraint equations.

In practice, the nominal ASV corresponding to the target source cannot be expressed in a compact mathematical form in term of array imperfections. In such case, new robust methods should be developed. In this chapter, we further extend the idea used in (Er & Ng, 1994) (Yu et al., 2004) (Zou et al., 2004) to design an adaptive beamformer robust against arbitrary ASV error. Since the output power of the standard Capon beamformer (SCB) is a function of the assumed nominal ASV, we can maximize the output power of the SCB with respect to all feasible ASVs instead of the modeled parameters of the ASV in (Er & Ng, 1994) (Yu et al., 2004) (Zou et al., 2004). Although nonzero scaling of ASV does not change the output signal-to-interference-plus-noise ratio (SINR) of the SCB, it introduces an arbitrary scale in the output power. To eliminate the ambiguity in the output power, we assume that the ASV has unit norm. If there are no other constraint on ASV, the design of the array processor can be simplified to a principal component analysis (PCA) or minor component analysis (MCA) problem (Jolliffe, 1986). Nevertheless, when the target signal is not the dominant one, such array beamformer may wrongly suppress the target signal and retrieve interference as the output signal. To solve this problem, we introduce an additional uncertainty constraint on the ASV. This uncertainty constraint of the ASV is also used in some robust methods (Vorobyov et al., 2003) (Stoica et al., 2003) (Li et al., 2003) (Li et al., 2004). It assumes that the feasible ASV is in an ellipsoid whose center is the nominal ASV. With this uncertainty constraint, the designed Capon beamformer is robust against arbitrary ASV error even with the existence of strong interferences.

Two types of robust Capon beamformer (RCB) are derived with and without the additional constraint on the norm of the ASV in optimization. Theoretical analysis shows that these RCBs can be generalized as the diagonal loading approach. The optimal output SINRs of the proposed RCBs are also derived. Unfortunately, the RCBs cannot achieve the optimal output SINR. Although the proposed RCBs cannot achieve the optimal output SINR, numerical experiments show that the RCBs demonstrate outstanding robustness against the ASV error and have relatively high output SINR.

## Beamformers Robust Against Arbitrary ASV Error

In practical applications, the true ASV $\mathbf{s}_0$ is always unknown or known but with some errors. If $\mathbf{s}_0$ deviates from the true one, target signal cancellation occurs and the array output power decreases. A method to overcome target signal cancellation is to search for an optimal ASVs, which results in maximal output power $\hat{p}$. We assume that the true ASV $\mathbf{s}_0$ belongs to the following uncertainty ellipsoid (Vorobyov et al., 2003) (Stoica et al., 2003) (Li et al., 2003):

$$\mathbf{s}_0 \in \left\{\mathbf{s} \mid (\mathbf{s}-\bar{\mathbf{s}}_0)^H \mathbf{D}^{-1}(\mathbf{s}-\bar{\mathbf{s}}_0) \le 1\right\} \tag{24}$$

where $\bar{\mathbf{s}}_0$ is the nominal ASV and $\mathbf{D}$is a positive definite matrix. If $\mathbf{D} = \varepsilon \mathbf{I}$, we have

$$\mathbf{s}_0 \in \left\{\mathbf{s} \middle| \, \|\mathbf{s}-\bar{\mathbf{s}}_0\|^2 \le \varepsilon\right\} \tag{25}$$

where ε is the uncertainty level. With the uncertainty constraint (25), robust beamformer can be formulated as

$$\begin{cases} \max\limits_{\mathbf{s}} \min\limits_{\mathbf{w}} \mathbf{w}^H \mathbf{R} \mathbf{w} \\ s.t. \ \mathbf{s}^H \mathbf{w} = 1 \\ \|\mathbf{s}-\bar{\mathbf{s}}_0\|^2 \le \varepsilon \end{cases} \tag{26}$$

This optimization problem can be solved in two steps. First, we fix $\mathbf{s}$ and search the minimal output power. Then we search for the maximal value of the minimal output power to all the feasible $\mathbf{s}$. For any given $\mathbf{s}$, the output power of the SCB is given in (19). Since $\mathbf{s}^H \mathbf{R}^{-1} \mathbf{s}$ is a scalar, maximizing $\frac{1}{\mathbf{s}^H \mathbf{R}^{-1} \mathbf{s}}$ is equivalent to minimize $\mathbf{s}^H \mathbf{R}^{-1} \mathbf{s}$. The optimization problem in (26) is simplified to

$$\begin{cases} \min\limits_{\mathbf{s}} \mathbf{s}^H \mathbf{R}^{-1} \mathbf{s} \\ s.t. \ \ \|\mathbf{s}-\bar{\mathbf{s}}_0\|^2 \le \varepsilon \end{cases} \tag{27}$$

The optimization problem (27) has similar mathematical form as the methods in (Stoica et al., 2003) (Li et al., 2003). It can be solved using the Lagrange multiplier methodology (Bazaraa & Shetty, 1979). The optimal solution is obtained on the boundary of the constraint set. Therefore (27) can be reformulated as

$$\begin{cases} \min\limits_{\mathbf{s}} \mathbf{s}^H \mathbf{R}^{-1} \mathbf{s} \\ s.t. \ \ \|\mathbf{s}-\bar{\mathbf{s}}_0\|^2 = \varepsilon \end{cases} \tag{28}$$

To exclude the trivial solution $\mathbf{s} = \mathbf{0}$ of (28), we assume that

$$\|\bar{\mathbf{s}}_0\|^2 \geq \varepsilon \tag{29}$$

Define a function

$$f = \mathbf{s}^H \mathbf{R}^{-1}\mathbf{s} + g(\|\mathbf{s} - \bar{\mathbf{s}}_0\|^2 - \varepsilon) \tag{30}$$

where $g$ is the Lagrange multiplier. The optimal vector $\hat{\mathbf{s}}$ is obtained by setting the differentiation of (30) with respect to $\mathbf{s}^*$ to zero,

$$\frac{\partial f}{\partial \mathbf{s}^*} = \mathbf{R}^{-1}\mathbf{s} + g(\mathbf{s} - \bar{\mathbf{s}}_0) = 0 \tag{31}$$

The above equation yields

$$\hat{\mathbf{s}} = (g^{-1}\mathbf{R}^{-1} + \mathbf{I})^{-1}\bar{\mathbf{s}}_0 \tag{32}$$

The Lagrange multiplier $g$ is the root of the constraint equation

$$\|\hat{\mathbf{s}} - \bar{\mathbf{s}}_0\|^2 = \|(\mathbf{I} + g\mathbf{R})^{-1}\bar{\mathbf{s}}_0\|^2 = \varepsilon \tag{33}$$

As shown in (12), the eigen-decomposition of $\mathbf{R}$ is

$$\mathbf{R} = \mathbf{U}\boldsymbol{\Sigma}\mathbf{U}^H \tag{34}$$

Substituting (34) into (33) and denoting $\mathbf{y} = \mathbf{U}\bar{\mathbf{s}}_0$, it yields

$$h(g) = \sum_{i=1}^{M} \frac{|y_i|^2}{(1 + \lambda_i g)^2} = \varepsilon \tag{35}$$

where $y_i$ is the $i$th element of the vector $\mathbf{y}$. Consider the derivative of $h(g)$ to $g$,

$$\frac{dh(g)}{dg} = -2\sum_{i=1}^{M} \frac{|y_i|^2 \lambda_i}{(1 + g\lambda_i)^3} < 0 \tag{36}$$

when $g \geq 0$. The function $h(g)$ is a monotonically decreasing function with respect to $g$. Moreover, $\lim_{g \to \infty} h(g) = 0$ and $h(0) = \|\mathbf{U}^H \bar{\mathbf{s}}_0\|^2 = \|\bar{\mathbf{s}}_0\|^2 > \varepsilon$. The equation (35) has a unique solution $\hat{g} \geq 0$. Substituting $\hat{g}$ into (32), it yields

$$\hat{\mathbf{s}} = (\hat{g}^{-1}\mathbf{R}^{-1} + \mathbf{I})^{-1}\bar{\mathbf{s}}_0 = \bar{\mathbf{s}}_0 - (\mathbf{I} + \hat{g}\mathbf{R})^{-1}\bar{\mathbf{s}}_0 \tag{37}$$

The corresponding optimal array weight is

$$\hat{\mathbf{w}} = \frac{\mathbf{R}^{-1}\hat{\mathbf{s}}}{\hat{\mathbf{s}}^H \mathbf{R}^{-1}\hat{\mathbf{s}}} \tag{38}$$

and the output power and SINR are

$$\hat{p} = \frac{\|\hat{\mathbf{s}}\|^2}{\hat{\mathbf{s}}^H \mathbf{R}^{-1} \hat{\mathbf{s}}}, \rho = \frac{\mathbf{w}_0^H \mathbf{R}_s \mathbf{w}_0}{\mathbf{w}_0^H \mathbf{R}_n \mathbf{w}_0} \tag{39}$$

where $\mathbf{R}_s$ and $\mathbf{R}_n$ are the sample covariance matrices of the target signal and noise, respectively. The numerator $\|\hat{\mathbf{s}}\|^2$ in $\hat{p}$ is used to eliminate the effect of norm of the ASV on the estimated power.

## Robust Beamformer with Multiple Constraints on Uncertainty of Steering Vector

In the above section, the optimal beamformer is derived ignoring the unit norm constraint on the ASV. The optimal ASV $\hat{\mathbf{s}}$ in (37) may not have unit norm. Its effect on estimated output power is eliminated by the normalization shown in (39). In this section, the unit norm constraint of the ASV is considered in the derivation of robust beamformer. From the following new derivation, a clear relationship between the proposed approach and the PCA/MCA based beamformer is shown.

In practice, the ASV $\mathbf{s}_0$ is always unknown or known but with some errors. If $\mathbf{s}_0$ deviates from the true one, target signal cancellation is inevitable. This results in decreasing of the output power (19). A solution to this problem is searching for an optimal ASV $\mathbf{s}$, which results in maximal output power $\hat{p}$. We also assume that the ASV has unit norm. Therefore, the robust beamformer can be formulated as

$$\begin{cases} \max\limits_{\mathbf{s}} \min\limits_{\mathbf{w}} \mathbf{w}^H \mathbf{R} \mathbf{w} \\ \quad s.t\ \mathbf{s}^H \mathbf{w} = 1 \\ \qquad \|\mathbf{s}\|^2 = 1 \end{cases} \tag{40}$$

This is equivalent to

$$\begin{cases} \min\limits_{\mathbf{s}} \mathbf{s}^H \mathbf{R}^{-1} \mathbf{s} \\ s.t.\ \ \|\mathbf{s}\|^2 = 1 \end{cases} \tag{41}$$

which becomes a PCA/MCA problem (Jolliffe, 1986). The optimal $\hat{\mathbf{s}}$ is the eigenvector corresponding to the largest eigenvalue of $\mathbf{R}$.

However, if the target signal is not the dominant one, this method leads to a wrong solution. Therefore, additional constraint must be incorporated in the optimization problem (40). We assume that the true ASV $\mathbf{s}_0$ belongs to an uncertainty set as shown in (25). With this uncertainty constraint on the ASV, the robust beamformer is constructed by maximizing the output power of the SCB when an imprecise knowledge of its steering vector $\mathbf{s}_0$ is available.

$$\begin{cases} \max\limits_{\mathbf{s}} \min\limits_{\mathbf{w}} \mathbf{w}^H \mathbf{R} \mathbf{w} \\ \quad s.t.\ \mathbf{s}^H \mathbf{w} = 1 \\ \qquad \|\mathbf{s}\|^2 = 1 \\ \quad \|\mathbf{s} - \bar{\mathbf{s}}_0\|^2 \le \varepsilon \end{cases} \tag{42}$$

This is equivalent to

$$\begin{cases} \min_{\mathbf{s}} \mathbf{s}^H \mathbf{R}^{-1} \mathbf{s} \\ s.t. \ \ \|\mathbf{s}\|^2 = 1 \\ \|\mathbf{s} - \overline{\mathbf{s}}_0\|^2 \leq \varepsilon \end{cases} \tag{43}$$

The optimization problem (43) can be solved as follows. According to the optimization theory (Bazaraa & Shetty, 1979), inequality constraint may be inactive during optimization. Otherwise, the optimal solution is on the boundary of the constraints. Hence, we define a phase shifted vector $\tilde{\mathbf{u}}_1$ of $\mathbf{u}_1$ as $\tilde{\mathbf{u}}_1 = \mathbf{u}_1 e^{j\phi}$, where $\mathbf{u}_1$ is the eigenvector of **R** corresponding to the largest eigenvalue. If

$$\max_{\phi}(\tilde{\mathbf{u}}_1^H \overline{\mathbf{s}}_0 + \overline{\mathbf{s}}_0^H \tilde{\mathbf{u}}_1) = 2\mathbf{u}_1^H \overline{\mathbf{s}}_0 e^{j\phi_{opt}} \geq 2 - \varepsilon, \phi_{opt} = -\arg(\mathbf{u}_1^H \overline{\mathbf{s}}_0) \tag{44}$$

the uncertainty constraint is inactive during optimization. The vector $\mathbf{u}_1$ is the optimal solution to (43). Otherwise, the optimal solution is obtained on the boundary of the constraints, i.e.,

$$\begin{cases} \min_{\mathbf{s}} \mathbf{s}^H \mathbf{R}^{-1} \mathbf{s} \\ s.t. \ \ \|\mathbf{s}\|^2 = 1 \\ \|\mathbf{s} - \overline{\mathbf{s}}_0\|^2 = \varepsilon \end{cases} \tag{45}$$

With assumption on unit norm on $\mathbf{s}, \overline{\mathbf{s}}_0$, the above optimization problem (45) is equivalent to

$$\begin{cases} \min_{\mathbf{s}} \mathbf{s}^H \mathbf{R}^{-1} \mathbf{s} \\ s.t. \ \ \|\mathbf{s}\|^2 = 1 \\ 2 - \mathbf{s}^H \overline{\mathbf{s}}_0 - \overline{\mathbf{s}}_0^H \mathbf{s} = \varepsilon \end{cases} \tag{46}$$

To solve the above problem, we define

$$f(\mathbf{s}, g_1, g_2) = \mathbf{s}^H \mathbf{R}^{-1} \mathbf{s} + g_1(\|\mathbf{s}\|^2 - 1) + g_2(\mathbf{s}^H \overline{\mathbf{s}}_0 + \overline{\mathbf{s}}_0^H \mathbf{s} - 2 + \varepsilon) \tag{47}$$

The minimization of (47) is obtained when

$$\frac{\partial f(\mathbf{s}, g_1, g_2)}{\partial \mathbf{s}^*} = \mathbf{R}^{-1}\mathbf{s} + g_1 \mathbf{s} + g_2 \overline{\mathbf{s}}_0 = 0 \tag{48}$$

It yields

$$\hat{\mathbf{s}} = -g_2(\mathbf{R}^{-1} + g_1 \mathbf{I})^{-1} \overline{\mathbf{s}}_0 \tag{49}$$

Substituting (49) into constraint $\|\mathbf{s} - \overline{\mathbf{s}}_0\|^2 = \varepsilon$, we obtain

$$\hat{g}_2 = \frac{2 - \varepsilon}{-2\overline{\mathbf{s}}_0^H (\mathbf{R}^{-1} + g_1 \mathbf{I})^{-1} \overline{\mathbf{s}}_0} = \frac{2 - \varepsilon}{-2\sum_{i=1}^{M} \frac{\lambda_i |\tilde{s}_i|^2}{\lambda_i g_1 + 1}} \tag{50}$$

where $\tilde{s}_i$ is the $i$th element of the vector $\tilde{\mathbf{s}}_0 = \mathbf{U}^H\bar{\mathbf{s}}_0$. The calculation of $g_2$ can be ignored when the beamformer is not used for power estimation because the scale $g_2$ does not influence the output SINR. Substitute (49) into the constraint $\|\mathbf{s}\|^2 = 1$, the optimal estimate $\hat{g}_1$ can be calculated from the following equation

$$\psi(g_1) = \frac{\bar{\mathbf{s}}_0^H(\mathbf{R}^{-1}+g_1\mathbf{I})^{-2}\bar{\mathbf{s}}_0}{(\bar{\mathbf{s}}_0^H(\mathbf{R}^{-1}+g_1\mathbf{I})^{-1}\bar{\mathbf{s}}_0)^2} = \frac{\sum_{i=1}^{M}|\tilde{s}_i|^2\left(\frac{\lambda_i}{\lambda_i g_1+1}\right)^2}{\left(\sum_{i=1}^{M}|\tilde{s}_i|^2\frac{\lambda_i}{\lambda_i g_1+1}\right)^2} = \frac{4}{(2-\varepsilon)^2} \tag{51}$$

It is shown in Appendix A that the above equation has a unique solution in the range $(-\lambda_1^{-1}, +\infty)$.

With the estimates $\hat{g}_1, \hat{g}_2$, the optimal solution $\hat{\mathbf{s}}$ is given by

$$\hat{\mathbf{s}} = \hat{g}_2(\mathbf{R}^{-1}+\hat{g}_1\mathbf{I})^{-1}\bar{\mathbf{s}}_0 \tag{52}$$

The corresponding optimal weight of the beamformer is given by

$$\hat{\mathbf{w}}_0 = \frac{\mathbf{R}^{-1}\hat{\mathbf{s}}}{\hat{\mathbf{s}}^H\mathbf{R}^{-1}\hat{\mathbf{s}}} \tag{53}$$

and the estimates of the signal power $\hat{p}$ and output SINR $\rho$ are given by

$$\hat{p} = \frac{1}{\hat{\mathbf{s}}^H\mathbf{R}^{-1}\hat{\mathbf{s}}}, \rho = \frac{\mathbf{w}_0^H\mathbf{R}\mathbf{w}_0}{\mathbf{w}_0^H(\mathbf{R}_i+\mathbf{R}_n)\mathbf{w}_0} \tag{54}$$

where $\mathbf{R}_s, \mathbf{R}_i$ and $\mathbf{R}_n$ are the covariance matrices of the target signal, interference and non-directional noise, respectively.

## Robust Beamformer with New Uncertainty Constraints

Numerical studies indicate that the RCBs have degraded output SINR when the uncertainty level is large. This is because the robustness of the RCBs is obtained by using large uncertainty level $\varepsilon$, which reduces the capability of the RCBs in noise and interference suppression. Hence, a compact uncertainty set is needed to guarantee the performance on noise and interference suppression. In this section, a new compact uncertainty set which has lower uncertainty level is proposed. The compact uncertainty constraint is obtained by replacing the nominal ASV in conventional uncertainty set (Li et al., 2003) (Stoica et al., 2003) with the projection of the nominal ASV onto the signal-plus-interference subspace. Using the new compact constraint, robust beamformers with higher output SINR is achievable, as verified by theoretical analysis and simulation results.

As shown in (32) and (52), the optimal ASV $\hat{\mathbf{s}}$ used for the RCB1 and the RCB2 are

$$\begin{aligned} &RCB1: \hat{\mathbf{s}} = (g^{-1}\mathbf{R}^{-1}+\mathbf{I})^{-1}\bar{\mathbf{s}}_0 \\ &RCB2: \hat{\mathbf{s}} = -g_2(\mathbf{R}^{-1}+g_1\mathbf{I})^{-1}\bar{\mathbf{s}}_0 \end{aligned} \tag{55}$$

where $g$ is calculated from (33), $g_1$ and $g_2$ are calculated from (50) and (51). From our studies, we find that the output SINR of the RCBs degrades with large uncertainty set. In this section, a new method to improve the output SINR of the RCB while maintaining the robustness against the ASV error is derived.

As shown in (16), the orthogonal bases $\mathbf{U}_s$ of the steering vectors $\mathbf{s}_k, k = 0, \cdots, K-1$, is obtained by extracting the eigenvectors corresponding to the largest $K$ eigenvalues. $\mathbf{U}_s$ spans a linear space H,

$$H = \{\mathbf{s} \mid \mathbf{s} = \mathbf{U}_s\mathbf{c}, \mathbf{c} \in \mathbb{C}^K\} \quad (56)$$

where $\mathbb{C}^K$ is K-dimensional complex vector space. For easy illustration, we assume that there are two signals. The space H is illustrated in 3-dimension as shown in Figure 2. Herein, the nominal ASV $\bar{\mathbf{s}}_0$ does not coincide with the actual one $\mathbf{s}_0$. The conventional RCBs can be regarded as searching for an optimal ASV in the uncertainty set B to maximize the output power.

Although the actual ASV $\mathbf{s}_0$ is unknown, it locates in the space H. With this property, a new nominal ASV $\hat{\bar{\mathbf{s}}}_0$ in H can be estimated to form a compact uncertainty set with a smaller uncertainty level. This new ASV $\hat{\bar{\mathbf{s}}}_0$ is designed as a vector in H and nearest to $\bar{\mathbf{s}}_0$. It can be expressed as

$$\hat{\bar{\mathbf{s}}}_0 = \mathbf{U}_s\hat{\bar{\mathbf{c}}} \quad (57)$$

where $\hat{\bar{\mathbf{c}}}$ is the solution of the following optimization problem

$$\hat{\bar{\mathbf{c}}} = \arg\min_{\mathbf{c}} \| \mathbf{U}_s\mathbf{c} - \bar{\mathbf{s}}_0 \|^2 \quad (58)$$

The optimal solution of $\hat{\bar{\mathbf{s}}}_0$ is

$$\hat{\bar{\mathbf{s}}}_0 = \mathbf{U}_s\mathbf{U}_s^H\bar{\mathbf{s}}_0 \quad (59)$$

Since $\hat{\bar{\mathbf{s}}}_0$ is the projection of $\bar{\mathbf{s}}_0$ onto the signal-plus-interferences subspace H, it is straight-forward to show that the distance between the estimated ASV $\hat{\bar{\mathbf{s}}}_0$ and the actual one $\mathbf{s}_0$ is shorter than that between $\bar{\mathbf{s}}_0$ and $\mathbf{s}_0$, i.e.,

$$\| \mathbf{s}_0 - \hat{\bar{\mathbf{s}}}_0 \|^2 \le \| \mathbf{s}_0 - \bar{\mathbf{s}}_0 \|^2 \quad (60)$$

*Figure 2. Geometry model of the uncertainty set*

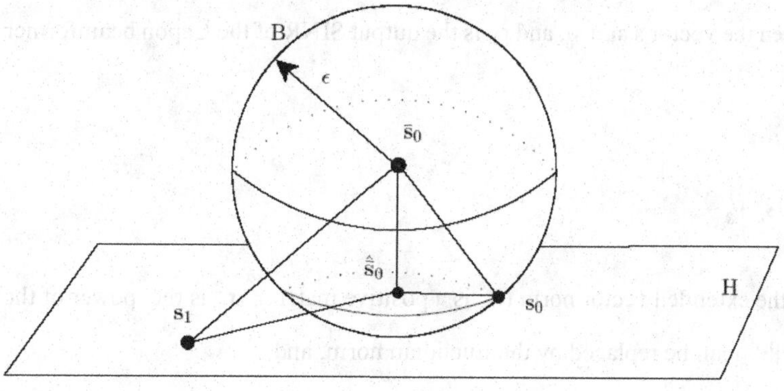

The new uncertainty constraint can be formulated as

$$\| \mathbf{s}_0 - \hat{\bar{\mathbf{s}}}_0 \|^2 \le \varepsilon' \tag{61}$$

where ε' is the new uncertainty level.

With the new uncertainty constraint (61), the proposed RCB is formulated as

$$\begin{cases} \min_{\mathbf{s}} \mathbf{s}^H \mathbf{R}^{-1} \mathbf{s} \\ s.t. \ \ \| \mathbf{s} - \hat{\bar{\mathbf{s}}}_0 \|^2 \le \varepsilon' \end{cases} \tag{62}$$

As we discussed, the optimal solution of the above problem is

$$\hat{\mathbf{s}} = (g^{-1}\mathbf{R}^{-1} + \mathbf{I})^{-1} \bar{\mathbf{s}}_0 \tag{63}$$

where the Lagrange multiplier $g$is the root of the constraint equation

$$\| \mathbf{s} - \bar{\mathbf{s}}_0 \|^2 = \| (\mathbf{I} + g\mathbf{R})^{-1} \bar{\mathbf{s}}_0 \|^2 = \varepsilon' \tag{64}$$

Additional norm constraint on **s** can also be included. Such kind of beamformer is called the projection based robust Capon beamformer (PRCB).

## Performance Analysis of the Proposed RCBs

In this section, the optimal output SINR of the proposed beamformer is derived. A complete performance analysis on the output SINR under general array imperfections represents a formidable analytical task. In this chapter, we assume that the array processor only has steering vector error. The theoretical covariance matrix is used in analysis. In such case, the performance degradation of the Capon beamformer is caused by the error in the nominal ASV. This assumption simplifies the performance analysis.

Before we derive the optimal output SINR of the proposed RCB, we first give the output SINR of the Capon beamformer in Lemma 1.

**Lemma 1:** Assume that the covariance matrices of the signal-of-interest (SOI) and interference/noise are $\mathbf{R}_s$ and $\mathbf{R}_n$, respectively. The covariance matrix of the array snapshot is $\mathbf{R} = \mathbf{R}_s + \mathbf{R}_n$. When the nominal ASV is given as **s**, and the true ASV is $\mathbf{s}_0$, the output SINR ρ of the Capon beamformer is given by

$$\rho = \frac{\rho_0 \cos^2(\theta)}{1 + \sin^2(\theta)\rho_0(\rho_0 + 2)}$$

where θ is the angle between the vector **s** and $\mathbf{s}_0$, and $\rho_0$ is the output SINR of the Capon beamformer when $\mathbf{s}_0$ is known, and

$$\cos^2(\theta) = \frac{|\mathbf{s}_0^H \mathbf{R}_n^{-1} \mathbf{s}|^2}{\| \mathbf{s}_0 \|_R^2 \| \mathbf{s} \|_R^2}$$

$$\rho_0 = \sigma_s^2 \mathbf{s}_0^H \mathbf{R}_n^{-1} \mathbf{s}_0 = \sigma_s^2 \| \mathbf{s}_0 \|_R^2$$

where $\| \mathbf{x} \|_R^2 = \mathbf{x}^H \mathbf{R}_n^{-1} \mathbf{x}$ is the extended vector norm ($\mathbf{R}_n$ is a positive matrix); $\sigma_s^2$ is the power of the SOI. If $\mathbf{R}_n = \sigma_n^2 \mathbf{I}$, the extended vector norm $\|\bullet\|_R$ can be replaced by the Euclidian norm, and

$$\cos^2(\theta) = \frac{|\mathbf{s}_0^H \mathbf{s}|^2}{\|\mathbf{s}_0\|_2^2 \|\mathbf{s}\|_2^2}$$

$$\rho_0 = \frac{\sigma_s^2}{\sigma_n^2} \|\mathbf{s}_0\|_2^2$$

**Proof.** Refer to (Yu, 2006a) (Yu, 2006b).

Lemma 1 indicates that the output SINR of the Capon beamformer is determined by the angle between the nominal ASV and the true ASV. Moreover, it is easy to find that the output SINR ρ is a monotonically increasing function of $\cos^2(\theta)$. From (55) and (63), we find that the proposed RCBs have similar mathematical form as the Capon beamformer except that the nominal vector $\overline{\mathbf{s}}_0$ is replaced by the estimated one, $\hat{\mathbf{s}}$. Therefore, the performance of the proposed RCBs can be analyzed via the angle between $\hat{\mathbf{s}}$ and $\mathbf{s}_0$. Different diagonal loading factors result in different $\hat{\mathbf{s}}$. The optimal diagonal loading factor should be selected to obtain the highest output SINR. In Lemma 2, we derive the optimal diagonal load factor and the output SINR of the RCB.

**Lemma 2:** Assume that the covariance matrix of the interference/noise is $\mathbf{R}_n$ and its eigen-decomposition is

$$\mathbf{R}_n = [\mathbf{U}_i \quad \mathbf{U}_n] \begin{bmatrix} \mathbf{\Sigma}_i & 0 \\ 0 & \mathbf{\Sigma}_n \end{bmatrix} [\mathbf{U}_i \quad \mathbf{U}_n]^H$$

where $\mathbf{U}_i$ and $\mathbf{U}_n$ are the eigenvector matrices which span the interference and noise subspaces respectively. $\mathbf{\Sigma}_i = diag\{\lambda_1, \lambda_2, \cdots, \lambda_K\}$ and $\mathbf{\Sigma}_n = \sigma_n^2 \mathbf{I}$ are the corresponding submatrices of the eigenvalue matrix. The nominal ASV is

$$\hat{\mathbf{s}} = (g^{-1}\mathbf{R}^{-1} + \mathbf{I})^{-1} \mathbf{s}_0$$

If $\lambda_i \geq \sigma_n^2, i = 1, \cdots, K$, the optimal output SINR $\rho_u$ of the output SINR is

$$\rho_u = \frac{\sigma_s^2 \|\mathbf{P}_{\mathbf{U}_n} \mathbf{s}_0\|^2}{\sigma_n^2}$$

which is achieved when

$$g_1 = \frac{-1}{\sigma_n^2 + \sigma_s^2 \|\mathbf{P}_{\mathbf{U}_n} \mathbf{s}_0\|^2}$$

provided that $\mathbf{s}_0^H \mathbf{P}_{\mathbf{U}_n} \mathbf{s}_0 \neq 0$. The matrix $\mathbf{P}_{\mathbf{U}_n} = \mathbf{U}_n \mathbf{U}_n^H$ is the projection matrix to the subspace spanned by $\mathbf{U}_n$. The power and ASV of the SOI are $\sigma_s^2$ and $\mathbf{s}_0$, respectively.

**Proof.** Refer to (Yu, 2006) and (Yu, 2006a; 2006b).

Lemma 2 indicates that the optimal output SINR of the proposed RCBs are achievable with negative diagonal loading factor. Since $\lambda_1 \geq \sigma_n^2 + \sigma_s^2 \|\mathbf{s}_0\|^2 \geq \sigma_n^2 + \sigma_s^2 \|\mathbf{P}_{\mathbf{U}_n} \mathbf{s}_0\|^2$, the optimal value of $g_1$ is not in $(\lambda_1^{-1}, +\infty)$, which is the solution range of the proposed RCBs. The proposed RCBs cannot achieve the highest output SINR. Nevertheless, the theoretical analysis in Lemma 3 and the simulation results in the next section will show that the proposed RCBs still have higher output SINRs than that of the others.

When there is only ASV error, a general conclusion on the output SINR of the Capon beamformer is given in Lemma 1. Lemma 1 indicates that the output SINR of the Capon beamformer is determined by the angle between the nominal and the true ASVs. Moreover, it is easy to show that the output SINR $\rho$ is a monotonically increasing function of $\cos^2(\theta)$.The PRCB and the RCB have similar mathematical form as the Capon beamformer except that the nominal vector **s** is replaced by the estimated one $\hat{\bar{\mathbf{s}}}$ or $\hat{\mathbf{s}}$. Therefore, the output SINR of the PRCB (RCB) can be analyzed via the angle between $\hat{\bar{\mathbf{s}}}$ ($\hat{\mathbf{s}}$) and $\mathbf{s}_0$.

**Lemma 3:** The ASVs used in the calculation of array optimal weight for the conventional RCBs and the PRCB are $\mathbf{s}_1$ and $\mathbf{s}_2$, respectively

$$\mathbf{s}_1 = (g_1^{-1}\mathbf{R}^{-1} + \mathbf{I})^{-1}\bar{\mathbf{s}}_0$$
$$\mathbf{s}_2 = (g_2^{-1}\mathbf{R}^{-1} + \mathbf{I})^{-1}\hat{\bar{\mathbf{s}}}_0$$

where the scales $g_1$ and $g_2$ are the optimal diagonal loading factors. Denoting

$$\cos^2(\theta_1) = \frac{|\mathbf{s}_0^H \mathbf{s}_1|^2}{\|\mathbf{s}_0\|_R^2 \|\mathbf{s}_1\|_R^2}$$
$$\cos^2(\theta_2) = \frac{|\mathbf{s}_0^H \mathbf{s}_2|^2}{\|\mathbf{s}_0\|_R^2 \|\mathbf{s}_2\|_R^2}$$

we have

$$\cos^2(\theta_1) \le \cos^2(\theta_2)$$

**Proof.** Refer to (Yu, 2006) and (Yu, 2006a; 2006b).

According to Lemma 1 and 3, it can be concluded that the output SINR $\rho_2$ of the PRCB is higher than that of the conventional RCB $\rho_1$, i.e.,

$$\rho_2 \ge \rho_1$$

## NUMERICAL EXPERIMENTS

### Performance Evaluation of RCBs without Projected ASV

In this section, some numerical simulations were carried out to evaluate the performance of the robust beamformers. A uniform linear array consisting of eight sensors with half-wavelength spacing is used to enhance the SOI in the presence of strong interferences as well as the uncertainty in ASV. There are two kinds of uncertainty under consideration: one is the well studied steering direction error; the other one is the random ASV error.

In the simulations, the estimate of signal power and SINR were the average of 200 Monte-Carlo experiments. The beam-patterns were obtained from one Monte-Carlo simulation. The non-directional noise was a spatially white Gaussian noise whose power is 0dB. The power of the SOI is $\sigma_0^2 = 10dB$, and the powers of the two interferences are $\sigma_1^2 = \sigma_2^2 = 20dB$. The DOA of the SOI is $\theta_0 = 0^0$. The DOAs of the two interferences are $\theta_1 = 60^0$ and $\theta_2 = 80^0$, respectively. The performance of the SCB is also demonstrated for the purpose of comparison.

In the first simulation, the array was assumed to have a steering direction error $\Delta = 3^0$. The covariance matrix was estimated with different number of snapshots. It is well known that the covariance matrix estimated using sample averaging method asymptotically approaches the true one. In the case only a small number of snapshots available, the estimated error in covariance matrix also affects the performance of beamformer. The result shown in Figure 3 indicates that the output powers of the RCBs are close to the true one even the number of snapshots is small. With increasing number of snapshots, the output SINRs are improved for the RCBs. However, for the SCB, due to steering direction error, the target signal is cancelled. Its output SINR remains a low level. From the beam-pattern shown in Figure 4, the difference between the beamformers is clear. The SCB has a null at the direction of target signal while the other beamformers still have high response gain at the direction of target signal.

*Figure 3. Output power and SINR versus the number of snapshots with steering error Δ = 3, ε = 0.12*

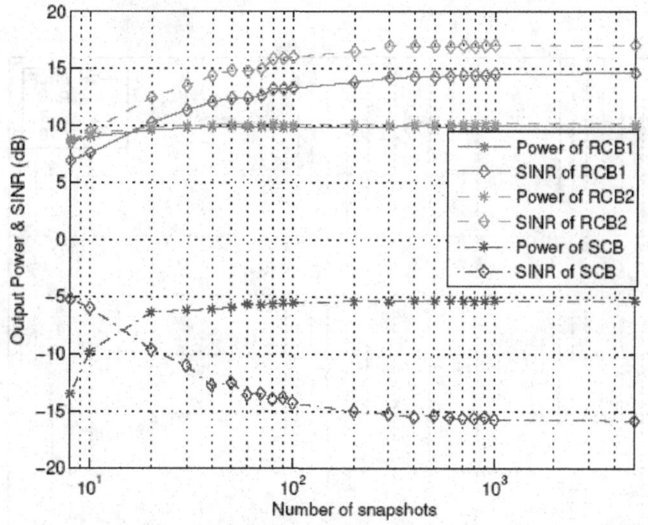


*Figure 4. Comparison of beampattern of RCB and SCB (The vertical dot lines indicate the direction of incident signals (Δ = 3, ε = 0.12)*

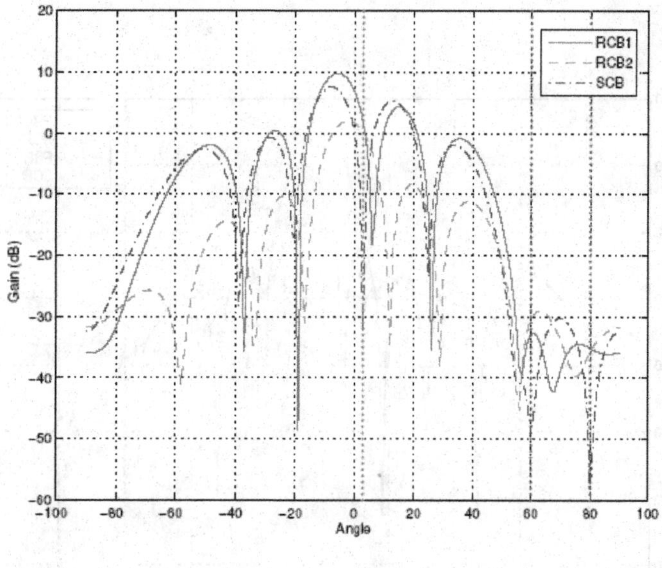


## Performance Evaluation of RCBs Using Projected ASV

In this section, some numerical experiments were carried out to evaluate the performance of the PRCB with a new compact constraint set. The performance of the SCB, the RCBs and the eigenspace-based beam-former (ESB) (Feldman & Griffiths, 1994) are also included for the purpose of performance comparison.

We assume that the array has steering direction error. The assumed direction of the SOI is $\theta_0 = 0^0$, but the actual one is Δ. In the first experiment, the direction and power pair of the SOI are $(6^0, 10dB)$. The direction and power pairs of the two interferences are $(60^0, 20dB)$ and $(80^0, 20dB)$, respectively. The output SINR of each beamformer versus

*Figure 5. Performance comparison of beamformers with 3 impinging sources (vertical dot lines stand for the directions of incident signals, Δ = 6, ε = 8.9, ε' = 0.1)*

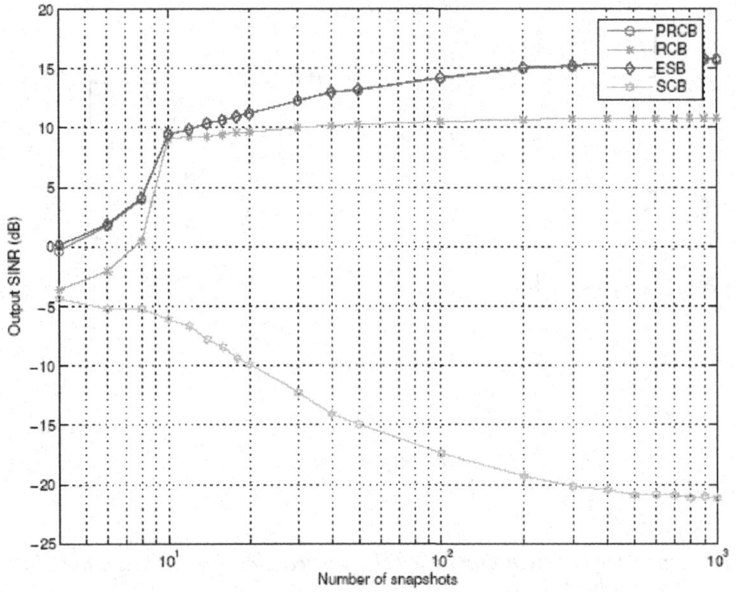


*(a) Comparison of output SINR of beamformers versus number of snapshots*

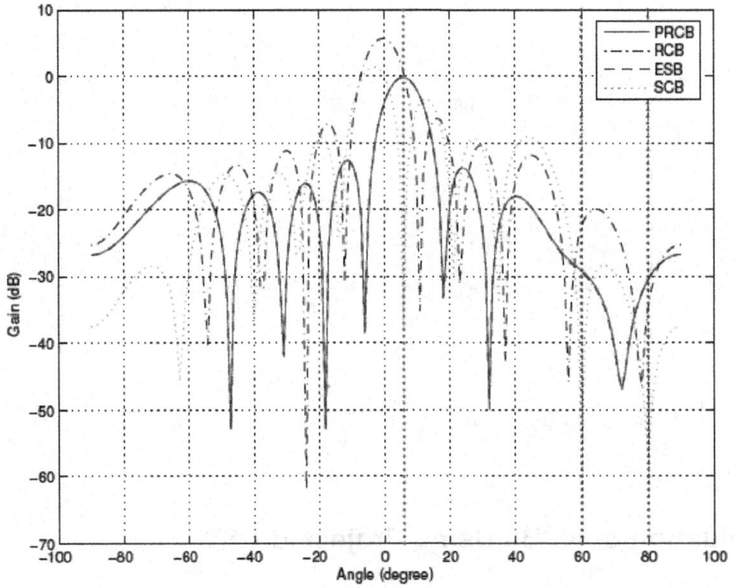


*(b) Comparison of the beampatterns (N = 100)*

the number of snapshots is shown to illustrate the beamformer performance because the covariance matrix is always estimated with a limited number of snapshots in practical applications.

From the results shown in Figure 5(a), it can be found that the PRCB, RCB and ESB all perform well at a small number of snapshots, *N*. Since the projected ASV has a high accuracy in this experiment, the PRCB and ESB have similar output SINR (The curves of PRCB and ESB in Figure 5 overlap). The SINR of the RCB is lower than that of the PRCB and ESB. The SCB completely fails. In Figure 5(b), it is found that the PRCB and ESB have similar beam-patterns. Although the response of the RCB to the SOI is similar to that of the PRCB and ESB, its response peaks at $\theta = 0^0$. This causes a larger noise gain of the RCB. Moreover, Fig. 5(b) implies that the interference suppression performance of the RCB is lower than that of the PRCB and ESB. All of these factors result in lower output SINR of the RCB.

In the next experiment, the interferences are moved close to the SOI, e.g., the direction and power pairs of the two interferences are changed to $(10^0, 20dB)$ and $(20^0, 20dB)$, respectively. The output SINR versus the number of snapshots *N* is shown in Figure 6(a), where the PRCB shows the highest output SINR. Since the directions of the interferences are close to that of the SOI in this experiment, the error in the projected ASV increases. It results in performance degradation of the ESB. On the other hand, the PRCB can tolerate the ASV error to some extent.

It is known that the performance of the ESB degrades when the dimension of the signal-plus-interference subspace is high. In this experiment, we evaluate the performance of the beamformers with a large number of interferences, i.e., the dimension of signal-plus-interference subspace is high. Considering six interferences whose direction and power pairs are ($60^0$, 10dB), ($80^0$, 20dB), ($-30^0$, 20dB), ($-50^0$, 10dB), ($-70^0$, 20dB) and ($-85^0$, 20dB), respectively. The results shown in Figure 7(a) clearly indicate that ESB has poor performance because of large error in the projected ASV due to the high dimension of the signal-plus-interference subspace. Meanwhile, the PRCB outperforms the other beamformers in output SINR because the PRCB also has capability to tolerate the ASV error.

It is also known that the performance of the ESB strongly depends on the accurate knowledge of the dimension of signal-plus-interference subspace. The following three experiments evaluate the performance of the beamformers when error exists in the estimation of the dimension of signal-plus-interference subspace. In these experiments, the direction and power pairs of the SOI and two interferences are ($6^0$, 10dB),($60^0$, 20dB) and ($80^0$, 20dB), respectively. The results in Figure 8 are obtained when the dimension of the subspace is overestimated as four. With this overestimated dimension parameter, the performance of the ESB seriously degrades, while the performance of PRCB is still better than those of the others. Two more experiments are carried out to evaluate the performances of these beamformers when the dimension parameter is underestimated as two. When the SOI is not the dominant signal, the results in Figure 9 show that both the PRCB and ESB fail. On the other hand, if the SOI becomes the dominant signal with power $\sigma_0^2 = 30dB$, these two beamformers still work, as can be found in Figure 10. From these experiments, we suggest that overestimation of the dimension of the signal-plus-interference subspace is needed for the PRCB to guarantee its robustness.

In the above simulations, it is assumed that the array has steering direction error only. In the following simulations, random error in the ASV is considered and the performance of beamformers is evaluated. The ASV errors in the simulations are generated as a random complex vector whose norm is 40% of the norm of the true ASV. Figure 11 shows that the PRCB and ESB have similar performance. The results in Figure 12 are obtained using 6 signals similar to the above experiments except that the direction of the SOI is changed to 0. The results in Figure 13 and Figure 14 are obtained when error exists in the estimate of the dimension of the signal-plus-interference subspace. Similar explanations of these results can be referred above. In summary, conclusion can be obtained that the PRCB is robust against arbitrary ASV error. Moreover, it has high output SINR regardless of the type of ASV error.

## SUMMARY

In this chapter, we introduce robust beamformers in two kinds of optimization methods: one is based on quadratic optimization with linear and nonlinear constraints; the other one is based on max-min optimization method. The contemporary robust beamformers are derived based on output power maximization subject to a uncertainty constraint on the nominal ASV. Theoretical performance analysis and comparison are presented for the robust Capon beamformers (RCBs). We proved that the RCBs cannot achieve the optimal output SINR. However, simulation results show that they have high robustness against arbitrary ASV error and relative high output SINR. The projection based RCB uses a new constraint by replacing the nominal ASV in the original constraint by the projected one into signal-plus-interference subspace. It provides much higher performance than the other RCBs. Simulations on various cases were carried out to show the effectiveness of all these RCBs. The theoretical analysis also indicates that there is still room for the improvement of the RCBs, which will be studied in future research.

*Figure 6. Performance comparison of beamformers with 3 impinging sources (vertical dot lines stand for the directions of incident signals Δ = 6, ε = 5.0, ε' = 4.0)*

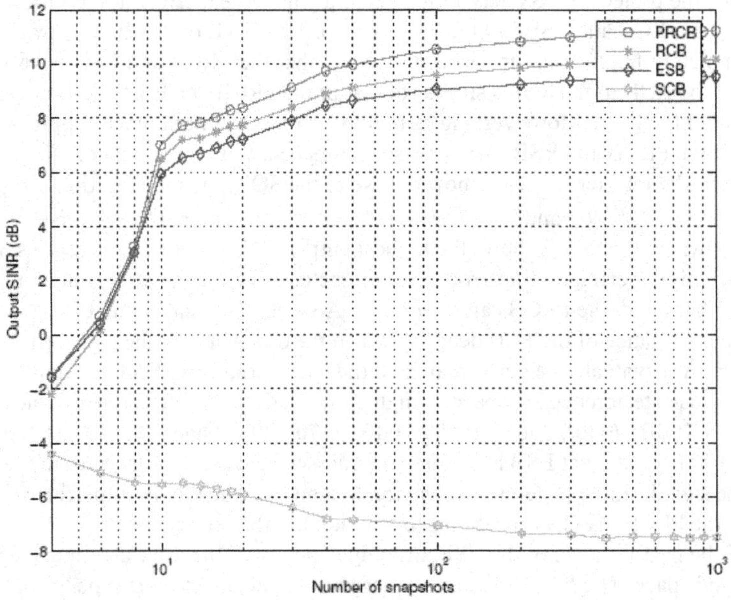


*(a) Comparison of output SINR of beamformers versus number of snapshots*

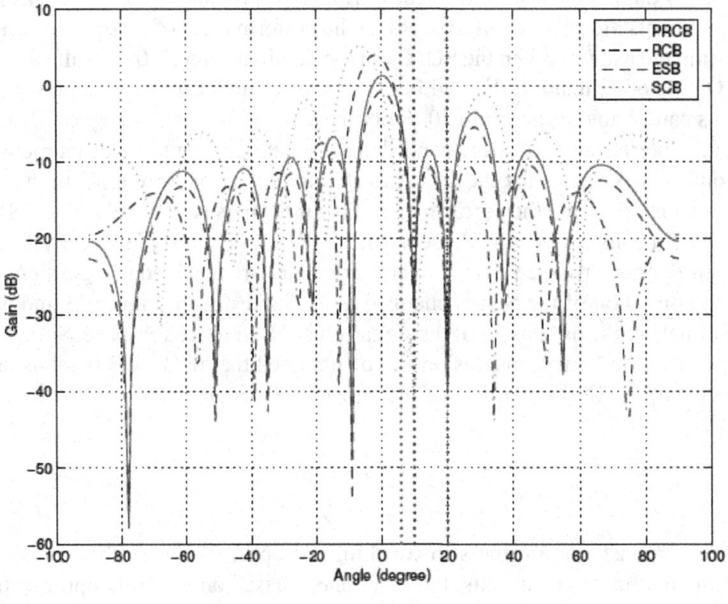


*(b) Comparison of the beampatterns (N = 100)*

*Figure 7. Performance comparison of beamformers with 6 impinging sources (vertical dot lines stand for the directions of incident signals, Δ = 6, ε = 7.0, ε' = 3.1)*

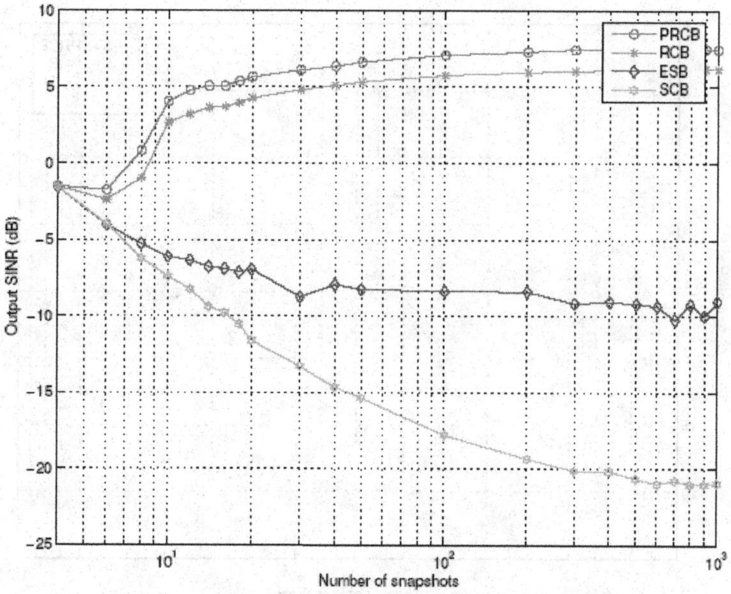


*(a) Comparison of output SINR of beamformers versus number of snapshots*

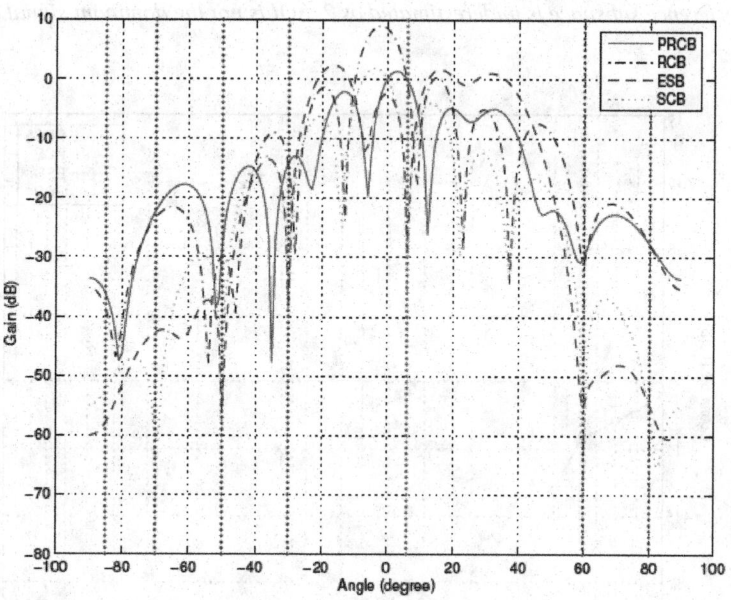


*(b) Comparison of the beampatterns*

*Figure 8. Comparison of output SINR of beamformers versus number of snapshots with 3 impinging sources (The dimension of signal-plus-interference subspace is overestimated as ε = 8.9, ε' = 4.0)*

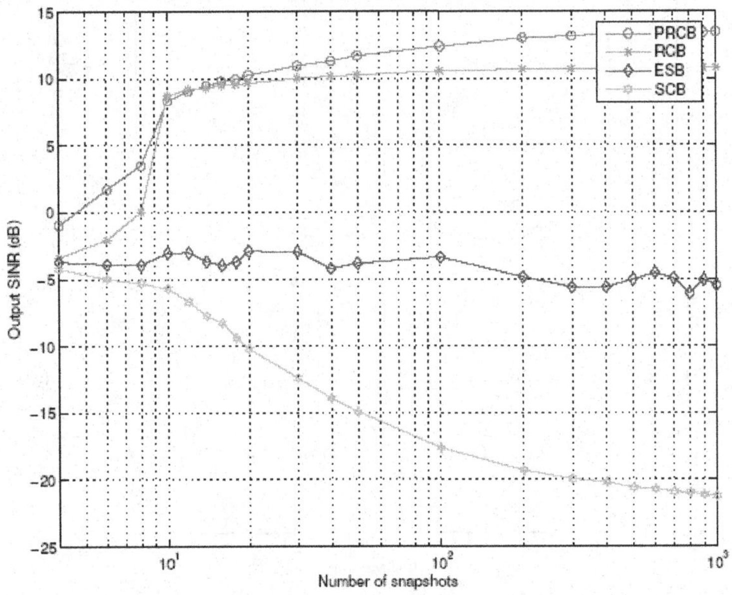


*Figure 9. Comparison of output SINR of beamformers versus number of snapshots with 3 impinging sources (The dimension of signal-plus-interference subspace is underestimated as 2. SOI is not the dominant signal ε = 9.0, ε' = 0.1)*

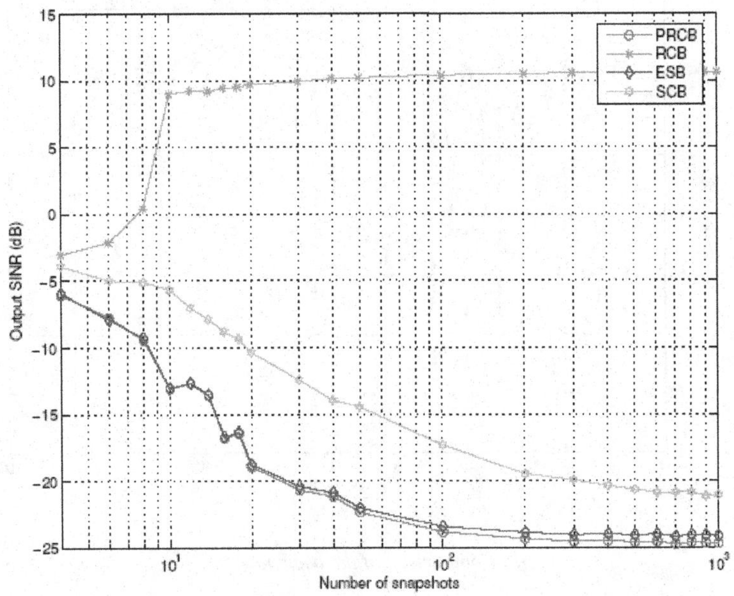

*Figure 10. Comparison of output SINR of beamformers versus number of snapshots with 3 impinging sources (The dimension of signal-plus-interference subspace is underestimated as 2. SOI is the dominant signal ε = 9.0, ε' = 8.0)*

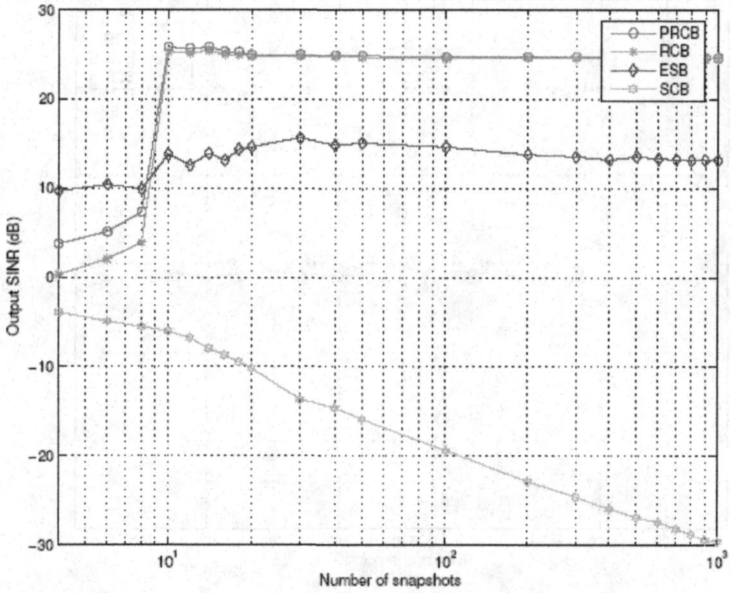


*Figure 11. Comparison of output SINR of beamformers with arbitrary ASV error (The parameters of target source and two interferences are ($0^0$,10dB),($60^0$,20dB) and ($80^0$,20dB), ε = 4.0, ε' = 0.1)*

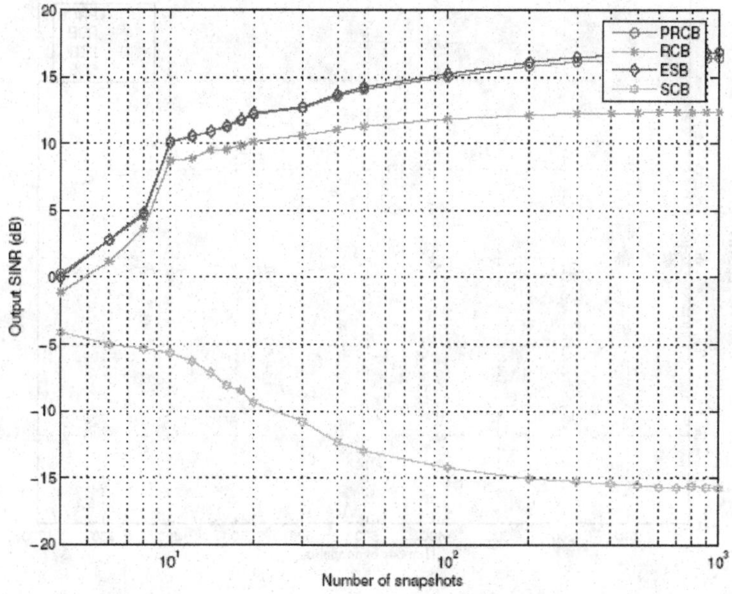

*Figure 12. Comparison of output SINR of beamformers with arbitrary ASV error with 7 impinging sources (ε = 4.0, ε' = 2.5)*

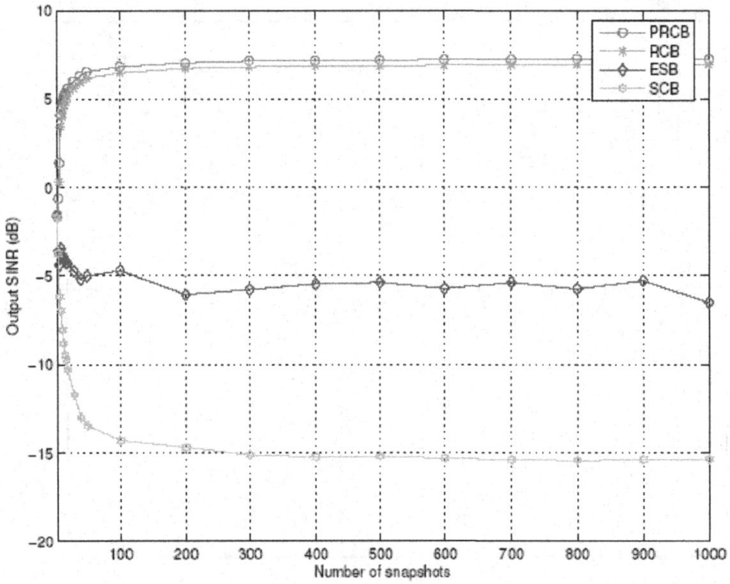


*Figure 13. Comparison of output SINR of beamformers versus number of snapshots with 3 impinging soruces (The dimension of signal-plus-interference subspace is overestimated as 4. ε = 6.0, ε' = 0.3)*

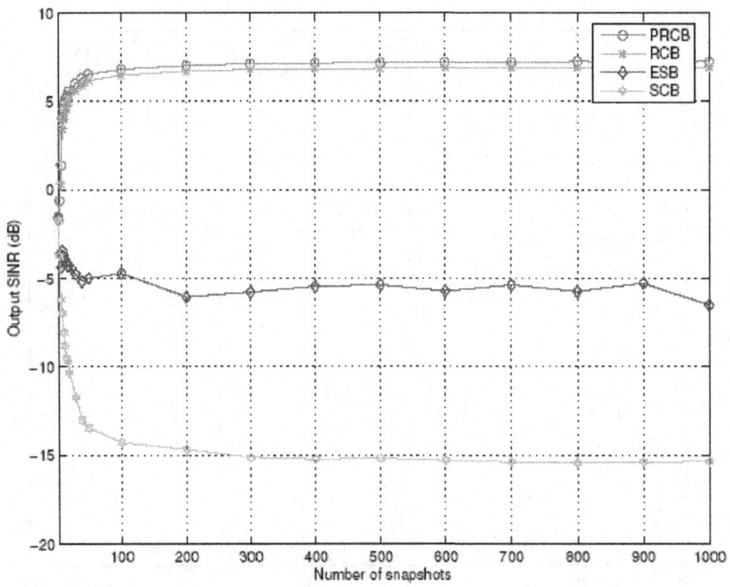

## ACKNOWLEDGMENT

The authors would like to thank the anonymous reviewers for their invaluable comments. Reprinted from Yu, Z.L. (2009). A robust minimum variance beamformer with new constraint on uncertainty of steering vector. In Handbook on Advancements in Smart Antenna Technologies for wireless Networks/Robust Adaptive Beamforming, (pp. 38-59); with permission from Elsevier.


## REFERENCES


Affes, S., & Grenier, Y. (1997). A signal subspace tracking algorithm for microphone array processing of speech. *IEEE Trans. Speech Audio Processing, 5*(5), 425-437.

Bazaraa, M. S., & Shetty, C. M. (1979). *Nonlinear programming: Theory and algorithms.* John Wiley & Sons.

Buckley, K. M. (1987). Spatial/spectral filtering with linearly constrained minimum variance beam-formers. *IEEE Trans. Acoust., Speech, Signal Processing, 35*(3), 249-266.

Buckley, K. M., & Griffiths, L. J. (1986). An adaptive generalized sidelobe canceller with derivative constraints. *IEEE Trans. Antennas Propagat., 34*(3), 311-319.

Capon, J. (1969). High-resolution frequency wavenumber spectrum analysis. In *Proc. IEEE* (Vol. 57).

Chang, L., & Yeh, C. C. (1993). Effect of pointing errors on the performance of the projection beamformer. *IEEE Trans. Antennas Propagat., 41*(8), 1045 -1056.

Claesson, I., & Nordholm, S. (1992). A spatial filtering approach to robust adaptive beamforming. *IEEE Trans. Antennas Propagat., 40*(9), 1093-1096.

Compton, R. (1988). Adaptive antennas: Concepts and performance. New Jersey: Prentice-Hall.

Cox, H. (1973). Resolving power and sensitivity to mismatch of optimum array processors. *J. Acoust. Soc. Am., 54*(3), 771-785.

Cox, H., Zeskind, R. M., & Owen, M. M. (1987). Robust adpative beamforming. *IEEE Trans. Acoust., Speech, Signal Processing, 35*(10), 1365-1376.

Cox, H., Zeskind, R. M., & Owen, M. M. (1988). Effects of amplitude and phase errors on linear predictive array processors. *IEEE Trans. Acoust., Speech, Signal Processing, 36*(1), 10-19.

Er, M. H. (1985). *Optimum antenna array processors with linear and quadratic constraints.* Doctoral dissertation, The University of Newcastle, New South Wales, Australia.

Er, M. H. (1988). Adaptive antenna array under directional and spatial derivative constraints. In *Proc. IEEE, 135*, 414-419.

Er, M. H., & Cantoni, A. (1983). Derivative constraints for broad-band element space antenna array processors. *IEEE Trans. Acoust., Speech, Signal Processing, 31*(6), 1378-1393.

Er, M. H., & Cantoni, A. (1986a). A new approach to the design of broad-band element space antenna array processors. *IEEE J. Ocean. Eng.*, 10, 231-240.

Er, M. H., & Cantoni, A. (1986b). A new set of linear constraints for broad-band time domain element space processor. *IEEE Trans. Antennas Propagat., 34*(3), 320-329.

Er, M. H., & Cantoni, A. (1986c). An unconstrained partitioned realization for derivative constrained broad-band antenna array processors. *IEEE Trans. Acoust. Speech Signal Processing, 34*(6), 1376-1379.

Er, M. H., & Cantoni, A. (1990). A unified approach to the design of robust narrow-band antenna array processor. *IEEE Trans. Antennas Propagat., 38*(1), 17-23.

Er, M. H., & Ng, B. C. (1994). A new approach to robust beamforming in the presence of steering vector errors. *IEEE Trans. Signal Processing, 42*(7), 1826-1829.

Feldman, D., & Griffiths, L. (1994). A projection approach for robust adaptive beamforming. *IEEE Trans. Signal Processing, 42*(4), 867-876.

Fudge, G. L. (1996). Spatial blocking filter derivative constraints for the generalized sidelobe canceller and MUSIC. *IEEE Trans. Signal Processing, 44*(1), 51-61.

Fudge, G. L., & Linebarger, D. A. (1994). A calibrated generalized sidelobe canceller for wideband beamforming. *IEEE Trans. Signal Processing, 42*(10), 2871 -2875.

Hudson, J. E. (1981). *Adaptive array principles.* Peter Peregrinus Ltd.

Jablon, N. K. (1986a). Adaptive beamforming with the generalized sidelobe canceller in the presence of array imperfection. *IEEE Trans. Antennas Propagat., 34*(8), 996-1012.

Jablon, N. K. (1986b). Steady state analysis of the generalized sidelobe canceller by adapitve noise cancelling techniques. *IEEE Trans. Antennas Propagat., 34*(3), 330-337.

Johnson, D. H., & Dudgeon, D. E. (1993). *Array signal processing: Concepts and techniques.* Prentice Hall, Inc.

Jolliffe, I. T. (1986). *Principal component analysis.* New York: Springer-Verlag.

Krim, H., & Viberg, M. (1996). Two decades of array signal processing research: the parametric approach. *IEEE Signal Processing Mag., 13*(4), 67-94.

Li, J., Stoica, P., & Wang, Z. (2003). On robust Capon beamforming and diagonal loading. *IEEE Trans. Signal Processing, 51*(7), 1702-1715.

Li, J., Stoica, P., & Wang, Z. (2004). Doubly constrained robust capon beamformer. *IEEE Trans. Signal Processing, 52*(9), 2407-2423.

Lorenz, R. G., & Boyd, S. P. (2005). Roubst minimum variance beamforming. *IEEE Trans. Signal Processing, 53*(5), 1684-1696.

Monzingo, R. A., & Miller, T. W. (1980). *Introduction to adaptive arrays.* New York: Wiley.

Naidu, P. S. (2001). *Sensor array signal processing.* CRC Press London.

Nunn, D. (1983). Performance assessments of a time-domain adaptive antenna processor in a broadband environment. In *Proceedings Inst. Elect. Eng.* F, H, (Vol. 130).

Stoica, P., Wang, Z., & Li, J. (2003). Robust Capon beamforming. *IEEE Signal Processing Lett., 10*(6), 172-175.

Takao, K., Fujita, M., & Nishi, T. (1976). An adaptive antenna array under directional constraint. *IEEE Trans. Antennas Propogat., 24*(5), 662-669.

Thng, I., Cantoni, A., & Leung, Y. H. (1993). Derivative constrained optimum broad-band antenna arrays. *IEEE Trans. Signal Processing, 41*(7), 2376-2388.

Thng, I., Cantoni, A., & Leung, Y. H. (1995). Constraints for maximally flat optimum broadband antenna arrays. *IEEE Trans. Signal Processing, 43*(6), 1334-1347.

Tsoulos, G. V. (2001). *Adaptive antennas for wireless communications.* New York: IEEE Press.

Van Veen, B. D., & Buckley, K. M. (1988). Beamforming: a versatile approach to spatial filtering. *IEEE ASSP Mag., 5*(2), 4-24.

Vorobyov, S. A., Gershman, A. B., & Luo, Z. Q. (2003). Robust adaptive beamforming using worst-case performance optimization: A solution to the signal mismatch problem. *IEEE Trans. Signal Processing, 51*(2), 313-324.

Vural, A. M. (1979). Effects of perturbation on the performance of optimum/adaptive arrays. *IEEE Trans. Aerosp. Electron. Syst., 15*(1), 76-87.

Yu, Z. L. (2006). *New techniques for robust beamformer design.* Doctoral dissertation, School of Electric and Electronic Engineering, Nayang Technological University, Singapore.

Yu, Z. L., & Er, M. H. (2006a). A robust capon beamformer against uncertainty of nominal steering vector. *EURASIP Journal on Applied Signal Processing*, 1-8.

Yu, Z. L., & Er, M. H. (2006b). A robust minimum variance beamformer with new constraint on uncertainty of steering vector. *Signal Processing, 86*(9), 2243-2254.

Yu, Z. L., Zou, Q., & Er, M. H. (2004). A new approach to robust beamforming against generalized phase errors. In *Proc. of IEEE 6th circuit and system symposium on emerging technologies* (Vol. 2, pp. 775-778). Shanghai, China.

Zhang, S., & Thng, I. (2002). Robust presteering derivative constraints for broadband antenna arrays. *IEEE Trans. Signal Processing, 50*(1), 1-10.

Zou, Q., Yu, Z. L., & Lin, Z. (2004). A robust algorithm for linearly constrained adaptive beam-forming. *IEEE Signal Processing Letters, 11*(1), 26-29.

## APPENDIX A

The diagonal loading factor $g_1$ is obtained by solving

$$\psi(g_1) = \frac{4}{(2-\varepsilon)^2}.$$

The derivative of $\psi$ to $g_1$ is given by

$$\frac{d\psi(g_1)}{dg_1} = \frac{\left(\sum_{i=1}^{M} |\tilde{s}_i|^2 \left(\frac{\lambda_i}{\lambda_i g_1+1}\right)^2\right)^2 - \left(\sum_{i=1}^{M} |\tilde{s}_i|^2 \left(\frac{\lambda_i}{\lambda_i g_1+1}\right)^3\right)\left(\sum_{i=1}^{M} |\tilde{s}_i|^2 \left(\frac{\lambda_i}{\lambda_i g_1+1}\right)\right)}{\left(\sum_{i=1}^{M} |\tilde{s}_i|^2 \frac{\lambda_i}{\lambda_i g_1+1}\right)^4}.$$

When $g_1 > -\lambda_1^{-1}$, the item $\frac{\lambda_i}{\lambda_i g_1+1} > 0$. Using Schwartz inequality, we have

$$\left(\sum_{i=1}^{M} |\tilde{s}_i|^2 \left(\frac{\lambda_i}{\lambda_i g_1+1}\right)^2\right)^2 - \left(\sum_{i=1}^{M} |\tilde{s}_i|^2 \left(\frac{\lambda_i}{\lambda_i g_1+1}\right)^3\right)\left(\sum_{i=1}^{M} |\tilde{s}_i|^2 \left(\frac{\lambda_i}{\lambda_i g_1+1}\right)\right) \le 0.$$

Hence $\frac{d\psi(g_1)}{dg_1} \le 0$, $\psi(g_1)$ is a monotonically decreasing function of $g_1$. Additionally,

$$\lim_{g_1 \to \infty} \psi(g_1) = 1,$$

$$\lim_{g_1 \to -\lambda_1^{-1}} \psi(g_1) = 1/|\tilde{s}_i|^2 \ge 1/|\mathbf{u}_1^H \bar{\mathbf{s}}_0|^2 \ge \frac{4}{(2-\varepsilon)^2}.$$

There is a unique root of (51) in the range $(-\lambda_1^{-1}, \infty)$.

# Chapter III
# Adaptive Beamforming Assisted Receiver

**Sheng Chen**
*University of Southampton, UK*

## ABSTRACT

*Adaptive beamforming is capable of separating user signals transmitted on the same carrier frequency, and thus provides a practical means of supporting multiusers in a space-division multiple-access scenario. Moreover, for the sake of further improving the achievable bandwidth efficiency, high-throughput quadrature amplitude modulation (QAM) schemes have become popular in numerous wireless network standards, notably, in the recent WiMax standard. This contribution focuses on the design of adaptive beamforming assisted detection for the employment in multiple-antenna aided multiuser systems that employ the high-order QAM signalling. Traditionally, the minimum mean square error (MMSE) design is regarded as the state-of-the-art for adaptive beamforming assisted receiver. However, the recent work (Chen et al., 2006) proposed a novel minimum symbol error rate (MSER) design for the beamforming assisted receiver, and it was demonstrated that this MSER design provides significant performance enhancement, in terms of achievable symbol error rate, over the standard MMSE design. This MSER beamforming design is developed fully in this contribution. In particular, an adaptive implementation of the MSER beamforming solution, referred to as the least symbol error rate algorithm, is investigated extensively. The proposed adaptive MSER beamforming scheme is evaluated in simulation, in comparison with the adaptive MMSE beamforming benchmark.*

## INTRODUCTION

The ever-increasing demand for mobile communication capacity has motivated the development of antenna array assisted spatial processing techniques (Winters *et al.*, 1994; Litva & Lo, 1996; Godara, 1997; Kohno, 1998; Winters, 1998; Petrus *et al.*, 1998; Tsoulos, 1999; Vandenameele *et al.*, 2001; Blogh & Hanzo, 2002; Soni *et al.*, 2002; Paulraj *et al.*, 2003; Paulraj *et al.*, 2004; Tse & Viswanath, 2005) in order to further improve the achievable spectral efficiency. A specific technique that has shown real promise in achieving substantial capacity enhancements is the use of adaptive beamforming with antenna arrays (Litva & Lo, 1996; Blogh & Hanzo, 2002). Through appropriately combining the signals received by the different elements of an antenna array, adaptive beamforming is capable of separating user signals transmitted on the same carrier frequency, provided that they are separated sufficiently in the angular or spatial domain. Adaptive beamforming technique thus provides a practical means of supporting multiusers in a space-division multiple-access scenario. For the sake of further improving the achievable bandwidth efficiency, high-throughput quadrature amplitude modulation (QAM) schemes (Hanzo *et al.*, 2004) have become popular in numerous wireless network standards. For example, the 16-QAM and 64-QAM schemes were adopted in the recent WiMax standard (IEEE 802.16). Classically, the beamforming process is carried out by minimising the mean square error (MSE) between the desired output and the actual array output, and this principle is rooted in the traditional beamforming employed in sonar and radar systems. An advantage of this minimum MSE (MMSE) beamforming design is that its adaptive implementation can readily be achieved using the well-known least mean square (LMS) algorithm, recursive least squares algorithm and many other adaptive schemes (Widrow *et al.*, 1967; Griffiths, 1969; Reed *et al.*, 1974; Widrow & Stearns, 1985; Ganz *et al.*, 1990; Haykin, 1996). For potential use in downlink adaptive beamforming receiver, we will only consider the stochastic gradient-based LMS algorithm in this study owing to the computational simplicity of this adaptive algorithm. The MMSE design has been regarded as the state-of-the-art for adaptive beamforming assisted receiver, despite of the fact that, for a communication system, it is the bit error rate (BER) or symbol error rate (SER) that really matters.

Ideally, the system design should be based directly on minimising the BER or SER, rather than the MSE. Adaptive beamforming design based directly on minimising the system's BER has been proposed for the binary phase shift keying (BPSK) modulation (Chen *et al.*, 2003; Wolfgang *et al.*, 2004; Garcia *et al.*, 2004; Liu & Yang, 2004; Chen *et al.*, 2005; Fan *et al.*, 2005; Ahmad, 2005) and quadrature phase shift keying (QPSK) modulation (Chen *et al.*, 2004; Chen *et al.*, 2005a). These studies have demonstrated that the adaptive minimum BER (MBER) beamforming design can significantly improve the system performance, in terms of achievable BER, over the conventional MMSE design. The MBER beamforming is the true state-of-the-art and it is more intelligent than the MMSE solution, since it directly optimises the system's BER performance, rather than minimising the MSE, where the latter strategy often turns out to be deficient in the rank-deficient situation when the number of the users supported exceeds the number of the receiver antennas. Thus, the adaptive MBER beamforming design has a larger user capacity than its adaptive MMSE counterpart. Simulation results also show that the MBER design is more robust in near-far situations than the MMSE design. For the system that employs high-order QAM signalling, it is computationally more attractive by minimising the system's SER. This has led to the adaptive minimum SER (MSER) beamforming design for QAM systems (Chen *et al.*, 2006). The present constribution expands the work by Chen *et al.*( 2006) and provides a detailed investigation for the adaptive MSER beamforming design for the generic multiple-antenna assisted multiuser system employing high-order QAM signalling.

The organisation of this contribution is as follows. Section 2 introduces the system model, which is used in Section 3 for studying the adaptive MMSE and MSER beamforming designs. Section 4 concentrates on investigating the achievable SER performance of the proposed adaptive MSER scheme in both the stationary and Rayleigh fading channels, using the adaptive MMSE scheme as a benchmark, while Section 5 presents the concluding remarks.

## SYSTEM MODEL

The system supports $S$ users, and each user transmits an $M$-QAM signal on the same carrier frequency of $\omega = 2\pi f$. For such a system, user separation can be achieved in the spatial or angular domain (Paulraj *et al.*, 2003; Tse & Viswanath, 2005) and the receiver is equipped with a linear antenna array consisting of $L$ uniformly spaced elements. Assume that the channel is narrow-band which does not induce intersymbol interference. Then the symbol-rate received signal samples can be expressed as

$$x_l(k) = \sum_{i=1}^{S} A_i b_i(k) e^{j\omega t_l(\theta_i)} + n_l(k) = \bar{x}_l(k) + n_l(k), \tag{1}$$

for $1 \le l \le L$, where $t_l(\theta_i)$ is the relative time delay at array element $l$ for source $i$ with $\theta_i$ being the direction of arrival for source $i$, $n_l(k)$ is a complex-valued Gaussian white noise with $E[|n_l(k)|^2] = 2\sigma_n^2$, $A_i$ is the narrow-band channel coefficient for user $i$, $\bar{x}_l(k)$ denotes the noiseless part of $x_l(k)$ and $b_i(k)$ is the $k$-th symbol of user $i$ which takes the value from the $M$-QAM symbol set

$$B \stackrel{\Delta}{=} \{b_{l,q} = u_l + ju_q, 1 \le l, q \le \sqrt{M}\} \tag{2}$$

with the real-part symbol $\Re[b_{l,q}] = u_l = 2l - \sqrt{M} - 1$ and the imaginary-part symbol $\Im[b_{l,q}] = u_q = 2q - \sqrt{M} - 1$. Without the loss of generality, assume that source 1 is the desired user and the rest of the sources are interfering users. The desired-user signal-to-noise ratio (SNR) is given by SNR $= |A_1|^2 \sigma_b^2 / 2\sigma_n^2$ and the desired signal-to-interferer $i$ ratio (SIR) is SIR$_i = A_1^2/A_i^2$, for $2 \le i \le S$, where $\sigma_b^2$ denotes the $M$-QAM symbol energy. The received signal vector $\mathbf{x}(k) = [x_1(k)\ x_2(k) \cdots x_L(k)]^T$ can be expressed as

$$\mathbf{x}(k) = \mathbf{Pb}(k) + \mathbf{n}(k) = \bar{\mathbf{x}}(k) + \mathbf{n}(k), \tag{3}$$

where $\mathbf{n}(k) = [n_1(k)\ n_2(k) \cdots n_L(k)]^T$, the system matrix $\mathbf{P} = [A_1\mathbf{s}_1\ A_2\mathbf{s}_2 \cdots A_S\mathbf{s}_S]$ with the steering vector for source $i$ given by $\mathbf{s}_i = [e^{j\omega t_1(\theta_i)}\ e^{j\omega t_2(\theta_i)} \cdots e^{j\omega t_L(\theta_i)}]^T$, and the transmitted QAM symbol vector $\mathbf{b}(k) = [b_1(k)\ b_2(k) \cdots b_S(k)]^T$.

Before it is proceeded further, the assumptions implied for the above system model are explained and justified. Although a linear antenna array structure with uniformly spaced elements is assumed, the approach is actually more general, and it is equally applicable to the generic narrow-band multiple-input multiple-output (MIMO) system (Paulraj *et al.*, 2003; Tse & Viswanath, 2005) modelled by $\mathbf{x}(k) = \mathbf{Pb}(k) + \mathbf{n}(k)$, where the $(l, i)$-th element of the channel matrix $\mathbf{P}$ represents the non-dispersive channel connecting the $i$-th transmit antenna to the $l$-th receive antenna. Except for the reference user's channel coefficient $A_1$ and steering vector $\mathbf{s}_1$, the receiver does not need to know the interfering users' channel coefficients $A_1$ and steering vector $\mathbf{s}_i$, $2 \le i \le S$. The adaptive beamforming approach considered is based on the so-called temporal reference technique, and during the training the reference user's transmitted symbols are available at the receiver for the adaptation purpose. The receiver, however, does not have access to the interfering users' data symbols. As will be explained latter, the first column $\mathbf{p}_1$ of the system matrix $\mathbf{P}$, corresponding to the desired user, is required at the receiver in order to detect the desired user's data symbols unbiasedly.

In the system model (3), the desired user and interfering signals are assumed to be symbol-synchronised. For the downlink scenario synchronous transmission of the users is guaranteed. By contrast, in an uplink scenario the differently delayed asynchronous signals of the users are no longer automatically synchronised. However, the quasi-synchronous operation of the system may be achieved with the aid of adaptive timing advance control, as in the global system of mobile (GSM) communications (Steele & Hanzo 1999). The GSM system has a timing-advance control accuracy of 0.25 bit duration. Since synchronous systems perform better than their asynchronous counterparts (Hwang & Hanzo 2003), the third-generation partnership research consortium (3GPP) is also considering the employment of timing-advance control in next-generation systems. In general, when the number of users is large, the users are asynchronous and the idealistic assumption of perfect power control is stipulated, the performance gain of the (symbol-rate) MSER solution over the MMSE beamformer may be expected to diminish, since the interference becomes nearly Gaussian at the symbol-rate samples. One way of maintaining the benefits of the MSER solution for asynchronous systems is to perform a joint MSER detection and synchronisation by sampling faster than the symbol rate. During each symbol period, several signal samples are taken and the receiver maintains several tentative MSER detectors. The detector having the smallest SER is chosen to perform symbol detection. In this study, symbol-rate synchronisation is assumed. For such a symbol-synchronised interference-limited QAM system the non-Gaussian nature of the interfering signals is effectively exploited by the MSER beamforming receiver, resulting in an improved SER performance.

A beamformer is employed at the receiver, whose soft output is given by

$$y(k) = \mathbf{w}^H \mathbf{x}(k) = \mathbf{w}^H (\bar{\mathbf{x}}(k) + \mathbf{n}(k)) = \bar{y}(k) + e(k), \tag{4}$$

where $\mathbf{w} = [w_1\ w_2 \cdots w_L]^T$ is the complex-valued beamformer weight vector and $e(k)$ is Gaussian distributed with zero mean and $E[|e(k)|^2] = 2\sigma_n^2 \mathbf{w}^H \mathbf{w}$. Define the combined system impulse response of the channel and beamformer as $\mathbf{w}^H \mathbf{P} = \mathbf{w}^H [\mathbf{p}_1\ \mathbf{p}_2 \cdots \mathbf{p}_S] = [c_1\ c_2 \cdots c_S]$. The beamformer's output can alternatively be expressed as

$$y(k) = c_1 b_1(k) + \sum_{i=2}^{S} c_i b_i(k) + e(k), \tag{5}$$

where the first term in the righthand side of equation is the desired user signal and the second term is the residual multiuser interference. Note that, in any detection scheme, the main tap $c_1$ must be known. That is, the desired user's channel and associated steering vector, namely $\mathbf{p}_1 = A_1\mathbf{s}_1$, must be known at the receiver. If this fact is overlooked, the decision will be biased (Cioffi *et al.*, 1995). Provided that $c_1 = c_{R_1} + jc_{I_1}$ satisfies $c_{R_1} > 0$ and $c_{I_1} = 0$, the symbol decision $\hat{b}_1(k) = \hat{b}_{R_1}(k) + j\hat{b}_{I_1}(k)$ can be made as

$$\hat{b}_{R_1}(k) = \begin{cases} u_1, & \text{if } y_R(k) \le c_{R_1}(u_1+1), \\ u_l, & \text{if } c_{R_1}(u_l-1) < y_R(k) \le c_{R_1}(u_l+1) \text{ for } 2 \le l \le \sqrt{M}-1, \\ u_{\sqrt{M}}, & \text{if } y_R(k) > c_{R_1}(u_{\sqrt{M}}-1), \end{cases} \tag{6}$$

$$\hat{b}_{I_1}(k) = \begin{cases} u_1, & \text{if } y_I(k) \le c_{R_1}(u_1+1), \\ u_q, & \text{if } c_{R_1}(u_q-1) < y_I(k) \le c_{R_1}(u_q+1) \text{ for } 2 \le q \le \sqrt{M}-1, \\ u_{\sqrt{M}}, & \text{if } y_I(k) > c_{R_1}(u_{\sqrt{M}}-1), \end{cases} \tag{7}$$

where $y(k) = y_R(k) + jy_I(k)$ and $\hat{b}_1(k)$ is the estimate for $b_1(k) = b_{R_1}(k) + jb_{I_1}(k)$. Fig. 1 depicts the decision thresholds associated with the decision $\hat{b}_1(k) = b_{l,q}$. In general, $c_1 = \mathbf{w}^H\mathbf{p}_1$ is complex-valued and the rotating operation

$$\mathbf{w}^{\text{new}} = \frac{c_1^{\text{old}}}{\left|c_1^{\text{old}}\right|}\mathbf{w}^{\text{old}} \tag{8}$$

can be used to make $c_1$ real and positive. This rotation is a linear operation and it does not change the system's SER.

*Figure 1. Decision thresholds associated with point $c_1 b_{l,q}$ assuming $c_{R_1} > 0$ and $c_{I_1} = 0$, and illustrations of symmetric distribution of $Y_{l,q}$ around $c_1 b_{l,q}$.*

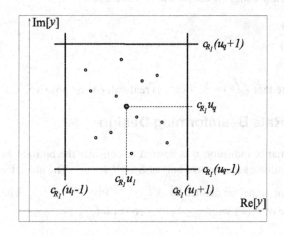

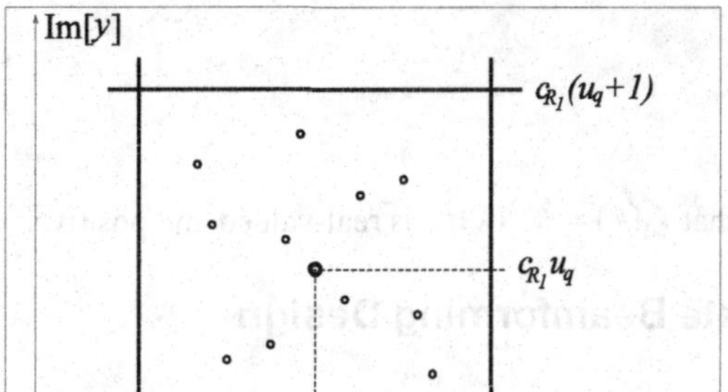

## ADAPTIVE BEAMFORMING ASSISTED RECEIVERS

Different beamforming designs derive the beamformer's weight vector $\mathbf{w}$ based on optimising different design criteria. The best-known design criterion is the MMSE criterion, while the novelty of this contribution is to optimise the beamformer's weight vector based on the MSER criterion. The MMSE and MSER designs are considered in this contribution.

### Minimum Mean Square Error Beamforming Design

The traditional design for the beamformer (4) is the MMSE solution, which minimises the MSE criterion $E[|b_1(k) - y(k)|^2]$, leading to the solution for the weight vector given by

$$\mathbf{w}_{\text{MMSE}} = \left( \mathbf{P}\mathbf{P}^H + \frac{2\sigma_n^2}{\sigma_b^2}\mathbf{I}_L \right)^{-1} \mathbf{p}_1, \tag{9}$$

where $\mathbf{I}_L$ denotes the $L \times L$ identity matrix. The MMSE beamforming design is computationally attractive, because it admits the closed-form solution given the second order statistics of the underlying system. Furthermore, it can be implemented adaptively using the classical LMS algorithm (Widrow & Stearns, 1985; Haykin, 1996). Given the current beamformer's output $y(k) = \hat{\mathbf{w}}^H(k)\mathbf{x}(k)$, the LMS algorithm modified for the adaptation of the QAM beamformer (4) is expressed as

$$\tilde{\mathbf{w}}(k+1) = \hat{\mathbf{w}}(k) + \mu\big(b_1(k) - y(k)\big)^* \mathbf{x}(k), \tag{10}$$

$$\tilde{c}_1(k+1) = \tilde{\mathbf{w}}^H(k+1)\hat{\mathbf{p}}_1, \tag{11}$$

$$\hat{\mathbf{w}}(k+1) = \frac{\tilde{c}_1(k+1)}{|\tilde{c}_1(k+1)|}\tilde{\mathbf{w}}(k+1), \tag{12}$$

where $\mu$ is a small positive step size, $\hat{\mathbf{p}}_1$ is an estimate of $\mathbf{p}_1$, and (11) and (12) implement the weight rotation operation. Given a training data block $\{b_1(k), \mathbf{x}(k)\}_{k=1}^N$, a block-based estimate of $\mathbf{p}_1$ is given by

$$\hat{\mathbf{p}}_1 = \frac{1}{N}\sum_{k=1}^{N} \frac{\mathbf{x}(k)}{b_1(k)}. \tag{13}$$

Alternatively, the receiver can track $\mathbf{p}_1$ using the simple moving average

$$\hat{\mathbf{p}}_1(k+1) = (1-\alpha)\hat{\mathbf{p}}_1(k) + \alpha\frac{\mathbf{x}(k)}{b_1(k)}, \tag{14}$$

where $0 < \alpha < 1$ is a step size. Note that $\hat{c}_1(k) = \hat{\mathbf{w}}^H(k)\hat{\mathbf{p}}_1$ is real-valued and positive.

### Minimum Symbol Error Rate Beamforming Design

Since the SER is the true performance indicator, it is desired to consider the optimal MSER Beamforming solution. Denote the $N_b = M^S$ number of legitimate sequences of $\mathbf{b}(k)$ as $\mathbf{b}_i$, $1 \le i \le N_b$. The noise-free part of the received signal, namely $\bar{\mathbf{x}}(k)$, takes values from the signal set defined by $X \stackrel{\Delta}{=} \{\bar{\mathbf{x}}_i = \mathbf{P}\mathbf{b}_i, 1 \le i \le N_b\}$. The set $X$ can be partitioned into $M$ subsets, depending on the value of $b_1(k)$ as $X_{l,q} \stackrel{\Delta}{=} \{\bar{\mathbf{x}}_i \in X : b_1(k) = b_{l,q}\}$, $1 \le l, q \le \sqrt{M}$. Similarly the noise-free part of the beamformer's output, namely $\overline{y}(k)$, takes values from the scalar set $Y \stackrel{\Delta}{=} \{\overline{y}_i = \mathbf{w}^H \overline{\mathbf{x}}_i, 1 \leq i \leq N_b\}$, and $Y$ can be divided into the $M$ subsets conditioned on $b_1(k)$

$$Y_{l,q} \stackrel{\Delta}{=} \{\overline{y}_i \in Y : b_1(k) = b_{l,q}\}, 1 \leq l, q \leq \sqrt{M}. \tag{15}$$

The following two lemmas summarise the properties of the signal subsets $Y_{l,q}$, $1 \leq l, q \leq \sqrt{M}$, which are useful in the derivation of the SER expression for the beamformer (4).

**Lemma 1:** The subsets $Y_{l,q}$, $1 \leq l, q \leq \sqrt{M}$, satisfy the shifting properties

$$Y_{l+1,q} = Y_{l,q} + 2c_1, 1 \leq l \leq \sqrt{M} - 1, \tag{16}$$

$$Y_{l,q+1} = Y_{l,q} + j2c_1, 1 \leq q \leq \sqrt{M} - 1, \tag{17}$$

$$Y_{l+1,q+1} = Y_{l,q} + (2 + j2)c_1, 1 \leq l, q \leq \sqrt{M} - 1. \tag{18}$$

**Proof.** Any point $\overline{y}_i^{(l+1,q)} \in Y_{l+1,q}$ can be expressed as

$$\overline{y}_i^{(l+1,q)} = \mathbf{w}^H \mathbf{P} \mathbf{b}_i^{(l+1,q)} = \mathbf{w}^H \mathbf{P} \left( \mathbf{b}_i^{(l,q)} + [2\, 0 \cdots 0]^T \right) = \overline{y}_i^{(l,q)} + 2c_1$$

where $\overline{y}_i^{(l,q)} \in Y_{l,q}$. This proves the shifting property (16). Proofs for the other two equations are similar.

**Lemma 2.** The points of $Y_{l,q}$ are distributed symmetrically around the symbol point $c_1 b_{l,q}$. This symmetric distribution is with respect to the two horizontal decision boundaries and the two vertical decision boundaries that separate $Y_{l,q}$ from the other subsets.

Lemma 2 is a direct consequence of symmetric distribution of the symbol constellation (2) and Lemma 1. This symmetric property is also illustrated in Figrue 1.

For the beamformer with weight vector **w**, denote

$$P_E(\mathbf{w}) = \text{Prob}\{\hat{b}_1(k) \neq b_1(k)\}, \tag{19}$$

$$P_{E_R}(\mathbf{w}) = \text{Prob}\{\hat{b}_{R_1}(k) \neq b_{R_1}(k)\}, \tag{20}$$

$$P_{E_I}(\mathbf{w}) = \text{Prob}\{\hat{b}_{I_1}(k) \neq b_{I_1}(k)\}. \tag{21}$$

$P_E(\mathbf{w})$ is the total SER, while $P_{E_R}(\mathbf{w})$ and $P_{E_I}(\mathbf{w})$ are the real-part and imaginary-part SERs, respectively. It is then easy to see that the SER is given by

$$P_E(\mathbf{w}) = P_{E_R}(\mathbf{w}) + P_{E_I}(\mathbf{w}) - P_{E_R}(\mathbf{w}) P_{E_I}(\mathbf{w}). \tag{22}$$

From the beamforming model (4) and the signal model (3), the conditional probability density function (PDF) of $y(k)$ given $b_1(k) = b_{l,q}$ is a Gaussian mixture (hence a non-Gaussian PDF) defined by

$$p(y \mid b_{l,q}) = \frac{1}{N_{sb} 2\pi\sigma_n^2 \mathbf{w}^H \mathbf{w}} \sum_{i=1}^{N_{sb}} e^{-\frac{|y-\bar{y}_i^{(l,q)}|^2}{2\sigma_n^2 \mathbf{w}^H \mathbf{w}}}, \quad (23)$$

where $N_{sb} = N_b/M$ is the size of $Y_{l,q}$, $\bar{y}_i^{(l,q)} = \bar{y}_{R_i}^{(l,q)} + j\bar{y}_{I_i}^{(l,q)} \in Y_{l,q}$, and $y = y_R + jy_I$. Noting that $c_1$ is real-valued and positive and taking into account the symmetric distribution of $Y_{l,q}$ (Lemma 2), for $2 \le l \le \sqrt{M}-1$, the conditional error probability of $\hat{b}_{R_1}(k) \neq u_l$ given $b_{R_1}(k) = u_l$ can be shown to be (Chen *et al.*, 2004a)

$$P_{E_R,l}(\mathbf{w}) = \frac{2}{N_{sb}} \sum_{i=1}^{N_{sb}} Q(g_{R_i}^{(l,q)}(\mathbf{w})), \quad (24)$$

where

$$Q(u) = \frac{1}{\sqrt{2\pi}} \int_u^{\infty} e^{-\frac{z^2}{2}} dz, \quad (25)$$

and

$$g_{R_i}^{(l,q)}(\mathbf{w}) = \frac{\bar{y}_{R_i}^{(l,q)} - c_{R_1}(u_l - 1)}{\sigma_n \sqrt{\mathbf{w}^H \mathbf{w}}}. \quad (26)$$

Further taking into account the shifting property (Lemma 1), it can be shown that

$$P_{E_R}(\mathbf{w}) = \gamma \frac{1}{N_{sb}} \sum_{i=1}^{N_{sb}} Q(g_{R_i}^{(l,q)}(\mathbf{w})), \quad (27)$$

where $\gamma = \frac{2\sqrt{M}-2}{\sqrt{M}}$. It is seen that $P_{E_R}$ can be evaluated using (real part of) any single subset $Y_{l,q}$. Similarly, $P_{E_I}$ can be evaluated using (imaginary part of) any single subset $Y_{l,q}$ as

$$P_{E_I}(\mathbf{w}) = \gamma \frac{1}{N_{sb}} \sum_{i=1}^{N_{sb}} Q(g_{I_i}^{(l,q)}(\mathbf{w})) \quad (28)$$

with

$$g_{I_i}^{(l,q)}(\mathbf{w}) = \frac{\bar{y}_{I_i}^{(l,q)} - c_{R_1}(u_q - 1)}{\sigma_n \sqrt{\mathbf{w}^H \mathbf{w}}}. \quad (29)$$

Note that the SER is invariant to a positive scaling of **w**.

The MSER solution $\mathbf{w}_{\text{MSER}}$ is defined as the weight vector that minimises the upper bound of the SER given by

$$P_{E_B}(\mathbf{w}) = P_{E_R}(\mathbf{w}) + P_{E_I}(\mathbf{w}), \quad (30)$$

that is,

$$\mathbf{w}_{\text{MSER}} = \arg\min_{\mathbf{w}} P_{E_B}(\mathbf{w}). \tag{31}$$

The solution obtained by minimising the upper bound (30) is practically equivalent to that of minimising $P_E(\mathbf{w})$, since the bound $P_E(\mathbf{w}) < P_{E_B}(\mathbf{w})$ is very tight, that is, $P_{E_B}(\mathbf{w})$ is very close to the true SER $P_E(\mathbf{w})$. Unlike the MMSE solution, the MSER solution does not admits a closed-form solution. However, the gradients of $P_{E_R}(\mathbf{w})$ and $P_{E_I}(\mathbf{w})$ with respect to $\mathbf{w}$ can be shown to be respectively

$$\nabla P_{E_R}(\mathbf{w}) = \frac{\gamma}{2N_{sb}\sqrt{2\pi}\sigma_n\sqrt{\mathbf{w}^H\mathbf{w}}} \sum_{i=1}^{N_{sb}} e^{-\frac{\left(\bar{y}_{R_i}^{(l,q)} - c_{R_1}(u_l-1)\right)^2}{2\sigma_n^2\mathbf{w}^H\mathbf{w}}}$$

$$\times\left(\frac{\bar{y}_{R_i}^{(l,q)} - c_{R_1}(u_l-1)}{\mathbf{w}^H\mathbf{w}}\mathbf{w} - \bar{\mathbf{x}}_i^{(l,q)} + (u_l-1)\mathbf{p}_1\right), \tag{32}$$

$$\nabla P_{E_I}(\mathbf{w}) = \frac{\gamma}{2N_{sb}\sqrt{2\pi}\sigma_n\sqrt{\mathbf{w}^H\mathbf{w}}} \sum_{i=1}^{N_{sb}} e^{-\frac{\left(\bar{y}_{I_i}^{(l,q)} - c_{R_1}(u_q-1)\right)^2}{2\sigma_n^2\mathbf{w}^H\mathbf{w}}}$$

$$\times\left(\frac{\bar{y}_{I_i}^{(l,q)} - c_{R_1}(u_q-1)}{\mathbf{w}^H\mathbf{w}}\mathbf{w} + j\bar{\mathbf{x}}_i^{(l,q)} + (u_q-1)\mathbf{p}_1\right), \tag{33}$$

where $\bar{\mathbf{x}}_i^{(l,q)} \in X_{l,q}$. The derivation of the gradients (32) and (33) follows the same procedure given in (Chen *et al.*, 2001; 2004a; 2005; 2006). With the gradient $\nabla P_{E_B}(\mathbf{w}) = \nabla P_{E_R}(\mathbf{w}) + \nabla P_{E_I}(\mathbf{w})$, the optimisation problem (31) can be solved iteratively using a gradient-based algorithm. Since the SER is invariant to a positive scaling of $\mathbf{w}$, it is computationally advantageous to normalise $\mathbf{w}$ to a unit-length vector $\mathbf{w}$ after every iteration, so that the gradients (32) and (33) are simplified to

$$\nabla P_{E_R}(\mathbf{w}) = \frac{\gamma}{2N_{sb}\sqrt{2\pi}\sigma_n} \sum_{i=1}^{N_{sb}} e^{-\frac{\left(\bar{y}_{R_i}^{(l,q)} - c_{R_1}(u_l-1)\right)^2}{2\sigma_n^2}}$$

$$\times\left(\left(\bar{y}_{R_i}^{(l,q)} - c_{R_1}(u_l-1)\right)\mathbf{w} - \bar{\mathbf{x}}_i^{(l,q)} + (u_l-1)\mathbf{p}_1\right) \tag{34}$$

and

$$\nabla P_{E_I}(\mathbf{w}) = \frac{\gamma}{2N_{sb}\sqrt{2\pi}\sigma_n} \sum_{i=1}^{N_{sb}} e^{-\frac{\left(\bar{y}_{I_i}^{(l,q)} - c_{R_1}(u_q-1)\right)^2}{2\sigma_n^2}}$$

$$\times\left(\left(\bar{y}_{I_i}^{(l,q)} - c_{R_1}(u_q-1)\right)\mathbf{w} + j\bar{\mathbf{x}}_i^{(l,q)} + (u_q-1)\mathbf{p}_1\right). \tag{35}$$

The following algorithm, which is a modified version of the simplified conjugate gradient algorithm (Bazaraa *et al.*, 1993; Chen *et al.*, 2001), provides an efficient means of finding an MSER solution.

- *Initialisation.* Choose a step size of $\mu > 0$ and a termination scalar of $\beta > 0$; given $\mathbf{w}(1)$ and $\mathbf{d}(1) = -\nabla P_{E_B}(\mathbf{w}(1))$; set the iteration index to $\iota = 1$.
- *Loop.* If $\nabla P_{E_B}(\mathbf{w}(\iota)) = \sqrt{(\nabla P_{E_B}(\mathbf{w}(\iota)))^H \nabla P_{E_B}(\mathbf{w}(\iota))} < \beta$: goto *Stop*. Else, $\tilde{\mathbf{w}}(\iota+1) = \mathbf{w}(\iota) + \mu\mathbf{d}(\iota)$,
- $c_1(\iota+1) = \tilde{\mathbf{w}}^H(\iota+1)\mathbf{p}_1$,
- $\bar{\mathbf{w}}(\iota+1) = \dfrac{c_1(\iota+1)}{|c_1(\iota+1)|}\tilde{\mathbf{w}}(\iota+1)$,
- $\mathbf{w}(\iota+1) = \dfrac{\bar{\mathbf{w}}(\iota+1)}{\|\bar{\mathbf{w}}(\iota+1)\|}$,
- $\phi_\iota = \dfrac{\|\nabla P_{E_B}(\mathbf{w}(\iota+1))\|^2}{\|\nabla P_{E_B}(\mathbf{w}(\iota))\|^2}$,
- $\mathbf{d}(\iota+1) = \phi_\iota \mathbf{d}(\iota) - \nabla P_{E_B}(\mathbf{w}(\iota+1))$,
- $\iota = \iota + 1$, goto *Loop.*
- *Stop.* $\mathbf{w}(\iota)$ is the solution.

At a minimum, $\|\nabla P_{E_B}(\mathbf{w})\| = 0$. Hence the termination scalar $\beta$ determines the accuracy of the solution obtained. The step size $\mu$ controls the rate of convergence. Typically, a much larger value of $\mu$ can be used compared to the steepest-descent gradient algorithm. As the SER surface $P_{E_B}(\mathbf{w})$ is highly nonlinear, occasionally the search direction **d** may no longer be a good approximation to the conjugate gradient direction or may even point to the "uphill" direction, when the iteration index becomes large. It is thus advisable to periodically reset **d** to the negative gradient in the above conjugate gradient algorithm. With this resetting mechanism, this conjugate gradient algorithm has been shown to converge fast to the theoretical MSER solution, typically in tens of iterations, in many simulation studies. Although in theory there is no guarantee that the above conjugate gradient algorithm can always find the global minimum point of the SER surface $P_{E_B}(\mathbf{w})$, in practice we have found that the algorithm works well and we have never observed any occurrence of the algorithm being trapped at some local minimum solution.

It is worth emphasising that there exist infinitely many global MSER solutions which forms an infinite half line in the beamforming weight space. This is because the SER is invariant to a positive scaling of **w**, i.e., the size of **w** does not matter (except for zero size). Thus, the SER surface has an infinitely long valley, and any point at the bottom of this valley is a true global MSER solution. For an illustration, see the simple example given in (Chen *et al.*, 2001). Once we restrict to the unit-length **w**, the MSER solution becomes unique. As alternatives to the simplified conjugate gradient algorithm, global optimisation search algorithms, such as the genetic algorithm (Goldberg, 1989; Man *et al.*, 1998) and adaptive simulated annealing (Ingber, 1993; Chen & Luk, 1999), can be used to obtain a global minimum solution of $P_{E_B}(\mathbf{w})$, at an expense of considerably increased computational requirements.

## Adaptive Minimum Symbol Error Rate Beamforming

In practice, the system matrix **P** is unknown (except its first column). Therefore adaptive implementation is required to realise the MSER beamforming. To adaptively implement the MMSE solution, the unknown second-order statistics can be estimated based on a block of training data. Furthermore, by considering a single-sample "estimate" of the MSE, the stochastic adaptive algorithm known as the LMS algorithm is derived. A similar adaptive implementation strategy can be adopted for adaptive MSER beamforming. The PDF $p(y)$ of $y(k)$ can be estimated using the Parzen window estimate (Parzen, 1962; Silverman, 1996; Bowman & Azzalini, 1997) based on a block of training data. This leads to an estimated SER for the beamformer. Minimising this estimated SER based on a gradient optimisation yields an approximated MSER solution. To derive a sample-by-sample adaptive algorithm, consider a single-sample "estimate" of $p(y)$

$$\tilde{p}(y,k) = \frac{1}{2\pi\rho_n^2} e^{-\frac{|y-y(k)|^2}{2\rho_n^2}} \tag{36}$$

and the corresponding one-sample SER "estimate" $\tilde{P}_{E_B}(\mathbf{w},k)$. The parameter $\rho_n$ is known as the kernel width. Using the instantaneous stochastic gradient of $\nabla\tilde{P}_{E_B}(\mathbf{w},k) = \nabla\tilde{P}_{E_R}(\mathbf{w},k) + \nabla\tilde{P}_{E_I}(\mathbf{w},k)$ with

$$\nabla\tilde{P}_{E_R}(\mathbf{w},k) = \frac{\gamma}{2\sqrt{2\pi}\,\rho_n} e^{-\frac{\left(y_R(k)-\hat{c}_{R_1}(k)(b_{R_1}(k)-1)\right)^2}{2\rho_n^2}} \times\left(-\mathbf{x}(k) + (b_{R_1}(k)-1)\hat{\mathbf{p}}_1\right) \tag{37}$$

and

$$\nabla\tilde{P}_{E_I}(\mathbf{w},k) = \frac{\gamma}{2\sqrt{2\pi}\,\rho_n} e^{-\frac{\left(y_I(k)-\hat{c}_{R_1}(k)(b_{I_1}(k)-1)\right)^2}{2\rho_n^2}} \times\left(j\mathbf{x}(k) + (b_{I_1}(k)-1)\hat{\mathbf{p}}_1\right) \tag{38}$$

gives rise to the stochastic gradient adaptive algorithm referred to as the least symbol error rate (LSER) algorithm

$$\tilde{\mathbf{w}}(k+1) = \hat{\mathbf{w}}(k) + \mu\left(-\nabla\tilde{P}_{E_B}(\hat{\mathbf{w}}(k),k)\right), \tag{39}$$

$$\tilde{c}_1(k+1) = \tilde{\mathbf{w}}^H(k+1)\hat{\mathbf{p}}_1, \tag{40}$$

$$\hat{\mathbf{w}}(k+1) = \frac{\tilde{c}_1(k+1)}{|\tilde{c}_1(k+1)|}\tilde{\mathbf{w}}(k+1). \tag{41}$$

The rotating operation (40) and (41) ensures that $\hat{c}_1(k) \triangleq \hat{\mathbf{w}}^H(k)\hat{\mathbf{p}}_1 = \hat{c}_{R_1}(k) + j\hat{c}_{I_1}(k)$ satisfies $\hat{c}_{R_1}(k) > 0$ and $\hat{c}_{I_1}(k) = 0$. The step size μ and the kernel width $\rho_n$ are the two algorithmic parameters that should be set appropriately in order to ensure an adequate performance in terms of convergence rate and steady-state SER misadjustment. Note that there is no need to normalise the weight vector after each updating. That is, the algorithm does not restrict to find the unit-length MSER solution. The estimate of $\mathbf{p}_1$ can be provided by either (13) or (14).

Theoretical proof for convergence of this LSER algorithm is very difficult if not impossible and it is still under investigation. However, it can be pointed out that this LSER algorithm belongs to the general stochastic gradient-based adaptive algorithm investigated by Sharma *et al.* (1996). Therefore, the results of local convergence analysis presented in (Sharma *et al.*, 1996) is applicable here. Our previous investigations (Chen *et al.*, 2004a; Chen *et al.*, 2006) have suggested that the LSER algorithm behaves well, has a reasonable convergence speed, and is consistently outperforms the LMS algorithm in terms of the achievable SER. Influence of the two algorithmic parameters of the LSER algorithm, namely μ and $\rho_n$, to the SER performance will be investigated in the following simulation.

## SIMULATION STUDY

The prototype system investigated consisted of four user sources and a three-element antenna array. Figure 2 shows the locations of the desired user source and the interfering user sources graphically, where the angular separation between the desired user and the interfering user 4 was $\theta < 65^\circ$. Note that the performance of a beamforming receiver mainly depends on the minimum angular separation between the desired user and the interfering users (in this case the minimum angular separation was θ), and whether or not the desired user is at the broadside of the antenna array is not too critical. As emphasised in Section 2, the column of the system matrix associated with the desired user, namely $\mathbf{p}_1$, must be

*Figure 2. Locations of the user sources with respect to the three-element linear array with λ /2 element spacing, λ being the wavelength, where* $\theta < 65^{\circ}$.

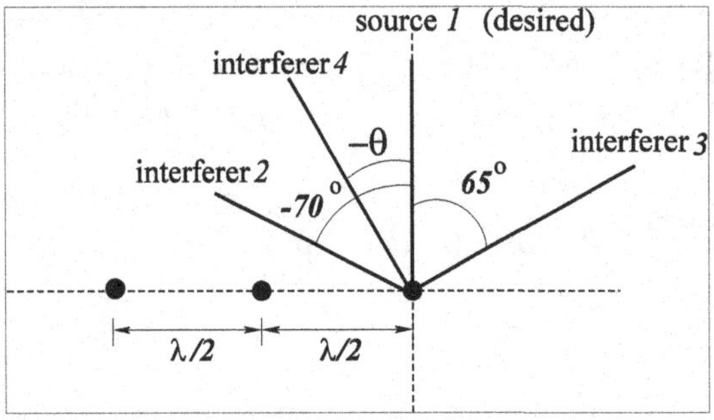


known in receiver. Usually, $\mathbf{p}_1$ can be estimated accurately during training using for example the block-based estimator (13) or the adaptive moving-average estimate (14). Thus, in the following simulation study, a perfect $\mathbf{p}_1$ was assumed at receiver. Hence, our attention was focused on the performance of the adaptive MMSE and MSER beamforming designs, rather than on the adaptive estimator for $\mathbf{p}_1$.

## Stationary System

The modulation scheme was 16-QAM and all the channels $A_i$, $1 \le i \le 4$, were time-invariant. Figure 3 compares the SER performance of the MSER beamforming solution to that of the MMSE beamforming solution under four different conditions: **(a)** the minimum angular separation between the desired user 1 and the interfering user 4 was $\theta = 32^{\circ}$, and all the four users had an equal signal power, i.e. $\text{SIR}_i = 0$ dB for $2 \le i \le 4$; **(b)** $\theta = 30^{\circ}$, and all the four users had an equal signal power; **(c)** $\theta = 28^{\circ}$, and all the four users had an equal signal power; and **(d)** $\theta = 30^{\circ}$, user 1 and user 2 had the same signal power but users 3 and 4 had 2 dB more power than users 1 and 2, i.e. $\text{SIR}_2 = 0$ dB and $\text{SIR}_3 = \text{SIR}_4 = -2$ dB. The MMSE beamformer was provided by the closed-form solution (9), while the MSER solution was obtained numerically using the simplified conjugate gradient algorithm.

For the case of equal user power with the minimum angular separation $\theta = 32^{\circ}$, the MSER beamforming solution had an SNR gain of 2 dB over the MMSE solution at the SER level of $10^3$, as can be seen from Figure 3 (a). When the minimum angular separation of the system was reduced to $\theta = 32^{\circ}$, as depicted in Figure 3 (b), the SNR gain of the MSER beamformer over the MMSE one was increased to 4 dB. With the minimum angular separation further reduced to $\theta = 32^{\circ}$, the MMSE beamforming solution became incapable of removing the interference and exhibited a high SER floor, as illustrated in Figure 3 (c). In contrast, the MSER beamformer remained capable of effectively removing the interference and achieving an adequate SER performance. By comparing Figure 3 (b) with Figure 3 (d), it can be seen that, with the minimum angular separation $\theta = 32^{\circ}$ and when facing stronger interfering users 3 and 4, the MMSE solution faltered while the MSER solution sufferred from very little degradation. This clearly demonstrated that the MSER beamformer is more robust in near-far situations than the MMSE beamformer.

The MSER solution is defined as the weight vector that minimises the upper bound SER $P_{E_B}(\mathbf{w}) = P_{E_R}(\mathbf{w}) + P_{E_I}(\mathbf{w})$, and in Section 2 it is pointed out that this is practically equivalent to minimise the true SER. The true SER is given by the sum of the inphase and quadrature components' error rates minus the appropriate correction term used for preventing the "double-counting" error-events as follows $P_E(\mathbf{w}) = P_{E_R}(\mathbf{w}) + P_{E_I}(\mathbf{w}) - P_{E_R}(\mathbf{w})P_{E_I}(\mathbf{w})$. The probability of simultaneous inphase and quadrature errors, which is represented by the term $P_{E_R}(\mathbf{w})P_{E_I}(\mathbf{w})$ tends to be quite low, unless the SNR is extremely low. More explicitly, the last term is typically orders of magnitude lower than the first two terms. Hence the bound $P_E(\mathbf{w}) < P_{E_B}(\mathbf{w})$ is very tight, i.e. $P_{E_B}(\mathbf{w})$ is very close to $P_E(\mathbf{w})$. In fact, $P_{E_B}(\mathbf{w})$ is almost indistinguishable from $P_E(\mathbf{w})$. This is not surprising, since the term $P_{E_R}(\mathbf{w})P_{E_I}(\mathbf{w})$ is negligible in comparison to the dominant term $P_{E_R}(\mathbf{w}) + P_{E_I}(\mathbf{w})$. For example, when $P_{E_R}(\mathbf{w})$ or $P_{E_I}(\mathbf{w})$ (they are symmetric) is of the order of $10^{-2}$, then $P_{E_R}(\mathbf{w})P_{E_I}(\mathbf{w})$ is of the order of $10^{-4}$, which constitutes an almost negligible factor. This is confirmed by the results of Figure 4, where both the true SER $P_E(\mathbf{w})$ and its upper bound $P_{E_B}(\mathbf{w})$ are plotted for the MMSE and MSER solutions under the channel conditions of $\theta = 30^{\circ}$ and equal user power.

*Figure 3. Desired user's symbol error rate performance comparison for the non-fading channel system employing the three-element array of Figure 2 to support four 16-QAM users*

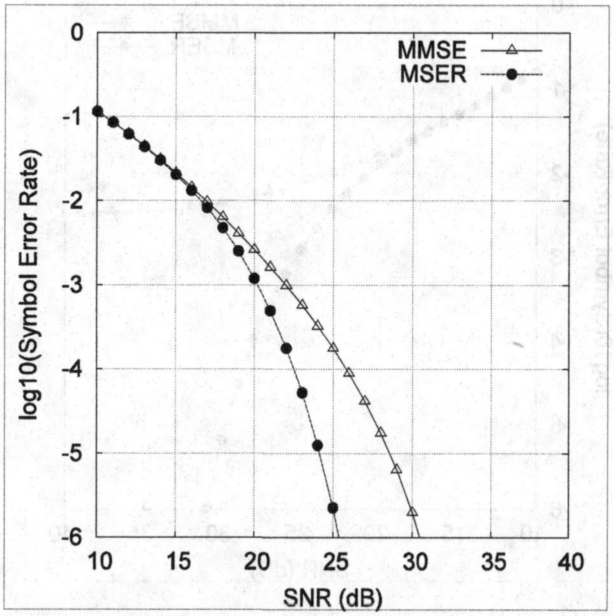


*(a)* $\theta = 32^{\circ}$, $SIR_i = 0$ *dB*, $2 \le i \le 4$

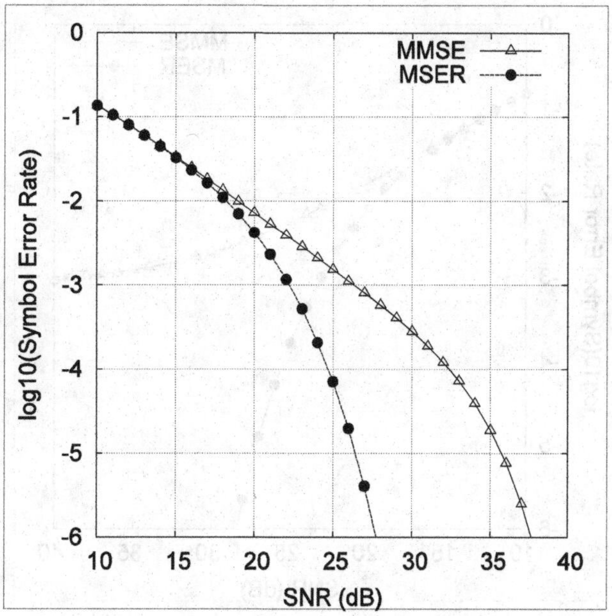


*continued on following page*

*Figure 3. continued*

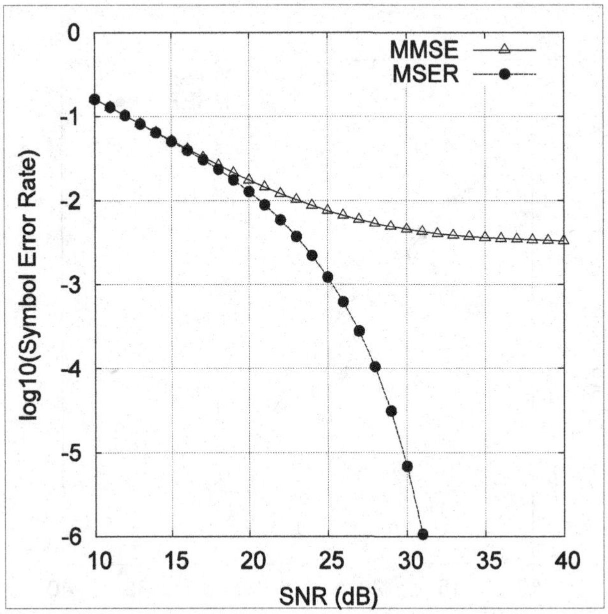


(c) $\theta = 28^{\circ}$, $SIR_i = 0\ dB,\ 2 \le i \le 4$

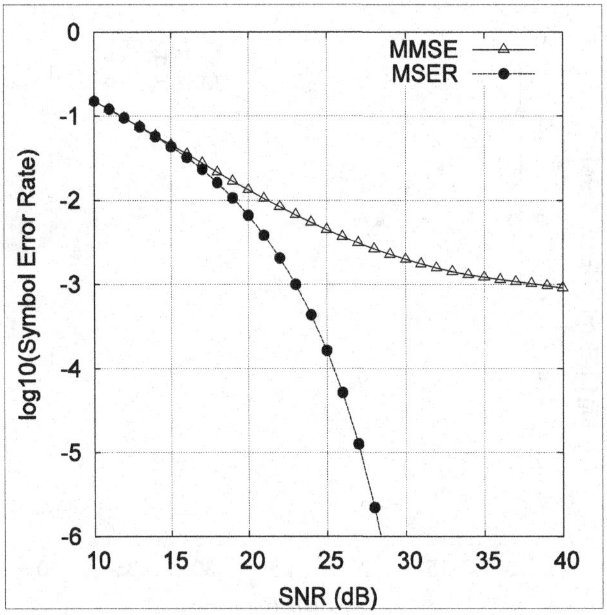


(d) $\theta = 30^{\circ}$, $SIR_2 = 0\ dB,\ SIR_3 = SIR_4 = -2\ dB$

Both the LMS and LSER based adaptive beamforming algorithms were next investigated using the system of the minimum angular separation $\theta = 30^o$, equal user power and SNR = 26 dB. Given $\hat{\mathbf{w}}(0) = [0.1 + j0.1 \;\; 0.1 - j0.01 \;\; 0.1 - j0.1]^T$ and the step size $\mu = 0.0005$, the learning curves of the LMS algorithm averaged over 20 different runs are plotted in Figure 5. There were two types of learning curves depicted in Figure 5, namely, the learning curve related to the training-based adaptation, when the desired user's transmitted symbol $b_1(k)$ was known to the receiver, and the learning curve related to the decision-directed (DD) adaptation, where from the sample $k = 250$ the beamformer's decision $\hat{b}_1(k)$ was used to substitute for $b_1(k)$. Similarly, the learning curves of the LSER algorithms under the same initial condition of $\hat{\mathbf{w}}(0)$ and given the step size 0.001 and the kernel width $\rho_n = \sigma_n$ are depicted in Figure 5, in comparison with those of the LMS algorithm. Lastly, the SER performance of both the LMS and LSER based beamformers are compared with those of the theoretic MMSE and MSER solutions in Figure 6 under the same condition of Figure 3 (b). The superiority of the adaptive LSER beamformer over the adaptive LMS beamformer is clearly demonstrated in Figure 6, where it can be seen that the performance of the adaptive LMS beamformer was notably deviated from its theoretic MMSE solution at high SNRs.

## Rayleigh Fading System

The modulation scheme was 64-QAM. Fading channels were simulated, where the magnitudes of $A_i$ for $1 \le i \le 4$ were Rayleigh processes with the normalised Doppler frequence $\overline{f}_D$ and each channel $A_i$ had the root mean power of $\sqrt{0.5} + j\sqrt{0.5}$. Thus the average $\text{SIR}_i = 0$ dB for $2 \le i \le 4$. Continuously fluctuating fading was used, which provided a different fading magnitude and phase for each transmitted symbol. The transmission frame structure consisted of 50 training symbols followed by 450 data symbols. Decision-directed adaptation was employed during data transmission, in which the adaptive beamforming detector's decision $\hat{b}_1(k)$ was used to substitute for $b_1(k)$. The SER of an adaptive beamforming detector was calculated using the 450 data symbols of the frame based on Monte Carlo simulation averaging over at least $2 \times 10^5$ frames, depending on the value of $\overline{f}_D$. Two initialisations were used for the adaptive LMS and LSER algorithms, where the initial weight vector $\hat{\mathbf{w}}(0)$ was initialised to either the MMSE solution (corresponding to the initial channel conditions) or $[0.1 + j0.0 \;\; 0.1 + j0.0 \;\; 0.1 + j0.0]^T$, and the performance were observed to be very similar for these two initialisations.

Given the minimum angular separation $\theta = 27^o$, Figure 7 compares the SER of the adaptive LSER beamformer with that of the LMS-based one, for the two normalised Doppler frequencies $\overline{f}_D = 10^{-4}$ and $10^{-3}$. It can be seen from Figure 7 that the SER performance of the adaptive LSER beamformer degraded only slightly when the fading rate increased from $\overline{f}_D = 10^{-4}$ to $10^{-3}$. This demonstrates that the LSER algorithm has an excellent tracking ability, capable of operating in fast fading conditions. The influence of the adaptive algorithm's parameters, the step size $\mu$ for the LMS algorithm, and the step size $\mu$ and kernel width $\rho_n$ for the LSER algorithm, were next investigated. Given $\overline{f}_D = 10^{-4}$, Figure 8 (a) show the influence of the adaptive algorithm's parameters, $\mu$ for the LMS algorithm, and $\mu$ and $\rho_n$ for the LSER algorithm, on the SER performance for a low average SNR value of 15 dB (Note that this was a 64-QAM system, and a SNR of 15 dB was relatively low), while Figure 8 (b) depicts the results for a high average SNR value of 30 dB. These results also explain why $\mu = 0.0002$ for the LMS algorithm and $\mu = 0.00005$ and $\rho_n = 4\sigma_n$ for the LSER algorithm were used in the simulation of Figure 7.

Lastly, the combining influence of the Rayleigh fading channels $A_i$, $1 \le i \le 4$, and the uniformly varying minimum angular seapration $\theta$ was investigated. Given the normalised Doppler frequency $\overline{f}_D = 10^{-3}$ and an average SNR of 25 dB, the minimum angular separation $\theta$ was varied in $[20^\circ, \; 50^\circ]$ and the SER performance of the LMS and LSER adaptive beamformers corresponding to each $\theta$ are depicted in Figure 9. It can be seen from Figure 9 that the performance of an adaptive beamformer depends on the combination of the channel coefficients $A_i$, $1 \le i \le 4$, and the value of $\theta$, and the adaptive LSER beamformer always performs better than the adaptive LMS beamformer. Finally, the average SER performance of the two adaptive beamformers over the uniformly distributed $\theta \in [20^\circ, \; 50^\circ]$ are plotted in Figure 10. Because certain combinations of the fading channels $A_i$, $1 \le i \le 4$, and the minimum angular separation $\theta$ led to catastrophic error rates, which dominated the average SER over $\theta \in [20^\circ, \; 50^\circ]$, the average SER curves exhibited high error floors, as can be seen in Figure 10. However, the adaptive LSER beamformer is seen to perform substantially better then the adaptive LMS beamformer.

To end the simulation experiment, we summarise the main simulation parameters in our simulation investigation in Table 1. In all the systems simulated, we observe that the MSER beamforming design significantly outperformed the MMSE beamforming design.

*Figure 4. Comparison of the true symbol error rate and its upper bound for the non-fading channel system employing the three-element array of Figure 2 with a minimum angular separation of* $\theta = 30^o$ *to support four equal-power 16-QAM users.*

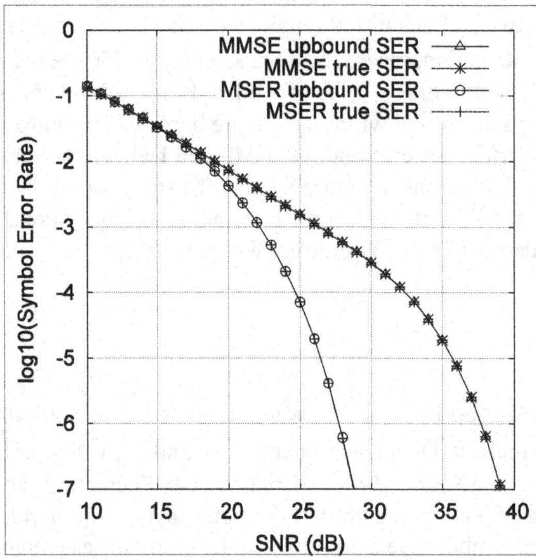


*Figure 5. Learning curves of the stochastic adaptive LMS and LSER algorithms averaged over 20 runs for the non-fading channel system employing the three-element array of Figure 2 with a minimum angular separation of* $\theta = 30^o$ *to support four equal-power 16-QAM users given SNR = 26 dB, where DD denotes decision-directed adaptation with* $\hat{b}_1(k)$ *substituting for* $b_1(k)$.

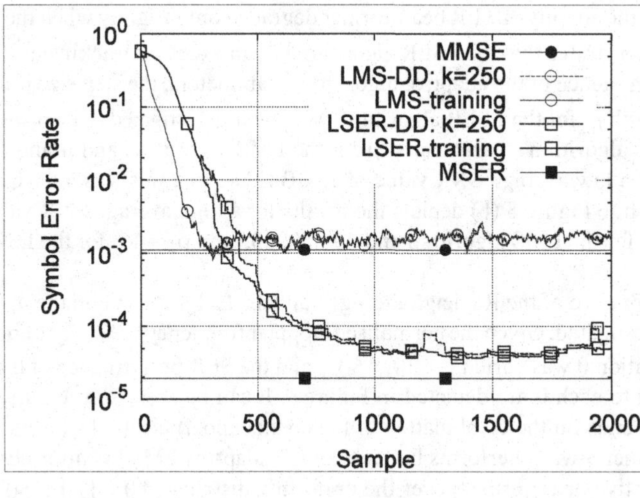

*Figure 6. Desired user's symbol error rate performance comparison for the non-fading channel system employing the three-element array of Figure 2 with a minimum angular separation of θ = 30º to support four equal-power 16-QAM users.*

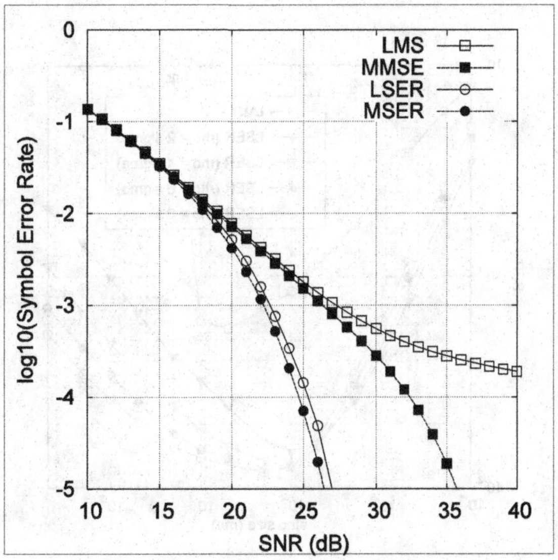


*Figure 7. Desired user's symbol error rate performance comparison for the fading channel systems of the two normalised Doppler frequencies* $\bar{f}_D = 10^{-4}$ *and* $10^{-3}$ *employing the three-element array of Figure 2 with a minimum angular separation of θ = 27º to support four 64-QAM users. The LMS algorithm has a step size μ = 0.0002, while the LSER algorithm has a step size μ = 0.00005 and a kernel width* $\rho_n = 4\sigma_n$.

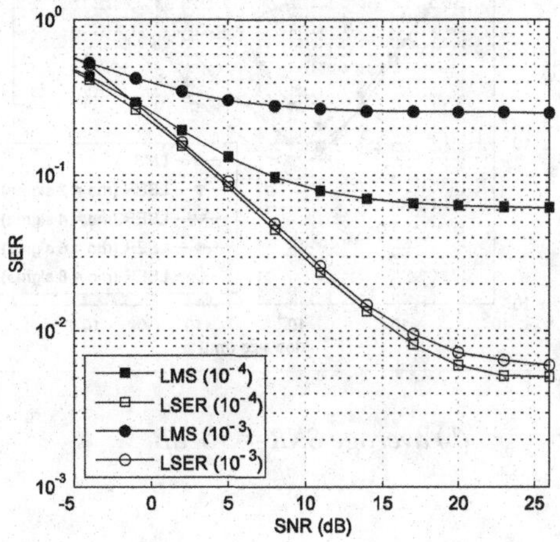

*Figure 8. Influence of the adaptive algorithm's parameters to the SER performance for the fading channel system employing the three-element array of Figure 2 to support four 64-QAM users, given* $\theta = 27^o$ *and* $\overline{f}_D = 10^{-4}$.

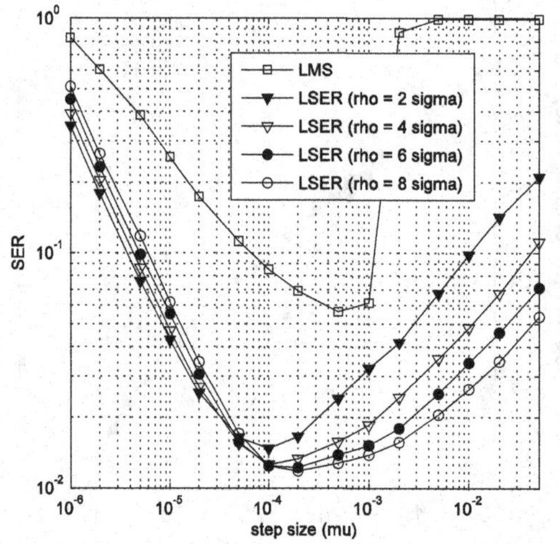


*(a) average* $SNR = 15$ *dB*

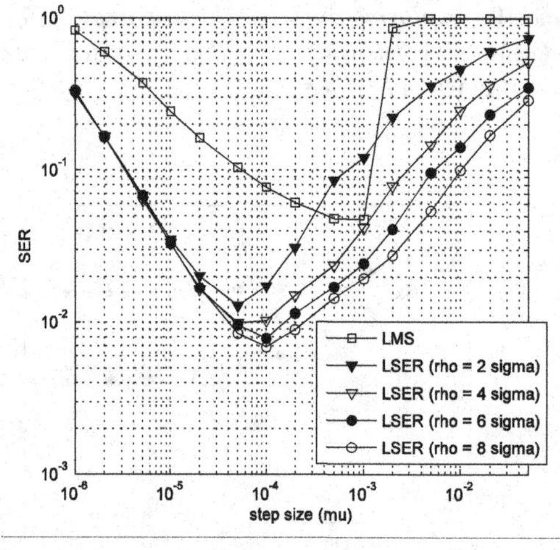


*(b) average* $SNR = 30$ *dB*

*Figure 9. Desired user's symbol error rate performance as a function of the minimum angular separation θ for the fading channel system of the normalised Doppler frequency $\bar{f}_D = 10^{-3}$ employing the three-element array of Figure 2 to support four 64-QAM users, given an average SNR of 25 dB. The LMS algorithm has a step size μ = 0.0002, while the LSER algorithm has a step size μ = 0.00005 and a kernel width $\rho_n = 4\sigma_n$.*

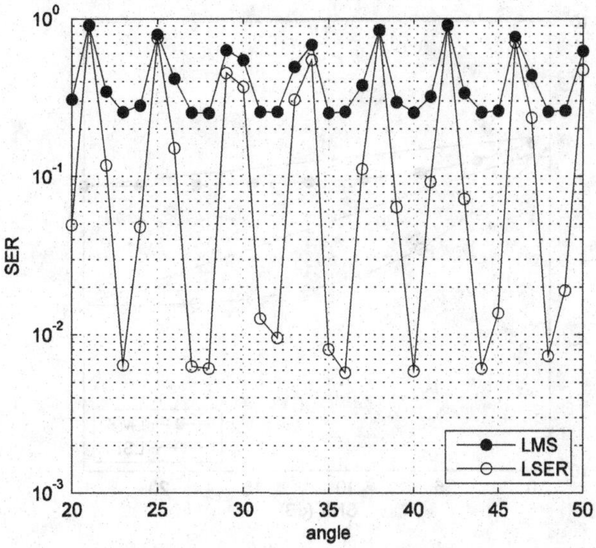


*Table 1. Main simulation parameters for three-element antenna-array system of Figure 2 to support four users*

| Channel taps | Minimum angular separation | Modulation | Beamforming algorithm |
|---|---|---|---|
| Fixed | $\theta = 32°, 30°, 28°$ | 16 QAM | MMSE and MSER |
| Fading | $\theta = 27°, \theta \in [20°, 50°]$ | 64 QAM | LMS and LSER |

## CONCLUSION

An adaptive MSER beamforming technique has been developed for multiple-antenna aided multiuser communication systems employing high-throughput QAM signalling. It has been demonstrated that the MSER beamforming design can provide significant performance enhancement, in terms of the achievable system's SER, over the standard MMSE design. It has also been demonstrated that the MSER beamforming design offers a higher user capacity and is more robust in the near-far senario, compared with the conventional MMSE beamforming design. An adaptive implementation of the MSER beamforming solution has been realised using the stochastic gradient adaptive algorithm known as the LSER technique. The simulation results presented in this study clearly show that the adaptive LSER beamforming is capable of operating successfully in fast fading conditions and it consistently outperforms the adaptive LMS beamforming benchmarker.

Since the discovery of turbo codes (Berrou & Glavieux, 1996), iterative detection (Hanzo *et al.*, 2002) has been applied to joint channel estimation and equalisation (Hanzo *et al.*, 2002a), multiuser detection (Hanzo *et al.*, 2003) and numerous other coded communication systems (Wang & Poor, 1999; Tüchler *et al.*, 2002; Tarable *et al.*, 2005). Most of the available literature discuss the MMSE based iterative receivers (Wang & Poor, 1999; Tüchler *et al.*, 2002; Tüchler *et al.*, 2002a; Li & Wang, 2005; Tarable *et al.*, 2005). It is however highly desired to consider the MBER based iterative

*Figure 10. Desired user's average symbol error rate performance comparison for the fading channel system of the normalised Doppler frequency* $\bar{f}_D = 10^{-3}$ *employing the three-element array of Figure 2 with the minimum angular separation* $\theta$ *uniformly distributed in* $[20^{\circ}, 50^{\circ}]$ *to support four 64-QAM users. The LMS algorithm has a step size* $\mu = 0.0002$*, while the LSER algorithm has a step size* $\mu = 0.00005$ *and a kernel width* $\rho_n = 4\sigma_n$.

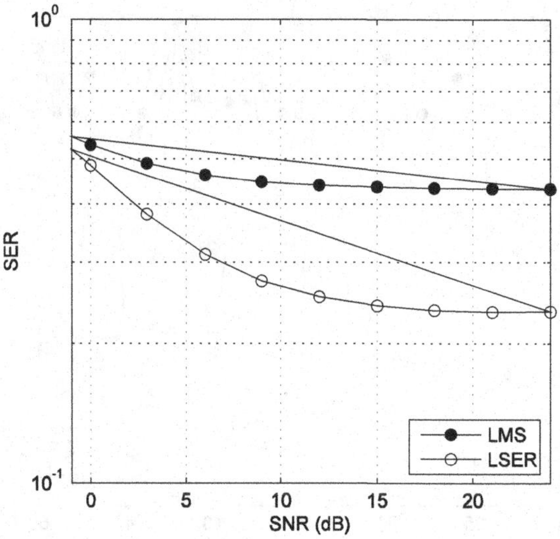


receivers, and the recent work (Tan *et al.*, 2006; Tan *et al.*, 2008) has studied turbo-detected MBER beamformer designs for BPSK and QPSK systems. Currently, the Communication Research Group at the University of Southampton is carrying out extensive investigation to design iterative MSER beamforming detection techniques for employment in the systems that adopt high-order QAM signalling.

The narrow-band MIMO model is considered in this study and beamforming is a spatial only processing technique. In order to deal with the generic frequency-selective MIMO system, space-time processing techniques should be employed. The recent work (Chen *et al.*, 2006a; Chen *et al.*, 2006b) has designed the novel MBER space-time equalisation for the space-division multiple-access induced MIMO system with BPSK modulation. Extension to the MSER space-time equalisation for the generic MIMO system employing high-throughput QAM modulation schemes is currently being conducted.

## ACKNOWLEDGMENT

The author acknowledges the contributions of professor L. Hanzo, Mr. A. Livingstone, and Miss H. Q. Du to the topic reported in this work.

## REFERENCES

Ahmad, N. N. (2005). *Minimum Bit Error Ratio Beamforming.* PhD Thesis, School of Electronics and Computer Sciences, University of Southampton, U.K.

Bazaraa, M. S., Sherali, H. D., & Shetty, C. M. (1993). *Nonlinear Programming: Theory and Algorithms.* New York: John Wiley.

Berrou, C., & Glavieux, A. (1996). Near optimum error correcting coding and decoding: turbo-codes. *IEEE Trans. Communications,* (10), 1261–1271.

Blogh, J. S., & Hanzo, L. (2002). *Third Generation Systems and Intelligent Wireless Networking – Smart Antenna and Adaptive Modulation.* Chichester, U.K.: John Wiley.

Bowman, A. W., & Azzalini, A. (1997). *Applied Smoothing Techniques for Data Analysis.* Oxford, U.K.: Oxford University Press.

Chen, S., & Luk, B. L. (1999). Adaptive simulated annealing for optimization in signal processing applications, *Signal Processing,* (1), 117–128.

Chen, S., Samingan, A. K., Mulgrew, B., & Hanzo, L. (2001.). Adaptive minimum-BER linear multiuser detection for DS-CDMA signals in multipath channels. *IEEE Trans. Signal Processing,* (6), 1240–1247.

Chen, S., Hanzo, L., & Ahmad, N. N. (2003). Adaptive minimum bit error rate beamforming assisted receiver for wireless communications. In *Proc. ICASSP 2003,* vol.IV, 640–643.

Chen, S., Hanzo, L., Ahmad, N. N., & Wolfgang, A. (2004). Adaptive minimum bit error rate beamforming assisted QPSK receiver. In *Proc. ICC 2004,* vol.6, 3389–3393.

Chen, S., Hanzo, L., & Mulgrew, B. (2004a). Adaptive minimum symbol-error-rate decision feedback equalization for multi-level pulse-amplitude modulation. *IEEE Trans. Signal Processing,* (7), 2092–2101.

Chen, S., Ahmad, N. N., & Hanzo, L. (2005). Adaptive minimum bit error rate beamforming. *IEEE Trans. Wireless Communications,* (2), 341–348.

Chen, S., Hanzo, L., Ahmad, N, N., & Wolfgang, A. (2005a). Adaptive minimum bit error rate beamforming assisted receiver for QPSK wireless communication. *Digital Signal Processing,* (6), 545–567.

Chen, S., Du, H. Q., & Hanzo, L. (2006). Adaptive minimum symbol error rate beamforming assisted receiver for quadrature amplitude modulation systems. In *Proc. VTC2006-Spring,* vol.5, 2236–2240.

Chen, S., Livingstone, A., & Hanzo, L. (2006a). Minimum bite-error rate design for space-time equalization-based multiuser detection. *IEEE Trans. Communications,* (5), 824–832.

Chen, S., Hanzo, L., & Livingstone, A. (2006b). MBER Space-time decision feedback equalization assisted multiuser detection for multiple antenna aided SDMA systems. *IEEE Trans. Signal Processing,* (8), 3090–3098.

Cioffi, J. M., Dudevoir, G. P., Eyuboglu, M. V., & Forney, G. D., Jr., (1995). MMSE decision-feedback equalizers and coding – Part I: equalization results. *IEEE Trans. Communications,* (10), 2582–2594.

Fan, L. Y., Zhang, H. B., & Chen, H. (2005). Minimum bit error rate beamforming for pre-FFT OFDM adaptive antenna array. In *Proc. VTC 2005-Fall,* vol.6, 359–363.

Ganz, M. W., Moses, R. L., & Wilson, S. L. (1990). Convergence of the SMI and the diagonally loaded SMI algorithms with weak interference (adaptive array). *IEEE Trans. Antennas and Propagation,* (3), 394–399.

Garcia, I. D. S., Marciano, J. J. S., Jr., & Cajote, R. D. (2004). Normalized adaptive minimum bit-error-rate beamformers. In *Proc. IEEE Region 10 TENCON Conf.,* vol.2, 625–628.

Godara, L. C. (1997). Applications of antenna arrays to mobile communications, Part I: Performance improvement, feasibility, and system considerations. *Proc. IEEE,* (7), 1031–1060.

Goldberg, D. E. (1989). *Genetic Algorithms in Search, Optimization and Machine Learning.* Reading, MA: Addison Wesley.

Griffiths, L. J. (1969). A simple adaptive algorithm for real-time processing in antenna arrays. *Proc. IEEE, 57,* 1696–1704.

Hanzo, L., Liew, T. H., & Yeap, B. L. (2002). *Turbo Coding, Turbo Equalisation and Space-Time Coding for Transmission over Fading Channels.* West Sussex, England: John Wiley and IEEE Press.

Hanzo, L., Wong, C. H., & Yee, M. S. (2002a). *Adaptive Wireless Transceiver: Turbo-Coded, Turbo-Equalised and Space-Time Coded TDMA, CDMA and OFDM Systems.* West Sussex, England: John Wiley and IEEE Press.

Hanzo, L., Yang, L. L., Kuan, E. L., & Yen, K. (2003). *Single- and Multi-Carrier DS-CDMA: Multi-User Detection, Space-Time Spreading, Synchronisation, Standards and Networking.* West Sussex, England: John Wiley and IEEE Press.

Hanzo, L., Ng, S. X., Keller, T., & Webb, W. (2004). *Quadrature Amplitude Modulation: From Basics to Adaptive Trellis-Coded, Turbo-Equalised and Space-Time Coded OFDM, CDMA and MC-CDMA Systems.* Chichester, U.K.: John Wiley and IEEE Press.

Haykin, S. (1996). *Adaptive Filter Theory.* 3rd edition, Upper Saddle River, NJ: Prentice-Hall.

Hwang, S. H., & Hanzo, L. (2003). Reverse-link performance of synchronous DS-CDMA systems in dispersive Rician multipath fading channels. *IEE Electronics Letters,* (23), 1682-1684.

IEEE 802.16 (2004). Air Interface for Fixed Broadband Wireless Access System, Section 8. PHY.

Ingber, L. (1993). Simulated annealing: practice versus theory. *Mathematical and Computer Modeling*, (11), 29–57.

Kohno, R. (1998). Spatial and temporal communication theory using adaptive antenna array. *IEEE Personal Communications*, (1), 28–35.

Li, K., & Wang, X. (2005). EXIT chart analysis of turbo multiuser detection. *IEEE Trans. Wireless Communications*, (1), 300–311.

Litva, J., & Lo, T. K. Y. (1996). *Digital Beamforming in Wireless Communications*. London: Artech House.

Liu, Y. H., & Yang, Y. H., (2004). Adaptive minimum bit error rate multitarget array algorithm. In *Proc. IEEE 6th CAS Symp. Emerging Technologies: Frontiers of Mobile and Wireless Communication*, vol.2, 745–748.

Man, K. F., Tang, K. S., & Kwong, S. (1998). *Genetic Algorithms: Concepts and Design*. London: Springer-Verlag.

Parzen, E. (19962). On estimation of a probability density function and mode. *The Annals of Mathematical Statistics*, *33*, 1066–1076.

Paulraj, A., Nabar, R., & Gore, D. (2003). *Introduction to Space-Time Wireless Communications*. Cambridge, U.K.: Cambridge University Press.

Paulraj, A. J., Gore, D. A., Nabar, R. U., & Bölcskei, H. (2004). An overview of MIMO communications – A key to gigabit wireless. *Proc. IEEE*, (2), 198–218.

Petrus, P., Ertel, R. B., & Reed, J. H. (1998). Capacity enhancement using adaptive arrays in an AMPS system. *IEEE Trans. Vehicular Technology*, (3), 717–727.

Reed, I. S., Mallett, J. D., & Brennan, L. E. (1974). Rapid convergence rate in adaptive arrays. *IEEE Trans. Aerospace and Electronic Systems*, 10, 853–863.

Sharma, R., Sethares, W. A., & Bucklew, J. A. (1996). Asymptotic analysis of stochastic gradient-based adaptive filtering algorithms with general cost functions. *IEEE Trans. Signal Processing*, (9), 2186–2194.

Silverman, B. W. (1996). *Density Estimation*. London: Chapman Hall.

Soni, R. A., Buehrer, R. M., & Benning, R. D. (2002). Intelligent antenna system for cdma2000. *IEEE Signal Processing Magazine*, (4), 54–67.

Steele, R., & Hanzo, L. (1999). *Mobile Radio Communications*. Piscataway, NJ: IEEE Press.

Tan, S., Xu, L., Chen, S., & Hanzo, L. (2006). Iterative soft interference cancellation aided minimum bit error rate uplink receiver beamforming. In *Proc. VTC2006-Spring*, vol.1, 17–21.

Tan, S., Chen, S., & Hanzo, L. (2008). On multi-user EXIT chart analysis aided turbo-detected MBER beamformer designs. *IEEE Trans. Wireless Communications*, *7*, 314-323.

Tarable, A., Montorsi, G., & Benedetto, S. (2005). A linear front end for iterative soft interference cancellation and decoding in coded CDMA. *IEEE Trans. Wireless Communications*, (2), 507–518.

Tse, D., & Viswanath, P. (2005). *Fundamentals of Wireless Communication*. Cambridge, U.K.: Cambridge University Press.

Tsoulos, G. V. (1999). Smart antennas for mobile communication systems: benefits and challenges. *IEE Electronics and Communications J.*, (2), 84–94.

Tüchler, M., Singer, A. C., & Koetter, R. (2002). Minimum mean squared error equalization using *a priori* information. *IEEE Trans. Signal Processing*, (3), 673–682.

Tüchler, M., Koetter, R., & Singer, A. C. (2002a). Turbo equalization: principle and new results. *IEEE Trans. Communications*, (5), 754-767.

Vandenameele, P., van Der Perre, L., & Engels, M. (2001). *Space Division Multiple Access for Wireless Local Area Networks*. Boston: Kluwer Academic Publishers.

Wang, X., & Poor, H. V. (1999). Iterative (turbo) soft interference cancellation and decoding for coded CDMA. *IEEE Trans. Communications*, (7), 1046–1060.

Widrow, B., Mantey, P. E., Griffiths, L. J., & Goode, B. B. (1967). Adaptive antenna systems. *Proc. IEEE*, *55*, 2143-2159.

Widrow, B., & Stearns, S. D. (1985). *Adaptive Signal Processing.* Englewood Cliffs, NJ: Prentice-Hall.

Winters, J. H., Salz, J., & Gitlin, R. D. (1994). The impact of antenna diversity on the capacity of wireless communication systems. *IEEE Trans. Communications*, (2), 1740–1751.

Winters, J. H. (1998). Smart antennas for wireless systems. *IEEE Personal Communications*, (1), 23–27.

Wolfgang, A., Ahmad, N. N., Chen, S., & Hanzo, L. (2004). Genetic algorithm assisted minimum bit error rate beamforming. In *Proc. VTC 2004-Spring*, 142–146.

# Chapter IV
# On the Employment of SMI Beamforming for Cochannel Interference Mitigation in Digital Radio

**Thomas Hunziker**
*University of Kassel, Germany*

## ABSTRACT

*Many common adaptive beamforming methods are based on a sample matrix inversion (SMI). The schemes can be applied in two ways. The sample covariance matrices are either computed over preambles, or the sample basis for the SMI and the target of the beamforming are identical. A vector space representation provides insight into the classic SMI-based beamforming variants, and enables elegant derivations of the well-known second-order statistical properties of the output signals. Moreover, the vector space representation is helpful in the definition of appropriate interfaces between beamfoming and soft-decision signal decoding in receivers aiming at adaptive cochannel interference mitigation. It turns out that the performance of standard receivers incorporating SMI-based beamforming on short signal intervals and decoding of BICM (bit-interleaved coded modulation) signals can be significantly improved by proper interface design.*

## INTRODUCTION

Cochannel interference (CCI) becomes a major performance limiting factor in today's growing variety and density of wireless links and networks. Cellular systems occupying licensed frequency bands may evade CCI by a smart channel reuse policy. But in emerging decentralized peer-to-peer networks an efficient management of the channel access with the guarantee of limited CCI is a complex task, especially if the peers have directional transmission and reception capabilities. And proactive interference control across different systems sharing an unlicensed band is even more difficult to realize. Receiver techniques aiming at *reactive* interference mitigation, on the other hand, do not require cooperation between transceivers or systems, and they are thus a more viable approach to limit outages in decentralized or heterogeneous networking scenarios.

Data streams are normally split up and conveyed in short frames from sender to receiver. In multi-hop networks the frames need to be lightweight in order to limit latency in links over multiple hops since the relaying peers can usually not receive and transmit simultaneously. Moreover, besides of the data frames a multitude of even shorter control frames conveying "Hello", "Request/Clear to Transmit", "Acknowledge" messages and others are exchanged. As a consequence, if a channel is shared without coordination the interference may fluctuate at a much higher rate than the actual channel gain does due to multipath fading. This necessitates interference mitigation techniques which can adapt to CCI characteristics within short signal periods.

Equipped with array antennas, receivers can suppress interference via beamforming, i.e., a weighting and combining of the signals from the multiple antennas. Classic beamforming methods include the minimum variance distortionless response (MVDR) beamformer, which maximizes the signal-to-interference-plus-noise ratio (SINR) under the constraint of undistorted desired signal, and the minimum mean squared error estimator. When the spatial signatures of the interfering signals are completely unknown, an adaptive beamforming becomes necessary. Many of the popular adaptive beamforming techniques, discussed in textbooks like (Monzingo et al, 1980; Van Trees, 2002), rely on an inversion of a sample covariance matrix (SCM). The methods presented in (Vorobyov et al, 2003; Feldman et al, 1994; Bell et al, 2000; Lorenz et al, 2005; Li et al, 2003) feature enhanced robustness to mismatches in the spatial signature of the desired signal and other uncertainties via a diagonal loading of the SCM or more elaborate arrangements. The properties of the output of classic SCM-based spatial filters are analyzed in (Richmond, 1996; Van Veen, 1991), exposing the performance degradation compared to ideal beamforming based on perfectly known CCI statistics.

The above referenced literature focuses on beamforming in general, with the aim to optimize the second-order statistics of the residual error in the filtered signals. Simply attaching such an optimized beamforming scheme to a standard baseband receiver does not necessarily result in a favourable architecture. In fact, optimal receivers perform interference mitigation and decoding jointly. Optimal maximum-likelihood signal decoding in the presence of CCI with unknown characteristics is discussed in (Hunziker et al, 2007), along with a suboptimal iterative procedure. Iterative solutions, which often clearly outperform their non-iterative equivalents, are also investigated in (Kuzminskiy et al, 2003; Swindlehurst et al, 1995; Biedka et al, 2000; Hunziker et al, 2004). On the other hand, joint beamforming and decoding as well as suboptimal iterative schemes do have the drawback of high complexity. Efficient algorithms are available for the sample matrix inversion (SMI), whereas iterative methods scale up receiver complexity by a factor two or more. In the following we shall thus restrict our attention to conventional baseband receiver architectures comprising an adaptive beamforming via an SMI, followed by the information decoding on the basis of the combined signal. Our concern is to amend the output of the beamforming subsystem such that it facilitates the subsequent soft-decision decoding.

## SYSTEM MODEL

In the following boldfaced lowercase characters are used for row and column vectors and boldfaced uppercase characters for matrices. The Hermitian transpose of $\mathbf{X}$ is written as $\mathbf{X}^{\mathrm{H}}$, $\mathbf{I}_K$ denotes the $K\times K$-identity matrix, and $[\mathbf{X}\ \mathbf{Y}]$ represents the horizontal concatenation of $\mathbf{X}$ and $\mathbf{Y}$. Furthermore, $\|\cdot\|$ denotes the 2-norm of a row/column vector.

Assume a staggered, block-wise transmission of an information-bearing signal over a narrow-band single-input/$N$-output channel. A block comprises $K$ data symbols. The channel gain is constant over many blocks and perfectly known, however, the characteristics of the CCI vary arbitrarily from block to block, and they are unknown. Following a proper sampling of the array signal at the symbol rate, the baseband receiver observes a block as the sequence $\mathbf{y}_1,\ldots,\mathbf{y}_K$ of complex $N\times 1$-vectors, where $\mathbf{Y} = [\mathbf{y}_1 \ \ldots \ \mathbf{y}_K]$ is given as

$$\mathbf{Y}=\mathbf{h}\mathbf{s}+\mathbf{W}\,. \tag{1}$$

The column vector $\mathbf{h}\in\mathbb{C}^N$ defines the signal attenuation at the $N$ receiver antennas, the $1\times K$-row vector $\mathbf{s} = [s_1 \ \ldots \ s_K]$ comprises the data symbols, and $\mathbf{W} = [\mathbf{w}_1 \ \ldots \ \mathbf{w}_K]$ includes the CCI and front end noise.

The interference may stem from an arbitrary number of unsynchronized sources. In many scenarios it is reasonable to model the composite CCI and noise as temporally white, spatially correlated Gaussian. Hence, the independent random vectors $\mathbf{w}_1,\ldots,\mathbf{w}_K$ are $CN(\mathbf{0},\mathbf{R})$, i.e., zero-mean circularly symmetric complex Gaussian with the covariance matrix $\mathbf{R}$. The matrix $\mathbf{R}$ depends on the radio channels between the sources of interference and the receiver. In environments with multipath signal propagation $\mathbf{R}$ is solely known to be Hermitian positive definite.

Preambles in the form of a number of leading zeros may facilitate the beamforming at the receiver end. Including a preamble of length $M$, the vector $\mathbf{s}$ comprises $M$ zeros and $K$-$M$ data symbols, i.e., $\mathbf{s} = [\mathbf{0}\ \mathbf{s}_{\mathrm{D}}]$. Likewise, $\mathbf{Y} = [\mathbf{Y}_{\mathrm{P}}\ \mathbf{Y}_{\mathrm{D}}]$, where $\mathbf{Y}_{\mathrm{P}}$ and $\mathbf{Y}_{\mathrm{D}}$ relate to the preamble and the data sections, respectively, of the block.

Figure 1 shows the structure of a typical receiver employing beamforming. Following the radio-frequency front ends and analog-digital (A/D) conversions, the signals from the $N$ antennas, represented by the rows of $\mathbf{Y}$, are linearly combined. The adaptive beamforming includes the computation of a proper row vector **f** with the weights and the signal combining into **fY**, and this procedure is assumed here to be carried out individually for every block. The weights are chosen on the basis of **Y** such that a favorable SINR results in **fY**. Implementing beamforming, demodulation, and decoding as separate blocks limits complexity as compared to optimal array signal decoding. Varying SINRs from block to block, due to fluctuating CCI, can be dealt with by forward error control (FEC) coding and interleaving over large numbers of blocks. Accomplishing log-likelihood ratio (LLR) computation, deinterleaving, and decoding in this order is appropriate for state-of-the-art bit-interleaved coded modulation (BICM). For generating BICM signals the transmitter employs an interleaving at bit-level between encoding and signal mapping. Under the assumption of ideal interleaving the signals in **s** can be modeled as uncorrelated random variables.

## BEAMFORMING VIA SAMPLE MATRIX INVERSION

In the ideal case where the covariance matrix **R** is perfectly known,

$$\mathbf{f}_{\mathrm{MVDR}} = (\mathbf{h}^{\mathrm{H}}\mathbf{R}^{-1}\mathbf{h})^{-1}\mathbf{h}^{\mathrm{H}}\mathbf{R}^{-1} \tag{2}$$

maximizes the SINR of the filtered signal $\mathbf{f}_{\mathrm{MVDR}}\mathbf{Y}$ subject to the constraint $\mathbf{f}_{\mathrm{MVDR}}\mathbf{h} = 1$. The spatial filter $\mathbf{f}_{\mathrm{MVDR}}$ is usually referred to as the MVDR (*minimum variance distortionless response*) beamformer. The resulting error in the filter output is given as

$$\mathbf{v} = \mathbf{f}_{\mathrm{MVDR}}\mathbf{W}. \tag{3}$$

From the characteristics of **W** it follows that the sample errors $\mathbf{f}_{\mathrm{MVDR}}\mathbf{w}_1, \ldots, \mathbf{f}_{\mathrm{MVDR}}\mathbf{w}_K$ are independent, complex normally distributed with the variance

$$\vartheta_{\mathrm{MVDR}} = \left(\mathbf{h}^{\mathrm{H}}\mathbf{R}^{-1}\mathbf{h}\right)^{-1}. \tag{4}$$

### Preamble-Based SMI Beamforming

The estimate $\hat{\mathbf{R}}_{\mathrm{P}} = M^{-1}\mathbf{Y}_{\mathrm{P}}\mathbf{Y}_{\mathrm{P}}^{\mathrm{H}}$ of the covariance matrix is termed *SCM*, representing the unconstrained maximum-likelihood estimate of **R** (Van Trees, 2002). Furthermore, the spatial filtering by

$$\mathbf{f}_{\mathrm{PSMI}} = \left(\mathbf{h}^{\mathrm{H}}\hat{\mathbf{R}}_{\mathrm{P}}^{-1}\mathbf{h}\right)^{-1}\mathbf{h}^{\mathrm{H}}\hat{\mathbf{R}}_{\mathrm{P}}^{-1} = \left(\mathbf{h}^{\mathrm{H}}\left(\mathbf{Y}_{\mathrm{P}}\mathbf{Y}_{\mathrm{P}}^{\mathrm{H}}\right)^{-1}\mathbf{h}\right)^{-1}\mathbf{h}^{\mathrm{H}}\left(\mathbf{Y}_{\mathrm{P}}\mathbf{Y}_{\mathrm{P}}^{\mathrm{H}}\right)^{-1} \tag{5}$$

is known as the *SMI* technique. The conditional variance of the sample errors $\mathbf{f}_{\mathrm{PSMI}}\mathbf{w}_{M+1}, \ldots, \mathbf{f}_{\mathrm{PSMI}}\mathbf{w}_K$ given $\mathbf{Y}_{\mathrm{p}}$ equals $\vartheta_{\mathrm{PSMI}} = \mathbf{f}_{\mathrm{PSMI}}\mathbf{R}\mathbf{f}_{\mathrm{PSMI}}^{\mathrm{H}}$. Both the beamformer $\mathbf{f}_{\mathrm{PSMI}}$ and $\vartheta_{\mathrm{PSMI}}$ are random quantities which depend on $\mathbf{Y}_{\mathrm{P}}$. If $M \geq N$, the

*Figure 1. Receiver model*

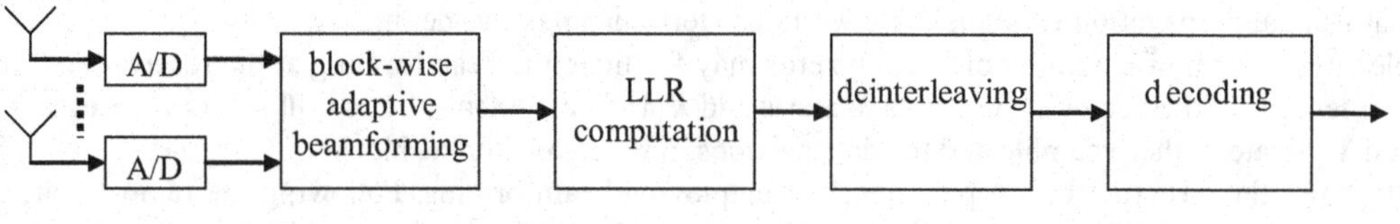

ratio $\vartheta_{\mathrm{MVDR}} / \vartheta_{\mathrm{PSMI}}$ is known to be beta distributed with the parameters $M-N+2$ and $N - 1$ (Reed, 1974). It follows that $E[\vartheta_{\mathrm{MVDR}} / \vartheta_{\mathrm{PSMI}}] = M/(M-N+1)$, where $E[\cdot]$ denotes the expectation, and that the unconditional mean squared sample error (MSSE) equals

$$\bar{\vartheta}_{\mathrm{PSMI}} = \frac{M}{M-N+1}\vartheta_{\mathrm{MVDR}} \tag{6}$$

provided that $K > M \geq N$. A comprehensive study on the statistical properties of the output of this beamforming scheme is found in (Richmond, 1996).

## Non-Preamble-Based SMI Beamforming

In the absence of a preamble, the SCM $\hat{\mathbf{R}} = K^{-1}\mathbf{Y}\mathbf{Y}^{\mathrm{H}}$ may be used as an estimate of $\mathbf{R}$, resulting in the spatial filter

$$\mathbf{f}_{\mathrm{DSMI}} = \left(\mathbf{h}^{\mathrm{H}}\hat{\mathbf{R}}^{-1}\mathbf{h}\right)^{-1}\mathbf{h}^{\mathrm{H}}\hat{\mathbf{R}}^{-1}. \tag{7}$$

In this approach, the sample basis for the SCM computation and the target of the beamformer are *identical*. Choosing $M = 0$ saves bandwidth, however, the presence of the desired signal in the sample basis degrades the accuracy of the SCM.

Under the assumptions of $K \geq N$ and the data symbols representing uncorrelated zero-mean random variables with 2nd moment

$$\varepsilon_{\mathrm{s}} = E\left[\left|s_k\right|^2\right],$$

the MSSE after this beamforming variant equals

$$\bar{\vartheta}_{\mathrm{DSMI}} = \frac{K-N+1}{K}\vartheta_{\mathrm{MVDR}} + \frac{N-1}{K}\varepsilon_{\mathrm{s}}. \tag{8}$$

The first and second terms on the right hand side in (8) may be regarded as due to the CCI (and noise) and due to the desired signal in the sample basis, respectively. Both MSSE terms were derived by Van Veen (1991), and an alternative way for their calculation is outlined below. We note that if $K$ is not significantly larger than $N$, the presence of the desired signal results in a large MSSE even in situations with minor interference. The dramatic effect of the desired signal in the sample basis has been widely known. Diagonal loading of the SCM has been shown to improve the beamforming performance (Carlson, 1988; Shahbazpanahi, 2003), however, the method is only effective for certain CCI constellations. In scenarios with "full load", where $N - 1$ interferers are similarly strong, any diagonal loading increases the sample error variance in the spatially filtered signal.

# VECTOR SPACE ILLUSTRATION

The expressions (5) and (7), containing inverse SCMs, do not provide much insight into the two beamforming schemes. In this section a vector space representation is discussed based on which further conclusions on the interference mitigation capabilities can be drawn. Let us first make the following definitions:

- The so-called *signal blocking matrix* $\mathbf{C} = [\mathbf{c}_1 \ldots \mathbf{c}_{N-1}]$ is a $N\times(N-1)$-matrix with orthonormal column vectors, i.e., $\mathbf{C}^{\mathrm{H}}\mathbf{C} = \mathbf{I}_{N-1}$, such that $\mathbf{h}^{\mathrm{H}}\mathbf{C} = \mathbf{0}$.
- B denotes the $(N-1)$-dimensional subspace of $\mathbb{C}^K$ that is spanned by the rows of $\mathbf{C}^{\mathrm{H}}\mathbf{W}$, i.e.,

$$\mathrm{B} = \mathrm{span}\left(\mathbf{c}_1^{\mathrm{H}}\mathbf{Y},\ldots,\mathbf{c}_{N-1}^{\mathrm{H}}\mathbf{Y}\right), \text{ and}$$

$\mathrm{B}^{\perp}$ represents the $(K\text{-}N\text{+}1)$-dimensional orthogonal subspace in $\mathbb{C}^K$.

- $\mathbf{P}_{\mathrm{B}}(\mathbf{x})$ stands for the orthogonal projection of $\mathbf{x}$ onto the subspace B, and

$$\mathbf{P}_{\mathrm{B}}^{\perp}(\mathbf{x}) = \mathbf{x} - \mathbf{P}_{\mathrm{B}}(\mathbf{x})$$

for the projection onto $\mathrm{B}^{\perp}$.

- $\mathbf{B} = [\mathbf{b}_1 \ldots \mathbf{b}_K]$ represent a $(N\text{-}1)\times K$-matrix with orthonormal rows spanning the subspace B. The rows of $\mathbf{B}$ can be obtained by applying the Gram-Schmidt orthogonalization algorithm to the rows of $\mathbf{C}^{\mathrm{H}}\mathbf{W}$.

## Non-Preamble-Based SMI Beamforming

Using the above notations, the signal $\mathbf{z} = \mathbf{f}_{\mathrm{DSMI}}\mathbf{Y}$ obtained by the *non-preamble-based* SMI beamforming variant can be expressed in the form (Hunziker et al, 2006)

$$\mathbf{z} = \mathbf{P}_{\mathrm{B}}^{\perp}\left(\mathbf{f}_{\mathrm{MVDR}}\mathbf{Y}\right). \tag{9}$$

That is, $\mathbf{z}$ may be seen as the result of an *ideal* filtering, projected onto the subspace $\mathrm{B}^{\perp}$. The signal $\mathbf{z}$ can further be written as

$$\mathbf{z} = \mathbf{P}_{\mathrm{B}}^{\perp}(\mathbf{s}) + \mathbf{e} \tag{10}$$

with $\mathbf{e} = \mathbf{P}_{\mathrm{B}}^{\perp}(\mathbf{v})$ the projection of the error signal (3) onto $\mathrm{B}^{\perp}$.

The following observations make the expressions (9) and (10) particularly practical.

- The subspace $\mathrm{B}^{\perp}$ is uniformly distributed on the Grassmann manifold $G(K,K\text{-}N\text{+}1)$.
- The random vector $\mathbf{v}$ is $CN\left(\mathbf{0},\vartheta_{\mathrm{MVDR}}\mathbf{I}_K\right)$.
- The subspace defining matrix $\mathbf{C}^{\mathrm{H}}\mathbf{W}$, the error vector $\mathbf{v}$, and the data vector $\mathbf{s}$ are independent.

The Grassmann manifold $G(K,K\text{-}N\text{+}1)$ represents the set of all $(K\text{-}N\text{+}1)$-dimensional subspaces of $\mathbb{C}^K$. The uniform distribution of $\mathrm{B}^{\perp}$ on $G(K,K\text{-}N\text{+}1)$ follows from the fact that $\mathbf{C}^{\mathrm{H}}\mathbf{W}$ does not change its distribution if post-multiplied by a constant unitary $K\times K$-matrix (Absil et al, 2005). And the independence of $\mathbf{C}^{\mathrm{H}}\mathbf{W}$ and $\mathbf{v}$ results from the independence of the jointly Gaussian distributed columns of $\mathbf{W}$ and $\mathbf{f}_{\mathrm{MVDR}}\mathbf{RC} = \mathbf{0}$, while the independence of $\mathbf{C}^{\mathrm{H}}\mathbf{W}$, $\mathbf{v}$, and $\mathbf{s}$ follows from the independence of $\mathbf{W}$ and $\mathbf{s}$.

In (10) we see that $\mathbf{z}$ contains no a posteriori information about the desired signal component in the subspace B. The signal part $\mathbf{P}_{\mathrm{B}}(\mathbf{s})$ is completely lost! Figure 2 shows the vector space with examples of the signals involved in the non-preamble-based interference mitigation procedure for the case $K = 3$, $N = 2$. The theoretical MVDR beamformer produces the signal $\mathbf{f}_{\mathrm{MVDR}}\mathbf{Y}$, which includes the error $\mathbf{v}$. The filtering by (7) leads to the signal $\mathbf{z}$, representing the sum of the projections of $\mathbf{s}$ and $\mathbf{v}$ onto $\mathrm{B}^{\perp}$. The error $(\mathbf{z} - \mathbf{s})$ comprises the error $\mathbf{e}$ due to the CCI (and noise) and $-\mathbf{P}_{\mathrm{B}}(\mathbf{s})$ resulting from the orthogonal projection onto $\mathrm{B}^{\perp}$. These two errors correspond to the two terms in (8). Considering the independence of $\mathbf{C}^{\mathrm{H}}\mathbf{W}$ and $\mathbf{v}$, the first term due to the CCI follows from the ratio defined by the dimensions of $\mathrm{B}^{\perp}$ and $\mathbb{C}^K$, while the second MSSE term due to the orthogonal projection amounts to $\varepsilon_s$ multiplied by the ratio defined by the dimensions of B and $\mathbb{C}^K$.

## Preamble-Based SMI Beamforming

A preamble of $M$ leading zeros constrains $\mathbf{s}$ to a $(K\text{-}M)$-dimensional subspace S of $\mathbb{C}^K$. Since $M \geq N$, the dimension of S is smaller than the dimension of $\mathrm{B}^{\perp}$. An obvious way for reconstructing $\mathbf{s}$ from $\mathbf{z}$ would be to choose the element in S for which the projection onto $\mathrm{B}^{\perp}$ comes closest to $\mathbf{z}$, i.e.,

$$\mathbf{z}_\mathrm{D} = \arg\min_{\mathbf{x}\in\mathbb{C}^{K-M}} \left\| \mathbf{P}_\mathrm{B}^{\perp}\left([\mathbf{0}\ \mathbf{x}]\right) - \mathbf{z} \right\|. \tag{11}$$

Interestingly, the signal $\mathbf{z}_\mathrm{D}$ corresponds to the result of the above described *preamble-based* SMI beamforming, that is, $\mathbf{z}_\mathrm{D} = \mathbf{f}_\mathrm{PSMI}\mathbf{Y}_\mathrm{D}$ (Hunziker et al., 2006).

Figure 3 illustrates the relationship between the signals (10) and (11) for the case $K = 3$, $N = 2$, $M$=2. The subspace S is here the one-dimensional space defined by the horizontal axis. The result $\mathbf{z}_\mathrm{D}$ of the preamble-based beamforming can be viewed as determined by **z**, which results from orthogonal projection of $\mathbf{f}_\mathrm{MVDR}\mathbf{Y}$ onto the randomly oriented subspace $\mathrm{B}^{\perp}$ as described above. As can be seen in Figure 3, the attempted reconstruction of $\mathbf{s}_\mathrm{D}$ from **z** suffers from severe noise enhancement when the angle between the subspaces $\mathrm{B}^{\perp}$ and S is large.

## BEAMFORMING-DECODING INTERFACE

The MSSE prior to demodulation and decoding does not directly determine the error rate performance at the decoder output. Particular attention needs to be given to the interface between beamforming and LLR computation. For facilitating accurate LLR computation as basis for soft-decision decoding, knowledge of the conditional distributions of the signals is required. If the sample errors are Gaussian distributed, SINRs need to be provided for calculating the LLRs of the bits encoded in the signals.

In conventional receivers the LLRs are directly computed on the basis of $\mathbf{f}_\mathrm{PSMI}\mathbf{Y}$ or $\mathbf{f}_\mathrm{DSMI}\mathbf{Y}$. Either $\varepsilon_\mathrm{s}\left(\mathbf{h}^\mathrm{H}\hat{\mathbf{R}}_\mathrm{P}^{-1}\mathbf{h}\right)$ or $\varepsilon_\mathrm{s}\left(\mathbf{h}^\mathrm{H}\hat{\mathbf{R}}^{-1}\mathbf{h}\right)$ may be employed as SINR estimates as they correspond to the SINR $\varepsilon_\mathrm{s}\left(\mathbf{h}^\mathrm{H}\mathbf{R}^{-1}\mathbf{h}\right)$ by the MVDR beamforming. More elaborate interfaces between beamforming and LLR computation can be formulated utilizing the results from the previous section, as discussed in the following.

### Non-Preamble-Based SMI Beamforming

The $k$th element of $\mathbf{z} = \mathbf{f}_\mathrm{DSMI}\mathbf{Y}$, in the following denoted as $z_k$, does not represent an unbiased estimate of the respective signal in **s**. To see this, write $z_k$ as

$$z_k = \left(\mathbf{P}_\mathrm{B}^{\perp}(\mathbf{u}_k)\mathbf{u}_k^\mathrm{H}\right)s_k + \sum_{\ell=1,\ldots,k-1,k+1,\ldots,K} \left(\mathbf{P}_\mathrm{B}^{\perp}(\mathbf{u}_\ell)\mathbf{u}_k^\mathrm{H}\right)s_\ell + \sum_{\ell=1}^{K}\left(\mathbf{P}_\mathrm{B}^{\perp}(\mathbf{u}_\ell)\mathbf{u}_k^\mathrm{H}\right)v_\ell \tag{12}$$

*Figure 2. Illustration of the non-preamble-based beamforming procedure in the vector signal space*

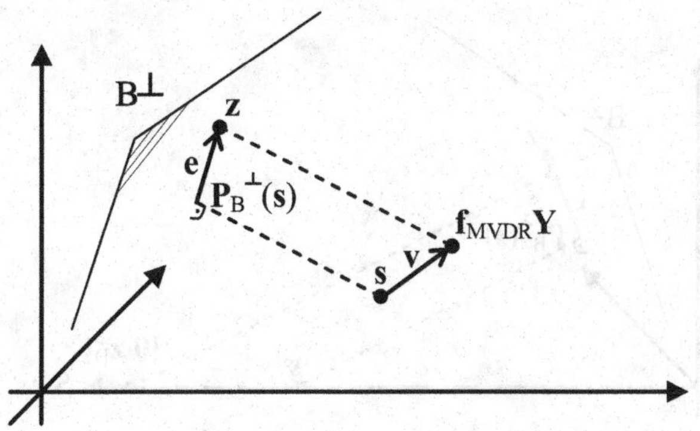

with $\mathbf{u}_k$ denoting the unit row vector containing only zeros except for a 1 at the $k$th position, and $v_k$ the $k$th element of $\mathbf{v} = \mathbf{f}_{\text{MVDR}}\mathbf{W}$. The expression (12) follows from (9) and (1). Obviously, $E[z_k \mid s_k] = (\mathbf{P}_{\text{B}}^{\perp}(\mathbf{u}_k)\mathbf{u}_k^{\text{H}})s_k$ as a consequence of the independence of $s_1,\ldots,s_K,v_1,\ldots,v_K$, where $(\mathbf{P}_{\text{B}}^{\perp}(\mathbf{u}_k)\mathbf{u}_k^{\text{H}}) \in [0,1]$. Moreover, $\varepsilon_{\text{s}}(\mathbf{h}^{\text{H}}\hat{\mathbf{R}}^{-1}\mathbf{h})$ may not be an accurate estimate of $\varepsilon_{\text{s}} / \vartheta_{\text{DSMI}}$ since the SCM $\hat{\mathbf{R}}$ is obtained from a sample space including the desired signal. Both the biased $z_k$ and an inaccurate SINR estimate may have a severe impact on the LLR computation in a conventional receiver.

Since the orientation of the subspace B is known at the receiver end, the factors $(\mathbf{P}_{\text{B}}^{\perp}(\mathbf{u}_\ell)\mathbf{u}_k^{\text{H}})$ appearing in (12) can be computed. Using the above introduced matrix $\mathbf{B} = [\mathbf{b}_1 \ldots \mathbf{b}_K]$ the factors can be expressed as

$$(\mathbf{P}_{\text{B}}^{\perp}(\mathbf{u}_\ell)\mathbf{u}_k^{\text{H}}) = \begin{cases} -\mathbf{b}_\ell^{\text{H}}\mathbf{b}_k, & \text{if } \ell \neq k \\ 1 - \|\mathbf{b}_k\|^2, & \text{if } \ell = k. \end{cases} \tag{13}$$

Multiplication of $z_k$ by $\left(1 - \|\mathbf{b}_k\|^2\right)^{-1}$ leads to an unbiased estimate of $s_k$, and the vector

$$\tilde{\mathbf{z}} = \left[ \left(1 - \|\mathbf{b}_1\|^2\right)^{-1} z_1 \ldots \left(1 - \|\mathbf{b}_K\|^2\right)^{-1} z_K \right] \tag{14}$$

represents a more appropriate basis for the LLR computation than $\mathbf{z}$. It further follows from (12) and (13) that the variance of the error term in the $k$th element of $\tilde{\mathbf{z}}$ equals

$$\vartheta_{\text{DX}}^{(k)} = \left(1 - \|\mathbf{b}_k\|^2\right)^{-2} \left( \|\mathbf{b}_k\|^2 \left(1 - \|\mathbf{b}_k\|^2\right)\varepsilon_{\text{s}} + \left(1 - \|\mathbf{b}_k\|^2\right)\vartheta_{\text{MVDR}} \right). \tag{15}$$

The sample-level SINRs $\varepsilon_{\text{s}}/\vartheta_{\text{DX}}^{(1)},\ldots,\varepsilon_{\text{s}}/\vartheta_{\text{DX}}^{(K)}$ were ideally used along with (14), however, there remains one unknown quantity in (15), namely $\vartheta_{\text{MVDR}}$. The known front end noise power $\vartheta_{\text{noise}}$ is a possible substitute for $\vartheta_{\text{MVDR}}$. Receivers computing LLRs on the basis of (14) and $\varepsilon_{\text{s}}/\hat{\vartheta}_{\text{DX}}^{(1)},\ldots,\varepsilon_{\text{s}}/\hat{\vartheta}_{\text{DX}}^{(K)}$, where

*Figure 3. Illustration of the preamble-based beamforming procedure in the vector signal space*

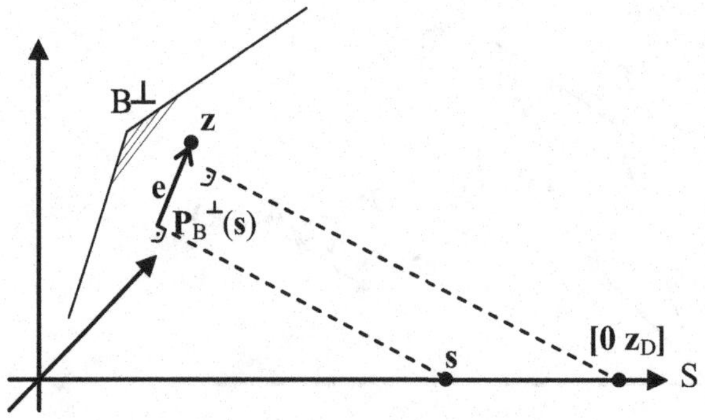

$$\varepsilon_s / \vartheta_{DX}^{(k)} = \varepsilon_s \left(1 - \|\mathbf{b}_k\|^2\right)\left(\|\mathbf{b}_k\|^2 \varepsilon_s + \vartheta_{noise}\right)^{-1}, \tag{16}$$

are investigated below.

## Preamble-Based SMI Beamforming

The preamble-based beamformer $\mathbf{f}_{PSMI}$ perfectly restores the signal $\mathbf{s}_D$ if $\mathbf{e} = \mathbf{0}$, as can be seen from (10) and (11). The effect of a non-zero $\mathbf{e}$ on a certain element of $\mathbf{z}_D$ depends on the orientation of $B^{\perp}$. In order to express the MSSE per each of the *K-M* samples composing $\mathbf{z}_D$, we define the matrix

$$\mathbf{T} = \begin{bmatrix} \mathbf{P}_B^{\perp}(\mathbf{u}_{M+1}) \\ \vdots \\ \mathbf{P}_B^{\perp}(\mathbf{u}_K) \end{bmatrix}. \tag{17}$$

The rows of $\mathbf{T}$ span the subspace $\{\mathbf{P}_B^{\perp}([\mathbf{0}\ \mathbf{x}]) : \mathbf{x} \in \mathbb{C}^{K-M}\}$ of $B^{\perp}$, which has *K-M* dimensions with probability 1. Using $\mathbf{T}$ the signal $\mathbf{z}$ defined in (10) can be expressed as $\mathbf{z} = \mathbf{s}_D\mathbf{T} + \mathbf{e}$. Furthermore, the mapping $\mathbf{z} \rightarrow \mathbf{z}_D$ defined by (11) can be expressed as $\mathbf{z}_D = \mathbf{z}\mathbf{T}^H(\mathbf{T}\mathbf{T}^H)^{-1}$. It follows that

$$\mathbf{z}_D = \mathbf{s}_D + \mathbf{v}\mathbf{T}^H\left(\mathbf{T}\mathbf{T}^H\right)^{-1} \tag{18}$$

since $\mathbf{e}\mathbf{T}^H = \mathbf{v}\mathbf{T}^H$. The random vector $\mathbf{v}$ is $CN(0, \vartheta_{MVDR}\ \mathbf{I}_K)$, hence, the sample-level MSSEs correspond to the diagonal of the matrix $\vartheta_{MVDR}(\mathbf{T}\mathbf{T}^H)^{-1}$. As easily seen, the matrix $\mathbf{T}$ is equal to the *K-M* lower rows of the matrix $\mathbf{I}_K - \mathbf{B}^H\mathbf{B}$, and $\mathbf{T}\mathbf{T}^H = \mathbf{I}_{K-M} - \mathbf{B}_D^H\mathbf{B}_D$, where the $(N\text{-}1)\times(K\text{-}M)$-matrix $\mathbf{B}_D$ is defined as the submatrix comprising the *K-M* rightmost columns of $\mathbf{B}$. Using some basic theorems from linear algebra, the $k$th diagonal element of $\vartheta_{MVDR}(\mathbf{T}\mathbf{T}^H)^{-1}$ can be expressed as

$$\vartheta_{PX}^{(k)} = \vartheta_{MVDR}\frac{\det\left(\mathbf{T}_{-k}\mathbf{T}_{-k}^H\right)}{\det\left(\mathbf{T}\mathbf{T}^H\right)} = \vartheta_{MVDR}\frac{\det\left(\mathbf{I}_{N-1} - \mathbf{B}_D\mathbf{B}_D^H + \mathbf{b}_{M+k}\mathbf{b}_{M+k}^H\right)}{\det\left(\mathbf{I}_{N-1} - \mathbf{B}_D\mathbf{B}_D^H\right)}, \tag{19}$$

where $\mathbf{T}_{-k}$ denotes the matrix resulting from deleting the $k$th row of $\mathbf{T}$. Note that instead of the inversion of the $(K\text{-}M)$-dimensional matrix $(\mathbf{T}\mathbf{T}^H)$ the evaluation of the expression on the right hand side of (19) only requires the computation of determinants of $(N\text{-}1)$-dimensional matrices.

Again, the MSSE $\vartheta_{MVDR}$ resulting from the ideal MVDR beamforming is unknown. A practical substitute may be $\left(\mathbf{h}^H\hat{\mathbf{R}}_P^{-1}\mathbf{h}\right)^{-1}$, leading to the sample-level SINR estimates $\varepsilon_s/\hat{\vartheta}_{PX}^{(1)}, \ldots, \varepsilon_s/\hat{\vartheta}_{PX}^{(K-M)}$, where

$$\varepsilon_s/\hat{\vartheta}_{PX}^{(k)} = \varepsilon_s\left(\mathbf{h}^H\hat{\mathbf{R}}_P^{-1}\mathbf{h}\right)\frac{\det\left(\mathbf{I}_{N-1} - \mathbf{B}_D\mathbf{B}_D^H\right)}{\det\left(\mathbf{I}_{N-1} - \mathbf{B}_D\mathbf{B}_D^H + \mathbf{b}_{M+k}\mathbf{b}_{M+k}^H\right)}, \qquad k \in \{1, \ldots, K-M\}. \tag{20}$$

Receivers computing the LLRs using the above SINR estimates are investigated in the following section.

## PERFORMANCE ANALYSIS

In this section we compare the error rate performance of receiver variants conforming to the architecture in Figure 1. In Monte-Carlo simulations, BICM signals are generated by encoding random bit sequences using the rate $R_c = ½$ convolutional encoder with the generators ($133_{oct}$, $171_{oct}$), followed by bitwise random interleaving and mapping onto signals from a quadrature amplitude modulation (QAM) signal set. The overall length of the information signals is 360 symbols, thus conveying 720 and 1080 bits of information in the cases of 16-QAM and 64-QAM, respectively (minus six tail bits in the convolutional encoding). After the mapping the signals are divided into blocks of length $K$-$M$, and preambles in the form of $M$ zeros are added to each block. The receiver observes $N = 4$ copies of the transmitted signal via four antennas, where the constant channel gain vector equals $\mathbf{h} = [1\ 1\ 1\ 1]^H$, and the array signal is subject to additive CCI stemming from $N$ - 1 sources as well as Gaussian front end noise. The independent vectors $\mathbf{g}_1$, $\mathbf{g}_2$, and $\mathbf{g}_3$, defining the single input/$N$-output channels between the three sources of interference and the receiver, are $CN(\mathbf{0},\mathbf{I}_N)$. Furthermore, the desired signal and the white Gaussian noise-like signals emitted by the interferers have equal mean power, while the front end noise power is down by 20 dB. The covariance matrix $\mathbf{R}$ follows as $\mathbf{R} = \varepsilon_s\left(\mathbf{g}_1\mathbf{g}_1^H + \mathbf{g}_2\mathbf{g}_2^H + \mathbf{g}_3\mathbf{g}_3^H\right) + 10^{-2}\varepsilon_s\mathbf{I}_4$. For each block the gain vectors $\mathbf{g}_1$, $\mathbf{g}_2$, and $\mathbf{g}_3$ are generated independently, being representative for scenarios with arbitrarily varying CCI characteristics from block to block. Large numbers of frame transmissions are simulated and the instances with erroneous output of the Viterbi decoder at the receiver end identified.

### Non-Preamble-Based SMI Beamforming

Figure 4 shows the observed frame error rates (FERs) at the decoder output of three receivers with different non-preamble-based beamforming and SINR estimation methods in the case of 16-QAM. Two conventional receivers let the LLR computation directly follow the classic beamforming (7), where one adopts the SMI-based value $\varepsilon_s\left(\mathbf{h}^H\hat{\mathbf{R}}^{-1}\mathbf{h}\right)$ and the other the signal-to-noise ratio (SNR) $\varepsilon_s / \vartheta_{noise}$ as SINR estimates. We find that calculating LLRs under the assumption of an SINR of $\varepsilon_s / \vartheta_{noise}$ leads to clearly better results, and conclude that $\varepsilon_s\left(\mathbf{h}^H\hat{\mathbf{R}}^{-1}\mathbf{h}\right)$ is an inadequate estimate of the SINR of the spatially filtered signal. The third receiver performs the LLR computation on the basis of (14) along with the sample-level SINRs $\varepsilon_s/\hat{\vartheta}_{DX}^{(1)}, \ldots, \varepsilon_s/\hat{\vartheta}_{DX}^{(K)}$. With this enhanced interface between beamforming and decoding the FERs can be improved at small block lengths. In the case of K = 24, for instance, the FER can be reduced from around 0.7 to around 0.4. At block lengths larger than 40 samples, however, no performance gains are achieved.

Figure 5 shows the FERs with the above two superior receivers in the case of 64-QAM. Longer blocks are necessary here in order to attain FERs significantly below 1. We observe a slight improvement by the enhanced interface at block lengths between 60 and 90 samples.

With either QAM scheme, occasional frame errors are encountered even at relatively large block lengths. The FERs can be decreased below $10^{-2}$ by more powerful FEC. The error rates in Figure 6 are obtained by employing the rate $R_c = 1/8$ convolutional encoder with the generators ($153_{oct}$, $111_{oct}$, $165_{oct}$, $173_{oct}$, $135_{oct}$, $135_{oct}$, $147_{oct}$, $137_{oct}$). The enhanced

*Figure 4. FER performance versus block length with four antennas, 16-QAM, and non-preamble-based beamforming*

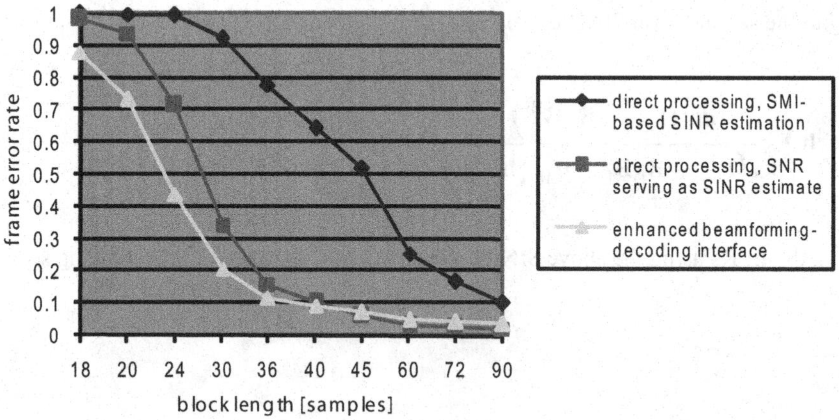

*Figure 5. FER performance versus block length with four antennas, 64-QAM, and non-preamble-based beamforming*

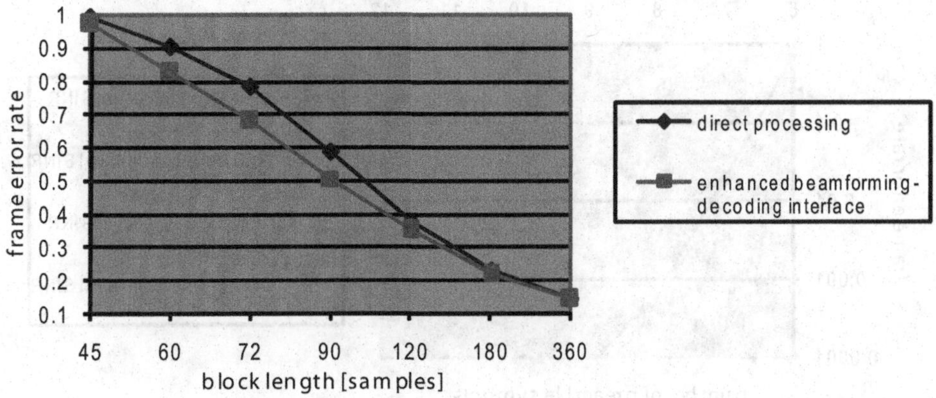


*Figure 6. FER performance versus block length with four antennas, rate 1/8 convolutional encoding, and non-preamble-based beamforming*

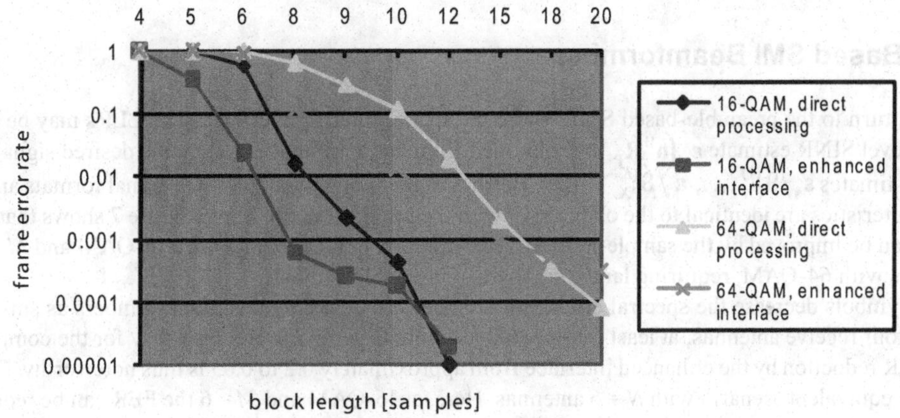


*Figure 7. FER performance versus number of preamble symbols with four antennas and preamble-based beamforming*

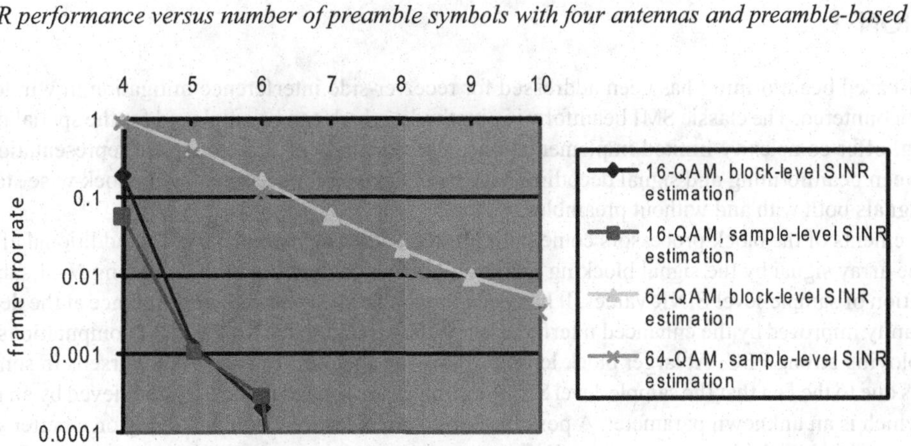

*Figure 8. FER performance versus number of preamble symbols with six antennas and preamble-based beamforming*

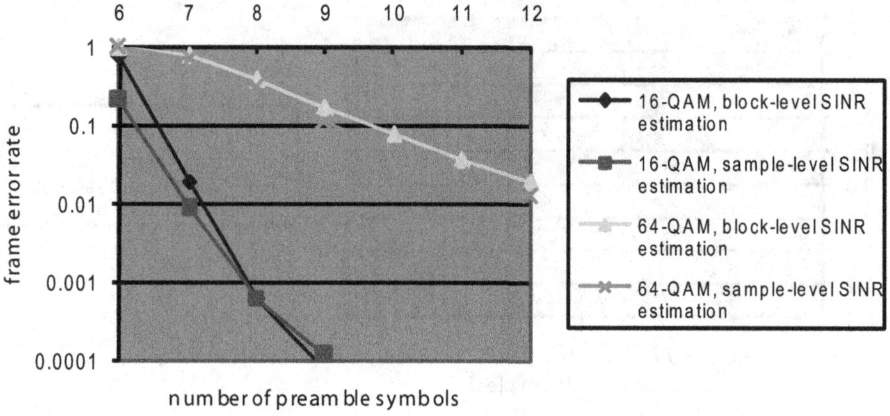


interface providing sample-level SINRs is clearly advantageous here, reducing the FERs by a factor 10 or more in some situations.

## Preamble-Based SMI Beamforming

We once again turn to the preamble-based SMI, where the spatial filtering yields $\mathbf{z}_D$. The LLRs may be computed using the block-level SINR estimate $\varepsilon_s\left(\mathbf{h}^H\hat{\mathbf{R}}_P^{-1}\mathbf{h}\right)$, obtained from an SCM unaffected by the desired signal, or using the sample-level estimates $\varepsilon_s/\hat{\vartheta}_{PX}^{(1)},\ldots,\varepsilon_s/\hat{\vartheta}_{PX}^{(K-M)}$ (20). Besides of including preambles, the signal formats and the channel and CCI characteristics are identical to the ones above with the rate $R_c = ½$ encoding. Figure 7 shows that the error rate performance can be improved by the sample-level SINR estimation in the scenario with 16-QAM and $M = 4$, as well as in the scenarios with 64-QAM, requiring larger numbers of preamble symbols.

Preamble symbols decrease the spectral efficiency, so there is a desire to keep their numbers as small as possible. In the case of four receive antennas, at least four preamble symbols per block are necessary for the computation of the SCMs. The FER reduction by the enhanced interface from approximately 0.2 to 0.05 is thus noteworthy. Figure 8 shows the FERs in an equivalent scenario with $N = 6$ antennas. Here, at 16-QAM and $M = 6$ the FER can be reduced from approximately 0.8 to 0.2, while minor improvements are also attainable in the case of 64-QAM.

## CONCLUSION

Adaptive SCM-based beamforming has been addressed for receiver-side interference mitigation in wireless networks where CCI is encountered. The classic SMI beamforming methods, which can be employed for the spatial filtering prior to the decoding, offer complexity-limited implementations. With the help of a vector space representation, enhanced interfaces between beamforming and signal decoding have been devised for scenarios with block-wise stationary CCI and transmit signals both with and without preambles.

The enhancements of the block processors come with limited additional complexity. The additional effort includes a filtering of the array signal by the signal blocking matrix, an orthogonalization of the resulting $N - 1$ sample vectors, and the calculation of sample-level SINR values. It has been found that the error rate performance at the decoder output can be significantly improved by the enhanced interfaces when the sample space for the SCM computations is small and also when employing strong FEC. At larger block lengths, however, the performance even worsens in some scenarios. This outcome is due to the fact that the sample-level SINR estimates are related to the SINR achieved by an ideal MVDR beamformer, which is an unknown parameter. A possible corrective measure would be to rely on a better substitute for this unknown quantity than the interference-free SNR.

## REFERENCES

Absil, P.-A., Edelman, A., & Koev, P. (2006). On the Largest Principal Angle between Random Subspaces. *Linear Algebra Appl.*, vol. 414, no. 1, pp 288-294, 2006.

Bell, K. L., Ephraim, Y., & Van Trees, H. L. (2000). A Bayesian Approach to Robust Adaptive Beamforming. *IEEE Trans. Sig. Processing*, vol. 48, no. 2, pp. 386-398.

Biedka, T. E., Reed, J. H., & Tranter, W. H. (2000). Mean Convergence Rate of a Decision Directed Adaptive Beamformer with Gaussian Interference. In *Proc. Sensor Array Multich. Sig. Processing Workshop:* (pp. 68-72), Cambridge, MA.

Carlson, B. D. (1988). Covariance Matrix Estimation Errors and Diagonal Loading in Adaptive Arrays. *IEEE Trans. Aerosp. Electron. Syst.*, vol. 24, no. 4, pp. 397-401.

Feldman, D. D., & Griffiths, L. J. (1994). A Projection Approach for Robust Adaptive Beamforming. *IEEE Trans. Sig. Processing*, vol. 42, no. 4, pp. 867-876.

Hunziker, T., Aono, T., & Ohira, T. (2004). An Iterative Beamforming and Decoding Procedure for Wireless Networks with Uncoordinated Channel Access. *IEEE Commun. Lett.*, vol. 8, no. 4, pp. 256-258.

Hunziker, T., & Taromaru, M. (2006). An Adaptive Beamforming Scheme for Enhanced Cochannel Interference Mitigation on Short Array Signal Intervals. In *Proc. IEEE Int. Conf. on Acoustics, Speech, and Signal Processing (ICASSP '06):* (pp. IV-1025-IV-1028), Toulouse, France.

Hunziker, T., Bordim, J. L., Taromaru, M., & Ohira, T. (2007). Maximum-Likelihood Array Signal Decoding in the Presence of Block-Wise Stationary Cochannel Interference. *IEEE Trans. Wireless Commun.*, vol. 6, no. 4, pp. 1476-1487.

Kuzminskiy, A. M. (2003). Iterative MIMO ML Detection with On-line Initialization Selection for Unknown Channels and Unstructured Asynchronous Interference. In *Proc. IEEE Vehicular Technology Conference (VTC 2003-Fall): Vol. 1* (pp. 617-621), Orlando, FL.

Li, J., Stioca, P., & Wang, Z. (2003). On Robust Cabon Beamforming and Diagonal Loading. *IEEE Trans. Sig. Processing*, vol. 51, no. 7, pp. 1702-1715, Jul., 2003.

Lorenz, R. G., & Boyd, P. (2005). Robust Minimum Variance Beamforming. *IEEE Trans. Sig. Processing*, vol. 53, no. 5, pp. 1684-1696.

Monzingo, R. A., & Miller, T. W. (1980). *Introduction to Adaptive Arrays*. Wiley, New York, NY.

Reed, I. S., Mallett, J. D., & Brennan, L. E. (1974). Rapid Convergence Rate in Adaptive Arrays. *IEEE Trans. Aerosp. Electron. Syst.*, vol. 10, no. 6, pp. 853-863.

Richmond, C. D. (1996). Derived PDF of Maximum Likelihood Signal Estimator which Employs an Estimated Noise Covariance. *IEEE Trans. Sig. Processing*, vol. 44, no. 2, pp. 305-315.

Shahbazpanahi, S., Gershman, A. B., Luo, Z.-Q., & Wong, K. M. (2003). Robust Adaptive Beamforming for General-Rank Signal Models. *IEEE Trans. Sig. Processing*, vol. 51, no. 9, pp. 2257-2269.

Swindlehurst, A. L., Daas, S., & Yang, J. (1995). Analysis of a Decision Directed Beamformer. *IEEE Trans. Sig. Processing*, vol. 43, no. 12, pp. 2920-2927.

Van Trees, H. L. (2002). *Optimum Array Processing*. Wiley, New York, NY.

Van Veen, B. D. (1991). Adaptive Convergence of Linearly Constrained Beamformers Based on the Sample Covariance Matrix. *IEEE Trans. Sig. Processing*, vol. 39, no. 6, pp. 1470-1473.

Vorobyov, S. A., Gershman, A. B., & Luo, Z.-Q. (2003). Robust Adaptive Beamforming Using Worst-Case Performance Optimization: A Solution to the Signal Mismatch Problem. *IEEE Trans. Sig. Processing*, vol. 51, no. 2, pp. 313-324.

# Chapter V
# Random Array Theory and Collaborative Beamforming

**Hideki Ochiai**
*Yokohama National University, Japan*

**Patrick Mitran**
*University of Waterloo, Canada*

**H. Vincent Poor**
*Princeton University, USA*

**Vahid Tarokh**
*Harvard University, USA*

## ABSTRACT

*In wireless sensor networks, the sensor nodes are often randomly situated, and each node is likely to be equipped with a single antenna. If these sensor nodes are able to synchronize, it is possible to beamform by considering sensor nodes as a random array of antennas. Using probabilistic arguments, it can be shown that random arrays formed by dispersive sensors can form nice beampatterns with a sharp main lobe and low sidelobe levels. This chapter reviews the probabilistic analysis of linear random arrays, which dates back to the early work of Y. T. Lo (1964), and then discusses recent work on the statistical analysis of two-dimensional random arrays originally derived in the framework of wireless sensor networks.*

## INTRODUCTION

Wireless sensor networks have recently attracted much attention in the communication engineering community (e.g., Yao, 1998). In many such networks, battery powered communicating nodes, each equipped with a single antenna, are distributed randomly. These nodes not only gather information but also communicate over a wireless channel. Some nodes may relay the received information to nearby nodes or directly transmit to the destination (or fusion center). In order to deliver information to the fusion center, provided neighboring nodes possess the same information and are able to synchronize, it is also possible for these nodes to beamform collaboratively. We refer to this kind of distributed beamforming as *collaborative beamforming* (Ochiai, 2005).

In wireless communications, beamforming enables an efficient implementation of space-division multiple access (SDMA), which has the potential to significantly increase communication rates over multiple access channels. SDMA by means of collaborative beamforming is also a powerful approach in the framework of wireless *ad hoc* sensor networks where several clusters operate asynchronously.

Let us consider the scenario illustrated in Figure 1, where the sensor nodes in two separate clusters A and B are communicating with respective distant fusion centers located in different directions. In this case, intra-cluster communication among sensor nodes for information sharing may be realized by low-cost short distance broadcast-type communication. Therefore, the main challenge is fair channel allocation for long distance communication links between the clusters and the fusion centers. Since synchronization among nodes in different clusters may not be easily established, random access is commonly used. With limited available spectrum resources, however, random access typically requires additional overhead such as collision detection. Furthermore, an increase in the number of communicating clusters considerably reduces the overall throughput.

On the other hand, with collaborative beamforming depicted in Figure 1, cognition of the other communicating clusters is not necessary, provided that the direction of the fusion center is different. Due to the random nature of *ad hoc* sensor networks, it is highly unlikely that the sensor nodes in the distinct clusters are transmitting in the same direction. Therefore, collaborative beamforming has the potential to become a low-cost SDMA implementation.

A question that arises in this scenario is whether or not collaborative sensors can form a nice beampattern. Since the distribution of the sensor nodes is typically random by nature, it is reasonable to consider arrays formed by the sensor nodes as random arrays and to treat them using probabilistic arguments.

Probabilistic analysis of antenna arrays dates back to the early work of Lo (1964), who first developed a comprehensive theory of linear random arrays. Using statistical arguments, Lo showed that a random array can form a nice average beampattern without the major grating lobes that are observed in a typical periodic array. He also derived the distribution of the beampattern based on a Gaussian approximation. Later, Steinberg (1972), Agrawal & Lo (1972), as well as Donvito & Kassam (1979) analyzed the distribution of the maximum of the sidelobe peaks associated with linear random arrays. An excellent overview and analytical treatment of linear random arrays can be found in (Steinberg, 1976).

With applications to wireless sensor networks in mind, in this chapter, we consider the beampatterns of two-dimensional (planar) phased arrays randomly distributed over a disk of a given radius. Much of the mathematical detail of this analysis can be found in our recent work (Ochiai, 2005) and here we focus mainly on the significance of the obtained results on the probabilistic distribution of such beampatterns. Upon studying this work, we notice that whereas the theories and optimizations of unequally-spaced arrays, both linear and planar, are prolific in the literature (e.g., Ishimaru, 1962; Bar-Ness, 1984; Leahy, 1991), there are relatively few publications on the subject of statistical properties of planar random arrays in the context of communication engineering applications (e.g., Fante, 1991). Nevertheless, the theory of two-dimensional random arrays has found diverse applications. Optimization issues of random arrays have been studied by Holm, Elgetun, & Dahl (1997) in the framework of ultrasound imaging. Kook, Davies, & Bolton (2002) analyze the statistical distribution of two dimensional microphone arrays. More recently, the statistical behavior of more complex arrays of random subarrays has been studied by Kerby & Bernhard (2006).

This chapter is organized as follows. The subsequent section is devoted to the description of assumptions and mathematical models in the framework of wireless sensor networks. We then study statistical properties of the beampatterns of random arrays uniformly distributed over a disk of a given radius. Specifically, we evalute the average beampattern,

*Figure 1. Collaborative beamforming in ad hoc wireless sensor networks*

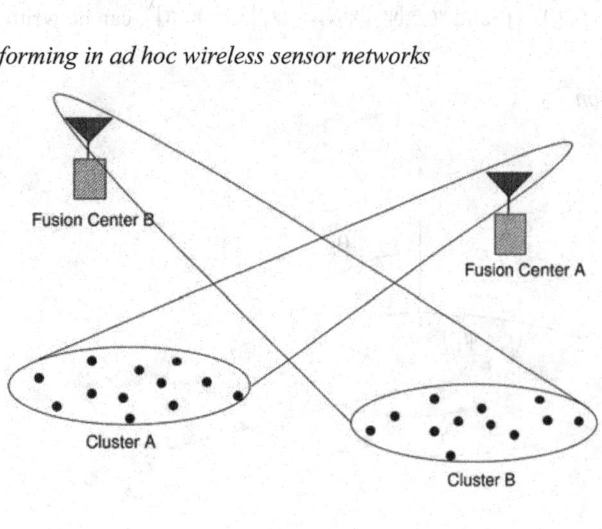

the distribution of sidelobes, and the distribution of the maximum sidelobe peak both theoretically and numerically. Finally, concluding remarks are given.

## SYSTEM MODEL AND BEAMPATTERN

The geometrical configuration of interest for the distributed nodes and destination is illustrated in Figure 2 where all the sensor nodes (or antenna elements) are assumed to be located on the $x-y$ plane, forming a planar array. For analytical convenience, we designate each node location in polar coordinates. The location of the $k$th node is thus denoted by $(r_k, \psi_k)$. The location of the destination is given in spherical coordinates by $(A, \phi_0, \theta_0)$. Following the standard notation in antenna theory (e.g., Balanis, 1997), the angle $\theta \in [0, \pi]$ denotes the elevation direction and the angle $\phi \in [-\pi, \pi]$ represents the azimuth direction. For analytical simplicity, we make the following assumptions.

1. Each antenna element is an ideal isotropic antenna.
2. All nodes transmit with identical energies, and the path losses of all nodes are also identical.
3. There is no reflection or scattering of the signal.
4. The antenna elements (sensor nodes) are sufficiently separated such that mutual coupling effects are negligible.
5. All the nodes are perfectly synchronized (or connected) such that no phase offset or jitter occurs among antenna elements.
6. The location of each antenna element is chosen randomly, following a uniform distribution within a disk of radius $R$.

In short, the above assumptions guarantee that the resulting array formed by the sensor nodes is an ideal phased array with each antenna element uniformly distributed over a planar disk of radius $R$.

Suppose we have $N$ sensor nodes within the disk and let $d_k(\phi, \theta)$ denote the Euclidean distance between the $k$th node and the reference location $(A, \phi, \theta)$, where $k \in \{1, 2,..., N\}$. It then follows that

$$d_k(\phi, \theta) = \sqrt{A^2 + r_k^2 - 2r_k A \sin\theta \cos(\phi - \psi_k)}. \quad (1)$$

If the initial phase of the node $k$ is set to

$$\Phi_k = -\frac{2\pi}{\lambda} d_k(\phi_0, \theta_0),$$

where $\lambda$ is the wavelength of the radio frequency (RF) carrier, the corresponding array factor, given the realization of node locations $r = [r_1, r_2, \ldots, r_N] \in [0, R]^N$ and $\psi = [\psi_1, \psi_2, \ldots, \psi_N] \in [-\pi, \pi]^N$, can be written as

*Figure 2. Definitions of notation*

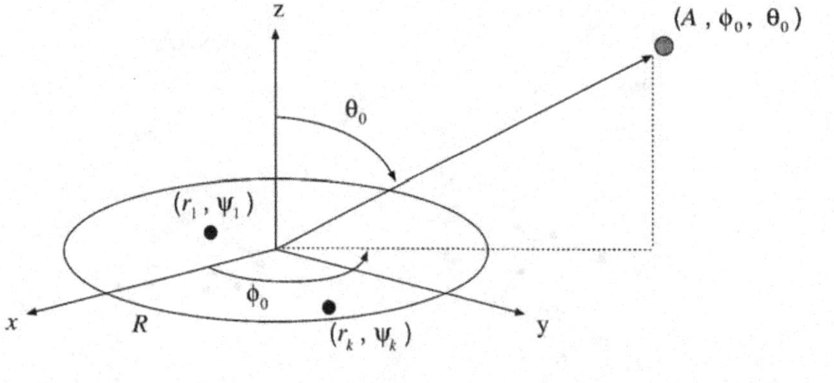

$$F(\phi, \theta | r, \psi) = \frac{1}{N}\sum_{k=1}^{N} e^{j\Phi_k} e^{j\frac{2\pi}{\lambda}d_k(\phi, \theta)} = \frac{1}{N}\sum_{k=1}^{N} e^{j\frac{2\pi}{\lambda}\{d_k(\phi, \theta) - d_k(\phi_0, \theta_0)\}}.$$

Since we are interested in the radiation pattern in the far-field region, assuming $A \gg r_k$, the distance $d_k(\phi, \theta)$ in (1) is approximated by

$$d_k(\phi, \theta) \approx A - r_k \sin\theta \cos(\phi - \psi_k).$$

The far-field radiation pattern is thus approximated by

$$F(\phi, \theta | r, \psi) \approx \frac{1}{N}\sum_{k=1}^{N} e^{j\frac{2\pi}{\lambda}r_k\{\sin\theta_0 \cos(\phi_0 - \psi_k) - \sin\theta \cos(\phi - \psi_k)\}} \triangleq \tilde{F}(\phi, \theta | r, \psi). \tag{2}$$

In order to make theoretical analysis tractable, we further assume that the target destination is located on the same plane as the antenna array, i.e., $\theta_0 = \pi / 2$. Without loss of generality, it is also assumed that $\phi_0 = 0$. The array factor of (2) can then be written as

$$\begin{aligned}\tilde{F}(\phi, \theta | r, \psi) &= \frac{1}{N}\sum_{k=1}^{N} e^{j\frac{2\pi}{\lambda}r_k\sqrt{(1-\sin\theta)^2 + 4\sin\theta \sin^2\frac{\phi}{2}}\,\sin\left(\psi_k - \arctan\left(\frac{1-\sin\theta\cos\phi}{\sin\theta\sin\phi}\right)\right)} \\ &= \frac{1}{N}\sum_{k=1}^{N} e^{j2\pi\frac{R}{\lambda}\sqrt{(1-\sin\theta)^2 + 4\sin\theta \sin^2\frac{\phi}{2}}\,\tilde{r}_k \sin\tilde{\psi}_k}\end{aligned} \tag{3}$$

where $\tilde{r}_k \triangleq r_k / R$ and $\tilde{\psi}_k \triangleq \psi_k - \arctan\left(\frac{1-\sin\theta\cos\phi}{\sin\theta\sin\phi}\right)$.

By assumption, the node locations $(r_k, \psi_k)$ follow a uniform distribution over the disk of radius $R$. Thus, the probability density functions (pdfs) of $r_k$ and $\psi_k$ are given by

$$\begin{aligned} f_{r_k}(r) &= \frac{2r}{R^2}, \quad 0 < r < R, \\ f_{\psi_k}(\psi) &= \frac{1}{2\pi}, \quad -\pi \le \psi < \pi. \end{aligned}$$

Referring to (3), we are interested in the pdf of the following compound random variable

$$z_k \triangleq \tilde{r}_k \sin\tilde{\psi}_k.$$

It can be readily shown that the pdf of this random variable is given by

$$f_{z_k}(z) = \frac{2}{\pi}\sqrt{1-z^2}, \quad -1 \le z \le 1.$$

The array factor of (3) can then be rewritten as

$$\tilde{F}(\phi, \theta | z) = \frac{1}{N}\sum_{k=1}^{N} e^{-j\alpha(\phi, \theta) z_k} \tag{4}$$

where

$$\alpha(\phi,\theta) \triangleq 2\pi\frac{R}{\lambda}\sqrt{(1-\sin\theta)^2+4\sin\theta\sin^2\frac{\phi}{2}}. \quad (5)$$

In the above expression, the term $R/\lambda$ represents the disk radius normalized by the wavelength and is one of the key parameters in the subsequent analysis.

Finally, the far-field beampattern can be defined as

$$\begin{aligned} P(\phi,\theta\,|\,z) &\triangleq \left|\tilde{F}(\phi,\theta\,|\,z)\right|^2 = \frac{1}{N^2}\sum_{k=1}^{N}\sum_{l=1}^{N} e^{-j\alpha(\phi,\theta)\left(z_k-z_l\right)} \\ &= \frac{1}{N}+\frac{1}{N^2}\sum_{k=1}^{N} e^{-j\alpha(\phi,\theta)\,z_k}\sum_{\substack{l=1\\ l\neq k}}^{N} e^{j\alpha(\phi,\theta)\,z_l}. \end{aligned} \quad (6)$$

Note that in (Ochiai, 2005) the elevation angle of observation is restricted to the target angle, i.e., only the case with $\theta = \pi/2$ is considered for simplicity. In this contribution, the parameter $\theta$ is retained.

## PROPERTIES OF BEAMPATTERN OF UNIFORMLY DISTRIBUTED RANDOM ARRAY

In what follows, several statistical properties of the beampattern formed by randomly distributed antenna elements are derived and numerically evaluated. We begin by studying the average beampattern of the random arrays.

### Average Far-Field Beampattern

Taking the statistical average of (6) with respect to $z$, we obtain

$$P_{av}(\phi,\theta) \triangleq E_z\left\{P(\phi,\theta\,|\,z)\right\} = \frac{1}{N}+\left(1-\frac{1}{N}\right)\left|\frac{2\,J_1\left(\alpha(\phi,\theta)\right)}{\alpha(\phi,\theta)}\right|^2 \quad (7)$$

where $E_z\{\cdot\}$ denotes expectation with respect to the random variables $z$ and $J_n(x)$ is the $n$th order Bessel function of the first kind, which is expressed (for a non-negative integer $n$) as

$$J_n(x) = \sum_{k=0}^{\infty}\frac{(-1)^k}{k!\,(n+k)!}\left(\frac{x}{2}\right)^{2k+n}.$$

The function $J_1(x)/x$ is oscillatory, but the local maxima of oscillation tend to decrease with increasing $x$. In (7), the first term represents the average power level of the sidelobe. The average level does not depend on any antenna element location and it is simply given by the inverse of the number of the antenna elements. On the other hand, the second term is the contribution of the main lobe factor. Numerical evaluation shows that the smallest value of $\alpha(\phi, \theta)$ that makes the second term zero is around 3.8317. From (5), the one-sided width of the main lobe (i.e., half of the null-to-null main lobe width observed in the average beampattern) in the azimuth direction (i.e., $\theta = 90^\circ$) can be estimated by

$$\Delta\phi = 2\arcsin\left(\frac{3.8317}{4\pi}\frac{\lambda}{R}\right) \approx 35^\circ \times \frac{\lambda}{R}. \quad (8)$$

The average beampattern of (7) with respect to azimuth angle $\phi$ is plotted in Figure 3 for several values of $R/\lambda$ with $N = 32$. As can be observed, the sidelobe approaches $1/N$ as the beam angle moves away from the target direction. In Figure 4, the main lobe beampattern is plotted with respect to elevation angle $\theta$. From Figures 3 and 4, it is apparent that the main lobe can be made sharper by increasing the disk radius $R$, but improving the sharpness in terms of elevation angle requires a much larger increase in the radius $R$. This observation stems from the fact that $\alpha(\phi, \theta)$ in (5) increases rapidly with $\phi$, but slowly with $\theta$. In fact, similar to the azimuthal case, from (5), the one-sided width of the main lobe in the elevation direction (with $\phi = 0^\circ$) can be estimated by

$$\Delta\theta = 2\arcsin\left(\sqrt{\frac{3.8317}{4\pi}\frac{\lambda}{R}}\right) \approx 63^\circ \times \sqrt{\frac{\lambda}{R}}.$$

## Distribution of Far-Field Beampattern of Random Arrays

Conditioned on $\phi$ and $\theta$, the array factor of the form (4) is a sum of bounded independent and identically distributed (i.i.d.) complex random variables. Therefore, by the strong law of large numbers for fixed $(\phi, \theta)$, the resulting beampattern should approach the ensemble average (7) with probability one as the number of antenna elements increases. However, in practice we are interested in finite (relatively small) values of $N$, and in this case the average beampattern does not necessarily represent the beampattern of any particular realization. For example, the average beampattern and those of two particular realizations of randomly generated antenna arrays are shown in Figure 5. The main lobes of the realization beampatterns closely match the average, but the sidelobes may fluctuate with a large dynamic range and their peaks often well exceed the average level.

Therefore, in practice, the statistical *distribution* of beampatterns and sidelobes in particular, is of importance. By approximating the beampattern sidelobes as a complex Gaussian process, Lo (1964) has derived the distribution of the beampattern in the case of linear random arrays. Using similar arguments, we now derive an approximate distribution for planar random arrays.

Considering that the array factor consists of a sum of $N$ statistically independent random variables, as $N$ increases, by the central limit theorem we may expect that the array factor in any given direction, except at the deterministic angle $\phi = 0$, approaches a complex Gaussian distribution. This approximation typically enables us to derive a simpler distribution formula. To this end, we write (4) as

$$\tilde{F}(\phi,\theta\,|\,\mathbf{z}) = \frac{1}{\sqrt{N}}(X - jY), \tag{9}$$

where

$$X \triangleq \frac{1}{\sqrt{N}}\sum_{k=1}^{N}\cos\left(z_k\alpha(\phi,\theta)\right),\; Y \triangleq \frac{1}{\sqrt{N}}\sum_{k=1}^{N}\sin\left(z_k\alpha(\phi,\theta)\right).$$

Since the $z_k$'s are i.i.d. random variables, as $N$ increases the distribution of $X$ and $Y$ will approach that of a Gaussian random variable with

$$\begin{aligned} E\{X\} &= \frac{2J_1(\alpha(\phi,\theta))}{\alpha(\phi,\theta)}\sqrt{N} \triangleq m_x, \\ \operatorname{Var}(X) &= \frac{1}{2}\left(1+\frac{J_1(2\alpha(\phi,\theta))}{\alpha(\phi,\theta)}\right) - \left(\frac{2J_1(\alpha(\phi,\theta))}{\alpha(\phi,\theta)}\right)^2 \triangleq \sigma_x^2, \\ E\{Y\} &= 0, \end{aligned} \tag{10}$$

*Figure 3. Average beampattern with different R/λ (N = 32, θ = 90°)*

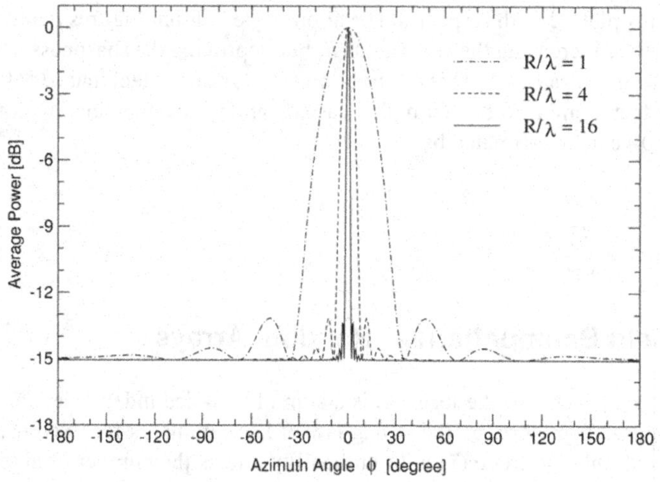


*Figure 4. Average beampattern with different R/λ (N = 32, φ = 0°)*

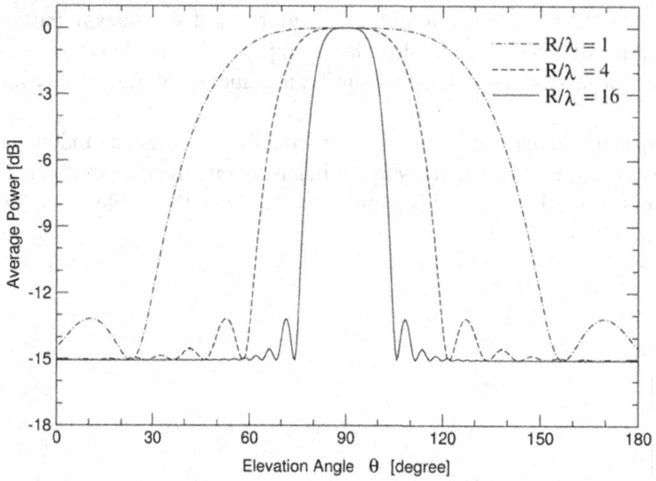


$$\mathrm{Var}(Y) \;=\; \frac{1}{2}\left(1-\frac{J_1(2\alpha(\phi,\theta))}{\alpha(\phi,\theta)}\right) \triangleq \sigma_y^2. \tag{11}$$

Note that $E\{XY\} = 0$, i.e., $X$ and $Y$ are orthogonal and thus statistically uncorrelated. Thus, for large $N$, the joint pdf of $X$ and $Y$ may be approximated by

$$f_{X,Y}(x,y) = \frac{1}{2\pi\sigma_x\sigma_y}\exp\left(-\frac{|x-m_x|^2}{2\sigma_x^2}-\frac{y^2}{2\sigma_y^2}\right).$$

For $\alpha(\phi,\theta) \gg 1$, the terms $J_1(2\alpha(\phi,\theta))/\alpha(\phi,\theta)$ and $\left(J_1(\alpha(\phi,\theta))/\alpha(\phi,\theta)\right)^2$ in the variance expressions (10) and (11) rapidly decrease and their contribution to the resulting variances becomes minor. Therefore, it is very likely that both variances are approximately equal in the sidelobe region. When this is the case, i.e., if $\sigma_x^2 \approx \sigma_y^2 \approx 1/2$, the distribution of the envelope, $R = \sqrt{X^2 + Y^2}$, follows a Nakagami-Rice distribution given by

$$f_R(r) = 2re^{-\left(r^2+m_x^2\right)} I_0(2m_x r)$$

where $I_n$ is the $n$th order modified Bessel function of the first kind, expressed (for a non-negative integer $n$) as

$$I_n(x) = \sum_{k=0}^{\infty} \frac{1}{k!\,(n+k)!}\left(\frac{x}{2}\right)^{2k+n}.$$

Let $P_0$ denote an arbitrary threshold sidelobe power level. From (9), the probability that the sidelobe power level of a beampattern with a specific angle does not exceed this threshold is given by

$$\begin{aligned}\Pr\left[P(\phi,\theta) < P_0\right] &= \Pr\left[\frac{1}{N}\left(X^2+Y^2\right) < P_0\right] = \Pr\left[R < \sqrt{NP_0}\right] \\ &= \int_0^{\sqrt{NP_0}} f_R(r)\,dr = 1 - Q\left(\sqrt{2}m_x, \sqrt{2NP_0}\right)\end{aligned} \tag{12}$$

where

$$Q(x,y) = \int_y^{\infty} t\,e^{-\frac{t^2+x^2}{2}} I_0(xt)\,dt$$

is the first-order Marcum-Q function.

If the reference angle is well away from the main lobe such that the mean $m_x$ can also be assumed to be zero, then the envelope follows a Rayleigh distribution and (12) can be further simplified to

$$\Pr\left[P(\phi,\theta) < P_0\right] = 1 - e^{-NP_0}, \tag{13}$$

which no longer depends on the observation angles $\theta$ and $\phi$.

*Figure 5. Average and realizations of beampattern (R/λ = 2, N = 32, θ = 90°)*

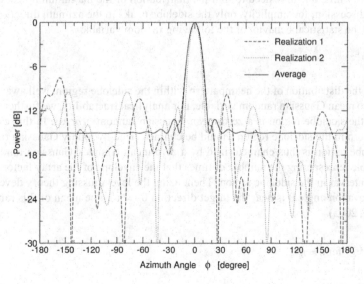

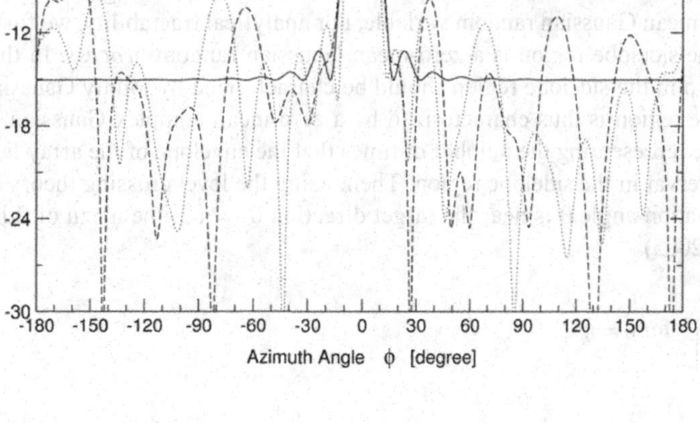

*Figure 6. CDF of beampattern with different values of reference azimuth angle $\phi$ ($R/\lambda = 2, N = 32, \theta = 90^o$); the vertical line represents the average sidelobe level 1/N*

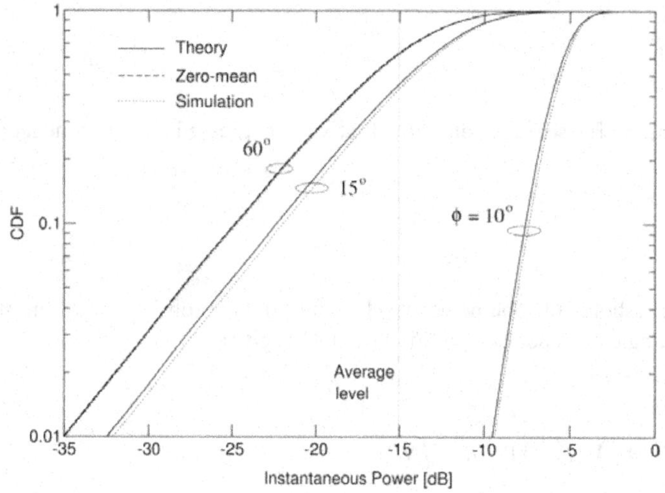


In Figure 6, the cumulative distribution function (CDF) of the beampattern (12) is plotted with $R/\lambda = 2$, $\theta = 90^\circ$, and $N = 32$ elements, along with those obtained by simulations. The CDF based on the zero-mean Gaussian approximation of (13) is also plotted. The azimuth angle $\phi$ is varied from $10^\circ$ to $60^\circ$. Note that from (8) the one-sided width of the main lobe $\Delta\phi$ in this case is approximately $18^\circ$. Therefore, the case with $\phi = 10^\circ$ corresponds to the main lobe region and thus the range of the distribution is relatively thin. This means that the main lobe is rather stable and does not vary much. On the other hand, as $\phi$ increases, the distribution approaches that of the zero-mean approximation (13). The above results suggest that except for the main lobe region, the zero mean Gaussian approximation is reasonable. This fact significantly simplifies further analysis of the statistical properties of the sidelobe region.

## Distribution of the Maximum of the Sidelobes

It is well known that unlike periodic or equally-spaced antenna arrays, many arrays with unequal spacing will yield no grating lobes for a large number of elements. This property is also preserved for random arrays (Lo, 1964; Steinberg, 1976), but in order to verify this, one may need to find the distribution of the maximum power of the whole sidelobe region. In the following discussion, for simplicity, only the sidelobe peaks in the azimuth direction will be considered. We are thus interested in the statistical behavior of the following random variable

$$\gamma(z) = \max_{\phi \in \text{ sidelobe region}} P(\phi, \theta \mid z).$$

So far, we have seen that the distribution of the beampattern within the sidelobe region, well away from main lobe, can be characterized by a zero mean Gaussian random variable. For analytical tractability, we further make the assumption that the beampattern in the sidelobe region is a zero-mean Gaussian random *process*. In this case, any two samples taken from the beampattern in the sidelobe region should be characterized by jointly Gaussian random variables. The array factor in the sidelobe region is thus characterized by a zero mean complex Gaussian random process. Let $\nu(a)$ denote the random variable representing the number of times that the envelope of the array factor crosses the threshold level $a$ upward per unit interval in the sidelobe region. Then, using the level crossing theory developed by Rice (1944, 1945), if the observed elevation angle $\theta$ is near the target direction $\theta_0 = \pi/2$, the mean of this random variable can be approximated by (Ochiai, 2005)

$$\nu(a) \approx 4\sqrt{\pi}(R/\lambda)\sqrt{\sin\theta}\, ae^{-a^2} \text{ for } \theta \approx \pi/2 \quad (14)$$

where $a$ corresponds to the envelope level normalized by the square root of the average power. Replacing the variable $a$ with the corresponding sidelobe reference level $\sqrt{NP_0}$ and using the argument of (Ochiai, 2001), the CDF of the maximum sidelobe power is approximated by

$$\Pr[\gamma(z) < P_0] \approx \exp\left\{-\nu\left(\sqrt{NP_0}\right)\right\}. \quad (15)$$

Figure 7 shows a comparison between simulation results and approximation (15) for $N = 32$ and 128 with several different values of $R/\lambda$. Due to a number of approximations involved, the theoretical curves do not match precisely with those of simulations, but they can at least serve for predicting the behavior of the maximum sidelobe peak. It is apparent from (14) and (15) that the maximum sidelobe peak level for a given value of CDF is inversely proportional to the number of antenna elements $N$. Therefore, at least in a statistical sense, increasing $N$ always contributes to the reduction of the

*Figure 7. Comparison of CDF of the sidelobe peaks ($\theta = 90^o$). Dotted curves: simulation, solid curves: theoretical approximation*

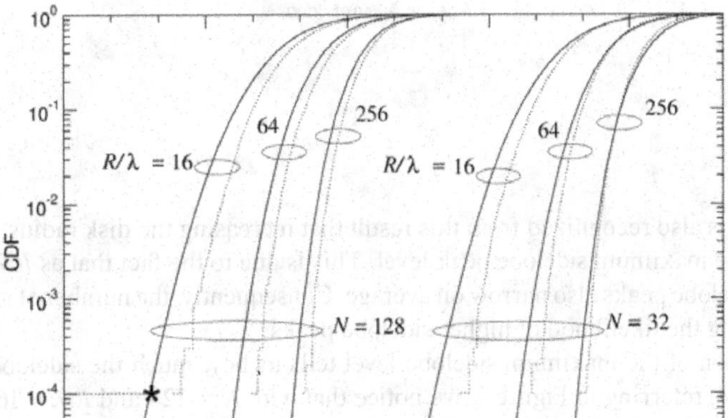


*Figure 8. Comparison of CCDF of the sidelobe peaks ($\theta = 90^o$). Dotted curves: simulation, solid curves: theoretical approximation*

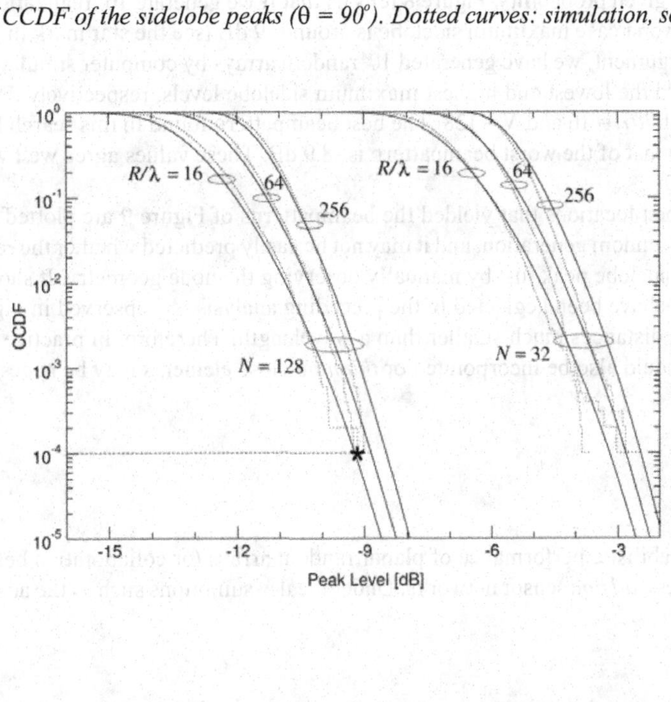

*Figure 9. Beampattern ($R/\lambda = 16$, $N = 128$, $\theta = 90^o$). Solid line: Best beampattern. Dotted line: Worst beampattern.*

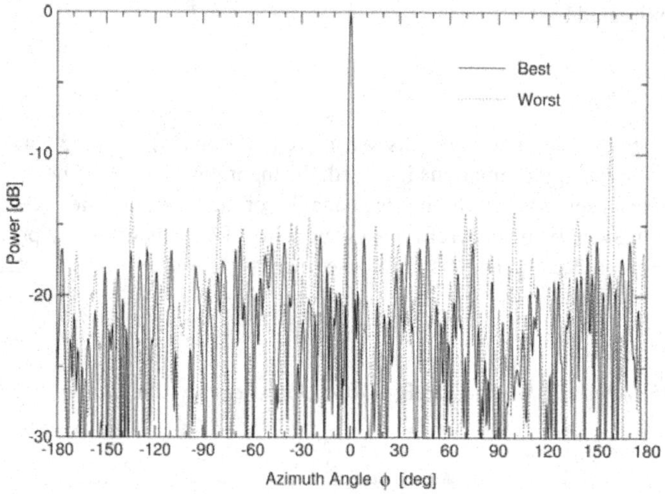


maximum sidelobe level. It is also recognized from this result that increasing the disk radius $R/\lambda$ for a given number of elements $N$ may increase the maximum sidelobe peak level. This is due to the fact that as $R/\lambda$ increases, the main lobe becomes narrow, but the sidelobe peaks also narrow on average. Consequently, the number of sidelobe peaks per sidelobe interval increases, increasing the likelihood of higher sidelobe peaks.

The statistical distribution of the maximum sidelobe level tells us how much the sidelobe peak can be reduced by random search. For example, referring to Figure 7, we notice that with $N = 128$ and $R/\lambda = 16$, the peak level is around -16 dB at $CDF = 10^{-4}$ (see the star mark in Figure 7). This means that if we randomly generate $10^4$ planar array patterns, then it is likely that the best pattern has a maximum sidelobe level near -16 dB.

Figure 8 shows the probability that the beampattern formed by a given realization exceeds a threshold, i.e., complementary CDF (CCDF), which is simply calculated by 1-*CDF*. The CCDF is useful when we do not have choice in node geometry (such as in wireless *ad hoc* sensor networks scenario) but yet wish to estimate the worst maximum sidelobe level that may occur with a given probability. Figure 8 tells us that if we generate $10^4$ realization of random arrays with $N = 128$ and $R/\lambda = 16$, the worst case maximum sidelobe is around -9 dB (see the star mark in Figure 8).

To confirm the above argument, we have generated $10^4$ random arrays by computer simulation and retained the best and worst patterns that yield the lowest and highest maximum sidelobe levels, respectively. Figure 9 shows the corresponding beampatterns with $R/\lambda = 16$ and $N = 128$. The best beampattern found in this search has a maximum sidelobe as low as -15.7 dB, whereas that of the worst beampattern is -8.9 dB. These values agree well with the predicted values based on the CDF and CCDF.

Finally, the best and worst locations that yielded the beampatterns of Figure 9 are plotted in Figure 10. These two examples are both based on random generation, and it may not be easily predicted whether the resulting beampattern has small (or large) maximum sidelobe peak just by manually observing the node geometry. It should also be remembered that mutual coupling effects have been neglected in the preceding analysis. As observed in Figure 10, many nodes are very closely located within distances much smaller than a wavelength. Therefore, in practice mutual coupling effects among antenna elements should also be incorporated or *thinning* close elements may be necessary before determining realistic beampatterns.

## CONCLUSION

We have analyzed the probabilistic performance of planar random arrays (or collaborative beamforming) with a view toward application to wireless *ad hoc* sensor networks. Under ideal assumptions such as the absence of mutual coupling

*Figure 10. The best and worst locations that formed the beampatterns of Figure 9; the units of both axes are normalized by the wavelength*

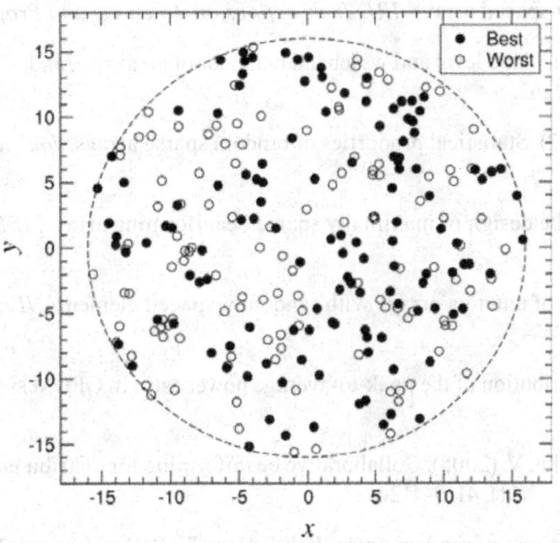


and perfect synchronization, it has been shown that such random arrays can form nice beampatterns with sharp main lobes and low sidelobe peaks. The sharpness of the main lobe is determined by the normalized disk radius $R/\lambda$, and the one-sided width of the main lobe in the azimuthal direction is roughly given by $35^\circ / (R/\lambda)$. Increasing the disk radius, however, also increases the probability of occurrence of higher sidelobe peaks. Therefore, there is a trade off between sharpness of the main lobe and maximum sidelobe level.

In (Ochiai, 2005) further details such as average directivity of planar random arrays and numerical methods for evaluating precise sidelobe distribution are discussed.

Throughout the chapter, ideal synchronization among sensor nodes is assumed. Considering the fact that beamforming nodes are distributed sparsely, and that low-cost sensor nodes are not capable of precise synchronization, the effect of imperfect synchronization should be critical in practice. This issue is further discussed in (Ochiai, 2005), where the impact of phase offsets among sensor nodes on the resulting beampatterns is analyzed in detail.

Further investigations on beampatterns under practical constraints are necessary before concluding the usefulness of random arrays. In particular, when the randomly allocated antenna elements are closely located, the effect of mutual coupling should be dominant. In this case, there is a potential for improving the beampattern by *thinning* closely located elements.

## REFERENCES

Agrawal, V. D., & Lo, Y. T. (1972). Mutual coupling in the phased arrays of randomly spaced antennas. *IEEE Transactions on Antennas and Propagation*, 20(3), 288-295.

Balanis, C. A. (1997). *Antenna Theory: Analysis and Design*. New York: Wiley.

Bar-Ness, Y., & Haimovich, A. M. (1984). Synthesis of random antenna array patterns with prescribed nulls. *IEEE Transactions on Antennas and Propagation*, 32(12), 1298-1307.

Donvito, M. B., & Kassam S. A. (1979). Characterization of the random array peak sidelobe. *IEEE Transactions on Antennas and Propagation*, 27(3), 379-385.

Fante, R. L., Robertshaw, G. A., & Zamoscianyk, S. (1991). Observation and explanation of an unusual feature of random arrays with a nearest-neighbor constraint. *IEEE Transactions on Antennas and Propagation*, 39(7), 1047-1049.

Holm, S., Elgetun, B., & Dahl, G. (1997). Properties of the beampattern of weight- and layout-optimized sparse arrays. *IEEE Transactions on Ultrasonics Ferroelectrics and Frequency Control*, 44(5), 983-991.

Ishimaru, A. (1962). Theory of unequally-spaced arrays. *IRE Transactions on Antennas and Propagation*, 10, 691-702.

Kerby, K. C., & Bernhard, J. T., (2006). Sidelobe level and wideband behavior of arrays of random subarrays. *IEEE Transactions on Antennas and Propagation*, 54(8), 2253-2262.

Kook, H., Davies, P., & Bolton, J. S. (2002). Statistical properties of random sparse arrays. *Journal of Sound and Vibration*, 255(5), 819-848.

Leahy, R. M., & Jeffs, B. D. (1991). On the design of maximally sparse beamforming arrays. *IEEE Transactions on Antennas and Propagation*, 39(8), 1178-1187.

Lo, Y. T. (1964). A mathematical theory of antenna arrays with randomly spaced elements. *IEEE Transactions on Antennas and Propagation*, 12, 257-268.

Ochiai, H., & Imai, H. (2001). On the distribution of the peak-to-average power ratio in OFDM signals. *IEEE Transactions on Communications*, 49(2), 282-289.

Ochiai, H., Mitran, P., Poor, H. V., & Tarokh, V. (2005). Collaborative beamforming for distributed wireless *ad hoc* sensor networks. *IEEE Transactions on Signal Processing*, 53(11), 4110-4124.

Rice, S. O. (1944,1945). Mathematical analysis of random noise. *Bell System Technical Journal*, 23/24, 282-332/46-156.

Steinberg, B. D. (1972). The peak sidelobe of the phased array having randomly located elements. *IEEE Transactions on Antennas and Propagation*, 20(2), 129-136.

Steinberg, B. D. (1976). *Principles of Aperture & Array System Design*. New York: Wiley.

Yao, K., Hudson, R. E., Reed, C. W., Chen, D. C., & Lorenzelli, F. (1998). Blind beamforming on a randomly distributed sensor array system. *IEEE Journal on Selected Areas in Communications*, 16(8), 1555-1567.

# Chapter VI
# Advanced Space-Time Block Codes and Low Complexity Near Optimal Detection for Future Wireless Networks

**W. H. Chin**
*Institute for Infocomm Research, Singapore*

**C. Yuen**
*Institute for Infocomm Research, Singapore*

## ABSTRACT

*Space-time block coding is a way of introducing multiplexing and diversity gain in wireless systems equipped with multiple antennas. There are several classes of codes tailored for different channel conditions. However, in almost all the cases, maximum likelihood detection is required to fully realize the diversity introduced. In this chapter, we present the fundamentals of space-time block coding, as well as introduce new codes with better performance. Additionally, we introduce the basic detection algorithms which can be used for detecting space-time block codes. Several low complexity pseudo-maximum likelihood algorithms will also be introduced and discussed.*

## INTRODUCTION

The use of multiple antennas at both transmitters and receivers has become increasingly common in recent years due to the higher capacities of such systems (Foschini, 1996; Telatar, 1995). A system with multiple transmit and multiple receive antennas, or more commonly known as multiple-input multiple-output (MIMO) systems, can provide two types of advantages (Zheng & Tse, 2003). Spatial multiplexing gains are obtained by sending different signals through the multiple Space-Time virtual channels created by the multiple transmit and receive antennas (Foschini, 1996). On the other hand, spatial diversity gains are obtained by transmitting or receiving copies of a signal through different antennas as a mean to combat fading and improve the performance of the system. Space-time coding is a way to achieve the above two advantages.

This chapter aims to deliver to readers a basic understanding of Space-Time Block Codes (STBCs), a special and important class of space-time coding. We will introduce the design and optimization of some advanced STBCs, which has better performance than prior arts. These high-rate STBCs can provide a transmit diversity gain of two and they can flexibly support either three or four transmit antennas with a code rate of between two to four. We show that by having good symbol dispersion property, the new codes lead to better performance, especially under spatially-correlated MIMO channels or with reduced number of transmit antennas. They also have better coding gain and lower decoding complexity than some existing high-rate STBCs at the same spectral efficiency.

Next, a tutorial of low complexity detection schemes will be delivered to introduce the readers to the detection schemes available for both STBC and non-STBC systems. The detection of symbols in a communications is a crucial link whereby we try to recover the transmitted symbols given certain information about the channel. Selecting an appropriate detection scheme would also make a big difference, especially in the performance and complexity of a communications system. We will start with basic detection schemes which can be applied not only to STBCs, but also to any MIMO system. We will also introduce the concept of tree search based detection, which have near maximum likelihood performance at a fraction of the computational cost. Additionally, list detection is also briefly covered to introduce the idea of soft detection to the readers.

The STBCs and detection schemes introduced in this chapter are the essential elements in future wireless system in order to provide a reliable, low cost and high throughput communication network.

## Signal Model

In this chapter, we consider a MIMO system with $N_{TX}$ transmit antennas and $N_{RX}$ receive antennas. Let **H** be the $N_{RX} \times N_{TX}$ channel gain matrix, which is assumed to remain unchanged across $P$ symbol periods. The $ij^{th}$ element of **H** is the channel coefficient for the path from the $j^{th}$ transmit antenna to the $i^{th}$ receive antenna. Let **C** be the $N_{TX} \times P$ transmitted STBC codeword, where $P$ is the code length. Then, the received $N_{RX} \times P$ signal **Y** can be written as

$$\mathbf{Y} = \mathbf{HC} + \mathbf{N} \tag{1}$$

where **N** is the additive white Gaussian noise. Hence if the codeword **C** can transmit $L$ complex symbols of information, the code rate of the STBC is defined as $R = L/P$. We will use bold big capital **M** to represent a matrix, bold small capital **v** to represent a vector, and italic small case capital $c$ for a variable.

The MIMO system is also assumed to be a coherent communication system. In other words, the transmitter has no information on the channel coefficients, whereas the receiver has the full information on the channel coefficients **H**. Since it is assumed that the receiver knows the **H**, the transmitted codeword can then be estimated with maximum-likelihood (ML) as follow:

$$\hat{\mathbf{C}} = \arg\min_{\mathbf{C} \in Ç} \operatorname{tr}\left(\{\mathbf{Y} - \mathbf{HC}\}^{\mathrm{H}} \{\mathbf{Y} - \mathbf{HC}\}\right) \tag{2}$$

where Ç represent the set of codewords.

Tarokh et al. (1998) further showed that the minimum rank of the codeword difference matrix quantifies the diversity gain, while the minimum product of the non-zero eigenvalues of the codeword distance matrix with the minimum rank quantifies the coding gain, of a STBC. They will be reviewed in the following. We assume that all the codewords have equal transmission probability and let $P(\mathbf{C} \to \mathbf{E})$ denote the probability that the codeword **C** is transmitted but the receiver decides erroneously in favor of another codeword **E**. This probability term is commonly called the pair-wise error probability (PEP). With ideal CSI, PEP is well approximated as follows (Tarokh et al., 1998):

$$P\left(\mathbf{C} \to \mathbf{E} \mid h_{i,k}, i = 1,2,\ldots,N_{\mathrm{TX}}; k = 1,2,\ldots,N_{\mathrm{RX}}\right) \le \exp\left(-d^2(\mathbf{C},\mathbf{E})\frac{E_s}{4N_o}\right) \tag{3}$$

where

$$d^2(\mathbf{C},\mathbf{E}) = \sum_{k=1}^{N_{\mathrm{RX}}} \sum_{p=1}^{P} \left| \sum_{i=1}^{N_{\mathrm{TX}}} h_{i,k}(c_{p,i} - e_{p,i}) \right|^2 \tag{4}$$

and $c_{p,i}$ and $e_{p,i}$ are the entries in the $p^{th}$ row and $i^{th}$ column of **C** and **E**.

To simplify the PEP expression, we first define two matrices, the first is the *codeword difference matrix* $\mathbf{B}^{\mathbf{CE}}$ of size $N_{TX} \times P$, which is defined as (Tarokh et al., 1998):

$$\mathbf{B}_{\mathbf{CE}} = \mathbf{C} - \mathbf{E} \tag{5}$$

and the second is the *codeword distance matrix* $\mathbf{A}_{\mathbf{CE}}$ of size $N_{TX} \times N_{TX}$, which is defined as :

$$\mathbf{A}_{\mathbf{CE}} = \mathbf{B}_{\mathbf{CE}} \mathbf{B}_{\mathbf{CE}}^{\mathrm{H}} \tag{6}$$

Furthermore, we also define $\lambda_1, \lambda_2, \ldots, \lambda_D$ as the non-zero eigenvalues of $\mathbf{A}_{\mathbf{CE}}$, where $D$ denotes the rank of $\mathbf{A}_{\mathbf{CE}}$, which is the same as the rank of $\mathbf{B}_{\mathbf{CE}}$. At high SNR, the upper bound of the PEP in (3) can be simplified as:

$$P(\mathbf{C} \to \mathbf{E}) \leq \left[ \left( \prod_{i=1}^{D} \lambda_i \right) \left( \frac{E_s}{4N_o} \right)^{-D} \right]^{-N_R} \tag{7}$$

From the above, we can define the following two quantities to account for the decoding performance of a STBC: (1) *Diversity Gain*: the minimum rank $D$ of the matrix $\mathbf{A}_{\mathbf{CE}}$ over all pairs of distinct codewords. It accounts for the slope of the bit error rate (BER) or block error rate (BLER) curve of the STBC. A STBC is said to achieve full diversity if $D = N_{TX}$. 2) *Coding Gain*: be the product of the non zero eigenvalue of the codeword distance matrix, it determines the left or right shift of the BER or BLER curve of the STBC.

Based on the above, a set of code design criteria for space-time codes, commonly called the Rank & Determinant Criteria, can be stated as follows (Tarokh et al., 1998): 1) *Rank Criterion*: To maximize the diversity gain of the STBC, maximize the minimum rank $D$ of the matrix $\mathbf{A}_{\mathbf{CE}}$ over all distinct pairs of codewords. Hence, in order to achieve the maximum diversity order, the matrix $\mathbf{A}_{\mathbf{CE}}$ has to be full rank, i.e., $D = N_{TX}$, for any codeword pair **C** and **E**. 2) *Determinant Criterion*: To maximize the coding gain of the STBC, maximize the minimum product of non-zero eigenvalues of the matrix $\mathbf{A}_{\mathbf{CE}}$ which has the minimum rank.

In addition, it is also important to normalize the power of a STBC (Hassibi & Hochwald, 2002). Assume every constellation symbol has an average power of 1. It is required that total average transmitted power of a STBC is normalized to $P \times N_{TX}$. In addition, we further restrict that every transmitted symbol has the same transmit power of $P \times N_{TX}/L$.

## Chapter Organization

The rest of this chapter will be partitioned into two major sections, with each section spanning approximately equal portions.

The next section on STBC will introduce the concept of STBC to the readers. In this section, the basic STBC will first be discussed, this is followed by description, design and optimization of more advanced STBCs, which have good dispersion properties and hence results in better performances than existing STBCs.

The third section will describe several low complexity pseudo-maximum likelihood detection schemes which can be applied to the detection of STBC. These include several tree search based detectors. A detailed description of the different low complexity detection algorithms will be included to introduce the readers to the finer workings of these algorithms.

Finally, we will conclude the chapter with a summary of the different codes and schemes described. Future challenges in the field will also be discussed.

## ADVANCED SPACE-TIME BLOCK CODE

Spatial Multiplexing (SM) can provide the highest possible bandwidth efficiency, but it has no transmit diversity gain and requires the number of received antennas to be greater than or equal to the number of transmit antennas (Paulraj &

Kailath, 1994). Hence Space-Time Block Codes (STBC) is proposed. One major class of STBC is Orthogonal Space-Time Block Code (O-STBC) (Alamouti, 1998); Tarokh et al., 1999), which can provide full transmit diversity and the lowest possible decoding complexity. However, O-STBC is not bandwidth efficient (as it has a code rate less than one) when more than two transmit antennas or more than one receive antenna are used (Sandhu & Paulraj, 2000).

The above two schemes, O-STBC achieves the optimal diversity gain and SM achieves the optimal multiplexing gain, a STBC that in between the above two schemes by achieving a tradeoff in diversity and multiplexing gain would be attractive. Therefore numerous STBCs with code rates greater than one and which provides different transmit diversity levels have been proposed (Texas Instruments, 2001; Hottinen et al., 2003; Uysal & Georghiades, 2002).

One example is the rate-two code D-STTD (Texas Instruments, 2001) as shown in (8) that can provide transmit diversity level of two. It is similar to transmit two Alamouti (Alamouti, 1998) in parallel. The code has been adopted in the 802.11n standard (IEEE, 2006). We will show that by considering the dispersion property of the code, one can design a STBC with better decoding performance without additional cost or increased in decoding complexity.

$$\mathbf{DSTTD} = \begin{bmatrix} x_1 & x_2 & x_3 & x_4 \\ -x_2^* & x_1^* & -x_4^* & x_3^* \end{bmatrix} \qquad (8)$$

In this section, we present a new high-rate STBC design that can provide transmit diversity of two which, due to its good symbol dispersion property, can support different number of transmit antennas more flexibly and has better decoding performance in correlated fading channels.

## New High Rate STBC

Let **A** and **B** be two Alamouti codes (Alamouti, 1998), **C** and **D** be two variants of Alamouti codes as shown below:

$$\mathbf{A} = \begin{bmatrix} x_1 & x_2 \\ -x_2^* & x_1^* \end{bmatrix}; \mathbf{B} = \begin{bmatrix} x_3 & x_4 \\ -x_4^* & x_3^* \end{bmatrix}; \mathbf{C} = \begin{bmatrix} x_5^* & x_6^* \\ x_6 & -x_5 \end{bmatrix}; \mathbf{D} = \begin{bmatrix} x_7^* & x_8^* \\ x_8 & -x_7 \end{bmatrix} \qquad (9)$$

Our high-rate STBC (Yuen et al., 2005) for four transmit antennas is:

$$\mathrm{C4} = [\mathbf{A}+\mathbf{B}+\mathbf{C}+\mathbf{D} \quad \mathbf{A}-\mathbf{B}+\mathbf{C}-\mathbf{D}]$$

$$= \begin{bmatrix} x_1+x_3+x_5^*+x_7^* & x_2+x_4+x_6^*+x_8^* & x_1-x_3+x_5^*-x_7^* & x_2-x_4+x_6^*-x_8^* \\ -x_2^*-x_4^*+x_6+x_8 & x_1^*+x_3^*-x_5-x_7 & -x_2^*+x_4^*+x_6-x_8 & x_1^*-x_3^*-x_5+x_7 \end{bmatrix} \qquad (10)$$

where $x_1$ to $x_8$ are the data symbols with arbitrary complex constellation. A column in **C4** represents the signals transmitted on a particular transmit antenna, while a row in **C4** represents the signals transmitted at a particular time slot. Since **C4** in (10) takes two symbol periods to transmit eight complex symbols, it has a code rate of four.

**C4** offers considerable flexibility in dimensioning: it can be made to support less transmit antennas by simply truncating its codeword columns, and its code rates can be adjusted by simply setting some codeword symbols to zero. In this section, we will study three such examples: rate-4 for four transmit antennas, rate-2 for four transmit antennas, and rate-2 for three transmit antennas. We will compare the STBC presented in this section with existing STBCs of similar dimensions and decoding complexity.

### Rate-Four Code for Four Transmit Antennas

We first optimize the decoding performance of **C4**. To optimize the decoding performance of the code **C4** in (10), we employ the multi-dimensional rotation technique proposed in (DaSilva & Sousa, 1997) to obtain:

$$C4_{Tx4} = \frac{1}{2}\begin{bmatrix} x_1 + x_3 + \hat{x}_5^* + \hat{x}_7^* & x_2 + x_4 + \hat{x}_6^* + \hat{x}_8^* & x_1 - x_3 + \hat{x}_5^* - \hat{x}_7^* & x_2 - x_4 + \hat{x}_6^* - \hat{x}_8^* \\ -x_2^* - x_4^* + \hat{x}_6 + \hat{x}_8 & x_1^* + x_3^* - \hat{x}_5 - \hat{x}_7 & -x_2^* + x_4^* + \hat{x}_6 - \hat{x}_8 & x_1^* - x_3^* - \hat{x}_5 + \hat{x}_7 \end{bmatrix} \quad (11)$$

where

$$\begin{bmatrix} \hat{x}_5^R & \hat{x}_5^I & \hat{x}_6^R & \hat{x}_6^I \end{bmatrix}^T = \left( \prod_{\substack{1 \le i \le 3, \\ i+1 \le k \le 4}} \mathbf{G}(i,k,\theta_{ik}) \right) \begin{bmatrix} x_5^R & x_5^I & x_6^R & x_6^I \end{bmatrix}^T \quad (12)$$

$$\begin{bmatrix} \hat{x}_7^R & \hat{x}_7^I & \hat{x}_8^R & \hat{x}_8^I \end{bmatrix}^T = \left( \prod_{\substack{1 \le i \le 3, \\ i+1 \le k \le 4}} \mathbf{G}(i,k,\theta_{ik}) \right) \begin{bmatrix} x_7^R & x_7^I & x_8^R & x_8^I \end{bmatrix}^T$$

and superscript R and I present the real and imaginary part of a symbol respectively. $\mathbf{G}(i, k, \theta_{ik})$ is a 4-by-4 matrix with entries at $(i, i)$ and $(k, k)$ equal to $\cos(\theta_{ik})$, entry at $(i, k)$ equals to $\sin(\theta_{ik})$, and entry at $(k, i)$ equals to $-\sin(\theta_{ik})$, one on the remaining diagonal positions and zero elsewhere. $\mathbf{G}(i, k, \theta_{ik})$ basically models a counter-clockwise rotation by degree with respect to the $(i, k)$ plane. For example, for $i = 2$, $k = 3$, the **G** matrix becomes:

$$\mathbf{G}(2,3,\theta_{23}) = \begin{bmatrix} 1 & 0 & 0 & 0 \\ 0 & \cos(\theta_{23}) & \sin(\theta_{23}) & 0 \\ 0 & -\sin(\theta_{23}) & \cos(\theta_{23}) & 0 \\ 0 & 0 & 0 & 1 \end{bmatrix} \quad (13)$$

The factor 1/2 in (11) is to ensure that the total transmission power is normalized such that it facilitates comparison with other STBC in the next section.

A set of optimum rotation angle to maximize the diversity and coding gain (Tarokh et al., 1998) of $\mathbf{C4}_{Tx4}$ for 4-QAM constellation has been found to be: $\theta_{12} = 78^0$, $\theta_{23} = -30^0$, $\theta_{34} = -23^0$, $\theta_{13} = -163^0$, $\theta_{14} = 149^0$, $\theta_{24} = -154^0$ through computer optimization. The transmit diversity level that can be provided by $\mathbf{C4}_{Tx4}$ is two. In addition, the ML decoding of $\mathbf{C4}_{Tx4}$ requires the joint detection of eight complex symbols (or a search space of $8^M$ in general, where $M$ is the constellation dimensionality).

## Rate-Two Code for Four Transmit Antennas

To obtain a lower code rate of two from **C4**, a new code $\mathbf{C2}_{Tx4}$ can be obtained by setting half of the symbols in **C4** (e.g., $x_5$ to $x_8$) to zero:

$$C2_{Tx4} = \frac{1}{\sqrt{2}}\begin{bmatrix} x_1 + x_3 & x_2 + x_4 & x_1 - x_3 & x_2 - x_4 \\ -x_2^* - x_4^* & x_1^* + x_3^* & -x_2^* + x_4^* & x_1^* - x_3^* \end{bmatrix} \quad (14)$$

The factor $1/\sqrt{2}$ of code **C4** in (10) is to ensure the total transmission power is normalized such that it facilitates comparison with other STBCs.

Again, we need to first investigate its performance optimization. The codeword distance matrix (Tarokh et al., 1998) of $\mathbf{C2}_{Tx4}$ is:

$$\mathbf{A}_{CE}(\mathbf{C2}_{Tx4}) = \frac{1}{2}\begin{pmatrix} w+v & 0 & x+y & z \\ 0 & w+v & -z^* & x-y \\ x-y & -z & w-v & 0 \\ z^* & x+y & 0 & w-v \end{pmatrix} \quad (15)$$

where $v = 2\mathrm{Re}(\Delta_2\Delta_4^* + \Delta_1\Delta_3^*)$, $w = \sum_{i=1}^{4}|\Delta_i|^2$, $x = \sum_{i=1}^{2}|\Delta_i|^2 - \sum_{i=3}^{4}|\Delta_i|^2$, $y = -2j\mathrm{Im}(\Delta_2\Delta_4^* + \Delta_1\Delta_3^*)$, $z = 2\left(\Delta_2\Delta_3^* - \Delta_4\Delta_1^*\right)$, and $\Delta_i$ is the possible error in the symbol $x_i$.

$\mathbf{A}_{CE}(\mathbf{C2}_{Tx4})$ has two zero eigenvalues (i.e., $\lambda_1 = \lambda_2 = 0$), and two non-zero eigenvalues that are equal:

$$\lambda_3 = \lambda_4 = 2(|\Delta_1|^2 + |\Delta_2|^2 + |\Delta_3|^2 + |\Delta_4|^2) \quad (16)$$

Since $\Delta_1$ to $\Delta_4$ cannot be all zero (or there won't be any codeword difference), $\lambda_3$ and $\lambda_4$ would never be zero. Also $\lambda_3$ and $\lambda_4$ depend on only $|\Delta_i|^2$, which invariant to constellation rotation, as a result, no performance optimization such as the one for **C4** can be performed to improve the decoding performance of $\mathbf{C2}_{Tx4}$, and the transmit diversity level that can be provided by $\mathbf{C2}_{Tx4}$ is two. In addition, the ML decoding of $\mathbf{C2}_{Tx4}$ requires the joint detection of four complex symbols (or a search space of $4^M$ in general, where $M$ is the constellation dimensionality).

## Rate-Two Code for Three Transmit Antennas

To use $\mathbf{C2}_{Tx4}$ for three transmit antennas, we may truncate one of its codeword columns, e.g. the last column, to obtain:

$$\mathbf{C2}_{Tx3} = \frac{1}{\sqrt{2}}\begin{bmatrix} x_1 + x_3 & x_2 + x_4 & x_1 - x_3 \\ -x_2^* - x_4^* & x_1^* + x_3^* & -x_2^* + x_4^* \end{bmatrix} \quad (17)$$

The codeword distance matrix of $\mathbf{C2}_{Tx3}$ is:

$$\mathbf{A}_{CE}(\mathbf{C2}_{Tx3}) = \frac{1}{2}\begin{pmatrix} w+v & 0 & x+y \\ 0 & w+v & -z^* \\ x-y & -z & w-v \end{pmatrix} \quad (18)$$

where $v, w, x, y, z$ are the same as in (15). It is found that $\mathbf{A}_{CE}(\mathbf{C2}_{Tx3})$ has one zero eigenvalue (i.e. $\lambda_1 = 0$). Hence its maximum rank is at most two. The other two eigenvalues of $\mathbf{A}_{CE}(\mathbf{C2}_{Tx3})$ are:

$$\lambda_2 = 2\left(|\Delta_1|^2 + |\Delta_2|^2 + |\Delta_3|^2 + |\Delta_4|^2\right)$$
$$\lambda_3 = |\Delta_1 + \Delta_3|^2 + |\Delta_2 + \Delta_4|^2 \quad (19)$$

Since $\Delta_1$ to $\Delta_4$ cannot be all zero (or there won't be any codeword difference), $\lambda_2$ would never be zero. However, $\lambda_3$ may become zero if $\Delta_1 = -\Delta_3$ or $\Delta_2 = -\Delta_4$, resulting in a loss of diversity gain. In order to maximize the diversity gain of $\mathbf{C2}_{Tx3}$, constellation rotation (CR) (Yuen et al., 2003) must be applied to ensure that $\lambda_3$ is non-zero. This is achieved by choosing $x_3$ from a different (i.e., rotated) constellation set than $x_1$, such that $\Delta_1 \neq -\Delta_3$; likewise for $x_4$ and $x_2$, such that $\Delta_2 \neq -\Delta_4$.

*Figure 1. Constellation rotation of 4-QAM symbol*

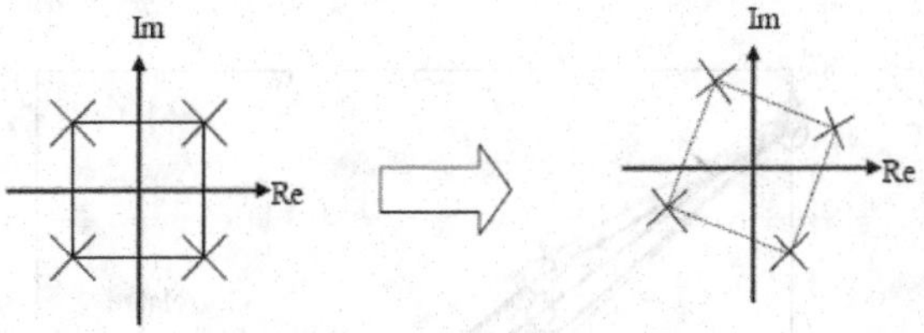


An example of rotation of 4-QAM constellation is shown in Figure 1. To maximize the coding gain of $\mathbf{C2}_{Tx3}$, the constellation rotation angles for $x_3$ and $x_4$ can be chosen such that the product of the non-zero eigenvalues of $\mathbf{A}_{CE}(\mathbf{C2}_{Tx3})$ (i.e., $\lambda_2 \lambda_3$) is maximized (Tarokh et al., 1998).

The best CR angles for $x_3$ and $x_4$ are found to be identical, with values 34 degrees, 11 degrees and 22.5 degrees for square-QAM (such as 4QAM), 4PSK and 8PSK constellations respectively. For the rest of this chapter, $\mathbf{C2}_{Tx3}$ with such optimized constellation rotations will be considered.

## Performance Comparisons

### Rate-Four Code for Four Transmit Antennas

A rate-four STBC for four transmit antennas requires at least four receive antennas, one of such, namely $\mathbf{QSTTD}_{Tx4}$, has been proposed in (Hottinen et al., 2003). It can provide a transmit diversity level of two and its ML decoding requires joint detection of eight complex symbols.

$$\mathbf{QSTTD}_{Tx4} = \frac{1}{\sqrt{2}}\begin{bmatrix} x_1 & x_2 & x_3 & x_4 \\ -x_2^* & x_1^* & -x_4^* & x_3^* \end{bmatrix} + \frac{1}{\sqrt{2}}\begin{bmatrix} 1 & 0 \\ 0 & -1 \end{bmatrix}\left[\begin{bmatrix} x_5 & x_6 \\ -x_6^* & x_5^* \end{bmatrix}\mathbf{U} \quad \begin{bmatrix} x_7 & x_8 \\ -x_8^* & x_7^* \end{bmatrix}\mathbf{U}\right] \tag{20}$$

where $\mathbf{U} = \frac{1}{\sqrt{7}}\begin{bmatrix} 1+j & 1+2j \\ -1+2j & 1-j \end{bmatrix}$ is the performance optimization factor.

Figure 2 compares the BER performance of $\mathbf{C4}_{TX4}$ and $\mathbf{QSTTD}_{TX4}$ with 4-QAM (spectral efficiency = 8 bps/Hz) for four transmit and four receive antennas under different spatial correlation. The results show that $\mathbf{QSTTD}_{TX4}$ and $\mathbf{C4}_{TX4}$ perform closely when the spatial channels are uncorrelated. However, when the first and second (or third and fourth) transmit antennas are correlated with correlated coefficients up to 0.9, the code $\mathbf{C4}_{TX4}$ is observed to have up to 0.8dB gain over $\mathbf{QSTTD}_{TX4}$. This is because $\mathbf{QSTT}_{TX4}$ transmits every symbol on only two of the four transmit antennas (as shown in (20)), whereas $\mathbf{C4}_{TX4}$ has each symbol transmitted on all four transmit antennas (as shown in (11)). Hence the code $\mathbf{C4}_{TX4}$ is more robust against spatial correlation than $\mathbf{QSTTD}_{TX4}$.

### Rate-Two Code for Four Transmit Antennas

A rate-two STBC for four transmit antennas requires at least two receive antennas, one of such, namely $\mathbf{DSTTD}_{Tx4}$ (which is repeated from (8)), has been proposed in (Texas Instruments, 2001). It can provide a transmit diversity level of two and its ML decoding requires joint detection of four complex symbols. It is obtained by transmitting two Alamouti STBCs (Alamouti, 1998) in parallel, as shown below:

*Figure 2. Effect of spatial correlation on the BER performance of 4-Tx 4-Rx STBC systems with spectral efficiency of 8bps/Hz*

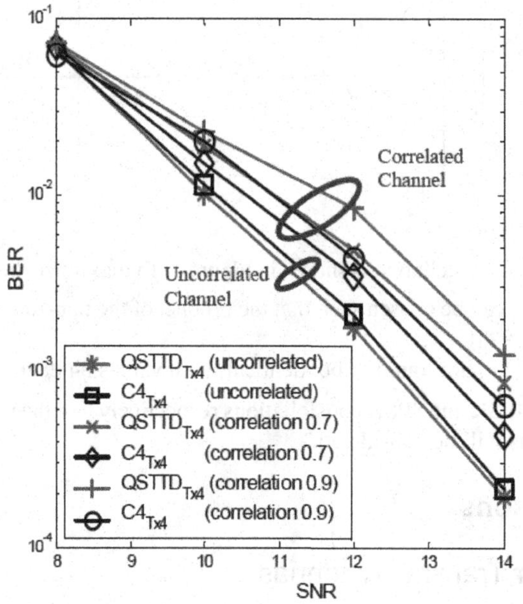


$$\mathbf{DSTTD}_{\mathrm{Tx4}} = \begin{bmatrix} x_1 & x_2 & x_3 & x_4 \\ -x_2^* & x_1^* & -x_4^* & x_3^* \end{bmatrix} \tag{21}$$

Comparing the codewords of $\mathbf{C2}_{\mathbf{Tx4}}$ in (14) and $\mathbf{DSTTD}_{\mathbf{Tx4}}$ in (21), it can be noted that every symbol ($x_1$ to $x_4$) in $\mathbf{C2}_{\mathbf{Tx4}}$ is transmitted on all the transmit antennas, whereas every symbol in $\mathbf{DSTTD}_{\mathbf{Tx4}}$ is transmitted only on two of the transmit antennas. Such symbol dispersion property is important in open-loop transmit diversity systems because, as a symbol is dispersed on more spatial channels, it has potentially better resistance against spatial channel correlation. The same property will also be shown to give $\mathbf{C2}_{\mathbf{Tx4}}$ more flexibility to support smaller number of transmit antennas, which cannot be achieved by $\mathbf{DSTTD}_{\mathbf{Tx4}}$.

The performance comparison of $\mathbf{DSTTD}_{\mathbf{Tx4}}$ and $\mathbf{C2}_{\mathbf{Tx4}}$ with 4-QAM and normalized total transmission power is shown in Figure 3 for a MIMO system with four transmit and two receive antennas (resultant spectral efficiency = 4 bps/Hz). Similar to the case of rate-four codes, $\mathbf{C2}_{\mathbf{Tx4}}$ performs better than $\mathbf{DSTTD}_{\mathbf{Tx4}}$ when the channel is correlated.

In this chapter, we aim to demonstrate how the dispersion properties of a STBC would affect the performance of a STBC without extra costs or complexity.

## Rate-Two Code for Three Transmit Antennas

Three STBCs for three transmit antennas and with rates 1, 3/2 and 4/3 have been proposed by Uysal & Georghiades (2002). In this chapter, they are named **UG1, UG2, UG3** respectively and their codewords are shown below. They need a joint detection of three or four symbols.

$$\mathbf{C}_{\mathrm{UG1}} = \begin{bmatrix} x_1 & x_2 & x_3 \\ x_2^* & -x_1^* & x_1 \\ -x_3^* & -x_3^* & x_2^* \end{bmatrix} \tag{22}$$

*Figure 3. Effect of spatial correlation on the BER performance of 4-Tx 2-Rx STBC systems with spectral efficiency of 4bps/Hz*

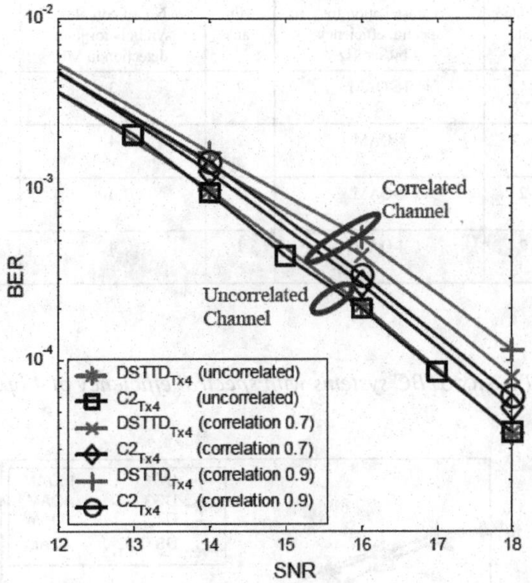


$$\mathbf{C}_{\mathbf{UG2}} = \begin{bmatrix} x_1 & x_3 & -x_2 \\ x_2 & -x_1^* & -x_3 \end{bmatrix} \tag{23}$$

$$\mathbf{C}_{\mathbf{UG3}} = \sqrt{\frac{9}{8}} \begin{bmatrix} x_1 & x_4 & \sqrt{\frac{2}{3}} x_3^* \\ x_2 & \sqrt{\frac{2}{3}} x_3^* & x_4^* \\ \sqrt{\frac{2}{3}} x_3 & -x_2^* & -x_1^* \end{bmatrix} \tag{24}$$

where the factor $\sqrt{9/8}$ and $\sqrt{2/3}$ in $\mathbf{C}_{\mathbf{UG3}}$ is to ensure that every symbols are transmitted in equal power and the total transmission power is the same as $\mathbf{C}_{\mathbf{UG1}}$ and $\mathbf{C}_{\mathbf{UG2}}$.

We apply a column truncation on $\mathbf{DSTTD}_{\mathbf{TX4}}$ to obtain a new STBC for three transmit antennas:

$$\mathbf{DSTTD}_{\mathbf{Tx3}} = \sqrt{\frac{3}{4}} \begin{bmatrix} x_1 & x_2 & \sqrt{2} x_3 \\ -x_2^* & x_1^* & -\sqrt{2} x_4^* \end{bmatrix} \tag{25}$$

Comparing (17) with (25), it can again be noted that all symbols in $\mathbf{C2}_{\mathbf{TX3}}$ are evenly distributed across all three transmit antennas, while the symbols in $\mathbf{DSTTD}_{\mathbf{TX3}}$ are transmitted on only one or two of the antennas.

*Table 1. Codes for comparisons for 3tx 2rx*

| Code | Code rate | Constellation for spectral efficiency 4 bit/sec/Hz | Min rank | No. of complex symbols for joint detection in ML | Decoding Search Space |
|---|---|---|---|---|---|
| **UG1** | 1 | 16-QAM | 2 | 3 | $16^3 = 4096$ |
| **UG3** | 4/3 | 8-QAM | 2 | 4 | $8^4 = 4096$ |
| **DSTTDTx3** | 2 | 4-QAM | 1 | 4 | $4^4 = 256$ |
| **C2Tx3** | 2 | 4-QAM | 1 | 4 | $4^4 = 256$ |

*Figure 4. BER performance of 3-Tx 2-Rx STBC systems with spectral efficiency of 4 bps/Hz*

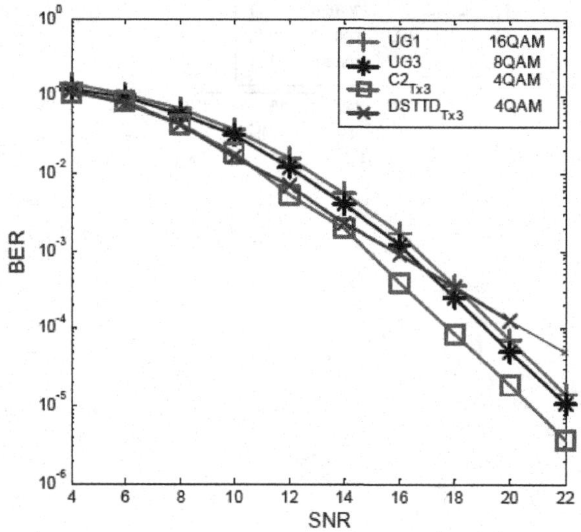


Figure 4 shows the BER performance of **UG1, UG3, C2**$_{TX3}$ and **DSTTD**$_{TX3}$ in a spatially-uncorrelated MIMO system with three transmit and two receive antennas. For comparison under the same spectral efficiency of 4 bps/Hz, **UG1** employs 16QAM, **UG3** employs 8QAM, **C2**$_{TX3}$ and **DSTTD**$_{TX3}$ employs 4QAM. The codes under comparisons are summarized in Table 1. **UG2** is not considered here because there is no standard constellation for it to achieve 4 bps/Hz. Figure 4 shows that **DSTTD**$_{TX3}$ performs the worst at high SNR, this is because some of its symbols are transmitted only on one transmit antenna and hence have no transmit diversity. On the other hand, the code **C2**$_{TX3}$ performs the best with a larger diversity gain over **DSTTD**$_{TX3}$ and a coding gain of up to 2 dB over **UG1** and **UG3**. The latter is because **C2**$_{TX3}$(code rate = 2) has a higher code rate than **UG1** (code rate = 1) and **UG3** (code rate = 4/3), hence **C2**$_{TX3}$ may use a smaller constellation dimension with larger Euclidean distances to support the same spectral efficiency. Similar conclusion can be drawn for **UG2**, whose code rate of 3/2 is also lower than that of **C2**$_{TX3}$.

Furthermore, as **C2**$_{TX3}$ needs to jointly decode four 4QAM symbols, it also has a much lower decoding complexity than **UG1** and **UG3**, which need to jointly decode three 16QAM symbols and four 8QAM symbols respectively.

In this section, we proposed a new rate-four STBC for four transmit antennas, and used it to derive new STBC with lower code rate of two by nulling some codeword symbols, or STBC for smaller number of transmit antennas by removing the last column of the codeword. Our new STBCs for four transmit antennas with rate two or four are shown to outperform the existing STBCs with same decoding complexity in spatially-correlated MIMO channels because our code distributes the code symbols more evenly across all the transmit antennas. For the same reason, our new STBC can be truncated for use with less transmit antennas without leaving some code symbols solely dependent on certain antennas. Furthermore, STBC for three transmit antennas presented in this section also has a higher code rate, better coding gain and lower decoding complexity than some of the existing STBCs proposed in the literature.

Also we would like to highlight that the gain in performance in correlated channel is obtained without extra transmission power or decoding complexity. Hence by properly designing a STBC with good dispersion property, extra advantages can be obtained with no extra cost.

However, to achieve those gain, maximum likelihood (ML) detection is required. Implementing ML in real time can be a challenging problem, hence an alternative detection algorithm that can achieve near ML decoding performance but with much lower complexity would be attractive. And this leads to the research in next section.

## LOW COMPLEXITY NEAR OPTIMAL DETECTION

In order to fully realize the diversity gains introduced by STBC, maximum likelihood detection (ML) is assumed at the receiver most of the time. However, such assumptions are impractical given the complexity of ML detectors. In such cases, low complexity near optimal detection methods are much more feasible as they are able to perform almost as well as ML detectors, but at a fraction of the computational complexity.

Several promising low complexity algorithms have been proposed in the last decade. These includes the Maximum Likelihood (ML) achieving lattice search based sphere decoder (Viterbo & Boutros, 1999), and the tree search based QR Decomposition-M (QRD-M) algorithm (Yue et al., 2003). The QRD-M algorithm is especially of interest as it achieves near-ML performance, while requiring comparatively low complexity. The algorithm also has the advantage of variable complexity depending on the number of branches it chooses to retain. In this chapter, we will also introduce the QRD-based Stack algorithm (Chin, 2005), which has even lower computational complexity than QRD-M.

In addition, we will introduce the list type version of the algorithms which will be required when soft decision decoding is required. We also introduce a method to further reduce the computational complexity of the list algorithms.

In the remaining portion of the chapter, we use the model in (Hassibi & Hochwald, 2002), where $\mathbf{H}$ is the equivalent channel and $\mathbf{C}$ is a vector of transmitted symbols.

### Basic Detection

We first introduce some linear detection techniques that can be used to detect STBC symbols at the receiver.

#### Least Squares

One of the most commonly used linear estimator is the least squares estimator. The estimator minimizes the least squares criterion,

$$\begin{aligned} J_{\text{LS}} &= \min_{\mathbf{C}} \| \mathbf{Y} - \mathbf{HC} \|^2 \\ &= \min_{\mathbf{C}} \left\{ (\mathbf{Y} - \mathbf{HC})^H (\mathbf{Y} - \mathbf{HC}) \right\} \\ &= \min_{\mathbf{C}} \left\{ \mathbf{Y}^H \mathbf{Y} - \mathbf{Y}^H \mathbf{HC} - \mathbf{C}^H \mathbf{H}^H \mathbf{Y} + \mathbf{C}^H \mathbf{H}^H \mathbf{HC}) \right\} \end{aligned} \tag{26}$$

The solution to (26) is well known to be the Moore-Penrose pseudoinverse of the matrix, $\mathbf{H}$,

$$\hat{\mathbf{C}} = \mathbf{H}^{\dagger} \mathbf{Y}, \tag{27}$$

where

$$\mathbf{H}^{\dagger} = \begin{cases} (\mathbf{H}^H\mathbf{H})^{-1}\mathbf{H}^H, & M < N \\ \mathbf{H}^{-1}, & M = N \\ \mathbf{H}^H(\mathbf{H}\mathbf{H}^H)^{-1}, & M > N. \end{cases} \tag{28}$$

Since we are only concerned with exact or over determined systems, the solution to our problem will be either of the first two solutions in (27), depending on the matrix dimensions.

Examining (27), we observe that each row of $\mathbf{H}^{\dagger}$ seeks to force all but one of the columns of $\mathbf{H}$ to be zero, *i.e.*,

$$(\mathbf{H}^{\dagger})_k \mathbf{H}_j \approx \begin{cases} 1, & k = j, \\ 0, & k \neq j \end{cases} \tag{29}$$

Considering this fact, the least squares estimator is analogous to the zero forcing equalizer, well known to the digital communications community. As a result, the least squares result is sometimes also referred to as zero forcing (ZF) (Wolniansky et al., 1998).

## Minimum Mean Square Error

While the least squares estimator is optimum in the least squares sense, it does not take into account the noise present. In instances where the received SNR on one particular antenna is much less than the average, the least squares estimator will attempt to enhance both the noise and the signal, resulting in unsatisfactory performance.

The minimum mean square estimator (MMSE) on the other hand, is a statistical estimator and considers the effects of the additive noise. The minimum mean square estimator works by minimizing the mean square error of the estimate,

$$J_{\text{MMSE}} = \min_{\mathbf{G}} E\left[ \| \mathbf{C} - \mathbf{G}\mathbf{Y} \|^2 \right], \tag{30}$$

where $\mathbf{G}$ is the matrix estimator of $\mathbf{C}$ from $\mathbf{Y}$.

Since,

$$\| \mathbf{A} \|^2 = \text{trace}\{\mathbf{A}\mathbf{A}^H\}, \tag{31}$$

we can minimize the trace of the covariance of the term $\mathbf{C} - \mathbf{G}\mathbf{Y}$ instead. Where

$$\begin{aligned} \text{cov}\{\mathbf{C} - \mathbf{G}\mathbf{Y}\} &= E[(\mathbf{C} - \mathbf{G}\mathbf{Y})(\mathbf{C} - \mathbf{G}\mathbf{Y})^H] \\ &= E[\mathbf{C}\mathbf{C}^H] - E[\mathbf{C}\mathbf{Y}^H]\mathbf{G}^H - \mathbf{G}E[\mathbf{Y}\mathbf{C}^H] + \mathbf{G}E[\mathbf{Y}\mathbf{Y}^H]\mathbf{G}^H \\ &= \mathbf{I} - \mathbf{H}^H\mathbf{G}^H - \mathbf{G}\mathbf{H} + \mathbf{G}(\mathbf{H}\mathbf{H}^H + \sigma\mathbf{I})\mathbf{G}^H \end{aligned} \tag{32}$$

The resulting estimator of $\mathbf{C}$ that minimizes the trace of (32) can be shown to be

$$\hat{\mathbf{C}} = \mathbf{H}^H(\mathbf{H}\mathbf{H}^H + \sigma^2\mathbf{I})^{-1}\mathbf{Y}. \tag{33}$$

## QRD Based Data Detection

Although the linear detectors have very low complexity, their performance is not good and is not able to extract the diversity introduced by STBCs. In this subsection, we introduce two low complexity pseudo maximum likelihood detectors which are able to perform very close to the maximum likelihood detector while only requiring a fraction of the computational cost. These two detectors are based the QR decomposition and tree-based search algorithms.

Consider a MIMO system as introduced in (1), to perform QRD-based detection, we perform QR decomposition on **H**, whereby we will obtain

$$\mathbf{H} = \mathbf{Q}\hat{\mathbf{R}}, \tag{34}$$

where **Q** is an $N_{RX} \times N_{RX}$ unitary matrix,

$$\hat{\mathbf{R}} = \begin{bmatrix} \mathbf{R} \\ 0_{(N_{RX}-N_{TX})\times N_{TX}} \end{bmatrix},$$

**R** is an $N_{TX} \times N_{TX}$ upper triangular matrix and $0_{(N_{RX}-N_{TX})\times N_{TX}}$ is a zero matrix of size $(N_{RX} - N_{TX}) \times N_{TX}$.

Noting that

$$\mathbf{Q}^H\mathbf{Q} = \mathbf{I}, \tag{35}$$

after pre-multiplying (1) by $\mathbf{Q}^H$, we can rewrite (1) as

$$\tilde{\mathbf{Y}} = \mathbf{RC} + \tilde{\mathbf{N}}, \tag{36}$$

where $\tilde{\mathbf{Y}}$ is the first $N_{TX}$ rows of $\mathbf{Q}^H\mathbf{Y}$ and $\tilde{\mathbf{N}}$ is the first $N_{TX}$ rows of $\mathbf{Q}^H\mathbf{N}$.

The resulting maximum likelihood criterion after QRD is

$$\min_{\mathbf{C}\in S} \left\| \tilde{\mathbf{Y}} - \mathbf{RC} \right\|^2, \tag{37}$$

where $S$ is the constellation set from which **C** is drawn.

The pre-multiplication of (1) by $\mathbf{Q}^H$ results in a tree-like structure with depth of $N_{TX}$ due to the property of the matrix **R**. Consequently, the tree search techniques such as the *M* algorithm and the stack algorithm can be applied to detect **C**.

## QRD-M

The QRD-M algorithm (Yue et al., 2003), is based on the classical *M*-algorithm (Anderson & Mohan, 1984). The concept is to apply the pre-multiplication step discussed above before applying the *M*-algorithm to detect the symbols in a sequential manner.

Starting from the last element of **C**, $x_{N_{TX}}$, the *M* algorithm calculates the metrics for all possible values of $x_{N_{TX}}$ (from constellation set *S* of size *C*) using

$$|\, \tilde{y}_{N_{TX}} - r_{N_{TX},N_{TX}} \hat{x}_{N_{TX}} \,|^2, \tag{38}$$

where $\tilde{y}_{N_{TX}}$ is the $N_{TX}$ element of $\tilde{\mathbf{Y}}$, and $r_{N_{TX},N_{TX}}$ is the $(N_{TX}, N_{TX})$ element of **R**. The metrics of these points or nodes are then ordered, and only *M* nodes with the smallest metrics are retained. These *M* nodes are subsequently extended with each node branching out to *C* nodes (possible values of $x_{N_{TX}-1}$) resulting in *MC* branches. Only *M* branches with the smallest branch metrics out of the *MC* branches are retained and the rest of the list are deleted. The same procedure is applied to the nodes of the next level, and this process is continued until a tree depth of $N_{TX}$ is reached.

The metrics of the branches are calculated by using the QRD reduced ML criterion of (37). For the tree depth of *i*, $1 \le i \le N_{TX}$, the metric for each branch is

*Figure 5. QRD-M algorithm*

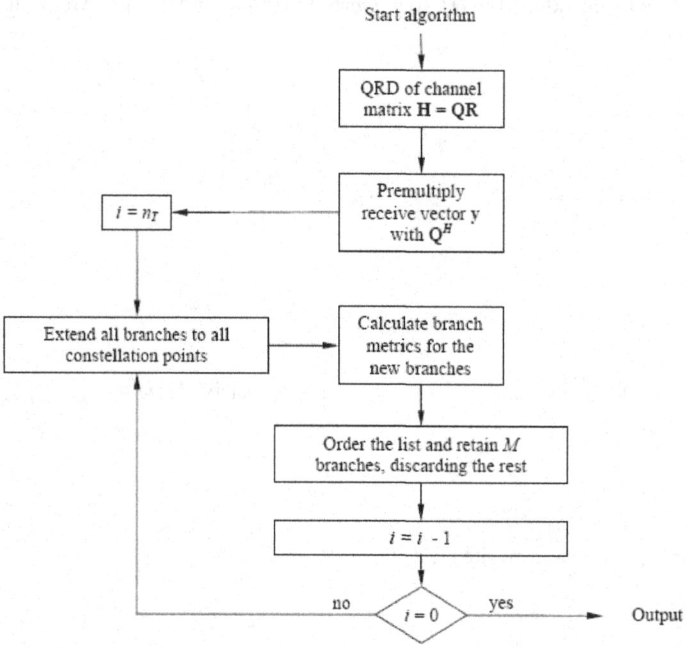


$$|\tilde{y}_{N_{TX}-i+1} - \mathbf{R}_{N_{TX}-i+1}\overline{\mathbf{x}}_i|^2, \quad (39)$$

where $\tilde{y}_i$ is the $i$ th element of $\tilde{\mathbf{y}}$, $\mathbf{R}_i$ is the $i$th row of $\mathbf{R}$, and $\overline{\mathbf{x}}_i$ is the vector of the appropriate nodes of the particular branch.

For the ease of understanding, the QRD-M algorithm can be summarized as follows:

1. Perform QRD on **H**
2. Premultiply **Y** with $\mathbf{Q}^H$
3. Extend all branches to $C$ nodes
4. Calculate the branch metrics using (39)
5. Order the branches according to their metrics, retaining only $M$ branches and discarding the rest
6. Move to next level and goto step 3

From the description of the algorithm, it can be seen that the algorithm extends all the branches in the list regardless of the value of the branch metric. This can result in unnecessary computational cost, which can be observed in 3.2.3. However, this property can prove to be useful in cases, such as list type decoding (Hochwald & ten Brink, 2003), which will be briefly touched on in 3.3, where a list of possible candidates are required for soft decision decoding. The flowchart for the QRD-M algorithm is shown in Figure 5.

## QRD-Stack

The QRD-Stack algorithm is different from QRD-M in the sense that it does not search all the nodes of the same tree depth before proceeding to the next depth.

The algorithm is based on the stack algorithm (Anderson & Mohan, 1984) and starts similarly as the QRD-M algorithm with the metric calculation of the $C$ possible nodes for $x_{N_{TX}}$. However, in contrast to the QRD-M algorithm, not all branches are extended subsequently. The stack algorithm only extends the branch with the smallest branch metric.

*Figure 6. QRD-Stack algorithm*

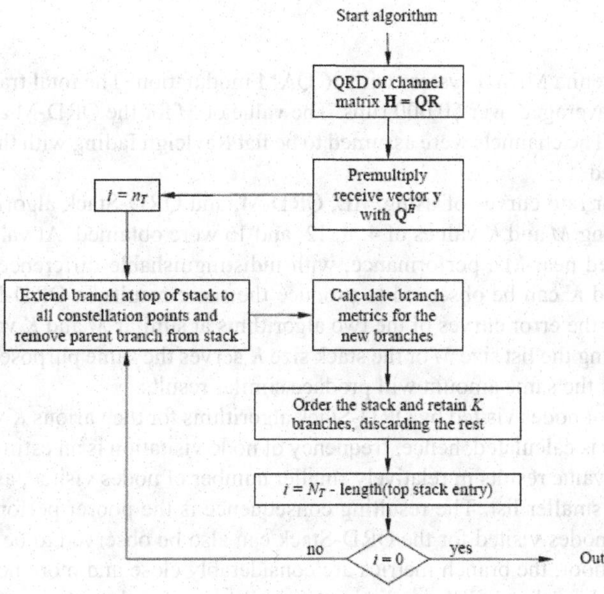


This branch is then replaced by the $C$ branches which it had spawned. The metrics of the branches, which are not necessarily of same depths, are ordered and only $K$ branches with the smallest metrics are retained. The algorithm continues until one of the branches reach the depth of $N_{TX}$. The branch metric calculation for QRD-Stack is the same as that of the QRD-M algorithm.

The advantage of the QRD-Stack algorithm over QRD-M algorithm is its selective extension of branches. This accounts for its lower computational complexity. This feature of algorithm may, however, also pose a problem if the size of the stack, $K$, is too small resulting in only a small part of the tree being searched. This situation can either be circumvented by using a larger $K$, or having the damage curtailed by permuting the rows of (1) such that more reliable symbols are shifted to the top of the tree.

While the QRD-Stack algorithm gains computationally from selecting only one single branch of smallest metric from which to extend, this property results in backtracking. As branch metrics from different levels are compared, in circumstances where there are no drastic differences in branch metrics, as in the case of low SNRs, the algorithm can frequently backtrack to a lower level. The undesirable property of multiple backtracking, is a resulting longer detection delay.

One other foreseeable shortfall of the QRD-Stack algorithm is that it will take a longer time as compared to QRD-M to collect full length candidates for list type soft decision decoding. The QRD-M algorithm has the advantage in this case as it has a guaranteed $MC$ full branches at the tree depth of $N_{TX}$ which it can use for its candidate list. The QRD-Stack algorithm, however, do not have that luxury, and would need to backtrack frequently to gather the candidate list. This could lead to a significant system delay.

For the ease of understanding, the QRD-Stack algorithm can be summarized as follows:

1. Perform QRD on $\mathbf{H}$
2. Premultiply $\mathbf{Y}$ with $\mathbf{Q}^H$
3. Extend branch on top of stack to $C$ nodes and remove the parent branch
4. Calculate the branch metrics of the new branches using (39)
5. Order the branches according to their metrics, retaining only $K$ branches and discarding the rest
6. Move to next level and goto step 3

The flowchart for the QRD-Stack algorithm is shown in Figure 6.

## Simulation Results

We consider a 4 transmit, 4 receive antenna MIMO system with 16 QAM modulation. The total transmit power is normalized to 1. Each simulation point is averaged over 50,000 runs. The value of $M$ for the QRD-M algorithm, and $K$ for QRD-Stack algorithm are both varied. The channels were assumed to be flat Rayleigh fading with the different transmit and receive antennas being uncorrelated.

In Figure 7, we present the bit error rate curves of using ML, QRD-M, and QRD-Stack algorithms. Performance curves of QRD-M and QRD-Stack using $M$ and $K$ values of 4, 8, 12, and 16 were obtained. At values of $M = K = 16$, QRD-M and QRD-Stack both achieved near-ML performance, with indistinguishable difference between the three curves. Subsequent reduction of $M$ and $K$ can be observed to produce the same results in QRD-M and QRD-Stack, with no observable difference between the error curves of the two algorithms at similar $M$ and $K$ values. The observed results are hardly surprising as restricting the list size $M$ or the stack size $K$ serves the same purpose in both algorithms. Consequently, reduction of $M$ and $K$ by the same amount will produce similar results.

In Figure 8, we present the number of nodes visited by QRD-Stack algorithms for the various $K$ values. Each node is a constellation point for which a metric is calculated, hence, frequency of node visitation is an estimate of the computational complexity involved. A small $K$ value results in relatively smaller number of nodes visited, as many of the nodes were discarded while maintaining the smaller list. The resulting consequence is the poorer performance as observed in the previous figure. The number of nodes visited for the QRD-Stack can also be observed to be a function of SNR. At lower SNRs, due to the high noise floor, the branch metrics are considerably close and more nodes are visited as a result. At high SNRs, a few dominant branches will have distinctly smaller metrics than the rest, and hence the tree search is restricted to a few branches, resulting in a smaller node count. It can be noted that at higher SNRs, the node count is similar regardless of $K$, since there is only one or two dominant branches and a higher $K$ value does little to increase the node count.

For QRD-M, as a fixed number of nodes are visited at each level, regardless of SNR. We can observe in Figure 9 that node counts of the QRD-M algorithm is not dependent on the SNR. The reduction of $M$ results in a reduction of the complexity of the algorithm. Comparing with the node count of the QRD-Stack in Figure 8, it can be observed that QRD-M has a higher node count than QRD-Stack at similar $M$ and $K$ values. This can be attributed to the nature of the

*Figure 7. BER of using QRD-M, QRD-Stack, and ML detection*

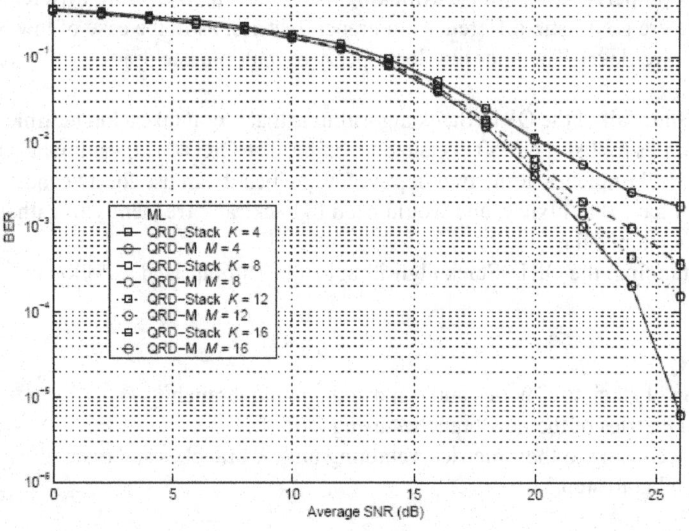

stack algorithm whereby only the branch with smallest branch metric is extended, whereas QRD-M indiscriminately extends all the branches in its list. This property of QRD-Stack, while desirable for its lower computational complexity, can also have the negative effect of having a higher detection delay as compared to QRD-M.

## List Detection

Since the invention of turbo codes, soft decision decoding has gained popularity and is increasingly replacing hard decision decoding in many communication systems. Soft decision decoding requires extrinsic information of the bits in order to maximize the *a posterior* probability (APP) of the bits. In cases where hard decisions are made, as in the case of maximum likelihood algorithms, this can still be done by constructing a list of candidates for which the maximum likelihood function has the lowest values (Hochwald & ten Brink, 2003). This subsection will introduce two algorithms which can produce this candidate list. The first is a variant of the QRD-Stack algorithm which is modified to produce a candidate list, while the other is an improved version of the List Stack algorithm which has significantly lower computational complexity.

### List Stack Algorithm

The Stack algorithm (Anderson & Mohan, 1984) starts from the last element of **C**, $x_{N_{TX}}$, and calculates the metrics for all possible values of $x_{N_{TX}}$ (from constellation set $S$ of size $C$). The metrics of these points or nodes are then ordered, and the branch with the smallest branch metric is extended to $C$ nodes (possible values of $x_{N_{TX}-1}$). The parent branch is replaced by the $C$ branches which it had spawned. The metrics of all the branches in the stack, which are not necessarily of same depths, are ordered and only $M$ branches with the smallest metrics are retained. The algorithm continues until one of the branches reach the length of $N_{TX}$. The metrics of the branches are calculated by using the QRD reduced ML criterion of (37).

For the calculation of the soft metrics, a single solution is insufficient. A list of possible candidates is required for the calculation of the bit metrics (Hochwald & ten Brink, 2003). To construct a candidate list for the Stack algorithm, full length branches are removed from the stack and placed on the list whenever they are obtained. The Stack algorithm will continue to operate on the remaining branches in the stack until the list has been filled.

*Figure 8. Average number of nodes visited by QRD-Stack for various values of **K** and SNR*

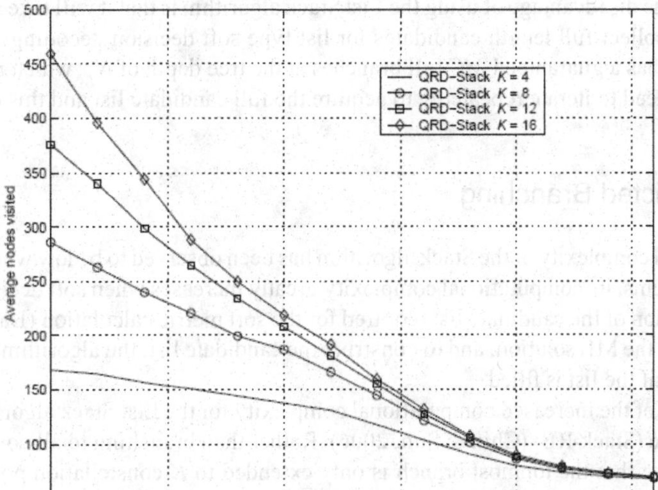

*Figure 9. Average number of nodes visited by QRD-M for various values of* $\mathbf{M}$ *and SNR*

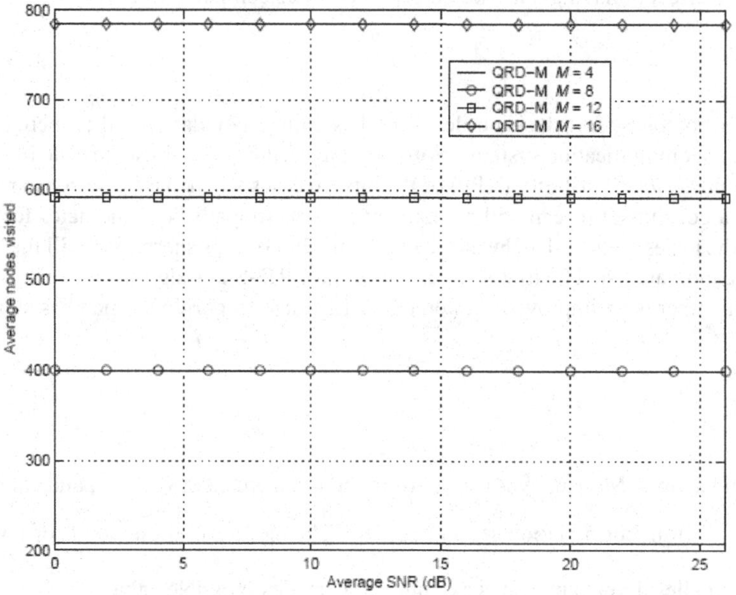


For the ease of understanding, the List Stack algorithm can be summarized as follows :

1. Perform QRD on $\mathbf{H}$
2. Premultiply $\mathbf{Y}$ with $\mathbf{Q}^H$
3. Extend branch on top of stack to $C$ nodes and remove the parent branch
4. If the new branches are fully extended, remove from stack and place in list.
5. If list is full, exit algorithm
6. Calculate the branch metrics of the new branches using (39)
7. Order the branches according to their metrics, retaining only $M$ branches and discarding the rest
8. Goto step 3

As mentioned earlier, one disadvantage of using the List Stack algorithm is that it will take a longer time as compared to the list $M$ algorithm to collect full length candidates for list type soft decision decoding. The $M$ algorithm has the advantage in this case as it has a guaranteed $MC$ full branches at the tree depth of $N_{TX}$ which it can use for its candidate list. The List Stack would need to iterate repeatedly to acquire the full candidate list and this could lead to a significant system delay.

## List Stack with Restricted Branching

Although the computational complexity of the Stack algorithm has been observed to be low when hard decision decoding is used and in uncoded systems, its computational complexity greatly increases when soft decision decoding is required. This is due to the construction of the candidate list required for the soft metric calculation (Baro et al., 2003). The Stack algorithm only searches for the ML solution, and to construct the candidate list, the algorithm have to backtrack repeatedly to extend branches until the list is filled.

To mitigate the problem of the increased computational complexity for the List Stack algorithm, we propose a simple restricted branching scheme (Stack-RB) (Chin & Sun, 2006). Rather than branching to all constellation points as in the Stack algorithm, we propose that the topmost branch is only extended to $K$ constellation points at each depth. The $K$ branches at each depth is selected according to their Euclidean distance from the zeroforcing (ZF) or MMSE solution, as illustrated in Figure 10.

The ZF solution can be obtained from (1) as

$$\hat{\mathbf{C}}_{ZF} = \mathbf{H}^{\dagger}\mathbf{Y} \quad (40)$$

and the MMSE solution is

$$\hat{\mathbf{C}}_{\text{MMSE}} = \mathbf{H}^{H}(\mathbf{H}\mathbf{H}^{H} + \sigma^{2}\mathbf{I})^{-1}\mathbf{Y}, \quad (41)$$

where $\mathbf{H}^{\dagger}$ is the Moore-Penrose pseudoinverse (Golub & van Loan, 1996) of **H**, and $\sigma^2$ is the variance of the noise.

The effect of the restricted branching is to only extend branches which are more likely to produce the ML solution and hence also more likely to contribute to the final candidate list. This is somewhat similar to the scheme proposed for the sphere decoder in Chan and Lee (2002).

The aim in constructing the candidate list is to choose a set of candidates for which (37) is smallest (Hochwald & ten Brink, 2003). In order to achieve lower computational complexity, the Stack-RB algorithm, through less branching, sacrificed some performance gain as compared to the List Stack algorithm, with the exception at $K = C$, when the two algorithms are the same. However, in doing so, the algorithm gained the flexibility of being able to adjust the computational complexity (through sacrificing performance) according to one's needs.

*Figure 10. Selection of closest K points in the list Stack-RB algorithm*

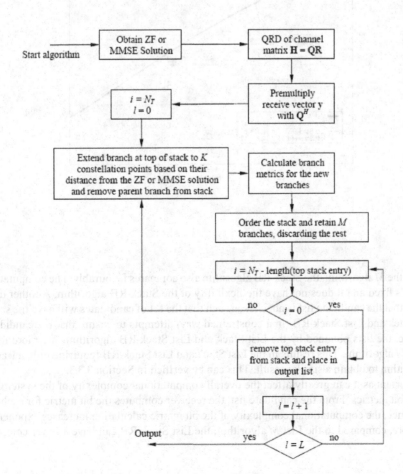

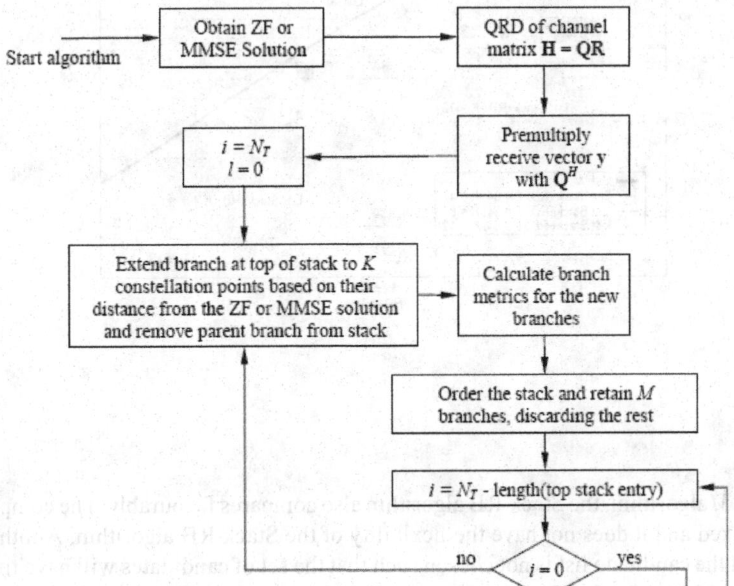

Figure 11. Flow diagram of the List Stack with Restricted Branching algorithm

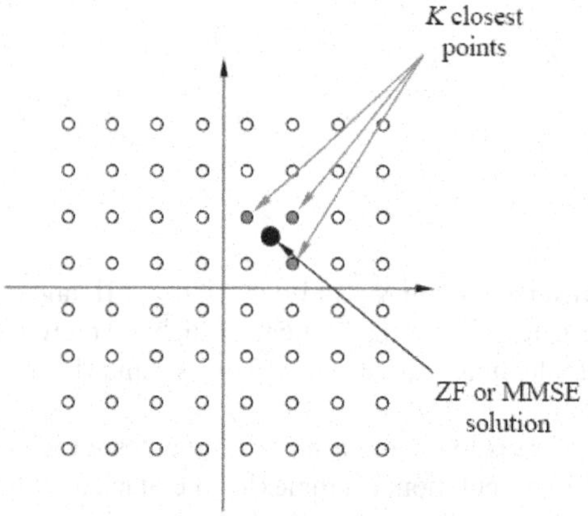


Figure 12. BLER of using List M with list size of 256 and List Stack-RB with list size of 64

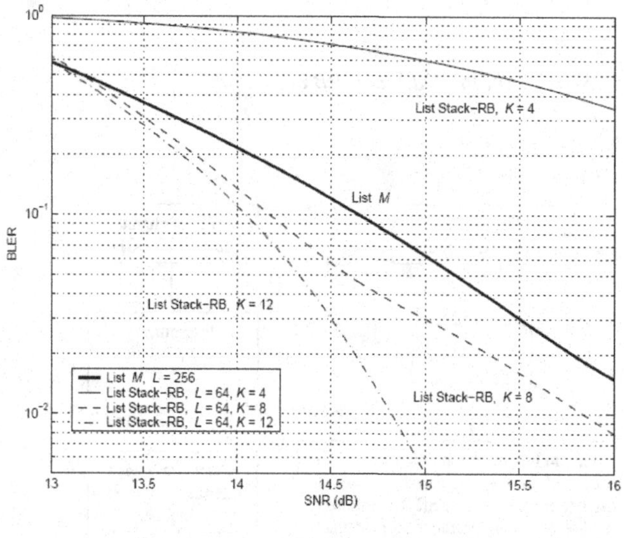


In comparison with the $M$ algorithm, the Stack-RB algorithm also compares favourably. The computational complexity of the $M$ algorithm is fixed and it does not have the flexibility of the Stack-RB algorithm. Another disadvantage of the List $M$ algorithm is that the candidate list is not chosen such that the set of candidates will have the smallest metrics. In contrast, the List Stack and List Stack-RB (in a constrained way) attempts to ensure that the candidate list has the smallest metrics. Hence, the lists obtained by the List Stack and List Stack-RB algorithms are more reliable than the one obtained from the $M$ algorithm. This results in the List Stack and List Stack-RB requiring only a fraction of the list size of the List $M$ algorithm to obtain a similar result. This can be verified in Section 3.3.3.

The list size is important as it can greatly affect the overall computational complexity of the system. This is due to the computation of the bit metrics. From the candidate list, the receiver computes the bit metric for each bit (Hochwald & ten Brink, 2003). Hence, the computational complexity of the bit metric calculation increases exponentially with the size of the list. Therefore, compared to the List $M$ algorithm, the List Stack-RB can have a lower computational complexity by

*Figure 13. Average number of nodes visited by the List M, and the List Stack-RB algorithms*

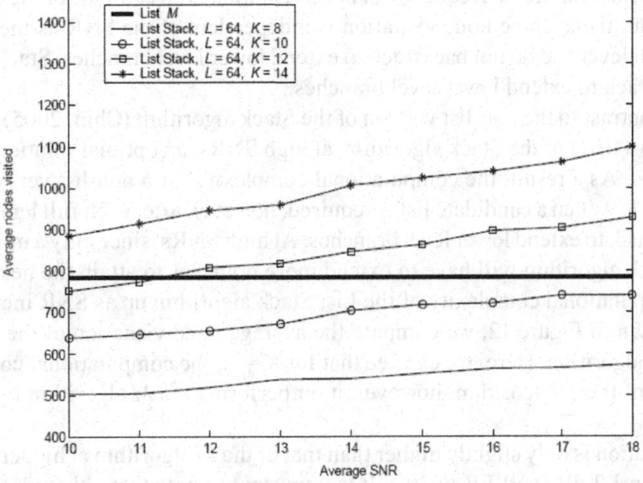


1. Reducing the branches to extend at each level,
2. Reducing the size of the candidate list by increasing the reliability of the list.

For the ease of understanding, the List Stack-RB algorithm can be summarized as follows:

1. Obtain the ZF/MMSE solution
2. Perform QRD on **H**
3. Premultiply Y with $\mathbf{Q}^H$
4. Extend branch on top of stack to $K$ nodes closest to the ZF/MMSE solution and remove the parent branch
5. If the new branches are full extended, remove from stack and place in list.
6. If list is full, exit algorithm
7. Calculate the branch metrics of the new branches using (39)
8. Order the branches according to their metrics, retaining only $M$ branches and discarding the rest.
9. Goto step 4

The flowchart of the proposed algorithm is shown in Figure 11.

## Simulation Results

In the simulation setup, we consider a 4 transmit, 4 receive antenna MIMO system with 16 QAM modulation. The total transmit power is normalized to 1. The number of branches to retain at each depth for both the List $M$ and Stack algorithms is 16. Each packet is 1/2 rate turbo coded with coded bit block length of 3072. The number of turbo iterations used for the simulations is 3. Each simulation point is averaged over 50,000 runs.

The list sizes used in the simulations are $L$ = 128 for the List $M$ algorithm and $L$ = 64 for the List Stack and List Stack-RB algorithms. This is to highlight the earlier discussion on list size and show the reliability of the stack algorithms' lists over the $M$ algorithm's list.

In Figure 12, we present the block error rate (BLER) curves of using $M$ and Stack-RB algorithms. For the Stack-RB algorithm, we used $K$ values of 4, 8, and 12. With half the list size, the Stack-RB algorithm with $K$ values of 8 and 12, still managed to perform better than the $M$ algorithm by approximately 0.3 dB and 1.2 dB respectively at BLER of $10^{-2}$. This stems from the fact that the list assembled by the Stack-RB algorithm is more reliable than the one produced by the

*M* algorithm. For $K = 4$, the number of constellation points being extended is too small, hence, the search space of the algorithm is too severely restricted to produce satisfactory results.

In Figure 13, we compare the number of nodes visited by the *M* and Stack-RB algorithms. Each node is a constellation point for which a metric is calculated, hence, frequency of node visitation is an estimate of the computational complexity involved. For the *M* algorithm, the average node visitation is independent of the SNR as the algorithm extends to the same number of nodes at each level and do not backtrack to extend lower level branches. Stack-based algorithms, on the other hand, frequently backtrack to extend lower level branches.

The results are in stark contrast to the non-list version of the Stack algorithm (Chin, 2005), where only a single solution is required as opposed to a list. For the Stack algorithm, at high SNRs, an optimal solution is obtained by traversing only a small number of nodes. As a result, the computational complexity for a non-list version of the Stack algorithm decreases with increasing SNR. When a candidate list is required, however, after each full length branch is obtained, the Stack algorithm has to backtrack to extend lower level branches. At high SNRs, since only a minimal number of branches have been extended, the Stack algorithm will have to extend more branches to attain the next full length branch. This consequently drives the computational complexity of the List Stack algorithm up as SNR increases.

For the performance shown in Figure 12, we compare the average node visitation of the Stack-RB algorithm with $K = 8$ and $K = 12$ and the *M* algorithm. Here, we can see that for $K = 8$, the computational complexity of the Stack-RB algorithm is lower than that of the *M* algorithm, however, it outperforms the *M* algorithm by approximately 0.3 dB at BLER of $10^{-2}$.

For $K = 12$, the node visitation is only slightly higher than that of the *M* algorithm at higher SNRs, but it outperforms the *M* algorithm by more than 1.2 dB at BLER of $10^{-2}$. It is important to note that although the computations required may be slightly higher for the list generation, with the smaller and more accurate list, the computations required for the soft metric calculation is immensely lower.

## CONCLUSION

In this chapter, an overview of STBC and several low complexity detection methods was given. We have introduced several STBC designs with good dispersion properties, and good performance in correlated channels. This has been achieved with no increase in transmission power or decoding complexity. We have also introduced several low complexity near optimal detection schemes. These schemes were shown to be able to closely match the performance of maximum likelihood detection while only requiring a small fraction of the computational cost. List type variants of the schemes were also discussed.

## REFERENCES


Alamouti, S. (1998). A simple transmit diversity technique for wireless communications. *IEEE JSAC*, *16*(8), 1451–1458.

Anderson, J., & Mohan, S. (1984). Sequential coding algorithms - a survey and cost analysis. *IEEE Trans. Comms*, *COM-32*(2), 169–176.

Baro, S., Hagenauer, J., & Witzke, M. (2003). Iterative detection of mimo transmission using a list sequential detector. In *Proc. ICC '03*, vol. 4, (pp. 11–15).

Chan, A. M., & Lee, I. (2002). A new reduced complexity sphere decoder for multiple antenna systems. In *Proc. ICC*, (pp. 460–464). New York City.

Chin, W. (2005). QRD-based tree search data detection for mimo communication systems. In *Proc. VTC 2005 Spring*, vol. 3, (pp. 1659–1662). Stockholm, Sweden.

Chin, W., & Sun, S. (2006). List stack detection with reduced search space for MIMO communication systems. In *Proc. VTC 2006 Spring*, vol. 4, (pp. 1713–1716). Melbourne, Australia.

DaSilva, V. M., & Sousa, E. (1997). Fading-resistant modulation using several transmitter antennas. *IEEE Trans. on Communications*, *45*(10), 1236–1244.

Foschini, G. (1996). Layered space-time architecture for wireless communication in a fading environment when using multi-element antennas. *Bell Labs Technical Journal*, (pp. 41–59).

Golub, G., & van Loan, C. (1996). *Matrix Computations.* Maryland: John Hopkins University Press, 3rd ed.

Hassibi, B., & Hochwald, B.M. (2002). High-rate codes that are linear in space and time. *IEEE Trans. on Information Theory, 48,* 1804–1824.

Hochwald, B., & ten Brink, S. (2003). Achieving near-capacity on a multiple-antenna channel. *IEEE Trans. Comms, 51*(3), 389–399.

Hottinen, A., Tirkkonen, O., & Wichman, R. (2003). *Multi-antenna Transceiver Techniques for 3G and Beyond.* John Wiley & Sons.

IEEE (2006). IEEE p802.11n/d001. *IEEE Standards Draft.*

Texas Instruments (2001). Double-sttd scheme for HSDPA systems with four transmit antennas: Link level simulation results. *TSG-RAN Working Group 1 #20.*

Paulraj, A., & Kailath, T. (1994). Increasing capacity in wireless broadcast systems using distributed transmission/directional reception (DTDR). *Tech. Rep. U.S. Patent #5345599.*

Sandhu, S., & Paulraj, A. (2000). Space-time block codes: A capacity perspective. *IEEE Comms Letters, 4*(12), 384–386.

Tarokh, V., Jafarkhani, H., & Calderbank, A. (1999). Space-time block codes from orthogonal designs. *IEEE Trans. Information Theory, 45*(5), 1456–1467.

Tarokh, V., Seshadri, N., & Calderbank, A. (1998). Space-time codes for high data rate wireless communication: Performance criterion and code construction. *IEEE Trans. Information Theory, 44*(2), 744–765.

Telatar, I. (1995). Capacity of multi-antenna gaussian channels. *Bell Labs Tech Memorandum.*

Uysal, M., & Georghiades, C. N. (2002). Non-orthogonal space-time block codes for 3TX antennas. *Electronics Letters, 38*(25), 1689–1691.

Viterbo, E., & Boutros, J. (1999). A universal lattice decoder for fading channels. *IEEE Trans. Information Theory, 45*(5), 1639–1642.

Wolniansky, P., Foschini, G., Golden, G., & Valenzuela, R. (1998). V-BLAST: An architecture for realizing very high data rates over the rich scattering wireless channel. In *Proc. IEEE ISSSE '98*, (pp. 295–300).

Yue, J., Kim, K., Gibson, J., & Iltis, R. (2003). Channel estimation and data detection for MIMO-OFDM systems. In *Proc. GLOBECOM 2003*, (pp. 581–585).

Yuen, C., Guan, Y., & Tjhung, T. (2003). Full-rate full-diversity stbc with constellation rotation. In *VTC 2003-Spring*, vol. 45, (pp. 296–300).

Yuen, C., Guan, Y., & Tjhung, T. (2005). New High-Rate STBC with Good Dispersion Property. In *PIMRC*, vol. 3, (pp. 1683–1687).

Zheng, L., & Tse, D. (2003). Diversity and multiplexing: A fundamental tradeoff in multiple-antenna channels. *IEEE Trans. Information Theory, 49*(5), 1073–1096.

# Chapter VII
# Space-Time Modulated Codes for MIMO Channels with Memory

**Xiang-Gen Xia**
*University of Delaware, USA*

**Genyuan Wang**
*Cisco Systems, USA*

**Pingyi Fan**
*Tsinghua University, China*

## ABSTRACT

*Modulated codes (MC) are error correction codes (ECC) defined on the complex field and therefore can be naturally combined with an intersymbol interference (ISI) channel. It has been previously proved that for any finite tap ISI channel there exist MC with coding gain comparing to the uncoded AWGN channel. In this chapter, we first consider space-time MC for memory channels, such as multiple transmit and receive antenna systems with ISI. Similar to MC for single antenna systems, the space-time MC can be also naturally combined with a multiple antenna system with ISI, which provides the convenience of the study. Some lower bounds on the capacities C and the information rates $I_{i.i.d}$ of the MC coded systems are presented. We also introduce an MC coded zero-forcing decision feedback equalizer (ZF-DFE) where the channel is assumed known at both the transmitter and the receiver. The optimal MC design based on the ZF-DFE are presented.*

## INTRODUCTION

Space-time coding for multiple transmit and receive antenna communication systems has recently attracted considerable attentions, see for example (Eittneben (1993), Winters, Salz, Gitlin (1994), & Telatar (1995), Foschini & Gans (1998), Tarokh, Seshadri, & Calderbank (1998), Tarokh, Naguib, Seshadri, & Calderbank (1999)), which is mainly because of the significant capacity increase from the diversities. Such studies include, for example, the capacity studies (Telatar (1995), Foschini & Gans (1998), Winters, Salz, Gitlin (1994)), space-time trellis coded modulation (TCM) schemes (Tarokh, Seshadri, & Calderbank (1998), Tarokh, Naguib, Seshadri, & Calderbank (1999)), and the combination of the space time coding and signal processing (Tarokh, Seshadri, & Calderbank (1998), Tarokh, Naguib, Seshadri, & Calderbank (1999)). Most studies for such systems so far are for memoryless channels that may fit slow fading environment well, where all the paths from different transmit and receive antennas are assumed constants and treated as independent random variables. A recent study on multiple transmit and receive antenna systems with memory can be found in (Ariyavisitakul, Winters, & Lee (1999)), where no space-time coding was considered.

In applications, MIMO-OFDM has been considered to be one of the best choices in the next generation of wireless communications. IEEE 802.16e has received some proposals on MIMO precoding with limited feedbacks (Kambourov (2006), Zhang et al (2004)). In MIMO-OFDMA systems, multiuser precoding with limited number of users has been proposed to increase the system capacity of users (Liu & Zhang (2007)). In the literature, there has been a lot of works focusing on the precoding techniques. The main reason is that if the wireless channels vary relatively slow and can be predicated or estimated by some methods in a relatively short time slot, the estimation and/or the prediction of the channel characteristics can be kept a relatively high accuracy when employing in the subsequent time slots. In this case, the precoding method can be used to reduce the effect of the multiuser interference and the channel fading, so that the system capacity can be greatly improved.

In this chapter, we are interested in multiple transmit and receive antenna channels with memory, where there are intersymbol interferences (ISI) for each pair of transmit and receive antennas. Note that, unlike those in the discussions in wireless ad hoc networks, we would not employ the statistics model to characterize the intersymbol interference but the algebraic model. Here we assume that all the ISI channels for all the different pairs are known at both the transmitter and the receiver. This assumption might be too strong for wireless communications. We have two reasons for such interest. The first reason is that the channel model here may fit some communication systems, such as multi-head and multi-track recording systems, such as (Soljanin & Georghiades (1995)), where there are ISI. The second reason is that, although the following study is based on the knowledge of all the ISI channels, the generalization of the study to unknown ISI channels might be possible in the WiMAX or some with fixed MIMO-OFDMA system.

For single antenna ISI channels, modulated codes (MC), i.e., error correction codes (ECC) defined on the complex field, have recently found useful in the ISI mitigation from both practical and theoretical perspectives, where the ISI is no longer distortion but diversity gain, see for example ((Xia, Xie, & Fan (1999), Xie & Xia (1998), Xia (1998), Xia (1999), Xia (1999, March), Fan & Xia (1999)). The main reason is that both MC encoding and the ISI arithmetic operations are defined on the complex field and therefore they can be combined naturally. The combination provides the convenience of the optimal study of the MC given an ISI channel. It has been proved that, for any finite tap ISI channel there always exist MC with coding gain comparing to the uncoded AWGN channel (Xia (1998)). An MC, however, does not have any coding gain in the AWGN channel (Xie & Xia (1998)). A general distance spectra calculation algorithm for a general MC was obtained in (Xie & Xia (1998)), which allows one to be able to search good MC given an ISI channel. Furthermore, due to the simplicity of the MC, it may be more convenient to update at the transmitter than the conventional ECC over finite fields for time-varying channels, in particular, when some suboptimal decoding algorithms, such as the decision feedback equalizer (DFE), are used at the receiver as we shall see later.

In this chapter, we generalize MC to space-time MC for multiple transmit and receive antenna ISI channels. Similar to the MC for single antenna ISI channels, the space-time MC can be naturally combined with the multiple antenna channels. By using the capacity formula of the multivariate channel with memory in (Brandenburg & Wyner (1974)), we first derive lower bounds of the capacities $C$ and the information rates $I_{i.i.d}$ for the MC coded systems, where $I_{i.i.d}$ is the i.i.d. information rates when the input is an i.i.d. source, see for example (Shamai, Ozarow, & Wyner (1991), Shamai & Laroia (1996)). As a property of the space-time MC, it is proved that for an N transmit and N receive antenna channel $\mathbb{H}(z)$ with memory and AWGN and for any rate $r$, $0 < r < 1$, there exist rate $r$ MC such that the MC coded systems have larger information rates $I_{i.i.d}$ than the system itself does, when the channel SNR is relatively low and the channel $\mathbb{H}(z)$ is not paraunitary, i.e., $\mathbb{H}(e^{j\theta})$ is not unitary. Notice that for a channel $\mathbb{H}(z)$, the condition that $\mathbb{H}(z)$ is not paraunitary holds almost surely. Another remark is that, when $N = 1$ this result is more general than the one obtained in (Fan & Xia (1999)) for MC coded single antenna systems, where only rate $1/P$ MC with $P \geq 2\Gamma - 1$ were constructed.

The MC studied here is basically a transmitter assisted ISI mitigation technique. There have been several approaches, such as the Tomlinson-Harashima (TH) precoding (Tomlinson (1971), Miyakawa & Harashima (1969)), the trellis precoding (Eyuboglu & Forney (1992)) and the transmitter assisted signal processing techniques and vector coding (Kasturia, Aslanis, & Cioffi (1996), Al-Dhahir & Cioffi (1996)). In the TH precoding and the most transmitter assisted signal processing techniques, the data rate is not expanded. Although in the trellis precoding the precoding is combined with the convolutional coding, it is not easy to analytically study the combination of the precoding and the ISI channel at all different channel SNR. In (Eyuboglu & Forney (1992)), there is some study on the combination when the channel SNR is high, which is based on the complete erasing of the ISI channel when the size of the signal constellation is large. The MC can be thought of as a transmitter assisted signal processing but with the data rate expansion, which is different from the existing pulse shaping filtering for shaping the channel spectrum from the following perspectives. First, the pulse shaping filter may not be easy to implement at the transmitter, in particular when the ISI channel has spectral nulls and the channel SNR is not high. Second, when the pulse shaping filter is long, the joint maximum likelihood sequence estimation (MLSE) with the ISI channel may be too complicated to implement at the receiver. In vector coding (Kasturia, Aslanis, & Cioffi (1996)), although the data rate is increased by inserting zeros to the input symbols to convert an ISI system into an ISI-free vector system, the vector size is usually large, which may cause some implementation problem. When the vector size is small, it may not be optimal. The coding in (Kasturia, Aslanis, & Cioffi (1996)) is after and separated from but related to the pulse shaping (signal processing) of the vector system. The MC is an alternative, which was studied from the conventional convolutional coding point of view, where the block size and memory size may not be large and therefore, the implementation may not be complicated. Spread-spectrum systems and orthogonal frequency division multiplexing (OFDM) systems can be thought of as block MC, which are, however, not optimally designed for a given ISI channel. The MC rate of the spread spectrum systems is usually low due to the spreading. Some recent related works of MC can be found in, for example, for single antenna systems (Xia (1997), Giannakis (1997), Scaglione, Barbarossa, & Giannakis (1999)), and for antenna arrays (Liu & Xia (1997)), where the name "filterbank precoders" were used. The study in (Scaglione, Barbarossa, & Giannakis (1999)) was similar to the study in (Al-Dhahir & Cioffi (1996)), where the optimal block (or constant) filterbank transceiver based on the maximal information rates was studied and it depends on a channel SNR. Notice that ECC matching to the partial response spectral null channels in magnetic recording systems can be found in (Karabed & Siegel (1991)), where the ECC are defined on finite fields and it is not easy to generalize to general ISI channels.

In the second half of this chapter, we propose the space-time MC coded zero-forcing decision feedback equalizer (ZF-DFE), which is similar to the one developed in (Xia (1999)) for single antenna ISI channels. The performance analysis and the optimal space-time MC design based on the ZF-DFE are then presented. Finally numerical simulations are presented to illustrate the theory. In the mean time, by using the optimal MC designed based on the above MC coded ZF-DFE, the information 3 rates $I_{i.i.d}$ of the MC coded channel are shown larger than the one of the channel itself, when the channel SNR are relatively low (when the channel SNR is below 6dB in the example).

This chapter is self-contained and organized as follows. In Section 2, we describe space-time MC and some key previously obtained properties of MC and then derive the capacity C and the information rate $I_{i.i.d}$ lower bounds and some properties for the MC coded multiple transmit and receive antenna channels with memory. In Section 3, we present the MC coded ZF-FE and its performance analysis and the optimal space-time MC design. In Section 4, we present simulation results.

## SPACE-TIME MODULATED CODES, CAPACITY AND INFORMATION RATES

Before going to space-time MC, let us first describe the channel model. Consider an N transmit antenna and M receive antenna channel with finite memory and AWGN, i.e.,

$$r_m(t) = \sum_{n=1}^{N}\sum_{k=0}^{\Gamma-1} h_{m,n}(k)s_n(t-k) + \eta_m(t), \quad 1 \le m \le M \tag{2.1}$$

where $s_n(t)$ is the information sequence at the $n$th transmit antenna, $r_m(t)$ is the received signal at the $m$th receive antenna, $h_{m,n}(k)$ is the ISI channel finite impulse response of length $\Gamma$ corresponding to the $n$th transmit antenna and the $m$th receive antenna, and $\eta_m(t)$ is the AWGN at the $m$th receive antenna. Let $H_{m,n}(z)$ denote the z-transform of $h_{m,n}(k)$ in terms of variable $k$. Let $\mathbb{H}(z)$ denote the following $M \times N$ matrix polynomial

$$\mathbb{H}(z) = \left(H_{m,n}(z)\right)_{1\le m\le M, 1\le n\le N}. \tag{2.2}$$

Then, channel (2.1) is an $N$ input and $M$ output system with transfer matrix function $H(z)$, i.e.,

$$\mathbb{R}(z) = \mathbb{H}(z)\mathbb{S}(z) + \eta(z), \tag{2.3}$$

where $\mathbb{R}(z) = (R_1(z), R_2(z), ..., R_m(z))^T$, $\mathbb{S}(z) = (S_1(z), S_2(z), ..., S_m(z))^T$ and $\eta(z) = (\eta_1(z), \; \eta_2(z), \; \cdots, \; \eta_M(z))^T$, $T$ stands for the transpose, and $R_m(z)$ $S_n(z)$ and $\eta_m(z)$ are the z-transforms of $r_m(t)$, $s_n(t)$ and $\eta_m(t)$, respectively. For convenience, in what follows the above channel transfer matrix function $\mathbb{H}(z)$ is normalized so that the channel itself does not have any gain:

$$\sum_{m=1}^{M}\sum_{n=1}^{N}\sum_{k}|h_{m,n}(k)|^2 = M. \qquad (2.4)$$

In the following, we want to encode the information sequences $s_n(t)$ by using a space-time MC given the channel $\mathbb{H}(z)$.

## Space-Time Modulated Codes

We first briefly describe MC studied in ((Xia, Xie, & Fan (1999), Xie & Xia (1998), Xia (1998), Xia (1999), Xia (1999, March)) and (Xia (2001), Xia, Wang, & Fan (2000)). For completeness, some of the key properties of MC and the combinations with ISI channels are described but without proofs, which are not used for the rest of this chapter.

### MC and Some Properties

A rate *K/L* modulated code (MC) encoder is an L by K polynomial matrix

$$\mathbb{G}(z) = \begin{bmatrix} g_1(z) & \cdots & g_{1K}(z) \\ \vdots & \vdots & \vdots \\ g_{L1}(z) & \cdots & g_K(z) \end{bmatrix}, \qquad (2.5)$$

where $g_k(z)$ is a polynomial of $z^{-1}$ with complex-valued coefficients.

Let *s(n)* be a binary information sequence and *x(n)* be the complex symbol sequence after the binary-to-complex symbol mapping of *s(n)*. Let *X(n)* be the *K* by 1 vector sequence of *x(n)* after the serial to parallel conversion. Their z transforms are *x(z)* and $\mathbb{X}(z)$, respectively. Then, the encoding of an MC is the following linear multiplication over the complex field:

$$\mathbb{Y}(z) = \mathbb{G}(z)\mathbb{X}(z),$$

where $\mathbb{Y}(z)$ is the z transform of the encoded *L* by 1 vector sequence. Notice that the MC encoding is after the binary-to-complex mapping as shown in Fig. 1, which is different from the conventional coded modulation.

Since, in the encoding of an MC, the coded signal mean power may be different from the information signal mean power. For convenience, an MC is normalized such that the mean power of the encoded signal *y(n)* is the same as the one of the information sequence *x(n)*. This can be achieved by normalizing the magnitude squared sum of all the coefficients of all the polynomials in $\mathbb{G}(z)$ as follows. Let

$$g_{nk}(z) = \sum_{l} g_{nk}(l)z^{-1}, \quad 1 \le n \le L, 1 \le k \le K. \qquad (2.6)$$

Then,

$$\sum_{n=1}^{L}\sum_{k=1}^{K}\sum_{l}|g_{nk}(l)|^2 = L. \qquad (2.7)$$

If an MC $\mathbb{G}(z)$ satisfies (2.7), it is called a normalized MC. In what follows, an MC at the transmitter always means a normalized MC unless otherwise specified.

For an MC $\mathbb{G}(z)$, its free distance can be defined as the minimum Euclidean distance between all two different encoded sequences. Comparing to an uncoded system in AWGN channel, the *coding gain* of a rate *K/L* MC with its free distance $d_{free}^2$ in AWGN channel is

$$\gamma = \frac{d_{free}^2 K}{L d_{\min}^2}, \qquad (2.8)$$

*Figure 1. Modulated code via coded modulation*

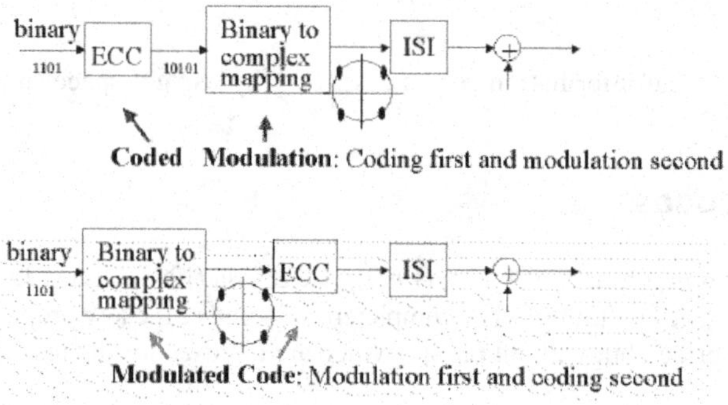


where $d_{\min}^2$ is the minimum distance between the complex symbols of the information sequence *x(n)*. The following result was obtained in (Xie & Xia (1998)).

Theorem 1. *A modulated code does not have any coding gain in an AWGN channel, i.e.,* $\gamma \le 1$ *in (2.8).*

Although an MC does not have any coding gain in the AWGN channel, it may be different in an ISI channel. Let *H(z)* be the z transform of an ISI channel with finite taps *h(n)*, for $\Gamma > 1$ and $h(0) \neq 0$ and $h(\Gamma - 1) \neq 0$ and the unit energy:

$$\sum_{n=0}^{\Gamma-1} |h(n)|^2 = 1. \tag{2.9}$$

Let $\mathbb{G}(z)$ be a normalized rate *K/L* MC. The combination of the MC $\mathbb{G}(z)$ and the ISI channel *H(z)* becomes another MC $\mathbb{C}(z)$:

$$\mathbb{C}(z) = \mathbb{H}(z)\mathbb{G}(z), \tag{2.10}$$

where $\mathbb{H}(z)$ is the blocked version of *H(z)* and has the form of the following *L* by L pseudo-circulant polynomial matrix (see [31]):

$$\mathbb{H}(z) = \begin{bmatrix} h_0(z) & z^{-1}h_{L-1}(z) & \cdots & z^{-1}h_1(z) \\ h_1(z) & h_0(z) & \cdots & z^{-1}h_2(z) \\ \vdots & \vdots & \vdots & \vdots \\ h_{L-2}(z) & h_{L-3}(z) & \cdots & z^{-1}h_{L-1}(z) \\ h_{L-1}(z) & h_{L-2}(z) & \cdots & h_0(z) \end{bmatrix}, \tag{2.11}$$

where $h_n(z)$ is the *n*th polyphase component of *H(z)*

$$h_n(z) = \sum_l h(Ll+n)z^{-l}, \quad 0 \le n \le L-1.$$

An important consequence of the above combined MC is its invertibility even when the ISI channel *H(z)* is spectral null, see (Xia (1997)). Although the encoding at the transmitter is based on the normalized MC $\mathbb{G}(z)$, the decoding is based on the combined MC $\mathbb{C}(z)$, where $\mathbb{G}(z)$ is over the ISI channel while $\mathbb{C}(z)$ is over the AWGN channel. The coding gain $\gamma_{ISI}$ of the MC $\mathbb{G}(z)$ at the transmitter in the ISI channel is defined as the coding gain of the combined MC $\mathbb{C}(z)$ in the AWGN channel, i.e.,

$$\gamma_{ISI} = \frac{d^2_{free,C}}{d^2_{\min}} \frac{K}{L}, \tag{2.12}$$

where $d_{free,C}$ is the free distance of the combined MC $\mathbb{C}(z)$. The above coding gain definition is compared to the ideal channel, i.e., the AWGN channel, and not the ISI channel itself, which is for convenience in the study. The total coding gain is illustrated in Figure 2, which is always above the coding gain $\gamma_{ISI}$ in our definition.

By the normalization, the mean power of the transmitted signal $y(n)$ is the same as the one of the information signal $x(n)$. An important observation is that the mean power of the received signal $q(n)$ may be different. It turns out that, by properly choosing an MC $\mathbb{G}(z)$, the mean power of the received signal $q(n)$ after the ISI channel may be greater than the information signal mean power. This is the key for the existence of a coding gain for an MC in an ISI channel, which is stated as follows. However, not any MC with increased power at the receiver may provide coding gain.

Theorem 2 *For any ISI channel $h(n)$, $0 \le n \le \Gamma - 1$, with $\Gamma > 1$ and $h(0) \ne 0$* and $h(\Gamma - 1) \ne 0$, *there exists a normalized modulated code $\mathbb{G}(z)$ that has a coding gain in the ISI channel, i.e., $\gamma_{ISI} > 1$. The coding gain is, however, upper bounded by the number of the ISI taps, $\Gamma$, i.e., $\gamma_{ISI} \le \Gamma$.*

These results were obtained in (Xie & Xia (1998), Xia (1998)).

## Space-Time MC

A space-time MC for $N$ transmit antennas is a rate $K/(NP)$ MC, where the $n$th block of block length $P$ of the MC encoded $NP \times 1$ vector is transmitted at the $n$th antenna as shown in Figure 3.

Similar to the combination of an MC and an ISI in the single antenna case (2.10)(2.11), which corresponds to $N = 1$ and $L = NP$ here, the space-time MC can be combined with the multiple transmit and receive antenna channel $\mathbb{H}(z)$ in (2.2) by blocking it with block size $P$. The combined MC and channel is

$$\mathcal{C}(z) = \mathcal{H}(z)\mathbb{G}(z), \tag{2.13}$$

where $\mathcal{H}(z)$ is the blocked version of $\mathbb{H}(z)$ and has the following pseudo-circulant form, see for example (Xia & Suter (1996), Vaidyanathan (1993))),

$$\mathcal{H}(z) = \begin{bmatrix} \mathbb{H}_0(z) & z^{-1}\mathbb{H}_{P-1}(z) & \cdots & z^{-1}\mathbb{H}_1(z) \\ \mathbb{H}_1(z) & \mathbb{H}_0(z) & \cdots & z^{-1}\mathbb{H}_2(z) \\ \vdots & \vdots & \vdots & \vdots \\ \mathbb{H}_{P-2}(z) & \mathbb{H}_{P-3}(z) & \cdots & z^{-1}\mathbb{H}_{P-1}(z) \\ \mathbb{H}_{P-1}(z) & \mathbb{H}_{P-2}(z) & \cdots & \mathbb{H}_0(z) \end{bmatrix}, \tag{2.14}$$

where $\mathbb{H}_p(z)$ is the $p$th polyphase component of $\mathbb{H}(z) = \sum_k H(k) z^{-k}$:

$$\mathbb{H}_p(z) = \sum_l H(Pl + p) z^{-l}, \quad 0 \le p \le P-1$$

Figure 2. Coding gain definition of an MC in an ISI channel

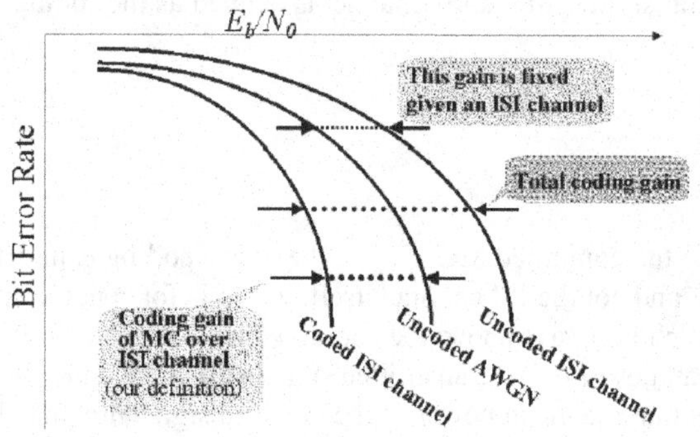


Figure 3. Space-time modulated code encoding

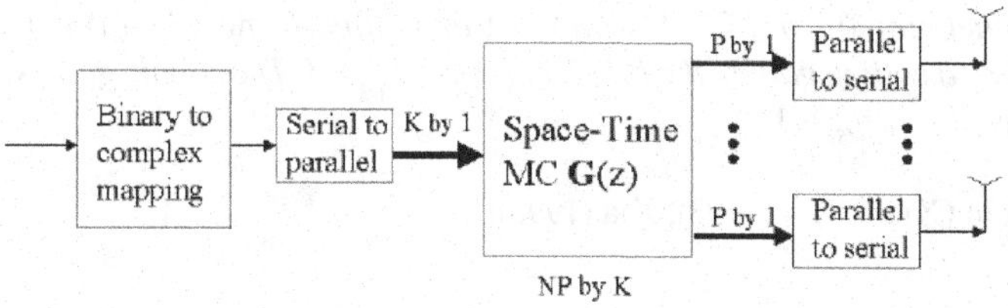


The space-time MC coded system in (2.1) and (2.3) after the blocking becomes

$$\mathcal{R}(z) = \mathcal{G}(z)\mathcal{X}(z) + \Xi(z), \quad (2.15)$$

where the size of $\mathcal{R}(z)$ is $MP \times 1$ and the $m$th receive antenna receives the $m$th block of block length $P$ in the vector $\mathcal{R}(z)$, the size of $\mathcal{G}(z)$ is $MP \times K$, the size of $\mathcal{X}(z)$ is $K \times 1$, and the size of $\Xi(z)$ is $MP \times 1$ and it is blocked from $\eta(z)$ in (2.3). All the components of $\Xi(z)$ are i.i.d. Gaussian with mean 0 and single sided power spectral density $N_0$. Since the rank of $\mathcal{G}(z)$ in (2.15) and (2.13) is at most min$\{MP, NP, K\}$ for the decodeability of the MC coded system (2.15), $K$ has to satisfy the following condition

$$K \leq \min\{MP, NP\}. \quad (2.16)$$

In the following we shall study the MC coded multi-input and multi-output (MIMO) system (2.15) under condition (2.16).

## Capacity and Information Rates of the Space-Time MC Coded MIMO Systems

The basic idea for the following capacity and information rate study is based on the capacity formula obtained by Brandenburg & Wyner (1974) for N-input and N-output systems (or multivariate channels) with memory and AWGN.

### Capacity and Information Rates of MIMO Systems without MC Encoding

We first want to study the capacity and the information rates of the original MIMO system (2.3). To do so, let us review the capacity and the information rates of a single antenna system with AWGN, i.e.,

$$Y(z) = H(z)X(z) + \eta(z) \tag{2.17}$$

The capacity, $C(E_s)$, of channel (2.17) is, see for example (Hirt & Massey (1998), Brandenburg & Wyner (1974)),

$$C(E_s) = \sup_x I(x,y) = \frac{1}{4\pi}\int_{-\pi}^{\pi} \max\left\{0, \log_2\left[\frac{2K_s\,|H(e^{j\theta})|^2}{N_0}\right]\right\}d\theta' \tag{2.18}$$

where

$$E_s = \frac{1}{2\pi}\int_{-\pi}^{\pi} \max\left\{0, K_s - \frac{N_0}{2}|H(e^{j\theta})|^{-2}\right\}d\theta, \tag{2.19}$$

where $K_s > 0$ is a constant parameter. Note that in the above capacity formula, the input $x$ may have any distribution. When the input $x$ is restricted to an i.i.d. source, the maximum mutual information is the information rate $I_{iid}$, see for example (Hirt & Massey (1998), Brandenburg & Wyner (1974), Shamai, Ozarow, & Wyner (1991), Shamai & Laroia (1996)). The information rate, $I_{iid}(E_S)$ of channel (2.17) is, see Hirt & Massey (1998),

$$I_{iid}(E_s) = \sup_{iid\ \ x} I(x,y) = \frac{1}{4\pi}\int_{-\pi}^{\pi} \log_2\left[1 + \frac{2E_s\,|H(e^{j\theta})|^2}{N_0}\right]d\theta, \tag{2.20}$$

which is achieved when the input $x$ is an i.i.d. Gaussian process. The above information rate determines an achievable reliable information rate when the standard random coding technique, such as the existing ECC defined on finite fields, is used. The information rate is also called, for example in (Shamai & Laroia (1996)), information capacity.

For N-input and N-output systems with memory and AWGN, Brandenburg and Wyner (Brandenburg & Wyner (1974)), derived the following similar formulas. Let an N-input and N-output system be

$$\mathcal{Y}(z) = \mathcal{P}(z)\mathcal{X}(z) + \Xi(z), \tag{2.21}$$

where all components in the noise $\Xi(n)$ are i.i.d. Gaussian and with the same statistics of $\eta(n)$ in (2.3). Then, the capacity of the system (2.21) is

$$C(S,N) = \frac{1}{4\pi}\sum_{k=1}^{N}\int_{-\pi}^{\pi} \max\left\{0, \log_2\left[\frac{2\lambda_k(\theta)K_S}{N_0}\right]\right\}d\theta \tag{2.22}$$

where

$$\frac{1}{2\pi}\sum_{k=1}^{N}\int_{-\pi}^{\pi} \max\left\{0, K_s - \frac{N_0}{2}\lambda_k^{-1}(\theta)\right\}d\theta = S \tag{2.23}$$

and $\lambda_k(\theta)$, $k = 1, 2, \ldots, N$, are the $N$ eigenvalues of matrix $P^+(e^{j\theta})P(e^{j\theta})$ for each $\theta$ with $-\pi \le \theta \le \pi$, the superscript $^+$ denotes the complex conjugate transpose of matrix $P(e^{j\theta})$, and $S$ denotes the mean norm squared, i.e., $\|X\|^2$, of $N \times 1$ vectors $X$ and is treated as the signal mean power of $N \times 1$ vectors. The information rate of the system (2.21) can be derived similar to (2.20):

$$I_{iid}(S,N) = \frac{1}{4\pi}\sum_{k=1}^{N}\int_{-\pi}^{\pi}\log_2\left[1+\frac{2S\lambda_k(\theta)}{NN_0}\right]d\theta, \quad (2.24)$$

where $\lambda_k(\theta)$, $k=1,2,\ldots,N$ are the same as in the capacity (2.22). Notice that the units of the capacity $C(S,N)$ and the information rates $I_{iid}(S,N)$ are bits per vector symbol of size $N$, i.e., bits per $N$ symbols. The signal mean power $S$ is also the mean power of vectors of size $N$, i.e., the mean power per $N$ symbols.

Let us now study the capacity and information rates of the MIMO system (2.3). The difficulty arises from that the number $N$ of the inputs may be different from the number $M$ of the outputs. When $M = N$, the above capacity formula (2.22) and the information rate formula (2.24) can be directly applied by replacing $\mathbb{P}(z)$ with $\mathbb{H}(z)$. By changing the units in (2.22) and (2.24) from per vectors to per symbols, **the capacity and the information rates of an $N$ transmit antenna and $N$ receive antenna system are**

$$C(E_s,N) = \frac{1}{4\pi N}\sum_{k=1}^{N}\int_{-\pi}^{\pi}\max\left\{0,\log_2\left[\frac{2\lambda_k(\theta)K_S}{N_0}\right]\right\}d\theta, \quad (2.25)$$

where

$$\frac{1}{2\pi}\sum_{k=1}^{N}\int_{-\pi}^{\pi}\max\left\{0,K_s-\frac{N_0}{2}\lambda_k^{-1}(\theta)\right\}d\theta = NE_s \quad (2.26)$$

and $\lambda_k(\theta)$, $k=1,2,\ldots,N$, are the $N$ eigenvalues of matrix $\mathbb{H}^+(e^{j\theta})\mathbb{H}(e^{j\theta})$ and $E_S$ denotes the mean symbol power. The information rate is

$$I_{iid}(E_s,N) = \frac{1}{4\pi N}\sum_{k=1}^{N}\int_{-\pi}^{\pi}\log_2\left[1+\frac{2E_s\lambda_k(\theta)}{N_0}\right]d\theta, \quad (2.27)$$

where $\lambda_K(\theta)$ are the same as in the capacity (2.25).

When the number of the receive antennas is not equal to the number of the transmit antennas, i.e., $M \neq N$, we may use the singular value decompositions of the MIMO system $\mathbb{H}(z)$ (2.3) and then convert it to a subsystem with $\min\{M,N\}$ inputs and $\min\{M,N\}$ outputs. By doing so, we have the following lower bound for the capacity and the information rates.

$$C(E_s,N,M) \geq \frac{1}{4\pi\min\{N,M\}}\sum_{k=1}^{\min\{N,M\}}\int_{-\pi}^{\pi}\max\left\{0,\log_2\left[\frac{2\lambda_k(\theta)K_S}{N_0}\right]\right\}d\theta, \quad (2.28)$$

where

$$\frac{1}{2\pi}\sum_{k=1}^{\min\{N,M\}}\int_{-\pi}^{\pi}\max\left\{0,K_s-\frac{N_0}{2}\lambda_k^{-1}(\theta)\right\}d\theta = \min\{N,M\}E_s \quad (2.29)$$

And $\lambda_k(\theta)$, $k=1,2,\ldots,\min\{N,M\}$ are the $\min\{N,M\}$ squared singular values of matrix $\mathbb{H}(e^{j\theta})$. The information rate is

$$I_{iid}(E_s,N,M) \geq \frac{1}{4\pi\min\{N,M\}}\sum_{k=1}^{\min\{N,M\}}\int_{-\pi}^{\pi}\log_2\left[1+\frac{2E_s\lambda_k(\theta)}{N_0}\right]d\theta' \quad (2.30)$$

where $\lambda_k(\theta)$, $k=1,2,\ldots,\min\{N,M\}$ are the same as in the capacity (2.28).

## Capacity ad Information Rates of the Space-Time MC Coded MIMO Systems

We next want to study the capacities and the information rates of the space-time MC coded MIMO systems (2.15). It is not hard to see that the MC coded MIMO system (2.15) is a $K$ input and $MP$ output system. Based on the decodeability condition (2.16), there are two cases for the parameters $K$, $NP$, $MP$: **Case (i)** when $K = MP \leq NP$ and **Case (ii)** when $K < MP$ and $K \leq NP$.

We first consider Case (i). Similar to (2.25)(2.27), **the capacity and the information rates of the space-time MC coded system in (2.15) when $K = MP \leq NP$ are**

$$C(E_s,K,NP,MP) = \frac{1}{4\pi NP}\sum_{k=1}^{K}\int_{-\pi}^{\pi}\max\left\{0,\log_2\left[\frac{2\lambda_k(\theta)K_S}{N_0}\right]\right\}d\theta, \tag{2.31}$$

where

$$\frac{1}{2\pi}\sum_{k=1}^{K}\int_{-\pi}^{\pi}\max\left\{0,K_s-\frac{N_0}{2}\lambda_k^{-1}(\theta)\right\}d\theta = KE_s \tag{2.32}$$

and $\lambda_k(\theta)$, $k=1,2,\ldots,K$, are the $K$ squared singular values of matrix $\mathcal{C}(e^{j\theta})$ in (2.13). The information rate is

$$I_{iid}(E_s,K,NP,MP) = \frac{1}{4\pi NP}\sum_{k=1}^{K}\int_{-\pi}^{\pi}\log_2\left[1+\frac{2E_s\lambda_k(\theta)}{N_0}\right]d\theta, \tag{2.33}$$

where $\lambda_k(\theta)$, $k=1,2,\ldots,K$ are the same as in the capacity (2.31). Notice that the data rate loss $K/(NP)$ in the above MC encoding has been taken into the account in the above formulas and otherwise the factor $1/(NP)$ in (2.31) and (2.33) would be $1/K$.

For Case (ii), similar to (2.28)(2.30), **the capacity and the information rates of the space-time MC coded system in (2.15) when $K < PM$ and $K \leq NP$ are lower bounded by**

$$C(E_s,K,NP,MP) \geq \frac{1}{4\pi NP}\sum_{k=1}^{K}\int_{-\pi}^{\pi}\max\left\{0,\log_2\left[\frac{2\lambda_k(\theta)K_S}{N_0}\right]\right\}d\theta, \tag{2.34}$$

where

$$\frac{1}{2\pi}\sum_{k=1}^{K}\int_{-\pi}^{\pi}\max\left\{0,K_s-\frac{N_0}{2}\lambda_k^{-1}(\theta)\right\}d\theta = KE_s \tag{2.35}$$

and $\lambda_k(\theta)$, $k=1,2,\ldots,K$ are the $K$ squared singular values of matrix $\mathcal{C}(e^{j\theta})$ in (2.13). The information rate is lower bounded by

$$I_{iid}(E_s,K,NP,MP) \geq \frac{1}{4\pi NP}\sum_{k=1}^{K}\int_{-\pi}^{\pi}\log_2\left[1+\frac{2E_s\lambda_k(\theta)}{N_0}\right]d\theta, \tag{2.36}$$

where $\lambda_k(\theta)$, $k=1,2,\ldots,K$ are the same as in the capacity (2.34).

We next want to show that there exists space-time MC such that the information rates of the MC coded system in (2.15) are larger than the ones of the original system in (2.3) when the channel

SNR is low and the number of transmit antennas is equal to the number of receive antennas, i.e., $M = N$. When $M \neq N$, similar arguments can be used to show the space-time MC existence with the larger information rate lower bound of the MC coded system over the information rate lower bound of the original system.

Before going to the results, we need to introduce two concepts on $N \times N$ polynomial matrix $\mathbb{H}(z)$. An $N \times N$ polynomial matrix $\mathbb{H}(z)$ is called paraunitary if and only if, see (Vaidyanathan (1993)),

$$\mathbb{H}(e^{j\theta})\mathbb{H}(e^{j\theta}) = dI_N, \quad -\pi \leq \theta \leq \pi, \tag{2.37}$$

where $d > 0$ is a constant. The above paraunitariness is a generalization of the unitariness for constant matrices. When $N > 1$, an $N \times N$ polynomial matrix $\mathbb{H}(z)$ is almost surely not paraunitary. When $N = 1$, a polynomial is paraunitary if and only if it is a single delay $dz^{-k_0}$, i.e., no ISI. An $N \times N$ polynomial matrix $\mathbb{H}(z)$ is called *pseudo-paraunitary* if and only if

$$\mathbb{H}(e^{j\theta}) = d(e^{j\theta})\mathbb{U}(e^{j\theta}), \quad -\pi \leq \theta \leq \pi, \tag{2.38}$$

where $d(e^{j\theta})$ is a scalar function of $e^{j\theta}$ and has at least two terms of $e^{jk\theta}$ for different $k$ and $\mathbb{U}(e^{j\theta})$ is an $N \times N$ unitary matrix for any $\theta$. For an $N \times N$ polynomial matrix with $N > 1$, it is almost surely not pseudo-paraunitary. However, for $N = 1$, any polynomial $H(z)$ is pseudo-paraunitary unless it is only a delay, i.e., $H(z) = dz^{-k_0}$, and in this case it is paraunitary.

**Lemma 1:** *If the blocked version $\mathcal{H}(z)$ in (2.14) with block size P of an $N \times N$ polynomial matrix $\mathbb{H}(z)$ is pseudo-paraunitary (or paraunitary), then $\mathbb{H}(z)$ is also pseudo-paraunitary (or parauni tary).*

***Proof.*** When $N = 1$, the blocked version $\mathcal{H}(z)$ can be diagonalized as follows, see Vaidyanathan (1993) or (2.5) in Xia (1997),

$$\mathcal{H}(z^P) = \left[W_P^* \mathcal{U}(z)\right]^+ \Lambda(z) W_P^* \mathcal{U}(z), \tag{2.39}$$

where $W_P = \frac{1}{\sqrt{P}}(w_P^{pq})_{0 \leq p,q \leq P-1}$, $w_P = \exp(-j2\pi / P)$

$\mathbb{U}(z) = diag\left(1, z^{-1}, \cdots, z^{-P+1}\right)$

and

$$\Lambda(z) = diag(\mathbb{H}(z), \mathbb{H}(zw_P), \cdots, \mathbb{H}(zw_P^{P-1})) \tag{2.40}$$

When $N > 1$, $\mathcal{H}(z^P)$ can be permuted both row-wisely and column-wisely such that each $P \times P$ submatrix of the permuted $\mathcal{H}(z^P)$ is pseudo-circulant and corresponds to the case when $N = 1$. Therefore, the diagonalization (2.39) can be used to each $P \times P$ submatrix of the permuted matrix of $\mathcal{H}(z^P)$. Then each diagonalized $P \times P$ submatrix of $\mathcal{H}(z^P)$ is permuted back. Let $P$ denote the permutation matrix. Then, $\mathcal{H}(z^P)$ has the following diagonalization

$$\mathcal{H}(z^P) = P^T diag\left(\left[W_P^* \mathbb{U}(z)\right]^+, \cdots, \left[W_P^* \mathbb{U}(z)\right]^+\right) P^T \Lambda(z) P^T diag\left(W_P^* \mathbb{U}(z), \cdots, W_P^* \mathbb{U}(z)\right) P^T \tag{2.41}$$

where $\Lambda(z)$ has the form (2.40). Since all matrices $W_P$, $\mathbb{U}(z)$, and $P$ are unitary, by the form of $\Lambda(z)$ in (2.40), $\mathcal{H}(z)$ is pseudo-paraunitary (or unitary) implies that $\mathbb{H}(z)$ is also pseudo-paraunitary (or unitary).

We are now ready to state and prove the following results.

**Theorem 3.** *Let $\mathbb{H}(z)$ be an $N \times N$ transfer polynomial matrix of an N transmit antenna and N receive antenna system with AWGN. If $\mathbb{H}(z)$ is not pseudo-paraunitary, then, for any $1 \leq K < NP$, there exists a rate K/(NP) space-time MC* *such that the information rates in (2.36) of the MC coded system (2.15) are larger than the information rates in (2.27) of the original system (2.3), when the channel SNR is sufficiently low.*

***Proof.*** For each $\theta$ with $-\pi \le \theta < \pi$ let $\mathcal{H}(e^{j\theta})$ in (2.14) have the following singular value decomposition

$$\mathcal{H}(e^{j\theta}) = \mathcal{U}(e^{j\theta})\Lambda(e^{j\theta})\mathcal{V}(e^{j\theta}), \tag{2.42}$$

where $U(e^{j\theta})$ and $\mathcal{V}(e^{j\theta})$ are both unitary and $\Lambda(e^{j\theta}) = diag(\overline{\lambda}_1(\theta), \cdots, \overline{\lambda}_{NP}(\theta))$ with

$$\overline{\lambda}_1(\theta) \ge \cdots \ge \overline{\lambda}_{NP}(\theta) \ge 0. \tag{2.43}$$

By the Parseval's equality, the channel normalization (2.4) with $M = N$ and the form (2.14) of $\mathcal{H}(z)$,

$$\frac{1}{2\pi}\int_{-\pi}^{\pi}\sum_{k=1}^{NP}\overline{\lambda}_k^2(\theta)d\theta = \frac{1}{2\pi}\int_{-\pi}^{\pi} trace(\mathcal{H}^+(e^{j\theta})\mathcal{H}(e^{j\theta}))d\theta = P\sum_{m=1}^{M}\sum_{n=1}^{N}\sum_{k}|h_{m,n}(k)|^2 = NP. \tag{2.44}$$

Let $\overline{\mathcal{G}}(e^{j\theta})$ be the following $NP \times K$ matrix

$$\overline{\mathcal{G}}(e^{j\theta}) = \mathcal{V}^+(e^{j\theta})\begin{bmatrix} diag(\sqrt{\frac{NP}{K}}, \cdots, \sqrt{\frac{NP}{K}}) \\ \mathbf{0}_{(NP-K)\times K} \end{bmatrix}_{NP\times K}. \tag{2.45}$$

Then, the $K$ squared singular values of matrix $\mathcal{H}(e^{j\theta})\overline{\mathcal{G}}(e^{j\theta})$ are

$$\overline{\lambda}_k^{(1)}(\theta) = \frac{NP}{K}\overline{\lambda}_k^2(\theta), \quad 1 \le k \le K. \tag{2.46}$$

We claim that, when $\mathbb{H}(z)$ is not pseudo-paraunitary, we have

$$\frac{1}{2\pi}\int_{-\pi}^{\pi}\sum_{k=1}^{K}\overline{\lambda}_k^2(\theta)d\theta > K. \tag{2.47}$$

In fact, if

$$\frac{1}{2\pi}\int_{-\pi}^{\pi}\sum_{k=1}^{K}\overline{\lambda}_k^2(\theta)d\theta \le K, \tag{2.48}$$

then, by (2.44) and (2.43) we have

$$\overline{\lambda}_1(\theta) = \cdots = \overline{\lambda}_{NP}(\theta) = \lambda(\theta) \text{ almost surely.} \tag{2.49}$$

By (2.42), we know that $\mathcal{H}(z)$ is pseudo-paraunitary. Therefore, by Lemma 1 $\mathbb{H}(z)$ is also pseudo paraunitary, which contradicts with the condition in Theorem 3.

Let $\lambda_n(\theta)$, $1 \le n \le N$, be the $N$ squared singular values of matrix $\mathbb{H}(e^{j\theta})$. By the normalization (2.4) of $\mathbb{H}(z)$ and the Parseval's equality, similar to (2.44) we have

$$\frac{1}{2\pi}\int_{-\pi}^{\pi}\sum_{k=1}^{N}\lambda_n(\theta)d\theta = N. \tag{2.50}$$

Let

$$\overline{\kappa} = \frac{\int_{-\pi}^{\pi} \sum_{k=1}^{K} \overline{\lambda}_k^{(1)}(\theta) d\theta}{P \int_{-\pi}^{\pi} \sum_{n=1}^{N} \lambda_n(\theta) d\theta}, \tag{2.51}$$

which only depends on the channel $\mathbb{H}(z)$. Therefore, by (2.46), (2.47), and (2.50), we have

$$\overline{\kappa} > 1. \tag{2.52}$$

If $\mathbb{V}(e^{j\theta})$ in the decomposition (2.42) is a polynomial matrix, i.e., each component in $\mathbb{V}(e^{j\theta})$ has only finite terms of $z^k$, then we claim that the rate $K/(NP)$ MC $\overline{\mathbb{G}}(e^{j\theta})$ in (2.45) is the MC $\overline{\mathbb{G}}(e^{j\theta})$ we wanted to construct for the proof. We next want to prove this claim. To do so, we consider the ratio $R_I(\gamma)$ of the information rate (2.36) for the MC coded system over the information rate (2.27) for the original system, where

$$\gamma = \frac{2E_s}{N_0}$$

By the lower bound in (2.36), the ratio $R_I(\gamma)$ is lower bounded by

$$R_I(\gamma) \geq \frac{\int_{-\pi}^{\pi} \sum_{k=1}^{K} \log_2(1+\gamma \overline{\lambda}_k^{(1)}(\theta)) d\theta}{P \int_{-\pi}^{\pi} \sum_{n=1}^{N} \log_2(1+\gamma \lambda_n(\theta)) d\theta}. \tag{2.53}$$

Thus,

$$\lim_{\gamma \to 0} R_I(\gamma) \geq \lim_{\gamma \to 0} \frac{\int_{-\pi}^{\pi} \sum_{k=1}^{K} \log_2(1+\gamma \overline{\lambda}_k^{(1)}(\theta)) d\theta}{P \int_{-\pi}^{\pi} \sum_{n=1}^{N} \log_2(1+\gamma \lambda_n(\theta)) d\theta}$$

$$\mathbb{H}(z^P) = \left[W_P^* \Psi(z)\right]^+ \Lambda(z) W_P^* \Psi(z)$$

$$\overset{step1}{=} \frac{\int_{-\pi}^{\pi} \sum_{k=1}^{K} \overline{\lambda}_k^{(1)}(\theta) d\theta}{P \int_{-\pi}^{\pi} \sum_{n=1}^{N} \lambda_n(\theta) d\theta} = \overline{\kappa} > 1 \tag{2.54}$$

where step 1 is from the L'H"opital's rule.

When $\mathbb{V}(z)$ has infinite terms of $z^{-k}$, it is truncated into a polynomial matrix $\mathbb{V}_1(z)$ in the way that it is close enough to $\mathbb{V}(z)$ and the corresponding MC $\mathbb{G}(z)$ defined similar to $\overline{\mathbb{G}}(z)$ in (2.45) by replacing $\mathbb{V}(z)$ with $\mathbb{V}_1(z)$ is also close enough to $\overline{\mathbb{G}}(z)$. Then, the corresponding squared singular values $\lambda_k^{(1)}(\theta)$ of $\mathbb{H}(z)\mathbb{G}(z)$ are also close to $\lambda_k^{(1)}(\theta)$ in (2.46) such that

$$\kappa = \frac{\int_{-\pi}^{\pi}\sum_{k=1}^{K}\lambda_k^{(1)}(\theta)d\theta}{P\int_{-\pi}^{\pi}\sum_{n=1}^{N}\lambda_n(\theta)d\theta} \approx \overline{\kappa}, \qquad and \quad \kappa > 1. \tag{2.55}$$

With the MC $\mathbb{G}(z)$ defined above, the corresponding information rate ratio $R_I(\gamma)$ is lower bounded by $\kappa > 1$ as the channel SNR $\gamma$ goes to 0, which is similar to (2.53)(2.54).

The above arguments prove that, when the channel SNR $\gamma$ is sufficiently small, the information rates of the MC coded system are larger than the ones of the original system. This proves Theorem 3. 2

From Theorem 3, it is known that, for almost all $N$ transmit antenna and $N$ receive antenna systems with $N > 1$ and any rate $K/(NP)$ with $K < NP$, there exist rate $K/(NP)$ space-time MC such that the MC coded systems have larger information rates than the ones of the original systems, when the channel SNR are small. Since when $N = 1$, i.e., a single antenna system, the ISI channel $H(z)$ with length $\Gamma > 1$ is always pseudo-paraunitary, Theorem 3 does not apply to the case when $N = 1$. In order to include this case, we have the following result.

**Theorem 4.** *Let*

$$\mathbb{H}(z) = \sum_{k=0}^{\Gamma-1} H(k)z^{-k}$$

*be an $N \times N$ transfer polynomial matrix of an N transmit antenna and N receive antenna system with AWGN. If $\mathbb{H}(z)$ is not paraunitary, then, for any $1 \le K < NP$ with $P > \Gamma$, there exists a rate K/(NP) space-time MC such that the information rates in (2.36) of the MC coded system (2.15) are larger than the information rates in (2.27) of the original system (2.3), when the channel SNR is sufficiently low.*

***Proof.*** By Theorem 3, we only need to prove Theorem 4 for the case when $\mathbb{H}(z)$ is pseudo paraunitary. In this case,

$$\mathbb{H}(e^{j\theta}) = d(e^{j\theta})\mathbb{U}(e^{j\theta})$$

where $d(e^{j\theta})$ is a scalar function of $e^{j\theta}$ and has at least two different terms of $e^{jk\theta}$ and $\mathbb{U}(e^{j\theta})$ is unitary. Since a unitary matrix multiplication does not change the information rates, by absorbing $\mathbb{U}(e^{j\theta})$ into the signal, the system $\mathbb{H}(z)$ is equivalently converted to $d(e^{j\theta})I_N$, which is equivalent to $N$ single antenna systems with the transfer functions $d(e^{j\theta})$. Since $\mathbb{H}(e^{j\theta})$ has length $\Gamma$, so does $d(e^{j\theta})$ due to the fact that

$$\mathbb{H}^+(e^{j\theta})\mathbb{H}(e^{j\theta}) = |d(e^{j\theta})|^2 I_N$$

Therefore, to prove Theorem 4, we only need to consider the case of single antenna systems, i.e., $N = 1$, with finite ISI. In this case, everything else but (2.47) in the proof of Theorem 3 directly applies here. This implies that we only need to prove (2.47) under the conditions that the single antenna system transfer function $H(z)$ of length $\Gamma$ with $\Gamma > 1$, and $P \ge \Gamma$. If (2.47) is false, i.e., (2.48) holds, then, similarly we have

$$\overline{\lambda}_1(\theta) = \cdots = \overline{\lambda}_P(\theta) = \lambda(\theta) \text{ almost surely.} \tag{2.56}$$

Going back to (2.42),

$$\mathcal{H}(e^{j\theta}) = \lambda(\theta)\mathbb{U}(e^{j\theta})\mathbb{V}(e^{j\theta}). \tag{2.57}$$

Since $\mathcal{H}(e^{j\theta})$ is pseudo-circulant, it has the diagonalization (2.39). By combining (2.57) and (2.39), we have

$$diag(H(e^{j\theta/P}), H(e^{j\theta/P}w_P), \cdots, H(e^{j\theta/P}w_P^{P-1})) = \lambda(\theta)\mathbb{W}(\theta),$$

where $\mathbb{W}(e^{j\theta})$ is unitary. Therefore,

$$|H(e^{j\theta/P}w_P)|^2 = \lambda^2(\theta), \qquad for \quad p = 0,1,\ldots,P-1. \tag{2.58}$$

By expanding (2.58) and setting $\theta = 0$, we have

$$\left|\sum_{k=0}^{\Gamma-1} h(k)\exp(jkp/P)\right|^2 = \lambda^2(0), \quad for \quad p = 0,1,\ldots,P-1. \tag{2.59}$$

Since $P \geq \Gamma$, (2.59) is possible only if the sequence $h(0), h(1), \ldots, h(\Gamma-1)$ has one nonzero element, which contradicts with the condition that $H(z)$ has at least two terms. This proves (2.47) and therefore Theorem 4 is proved. 2

From the above proof, the following corollary is immediate.

**Corollary 1.** *For a single antenna system with transfer function $H(z)$ of length $\Gamma > 1$ and AWGN, there always exist rate K/P MC with $1 \leq K < P \geq \Gamma$ such that the MC coded systems have larger information rates than the original system does, when the channel SNR is small.*

The above result is a generalization of the one obtained in (Fan & Xia (1999)), where rate $1/P$ with $P \geq 2\Gamma - 1$ MC were constructed. Examples shall be presented later to illustrate the above results. One thing should be emphasized here is that, in all the above proofs of the information rate increase, the condition $K < NP$, i.e., the data rate increase of the MC encoding, ensures (2.47) and therefore $\kappa > 1$ in (2.55) or $\overline{\kappa} > 1$ in (2.52).

## SPACE-TIME MC CODED ZERO-FORCING DECISION FEEDBACK EQUALIZER

There are several decoding schemes for the space-time MC coded multiple transmit and receive antenna systems discussed in the preceding sections, such as the joint maximum likelihood sequence estimation (MLSE) decoding, the minimum mean squared error (MMSE) decoding, and the joint zero-forcing decision feedback equalizer (ZF-DFE) decoding. Different decoding algorithms also give different criterions for the optimal MC design at the transmitter. For single antenna systems, the joint MLSE decoding was studied in (Xie & Xia (1998)), where a general distance spectra calculation algorithm was obtained for a general MC. Note that there is no general distance spectra calculation algorithm for a general trellis coded modulation (TCM). The distance spectra calculation algorithm for a general MC allows us to be able to search good MC to match an arbitrarily given ISI channel. The joint MLSE decoding, however, has the highest computational complexity, which becomes even more prohibitive for multiple transmit and receive antenna systems than single antenna systems do. The MMSE is computationally simple but the performance sometimes may not be acceptable, in particular for spectral-null channels. The ZF-DFE has not only good performance but also low computational complexity. In the following, we propose a space-time MC coded ZF-DFE, which is similar to the one for single antenna systems studied in (Xia (1999)). The difference here is that the channel $\mathbb{H}(z)$ is no longer a polynomial but a polynomial matrix that may be non-squared. For the completeness, we describe it in details as follows.

### MC Coded ZF-DFE and Performance Analysis

Consider an $N$ transmit antenna and $M$ receive antenna system with $M \times N$ transfer polynomial matrix $\mathbb{H}(z)$. For a space-time rate $K/(NP)$ MC $\mathbb{G}(z)$, $K < NP$ is always assumed in what follows. The block diagram for the space-time MC coded ZF-DFE is shown in Figure 4, where $I_K$ is the $K$ by $K$ identity matrix, and the $K$ by 1 vector decision takes the best $K$ by 1 vector of all the possible $K$ by 1 information symbol vectors, and $\eta$ is the channel additive white Gaussian noise with zero mean and variance $\sigma_\eta^2 = N_0/2$, and $\mathcal{H}(z)$ is the blocked version of $\mathbb{H}(z)$ and has size $MP \times NP$.

The matrix multiplier $\mathcal{D}(z)$ at the receiver in Figure 4 for the MC coded ZF-DFE converts the nonsquare polynomial matrix $\mathcal{C}(z)$ of the combination into a square polynomial matrix so that the DFE can be implemented as shown in Fig. 4. It is usually the case that the higher of the order of the ISI channel to equalize is, the worse of the DFE performance is. To make the order of the overall system $\mathcal{F}(z)$ after the matrix multiplier as low as possible, where

$$\mathcal{F}(z) \overset{\Delta}{=} \mathcal{D}(z)\mathcal{C}(z) = \mathcal{D}(z)\mathcal{H}(z)\mathbb{G}(z) \tag{3.1}$$

*Figure 4. Space-time MC coded zero-forcing decision feedback equalizer*

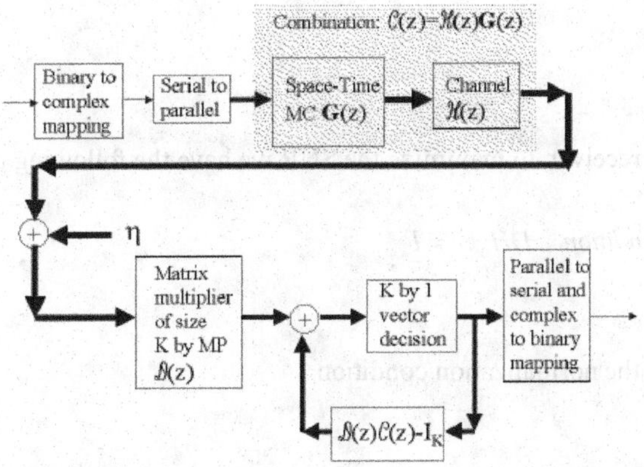


and $\mathcal{H}(z)$ is as (2.14), the matrix multiplier D(z) simply takes a *K* by *PM* constant matrix. It also suggests that the MC $\mathcal{G}(z)$ takes a block code, i.e., $\mathcal{G}(z)$ is an *PM* by *K* constant matrix. We next want to study the MC design rule for the above ZF-DFE. Consider an *PN* by *K* block MC $\mathcal{G}(z)=\mathcal{G}$ and a constant *K* by *PN* matrix multiplier $\mathcal{D}(z)=\mathcal{D}$. The combined MC of the channel and the transmitter MC $\mathcal{G}$ becomes

$$\mathcal{C}(z)=\mathcal{H}_0\mathcal{G}+\mathcal{H}_1\mathcal{G}z^{-1}+\cdots+\mathcal{H}_{\Gamma_1}\mathcal{G}z^{-\Gamma_1}, \tag{3.2}$$

where $\mathcal{H}(z)=\sum_{k=0}^{\Gamma_1}\mathcal{H}_k z^{-k}$ and

$$\mathcal{H}_0=\begin{bmatrix} H(0) & 0 & \cdots & 0 \\ H(1) & H(0) & \cdots & 0 \\ \vdots & \vdots & \vdots & \vdots \\ H(P-1) & H(P-2) & \cdots & H(0) \end{bmatrix}_{MP\times NP}, \tag{3.3}$$

where $H(0)\ldots,H(P-1)$ are $M\times N$ constant coefficient matrices in the channel

$$\mathbb{H}(z)=\sum_{k=0}^{\Gamma-1}H(k)z^{-k}.$$

From the feedback loop in the ZF-DFE in Figure 4, we want to have $\mathcal{D}\mathcal{H}_0\mathcal{G}=I_K$, i.e., the feedback does not depend on the current vector. Therefore, the matrix multiplier $\mathcal{D}(z)=D$ is the right inverse (pseudo inverse), $(\mathcal{H}_0\mathcal{G})^{-1}$, of the $MP\times K$ constant matrix $\mathcal{H}_0\mathcal{G}$. This also implies that the MC coded matrix $\mathcal{H}_0\mathcal{G}$ should have full column rank. In other words, the $MP\times NP$ matrix $\mathcal{H}_0$ should have rank at least *K*. For an arbitrarily given $\mathbb{H}(z)$, the matrix $\mathcal{H}_0$ has full rank almost surely. Without loss of generality, in what follows we always assume that $\mathcal{H}_0$ has full rank. As we shall see later on the MC design, it is only needed that the rank of $\mathcal{H}_0$ is at least *K*.

After the matrix multiplier of $\mathcal{D}(z)=\mathcal{D}=(d_{ij})_{K\times MP}$ at the receiver, the mean power of the multiplied noise $\bar{\eta}$ of $\eta$ becomes

$$\sigma_{\bar{\eta}}^2=\frac{\sum_{i=1}^{K}\sum_{j=1}^{MP}|d_{ij}|^2}{K}\sigma_\eta^2=\frac{\sum_{i=1}^{K}\sum_{j=1}^{MP}|d_{ij}|^2}{2K}N_0. \tag{3.4}$$

By the normalization condition of the MC $\mathcal{G}$, the mean transmitted signal power is still $\sigma_x^2$. Similar to the conventional ZF-DFE for invertible ISI channel, see, for example, (Forney & Eyuboglu (1991)), the signal-to-noise ratio (SNR) after the MC coded ZF-DFE for the invertible $\mathcal{C}(z)$ is

$$SNR = \frac{\sigma_x^2}{\sigma_{\bar{\eta}}^2} = \frac{2K}{\sum_{i=1}^{K}\sum_{j=1}^{MP} |d_{ij}|^2 N_0} \sigma_x^2. \tag{3.5}$$

Based on this SNR analysis at the receiver, to maximize the SNR we have the following **optimal MC design rule**:

$$\min_{\mathcal{G}} \sum_{i=1}^{K}\sum_{j=1}^{MP} |d_{ij}|^2 \quad under \quad the \quad condition \quad \mathcal{D}\mathcal{H}_0\mathcal{G} = I_K, \tag{3.6}$$

where the MC $\mathcal{G} = (g_{ij})$ satisfies the normalization condition

$$\sum_{i=1}^{NP}\sum_{j=1}^{K} |g_{ij}|^2 = NP. \tag{3.7}$$

Let the singular value decomposition of the matrix $\mathcal{H}_0\mathcal{G}$ be

$$\mathcal{U}_l \mathcal{V} \mathcal{U}_r = \mathcal{H}_0 \mathcal{G}, \tag{3.8}$$

where $\mathcal{U}_l$ and $\mathcal{U}_r$ are $MP \times NP$ and $K \times K$ unitary matrices, respectively, and

$$\mathcal{V} = \begin{pmatrix} diag(\lambda_1, \cdots, \lambda_K) \\ 0_{(MP-K)\times K} \end{pmatrix} \tag{3.9}$$

and $\lambda_i \quad for \quad i = 1, 2, \ldots, K$ are the singular values of the matrix $\mathcal{H}_0\mathcal{G}$. Then the matrix multiplier $\mathcal{D}$ is

$$\mathcal{D}(z) = \mathcal{D} = \mathcal{U}_r^+ \mathcal{V}^{-1} \mathcal{U}_l^+, \tag{3.10}$$

where

$$\mathcal{V}^{-1} = (diag(1/\lambda_1, \cdots, 1/\lambda_K), 0_{K\times(MP-K)}). \tag{3.11}$$

Thus, the total energy of the matrix $\mathcal{D}$ is

$$\sum_{i=1}^{K}\sum_{j=1}^{MP} |d_{ij}|^2 = \sum_{i=1}^{K} \frac{1}{\lambda_i^2}. \tag{3.12}$$

Therefore, using the elementary inequality on the right hand side of (3.12) we have

$$\sum_{i=1}^{K}\sum_{j=1}^{MP} |d_{ij}|^2 \geq K \left( \prod_{i=1}^{K} \frac{1}{\lambda_i^2} \right)^{1/K}, \tag{3.13}$$

where the equality (the minimum) is reached if and only if

$$\lambda_1 = \lambda_2 = \cdots = \lambda_K = \lambda. \tag{3.14}$$

The optimality condition (3.14) is the one to design the MC $\mathcal{G}$ that whitens the matrix $\mathcal{H}_0$ generated from the ISI channel. In the next subsection, we propose a method to design such MC $\mathcal{G}$ given an $\mathcal{H}_0$ in (3.3).

We next want to study the performance of the MC coded ZF-DFE in Figure 4, i.e., the error probability. Let us consider the vector decision block in Figure 4. For a general MC $\mathcal{G}$ at the transmitter and the matrix multiplier $\mathcal{D}$ with the form in (3.10), each $K \times 1$ multiplied noise vector $\eta$ for a fixed time may be colored when $K \times 1$. In this case, the vector decision is necessary for the optimal detection. If the MC $\mathcal{G}$ whitens $\mathcal{H}_0$ , i.e., the condition (3.14) holds, then it is not hard to see that each $K \times 1$ multiplied noise vector $\eta$ for a fixed time is white too. Thus, the vector decision in Figure 4 can be reduced to the symbol-by-symbol detection as shown in Figure 5.

Assume that the condition (3.14) for the MC encoding holds, which is always possible to design as we shall see later. In this case,

$$\sum_{i=1}^{K}\sum_{j=1}^{MP}| d_{ij} |^2 = \frac{K}{\lambda^2}$$

Let $P_s(\gamma_s)$ denote the symbol error probability at the symbol SNR $\gamma_s$ for the binary-to-complex symbol mapping used at the transmitter in Figure 4. For convenience, in what follows we only consider the BPSK binary-to-complex symbol mapping. In this case, the symbol error probability is $P_s(\gamma_s) = Q(\sqrt{2\gamma_s})$, where $\gamma_s$ is the SNR before the decision block in Figure 4. Using the SNR (3.5), the corresponding $\gamma_s$ is

$$\gamma_s = \frac{\sigma_x^2}{2\sigma_{\bar{\eta}}^2} = \frac{2K\sigma_x^2}{\sum_{i=1}^{K}\sum_{j=1}^{MP}| d_{ij} |^2 N_0} = \frac{\lambda^2\sigma_x^2}{N_0} = \frac{\lambda^2 K}{NP}\frac{E_b}{N_0}, \tag{3.15}$$

where $E_b$ is the total energy of all the transmit antennas per bit. Then, the bit error rate (BER) for the MC coded ZF-DFE at the $E_b/N_0$ is

$$BER = P_s(\gamma_s) = Q(\sqrt{2\frac{E_b}{N_0}\gamma}), \tag{3.16}$$

where $\gamma$ is the coding gain as follows, which is based on the joint ZF-DFE decoding and compared to the uncoded BPSK in AWGN channel:

$$\gamma = \frac{\lambda^2 K}{NP}, \tag{3.17}$$

where $\lambda$ is defined in (3.14).

## The Optimal MC Design

In this subsection, we present the optimal MC design such that the optimality condition (3.14) is satisfied. Consider the singular value decomposition of $\mathcal{H}_0^+\mathcal{H}_0$ :

$$\mathcal{H}_0^+\mathcal{H}_0 = W^+\mathcal{A}W, \tag{3.18}$$

where $W$ is an $MP \times NP$ unitary matrix and

$$\mathcal{A} = \begin{cases} diag(\xi_1^2,\cdots,\xi_{NP}^2), & when \quad MP \geq NP \\ diag(\xi_1^2,\cdots,\xi_{MP}^2,0,\cdots,0) & when \quad MP < NP \end{cases}, \tag{3.19}$$

where

$$\xi_1 \geq \cdots \geq \xi_{\min\{MP,NP\}} > 0 \tag{3.20}$$

are the min{$NP$, $MP$} singular values of $\mathcal{H}_0$. Using the singular value decomposition (3.8) of $\mathcal{H}_0\mathcal{G}$, we have

$$\mathcal{U}_r\mathcal{G}^+W^+\mathcal{A}W\mathcal{G}\mathcal{U}_r^+ = diag(\lambda_1^2,\cdots,\lambda_K^2). \tag{3.21}$$

Define

$$\overline{\mathcal{G}} = W\mathcal{G}\mathcal{U}_r^+$$

Then

$$\overline{\mathcal{G}}^+\mathcal{A}\overline{\mathcal{G}} = diag(\lambda_1^2,\cdots,\lambda_K^2) \overset{step1}{=} \lambda I_K, \tag{3.22}$$

where step 1 is from the optimal criterion (3.14). Let

$$\lambda = \frac{NP}{\sum_{i=1}^{K}\xi_i^{-2}} \tag{3.23}$$

and

$$\overline{\mathcal{G}} = \begin{pmatrix} diag(\lambda/\xi_1,\cdots,\lambda/\xi_K) \\ 0_{(NP-K)\times K} \end{pmatrix}. \tag{3.24}$$

By (3.23), it is not hard to see that $\overline{\mathcal{G}}$ is normalized. Since both $W$ and $\mathcal{U}_r$ are unitary, by (3.21) the MC $\mathcal{G}$ is also normalized. Clearly, $\mathcal{G}$ satisfies the optimality condition (3.14), which is therefore optimal. Note that only the inverses of the $K$ largest singular values of $\mathcal{H}_0$ are needed, the rank of $\mathcal{H}_0$ only needs to be $K$ or above.

Going back to (3.18)(3.22), the optimal normalized MC $\mathcal{G}$ is

*Figure 5. Space-time MC coded zero-forcing decision feedback equalizer with optimal MC*

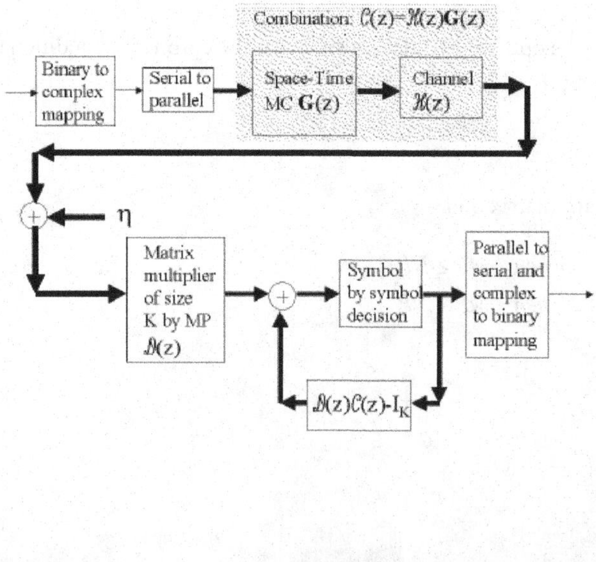

$$\mathcal{G}_{opt} = W^{+}\overline{\mathcal{G}}\mathcal{U}_r, \quad (3.25)$$

where $\mathcal{U}_r$ is an arbitrary $K \times K$ unitary matrix, $W$ is the $NP \times NP$ unitary matrix defined in (3.18), and $\overline{\mathcal{G}}$ is defined in (3.23)(3.24).

**Theorem 5.** *Given an N transmit and M receive antenna channel $\mathbb{H}(z)$, the optimal normalized rate K/(NP) modulated code $\mathcal{G}$ for the space-time MC coded zero-forcing decision feedback equalizer in Figure 2 is given in (3.25).*

Using the optimal $\lambda$ in (3.23) and the optimal coding gain formula in (3.17) for the BPSK signaling, we have the following optimal coding gain using the optimal rate *K/(NP)* MC $\mathcal{G}_{opt}$ in (3.25) for a given channel:

$$\gamma_{opt} = \frac{K}{\sum_{i=1}^{K} \xi_i^{-2}}, \quad (3.26)$$

where $\xi_i, i = 1, 2, \ldots, K$ are the first $K$ largest singular values of $\mathcal{H}_0$. One might want to ask when the above coding gain $\gamma_{opt} > 1$. We have the following simple result.

**Theorem 6.** *When $K = 1$ and $P \geq \Gamma$ in the MC $\mathcal{G}_{opt}$ in (3.25), if $M > N$, i.e., the number of the receive antennas is greater than the number of the transmit antennas, then the corresponding coding gain $\gamma_{opt} > 1$.*

***Proof:*** To prove this theorem, by (3.26) we need to prove that the first singular value $\xi_1$ of $\mathcal{H}_0$ is greater than 1, i.e., $\xi_1 > 1$. Assume that all the singular values of $\mathcal{H}_0$ are less than 1. We want to derive a contradiction. Consider the following matrix sequence for $l = 1, 2, \ldots$

$$A_l \stackrel{\Delta}{=} \left[\mathcal{H}_0^{+}\mathcal{H}_0\right]^{2^l}. \quad (3.27)$$

By (3.18),

$$A_l = W^{+} diag(\xi_1^{2^{l+1}}, \cdots, \xi_{NP}^{2^{l+1}})W. \quad (3.28)$$

By the assumption that $\xi_k \leq 1$ for all $k$, we have that all the magnitudes of the elements in matrix $A_l$ are bounded by $NP$.

(3.29)

We now want to see the direct expansion of $A_l$ in (3.27). Let

$$H = \sum_{k=0}^{P-1} H^{+}(k)H(k). \quad (3.30)$$

It is not hard to see that the left upper corner $N \times N$ submatrix of $A_l$ is always

$$C_l \stackrel{\Delta}{=} (H)^{2^l} + B_l, \quad (3.31)$$

where $B_l$ is a nonnegative definite matrix. Since $P \geq \Gamma$, by the normalization condition in (2.4) on $\mathbb{H}(z)$ and the condition $M > N$, we have

$$\sum_{n=1}^{N} \text{I}_n = trace(H) = M > N, \quad (3.32)$$

where $\lambda_1 \geq \cdots \geq \lambda_N \geq 0$ are the eigenvalues of $H$. Clearly, $\lambda_1 > 1$ by (3.32). Since $B_l$ is nonnegative definite, we have $trace(B_l) \geq 0$. Therefore,

$$trace(C_l) = trace(H^{2^l}) + trace(B_l) \geq \lambda_1^{2^l} \to \infty, \quad when \quad l \to \infty. \tag{3.33}$$

This implies that some elements in matrix $A_l$ go to 1 as $l$ goes to $\infty$, which contradicts with (3.29). This proves Theorem 6. **q.e.d.**

Although in Theorem 6 the condition $M > N$ is required, the result in Theorem 6 still holds when $M = N$ and the channel matrix $\mathbb{H}(z)$ is not paraunitary. Since the proof in this case is notationally tedious, we omit it here. We next want to see some numerical simulation results.

## NUMERICAL RESULTS

In this section, we want to see some simulation results on both MC coded ZF-DFE and the capacity and the information rates of the MC coded systems and the systems themselves. In the following simulations, the number of transmit and receive antennas are both 2, i.e., $M = N = 2$. The block size in the space-time MC in Figure 3 is $P = 3$. The space-time MC code rate is 2/6, i.e., $K = 2$. In this case, the rate for each transmit antenna is 2/3. The BPSK is used for all the simulations and no additional coding is used before the MC encoding.

We consider two different multiple transmit and receive antenna channels with memory and AWGN: Channel A $\mathbb{H}_A(z)$ and Channel B $\mathbb{H}_B(z)$.

Channel A $\mathbb{H}_A(z)$ has length $\Gamma = 3$ and its 3 coefficient matrices are:

$$H(0) = H(1) = H(2) = \begin{bmatrix} 0.4762 & 0.4286 \\ -0.3810 & 0.3333 \end{bmatrix}, \tag{4.1}$$

which is a spectral-null channel because $\mathbb{H}_A(z) = (1 + z^{-1} + z^{-2})H(0)$. The optimal rate 2/6 space-time MC $\mathbb{G}_{opt}$ in (3.25) for this channel is

*Figure 6. Channel A: Capacities C and information rates $I_{iid}$ for the MC coded and uncoded channels*

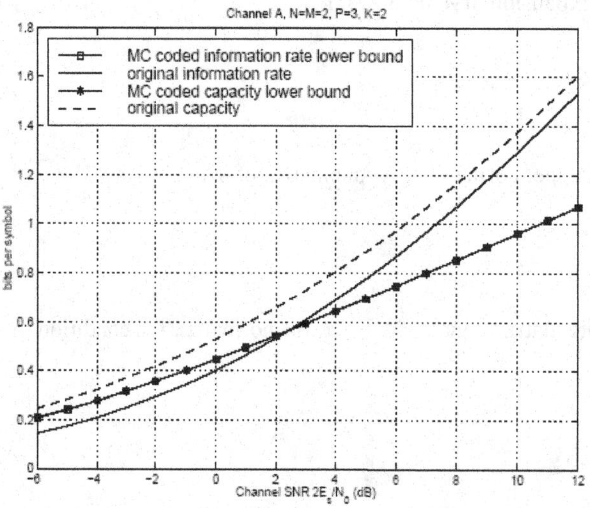

$$\mathcal{G}_{opt} = \begin{bmatrix} 0.9350 & -0.7529 \\ 0.5778 & 1.2182 \\ 0.7498 & -0.6038 \\ 0.4634 & 0.9769 \\ 0.4161 & -0.3351 \\ 0.2572 & 0.5421 \end{bmatrix}. \tag{4.2}$$

The optimal coding gain with this MC is $\gamma_{opt} = 1.96dB$. Figure 6 shows the capacities and information rates $I_{iid}$ of the MC coded/uncoded systems. The solid line shows the original channel information rates $I_{iid}$ (2.27) with $N = 2$ while the solid line marked by 2 shows the lower bound (2.36) of the information rates $I_{iid}$ of the space-time MC coded channel using the optimal MC in (4.2). One can see that the information rates $I_{iid}$ of the MC coded channel are above the ones of the original channel when the channel SNR below about 2.5 *dB*. The dashed line shows the original channel capacity (2.25) while the solid line marked by * shows the capacity lower bound (2.34) of the MC coded channel. Figure 7 shows the BER performance comparison. The solid line shows the theoretical BER vs. $E_b / N_0$ curve of the MC coded ZF-DFE and the solid line marked by $\times$ shows the simulation results. The dashed line shows the BER vs. $E_b / N_0$ curve of the uncoded BPSK over the AWGN channel (i.e., the ideal single antenna channel). One can clearly see the coding gain of the MC. Since the rate for each antenna in this case is 2/3 and all the ISI channels of all transmit and receive antenna pairs are the same, basically $1 + z^{-1} + z^{-2}$, it is possible to compare it with its single antenna system with the ISI channel $0.5774(1 + z^{-1} + z^{-2})$. The rate of the MC for the single antenna system is 2/3 comparing to Channel A. In this case, the optimal coding gain using the ZF-DFE developed in (Xia (1999)) is -4.20 *dB*, i.e., coding loss. Comparing to the coding gain 1.96 *dB*, the MC coded multiple transmit and receive antenna channel significantly outperforms the corresponding single antenna channel.

*Figure 7. Channel A: BER performance for the MC coded channel using the joint ZF-DFE, where the theoretical coding gain is 1.96dB*

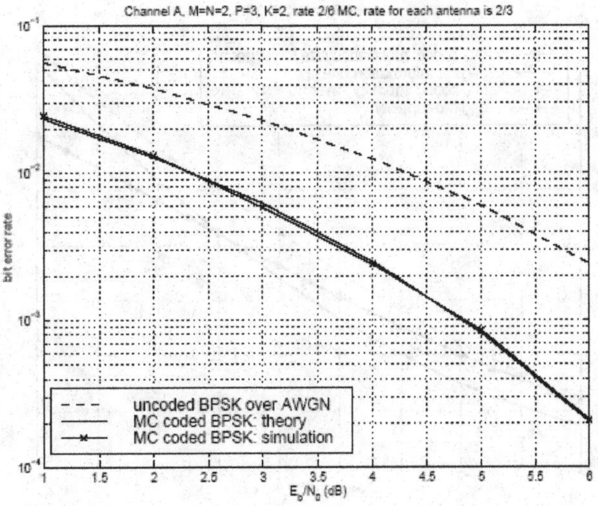

Channel B $\mathbb{H}_B(z)$ has length $\Gamma = 5$ and its 5 coefficient matrices are

$$H(0) = \begin{bmatrix} 0.8111 & 0.5469 \\ 0.7117 & 0.6691 \end{bmatrix} \quad H(1) = \begin{bmatrix} -0.1459 & -0.0136 \\ -0.0880 & -0.0507 \end{bmatrix} \quad H(2) = \begin{bmatrix} -0.0183 & -0.1676 \\ -0.1263 & 0.1129 \end{bmatrix}$$

$$H(3) = \begin{bmatrix} 0.0154 & -0.0152 \\ -0.0457 & 0.0323 \end{bmatrix} \quad H(4) = \begin{bmatrix} -0.0620 & -0.0617 \\ -0.0228 & -0.0970 \end{bmatrix}. \tag{4.3}$$

This channel is randomly chosen. The optimal rate 2=6 space-time MC

$$G_{opt,B} = \begin{bmatrix} 0.6276 & -0.9998 \\ 0.4449 & -0.7658 \\ -1.0787 & -0.0931 \\ -0.8155 & 0.0032 \\ 0.5606 & 0.9498 \\ 0.4488 & 0.7529 \end{bmatrix}. \tag{4.4}$$

The optimal coding gain with this MC is $\gamma_{opt} = 3.24dB$. Similar to Channel A, Figure 8 shows the capacities C and information rates $I_{iid}$ of the MC coded/uncoded systems and Figure 9 shows the BER performance comparison. From Figure 8, one can see that the information rates $I_{iid}$ of the MC coded channel are above the ones of the original channel when the channel SNR below about 6 *dB*.

*Figure 8. Channel B: Capacities C and information rates $I_{iid}$ for the MC coded and uncoded channels*

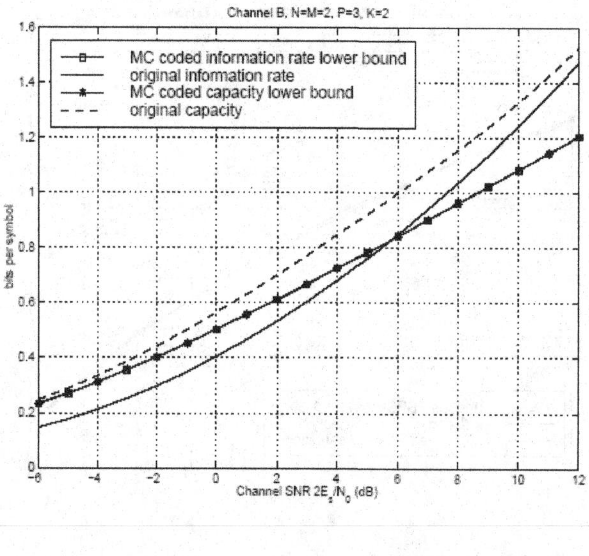

*Figure 9. Channel B: BER performance for the MC coded channel using the joint ZF-DFE, where the theoretical coding gain is 3.24dB*

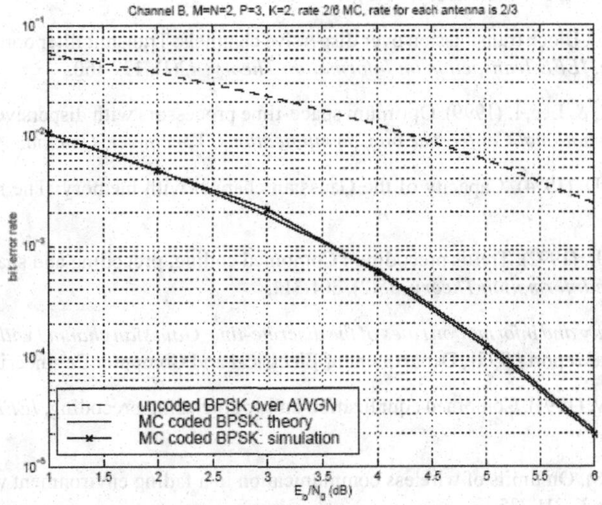


From both Figure 6 and Figure 8, the capacity lower bound curves almost coincide with the information rate lower bound curves of the MC coded channels. It is, however, not always the case for any MC.

On the implementation of the proposed method, one should first estimate the channel characteristics (the channel coefficients) based on some known methods such as pilot tone or pseudo random sequence. Then, one may employ the estimated channel coefficients to calculate the corresponding maximum coding gain that can be possibly obtained. If the maximum coding gain is higher than the expected one paid by the cost of the corresponding complexity, one can employ the proposed precoding method, otherwise, one may select the nonencoding method. In fact, such an encoding method will not always get much high coding gain. In some cases, it only will provide a little coding gain. It is totally depending on the channel characteristics.

## CONCLUSION

In this chapter, we proposed space-time MC for multiple transmit and receive antenna channels with memory, where MC are defined on the complex field. We have shown previously that MC may have coding gain in ISI channels comparing to uncoded AWGN channels. In this chapter, we derived some capacity $C$ and information rate $I_{iid}$ lower bounds for the space-time MC coded channels. We proved that, for almost any $N$ transmit antenna and $N$ receive antenna channel with finite memory and for any rate $0 < r < 1$, there exists a rate r space-time MC such that the information rates $I_{iid}$ of the MC coded channel are larger than the ones of the channel itself, when the channel SNR is low, say for example below 6dB in one of our numerical examples. This basically tells us that, the achievable reliable information rates of the MC coded channels based on the standard random coding techniques are larger than the ones of the channel themselves when the channel SNR is relatively low.

In the second half of this chapter, we proposed a joint decoding method of the space-time MC encoded channel, i.e., the joint ZF-DFE. The performance analysis and the optimal space-time MC design were obtained when an $M \times N$ channel $\mathbb{H}(z)$ is given. Finally, numerical simulations were presented to illustrate the theory, where coding gains are achieved when the joint ZF-DFE is used. The information rates $I_{iid}$ of the MC coded channels using the optimal MC designed based on the joint ZF-DFE were shown larger than the ones of the original channels when the channel SNR is below 2.2 *dB* for one channel and 6 *dB* for the other channel.

## REFERENCES


A. Scaglione, S. Barbarossa, and G. B. Giannakis (1999). Filterbank transceivers optimizing infor mation rate in block transmissions over dispersive channels," *IEEE Trans. Inform. Theory, 45*(3), 1019-1032.

Al-Dhahir, N., & Cioffi, J. M. (1996). Block transmission over dispersive channels: Transmit filter optimization and Realization, and MMSE-DFE receiver performance, *IEEE Transaction on Information Theory, 42*(1), 137-160.

Ariyavisitakul, S. L., & Winters, J., & Lee, I. (1999). Optimum space-time processors with dispersive interference-unified analysis and required filter span, *IEEE International Conference on Communications, Vancouver, Canada.*

Brandenburg, L. H. & Wyner, A. D. (1974). Capacity of the Gaussian channel with memory: The multivariate case, *Bell System Technical Journal, 53*(5), 745-778.

Eyuboglu, M. V., & Forney, G. D. Jr. (1992). Trellis precoding : Combined coding, precoding and shaping for intersymbol interference channel, *IEEE Transaction on Information Theory, 38*(2), 301-314.

Fan, P., & Xia, X.-G. (1999). *Capacity and information rates of the discrete-time Gaussian channel with intersymbol interference and modulated code encoding*, technical report #9951, Department of Electrical and Computer Engineering, University of Delaware.

Forney, G. D. Jr., & Eyuboglu, M. V. (1991). Combined equalization and coding using precoding, *IEEE Communications Magazine*, 25-34.

Foschini, G. J., & Gans, M. J. (1998). On limits of wireless communication in a fading environment when using multiple antennas, *Wireless Personal Communication, 6,* 311-335.

Giannakis, G. B. (1997). Filterbanks for blind channel identification and equalization, *IEEE Signal Processing Letters, 4.*

Hirt, W. & Massey, J. L. (1998). Capacity of the discrete-time Gaussian channel with intersymbol interference, *IEEE Transaction on Information Theory, 34*(3), 380-388.

Kambourov, L. (2006). *MIMO Aspects in 802.16e WiMAX OFDMA*, WiMAX Torial, Vienna.

Karabed, R., & Siegel, P. H. (1991). Matched spectral-null codes for partial response channels, *IEEE Transaction on Information Theory, 37,* 818-855.

Kasturia, S., & Aslanis, J. T. & Cioffi, J. M. (1990). Vector coding for partial response channels, *IEEE Transaction on Information Theory, 36,* 741-762.

Liu, H., & Xia, X.-G. (1997). Precoding for undersampled antenna array receiver systems, *Proceedings of the 28th Annual Asilomar Conference on Signals, Systems, and Computers*, Pacific Grove, California, (pp.1043-1047).

Liu, Y. & Zhang, H. L. (2007). Precoding for multiuser MIMO-OFDMA downlink with limited feedback, *Journal of Xi'dian University, 1.*

Miyakawa, H. & Harashima, H. (1969). A method of code conversion for a digital communication channel with intersymbol interference, *Transactions Of the Institute of Electronics and Communication Engineers of Japan, 52-A*, 272-273.

Shamai, S., & Laroia, R. (1996). The intersymbol interference channel: Lower bounds on capacity and channel precoding loss, *IEEE Transaction on Information Theory, 42*(5), 1388-1404.

Shamai, S., & Ozarow, L. H. & Wyner, A. D. (1991). Information rates for a discrete-time Gaussian channel with intersymbol interference and stationary inputs, *IEEE Transaction on Information Theory, 37*(6), 1527-1539.

Soljanin, E., & Georghiades, C. N. (1995). Coding for two-head recording systems, *IEEE Transaction on Information Theory, 41,* 747-755.

Tarokh, V., & Naguib, A., & Seshadri, N., & Calderbank, A. R. (1999). Combined array processing and space-time coding, *IEEE Transaction on Information Theory, 45,* 1121-1128.

Tarokh, V., & Seshadri, N., & Calderbank, A. R. (1998). Space-time codes for high data rate wireless communication: Performance analysis and code construction, *IEEE Transaction on Information Theory, 44,* 744-765.

Telatar, E. (1995). Capacity of multiantenna Gaussian channels," *Internal AT&T Bell Labs Technical Memorandum.*

Tomlinson, M. (1971). New automatic equalizer employing modulo arithmetic, *Electronics Letters, 7,* 138-139.

Vaidyanathan, P. P. (1993). *Multirate Systems and Filter Banks*, Prentice Hall, Englewood Cliffs, New Jersey.

Winters, J., & Salz J., & Gitlin, R. D. (1994). The impact of antenna diversity on the capacity of wireless communication systems, *IEEE Transaction on Communication, 42,* 1740-1751.

Wittneben, A. (1993). Base station modulation diversity for signal simulcast, *Vehicular Technology Conference Proceedings* (pp. 505-511).

Xia, X.-G. (1997). New precoding for intersymbol interference cancellation using nonmaximally decimated multirate filterbanks with ideal FIR equalizers, *IEEE Transaction on Signal Processing, 45*, 2431-2441.

Xia, X.-G. (1998). Coding gain for modulated codes combined with ISI channels, preprint.

Xia, X.-G. (1999). *A new coded zero-forcing decision feedback equalizer using modulated codes*, Proceedings of Conference on Information Sciences and Systems, The Johns Hopkins University, Baltimore.

Xia, X.-G. (1999). *Modulated-coded zero-forcing decision feedback equalizer: performance analysis and optimal modulated code design*, technical report #9852, Department of Electronical and Computer Engineering, University of Delaware, May 1998, also in Proc. MILCOM'99, Atlantic City, New Jersey, Oct. 31-Nov. 3, 1999.

Xia, X.-G. (2001). *Modulated Coding for Intersymbol Interference Channels*, Marcel Dekker, New York.

Xia, X.-G., & Fan, P., & Xie, Q. (1999). A new coding scheme for ISI channels: Modulated codes, *Procedings of IEEE International Conference on Communications*, Vancouver, Canada.

Xia, X.-G., & Suter, B. W. (1996). Multirate filterbanks with block sampling, *IEEE Transaction on Signal Processing, 44*, 484-496.

Xia, X.-G., & Wang, G., & Fan, P. (2000). *Space-time modulated codes for memory channels: Capacity and information rates, zero-forcing decision feedback equalizer,* the Proceedings of IEEE Sensor Array and Multichannel Signal Processing Workshop (SAM), Cambridge, MA, USA, March 16-17, (pp. 183-187).

Xie, Q., & Xia, X.-G. (1998). *Modulated codes with ISI channels*, technical report #9831, Department of Electrical and Computer Engineering, University of Delaware.

Zhang, J., & Zhang, H., & Waes, N. V., & Reid, A. V., & Stolpman. (2004). *Closed-loop MIMO precoding with limited feedback*, IEEE 802.16 broadband wireless access working group.

# Chapter VIII
# Blind Channel Estimation in Space-Time Block Coded Systems

**Javier Vía**
*University of Cantabria, Spain*

**Ignacio Santamaría**
*University of Cantabria, Spain*

**Jesús Ibáñez**
*University of Cantabria, Spain*

## ABSTRACT

*This chapter analyzes the problem of blind channel estimation under Space-Time Block Coded transmissions. In particular, a new blind channel estimation technique for a general class of space-time block codes is proposed. The method is solely based on the second-order statistics of the observations, and its computational complexity reduces to the extraction of the main eigenvector of a generalized eigenvalue problem. Additionally, the identifiability conditions associated to the blind channel estimation problem are analyzed, which is exploited to propose a new transmission technique based on the idea of code diversity or combination of different codes. This technique resolves the ambiguities in most of the practical cases, and it can be reduced to a non-redundant precoding consisting in a single set of rotations or permutations of the transmit antennas. Finally, the performance of the proposed techniques is illustrated by means of several simulation examples.*

## INTRODUCTION

In the last ten years, since the well known work of Alamouti (1998), and the later generalization by Tarokh et al. (1999), space-time block coding (STBC) has emerged as a promising technique to exploit the spatial diversity in multiple-input multiple-output (MIMO) communication systems. A common assumption for most of the STBCs is that perfect channel state information (CSI) is available at the receiver, which has motivated an increasing interest in blind techniques (Ammar & Ding, 2006, 2007, Larsson et al., 2003; Ma et al., 2006; Shahbazpanahi et al., 2006; Stoica & Ganesan, 2003; Swindlehurst & Leus, 2002). The main advantage of blind approaches resides in their ability to avoid the penalty in bandwidth efficiency or signal to noise ratio (SNR) associated, respectively, to training based techniques (Hassibi & Hochwald, 2003; Naguib et al., 1998; Pohl et al., 2005), or differential schemes (Ganesan & Stoica, 2002; Hochwald & Sweldens, 2000; Hughes, 2000; Jafarkhani & Tarokh, 2001; Tarokh & Jafarkhani, 2000; Zhu & Jafarkhani, 2005). On the other hand, these advantages come at the cost of an increase in both computational complexity and latency, which can be seen as a direct consequence of the common assumption about the coherence time of the MIMO channel.

Blind channel estimation or blind decoding techniques can be divided into two groups depending on whether they exploit the higher-order statistics (HOS) or the second-order statistics (SOS) of the signals. The main advantage of SOS-based approaches consists in their reduced computational complexity and independency of the specific signal constellation. Unfortunately, most of the blind techniques have been proposed for the particular case of orthogonal STBCs (OSTBCs) (Ammar & Ding, 2006; Larsson et al., 2003; Ma et al., 2006; Shahbazpanahi et al., 2006; Stoica & Ganesan, 2003), and the number of methods for more general settings is rather scarce (Shahbazpanahi et al., 2006; Swindlehurst, 2002; Swindlehurst & Leus, 2002). Furthermore, it can be easily proven that some of these techniques are affected by additional indeterminacies to those associated to the blind channel estimation problem.

In this chapter, the blind channel estimation problem is formulated for a general class of STBCs, and a new SOS-based technique is proposed. The method reduces to the extraction of the main eigenvector of a generalized eigenvalue problem (GEV), it does not introduce additional indeterminacies to those of the blind channel estimation problem, and it can be easily extended to multiuser settings. Additionally, we provide an identifiability analysis for the general STBC case, where some intuitive necessary conditions are obtained, and in the particular OSTBC case, we present several sufficient conditions for blind channel identifiability, which shed some light into previous numerical results obtained by other authors. Finally, we propose several techniques for the solution of the indeterminacies. On one hand, in the OSTBC case the ambiguities can be easily avoided by exploiting the HOS, the correlation properties of the sources, or by slightly reducing the transmission rate. On the other hand, we propose a new technique for the general STBC case. The proposed method is based on the general idea of code diversity, which consists in combining different STBCs. However, it can be reduced to a non-redundant precoding consisting in a single rotation or permutation of the transmit antennas, which comes at virtually no computational expense at the transmitter. Unlike previous approaches, the code diversity technique is able to avoid the ambiguities in most of the cases without any penalty in terms of transmission rate nor capacity.

The structure of this chapter is as follows: The STBC data model and some code examples are introduced in Section 2. A brief review of some previously proposed blind channel estimation techniques is presented in Section 3. In Section 4 the new blind channel estimation criterion is presented, and in Section 5 the identifiability conditions are analyzed. In Section 6 we propose several techniques for the solution of the ambiguities. Finally, the performance of the proposed techniques is illustrated in Section 7 by means of several simulation examples, and the main conclusions are summarized in Section 8.

## SPACE-TIME BLOCK CODING DATA MODEL

### Notation

Throughout this chapter we will use bold-faced upper case letters to denote matrices, e.g., $\mathbf{X}$ with elements $x_{i,j}$; bold-faced lower case letters for column vector, e.g., $\mathbf{x}$, and light-faced lower case letters for scalar quantities. The superscripts $(\cdot)^T$, $(\cdot)^H$ and $(\cdot)^*$ denote transpose, Hermitian and complex conjugate, respectively. The real and imaginary parts will be denoted as $\Re(\cdot)$ and $\Im(\cdot)$, and superscript $\hat{(\cdot)}$ will denote estimated matrices, vectors or scalars. The trace, range (or column space) and Frobenius norm of matrix $\mathbf{A}$ will be denoted as Tr($\mathbf{A}$), range($\mathbf{A}$) and $\|\mathbf{A}\|$, respectively. Finally, the identity and zero matrices of the required dimensions will be denoted as $\mathbf{I}$ and $\mathbf{0}$, $E[\cdot]$ will denote the expectation operator, and $\lceil q \rceil$ will denote the smallest integer greater or equal than $q$.

We will consider a flat fading MIMO system with $n_T$ transmit and $n_R$ receive antennas. The $n_T \times n_R$ complex channel matrix is

$$\mathbf{H} = \left[\mathbf{h}_1 \cdots \mathbf{h}_{n_R}\right] = \begin{bmatrix} h_{1,1} & \cdots & h_{1,n_R} \\ \vdots & \ddots & \vdots \\ h_{n_T,1} & \cdots & h_{n_T,n_R} \end{bmatrix},$$

where $h_{i,j}$ denotes the channel response between the $i$-th transmit and the $j$-th receive antennas, and $\mathbf{h}_j$ contains the channel response associated to the $j$-th receive antenna. The complex noise at the receive antennas is considered both spatially and temporally white with variance $\sigma^2$.

## STBC Data Model

Let us start by considering the single-user case and a linear space-time block code (STBC) transmitting $M$ symbols during $L$ time slots with $n_T$ antennas at the transmitter side. The transmission rate is defined as $R = M/L$, and the number of real symbols $M'$ transmitted in each block is

$$M' = \begin{cases} M & \text{for real codes,} \\ 2M & \text{for complex codes.} \end{cases}$$

For a STBC, the $n$-th block of data can be expressed as

$$\mathbf{S}(\mathbf{s}[n]) = \sum_{k=1}^{M'} \mathbf{C}_k s_k[n],$$

where $\mathbf{s}[n] = [s_1[n], \ldots, s_{M'}[n]]^T$ contains the $M'$ real information symbols transmitted in the $n$-th block, and $\mathbf{C}_k \in \mathbb{C}^{L \times n_T}$, $k = 1, \ldots, M'$, are the code matrices. In the case of real STBCs, the transmitted matrix $\mathbf{S}(\mathbf{s}[n])$ and the code matrices $\mathbf{C}_k$ are real.

The complex signal at the $j$-th receive antenna can be written as

$$\mathbf{y}_j[n] = \mathbf{S}(\mathbf{s}[n])\mathbf{h}_j + \mathbf{n}_j[n] = \sum_{k=1}^{M'} \mathbf{w}_k(\mathbf{h}_j) s_k[n] + \mathbf{n}_j[n],$$

where $\mathbf{n}_j[n]$ is the white complex noise with variance $\sigma^2$, and

$$\mathbf{w}_k(\mathbf{h}_j) = \mathbf{C}_k \mathbf{h}_j, \qquad k = 1, \ldots, M',$$

represent the composite effect of the MIMO channel and the STBC code.

Defining now the real vectors $\tilde{\mathbf{w}}_k(\mathbf{h}_j) = \left[\Re(\mathbf{w}_k(\mathbf{h}_j))^T, \Im(\mathbf{w}_k(\mathbf{h}_j))^T\right]^T$ and the extended code matrices

$$\tilde{\mathbf{C}}_k = \begin{bmatrix} \Re(\mathbf{C}_k) & -\Im(\mathbf{C}_k) \\ \Im(\mathbf{C}_k) & \Re(\mathbf{C}_k) \end{bmatrix},$$

we can write

$$\tilde{\mathbf{w}}_k(\mathbf{h}_j) = \tilde{\mathbf{C}}_k \tilde{\mathbf{h}}_j,$$

with $\tilde{\mathbf{h}}_j = \left[\Re(\mathbf{h}_j)^T, \Im(\mathbf{h}_j)^T\right]^T$. Thus, defining the real vectors $\tilde{\mathbf{y}}_j[n] = [\Re(\mathbf{y}_j[n])^T, \Im(\mathbf{y}_j[n])^T]^T$ and $\tilde{\mathbf{n}}_j[n] = [\Re(\mathbf{n}_j[n])^T, \Im(\mathbf{n}_j[n])^T]^T$, we obtain the real signal model

$$\tilde{\mathbf{y}}_j[n] = \sum_{k=1}^{M'} \tilde{\mathbf{w}}_k(\mathbf{h}_j) s_k[n] + \tilde{\mathbf{n}}_j[n] = \tilde{\mathbf{W}}(\mathbf{h}_j)\mathbf{s}[n] + \tilde{\mathbf{n}}_j[n],$$

where $\tilde{\mathbf{W}}(\mathbf{h}_j) = \left[\tilde{\mathbf{w}}_1(\mathbf{h}_j) \cdots \tilde{\mathbf{w}}_{M'}(\mathbf{h}_j)\right]$.

Finally, stacking all the received signals into $\tilde{\mathbf{y}}[n]=\left[\tilde{\mathbf{y}}_1^T[n],\ldots,\tilde{\mathbf{y}}_{n_R}^T[n]\right]^T$, we can write

$$\tilde{\mathbf{y}}[n]=\tilde{\mathbf{W}}(\mathbf{H})\mathbf{s}[n]+\tilde{\mathbf{n}}[n], \quad (1)$$

where $\tilde{\mathbf{W}}(\mathbf{H})=\left[\tilde{\mathbf{W}}^T(\mathbf{h}_1)\cdots\tilde{\mathbf{W}}^T(\mathbf{h}_{n_R})\right]^T$ is the equivalent channel, and $\tilde{\mathbf{n}}[n]$ is defined analogously to $\tilde{\mathbf{y}}[n]$.

## Decoding with Perfect Channel State Information

If $\mathbf{H}$ is known at the receiver, and assuming a Gaussian distribution for the noise, the coherent maximum likelihood (ML) decoder amounts to minimizing the following criterion

$$\underset{\hat{\mathbf{s}}[n]}{\operatorname{argmin}}\left\|\tilde{\mathbf{y}}[n]-\tilde{\mathbf{W}}(\mathbf{H})\hat{\mathbf{s}}[n]\right\|^2,$$

subject to the constraint that the elements of $\hat{\mathbf{s}}[n]$ belong to some finite alphabet. In general, this is a NP-hard problem and optimal algorithms to solve it, such as *sphere decoding*, can be computationally expensive (Damen et al., 2000; Fincke & Pohst, 1985; Gesbert et al., 2003; Jaldén et al., 2003). However, for certain codes such as OSTBCs (Tarokh et al., 1999) or quasi-orthogonal STBCs (QSTBCs) (Jafarkhani, 2005; Sezgin & Oechtering, 2008), the code-channel matrix $\tilde{\mathbf{W}}(\mathbf{H})$ satisfies certain orthogonality properties which simplify the ML decoding process. Finally, an alternative receiver with a reduced computational cost can be obtained from the direct application of the linear minimum mean square error (LMMSE) criterion, which yields

$$\hat{\mathbf{s}}[n]=\mathbf{R}_s\tilde{\mathbf{W}}^T(\mathbf{H})\left(\tilde{\mathbf{W}}(\mathbf{H})\mathbf{R}_s\tilde{\mathbf{W}}^T(\mathbf{H})+\frac{\sigma^2}{2}\mathbf{I}\right)^{-1}\tilde{\mathbf{y}}[n], \quad (2)$$

where $\mathbf{R}_s=E\left[\mathbf{s}[n]\mathbf{s}^T[n]\right]$ is the correlation matrix of the information symbols.

## Some Common STBCs

In this subsection we summarize the main properties of some of the most common STBCs.

### Orthogonal Space-Time Block Codes (OSTBCs)

This class of codes generalizes the Alamouti scheme (Tarokh et al., 1999) for $n_T>2$. The main OSTBC characteristic is the following:

$$\mathbf{S}^H(\mathbf{s}[n])\mathbf{S}(\mathbf{s}[n])=\|\mathbf{s}[n]\|^2\mathbf{I},\ \forall\mathbf{s}[n], \quad \Leftrightarrow \quad \tilde{\mathbf{W}}^T(\mathbf{H})\tilde{\mathbf{W}}(\mathbf{H})=\|\mathbf{H}\|^2\mathbf{I},\ \forall\mathbf{H}.$$

This orthogonality property ensures that the code exploits the diversity of the MIMO channel, and it reduces the complexity of the ML receiver to a matched filter followed by a symbol by symbol decoder. The main drawback of OSTBCs is due to the strict restrictions imposed by the orthogonality conditions. In particular, these restrictions imply a limitation in the achievable transmission rate, which also translates into a penalty in the capacity of the MIMO-OSTBC system. The conditions in the code matrices are the following

$$\mathbf{C}_k^H\mathbf{C}_l=\begin{cases}\mathbf{I} & k=l,\\ -\mathbf{C}_l^H\mathbf{C}_k & k\neq l,\end{cases} \qquad k,l=1,\ldots,M',$$

and the maximum achievable transmission rates are (Liang, 2003)

$$R = \begin{cases} \frac{1}{2} + \frac{1}{2\lceil \frac{n_T}{2} \rceil} & \text{for complex codes,} \\ 1 & \text{for real codes.} \end{cases}$$

The most popular OSTBC is the Alamouti code, whose transmission matrix is

$$\mathbf{S}(\mathbf{s}[n]) = \begin{bmatrix} d_1[n] & d_2[n] \\ -d_2^*[n] & d_1^*[n] \end{bmatrix}, \quad (3)$$

where $d_1[n]$, $d_2[n]$ represent the complex information symbols. Another popular complex OSTBC, achieving the maximum possible rate ($R$ = 3/4) for $n_T$ = 4 is

$$\mathbf{S}(\mathbf{s}[n]) = \begin{bmatrix} d_1[n] & 0 & -d_2^*[n] & d_3^*[n] \\ 0 & d_1[n] & -d_3[n] & -d_2[n] \\ d_2[n] & d_3^*[n] & d_1^*[n] & 0 \\ -d_3[n] & d_2^*[n] & 0 & d_1^*[n] \end{bmatrix}. \quad (4)$$

Finally, the following code provides an example of real OSTBC with $R$ = 1 and $n_T = M = L = 4$ (Larsson and Stoica, 2003)

$$\mathbf{S}(\mathbf{s}[n]) = \begin{bmatrix} s_1[n] & s_2[n] & -s_3[n] & -s_4[n] \\ -s_2[n] & s_1[n] & s_4[n] & -s_3[n] \\ s_3[n] & -s_4[n] & s_1[n] & -s_2[n] \\ s_4[n] & s_3[n] & s_2[n] & s_1[n] \end{bmatrix}. \quad (5)$$

## Quasi-Orthogonal Space-Time Block Codes (QSTBCs)

Quasi-orthogonal STBCs (QSTBCs) (Jafarkhani, 2005; Sezgin, 2005; Sezgin & Oechtering, 2008) constitute an alternative to OSTBCs, which provides transmission rate $R$ = 1 for any number of transmit antennas $n_T$, establishing a trade-off between the complexity of the optimum decoder and the system diversity (Sezgin, 2005; Sezgin & Oechtering, 2008).

Considering an even number of transmit antennas, the QSTBC transmission matrix is recursively obtained as (Sezgin, 2005, Sezgin and Oechtering, 2008)

$$\mathbf{S}_{n_T}(\mathbf{s}_{n_T}[n]) = \begin{bmatrix} \mathbf{S}_{\frac{n_T}{2}}(\mathbf{s}_{\frac{n_T}{2},1}[n]) & \mathbf{S}_{\frac{n_T}{2}}(\mathbf{s}_{\frac{n_T}{2},2}[n]) \\ \mathbf{S}_{\frac{n_T}{2}}(\mathbf{s}_{\frac{n_T}{2},2}[n])\mathbf{\Lambda} & -\mathbf{S}_{\frac{n_T}{2}}(\mathbf{s}_{\frac{n_T}{2},1}[n])\mathbf{\Lambda} \end{bmatrix},$$

where

$\mathbf{S}_{\frac{n_T}{2}}(\mathbf{s}_{\frac{n_T}{2},1}[n])$ and $\mathbf{S}_{\frac{n_T}{2}}(\mathbf{s}_{\frac{n_T}{2},2}[n])$

are two QSTBC matrices for $n_T/2$ transmit antennas, and $\Lambda$ is a diagonal matrix with $(-1)^{k-1}$ in its $k$-th position. The basic QSTBC block is given by an OSTBC matrix (usually the Alamouti code), which implies certain orthogonality properties in the equivalent channel $\tilde{\mathbf{W}}(\mathbf{H})$, simplifying the decoding process. Finally, by eliminating columns of a QSTBC transmission matrix, we can obtain designs for any number of transmit antennas.

### Trace-Orthogonal Space-Time Block Codes (TOSTBCs)

Recently, trace-orthogonal STBCs (TOSTBCs) have been proposed as an example of linear dispersion codes (LDCs) with full rate ($R = n_T$), diversity, and ergodic capacity (Barbarossa, 2005, Fasano and Barbarossa, 2006, Zhang et al., 2005, 2007). Although the complexity of the optimum ML receiver is prohibitive, an attractive property of TOSTBCs consists in their optimality, in terms of bit error rate (BER), when the LMMSE receiver is employed. The TOSTBC code matrices must satisfy

$$\mathbf{C}_k^H \mathbf{C}_k = \mathbf{I},$$
$$\Re\left\{\mathrm{Tr}\left(\mathbf{C}_k^H \mathbf{C}_l\right)\right\} = n_T \delta(k-l).$$

## REVIEW OF PREVIOUS BLIND DECODING APPROACHES

Usually, the channel knowledge at the receiver is obtained by means of a training sequence (Hassibi and Hochwald, 2003, Naguib et al., 1998, Pohl et al., 2005). Specifically, assuming that $N_{tr}$ orthogonal blocks ($\mathbf{S}_{tr}^H[n]\mathbf{S}_{tr}[n] = \mathbf{I}$) are transmitted during the training phase, the received signal is

$$\mathbf{Y}_{tr}[n] = \mathbf{S}_{tr}[n]\mathbf{H} + \mathbf{N}[n], \qquad n = 0, \ldots, N_{tr} - 1,$$

where $\mathbf{Y}_{tr}[n] \in \mathbb{C}^{L \times n_R}$ and $\mathbf{N}[n] \in \mathbb{C}^{L \times n_R}$ are the received and complex noise matrices, respectively. Therefore, the estimated channel matrix is obtained by means of the least squares method (LS) as

$$\hat{\mathbf{H}}_{tr} = \frac{1}{N_{tr}} \sum_{n=0}^{N_{tr}-1} \mathbf{S}_{tr}^H[n]\mathbf{Y}_{tr}[n].$$

Obviously, the transmission of pilot symbols translates into a penalty in the bandwidth efficiency, which has motivated the development of several schemes to recover the signals, or the channel, without transmitting a training sequence. The main approaches are summarized in this section.

### Differential and Unitary Schemes

A first solution to the penalty in bandwidth efficiency consists in the use of differential techniques (Ganesan and Stoica, 2002, Hochwald and Sweldens, 2000, Hughes, 2000, Jafarkhani and Tarokh, 2001, Tarokh and Jafarkhani, 2000, Zhu and Jafarkhani, 2005). Basically, these schemes encode the information symbols in the difference between two consecutive STBC blocks, where the channel is assumed to remain constant. The encoding and decoding processes are very simple, and the receiver only needs to apply a conventional decoder assuming that the equivalent MIMO channel is given by the observations associated to the preceding block (Larsson and Stoica, 2003). Unfortunately, these schemes are limited to OSTBCs and QSTBCs, and they incur in a penalty in SNR of 3dB (OSTBC) or higher (QSTBC).

On the other hand, in (Hochwald and Marzetta, 2000, Hochwald et al., 2000) the authors have proposed a unitary space-time modulation, which does not require channel knowledge at the receiver. However, the penalty in performance of these schemes is about 2-4dB, and the receiver complexity increases exponentially with the number of points in the unitary space-time constellation.

## Blind Channel Estimation Techniques

### Blind Maximum Likelihood Receiver

When the channel is not known at the receiver, the information symbols can be recovered by means of the blind maximum likelihood (ML) receiver, which, assuming a Gaussian distribution for the noise and N available blocks at the receiver, amounts to minimizing

$$L(\hat{\mathbf{H}}, \hat{\mathbf{s}}[n]) = \sum_{n=0}^{N-1} \left\| \tilde{\mathbf{y}}[n] - \tilde{\mathbf{W}}(\hat{\mathbf{H}})\hat{\mathbf{s}}[n] \right\|^2, \tag{6}$$

subject to the constraint that the estimated symbols $\hat{\mathbf{s}}[n]$ belong to some finite alphabet. Unfortunately, this is a fairly difficult problem even in the case of matrices $\tilde{\mathbf{W}}(\hat{\mathbf{H}})$ with special properties, which is due to the fact that all the possible information symbol sequences have to be considered.

Interestingly, Ma et al. (2006) have proposed two approaches to the ML solution for the particular case of OSTBCs and BPSK or QPSK constellations. Specifically, the authors propose a suboptimum technique based on a semidefinite relaxation approach, and an optimal solution based on the sphere decoder. However, for a moderate number of observations, the computational complexity of these techniques remains relatively high.

### Subspace-Based Techniques

In general, the subspace techniques are based on a relaxation of the finite alphabet constraint, which permits the derivation of blind channel estimation algorithms independent of the specific signal constellation. Unfortunately, most of the algorithms have been proposed for the particular case of single-user systems and OSTBC codes. The main techniques are the following:

- In (Larsson et al., 2003, Stoica and Ganesan, 2003) the authors propose blind and semiblind channel estimation techniques based on alternating minimizations over the channel and signal estimates. The main drawback of these approaches consists in their iterative nature and their restrictive identifiability conditions. Specifically, the method in (Stoica and Ganesan, 2003) requires $M \leq n_T$, which is a condition associated to the specific technique, and not to the blind channel estimation problem.
- Recently, Ammar and Ding (2006) have proposed a method solely based on the SOS of the observations. The technique is based on the extraction of the noise subspace associated to the observations, and the minimization of the projection of the equivalent complex channel onto that subspace. Unfortunately, this technique also introduces additional ambiguities to those associated to the blind channel estimation problem.
- The technique proposed by Shahbazpanahi et al. (2005) is equivalent to the relaxed blind ML decoder, and it is based on the following maximization problem

  $$\underset{\hat{\mathbf{H}}}{\operatorname{argmax}} \operatorname{Tr}\left(\tilde{\mathbf{W}}^T(\hat{\mathbf{H}})\hat{\mathbf{R}}_{\tilde{y}}\tilde{\mathbf{W}}(\hat{\mathbf{H}})\right), \tag{7}$$

  where

  $$\hat{\mathbf{R}}_{\tilde{y}} = \sum_{n=0}^{N-1} \tilde{\mathbf{y}}[n]\tilde{\mathbf{y}}^T[n]$$

  can be seen as an estimate of the correlation matrix of $\tilde{\mathbf{y}}[n]$. Due to the orthogonality properties of $\tilde{\mathbf{W}}(\hat{\mathbf{H}})$, the solution of the above criterion is given by a single eigenvalue problem. Furthermore, in (Vía & Santamaría, 2008) it has been proven that this technique does not introduce additional ambiguities to those associated to the problem of blind channel estimation from SOS.

Unfortunately, the number of blind techniques for more general settings is rather scarce. The three main approaches are:

- In (Swindlehurst, 2002, Swindlehurst and Leus, 2002) the authors propose a blind receiver for a general class of STBCs, which include the multiuser case. However, the proposed receiver is affected by additional ambiguities to those of the blind channel estimation problem. For instance, it is not able to recover the sources when $n_T = L$.
- Ammar and Ding (2007) have extended the technique in (Ammar and Ding, 2006) to the case of general STBCs. The main drawback is again related to the identifiability problems, which are solved by means of a short training sequence, i.e., it is a semiblind approach.
- In the multiuser OSTBC case, two different blind channel estimation algorithms have been proposed in (Shahbazpanahi et al., 2006). The methods are based on the MUSIC (Schmidt, 1986) and Capon (Capon, 1969) techniques, and they minimize the projection of $\tilde{\mathbf{W}}(\hat{\mathbf{H}})$ onto the noise subspace associated to $\tilde{\mathbf{y}}[n]$. Unfortunately, the assumption of users with a common OSTBC introduces indeterminacies, which are solved by means of some pilot symbols.

## PROPOSED BLIND CHANNEL ESTIMATION CRITERION

In this section a new blind channel estimation criterion is proposed. The technique is inspired by the relaxed blind ML receiver, it can be applied to a very general class of STBCs, and it is easily extended to multiuser settings. The proposed criterion reduces to the extraction of the principal eigenvector of a generalized eigenvalue problem (GEV), and it can be interpreted as a deterministic approach, i.e., in the absence of noise it exactly recovers the channel, up to a real scalar, within a finite number of observations. Let us start by introducing the main assumptions of the proposed technique.

### Main Conditions

**Condition 1 (Number of available blocks)** *The MIMO channel is flat fading and constant during a period of* $N \geq M'$ *transmission blocks.*

**Condition 2 (Input signals)** *The input* $\mathbf{S}_N = [\mathbf{s}[0] \cdots \mathbf{s}[N-1]]$ *is persistently exciting, i.e.,* $\mathbf{S}_N$ *is full row rank.*

**Condition 3 (Equivalent Channel)** *The equivalent channel* $\tilde{\mathbf{W}}(\mathbf{H})$ *is full column rank.*

**Condition 4 (Rate and number of antennas)** *The number of transmit and receive antennas satisfy*

$$n_T, n_R > \begin{cases} R & \text{for complex codes,} \\ R/2 & \text{for real codes.} \end{cases}$$

The above conditions are easily satisfied and their main implications are:

- Conditions 1 and 2 establish mild assumptions on the coherence time of the channel and the correlation properties of the inputs.
- Condition 3 is a desirable property for any STBC. In particular, we must note that if $\tilde{\mathbf{W}}(\mathbf{H})$ is not full column rank, any information vector $\mathbf{s}[n] + \mathbf{z}[n]$, with $\mathbf{z}[n]$ belonging to the null subspace of $\tilde{\mathbf{W}}(\mathbf{H})$, provides the same observations $\tilde{\mathbf{y}}[n]$ as $\mathbf{s}[n]$. On the other hand, for the most common STBCs, the channels $\mathbf{H}$ providing rank deficient matrices $\tilde{\mathbf{W}}(\mathbf{H})$ form a set of measure zero.
- As will be shown later, Condition 4 is a necessary condition for the blind identifiability of the channel from SOS, which implies that full rate codes ($R = n_T$), such as TOSTBCs, do not allow the blind extraction of the channel.

### Proposed Criterion

The proposed criterion is inspired by the relaxed blind maximum likelihood (ML) receiver. Thus, solving for $\hat{\mathbf{s}}[n]$, the minimization problem in (6) can be rewritten in terms of $\hat{\mathbf{H}}$ as

$$\underset{\hat{\mathbf{H}}}{\operatorname{argmax}} \operatorname{Tr}\left(\tilde{\mathbf{U}}^T(\hat{\mathbf{H}})\hat{\mathbf{R}}_{\tilde{\mathbf{y}}}\tilde{\mathbf{U}}(\hat{\mathbf{H}})\right), \tag{8}$$

where $\tilde{\mathbf{U}}(\hat{\mathbf{H}})$ is an orthogonal basis for the subspace spanned by the columns of $\tilde{\mathbf{W}}(\hat{\mathbf{H}})$.

Obviously, the relaxation of the finite alphabet property translates into a real scalar ambiguity, and a slight performance degradation with respect to the blind ML decoder. However, this relaxation permits the obtention of a channel estimation algorithm independent of the specific symbol constellation and, as will be shown later, for a sufficiently large number of blocks $N$, the performance of the receiver based on the estimated channel is very close to that of the receiver with perfect channel state information (CSI). Finally, the scalar ambiguity introduced by the relaxation of the finite alphabet constraint is a common ambiguity to all the blind techniques solely based on SOS, and it can be easily resolved in a later step. Thus, from now on we will consider $\|\tilde{\mathbf{W}}(\hat{\mathbf{H}})\| = \|\tilde{\mathbf{W}}(\mathbf{H})\| = 1$.

Unfortunately, excluding the OSTBC case, the dependence of $\tilde{\mathbf{U}}(\hat{\mathbf{H}})$ with $\hat{\mathbf{H}}$ is not trivial, and (8) can not be easily solved. However, the optimization problem in (8) can be interpreted as the maximization of the correlation between the observations and the whitened version of $\tilde{\mathbf{W}}(\hat{\mathbf{H}})$. Thus, defining $\mathbf{\Phi}_{\tilde{\mathbf{y}}}$ as the prewhitened and rank-reduced (with rank $M'$) version of $\hat{\mathbf{R}}_{\tilde{\mathbf{y}}}$, we propose the following alternative criterion

$$\underset{\hat{\mathbf{H}}}{\operatorname{argmax}} \operatorname{Tr}\left(\tilde{\mathbf{W}}^T(\hat{\mathbf{H}})\mathbf{\Phi}_{\tilde{\mathbf{y}}}\tilde{\mathbf{W}}(\hat{\mathbf{H}})\right), \qquad \text{subject to} \qquad \|\tilde{\mathbf{W}}(\hat{\mathbf{H}})\| = 1, \tag{9}$$

i.e., we propose to maximize the correlation between the equivalent channel $\tilde{\mathbf{W}}(\hat{\mathbf{H}})$ and the prewhitened and rank-reduced version of the observations.

In the absence of noise, or in the asymptotic case of $N \to \infty$, the theoretical value of $\mathbf{\Phi}_{\tilde{\mathbf{y}}}$ is

$$\mathbf{\Phi}_{\tilde{\mathbf{y}}} = \tilde{\mathbf{U}}(\mathbf{H})\tilde{\mathbf{U}}^T(\mathbf{H}),$$

and it can be easily proven that the solutions to the blind channel estimation criterion in (9) satisfy

$$\operatorname{range}\left(\tilde{\mathbf{W}}(\hat{\mathbf{H}})\right) = \operatorname{range}\left(\tilde{\mathbf{W}}(\mathbf{H})\right), \tag{10}$$

and are also solutions of the relaxed blind ML decoder (8). This implies that, unlike other previously proposed techniques (Ammar and Ding, 2007, Stoica and Ganesan, 2003, Swindlehurst and Leus, 2002), the proposed method does not introduce additional ambiguities to those associated to the blind channel estimation problem.

## Algorithm Implementation

Unlike $\tilde{\mathbf{U}}(\hat{\mathbf{H}})$, the relationship between $\hat{\mathbf{H}}$ and $\tilde{\mathbf{W}}(\hat{\mathbf{H}})$ is trivial. Specifically, defining the vectorized MIMO channel $\tilde{\mathbf{h}} = \left[\tilde{\mathbf{h}}_1^T, \ldots, \tilde{\mathbf{h}}_{n_R}^T\right]^T$, and the $M'$ block diagonal matrices $\tilde{\mathbf{D}}_k \in \mathbb{R}^{2Ln_R \times 2n_T n_R}$ as

$$\tilde{\mathbf{D}}_k = \begin{bmatrix} \tilde{\mathbf{C}}_k & \cdots & \mathbf{0} \\ \vdots & \ddots & \vdots \\ \mathbf{0} & \cdots & \tilde{\mathbf{C}}_k \end{bmatrix}, \qquad k = 1, \ldots, M', \tag{11}$$

it is easy to see that the $k$-th column of $\tilde{\mathbf{W}}(\mathbf{H})$ is given by $\tilde{\mathbf{D}}_k\tilde{\mathbf{h}}$. Therefore, (9) can be rewritten as

$$\underset{\hat{\mathbf{H}}}{\operatorname{argmax}}\, \hat{\tilde{\mathbf{h}}}^T\mathbf{\Theta}\hat{\tilde{\mathbf{h}}}, \qquad \text{subject to} \qquad \hat{\tilde{\mathbf{h}}}^T\mathbf{\Psi}\hat{\tilde{\mathbf{h}}} = 1, \tag{12}$$

where $\mathbf{\Theta} \in \mathbb{R}^{2n_T n_R \times 2n_T n_R}$ is a modified correlation matrix given by

$$\mathbf{\Theta} = \sum_{k=1}^{M'} \tilde{\mathbf{D}}_k^T \mathbf{\Phi}_{\tilde{\mathbf{y}}} \tilde{\mathbf{D}}_k,$$

and $\mathbf{\Psi} \in \mathbb{R}^{2n_T n_R \times 2n_T n_R}$ is

$$\mathbf{\Psi} = \sum_{k=1}^{M'} \tilde{\mathbf{D}}_k^T \tilde{\mathbf{D}}_k.$$

Finally, the solutions $\hat{\tilde{\mathbf{h}}}$ of (12) can be easily obtained as the eigenvectors associated to the largest eigenvalue β of the following generalized eigenvalue problem (GEV)

$$\mathbf{\Theta}\hat{\tilde{\mathbf{h}}} = \beta \mathbf{\Psi} \hat{\tilde{\mathbf{h}}}. \tag{13}$$

## Generalization to Multiuser Settings

The proposed blind channel estimation technique can be extended to the multiuser case in a straightforward manner. Specifically, we consider the uplink MIMO channel with $U$ synchronous multi-antenna transmitters and one multi-antenna receiver. Each user $u$ employs a STBC $C_u$, with $n_T(C_u)$ transmit antennas, $M(C_u)$ information symbols, and a common block length of $L$ time slots. Therefore, the set of $U$ codes can be considered as a composite code with $n_T = \sum_{u=1}^{U} n_T(C_u)$ transmit antennas, $M = \sum_{u=1}^{U} M(C_u)$ information symbols, and transmission rate $R = \sum_{u=1}^{U} R(C_u)$. Furthermore, we must note that the assumption of a common block length $L$ is not restrictive. In the case of different lengths $L(C_u)$, the concatenation of several blocks can be considered as a composite code, and a common length $L$ can be defined as the least common multiple of $L(C_1),\ldots,L(C_U)$.

Taking into account the previous definitions, it is easy to see that the data model given by eq. (1) holds, with

$$\mathbf{H} = \left[\mathbf{H}_1^T \cdots \mathbf{H}_U^T\right]^T$$

$$\tilde{\mathbf{W}}(\mathbf{H}) = \left[\tilde{\mathbf{W}}(\mathbf{H}_1, C_1) \cdots \tilde{\mathbf{W}}(\mathbf{H}_U, C_U)\right],$$

$$\mathbf{s}[n] = \left[\mathbf{s}_1^T[n], \ldots, \mathbf{s}_U^T[n]\right]^T,$$

and where $\mathbf{H}_u$ and $\mathbf{s}_u[n]$ denote, respectively, the MIMO channel and the vector with the $M'(C_u)$ real information symbols for the $u$-th user. Thus, the MIMO channel $\tilde{\mathbf{h}}_u$ associated to the $u$-th user can be estimated by means of the following maximization problem

$$\underset{\hat{\tilde{\mathbf{h}}}_u}{\operatorname{argmax}}\ \hat{\tilde{\mathbf{h}}}_u^T \mathbf{\Theta}(C_u) \hat{\tilde{\mathbf{h}}}_u, \qquad \text{subject to} \qquad \hat{\tilde{\mathbf{h}}}_u^T \mathbf{\Psi}(C_u) \hat{\tilde{\mathbf{h}}}_u = 1,$$

where

$$\mathbf{\Theta}(C_u) = \sum_{k=1}^{M'(C_u)} \tilde{\mathbf{D}}_k^T(C_u) \mathbf{\Phi}_{\tilde{\mathbf{y}}} \tilde{\mathbf{D}}_k(C_u), \qquad \mathbf{\Psi}(C_u) = \sum_{k=1}^{M'(C_u)} \tilde{\mathbf{D}}_k^T(C_u) \tilde{\mathbf{D}}_k(C_u),$$

and $\tilde{\mathbf{D}}_k(C_u)$ are defined as in eq. (11). Finally, the channel estimate $\hat{\tilde{\mathbf{h}}}_u$ for the $u$-th user is given by the eigenvector associated to the largest eigenvalue $\beta_u$ of the following GEV

$$\mathbf{\Theta}(C_u) \hat{\tilde{\mathbf{h}}}_u = \beta_u \mathbf{\Psi}(C_u) \hat{\tilde{\mathbf{h}}}_u,$$

and, analogously to the single-user case, this introduces an unavoidable real scale factor for each user. Finally, it is easy to prove that, if several users employ a common STBC, then there exists an additional indeterminacy problem given by the linear combinations of the associated MIMO channels.

## Algorithm Properties and Relationship with Other Techniques

In this section we have presented a new SOS-based blind channel estimation technique for STBC systems. The proposed algorithm is inspired by the relaxed blind ML receiver, and it is easily generalized to the multiuser case. Interestingly, for a large number of STBCs, including OSTBCs and QSTBCs, the code matrices satisfy, for some constant $c$

$$\mathbf{\Psi} = \sum_{k=1}^{M'} \mathbf{C}_k^H \mathbf{C}_k = c^2 \mathbf{I}, \qquad \Leftrightarrow \qquad \left\| \tilde{\mathbf{W}}(\mathbf{H}) \right\| = c \left\| \mathbf{H} \right\|, \qquad \forall \mathbf{H}.$$

In these cases, the proposed criterion reduces to the extraction of the eigenvector associated to the largest eigenvalue of $\mathbf{\Theta}$, and it is theoretically equivalent to the MUSIC-based (Schmidt, 1986) technique presented in (Shahbazpanahi et al., 2006) for the multiuser OSTBC case. Specifically, the method in (Shahbazpanahi et al., 2006) amounts to minimizing the energy of the projection of $\tilde{\mathbf{W}}(\hat{\mathbf{H}})$ onto the noise subspace associated to $\hat{\mathbf{R}}_{\tilde{\mathbf{y}}}$, whose dimension depends on the number of receive antennas. However, the proposed technique has been derived from a different point of view, and it can be reformulated as a principal component analysis (PCA) problem, which can be exploited to obtain adaptive algorithms for blind channel estimation (Vía et al., 2006a, 2006c). Specifically, the prewhitening step can be performed by means of the adaptive principal component extraction (APEX) algorithm (Diamantaras and Kung, 1996), and the final channel estimate can be extracted by direct application of the Oja's rule (Oja, 1992).

Finally, in the single-user OSTBC case, the solutions of (8) can be directly obtained, and the prewhitening step is not necessary, i.e., $\mathbf{\Phi}_{\tilde{\mathbf{y}}}$ can be replaced by $\hat{\mathbf{R}}_{\tilde{\mathbf{y}}}$, obtaining the blind channel estimation technique proposed in (Shahbazpanahi et al., 2005). Therefore, the proposed criterion can be seen as a generalization of the technique in (Shahbazpanahi et al., 2005) for a wider class of STBCs and multiuser settings.

## IDENTIFIABILITY ANALYSIS

In the previous section we have proposed a new blind channel estimation criterion, whose theoretical solutions are given by eq. (10). From a practical point of view, the existence of spurious solutions, i.e., MIMO channels $\hat{\mathbf{H}} \neq \pm\mathbf{H}$ maximizing (9), is translated into a multiplicity $P > 1$ of the largest eigenvalue of the GEV in (13). In this section the indeterminacy problems are illustrated by means of some numerical examples, and some identifiability conditions are presented, with special interest in the case of OSTBCs.

## Numerical Examples of the Ambiguity Problems

In order to illustrate the ambiguity problems we present some simulation results. All the examples have been repeated for random Rayleigh distributed channels $\mathbf{H}$. As pointed out in (Shahbazpanahi et al., 2005) for the case of OSTBCs, we have observed that the multiplicity $P$ of the largest eigenvalue of (13) depends on the specific STBC and the number $n_R$ of receive antennas, but not on the specific channel realization.

The first set of examples is illustrated in Table 1, which is a partial reproduction of the table presented in (Shahbazpanahi et al., 2005). Here, we can see that, for most of the common OSTBCs and $n_R > 1$, the multiplicity of the largest eigenvalue of (13) is $P - 1$, i.e., the MIMO channel can be recovered, up to a real scalar, by means of the proposed method.

In the second set of examples, the rate one QSTBCs have been analyzed. The codes have been recursively obtained using the Alamouti code as a basic block, which provides designs for a number of transmit antennas $n_T$ power of two. For different $n_T$ values, the codes have been obtained by removing some of the transmit antennas from the design for the smallest power of two greater than $n_T$. Table 2 shows the multiplicity $P$ in this case, which in general decreases with $n_R$ up to a certain value (the rightmost multiplicity shown in the table). As can be seen, the QSTBCs do not allow the unambiguous recovery of the channel regardless of the number of receive antennas $n_R$. In the next section we present a new transmission technique which enables the blind channel recovery, without reducing the transmission rate nor the system capacity, by means of a slight modification of the code.

*Table 1. Identifiability characteristics for some of the most common OSTBCs*

| Constellation | $n_T$ | $M$ | $L$ | $R = M/L$ | Design | $P_{n_R=1}$ | $P_{n_R>1}$ |
|---|---|---|---|---|---|---|---|
| real | 2 | 2 | 2 | 1 | Alamouti | 2 | 2 |
| real | 3 | 4 | 4 | 1 | gen. ort | 2 | 1 |
| real | 4 | 4 | 4 | 1 | gen. ort | 4 | 4 |
| real | 5 | 8 | 8 | 1 | gen. ort | 2 | 1 |
| real | 6 | 8 | 8 | 1 | gen. ort | 2 | 1 |
| real | 7 | 8 | 8 | 1 | gen. ort | 2 | 1 |
| real | 8 | 8 | 8 | 1 | gen. ort | 2 | 1 |
| complex | 2 | 2 | 2 | 1 | Alamouti | 4 | 4 |
| complex | 3 | 4 | 8 | 1/2 | gen. ort | 2 | 1 |
| complex | 4 | 4 | 8 | 1/2 | gen. ort | 4 | 4 |
| complex | 5 | 8 | 16 | 1/2 | gen. ort | 2 | 1 |
| complex | 6 | 8 | 16 | 1/2 | gen. ort | 2 | 1 |
| complex | 7 | 8 | 16 | 1/2 | gen. ort | 2 | 1 |
| complex | 8 | 8 | 16 | 1/2 | gen. ort | 2 | 1 |
| complex | 3 | 3 | 4 | 3/4 | amicable | 2 | 1 |
| complex | 4 | 3 | 4 | 3/4 | amicable | 2 | 1 |
| complex | 5 | 4 | 8 | 1/2 | amicable | 1 | 1 |
| complex | 6 | 4 | 8 | 1/2 | amicable | 1 | 1 |
| complex | 7 | 4 | 8 | 1/2 | amicable | 1 | 1 |
| complex | 8 | 4 | 8 | 1/2 | amicable | 1 | 1 |

## Necessary Identifiability Conditions for General STBCs

In order to obtain some identifiability results we will start by rewriting the ambiguity condition in eq. (10) as

$$\text{range}\left(\tilde{\mathbf{W}}(\hat{\mathbf{H}})\right) = \text{range}\left(\tilde{\mathbf{W}}(\mathbf{H})\right), \quad \Leftrightarrow \quad \tilde{\mathbf{W}}(\hat{\mathbf{H}}) = \tilde{\mathbf{W}}(\mathbf{H})\mathbf{A}, \quad \Leftrightarrow \quad \mathbf{S}(\mathbf{s}[n])\mathbf{H} = \mathbf{S}(\hat{\mathbf{s}}[n])\hat{\mathbf{H}},$$

where $\mathbf{A} \neq \pm\mathbf{I}$ is a nonsingular matrix and $\hat{\mathbf{s}}[n] = \mathbf{A}^{-1}\mathbf{s}[n]$ is a spurious information vector.

Firstly, we must observe that $\tilde{\mathbf{W}}(\mathbf{H})$ is a $2Ln_R \times M'$ matrix. Thus, taking Condition 3 into account, we can conclude that a necessary identifiability condition is given by

$$n_R > \frac{M'}{2L}, \quad \Leftrightarrow \quad n_R > \begin{cases} R & \text{for complex codes,} \\ R/2 & \text{for real codes,} \end{cases}$$

and if it is not satisfied, the multiplicity of the largest eigenvalue of (13) is $P = 2n_T n_R$. This result explains the multiplicity $P = 2n_T$ observed in Table 2 for QSTBCs with $n_R = R = 1$.

Finally, given any full-rank MIMO channel $\mathbf{H}$ with $n_R \geq n_T$, it is easy to prove (Vía & Santamaría, 2008) that, if the channel can not be unambiguously identified, then all the channels are affected by the ambiguity problem. Therefore, the identifiability analysis can be reduced to $n_R \leq n_T$, which justifies the assumption in Condition 4.

## Sufficient Identifiability Conditions for OSTBCs

Although we have obtained some intuitive necessary identifiability conditions, in a general case, the identifiability analysis is a very difficult task (Ammar and Ding, 2007). Fortunately, in the case of OSTBCs, the structure of the equivalent channel matrix $\tilde{\mathbf{W}}(\mathbf{H})$ and the ambiguity matrix **A** can be exploited to obtain some sufficient identifiability conditions. Here we summarize the main results presented in (Vía & Santamaría, 2008).

- Firstly, it can be easily proven that, if the number of real information symbols $M'$ is odd, then the channel can be unambiguously extracted regardless of the number of receive antennas.
- As pointed out in the previous subsection, if for a full row rank channel **H** ($n_R \geq n_T$) there exists an ambiguity problem, then the code does not allow the blind channel recovery for any MIMO channel (*non-identifiable OSTBCs*). Conversely, if the code is *identifiable*, any full row rank MIMO channel can be unambiguously extracted.
- All the real OSTBCs with an odd number of transmit antennas $n_T$ are identifiable.
- All the OSTBCs with a transmission rate $\frac{M'}{L} > \frac{2}{\lceil n_T/2 \rceil}$ are identifiable.

The above results ensure the blind channel identifiability for most of the OSTBCs and full row rank MIMO channels **H**. However, from Table 1 we can see that the full row rank property is not necessary. Thus, assuming a Gaussian distributed random channel with non-singular covariance matrix, we can obtain the following results (Vía & Santamaría, 2008):

- Consider an identifiable OSTBC and a MIMO channel providing an ambiguity with multiplicity $P > 1$. Then, the addition of a new receive antenna will decrease the multiplicity with probability one.
- The impact of the previous result is increased by the observation, through numerical examples (see Table 1), that the multiplicity associated to identifiable OSTBCs and multiple-input single-output (MISO) channels ($n_R = 1$) is $P \leq 2$ with probability one. This allows us to state that, for $n_R > 1$ and an identifiable OSTBC, the channel can be unambiguously recovered with probability one.

These new results shed some light into Table 1. Furthermore, we must note that the threshold in the transmission rate provides a tight condition, which allows us to state that the only non-identifiable OSTBCs with practical interest are the Alamouti codes and the real design with $n_T = M = L = 4$ given in eq. (5) (Vía & Santamaría, 2008).

# SOLUTION OF THE CHANNEL INDETERMINACIES

In the previous section we have illustrated the identifiability problems in the particular OSTBC and QSTBC cases. In general, the ambiguity problem implies that the true channel belongs to a $P$-dimensional subspace defined by the principal eigenvectors of (13), which suggest the idea of extracting the channel, from that reduced subspace, by means of a short training sequence (Ammar and Ding, 2007, Shahbazpanahi et al., 2006). However, in order to avoid the penalty in bandwidth efficiency, several blind techniques can be applied. In this section, the main approaches for single-user OSTBC transmissions are summarized, and we propose a new transmission technique for the resolution of the ambiguities in a general case. The proposed technique is based on the combination of different STBCs, which we refer to as code diversity, and it can be reduced to a non-redundant precoding consisting in a single rotation or permutation of the transmit antennas. Unlike other previously proposed approaches (Ma, 2007, Shahbazpanahi et al., 2006, 2005, Vía & Santamaría, 2008, Vía et al., 2006b), the transmission technique is able to resolve many of the ambiguity problems without reducing the transmission rate nor the capacity of the MIMO-STBC system.

## Several Techniques for the Particular OSTBC Case

Several techniques have been proposed for the resolution of the ambiguities in the particular case of OSTBCs. Here we summarize the three main approaches.

*Table 2. Identifiability characteristics for QSTBCs*

| $n_T$ | $P_{n_R=1}$ | $P_{n_R=2}$ | $P_{n_R=3}$ | $P_{n_R=4}$ | $P_{n_R=5}$ | $P_{n_R=6}$ | $P_{n_R=7}$ | ... | $P_{n_R=15}$ |
|---|---|---|---|---|---|---|---|---|---|
| 2 | 4 | | | | | | | | |
| 3 | 6 | 4 | 2 | | | | | | |
| 4 | 8 | | | | | | | | |
| 5 | 10 | 4 | 2 | | | | | | |
| 6 | 12 | 8 | 4 | | | | | | |
| 7 | 14 | 12 | 10 | 8 | 6 | 4 | 2 | | |
| 8 | 16 | | | | | | | | |
| 9 | 18 | 4 | 2 | | | | | | |
| 10 | 20 | 8 | 4 | | | | | | |
| 11 | 22 | 12 | 2 | | | | | | |
| 12 | 24 | 16 | 8 | | | | | | |
| 13 | 26 | 20 | 14 | 8 | 2 | | | | |
| 14 | 28 | 24 | 20 | 16 | 12 | 8 | 4 | | |
| 15 | 30 | 28 | 26 | 24 | 22 | 20 | 18 | $32-2_{n_R}$ | 2 |
| 16 | 32 | | | | | | | | |

## Combination of SOS and HOS Approaches

In (Vía et al., 2006a) we have proposed a technique based on the combination of SOS and higher-order statistics (HOS) approaches. The main idea is based on the extraction of the $P$-dimensional subspace containing the channel by means of a SOS technique. In particular, in (Vía et al., 2006a) we have applied the adaptive principal component extraction algorithm (APEX) (Diamantaras and Kung, 1996). In a second stage, the channel is extracted from that reduced subspace by exploiting the finite alphabet properties or the HOS of the sources, for instance, by means of the constant modulus algorithm (CMA) (Godard, 1980). Thus, the combined technique can be seen from two different points of view: On one hand, the HOS-based technique is used to solve the ambiguity problem, which only affects to the SOS-based techniques. On the other hand, the SOS-based technique can be seen as a preprocessing step, which reduces the complexity of the problem to be solved by means of HOS approaches, obtaining better results than those provided by direct application of HOS techniques (Vía et al., 2006a).

## Linear Precoding and Correlation Matching

The blind channel estimation criterion presented in this chapter is a deterministic technique which does not make any assumption about the correlation properties of the information symbols. However, this information could be used to solve the indeterminacy problems in a large number of cases. This idea has been exploited in (Shahbazpanahi et al., 2005, Vía et al., 2006c) to obtain a straightforward modification of the criterion in (7). Furthermore, it is easy to prove that a sufficient condition for the identifiability of the channel up to a sign change is given by the existence of an eigenvalue with multiplicity one in the correlation matrix $\mathbf{R}_s$ of the information symbols $\mathbf{s}[n]$. This dispersion in the eigenvalues can be imposed by means of a linear precoding step, but it is also present in some specific OSTBCs designed for minimizing the peak to average power ratio (Tran et al., 2006). Unfortunately, this technique is not deterministic, which translates into a noise floor due to the difference between the theoretical correlation matrices and their finite sample estimate. Additionally, it can be proven that the indeterminacies can not be solved in the case of uncorrelated signals, and the transmission of correlated sources incurs in a penalty in the capacity of the MIMO-OSTBC system.

## A Technique Based on Rate-Reduction

In (Vía & Santamaría, 2008, Vía et al., 2006b) we have proposed a new technique based on one of the identifiability results for the OSTBC case. Specifically, the main idea consists in the elimination of one real information symbol (the real or imaginary part of a complex symbol) in one of each $B$ OSTBC blocks, which ensures the blind identifiability of the channel up to a real scalar. Interestingly, this technique can be seen as a particular case of semiblind or linear precoding approaches. On one hand, the method can be interpreted as a semiblind technique including a *zero value* pilot symbol. Thus, unlike classical semiblind approaches, the *pilots* do not require power, which can be used in the transmission of the information symbols. On the other hand, the technique can be considered as a limiting case of linear precoding. However, unlike linear precoding techniques, this scheme is deterministic, and therefore it is not affected by the noise floor due to the estimation of the correlation matrices. Obviously, the main drawback of the rate-reduction approach consists in a slight penalty in the bandwidth efficiency.

# New Techniques for a General Class of STBCs

## Code Diversity

Here we introduce the idea of code diversity for blind channel estimation. Analogously to the case of time, frequency or spatial diversity, where one or several signals are observed through *different channels*, the proposed technique relies on the *observation* of the MIMO channel *through* different STBCs, which provides the sufficient information to unambiguously identify the channel.

From the identifiability discussion in the previous section we know that the true channel $\tilde{\mathbf{h}}$ belongs to the subspace $\mathbf{G}(\mathbf{H},C) \in \mathbb{R}^{2n_T n_R \times P(C)}$ spanned by the $P(C)$ principal eigenvectors of (13), where we have explicitly included the dependency on the channel $\mathbf{H}$ and the code $C$, i.e.,

$$\tilde{\mathbf{h}} \in \text{range}\left(\mathbf{G}(\mathbf{H},C)\right).$$

Let us now consider $K$ different codes $C_k$, $k = 1,\dots,K$, with $n_T$ transmit antennas and transmitting $M(C_k)$ information symbols in $L(C_k)$ uses of the channel. Then, it is obvious that

$$\tilde{\mathbf{h}} \in \left\{\text{range}\left(\mathbf{G}(\mathbf{H},C_1)\right) \cap \cdots \cap \text{range}\left(\mathbf{G}(\mathbf{H},C_K)\right)\right\},$$

i.e., the true channel belongs to the intersection of the $K$ different subspaces, of size $P(C_k)$, defined by the matrices $\mathbf{G}(\mathbf{H},C_k)$. However, in a general case, there is no reason to think that the rank of such intersection will be larger than one, i.e., the spurious solutions to the blind channel estimation problem for code $C_k$ do not necessarily maximize the criterion (9) when a different code $C_l (l \neq k)$ is used.

The proposed technique is based on the previous idea. Assuming that the MIMO channel remains constant during a large enough interval, the first $M(C_1)$ information symbols are transmitted during the first $L(C_1)$ time slots using $C_1$. In the following $L(C_2)$ channel uses, $M(C_2)$ new information symbols are transmitted by means of $C_2$, and the same procedure is used with the $K$ STBCs. Thus, with obvious definitions of $\tilde{\mathbf{W}}(\mathbf{H},C_k)$, $\tilde{\mathbf{y}}[n,k]$, $\tilde{\mathbf{n}}[n,k]$ and $\mathbf{s}[n,k]$, the data model in (1) remains valid, where now

$$\tilde{\mathbf{y}}[n] = \left[\tilde{\mathbf{y}}^T[n,1],\dots,\tilde{\mathbf{y}}^T[n,K]\right]^T,$$

$$\tilde{\mathbf{n}}[n] = \left[\tilde{\mathbf{n}}^T[n,1],\dots,\tilde{\mathbf{n}}^T[n,K]\right]^T,$$

$$\mathbf{s}[n] = \left[\mathbf{s}^T[n,1],\dots,\mathbf{s}^T[n,K]\right]^T,$$

and

*Table 3. Application of the non-redundant precoding technique; QSTBC codes and K = 4 random rotations*

| $n_T$ | $P_{n_R=1}$ | $P_{n_R=2}$ |
|---|---|---|
| 2 | 4 | 4 |
| 3 | 6 | 1 |
| 4 | 8 | 1 |
| 5 | 10 | 1 |
| 6 | 12 | 1 |
| 7 | 14 | 1 |
| 8 | 16 | 1 |
| 9 | 18 | 1 |
| 10 | 20 | 1 |
| 11 | 22 | 1 |
| 12 | 24 | 1 |
| 13 | 26 | 1 |
| 14 | 28 | 1 |
| 15 | 30 | 1 |
| 16 | 32 | 1 |

$$\tilde{\mathbf{W}}(\mathbf{H})=\begin{bmatrix}\tilde{\mathbf{W}}(\mathbf{H},C_1) & \cdots & \mathbf{0}\\ \vdots & \ddots & \vdots\\ \mathbf{0} & \cdots & \tilde{\mathbf{W}}(\mathbf{H},C_K)\end{bmatrix}.$$

Following the derivation of the proposed blind channel estimation technique, the final channel estimation criterion amounts to maximize

$$\underset{\hat{\mathbf{h}}}{\operatorname{argmax}}\sum_{k=1}^{K}\hat{\mathbf{h}}^T\boldsymbol{\Theta}(C_k)\hat{\mathbf{h}}, \qquad \text{subject to} \qquad \sum_{k=1}^{K}\hat{\hat{\mathbf{h}}}^T\boldsymbol{\Psi}(C_k)\hat{\hat{\mathbf{h}}}, \tag{14}$$

where

$$\boldsymbol{\Theta}(C_k)=\sum_{l=1}^{M'(C_k)}\tilde{\mathbf{D}}_l^T(C_k)\boldsymbol{\Phi}_{\tilde{\mathbf{y}}}(C_k)\tilde{\mathbf{D}}_l(C_k), \qquad \boldsymbol{\Psi}(C_k)=\sum_{l=1}^{M'(C_k)}\tilde{\mathbf{D}}_l^T(C_k)\tilde{\mathbf{D}}_l(C_k),$$

$\boldsymbol{\Phi}_{\tilde{\mathbf{y}}}(C_k)$ is the rank-$M'(C_k)$ and whitened version of $\hat{\mathbf{R}}_{\tilde{\mathbf{y}}}(C_k)=\sum_{n=0}^{N/K-1}\tilde{\mathbf{y}}[n,k]\tilde{\mathbf{y}}^T[n,k]$, and $\tilde{\mathbf{D}}_l(C_k)$ are the block diagonal matrices for code $C_k$, which are defined analogously to (11).

From (14) it is obvious that the channel estimate $\hat{\hat{\mathbf{h}}}$ can be obtained as the eigenvector associated to the largest eigenvalue β of the following GEV

$$\boldsymbol{\Theta}\hat{\hat{\mathbf{h}}}=\beta\boldsymbol{\Psi}\hat{\hat{\mathbf{h}}},$$

where

$$\mathbf{\Theta} = \sum_{k=1}^{K} \mathbf{\Theta}(C_k), \qquad \text{and} \qquad \mathbf{\Psi} = \sum_{k=1}^{K} \mathbf{\Psi}(C_k).$$

Furthermore, since the true channel $\tilde{\mathbf{h}}$ maximizes the $K$ factors $\tilde{\mathbf{h}}^T \mathbf{\Theta}(C_k)\tilde{\mathbf{h}}$ simultaneously, it is easy to see that all the solutions to (14) belong to the intersection of the subspaces spanned by $\mathbf{G}(\mathbf{H}, C_k)$, for $k$ = 1, ..., $K$.

It is also interesting to point out that the set of $K$ code matrices can be interpreted as a composite code with a larger delay

$$L = \sum_{k=1}^{K} L(C_k),$$

which resembles the idea of designing codes with $L \gg n_T$ in order to improve the bit error rate (BER) (Fasano and Barbarossa, 2006, Henkel, 2005). Finally, we must take into account that the total number of time slots used in the blind channel estimation process has to be distributed among the $K$ different STBCs. This implies a reduction of the effective number of blocks for the estimation of the $K$ correlation matrices $\hat{\mathbf{R}}_{\tilde{\mathbf{y}}}(C_k)$. Therefore, there exists a tradeoff between the identifiability properties of the composite code, which are improved by increasing $K$, and the accuracy of the estimates $\hat{\mathbf{R}}_{\tilde{\mathbf{y}}}(C_k)$, which is degraded with increasing $K$.

## A Particular Solution: Non-Redundant Precoding

The proposed technique for the solution of the ambiguities exploits the general idea of code diversity, which raises a great number of questions related to the analysis of the identifiability conditions, or the best combination of codes. Here, we propose a particular code combination strategy, which is based on a single STBC, and achieves the code diversity by means of a non-redundant precoding consisting in the rotation of the transmit antennas.

Let us consider $K$ different unitary matrices $\mathbf{Q}_k$, $k$ = 1, ..., $K$, and one STBC code $C$ with transmission matrix $\mathbf{S}(\mathbf{s}[n], C)$. Then, we can define the following transmission matrices

$$\mathbf{S}(\mathbf{s}[n], C_k) = \mathbf{S}(\mathbf{s}[n], C)\mathbf{Q}_k, \qquad k = 1, \ldots, K,$$

which are associated to $K$ different codes $C_k$. Thus, the code diversity is obtained by rotating the transmission matrix of one STBC and, since the effect of the rotations can be considered as part of the channel, the proposed strategy preserves the code properties. Finally, we must point out that the matrices $\mathbf{Q}_k$ can be chosen as permutation matrices, which does not increase the complexity of the transmitter and preserves the power properties associated to each transmit antenna.

## Practical Examples

Let us illustrate the performance of the proposed non-redundant precoding technique by means of a numerical example. Table 3 shows the identifiability properties for QSTBCs when the non-redundant precoding technique is applied. The results have been obtained by random generations of $K$ = 4 unitary matrices, and Rayleigh MIMO channels. As can be seen, most of the ambiguities in Table 2 have been resolved, with the only exceptions of the MISO cases ($n_R$ = 1) and the Alamouti code ($n_T$ = 2). The ambiguity in the MISO cases is explained by the necessary identifiability condition $n_R > R$, which is still violated. Finally, taking into account the structure of the Alamouti transmission matrix given in eq. (3), it is easy to prove that, for any unitary matrix $\mathbf{Q}$ of the same form, $\mathbf{S}(\mathbf{s}[n], C)\mathbf{Q}_k\mathbf{Q}$ preserves the structure of $\mathbf{S}(\mathbf{s}[n], C)\mathbf{Q}_k$, for $k$ = 1, ..., $K$, which implies that the channel can not be unambiguously identified.

## Code Diversity and Non-Redundant Precoding in Multiuser Systems

Although we have considered the single-user case for notational simplicity, the ideas of code diversity and non-redundant precoding can be directly applied in the multiuser case. Furthermore, the non-redundant precoding allows us to avoid the ambiguities associated to several users with a common STBC. The idea consists in obtaining, by means of rotations, a virtually different STBC for each user. Thus, the non-redundant precoding is applied *vertically* to avoid the intra-user ambiguities, and *horizontally* to solve the inter-user indeterminacies. This transmission strategy is illustrated in the next section by means of some simulation examples.

## SIMULATION RESULTS

In this section, the performance of the blind channel estimation technique and the non-redundant precoding is illustrated by means of some numerical examples. All the results have been obtained by averaging 5000 independent experiments, where the MIMO channel $\mathbf{H}_u$ for each user has been generated as a Rayleigh channel with unit-variance elements. The signal to noise ratio (SNR) for the $u$-th user is defined as

$$\text{SNR}_u = 10\log_{10}\frac{\sigma_u^2}{\sigma^2},$$

where $\sigma^2$ is the variance of the white Gaussian noise, and $\sigma_u^2$ is the energy per time slot transmitted by the $n_T(C_u)$ transmit antennas of the $u$-th user. The i.i.d information symbols belong to a quadrature phase shift keying (QPSK) constellation, and the receivers have been designed based on the LMMSE criterion (2) and a hard decoder, which in the case of single-user OSTBC transmissions is equivalent to the ML receiver.

In order to avoid the ambiguity problems, the non-redundant precoding technique with $K = 4$ permutations has been applied. Specifically, the permutations of the transmit antennas are based on the following matrices:

$$\mathbf{Q}_1(4) = \begin{bmatrix} 1 & 0 & 0 & 0 \\ 0 & 1 & 0 & 0 \\ 0 & 0 & 1 & 0 \\ 0 & 0 & 0 & 1 \end{bmatrix}, \quad \mathbf{Q}_2(4) = \begin{bmatrix} 0 & 0 & 0 & 1 \\ 1 & 0 & 0 & 0 \\ 0 & 0 & 1 & 0 \\ 0 & 1 & 0 & 0 \end{bmatrix},$$

$$\mathbf{Q}_3(4) = \begin{bmatrix} 0 & 0 & 0 & 1 \\ 0 & 0 & 1 & 0 \\ 0 & 1 & 0 & 0 \\ 1 & 0 & 0 & 0 \end{bmatrix}, \quad \mathbf{Q}_4(4) = \begin{bmatrix} 0 & 0 & 0 & 1 \\ 1 & 0 & 0 & 0 \\ 0 & 1 & 0 & 0 \\ 0 & 0 & 1 & 0 \end{bmatrix}.$$

### Single-User Simulations

In the first set of examples, the real OSTBC design with $n_T = M = L = 4$ given in eq. (5) has been evaluated. As shown in Table 1, this code does not allow the direct blind channel recovery for any number of receive antennas, which in this case is $n_R = 4$. The non-redundant precoding technique has been applied with the permutation matrices $\mathbf{Q}_1(4),\ldots,\mathbf{Q}_4(4)$, i.e., the precoding technique is limited to a single set of permutations of the transmit antennas, which comes at virtually no computational expense at the transmitter. Figures 1 and 2 show the MSE in the channel estimate and the bit error rate (BER) after decoding for different numbers $N$ of available blocks at the receiver. As can be seen, the proposed technique provides very accurate channel estimates, and the performance loss with respect to the coherent receiver, which is included for comparison purposes, is much lower than the 3-dB loss associated to the differential receivers (Ganesan & Stoica, 2002; Hochwald & Sweldens, 2000; Hughes, 2000;Jafarkhani & Tarokh, 2001; Tarokh & Jafarkhani, 2000). A comparison with the least squares training method can be seen in Figure 3, where the two channel estimation techniques and the coherent receiver have been evaluated for three different SNR values. As can be seen, the BER performance obtained by the proposed method is equivalent to that obtained with a training based approach and some number $N_{tr} < N$ of training blocks. However, the blind channel estimation technique avoids the loss in bandwidth efficiency. Specifically, we can see that for $N = 100$ available blocks at the receiver, the proposed method outperforms the training technique with $N_{tr} = 10$, not only in bandwidth efficiency, but also in terms of BER.

In order to illustrate the advantage of the blind approach we have considered a scenario with fixed size packets of $N$ OSTBC blocks. Specifically, the ergodic capacity of the binary symmetric channel (BSC) seen by the higher layer is obtained as

*Figure 1. Channel estimate MSE of the proposed method with K=4 code permutations. Real OSTBC design with* $n_T = n_R = M = L = 4$

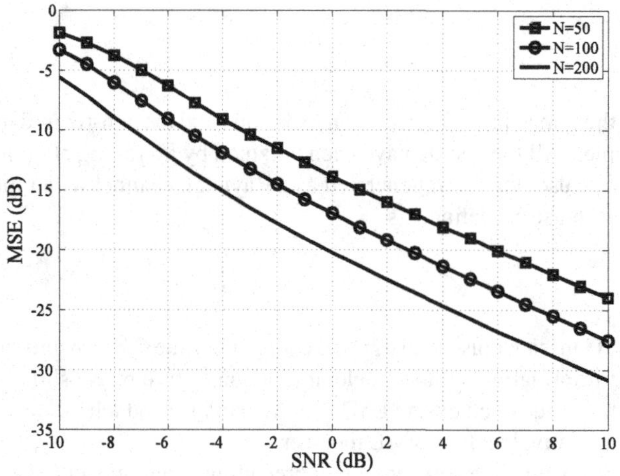


*Figure 2. Bit Error Rate (BER) for the ML receiver with perfect and estimated CSI. Real OSTBC design with* $n_T = n_R = M = L = 4$. $K = 4$ *code permutations.*

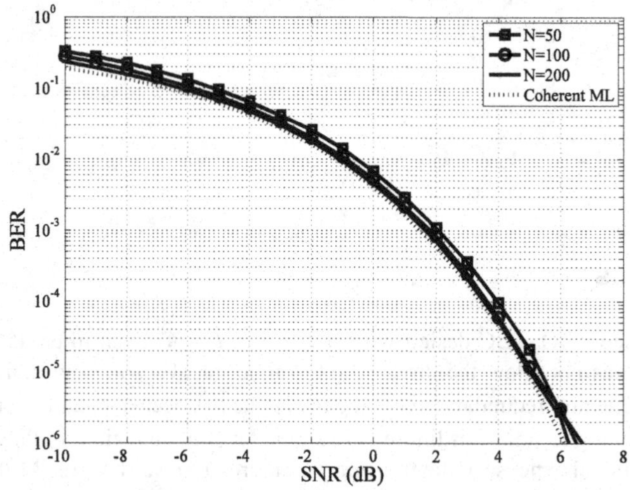


$$C = \eta(1 - H_{BSC}),$$

where $H_{BSC} = -(1-p)\log_2(1-p) - p\log_2(p)$ is the entropy of the BSC, $p$ is the BER shown in Figure 3, and $\eta = \frac{N-N_{tr}}{N}$ is the bandwidth efficiency. Figure 4 shows the capacity as a function of the packet size for the different techniques, where we can see that the capacity of the training approach is dominated by the penalty in bandwidth efficiency, whereas the performance of the proposed technique is close to that of the coherent receiver, which comes at the expense of a slight increase in the computational complexity and latency.

In the second set of examples, the Alamouti-based QSTBC design for $n_T = M = L = 8$ and $n_R = 4$ has been evaluated. Analogously to the previous examples, this code does not allow the direct blind channel identification based only on SOS (see Table 2). The $K = 4$ permutation matrices have been defined as

*Figure 3. Bit Error Rate (BER) versus number N of observation blocks for the proposed method (left) or number $N_{tr}$ of training blocks (right). Real OSTBC design with $n_T = n_R = M = L = 4$. K = 4 code permutations.*

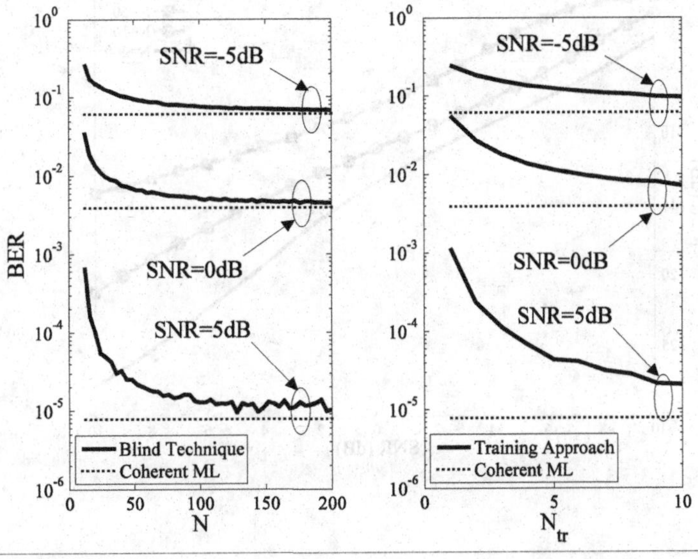


*Figure 4. Capacity of the binary symmetric channel after decoding. Real OSTBC design with $n_T = n_R = M = L = 4$. K = 4 code permutations.*

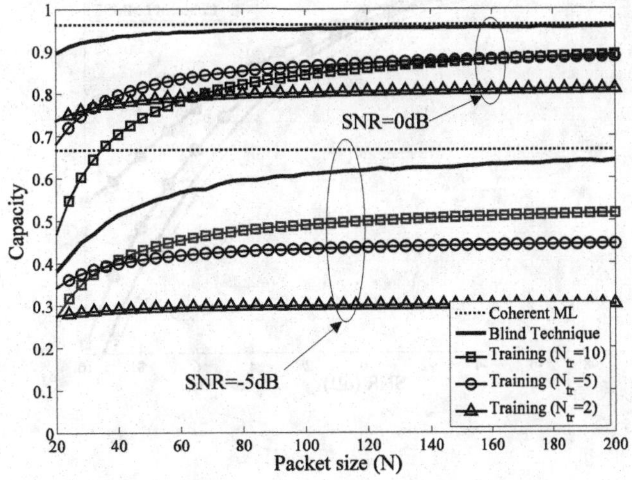

*Figure 5. Channel estimate MSE of the proposed method with K = 4 code permutations. QSTBC with* $n_T = M = L = 8$ *and* $n_R = 4$.

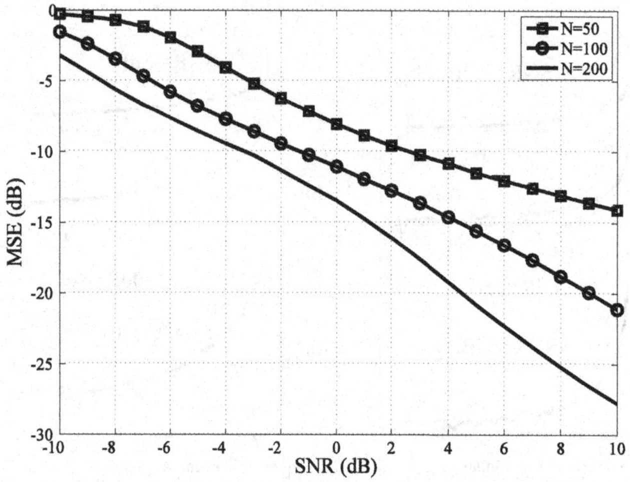


*Figure 6. Bit Error Rate (BER) for the LMMSE receiver with perfect and estimated CSI. QSTBC with* $n_T = M = L = 8$ *and* $n_R = 4$. *K* = 4 *code permutations.*

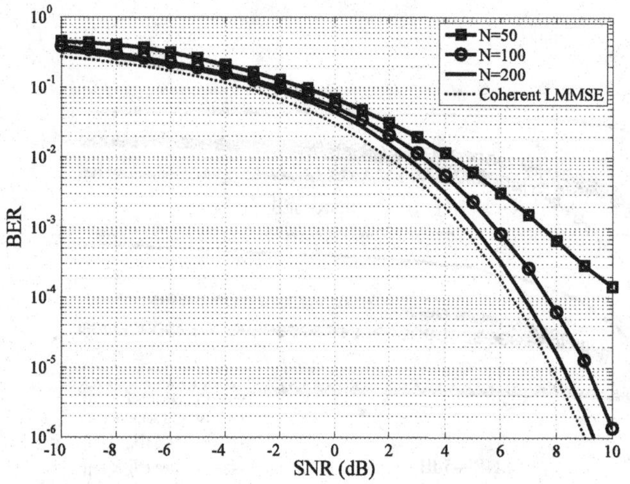

*Figure 7. Bit Error Rate (BER) versus number N of observation blocks for the proposed method (left) or number $N_{tr}$ of training blocks (right). QSTBC with $n_T = M = L = 8$ and $n_R = 4$. K = 4 code permutations.*

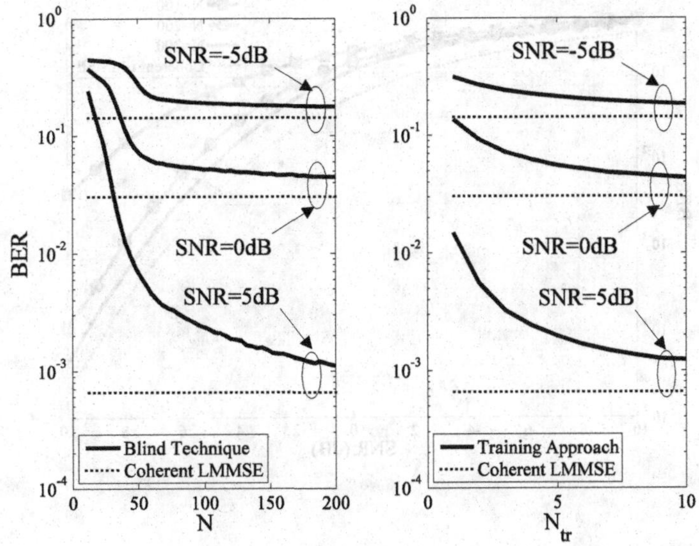


*Figure 8. Bit Error Rate (BER) versus SNR. Example with U = 2 users and R = 3/4 OSTBC. $n_T = L = 4$, M = 3, $n_R$ = 4*

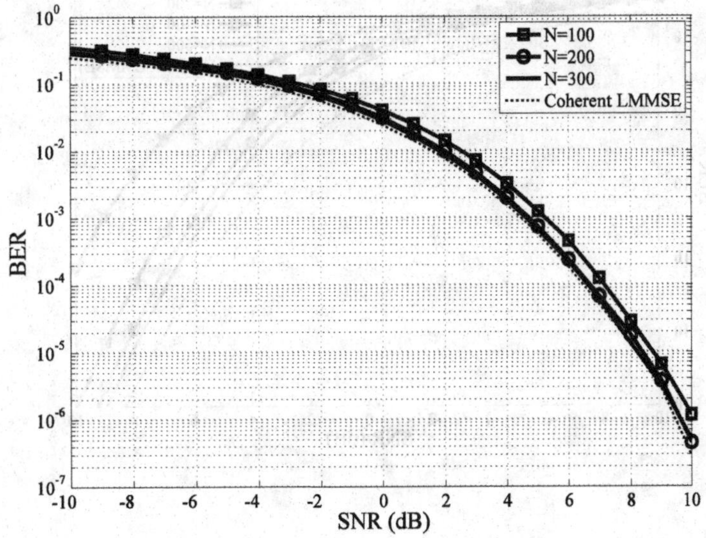

*Figure 9. Bit Error Rate (BER) versus SNR. Example with U = 2 users and R = 1 QSTBC.* $n_T = L = M = 4$, $n_R = 4$.

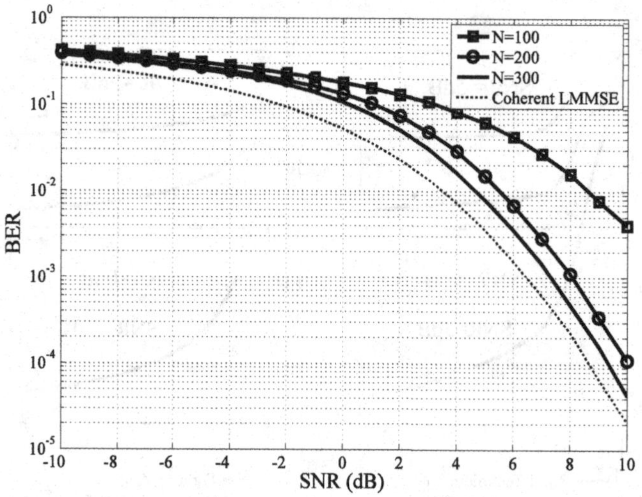


*Figure 10. Bit Error Rate (BER) versus SNR. Example with U = 4 users and R = 1; QSTBC.* $n_T = L = M = 4$, $n_R = 8$

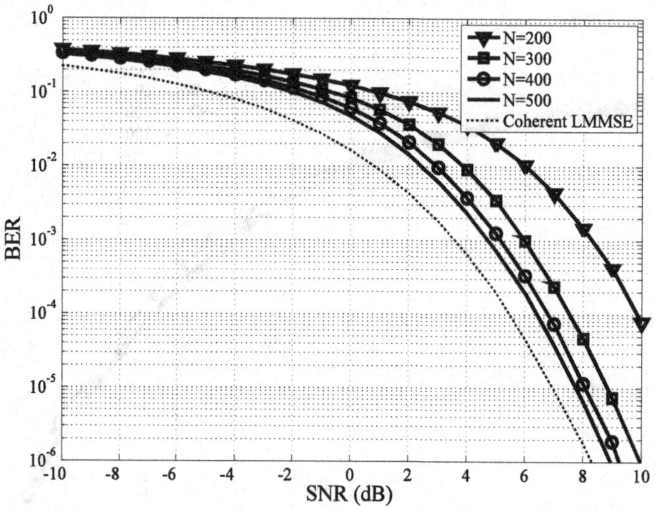


$$\mathbf{Q}_1(8) = \begin{bmatrix} \mathbf{Q}_1(4) & 0 \\ 0 & \mathbf{Q}_2(4) \end{bmatrix}, \quad \mathbf{Q}_2(8) = \begin{bmatrix} \mathbf{Q}_3(4) & 0 \\ 0 & \mathbf{Q}_4(4) \end{bmatrix},$$

$$\mathbf{Q}_3(8) = \begin{bmatrix} \mathbf{Q}_1(4) & 0 \\ 0 & \mathbf{Q}_3(4) \end{bmatrix}, \quad \mathbf{Q}_4(8) = \begin{bmatrix} \mathbf{Q}_2(4) & 0 \\ 0 & \mathbf{Q}_4(4) \end{bmatrix},$$

and the simulation results are shown in Figures 5-7. As can be seen, the results are not as accurate as in the OSTBC examples, which is due to the fact that this is a more complicated problem (non-orthogonal transmissions, higher $n_T$ and transmission rate). However, for a sufficiently large $N$ the proposed technique outperforms the pilot-based approach, and its performance degradation with respect to the coherent receiver is lower than the minimal loss (3-dB) associated to the QSTBC differential technique proposed in (Zhu and Jafarkhani, 2005).

## Multiuser Simulations

In this subsection three different multiuser examples are presented. In all the examples the users transmit with the same power, i.e., the SNR is equal for all the users, and they share a common STBC with $n_T = 4$ transmit antennas, which is modified by means of the non-redundant precoding technique. Specifically, the transmission matrices for the first 4 blocks of $U = 4$ different users are given by

$$\mathbf{S}_1 = \begin{bmatrix} \mathbf{S}(\mathbf{s}_1[0])\mathbf{Q}_1(4) \\ \mathbf{S}(\mathbf{s}_1[1])\mathbf{Q}_2(4) \\ \mathbf{S}(\mathbf{s}_1[2])\mathbf{Q}_3(4) \\ \mathbf{S}(\mathbf{s}_1[3])\mathbf{Q}_4(4) \end{bmatrix}, \quad \mathbf{S}_2 = \begin{bmatrix} \mathbf{S}(\mathbf{s}_2[0])\mathbf{Q}_2(4) \\ \mathbf{S}(\mathbf{s}_2[1])\mathbf{Q}_3(4) \\ \mathbf{S}(\mathbf{s}_2[2])\mathbf{Q}_4(4) \\ \mathbf{S}(\mathbf{s}_2[3])\mathbf{Q}_1(4) \end{bmatrix},$$

$$\mathbf{S}_3 = \begin{bmatrix} \mathbf{S}(\mathbf{s}_3[0])\mathbf{Q}_3(4) \\ \mathbf{S}(\mathbf{s}_3[1])\mathbf{Q}_4(4) \\ \mathbf{S}(\mathbf{s}_3[2])\mathbf{Q}_1(4) \\ \mathbf{S}(\mathbf{s}_3[3])\mathbf{Q}_2(4) \end{bmatrix}, \quad \mathbf{S}_4 = \begin{bmatrix} \mathbf{S}(\mathbf{s}_4[0])\mathbf{Q}_4(4) \\ \mathbf{S}(\mathbf{s}_4[1])\mathbf{Q}_1(4) \\ \mathbf{S}(\mathbf{s}_4[2])\mathbf{Q}_2(4) \\ \mathbf{S}(\mathbf{s}_4[3])\mathbf{Q}_3(4) \end{bmatrix},$$

i.e., all the users employ the same STBC and the same permutation matrices, but the permutations are shifted in time. With this scheme, each user has a virtually different STBC and the ambiguity problems are avoided.

In the first example, $U = 2$ different users transmit with the $R = 3/4$ ($n_T = L = 4$, $M = 3$) amicable-design OSTBC given in eq. (4) (see also Table 1), and the receiver is equipped with $n_R = 4$ receive antennas. We must note that this code is the same used in the simulation examples in (Shahbazpanahi et al., 2006), where the channel ambiguity is resolved by means of some pilot symbols. Figure 8 shows the BER as a function of the SNR for different values of $N$. As can be seen, for $N \geq 100$, the performance loss with respect to the coherent receiver is lower than 1dB.

A similar scenario is considered in the second example, where $U = 2$ users transmit with the Alamouti-based QSTBC for $n_T = L = M = 4$, and the number of receive antennas is $n_R = 4$. Figure 9 shows the BER after decoding, where we can see that, in order to obtain accurate estimates, the number $N$ of blocks at the receiver must be higher than in the OSTBC case. As pointed out before, this difference can be seen as a consequence of the higher complexity of the code.

In the final example, the QSTBC code with $n_T = L = M = 4$ is shared by $U = 4$ users, and the receiver is equipped with $n_R = 8$ receive antennas. Figure 10 shows the simulation results, which allow us to conclude that, for a sufficiently large number of available blocks $N$ (which depends on the problem complexity), the performance of the proposed technique is close to that of the coherent receiver.

# CONCLUSION

In this chapter the problem of blind channel estimation in general space-time block coded (STBC) systems has been analyzed, and a new blind channel estimation criterion has been proposed. The technique reduces to the extraction of the main eigenvector of a generalized eigenvalue (GEV) problem, and it can be easily extended to the multiuser case. The proposed method does not introduce additional ambiguities to those associated to the blind channel estimation problem, and it can be seen as a deterministic approach, i.e., in the absence of noise it is able to exactly recover the channel, up to a real scalar, within a finite number of observations.

Additionally, the problem of blind channel identifiability from second-order statistics (SOS) has been studied, obtaining several intuitive results for the general STBC case, and presenting some sufficient conditions for the case of orthogonal STBCs (OSTBCs). The identifiability analysis has also been exploited to propose a new transmission technique, which is able to avoid the indeterminacies in a large number of situations. The general idea is based on the combination of different

STBCs (*code diversity*) but it can be reduced to a non-redundant precoding consisting in a single rotation or permutation of the transmit antennas. Finally, the proposed techniques have been evaluated by means of several numerical examples, which show that the performance of the blind approaches is close to that of the coherent receiver.

Future research lines include an in-deep study of the blind channel identifiability conditions for a general class of STBCs, as well as the identifiability properties associated to the code diversity and non-redundant precoding techniques. This study should provide answers to questions such as: How many rotations or code combinations are necessary to avoid the ambiguity problems? or, What is the best selection of rotation/permutation matrices?

## REFERENCES

Alamouti, S. (1998). A simple transmit diversity technique for wireless communications. *IEEE Journal on Selected Areas in Communications*, 45(9):1451–1458.

Ammar, N. & Ding, Z. (2006). Channel identifiability under orthogonal space-time coded modulations without training. *IEEE Transactions on Wireless Communications*, 5(5):1003–1013.

Ammar, N. & Ding, Z. (2007). Blind channel identifiability for generic linear space-time block codes. *IEEE Transactions on Signal Processing*, 55(1):202–217.

Barbarossa, S. (2005). *Multiantenna Wireless Communication Systems*. Artech House Publishers.

Capon, J. (1969). High-resolution frequency-wavenumber spectrum analysis. *Proceedings of the IEEE*, 57(8):1408–1418.

Damen, O., Chkeif, A., & Belfiore, J. (2000). Lattice code decoder for space-time codes. *IEEE Communications Letters*, 4(5):161–163.

Diamantaras, K. I. & Kung, S. Y. (1996). *Principal Component Neural Networks, Theory and Applications*. John Wiley & Sons, Inc., New York, NY, USA.

Fasano, A. & Barbarossa, S. (2006). Information lossless full-rate full-diversity trace-orthogonal space-time codes. In *IEEE Workshop on Signal Processing Advances in Wireless Communications (SPAWC 2006)*, Cannes, France.

Fincke, U. & Pohst, M. (1985). Improved methods for calculating vectors of short length in a lattice, including a complexity analysis. *Mathematics of Computation*, 44:463–471.

Ganesan, G. & Stoica, P. (2002). Differential modulation using space-time block codes. *IEEE Signal Processing Letters*, 9(2):57–60.

Gesbert, D., Shafi, M., Shiu, D.-S., Smith, P. J., & Naguib, A. (2003). From theory to practice: an overview of MIMO space-time coded wireless systems. *IEEE Journal on Selected Areas in Communications*, 21(3):281–302.

Godard, D. (1980). Self-recovering equalization and carrier tracking in two-dimensional data communication systems. *IEEE Transactions on Communications*, 28(11):1867–1875.

Hassibi, B. & Hochwald, B. M. (2003). How much training is needed in multiple-antenna wireless links? *IEEE Transactions on Information Theory*, 49(4):951–963.

Henkel, O. (2005). Sphere packing bounds in the Grassmann and Stiefel manifolds. *IEEE Transactions on Information Theory*, 51(10):3445–3456.

Hochwald, B. & Marzetta, T. (2000). Unitary space-time modulation for multiple-antenna communications in Rayleigh flat fading. *IEEE Transactions on Information Theory*, 46(2):543–564.

Hochwald, B., Marzetta, T., Richardson, T., Sweldens, W., & Urbanke, R. (2000). Systematic design of unitary space-time constellations. *IEEE Transactions on Information Theory*, 46(6):1962–1973.

Hochwald, B. & Sweldens, W. (2000). Differential unitary space-time modulation. *IEEE Transactions on Communications*, 48(12):2041–2052.

Hughes, B. L. (2000). Differential space-time modulation. *IEEE Transactions on Information Theory*, 46(7):2567–2578.

Jafarkhani, H. (2005). *Space-Time Coding: Theory and Practice*. Cambridge University Press.

Jafarkhani, H. & Tarokh, V. (2001). Multiple transmit antenna differential detection from generalized orthogonal designs. *IEEE*

*Transactions on Information Theory*, 47(6):2626–2631.

Jaldén, J., Martin, C., & Ottersten, B. (2003). Semidefinite programming for detection in linear systems – optimality conditions and space-time decoding. In *IEEE International Conference on Acoustics, Speech, and Signal Processing (ICASSP 2003)*, volume 4, pages 9–12.

Larsson, E., Stoica, P., & Li, J. (2003). Orthogonal space-time block codes: Maximum likelihood detection for unknown channels and unstructured interferences. *IEEE Transactions on Signal Processing*, 51(2):362–372.

Larsson, E. G. & Stoica, P. (2003). *Space-Time Block Coding for Wireless Communications*. Cambridge University Press, New York, USA.

Liang, X.-B. (2003). Orthogonal designs with maximal rates. *IEEE Transactions on Information Theory*, 49(10):2468–2503.

Ma, W.-K. (2007). Blind ML detection of orthogonal space-time block codes: Identifiability and code construction. *IEEE Transactions on Signal Processing*, 55(7):3312–3324.

Ma, W. K., Vo, B. N., Davidson, T. N., & Ching, P. C. (2006). Blind ML detection of orthogonal space-time block codes: Efficient high-performance implementations. *IEEE Transactions on Signal Processing*, 54(2):738–751.

Naguib, A. F., Tarokh, V., Seshadri, N., & Calderbank, A. R. (1998). A space-time coding modem for high-data-rate wireless communications. *IEEE Journal on Selected Areas in Communications*, 16(8):1459–1478.

Oja, E. (1992). Principal components, minor components, and linear neural networks. *Neural Networks*, 5(6):927–935.

Pohl, V., Nguyen, P., Jungnickel, V., & von Helmolt, C. (2005). Continuous flat fading MIMO channels: Achievable rate and the optimal length of the training and data phase. *IEEE Transactions on Wireless Communications*, 4(4):1889–1900.

Schmidt, R. (1986). Multiple emitter location and signal parameter estimation. *IEEE Transactions on Antennas Propagation*, 34(3):276–280.

Sezgin, A. (2005). *Space-Time Codes for MIMO Systems: Quasi-Orthogonal Design and Concatenation*. PhD thesis, University of Technology, Berlin, Germany.

Sezgin, A., & Oechtering, T. (2008). Complete characterization of the equivalent MIMO channel for quasi-orthogonal space-time codes. To appear in *IEEE Transactions on Information Theory*.

Shahbazpanahi, S., Gershman, A. B., & Giannakis, G. B. (2006). Semiblind multiuser MIMO channel estimation using Capon and MUSIC techniques. *IEEE Transactions on Signal Processing*, 54(9):3581–3591.

Shahbazpanahi, S., Gershman, A. B., & Manton, J. H. (2005). Closed-form blind MIMO channel estimation for orthogonal space-time block codes. *IEEE Transactions on Signal Processing*, 53(12):4506–4517.

Stoica, P. & Ganesan, G. (2003). Space–time block codes: Trained, blind, and semi–blind detection. *Digital Signal Processing*, 13:93–105.

Swindlehurst, A. L. (2002). Blind and semi-blind equalization for generalized space-time precoding. In *IEEE International Conference on Acoustics, Speech and Signal Processing (ICASSP 2002)*, Orlando, FL.

Swindlehurst, L. & Leus, G. (2002). Blind and semi-blind equalization for generalized space-time block codes. *IEEE Transactions on Signal Processing*, 50(10):2489–2498.

Tarokh, V. & Jafarkhani, H. (2000). A differential detection scheme for transmit diversity. *IEEE Journal on Selected Areas in Communications*, 18(7):1169–1174.

Tarokh, V., Jafarkhani, H., & Calderbank, A. R. (1999). Space-time block codes from orthogonal designs. *IEEE Transactions on Information Theory*, 45(5):1456–1467.

Tran, L. C., Wysocki, T. A., Mertins, A., & Seberry, J. (2006). *Complex Orthogonal Space-Time Processing in Wireless Communications*. Springer-Verlag New York, Inc., Secaucus, NJ, USA.

Vía, J., & Santamaría, I. (2008). On the blind identifiability of orthogonal space-time block codes from second order statistics. *IEEE Transactions on Information Theory, 54*(2), 709-722.

Vía, J. & Santamaría, I. (2007b). Some results on the blind identifiability of orthogonal space-time block codes from second order statistics. In *IEEE International Conference on Acoustic, Speech, and Signal Processing (ICASSP 2007)*, Honolulu, Hawaii, USA.

Vía, J., Santamaría, I., & Pérez, J. (2006a). Blind identification of MIMO-OSTBC channels combining second and higher order statistics. In *European Signal Processing Conference (EUSIPCO 2006)*, Florence, Italy.

Vía, J., Santamaría, I., & Pérez, J. (2006b). A sufficient condition for blind identifiability of MIMO-OSTBC channels based on second order statistics. In *IEEE Workshop on Signal Processing Advances in Wireless Communications (SPAWC 2006)*, Cannes, France.

Vía, J., Santamaría, I., Pérez, J., & Ramírez, D. (2006c). Blind decoding of MISO-OSTBC systems based on principal component analysis. In *IEEE International Conference on Acoustic, Speech, and Signal Processing (ICASSP 2006)*, Toulouse, France.

Zhang, J. K., Liu, J., & Wong, K. M. (2005). Trace-orthogonal full diversity cyclotomic space-time codes. In *Space-Time processing for MIMO communications*. Wiley.

Zhang, J. K., Liu, J., & Wong, K. M. (2007). Trace-orthonormal full-diversity cyclotomic space-time codes. *IEEE Transactions on Signal Prcessing*, 55(2):618–630.

Zhu, Y. & Jafarkhani, H. (2005). Differential modulation based on quasi-orthogonal codes. *IEEE Transactions on Wireless Communications*, 4(6):3005–3017.

# Chapter IX
# Fast Beamforming of Compact Array Antenna

**Chen Sun**
*ATR Wave Engineering Laboratories, Japan*

**Takashi Ohira**
*Toyohashi University of Technology, Japan*

**Makoto Taromaru**
*ATR Wave Engineering Laboratories, Japan*

**Nemai Chandra Karmakar**
*Monash University, Australia*

**Akifumi Hirata**
*Kyocera Corporation, Japan*

## ABSTRACT

*In this chapter, we describe a compact array antenna. Beamforming is achieved by tuning the load reactances at parasitic elements surrounding the active central element. The existing beam forming algorithms for this reactively controlled parasitic array antennas require long training time. In comparison with these algorithms, a faster beamforming algorithm, based on simultaneous perturbation stochastic approximation (SPSA) theory with a maximum cross-correlation coefficient (MCCC) criterion, is proposed in this chapter. The simulation results validate the algorithm. In an environment where the signal-to-interference ratio (SIR) is 0 dB, the algorithm converges within 50 iterations and achieves an output SINR of 10 dB. With the fast beamforming ability and its low power consumption attribute, the antenna makes the mass deployment of smart antenna technologies practical. To give a comparison of the beamforming algorithm with one of the standard beamforming algorithms for a digital beamforming (DBF) antenna array, we compare the proposed algorithm with the least mean square (LMS) beamforming algorithm. Since the parasitic array antenna is in nature an analog antenna, it cannot suppress correlated* **interference***. Here, we assume that the interferences are uncorrelated.*

## INTRODUCTION

The evolution of wireless communications systems requires new technologies to support better quality communications, new services and applications. Smart antennas have become a hot topic of research. With a **smart antenna** directive beam patterns can be steered toward the desired signal and deep nulls can be formed toward the interference, thus spatial filtering is realized. This brings the benefits such as lower power transmission, higher spectrum efficiency, better link quality and higher system capacity (Godara, 1997a; Winters, 1998; Tsoulos, 1999; Boukalov, 2000; Jana, 2000; Friodigh, 2001; Ogawa, 2001; Bhobe, 2001; Soni, 2002; Blogh, 2002; Bellofiore, 2002a; Bellofiore, 2002b; Diggavi, 2004).

Various **beamforming** and direction of arrival (DOA) estimation algorithms have been designed (Widrow, 1967; Van-Veen, 1988; Litva, 1996; Godara, 1997b; Anderson, 1999; Lehne, 1999; Boukalov, 2000; Janaswamy, 2001; Rappaport, 2002; Blogh, 2002; Bellofiore, 2002b). The simulation and experiments carried out by many researchers have shown the abilities of these algorithms (Anderson, 1996a; Anderson, 1996b, Winters, 1997; Tsoulos, 1997; Boukalov, 2000). Most of these algorithms are designed based on the digital beamforming (DBF) antenna arrays. Signals received by individual antenna elements are down-converted into baseband signals. These signals are digitized and fed into digital signal processing (DSP) chip where the algorithms reside in. However, radio-frequency (RF) circuit branches connected to the array elements, analog-to-digital converters (ADCs) and the baseband DSP chip consume a considerable amount of dc power. Furthermore, each channel connected to the array sensor has the same structure, so the cost of fabrication increases with the number of array elements (Ohira, 2000; Boukalov, 2000; Thiel, 2001). Thanks to the recent development of GaAs monolithic microwave integrated circuit (MMIC) technologies, the beamformer could be integrated into a single chip at the RF front end such as MBF (Ohira, 1997), instead of the baseband. The advantages are the reduced quantization errors and the increased dynamic range. However, their costs of fabrication still limit the range of implementations. All these factors make DBF and microwave beamforming (MBF) antennas unsuitable for low power consumption and low cost systems and thus hinder the mass applications of the **smart antenna** technologies. For example, it could be too costly to equip DBF antenna arrays at battery powered lap-tops or mobile computing terminals within a wireless network.

As the wireless communications become more ubiquitous, the demands for smart antennas expend. Smart antennas are now not only required to be installed at base stations (BSs) but also strongly desired to be installed at mobile stations (MSs) such as vehicles, laptops, and even commercial mobile phones. The challenges emerge. Due to the size and power limit of the MSs, the **smart antennas** are required to be low power consumption, small in size and low cost so that it is affordable to mobile users. With this goal in mind, researchers are now investigating compact array antennas such as the **parasitic array** antennas (Black, 1973; Gueguen, 1974; Himmel, 1978; Harrington, 1978; Milne, 1985; Thiel, 1996; Sibille, 1997; Preston, 1997; Preston, 1998; Preston, 1999; Vaughan, 1999; Ohira, 2000a; Ohira, 2000b; Svantesseon, 2001; Thiel, 2001; Sun, 2002a; Varlamos, 2004; Sun, 2004). The antenna normally has one RF port, therefore, the size and power consumption can be significantly reduced (Ohira, 2000b; Thiel, 2001).

Various forms of switched **parasitic array** antennas have been documented (Himmel, 1978; Black, 1973; Gueguen, 1974; Milne, 1985; Thiel, 1996; Sibille, 1997; Preston, 1998; Preston, 1999; Vaughan, 1999; Svantesseon, 2001; Varlamos, 2004). Beam steering of these antennas are achieved by switching ON and OFF parasitic elements, or switching the position of the active element. However, beam patterns can only be steered to a predefined set of directions in a way similar to that for switched beam antennas. This limits the performance, especially for the applications scenarios where desired signals impinging from different directions due to multipath reflection.

Another category of **parasitic array** antennas, reactively controlled directive array antennas, circumvents the problem with switched parasitic array antennas. The first work was presented in (Harrington, 1978). The antenna has one central element connected to the sole RF port and a number of surrounding parasitic elements form the array. Beam steering is achieved by tuning the load reactances at parasitic elements surrounding the central active element. Each parasitic element is loaded with tuneable reactance.

Recently, the adaptive **beamforming** with reactively controlled directive array antennas has been studied (Ohira, 2000a; Ohira, 2000b; Gyoda, 2000; Komatsuzaki, 2000; Cheng, 2001a; Cheng, 2001b; Shishkov, 2001; Shishkov, 2002; Sun, 2002b; Ohira, 2002; Cheng, 2002; Hirata, 2002; Sun, 2003; Sun, 2004). The antenna is named as "electronically steerable parasitic array radiator (**Espar**)". In (Gyoda, 2000), a random search method is applied to optimize the reactance values. A Hamiltonian approach is presented by Komatsuzaki, (2000). It projects an optimization problem on the motion of a particle in an $M$-dimensional space, by making the objective function and the unknown variables correspond to the potential energy and the coordinates, respectively. However, the optimum result is only achieved through a large number of calculation cycles. The steepest gradient algorithm (SGA) is applied in (Cheng, 2001a; Cheng, 2001b). The algorithm maximizes the cross correlation coefficient (CCC) between the received signal and the training signal, using steepest gradient method based on sequential perturbation (Moon, 2001). To cope with the noise effect, maladjustment and uncertainty in estimating the objective function, another beamforming algorithm is proposed in (Shishkov, 2001; Shishkov, 2002). The algorithm is based on the stochastic approximation theory. A blind adaptive algorithm maximizes the higher-order moment using steepest gradient method is presented in (Ohira, 2002). The algorithm does not require the **training signals**.

However, the above algorithms are not efficient enough for real time implementations in the wireless communications systems, where a fast beam tuning ability is required.

In this chapter a fast beamforming algorithm based on the **simultaneous perturbation stochastic approximation** (**SPSA**) theory (Spall, 1998) is proposed. With an analytical model, the performance is examined with simulations. The experiment result can be found in (Sun, 2004). Our study proves the adaptive beamforming ability of the Espar antennas and demonstrates its suitability for mass commercial implementations of smart antenna technologies to wireless networks.

## BEAMFORMING ARCHITECTURES AND THEORY

### A. Antenna Configuration

Figure 1 shows the configuration of a 7-element Espar antenna. One active central element (monopole) is surrounded by 6 parasitic elements that are uniformly placed on a circular grid of radius $R$ on a circular grounded baseplate. The length of each monopole $L$ and the radius $R$ of the circular grid are one-quarter wavelength ($\lambda/4$) of the transmitting RF signal at 2.484 GHz. According to Harrington (1978), large spacing among elements (for example $\lambda/2$) leads to reduced gain and significant back lobes. Moreover, further reduction of the inter-element spacing to $\lambda/8$ leads to the super-gain array. Therefore, a realistic design for spacing of $\lambda/4$ is chosen. The baseplate transforms monopoles with their images to dipoles with a length of $2L$. The central monopole is connected to a RF receiver and each parasitic monopole is loaded with a tuneable varactor.

The working principle of **Espar** antennas is different from that of DBF array antennas. We first assume that the antenna is working in the transmitting mode. The antenna generates a directional beam based on tuning load reactances ($x_1, x_2, \ldots, x_6$) on the parasitic monopoles. Signals received or transmitted from the central RF port excite the parasitic monopoles with substantial induced mutual currents on them. The currents and the voltages on the monopoles are denoted by vectors **i** and **v**,

$$\mathrm{i} = [i_0\ i_1\ i_2\ i_3\ i_4\ i_5\ i_6]^T, \tag{1}$$

$$\mathbf{v} = [v_0\ v_1\ v_2\ v_3\ v_4\ v_5\ v_6]^T. \tag{2}$$

*Figure 1. Configuration of a 7-element Espar antenna. The length of each monopole L and the radius R of the circular grid are one-quarter wavelength of the transmitting RF signal at 2.484 GHz.*

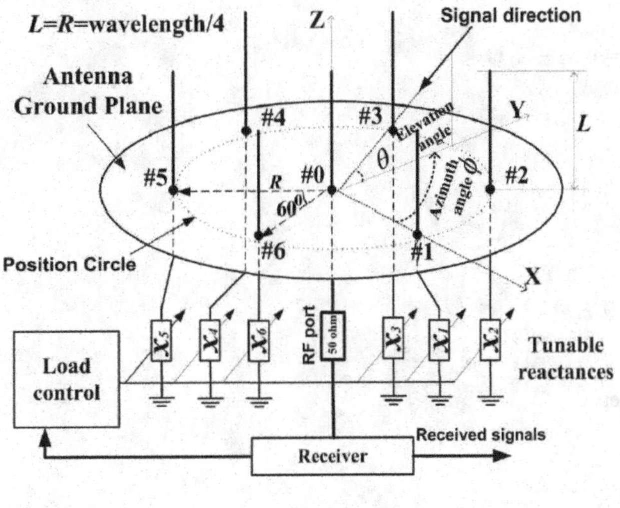

The superscript $T$ represents the transpose. $i_0$ and $v_0$ represent the current and the voltage on the active central element (#0), respectively. Mutual admittances are represented by the matrix $\mathbf{Y}_C$ with each entity $y_{ij}$ denoting the mutual admittance between the $i$th element and the $j$th element. The induced currents are represented with mutual admittances

$$\mathbf{i} = \mathbf{Y}_C \mathbf{v} = \begin{bmatrix} y_{00} & y_{01} & y_{02} & y_{03} & y_{04} & y_{05} & y_{06} \\ y_{10} & y_{11} & y_{12} & y_{13} & y_{14} & y_{15} & y_{16} \\ y_{20} & y_{21} & y_{22} & y_{23} & y_{24} & y_{25} & y_{26} \\ y_{30} & y_{31} & y_{32} & y_{33} & y_{34} & y_{35} & y_{36} \\ y_{40} & y_{41} & y_{42} & y_{43} & y_{44} & y_{45} & y_{46} \\ y_{50} & y_{51} & y_{52} & y_{53} & y_{54} & y_{55} & y_{56} \\ y_{60} & y_{61} & y_{62} & y_{63} & y_{64} & y_{65} & y_{66} \end{bmatrix} \times \begin{bmatrix} v_0 \\ v_1 \\ v_2 \\ v_3 \\ v_4 \\ v_5 \\ v_6 \end{bmatrix}. \tag{3}$$

Because of the symmetric structure of the Espar antenna, the admittances are

$$y_{11} = y_{22} = y_{33} = y_{44} = y_{55} = y_{66},$$
$$y_{01} = y_{02} = y_{03} = y_{04} = y_{05} = y_{06},$$
$$y_{12} = y_{23} = y_{34} = y_{45} = y_{56} = y_{61},$$
$$y_{13} = y_{24} = y_{35} = y_{46} = y_{51} = y_{62},$$
$$y_{14} = y_{25} = y_{36},$$

Therefore, the admittance matrix $\mathbf{Y}_C$ is described with only 6 independent parameters.

These values are obtained with the Numerical Electromagnetic Code (NEC). The voltages on the active central monopole and the $m$th parasitic monopole are represented by (4) and (5), respectively,

$$v_0 = v_s - z_0 i_0 , \tag{4}$$

and

$$v_m = -jx_m i_m , \tag{5}$$

where $m = 1, \ldots, 6$. $z_0$ is the characteristic impedance of 50 Ω at the RF port. $v_s$ represents the transmitted voltage signal source with the amplitude and the phase from the driving RF port at the central element. We write equations (4) and (5) into a matrix form as

$$\mathbf{v} = [v_s \ 0\,0\,0\,0\,0\,0]^T - \mathbf{X}_L \mathbf{i} = v_s \mathbf{u}_1 - \mathbf{X}_L \mathbf{i}, \tag{6}$$

and define

$$\mathbf{X} = \mathrm{diag}[50, \quad jx_1, \quad \ldots, \quad jx_6], \tag{7}$$

where

$$\mathbf{u}_1 = [1 \quad 0 \quad 0 \quad 0 \quad 0 \quad 0 \quad 0]^T. \tag{8}$$

Combining (3) and (6), we get

$$\mathbf{i} = \mathbf{Y}_C \mathbf{v} = \mathbf{Y}_C (v_s \mathbf{u}_1 - \mathbf{X}_L \mathbf{i}). \tag{9}$$

After a simple mathematical manipulation, we can obtain

$$\mathbf{i} = v_s (\mathbf{Y}_C^{-1} + \mathbf{X}_L)^{-1} \mathbf{u}_1 = v_s \mathbf{w}_{ESP}, \tag{10}$$

where $(\mathbf{Y}_C^{-1} + \mathbf{X}_L)^{-1}\mathbf{u}_1$ is denoted by $\mathbf{w}_{ESP}$, a 7-by-1 dimensional vector, called as "**Espar** equivalent weight vector". The far-field radiation pattern is the superposition of all monopoles' radiation patterns (Balanis, 1997). Therefore, the far-field current signal in the azimuthal direction $\phi$ with its amplitude and the phase is represented as

$$y_{far}(\phi) = \mathbf{i}^T \boldsymbol{\alpha}(\phi) = \mathbf{w}_{ESP} \boldsymbol{\alpha}(\phi) v_s. \tag{11}$$

The steering vector $\boldsymbol{\alpha}(\phi)$ is defined based on the array geometry (see Figure 1) as

$$\boldsymbol{\alpha}(\phi) = \left[1 \;\; e^{j\frac{\pi}{2}\cos(\phi)} \;\; e^{j\frac{\pi}{2}\cos\left(\phi - \frac{\pi}{3}\right)} \;\; e^{j\frac{\pi}{2}\cos\left(\phi - \frac{2\pi}{3}\right)} \;\; e^{j\frac{\pi}{2}\cos(\phi - \pi)} \;\; e^{j\frac{\pi}{2}\cos\left(\phi - \frac{4\pi}{3}\right)} \;\; e^{j\frac{\pi}{2}\cos\left(\phi - \frac{5\pi}{3}\right)}\right]^T. \tag{12}$$

According to the reciprocity theory for radiation patterns (Balanis, 1997), if the antenna is working in the receiving mode, the voltage signals *y*(t) at the RF port are

$$y(t) = \mathbf{w}_{ESP}^T \boldsymbol{\alpha}(\phi) s(t). \tag{13}$$

where, *s*(*t*) represents the far-field incident current waves with the amplitude and the phase in the azimuthal direction $\phi$.

## B. Calculation of Mutual Admittance Matrix

In the previous section, we have shown that **beamforming** of an Espar antenna is achieved based on strong **mutual coupling** among neighbouring elements (one active centre monopoles and 6 parasitic monopoles loaded with tuneable reactance).

The calculation of the mutual impedance matrix $\mathbf{Y}_C$ is achieved with Numerical Electromagnetic Code, i.e. NEC simulator. NEC is a moment method code for analyzing the interaction of electromagnetic waves with arbitrary structures. The NEC simulator we used is from Poynting Software (Pty) Ltd. It is an object-oriented software, which uses an electric-field integral equation (EFIE) to model the electromagnetic response of general structures. Surface and elements are modelled with wire segments. In Figure 2, each monopole is modelled with 21 wire segments. All the monopoles are loaded with 50 Ω impedance at the centre segment. By exciting one monopole at the centre segment with a unitary voltage, we can get the currents at the centre segments of all monopoles. Since the exciting voltage is unitary, the currents on the neighbouring monopoles are the mutual admittance and the current on the excited monopole is the self admittance. The values are listed below.

$y_{00} = 0.00023344 - j0.0066918$
$y_{10} = -0.0001816 + j0.00242469$
$y_{11} = 0.00300676 - \mathrm{j}0.0044176$
$y_{21} = 0.00097427 + j0.00299986$
$y_{31} = -0.0003066 - j0.0003067$
$y_{41} = -0.0000973 - j0.0001353$

*Figure 2. Structure of a seven-element Espar antenna on an infinite ground plane in NEC simulator. The ground plane transforms the monopoles with their image into λ/2 dipoles. The inter-element spacing is λ/4 at 2.484 GHz.*

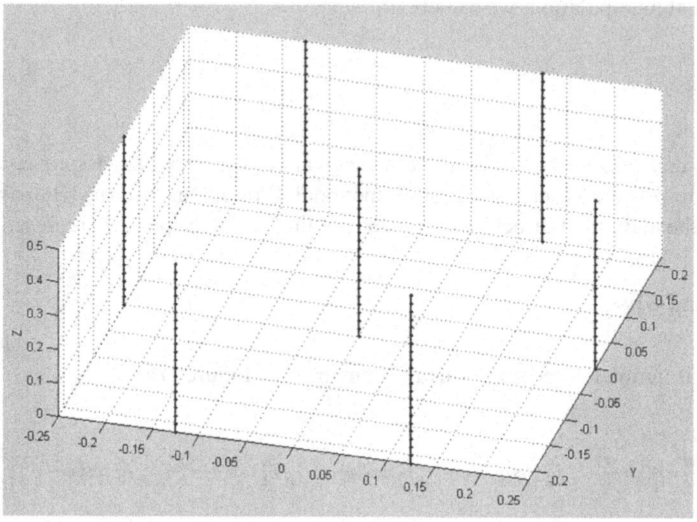


## C. Espar Antenna Beam Pattern

In the previous two subsections, we have described the theoretical foundation of Espar antennas, and explained how to calculate the mutual impedance matrix of an Espar antenna using NEC simulator. In this subsection, we investigate the beam pattern generated by an Espar antenna.

Firstly, we compared the beam pattern generated using MATLAB© code with equation (13) with the pattern generated using NEC simulator. The reactance values $x_1$, $x_2$, $x_3$, $x_4$, $x_5$ and $x_6$ are set to -90 Ω, 0 Ω, 0 Ω, 0 Ω, 0 Ω and 0 Ω, respectively. This produces a sector beam pattern. The three dimensional (3D) current distributions generated using the NEC simulator is shown in Figure 3. The colour bar indicates the current amplitude. The 2D radiation pattern in the azimuthal plane ($\theta = 90°$) is compared the radiation pattern calculated from MATLAB© code using the analytical model given in (13). As shown in Figure 4, the close match between the two patterns verifies (13).

So far, we have simulated beam patterns of a 7-element Espar antenna with quarter wavelength inter-element spacing. Here, we keep the length of the antenna elements and examine the influence of inter-element spacing to the beam forming performance. From the previously published study (Harrington, 1978), it is understood that a larger inter-element spacing (e.g., λ/2) leads to less beam pattern controllability by loaded reactances. From Figure 5, it can be observed that pattern 1 has a larger back lobe than pattern 2 for which the inter-element spacing is λ/4. Further decreasing of the inter-element spacing leads to more directive patterns (see pattern 3). However, this tends to be a supergain excitation of the array (Balanis, 1997), which means the optimum design is very sensitive to frequency and design variations. This is also illustrated in Figure 6. As the inter-element spacing decreases, the pattern gain becomes more sensitive to the frequency.

## D. Adaptive Beamforming Theory

Examining $\mathbf{w}_{ESP}$ in (13), we found that it is dependent on the reactance at each parasitic monopole. Desired beam patterns could be formed by tuning the reactances ($x_1$, $x_2$, ..., $x_6$). This is achieved by changing the control voltages ($V_{cc1}$, $V_{cc2}$, ..., $V_{cc6}$) of the diodes, which are connected to parasitic elements. Note that (13) has the same form as the beamforming equation in array processing literatures (Widrow, 1967; Johnson, 1993). Signals at the antenna receiver are the sum of the weighted signal samples at individual antenna sensor elements.

For **beamforming** based on conventional DBF array antennas, signal samples are obtained at the baseband, and the desired output is the linear combination of these samples. If the objective function is the mean square error (MSE) between the **training signals** and the combined output, the **beamforming** is a Wiener filtering problem. The optimum

*Figure 3. The current distributions on the monopoles of a 7-element Espar antenna generated using NEC simulator. The ground plane transforms the monopoles with their image into λ/2 dipoles. The inter-element spacing is λ/4. The reactance values x1, x2, x3, x4, x5 and x6 are set to -90 Ω, 0 Ω, 0 Ω, 0 Ω, 0 Ω and 0 Ω, respectively*

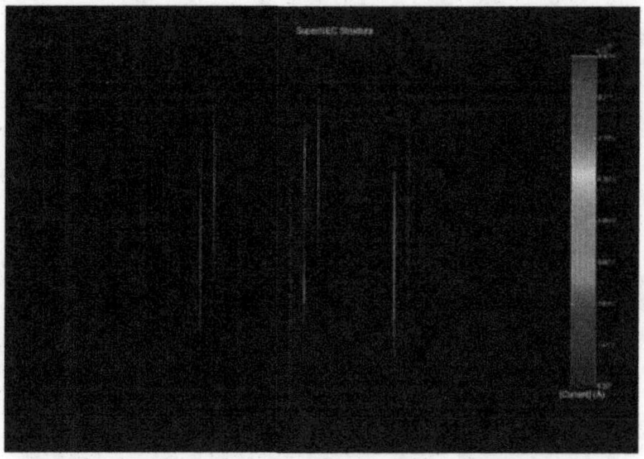

*Figure 4. The radiation patterns of a 7-element Espar antenna generated using NEC simulator and MATLAB© code. The inter-element spacing is λ/4. The reactance values x1, x2, x3, x4, x5 and x6 are set to -90 Ω, 0 Ω, 0 Ω, 0 Ω, 0 Ω and 0 Ω, respectively.*

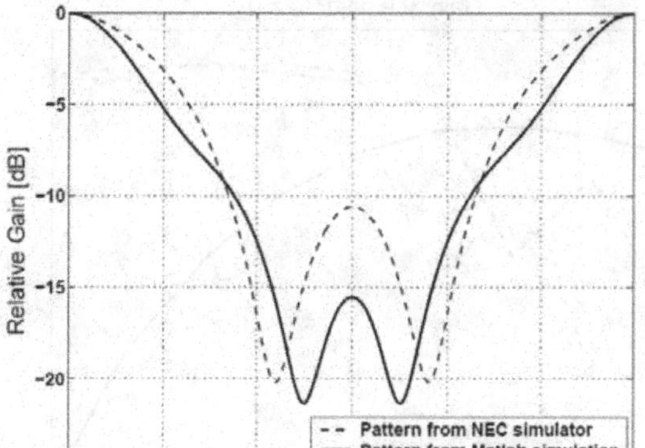


solution of the combining weight is obtained by solving the Wiener-Hopf equation (Widrow, 1967; Haykin, 1902). To reduce the computational complexity, the (least mean squre) LMS algorithm is employed to obtain the optimum solution through iteration using the gradient of the MSE objective function which is in a linear quadratic form (Haykin, 2002). On the contrary, the Espar antenna has only one RF port. Signals impinging on the antenna elements are combined together due to the strong **mutual coupling**. The baseband receiver receives the combined signals from the sole RF port. This combination is a nonlinear function of the loading reactance as shown in the above equations.

Furthermore, for the DBF array antennas, the optimum *weight* obtained is multiplied with the signal sample vector digitally at the baseband. While for Espar antennas, the *weight* is tuned by changing the reactance values at antenna elements based on a nonlinear relationship between the control voltages and the beam pattern. Therefore, $\mathbf{w}_{ESP}$ defined in (10) is called "equivalent weight vector". This unique property of the Espar antenna initiates our design of the adaptive beamforming algorithm based on a numerical iterative manner as for a nonlinear filtering problem.

*Figure 5. The radiation patterns of a 7-element Espar antenna generated using NEC simulator correspond to different inter-element spacing. The reactance values x1, x2, x3, x4, x5 and x6 are set to -90 Ω, 0 Ω, 0 Ω, 0 Ω, 0 Ω and 0 Ω, respectively.*

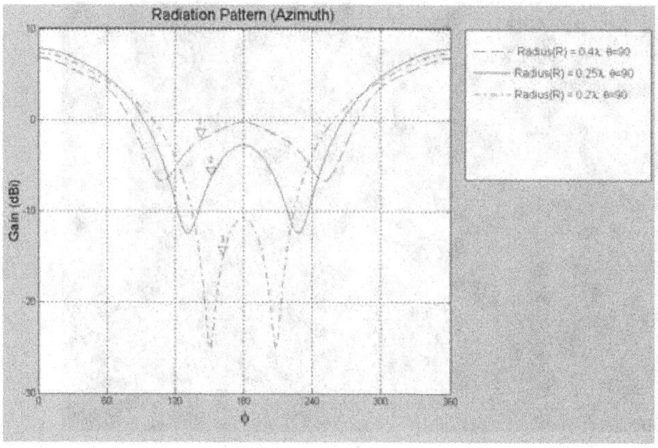


*Figure 6. The radiation pattern gains of a 7-element Espar antenna generated using NEC simulator correspond to different frequencies. The reactance values x1, x2, x3, x4, x5 and x6 are set to -90 Ω, 0 Ω, 0 Ω, 0 Ω, 0 Ω and 0 Ω, respectively.*

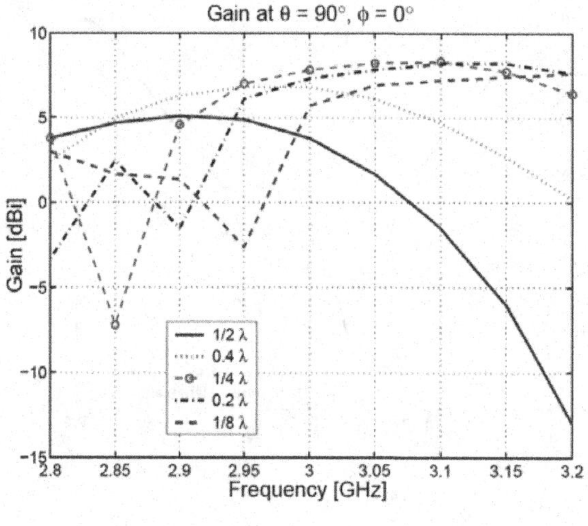


## ADAPTIVE ALGORITHM

As aforementioned, the optimum weight has to be obtained numerically. The objective function of the iteration process is the normalized mean square error (NMSE) between the received signal and the desired signal (training signal). It is simply represented by $1-|\rho_{yr}|^2$, where $\rho_{yr}$ is the cross correlation coefficient (CCC) between the desired and the received signals (Shishkov, 2001). Here, $\rho_{yr}$ is expressed as

$$\rho_{yr} = \mathrm{E}\left[y(t)r(t)^*\right] \Big/ \sqrt{\mathrm{E}\left[y(t)y(t)^*\right]\mathrm{E}\left[r(t)r(t)^*\right]}, \tag{14}$$

where $r(t)$ is the training signal. Since the objective function is based on the CCC, and the goal of the iteration is to maximize the CCC thus minimizing the NMSE, we also call this algorithm a maximum CCC (MCCC) algorithm. The output signal-to-interference-plus-noise ratio (SINR) is approximately represented by $|\rho_{yr}|^2/(1-|\rho_{yr}|^2)$. In this optimization problem, the system parameters are the reactance values, which are controlled by the reversely biased (RB) voltages ($V_{cc1}$, $V_{cc2}$, ..., $V_{cc6}$) at tuneable varactors. The desired beam pattern is obtained by changing the reactance values through iterations.

The reported algorithm (Cheng, 2001b) is based on the SGA (Moon, 2001), a classical deterministic optimization method. It assumes that one knows exactly the loss function (and its derivatives if relevant) and that information is used to calculate the gradient of the optimization search route in a deterministic manner at each step of the iteration. The process perturbs each parameter of the system; measures the loss function; and determines the gradient with respect to each individual parameter. This is not computationally efficient if the system has a considerable amount of parameters. As for the 7-element Espar antenna, there are 6 parameters (load reactances at six parasitic elements). This algorithm hinders us to implement more parasitic elements to improve the performance and impedes the reduction of the iteration process. Furthermore, we do not have the perfect information about the loss function, because the measurement of the loss function is quite *noisy*. This *noise* denoted by $\varepsilon$ in following equations is not added white Gaussian noise (AWGN). Rather, it is the randomness and the inevitable error in observing the loss function (Kushner, 1997). For example, we cannot compute the mean square error. In practice, it is a squared error over a finite length of observations. In our algorithm the observation length of the loss function is 50 data bits. Therefore, the optimization process of the Espar antenna should be modelled more properly as the stochastic optimization problem, especially in the Kiefer-Wolfowitz setting (Kushner), in which one uses only noisy measurements of the objective function to estimate the gradient of the objective function.

## A. Signal Model and Assumptions

To model the propagation environment and simulate the adaptive beamforming of an Espar antenna, we assume a line-of-sight (LOS) propagation environment, and the multipath components are not considered. These assumptions give a direct insight into the beamforming ability of the Espar antenna.

Signals are assumed to come in the azimuth plane. When there are total $K$ signals impinging on the antenna, the received RF signal $y(t)$ is the sum of all the RF signals from individual signal sources

$$y(t)=\sum_{k=1}^{K}\mathbf{w}_{ESP}^{T}\boldsymbol{\alpha}(\phi_k)s_k(t)+n(t), \tag{15}$$

where $\phi_k$ is the azimuthal angle of the $k$th impinging signal. And, $n(t)$ is the AWGN component with power $\sigma^2$. Also, $s_k(t)$ denotes the $k$th signal wave impinging on the antenna array with unit power. Here we define a scalar $\alpha(\phi_k)$ as $\mathbf{w}_{ESP}^{T}\boldsymbol{\alpha}(\phi_k)$.

## B. Optimization Procedure

The objective function is the NMSE. It is represented as

$$L(X(n))=1-|\rho_{yr}|^2, \tag{16}$$

where $X(n)$ is the reactance values ($x_1, x_2, \ldots, x_6$) at the $n$th iteration stage. And, $\rho_{yr}$ is the CCC between the received signal and the training signal. The estimated gradient of the optimization problem is

$$\hat{g}_n=\left(\hat{L}_n^{(+)}-\hat{L}_n^{(-)}\right)\Big/2c(n)\Delta_n, \tag{17}$$

where ^ denotes estimated values. $\hat{L}_n^{(\pm)}$ is the estimated value of the objective function with the parameters (6 reactance values) simultaneously perturbed to the positive and negative directions, respectively, at the $n$th iteration,

$$\hat{L}_n^{(\pm)}=L\left(X(n)\pm c(n)\Delta_n\right)\pm\varepsilon^{(\pm)}, \tag{18}$$

$$\Delta_n=\left[\Delta_n^{(1)}\quad\Delta_n^{(2)}\quad\cdots\quad\Delta_n^{(6)}\right]. \tag{19}$$

Each entity is selected based on the Bernoulli distribution. Their values are either 1 or -1 with a probability of 1/2 (Spall, 1998). $c(n)$ is the perturbation size at the $n$th iteration

$$c(n)=C/(n+1)^{\gamma}, \tag{20}$$

and $C$ is a constant. $\varepsilon$ is the *noise* component as aforementioned. Please note that, although the observation of the loss function is a scalar, $g_n$ is a vector with 6 entities. New reactance values are updated as

$$X(n+1)=X(n)-a(n)\hat{g}_n, \tag{21}$$

where $a(n)$ is the updating step size, $a$ and $A$ are the constants. $a$ and $\gamma$ are nonnegative coefficients

$$a(n)=a/(A+n+1)^{\alpha}. \tag{22}$$

Please note that only two measurements of the loss function are needed to get the gradient $g_n$. This is obviously because all the parameters are always perturbed together instead of one by one as in the steepest gradient optimization process (Cheng, 2001b). The step size goes to zero as n → ∞. This characteristic allows an average of the *noise* in the long run

(Kushner, 1997). And, $c(n)$ follows $\sum_{n=0}^{\infty} c(n)=\infty$ to satisfy the convergence condition.

## C. Selection of Parameters

In this study, we select the values of $a$, $\gamma$, C, and $a$ from (Spall, 1998). It is recommended to choose $a$ and $\gamma$ as 0.602 and 0.101, respectively (practically effective and theoretically valid). $C$ is set at the level approximately equal to the standard deviation of the measured noise in the objective function. $a$ is selected together with $A$ while satisfying that the magnitude of the gradient times $a/(A+1)^{a}$ approximately equals to the smaller perturbation size of the parameters in early iterations. In our design, the parameters are the control voltages ($V_{cc1}$, $V_{cc2,\ldots,}$ $V_{cc6}$) at the tunable reactances. We selected 50 as the maximum number of iterations in the algorithm, because it is not too long, but is sufficient for the algorithm to converge.

Although the selection of parameters is based on the guideline in (Spall, 1998), in the experiment we cannot directly use the parameters that are optimal to simulation environment as differences between two systems exist. The practical parameters have to be selected through trials and errors. In our system, we set $a = 8002000$, $C = 280$ and $A = 100$.

# SIMULATION RESULTS

## A. Speed of Convergence

In the simulation the DOA of the desired signal is 0°. The length of each data sample block $P$ is 50 bits. The total number of iterations $N$ is 50. For the proposed algorithm based on **SPSA**, the total length of training signals required is $P \times 2 \times N$ (Spall, 1998). Based on the varactor catalog data (Diode-datasheet), we model the relationship between the digital control voltage $V_{cci}$ and the reactance $x_i$ ($i = 1, \ldots, 6$) with a linear equation

$$x_i = -0.0217V_{cci} - 49.21. \tag{23}$$

First, we study the convergence of the algorithm. The interference is impinging on the antenna from 90°. We set the signal-to-noise ratio (SNR) 30 dB and vary the signal-to-noise ratio (SIR) from -5 dB to 5 dB. Secondly, we vary the SNR keeping the SIR at 0 dB. The learning curves of CCCs for different settings are plotted in Figure 7.

When we fix the SNR at 30 dB and vary the SIR from 5 dB to -5 dB, the algorithm converges to a point where the output SINR is more than 15 dB. At the beginning of the iteration, the value of $\rho_{yd}$ is influenced by the input SIR. When the SIR is kept at 0 dB and lower the SNR to 20 dB, $\rho_{yd}$ still reaches more than 0.95 (solid line with plus signs), which

*Figure 7. CCC between the received signal and the training signal in different environment settings for the proposed Espar SPSA beamforming algorithm. The azimuthal DOAs of one desired signal and one interference signal are 0° and 90°, respectively.*

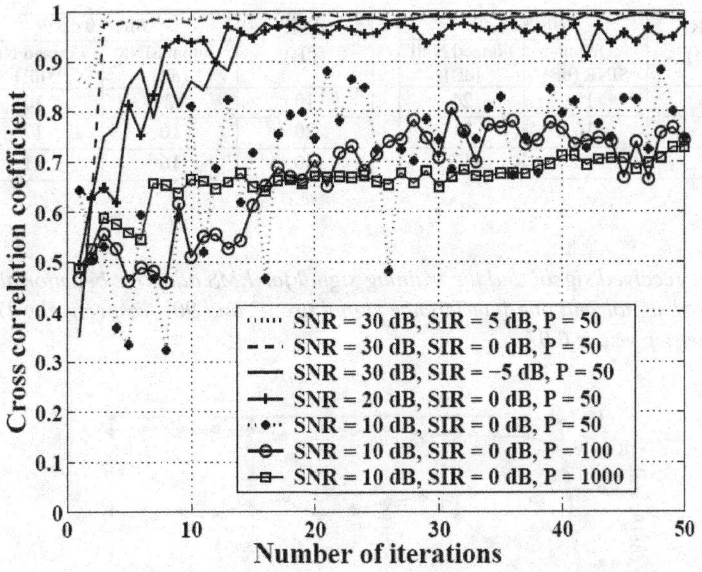


corresponds to an output SINR of 10 dB. However, the algorithm could not converge when SNR ≤ 10 dB. In a high noise level environment (SNR ≤ 10 dB), it is difficult to estimate correctly the CCC between the training signals and the received signals with only 50 bit samples. To ensure a correct calculation of the CCC, the sample length *P* needs to be increased. In Figure 7 the CCCs with 100-bit and 1,000-bit block samples are plotted, respectively.

We also note that the performance of the algorithm varies at each time. This is because of the randomness of the CCC between the shot training signal and the signals transmitted by the interference. Their correlation is varying randomly rather than being zero. This phenomenon is also observed in other experiments and field tests (Mogensen, 1997; TSUNAMI-98, Wennstrom, 1999). Averaged results for different environment settings are listed in Table 1. Simulation shows that 50 iterations are more than sufficient for the algorithm to converge.

## B. Comparison with Standard LMS Beamforming

To give a comparison on the convergence of the proposed beamforming algorithm for reactively controlled directive array with the standard LMS beamforming algorithm based on DBF array antennas of the same array geometry of the 7-element Espar antenna, we plot the learning curve of the LMS algorithm in Figure 8. The mutual coupling of the array antenna is also considered. The length of each data sample block is 50 bits. The step size is 0.07. From the figure we can see that the LMS beamforming algorithm requires comparatively less iterations to achieve the same CCC (the same output SINR). However, the penalties are the storing of seven times the data as for the Espar beamforming algorithm and the increase of the computational complexity, needless to say the complexity of the circuit and the power consumption as aforementioned.

## C. Null Forming Ability

To justify the null forming ability of the proposed algorithm, the formed beam patterns are also plotted in Figure 9. We observe that the algorithm still forms deep nulls even when the interference level increases. This is because the CCC between the received signals and the training signals is mostly influenced by the **interference** instead of the AWGN. While the MCCC based algorithm is searching to maximize the CCC, the interference is getting suppressed.

*Table 1. Averaged simulation results (averaged over 10 times of executing the program) for different environment settings. P = 50. N = 50. DOAs of one desired signal and one interference are 0° and 90° in the Azimuth Plane, respectively. (Results in the shaded region is obtained when P = 1,000.)*

| SIR (dB) | SNR = 30 dB | |
|---|---|---|
| | Output SINR (dB) | Formed Null (dB) |
| -5 | 15 | 24 |
| 0 | 16 | 21 |
| 5 | 18 | 17 |

| SNR (dB) | SIR = 0 dB | |
|---|---|---|
| | Output SINR (dB) | Formed Null (dB) |
| 10 | 2 | 16 |
| 20 | 10 | 17 |
| 30 | 16 | 21 |

*Figure 8. CCC between the received signal and the training signal for LMS adaptive beamforming algorithm. The azimuthal DOAs of one desired signal and one interference signal are 0° and 90°, respectively. The length of each data sample block is 50 bits. The step size is 0.07.*

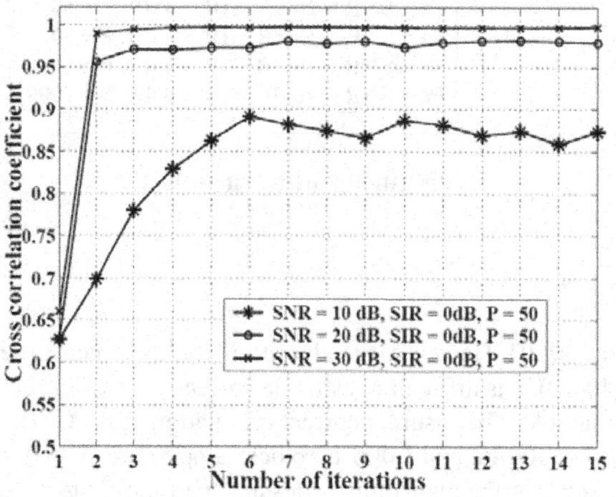


*Figure 9. Averaged (P = 1,000) simulated beam patterns (normalized gain in dB) in the azimuth plane for different SIRs. SNR = 30 dB.*

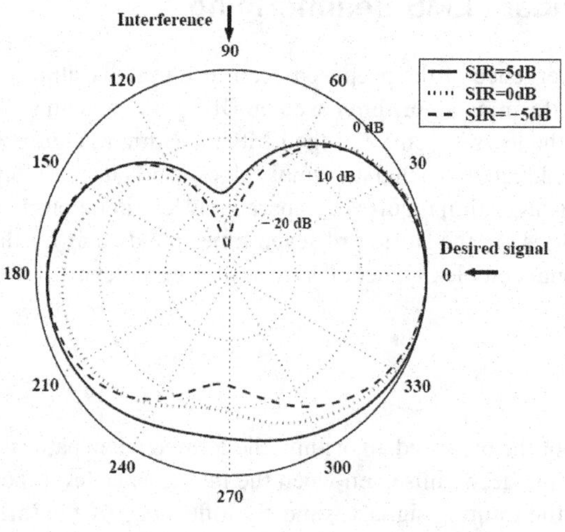

*Table 2. Simulation results (averaged over 10 times of executing the program) for different DOAs in the Azimuth Plane of the interference signal. DOAs of the signal of interest is 0°.*

| Interferer's direction in the azimuth plane | | 90° | 180° | 225° | 315° |
|---|---|---|---|---|---|
| SNR 20 dB | Output SINR (dB) | 10 | 8 | 9 | 6 |
| | Formed Null (dB) | 17 | 15 | 16 | 11 |
| SNR 30 dB | Output SINR (dB) | 18 | 11 | 14 | 11 |
| | Formed Null (dB) | 22 | 14 | 18 | 13 |

*Figure 10. Spatial correlation coefficients at the Espar antenna between the desired signal ($\phi_1$ = 0°) and the interference signal with respect to different azimuthal angles ($\phi_1$ is from 0° to 360°).*

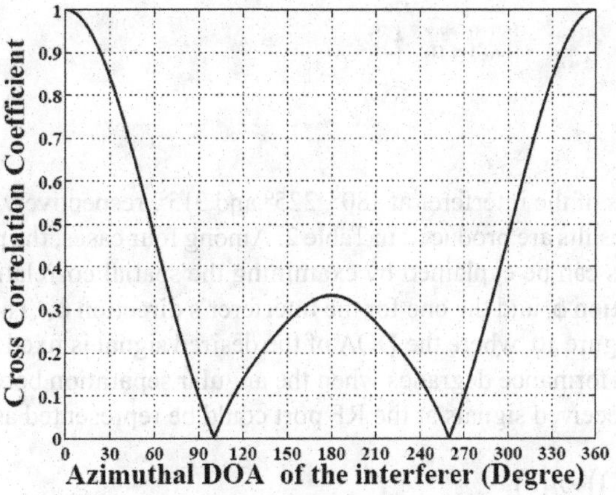


*Table 3. Simulation results (averaged over 10 times of executing the program) where there are two interferers. Case I: Interferers' directions are 90° and 270°. Case II: Interferers' directions are 225° and 270°. DOA of the desired signal is 0°.*

| | SNR (dB) | Formed null at interferer 1's direction (dB) | Formed null at interferer 2's direction (dB) | Output SINR (dB) |
|---|---|---|---|---|
| Case I | 20 | 15 | 13 | 9 |
| | 30 | 16 | 16 | 15 |
| Case II | 20 | 14 | 6.5 | 8 |
| | 30 | 17 | 10 | 12 |

*Figure 11. Simulated beam patterns (normalized gain in dB, P = 1,000) in the azimuth plane when there are two interferers. SIR = 0 dB. SNR = 30 dB. N = 50. Case I: Interferers' directions are 90° and 270°. Case II: Interferers' directions are 225° and 270°.*

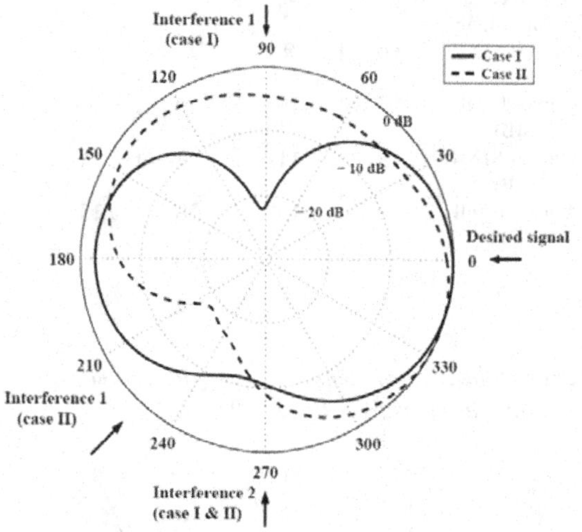


We also set the azimuthal DOAs of the interferer at 180°, 225° and 315°, respectively, while keeping the DOA of the desired signal at 0°. The averaged results are produced in Table 2. Among four cases, the performances are the best when the interferer is at 90° or 225°. This can be explained by examining the spatial correlation between the Espar steering vector for the desired signal's direction $\phi_1$ and the one for the interferer's direction $\phi_2$. The spatial correlation coefficient at the Espar antenna is plotted in Figure 10, where the DOA of the desired signal is fixed at 0°, the DOA of the interferer is changed from 0° to 360°. The performance degrades when the angular separation between the desired signal and the interference is getting small. The received signals at the RF port could be represented as:

$$y(t) = \mathbf{w}_{ESP}^T \left( s_1(t)\boldsymbol{\alpha}(\phi_1) + s_2(t)\boldsymbol{\alpha}(\phi_2) \right) + n(t). \quad (24)$$

Analogous to the duplicate channel multi-user division in the SDMA scheme, the less spatial correlation between the desired signal and the interference signal is, the better the null forming is to the interference.

Finally we increase the number of interferers to two. The powers of the interference signals are equally set to be half of the desired signal's power. Therefore, the SIR is kept 0 dB. The SNR is set to be 30 dB. Directions of two interferers are 90° and 270° in the first case, and 225° and 270° in the second case. Averaged results over 10 times of executing the algorithm are listed in Table 3. Although the performances with respect to the output SINR and the formed null degrade around 2 dB comparing to the same condition with one interferer, about 10 dB nulls are still formed at the interferers' directions. Beam patterns for the two cases when the SNR is 30 dB are plotted in Figure 11. As it can be seen in the figure for case II, more than 10 dB null is formed at interference 1 which is at 225°, and about 9 dB null is formed at interference 2 which is at 225°.

## CONCLUSION

Based on the knowledge gained from the thorough literature review, we notice that **parasitic array** antennas are suitable for power-consumption limited mobile terminal applications. Two categories of parasitic array antennas are examined. In comparison with switched parasitic array antennas, reactively controlled directive array can realize a smoother beam pattern steering, thus achieving a better **interference** suppression performance. Therefore, we develop a fast beamforming for a reactively controlled directive array antenna named "**Espar**." The proposed **beamforming** technique, maximizes the CCC between desired signals and training signals based on the **SPSA** theory and adaptively adjusts the tuneable reactance at each parasitic element to form a beam pattern that achieves efficient interference suppression. Simulation studies are carried out to examine the beamforming ability, convergence, and interference suppression of the proposed technique. Based on the simulation studies in this chapter, the following conclusions can be made.

1. The reactively controlled directive array antenna is an analog adaptive antenna. Adaptive beamforming is achieved by adaptively adjusting the load reactance of parasitic elements. The problem of beamforming cannot be modelled as a linear filtering problem as for DBF array antennas. Instead of solving the linear equation to obtain the Wiener solution, iterative approach is employed to achieve the optimum solution.
2. In DBF array antennas, mutual coupling influences the beam steering ability and degrades the beamforming and interference suppression performances. On the contrary, the reactive controlled directive array antennas take the advantage of strong **mutual coupling** between neighbouring elements to achieve beamforming.
3. We also compare the convergence speeds between the adaptive algorithm for a 7-element Espar antenna and the standard LMS beamforming algorithm using a 7-element circular array antenna with the same geometry as that of the Espar antenna. Results reveal that Espar antennas converge slower than DBF array antennas. However, the Espar antenna store much less signal datum and has much lower power consumption and system complexity. The sufficient beamforming ability and low power consumption feature make the Espar antenna a strong candidate for mobile terminal applications.
4. The beamforming algorithm suppresses the **interference** by maximizing the CCC between the desired signal and the training signal. It is more efficient to combat against the **interference** than to the noise. Therefore, the algorithm is suitable and efficient for the interference dominant environment, for example in ad hoc network applications where the link quality and throughput at a node are mainly limited by the interferences from the neighbouring nodes (mobile terminals). In a moderate condition (SNR is around 20 to 30 dB), the algorithm could achieve an output SINR of around 10 dB within a few iterations. Thus, a sound link [bit error rate (BER) $\leq 10^{-3}$] could be established quickly. When the SNR is lower than 10 dB, the method breaks down.
5. When the DOA of the impinging interference signal is close to that of the desired signal, the null forming ability degrades. This is due to the high spatial correlation between the desired signal and the interference signal.

## REFERENCES

Anderson, S., Forssen, U., Karlsson, J., Witzschel, T., Fischer, P., & Krug, A. (1996a, May), *Ericsson/mannesmann GSM field-trials with adaptive antennas*, Paper presented at the IEEE 47th Vehicular Technology Conference.

Anderson, S., Forssen, U., Karlsson, J., Witzschel, T., Fischer, P., & Krug, A. (1996b, October), *Ericsson/Mannesmann GSM field-trials with adaptive antennas*, Paper presented at the IEEE Colloquium on Advanced TMP Techniques and Applications,

Anderson, S., Hagerman, B., Dam, H., Forssen, U., Karlsson, J., Kronestedt, F., Mazur, S., & Molnar, K. J. (1999), Adaptive antennas for GSM and TMP systems, *IEEE Personal Communications Magazine*, 6(3), 74-86.

Balanis, C. A. (1997), *Antenna Theory: Analysis and Design*, John Wiley & Sons, Inc.

Bellofiore, S., Balanis, C. A., Foutz, J., & Spanias, A. S. (2002a), Smart-antenna systems for mobile communication networks, part 1. overview and antenna design, *IEEE Antennas and Propagation Magazine*, 44(3), 145-154.

Bellofiore, S., Foutz, J., Balanis, C. A., & Spanias, A. S. (2002b), Smart-antenna systems for mobile communication networks. part 2. beamforming and network throughput, *IEEE Antennas Propagation Magazine*, 44(4), 106-114.

Bhobe, A. U., & Perini, P. L. (2001), An overview of smart antenna technology for wireless communication, *IEEE Aerospace Conference*, vol. 2, Big Sky, USA, Mar. 10-17, 875-883.

Black, S. H., & Formeister, R. B. (1973), *Direction finding system*, U.S. Patent No. 3725938.

Blogh, J. S., & Hanzo, L. (2002), *Third-Generation Systems and Intelligent Wireless Networking: Smart Antennas and Adaptive Modulation*, John Wiley & Sons, Inc.

Boukalov, A. O., & Häggman, S. G. (2000), System aspects of smart-antenna technology in cellular wireless communications—an overview, *IEEE Transactions on Microwave Theory Techniques*, 48(6), 919-929.

Cheng, J., Kamiya, Y., & Ohira, T. (2001a), Adaptive beamforming of ESPAR antenna based on steepest gradient algorithm, *IEICE Transactions on Communications*, E84-B(7), 1790--1800.

Cheng, J., Kamiya, Y., & Ohira, T. (2001b, May), *Adaptive beamforming of ESPAR antenna using sequential perturbation*, Paper presented at the IEEE MTT-S International Microwave Symposium, Phoenix, AZ.

Cheng, J., Iigusa, K., Hashiguchi, M., & Ohira, T. (2002, November), *Blind aerial beamforming based on a higher-order maximum moment criterion (part II: Experiments)*, Paper presented at the Asia-Pacific Microwave Conference, Kyoto, Japan.

Diggavi, S. N., Al-Dhahir, N., Stamoulis, A., & Calderbank, A. R. (2004), Great expectations: The value of spatial diversity in wireless networks, *Proceedings of the IEEE*, 99(2), 219-270.

Diode-datasheet, Data retrieved from http://www.chipdocs.com.

Frodigh, M, Parkvall, S., Roobol, C., Johansson, P., & Larsson, P. (2001), Future-generation wireless networks, *IEEE Personal Communications Magazine*, 8(5), 10-17.

Godara, L. C. (1997a), Applications of antenna arrays to mobile communications, part I: Performance improvement, feasibility, and system considerations, *Proceedings of the IEEE*, 85(7), 1029-1030.

Godara, L. C. (1997b), Application of antenna arrays to mobile communications, part II: Beam-forming and direction-of-arrival considerations, *Proceedings of the IEEE*, 85(8), 1195-1245.

Gueguen, M. (1974), *Electronically step-by-step rotated directive radiation beam antenna*, U.S. Patent No. 3856799.

Gyoda, K., & Ohira, T. (2000, July), *Design of electronically steerable passive array radiator (ESPAR) antennas*, Paper presented at the IEEE International Symposium on Antennas and Propagation, Kyoto, Japan.

Harrington, R. F. (1978), Reactively controlled directive arrays, *IEEE Transactions on Antennas and Propagation*, AP-26(3), 390-395.

Haykin, S. (2002), *Adaptive Filter Theory,* Prentice Hall PTR.

Himmel, L., Dodington, S. H., & Parker, E. G. (1978), *Electronically controlled antenna system*, U.S. Patent NO. 3560978.

Hirata A., & Ohira, T. (2002, November), *Spotted null forming of electronically steerable parasitic array radiator antennas in indoor multipath propagation*, Paper presented at the Asia-Pacific Microwave Conference, Kyoto, Japan.

Jana, R., & Dey, S. (2000), 3G wireless capacity optimization for widely spaced antenna arrays, *IEEE Personal Communications Magazine*, 7(6), 32-35.

Janaswamy, R. (2001), *Radiowave Propagation and Smart Antennas for Wireless Communications*, Kluwer.

Johnson D. H., & Dudgeon, D. E., *Array Signal Processing: Concepts and Techniques*, Prentice Hall PR.

Komatsuzaki, A., Saito, S., Gyoda, K., & Ohira, T. (2000, December), *Hamiltonian approach to reactance optimization in ESPAR antennas*, Paper presented at the Asia-Pacific Microwave Conference, Ohio.

Kushner, H. J., & Yin, G. G. (1997), *Stochastic Approximation Algorithms and Applications*. Springer.

Lehne, P. H., & Pettersen, M. (1999), An overview of smart antenna technology for mobile communications systems, *IEEE Communications Surveys*, 2(4),2-13.

Litva, J. (1996), *Digital Beamforming in Wireless Communications*, Artech House, Inc.

Milne, R. (1985, June), *A small adaptive array antenna for mobile communications*, Paper presented at the IEEE International Symposium on Antennas and Propagation.

Mogensen, P. E., Pedersen, K. I., Espensen, P. L., Fleury, B., Frederiksen, F., Olesen, K., & Larsen, S. L. (1997, May), *Preliminary measurement results from an adaptive antenna array testbed for GSM/UMTS*, Paper presented at the IEEE 47th Vehicular Technology Conference.

Moon, T., & Stirling, W. (2001), *Mathematical Methods and Algorithms for Signal Processing*, Prentice Hall.

Ogawa, Y., & Ohgane, T. (2001), Advances in adaptive antenna technologies in Japan, *IEICE Transactions on Communications*, 84B(7), 1704-1712.

Ohira, T., Suzuki, Y., Ogawa, H., & Kamitsuna, K. (1997), Megalithic microwave signal processing for phased-array beamforming and steering, *IEEE Transactions on Microwave Theory and Techniques*, 45(12), 2324-2332.

Ohira, T., & Gyoda, K. (2000a, May), *Electronically steerable passive array radiator antennas for low-cost analog adaptive beamforming*, Paper presented at the IEEE Conference on Phased Array Systems and Technology, CA.

Ohira, T. (2000b, December), *Adaptive array antenna beamforming architectures as viewed by a microwave circuit designer*, Paper presented at the IEEE Asia-Pacific Microwave Conference, Sydney.

Ohira, T. (2002, November), *Blind aerial beamforming based on a higher-order maximum moment criterion (part I: Theory)*, Paper presented at the Asia-Pacific Microwave Conference, Kyoto, Japan.

Preston, S. L., Thiel, D. V., Lu, J. W., O'Keefe, S. G., & Bird, T. S. (1997), Electronic beam steering using switched parasitic patch elements, *IEE Electronics Letters*, (33)1, 7-8.

Preston, S. L., Thiel, D. V., Smith, T. A., O'Keefe, S. G., & Lu, J. W. (1998), Base-station tracking in mobile communications using a switched parasitic antenna array, *IEEE Transactions on Antennas and Propagation*, 46(6), 841-844.

Preston, S. L., Thiel, D. V., & Lu, J. W. (1999), A multibeam antenna using switched parasitic and switched active elements for space-division multiple access applications, *IEICE Transactions on Electronics*, E82-C(7), 1202-1210.

Rappaport, T. S. (2002), *Wireless Communications: Principles and Practice*, Prentice Hall PTR.

Shishkov, B., & Ohira, T. (2001, December), *Adaptive beamforming of ESPAR antenna based on stochastic approximation theory*, Paper presented at the IEEE Asia-Pacific Microwave Conference.

Shishkov, B., Cheng, J., & Ohira, T. (2002), Adaptive control algorithm of ESPAR antenna based on stochastic approximation theory, *IEICE Transactions on Communications*, E85-B(4), 802-811.

Sibille, A., Roblin, C., & Poncelet, G. (1997), Circular switched monopole arrays for beam steering wireless communications, *IEE Electronics Letters*, 33(7), 551-552.

Soni, R. A., Buehrer, R. M., & Benning, R. D. (2002), Intelligent antenna system for cdma2000, *IEEE Signal Processing Magazine*, 19(4), 54-67.

Spall, J. C. (1998), Implementation of the simultaneous perturbation algorithm for stochastic optimization, *IEEE Transactions on Aerospace and Electronic Systems*, 34(3), 817-823.

Sun, C., & Karmakar, N. C. (2002a, November), *A DOA estimation technique based on a single-port smart antenna for position location services*, Paper presented at the Asia-Pacific Microwave Conference, Kyoto, Japan.

Sun, C., & Karmakar, N. C. (2002b, November), *Adaptive beamforming of ESPAR antenna based on simultaneous perturbation stochastic approximation theory*, Paper presented at the Asia-Pacific Microwave Conference, Kyoto, Japan.

Sun, C., Hirata, A., Ohira, T., & Karmakar, N. C. (2003, October), *Experimental study of a fast beamforming algorithm for ESPAR antennas*, Paper presented at the European Conference on Wireless Technologies, Munich.

Sun, C. Hirata, A., Ohira, T., & Karmakar, N. C. (2004), Fast beamforming of electronically steerable parasitic array radiator antennas: Theory and experiment, *IEEE Transactions on Antennas and Propagation*, 52(7), 1819-1832.

Svantesson, T., & Wennstrom, M. (2001, August), *High-resolution direction finding using a switched parasitic antenna*, Paper presented at the IEEE 11th Signal Processing Workshop.

Thiel, D. V., O'Keefe, S., & Lu, J. W. (1996, July), *Electronic beam steering in wire and patch antenna systems using switched parasitic elements*, Paper presented at the IEEE International Symposium on Antennas and Propagation.

Thiel D. V., & Smith, S. (2001), *Switched Parasitic Antennas for Cellular Communications*. Artech House, Inc.

Tsoulos, G. V., Beach, M., & McGeehan, J. (1997), Wireless personal communications for the 21st century: European technological advances in adaptive antennas, *IEEE Communications Magazine*, 35(9), 102-109.

Tsoulos, G. V. (1999), Smart antennas for mobile communication systems: benefits and challenges, *Electronics & Communication Engineering Journal*, 11(2), 84-94.

Howard, P., Simmonds, C. M., Darwood, P., Beach, M. A., Arnott, R., Cesbron, F., & Newman, M. (1998), Adaptive antenna performance in mobile systems (Technical Report 98-0797), ERA Technology Ltd.

Van-Veen, B. D., & Buckley, K. M. (1988), Beamforming: A versatile approach to spatial filtering, *IEEE Acoustics, Speech and Signal Processing*, 5(2), 4-24.

Varlamos, P. K., & Capsalis, C. N. (2004), Direction-of-arrival estimation DoA using switched parasitic planar arrays and the method of genetic algorithms, *Wireless Personal Communications*, 28(1), 59-75.

Vaughan, R. (1999), Switched parasitic elements for antenna diversity, *IEEE Transactions on Antennas and Propagation*, 47(2), 399-405.

Wennstrom, M. (1999), Smart antenna implementation issues for wireless communications," doctorial dissertation, Uppsala University, Sweden.

Widrow, B, Mantey, P. E., Griffiths, L. J., & Goode, B. B. (1967), Adaptive antenna systems, *IEEE Transactions on Antennas and Propagation*, 55(12), 2143-2159.

Winters, J. H. (1998), Smart antennas for wireless systems, *IEEE Personal Communications Magazine*, 5(1), 23-27.

Winters, J. H., & Golden, G. D. (2007, July), *Adaptive antenna applique field test*, Paper presented at the 4th Workshop on Smart Antennas in Wireless Mobile Communications, Stanford, CA.

# Chapter X Direction of Arrival Estimation with Compact Array Antennas: A Reactance Switching Approach

**Eddy Taillefer**
*Doshisha University Miyakodani 1-3, Japan*

**Jun Cheng**
*Doshisha University Miyakodani 1-3, Japan*

**Takashi Ohira**
*Toyohashi University of Technology Toyohashi, Japan*

## ABSTRACT

*This chapter presents direction of arrival (DoA) estimation with a compact array antenna using methods based on reactance switching. The compact array is the single-port electronically steerable parasitic array radiator (Espar) antenna. The antenna beam pattern is controlled though parasitic elements loaded with reactances. DoA estimation using an Espar antenna is proposed with the power pattern cross correlation (PPCC), reactance-domain (RD) multiple signal classification (MUSIC), and, RD estimation of signal parameters via rotational invariance techniques (ESPRIT) algorithms. The three methods exploit the reactance diversity provided by an Espar antenna to correlate different antenna output signals measured at different times and for different reactance values. The authors hope that this chapter allows the researchers to appreciate the issues that may be encountered in the implementation of direction-finding application with a single-port compact array like the Espar antenna.*

## INTRODUCTION

The efficient use of direction-of-arrival (DoA) estimation techniques with smart antennas is an important research topic in wireless systems like *ad hoc* networks (Alexiou, 2004) or for the design of a DoA finder.

As a kind of smart antenna, switched parasitic antennas were proposed for cellular communications (Vaughan, 1999; Scott, 1999; Almhdie, 2000; Svantesson, 2002; Thiel, 2002). Compared to a conventional array antenna, which needs as many active radio receivers as antenna elements, a switched parasitic array antenna needs only a single active radio receiver. Therefore, the use of a switched parasitic antenna at the user side is an interesting alternative, since it can provide a small size, low cost and low power consumption for the receiver part. The switched parasitic antenna forms beams by using passive antenna elements that serve as reflectors when shorted to ground. Thus, a fixed number of directional patterns can be achieved by switching the short-circuits of the passive elements using p.i.n. diodes (Thiel, 2002). Switched parasitic antennas are known to improve the communication capacity in wireless communication systems (Almhdie, 2000), to perform high-resolution DoA estimation (Svantesson, 2002), such as that for personal locating services, and to provide antenna diversity (Vaughan, 1999; Scott, 1999) for adaptive communication systems.

The electronically steerable parasitic array radiator (Espar antenna), a kind of reactively controlled antenna (Harrington, 1978), was first proposed for low-cost user terminal applications (Ohira, 2000). Compared with simple switched parasitic antennas, the Espar antenna exhibits greater steerability control by means of its electronically controllable reactances and a more complex system for controlling the reactances (Thiel, 2004). Indeed, the parasitic element is connected to the ground by means of a reactance made with a reverse-bias varactor diode that can be controlled through loaded voltage. Thus, as a function of the reactance value, the parasitic element can variably act as a reflector or a radiator (Ohira, 2004, pp. 184-204). The continuous variability of the loaded reactance of the Espar antenna makes it more flexible than switched parasitic antennas because the number of possible directional patterns becomes greater (Thiel, 2004). Such a feature could be successfully employed, for example, in adaptive control processes involving beam and null forming, where the radiation pattern of an Espar antenna with a beampointing direction in the desired signal direction and nulls in the interference directions is obtained by optimizing the loaded reactances (Sun, 2004).

For DoA estimation applications with an Espar antenna, an earlier method was proposed to develop a hand-held microwave DoA finder for locating transmitters (Ohira, 2001), for example, after an avalanche (Furuhi, 2002). This method was based on a simple algorithm that switches twelve directional beam patterns by means of reactances and then chooses as the DoA estimate the beampointing direction of the pattern that provides the highest antenna output amplitude gain. Therefore, this method could only provide estimation for one impinging signal with a coarse precision of (Ohira, 2001; Furuhi, 2002). An alternative that used antenna output amplitude gain with pre-measured directional patterns was proposed for high-precision DoA estimation of one impinging signal (Taillefer, 2005a). The method is called Power Pattern Cross Correlation (PPCC). As its name suggests, it is based on the correlation between pre-measured power radiation patterns and power outputs of the antenna.

As a further step, to use more sophisticated algorithms that provide high-resolution and high-precision DoA estimation of multiple signals, and which are available in conventional array processing, a correlation matrix of the antenna elements output is required. In the case of the Espar antenna, since the beamforming is performed in the analog domain, only one output port is observable. However, by using a vector composed of this output-port complex gain, measured sequentially for different directional patterns, a technique called the *reactance-domain* (RD) technique could be adopted to create a correlation matrix for the Espar antenna. Consequently, based on this technique, the multiple signal classification (MUSIC) subspace DoA estimator (Schmidt, 1986) was proposed (Plapous, 2004) and experimentally verified (Taillefer, 2003) with a 7-element Espar antenna. In (Taillefer, 2003), the RD technique was developed by transmitting the same information as many times as the number of used directional patterns. Although such a data transmission scheme decreases the transmission rate, it is still sufficient in applications such as terminal position location, a hand-held DoA finder, or when the DoA estimation is needed from time to time for such tasks as forming a node position location table in an *ad-hoc* network where the nodes do not frequently relocate. Another interesting possibility for obtaining the RD output information without decreasing the transmission rate is to sample the received signal with different radiation patterns. This technique of oversampling is common in many communication systems, but here it needs to be considered as spatio-temporal oversampling, since beam pattern switching implies spatial diversity. Furthermore, more work is needed to clarify the practical aspects of using RD with an oversampling strategy.

The MUSIC algorithm employed in the RD-MUSIC for Espar antennas requires two stages: singular value decomposition (SVD) of the RD correlation matrix and a search over parameter space. This last stage implies a general optimization procedure, which becomes computationally costly when the DoA estimate must have high precision. It is known that compared with the MUSIC algorithm, the ESPRIT (Estimation of signal parameters *via* rotational invariance techniques) algorithm dramatically reduces the computational cost of the DoA estimator, since the second stage just consists of direct calculations that need only SVD (Roy, 1989). However, ESPRIT does not perform as well as MUSIC in terms of DoA estimation precision.

The RD-ESPRIT algorithm was first presented in (Taillefer, 2004), for high-precision, low-computational cost and low-power-consumption DoA finding applications with a hexagonally shaped 7-element Espar antenna. However, the proposed method made used of an intermediate correlation matrix computation and an additional computation SVD, which thus increased the number of computed SVD. An improved method and a performance study was then proposed in (Taillefer, 2005b) that needs no more that additional computation.

The present chapter presents DoA estimation with a Espar antenna using the PPCC algorithm and RD-based method, in the experimental point of view. The Espar antenna signal model and the employed DoA estimation algorithm are first explained. Then the PPCC algorithm, the RD-MUSIC and the RD-ESPRIT algorithm are presented. Some results of experiments conducted in an anechoic chamber, illustrate the DoA estimation capability of the single-port Espar antenna.

## ESPAR ANTENNA SIGNAL MODEL

The Espar antenna is a reactively controlled array antenna that consists of an active monopole element located at the center of a ground plane and connected to the receiver. The active monopole element is surrounded by $M$ parasitic controllable monopole elements equally spaced around a circle. Each element is a quarter-wavelength long, and the inter-element separation is $d$. Each parasitic element is loaded by an adjustable reactance (Vaughan, 1999; Scott, 1999; Ohira, 2004; Schlub, 2004).

Figure 1 shows an $(M + 1)$-Espar antenna with $M = 6$, inter-element spacing $d$, and working wavelength $\lambda$. A small space (*i.e.* $d < \lambda/2$) between elements involves mutual coupling between the elements and thus allows antenna beamforming; a small $d$ also allows reduction of the total antenna size. The model formulation of the beamformer *equivalent radio-frequency (RF) weight vector* is given by (Ohira, 2004, pp. 184-204):

$$\mathbf{w} = 2z_s(\mathbf{Z}+\mathbf{X})^{-1}\mathbf{u}_0 \tag{1}$$

where $z_s$ is the receiver's input impedance, $\mathbf{z}$ is an $(M+1)\times(M+1)$-sized mutual impedance matrix, and $\mathbf{u}_0$ is an $(M+1)$-dimensional vector taken as $[1,0,\ldots,0]^T$. $\mathbf{X} = \mathrm{diag}\{z_s, j\mathbf{x}^T\}$ is called the reactance matrix with $\mathbf{x} \equiv [x_1,\ldots,x_M]$ as the vector containing the reactance values applied to the parasitic elements.

The response of the antenna to a unit wavefront from direction $\theta_q$ is modeled by the complex $(M + 1)$-dimensional steering vector

$$\mathbf{a}(\theta_q) = [1, \exp(j\psi_1(\theta_q)), \ldots, \exp(j\psi_M(\theta_q))]^T \tag{2}$$

where

$$\psi_m(\theta_q) = \frac{2\pi d}{\lambda}\cos\left(\theta_q - 2\pi\frac{m-1}{M}\right) \tag{3}$$

and $m = 1, \ldots, M$.

The $Q$ complex signal components at time $t$, $s_q(t)$, $q = 1, \ldots, Q$, are collected in the vector $\mathbf{s}(t) = [s_1(t), s_2(t), \ldots, s_Q(t)]^T$. The Espar antenna output $y(t)$ is then expressed as

$$\begin{aligned} y(t) &= \mathbf{w}^T[\mathbf{a}(\theta_1), \ldots, \mathbf{a}(\theta_Q)]\mathbf{s}(t) + n(t) \\ &= \mathbf{w}^T\mathbf{A}\mathbf{s}(t) + n(t) \end{aligned} \tag{4}$$

where $n(t)$ is a complex white Gaussian noise component. Changing the value of the reactances allows us to control $\mathbf{w}$ and thus the antenna radiation pattern.

Due to its single port design, the Espar antenna seems only dedicated to applications, like beamforming, requiring adaptive control. However the ability to control $\mathbf{w}$ through the fed reactances still allows DoA estimation by exploiting the reactance diversity to recreate output diversity.

## DoA ESTIMATION WITH ESPAR ANTENNAS

In this section the DoA estimation problem is considered with adaptive arrays antennas like the Espar antenna. In array processing, many methods have been proposed for direction-finding using conventionnal adaptive array antennas

*Figure 1. Diagram of the 7-element Espar antenna*

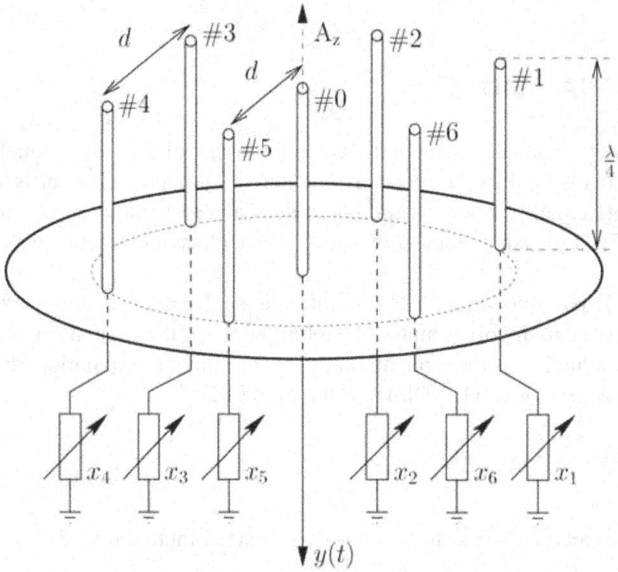


(Krim, 1996; Jhonson, 1993). However, these methods required that all the array antenna element output to be observable and measurable, which is not the case of the single output multi-element Espar antenna. Since wireless communication application systems may require high-performance DoA estimation capabilities of the receiver, efforts for developping high-performance DoA finder for Espar antenna have to be made, despite of the difficult hardware limitation. In the folowing three DoA estimation methods based on reactances switching is explained. The first method used a cross correlation approach for performing high-precision DoA estimation, whereas the two others methods employ the RD approach applied to the well known MUSIC and ESPRIT algorithms (Schmidt, 1986; Roy, 1989). The cross correlation approach has the advantage of a low computationnal complexity and robustness against impinging signal phase fluctuation. However this method is limitated to one signal estimation. The RD approach allows to apply most of the DoA estimation algorithms avalaible for conventional adaptive array antennas.

## Power Pattern Cross Correlation (PPCC) Algorithm

This section introduces a DoA estimation method for Espar-like antenna that can provide diverse directive beam patterns. As the section name suggests, the method based on the computation of the correlation between *N pre-measured* power radiation patterns and *N* power outputs of the antenna measured at each estimation times. A suitable selection of pattern shapes, which was derived directly from the principle, shows that with four single-peaked directive patterns, the method can efficiently achieve DoA estimation of an unknown signal. Moreover, since only the amplitude of the power output is used the method exhibits robustness against arrival signal data phase fluctuation. Another advantage is low computational cost allowing the PPCC algorithm to be employed in many applications requiring direction-finding. However, *N* pattern data need to be measured and saved in order to practically apply the method, which can be costly in terms of measurement equipment.

## PPCC Principle

When considering one signal impinging on the antenna at the unknown DoA $\theta_{sig}$, the PPCC principle is explained as follows: for a given set of *N* antenna power patterns corresponding to a set of *N* reactance vectors $\{\mathbf{x}^{(1)}, \mathbf{x}^{(2)}, \ldots, \mathbf{x}^{(N)}\}$, the correlation coefficient between the output power of the antenna for each corresponding reactance vector $\{Y(\mathbf{x}^{(1)}),$

$Y(\mathbf{x}^{(2)}), \ldots, Y(\mathbf{x}^{(N)})\}$ and the antenna power pattern set is the highest at the signal DoA angle. Here, according to (1) and (4), the antenna output power can be modelized by

$$Y(\mathbf{x}^{(k)}) = \mathrm{E}\left[|s(t)|^2\right]\mathbf{w}_k^{\mathrm{T}}\mathbf{a}(\theta_{\mathrm{sig}})\mathbf{a}^{\mathrm{H}}(\theta_{\mathrm{sig}})\mathbf{w}_k^* + \mathrm{E}\left[|n(t)|^2\right], \quad (5)$$

where

$$\mathbf{w}_k = 2z_s\left(\mathbf{Z} + \mathrm{diag}\{z_s, j(\mathbf{x}^{(k)})^{\mathrm{T}}\}\right)^{-1}\mathbf{u}_0. \quad (6)$$

The part $\mathbf{w}_k^{\mathrm{T}}\mathbf{a}(\theta_{\mathrm{sig}})\mathbf{a}^{\mathrm{H}}(\theta_{\mathrm{sig}})\mathbf{w}_k^*$ in (5) represents the power of the antenna radiation pattern toward $\theta_{\mathrm{sig}}$ for antenna parasitic elements load with reactance values $\mathbf{x}^{(k)} = \left[x_1^{(k)}, x_2^{(k)}, \ldots, x_6^{(k)}\right]$.

In a practical case of the DoA estimation of one modulated data signal (*e.g.*, BPSK, QPSK, $2^q$-QAM), the method follows these steps:

1. Choose $N$ different sets of reactances, $\{\mathbf{x}^{(1)}, \mathbf{x}^{(2)}, \ldots, \mathbf{x}^{(N)}\}$. Measure the antenna power pattern of each set. The antenna power pattern value at angle q corresponding to the $i$-th set is denoted $P(\mathbf{x}^{(i)}, \theta)$. Note that the first step is performed only one time.
2. For each set of reactances, $\{\mathbf{x}^{(1)}, \mathbf{x}^{(2)}, \ldots, \mathbf{x}^{(N)}\}$, measure the corresponding antenna output power $\{Y(\mathbf{x}^{(1)}), Y(\mathbf{x}^{(2)}), \ldots, Y(\mathbf{x}^{(N)})\}$.
3. For $\theta$ from 0º to 360 º, compute the correlation coefficient $\Gamma(\theta)$ between the measured power patterns and the measured antenna output:

$$\Gamma(\theta) = \frac{\sum_{n=1}^{N} P(\mathbf{x}^{(n)}, \theta) Y(\mathbf{x}^{(n)})}{\sqrt{\sum_{n=1}^{N} P(\mathbf{x}^{(n)}, \theta)^2 \sum_{n=1}^{N} Y(\mathbf{x}^{(n)})^2}} \quad (7)$$

4. The DoA estimate, $\hat{\theta}_{\mathrm{sig}}$ will correspond to the highest value of the function $\Gamma$.

By making a *pre*-decision on the search range of the maximum value of $\Gamma$, it is possible to practically speed up the estimation calculation. Indeed, employing an adaptive beamforming algorithm (Shiskov, 2001; Ohira, 2004, pp. 184-204) , it is possible to provide a set of reactances corresponding to a directive beam pattern for each of the $Q$ regular directions of the azimuth plane (*e.g.*, for $Q$=6, form beams in 0º, 60º,. . ., and 300º). Then, within this $Q$ *pre*-calculated reactance set, we look for the angle $\theta_C$ providing the highest gain value. The search range of the maximum of $\Gamma$ will be bound by the angle before $\theta_C$ and the one after $\theta_C$.

## Basic Theory on the PPCC Method

In this section we make a mathematical consideration of the PPCC method as a DoA estimation algorithm.

Let $P_k(\theta) = P(\mathbf{x}^{(k)}, \theta)$, and $Y_k = Y(\mathbf{x}^{(k)})$. We assume that $\forall k = 1, \ldots, N$, $\forall \theta \in [0, 2\pi[$, $P_k$ is continuous, and $P_k(\theta) > 0$. Then, the normalized versions of the patterns and the antenna outputs for $\forall k = 1, \ldots, N$ become:

$$\overline{P}_k(\theta) = \frac{P_k(\theta)}{\sqrt{\sum_{n=1}^{N} P_n(\theta)^2}}, \quad \text{and} \quad \overline{Y}_k = \frac{Y_k}{\sqrt{\sum_{n=1}^{N} Y_n^2}} \quad (8)$$

*Table 1.*

| | $\overline{P}_1(\theta)$ | $\overline{P}_2(\theta)$ | $\overline{P}_3(\theta)$ | $\overline{P}_4(\theta)$ |
|---|---|---|---|---|
| Formulation | $\frac{\cos(\theta)+1}{2}$ | $\frac{-\cos(\theta)+1}{2}$ | $\frac{\sin(\theta)+1}{2}$ | $\frac{-\sin(\theta)+1}{2}$ |
| Beampointing angle | $\pi$ | 0 | $\frac{3\pi}{2}$ | $\frac{\pi}{2}$ |

Thus (9) is rewritten

$$\Gamma(\theta)=\sum_{n=1}^{N}\overline{P}_k(\theta)\overline{Y}_k \tag{9}$$

Obviously, $\Gamma$ is continuous and $\forall\theta, 0<\Gamma(\theta)<1$. Consequently, at least one minimum and one maximum exist.

Estimator condition: *To be an angle estimator, the maximum of* $\Gamma(\theta)$ *must be unique (i.e., in the ideal noise-free case, the solution* $\theta$ *of* $\Gamma(\theta)=1$ *must be unique).*

$\Gamma$ admits an extremum when its derivative verifies:

$$\frac{\partial\Gamma(\theta)}{\partial\theta}=\sum_{n=1}^{N}\frac{\partial\overline{P}_k(\theta)}{\partial\theta}\overline{Y}_k=0 \tag{10}$$

To fulfill the "estimator condition", the derivative should not null independently of $\overline{Y}_k$.

An analysis of (10) leads us to the following conditions for the most appropriate shape of the patterns $\overline{P}_k$:

- Case $N = 2$: the "estimator condition" is fulfilled if the curves of $\overline{P}_1$ and $\overline{P}_2$ have fully opposite slopes.
- Case $N > 2$: the "estimator condition" will depend on the effect of each couple of patterns in the $N$ patterns. So, this concurs with Case $N = 2$. The resulting performance will be a combination of the performance of each couple.

Let these results be illustrated with directive beam patterns in the case of $N = 2,4$.

The directive pattern formulations and the beampointing angles are indicated in Table 1.

In case $N = 2$ ($\overline{P}_{k=1,2}$), the derivative of (9) nulls without the help of $\{\overline{Y}_{k=1,2}\}$, in zero and $\pi$. Moreover, due to the symmetry of the patterns, the estimator always admits two maxima with symmetry in relation to the axis of $\theta - \pi$. In this case, the estimator condition is fulfilled only if half of the sector is considered.

In case $N = 4$ ($\overline{P}_{k=1...4}$), the derivative never nulls independently of $\{\overline{Y}_{k=1...4}\}$. Consequently, for a given quadruple $\overline{Y}_1$, $\overline{Y}_2$, $\overline{Y}_3$, $\overline{Y}_4$, (9) will always admit a unique maximum. The two couples of patterns (1, 2 and 3, 4) have opposite qualities, so the angles that provide a bad estimation for one pair of patterns will not do so for the other pair.

From these examples using single-peaked directive beam patterns, it can be said that the estimations can be improved by increasing the number of patterns used.

## RD-Based Algorithms: RD-MUSIC and RD-ESPRIT

### RD Correlation Matrix

In a conventional array antenna, the correlation matrix is created by measuring the signals on each element of the antenna. For a single-port output like the Espar antenna, the spatial diversity of a conventional array antenna is recreated by periodically changing the reactance values while measuring the antenna output.

First, the RD complex output vector, $\mathbf{y}(t) = [y(t_1), y(t_2), ..., y(t_N)]^{\mathrm{T}}$, is formed by choosing $N$ different sets of reactance values $\mathbf{x}^{(1)}, \mathbf{x}^{(2)}, ..., \mathbf{x}^{(N)}$, then, for each reactance set $\mathbf{x}^{(m)}$, by getting the output $y(t_m)$ of the antenna, ($n$ = 1, ..., $N$):

$$\mathbf{y} = \begin{bmatrix} \mathbf{w}_1^{\mathrm{T}}\mathbf{A}\mathbf{s}(t_1) \\ \mathbf{w}_2^{\mathrm{T}}\mathbf{A}\mathbf{s}(t_2) \\ \vdots \\ \mathbf{w}_N^{\mathrm{T}}\mathbf{A}\mathbf{s}(t_N) \end{bmatrix} + \begin{bmatrix} n(t_1) \\ n(t_2) \\ \vdots \\ n(t_N) \end{bmatrix} \tag{11}$$

It is assumed that the same signals are repeated $N$ times, *i.e.*, for $n$ = 1, ..., $N$, $\mathbf{s}(t_m) = \mathbf{s}$, thus $\mathbf{y}$ can be rewritten as:

$$\mathbf{y} = \begin{bmatrix} \mathbf{w}_1^{\mathrm{T}} \\ \mathbf{w}_2^{\mathrm{T}} \\ \vdots \\ \mathbf{w}_N^{\mathrm{T}} \end{bmatrix} \mathbf{A}\mathbf{s} + \begin{bmatrix} n(t_1) \\ n(t_2) \\ \vdots \\ n(t_N) \end{bmatrix} = \mathbf{W}^{\mathrm{T}}\mathbf{A}\mathbf{s} + \mathbf{n} \tag{12}$$

where $\mathbf{w}^{\mathrm{T}}$ is called the RF equivalent weight matrix or RD weight matrix.

Then, assuming that the noise for different times and the incoming signals are not correlated with each other, the RD correlation matrix, $\mathbf{R}_{yy}$, thus has the following structure:

$$\mathbf{R}_{yy} = \mathrm{E}\left[\mathbf{y}\mathbf{y}^{\mathrm{H}}\right] = \mathbf{W}^{\mathrm{T}}\mathbf{A}\mathbf{P}_s\mathbf{A}^{\mathrm{H}}\mathbf{W}^{*} + \mathbf{\Sigma} \tag{13}$$

where $\mathbf{P}_s = \mathrm{E}\left[\mathbf{s}\mathbf{s}^{\mathrm{H}}\right]$ is the signals' correlation matrix, and $\mathbf{\Sigma} = \mathrm{E}\left[\mathbf{n}\mathbf{n}^{\mathrm{H}}\right]$.

In practice, instead of (13), a sample RD correlation matrix $\hat{\mathbf{R}}_{yy}$, based on $N \times K$ observations (snapshots) of the antenna output $y_n[k]$ ($n$ = 1, ..., $N$), is computed as:

$$\hat{\mathbf{R}}_{yy} = \frac{1}{K}\sum_{k=1}^{K}\left[\begin{bmatrix} y_1[k] \\ y_2[k] \\ \vdots \\ y_N[k] \end{bmatrix}\begin{bmatrix} y_1^*[k] & y_2^*[k] & \cdots & y_N^*[k] \end{bmatrix}\right] \tag{14}$$

## Signal Subspace Computation

The eigen decomposition of the RD correlation matrix estimate, $\hat{\mathbf{R}}_{yy}$, has the following form:

$$\hat{\mathbf{R}}_{yy} = \mathbf{E}_s\mathbf{\Lambda}_s\mathbf{E}_s^{\mathrm{H}} + \mathbf{E}_n\mathbf{\Lambda}_n\mathbf{E}_n^{\mathrm{H}} \tag{15}$$

where $\mathbf{E}_s = \left[\mathbf{e}_1, \ldots, \mathbf{e}_Q\right]$, $\mathbf{E}_n = \left[\mathbf{e}_{Q+1}, \ldots, \mathbf{e}_N\right]$, $\lambda_1 \geq \cdots \geq \lambda_N$, $\mathbf{\Lambda}_s = \mathrm{diag}\{\lambda_1, \ldots, \lambda_Q\}$, and $\mathbf{\Lambda}_n = \mathrm{diag}\{\lambda_{Q+1}, \ldots, \lambda_N\}$.

In the case of conventional array antennas, the space spanned by the columns of $\mathbf{E}_s$ is often referred to as the signal subspace, and $\mathbf{E}_n$ is called the noise subspace. However, in the case of RD signal processing, the signal subspace correlation is performed by means of the RD weight matrix $\mathbf{w}^{\mathrm{T}}$, as can be seen in (13).

It must be noted here, as a required condition, that the number of emitters, $Q$, and the RF equivalent weight matrix $W$ have to be known or estimated as a preliminary step. An estimate of the number of sources $Q$ can be obtained by using AIC or MDL (Wax, 1985); the matrix $\mathbf{w}^T$ can be estimated by using an array manifold calibration technique, which is explained in Section. It is assumed in the following that $Q$ and $\mathbf{w}^T$ are known.

## RD-MUSIC Algorithm

According to the RD technique the antenna steering vector $\mathbf{a}(\theta)$ will result to a modified steering vector $\mathbf{W}^T\mathbf{a}(\theta)$ (Plapous, 2004). The RD-MUSIC algorithm thus consists in computing the following modified MUSIC DoA spectrum:

$$P_{MUSIC}^{RD}(\theta) = \frac{1}{\mathbf{W}^T\mathbf{a}(\theta)\mathbf{E}_n\mathbf{E}_n^H\mathbf{a}^H(\theta)\mathbf{W}^*} \quad (16)$$

for $0^\circ \leq \theta \leq 360^\circ$.

The DoA estimates $\hat{\theta}_1, \hat{\theta}_2, \ldots, \hat{\theta}_Q$ correspond to the values of $\theta$ at the peaks of $P_{MUSIC}^{RD}(\theta)$. In the aim at practically using of this method in direction-finding applications, the problem is how to compute the RD-MUSIC spectrum while the matrix $\mathbf{W}^T$ is a required unknown constant. In the following, two practical methods for estimating the RD steering vector, $\mathbf{W}^T\mathbf{a}(\theta)$, are explained (Taillefer, 2003). The two methods differ in the way the RD steering vector is obtained. In addition, a third method, based on antenna calibration, allowing direct estimation of $\mathbf{W}^T$ is explained (Hirata, 2003). The first two methods can be used only in the computation of the RD-MUSIC spectrum, whereas the last one can also be used in the RD-ESPRIT algorithm which is presented latter.

## Equivalent Weight Vector (EWV) Method

The RD steering vector, $\mathbf{W}^T\mathbf{a}(\theta)$, is computed by simply use the formulation of the current vector given in (1). Precisely, for each set of reactance $\mathbf{x}^{(l)} = [x_1^{(l)}, \ldots, x_N^{(l)}]$ the corresponding RF equivalent weight vector $\mathbf{w}^{(l)}$ is calculated from (1). The matrix $\mathbf{W}^T$, used for computing the RD-MUSIC spectrum, is formed after repeating the process for the $N$ sets of reactances.

Notice that in this method, the computation of, the RD-MUSIC spectrum is led by the estimation of the impedance matrix $\mathbf{Z}$ and the relationship used to obtain the equivalent reactances. So, these parameters have to be calibrated.

This critical problem of calibrating the impedance parameters used in the Espar antenna output model can be avoided by using a more direct method.

## Direct Measurement (DM) Method

In the previous method, for a given Espar antenna we need to estimate the matrix $\mathbf{Z}$ and the relation between voltage and reactance. The accuracy of the obtained MUSIC-Espar spectrum strongly depends on these estimations. In the DM method, the DoA spectrum is calculated *only* using *directly measured data*.

The method is based on the idea that for one constant impinging signal $u(t) = K = 1$ arriving toward $\theta$ in a noise-free assumption, and for a given reactance set $\mathbf{x}$, the Espar antenna output is expressed by

$$y(t) = \mathbf{w}^T\mathbf{a}(\theta)u(t) = \mathbf{w}^T\mathbf{a}(\theta) \quad (17)$$

which can be estimated by measuring the phase and power outputs of the antenna.

The current-steering vector data $\mathbf{W}^T\mathbf{a}(\theta) = [\mathbf{w}_1^T\mathbf{a}(\theta), \mathbf{w}_2^T\mathbf{a}(\theta), \ldots, \mathbf{w}_N^T\mathbf{a}(\theta)]^T$ for a given $\theta$, is obtained by measuring the power and phase of the antenna output for each of the $N$ sets of reactances. Then, the RD-MUSIC spectrum in (16) is computed using this data.

Notice that for a given number, $N$, of sets of reactances and, for example, angles $\theta = 0^\circ, 1^\circ, \ldots, 359^\circ$, $P_{MUSIC}^{RD}(\theta)$ is formed with $\mathbf{E}_n$ and the measure of $N$ patterns having 360 points each. Also, to perform several estimations of the same impinging signal at different DoAs, the $N$ patterns are measured only one time.

## Calibration of $\mathbf{w}^T$

Because of the unknown element gain, phase, mutual coupling errors, and varactor flaws of the real antenna that could not be considered in the antenna model, the $\mathbf{W}^T$ parameter directly calculated from the analytic model in (1) may be not suitable for practical DoA estimation (Taillefer, 2003). In this case, a calibrated estimate of $\mathbf{W}^T$ should be considered.

The chosen calibration method is based on an array manifold calibration (Pierre, 1991; See, 1994). In the case of the Espar antenna, this calibration procedure follows these steps (Hirata, 2003):

1. Choose $N$ sets of reactances, $\mathbf{x}^{(1)},\mathbf{x}^{(2)},\ldots,\mathbf{x}^{(N)}$.
2. Choose $P$ different values for angles $\theta_p, p=1,\ldots,P$, with $0^\circ \le \theta_p < 360^\circ$, and $P >> N$.
3. For one signal impinging the antenna toward $\theta_p$ form $\hat{\mathbf{R}}_{yy}$ from (14). Then, Eigen decompose $\hat{\mathbf{R}}_{yy}$ to obtain the eigenvector $\mathbf{e}_1^{(p)}$ corresponding to the highest eigenvalue $\lambda_1^{(p)}$. Carry out the same procedure for all $\theta_p$ with $p$ from 1 to $P$ and form the matrix $\mathbf{E}_c = \left[\mathbf{e}_1^{(1)},\mathbf{e}_1^{(2)},\ldots,\mathbf{e}_1^{(P)}\right]$, which collects all of the computed eigenvectors.
4. Form the matrix $\mathbf{A}_c = [\mathbf{a}(\theta_1),\ldots,\mathbf{a}(\theta_P)]$, which contains the steering vector for each of the chosen $\theta_p$.
5. Finally, the RF weight vector estimate, $\hat{\mathbf{W}}^T$, is given by the total least square solution of $\hat{\mathbf{W}}^T\mathbf{A}_c = \mathbf{E}_c$

Notice that for better performance the chose $N$ sets of reactances in the calibration procedure, should be the same that the reactance sets used during the DoA estimation.

## RD-ESPRIT Algorithm

The ESPRIT algorithm fundamentally requires that the signal subspace matrix be computed from the elements' correlation matrix and that translational invariance be designed into the array element geometry. Consequently the RD subspace matrix, $\mathbf{E}_s$ needs to be processed before employing any subspace-based method that requires the element signal subspace. The element signal subspace can thus be obtained as follows:

$$\tilde{\mathbf{E}}_s = \left(\mathbf{W}^T\right)^{-1}\mathbf{E}_s \qquad (18)$$

Thanks to the hexagonal shape of the 7-element Espar antenna, three configurations showing translational invariance could be heuristically found. The configurations are shown in Figure 2, where $\vec{\Delta}$ characterizes the translational invariance between the two subarrays.

The spatial delay between subarrays 1 and 2 due to displacement invariance $\vec{\Delta}$ leads to a phase delay $\phi_q$ on the coming signal $s_q(t)$. This phase delay is expressed as:

$$\phi_q = \exp\left(j\left(\frac{\omega_0\Delta}{c}\right)\sin\left(\theta_q - \theta_r\right)\right), q = 1,2,\ldots,Q \qquad (19)$$

where $\omega_0 = 2\pi f_0$ with $f_0$ is the antenna working frequency, $\Delta = \|\vec{\Delta}\| = d$, $c$ is the speed of light, and $\theta_r$ is a constant angle value that depends on the translational invariance axis.

Therefore, the total-least-square (TLS) ESPRIT algorithm applied to one of the three configurations can be summarized as follows (Roy, 1989):

1. Decompose $\tilde{\mathbf{E}}_s$ into $\tilde{\mathbf{E}}_{s1}$ and $\tilde{\mathbf{E}}_{s2}$ by using the selection matrices $\mathbf{J}_1$ and $\mathbf{J}_2$, which pick up the elements of subarrays 1 and 2, respectively:

$$\tilde{\mathbf{E}}_s \equiv \begin{bmatrix}\tilde{\mathbf{E}}_{s1}\\ \tilde{\mathbf{E}}_{s2}\end{bmatrix} = \begin{bmatrix}\mathbf{J}_1\tilde{\mathbf{E}}_s\\ \mathbf{J}_2\tilde{\mathbf{E}}_s\end{bmatrix} \qquad (20)$$

*Figure 2. Translational invariance configurations designed into the 7-element Espar antenna. Subarrays 1 and 2 are drawn with dotted and dashed lines, respectively.*

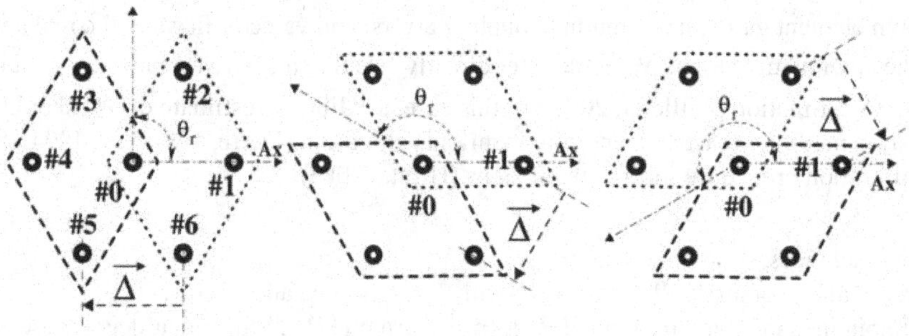


2. Eigen decompose (with eigenvalues in decrementing order)

$$\begin{bmatrix} \tilde{\mathbf{E}}_{s1}^{H} \\ \tilde{\mathbf{E}}_{s2}^{H} \end{bmatrix} \begin{bmatrix} \tilde{\mathbf{E}}_{s1} & \tilde{\mathbf{E}}_{s2} \end{bmatrix} \quad (21)$$

to obtain its eigenvectors, and then form the matrix **E** containing these eigenvectors.

3. Decompose **E** to form the $Q \times Q$-sized matrices $\mathbf{E}_{12}$ and $\mathbf{E}_{22}$ as follows:

$$\mathbf{E} \equiv \begin{bmatrix} \mathbf{E}_{11} & \mathbf{E}_{12} \\ \mathbf{E}_{21} & \mathbf{E}_{22} \end{bmatrix} \quad (22)$$

4. Calculate the eigenvalues $\hat{\phi}_q$ (also called phase factor estimates) of $\mathbf{\Psi} = -\mathbf{E}_{12}\mathbf{E}_{22}^{-1}$, $q = 1,\ldots,Q$.
5. Calculate the DoA estimates from

$$\hat{\theta}_q = \arcsin\left(\frac{c}{\omega_0 \Delta}\arctan\left(\frac{\Im m(\hat{\phi}_q)}{\Re e(\hat{\phi}_q)}\right)\right) + \theta_r \quad (23)$$

Here, it should be noted that generally speaking the RD technique can be performed with more or less than $M + 1$ reactance sets, as in (Taillefer, 2003; Plapous, 2004) where it is performed with $M$ reactance sets. However, in the case of the ESPRIT algorithm which requires that a translational invariance be designed into the array element geometry, the RD correlation matrix must have the same dimension as the array element number.

## EXPERIMENTS WITH A 7-ELEMENTS ESPAR ANTENNA

Experimental verifications of the DoA estimation capabilities of the Espar antenna has been conducted in an anechoic chamber. The experimental settings is the same as explained in (Cheng, in press).

### Experimental Results

All of the DoA estimation methods required radiation pattern data, an example of measured radiation patterns is given in Figure 3. As shown in the previous section, the PPCC method has best performance when used with single-peaked directive beam patterns. However due to the design limitation of the Espar antenna, single-peaked beam patterns could not be obtained.

*Figure 3. Examples of six measured directive beam patterns amplitude beampointing toward* $0^\circ$, $60^\circ$, ..., *and* $360^\circ$

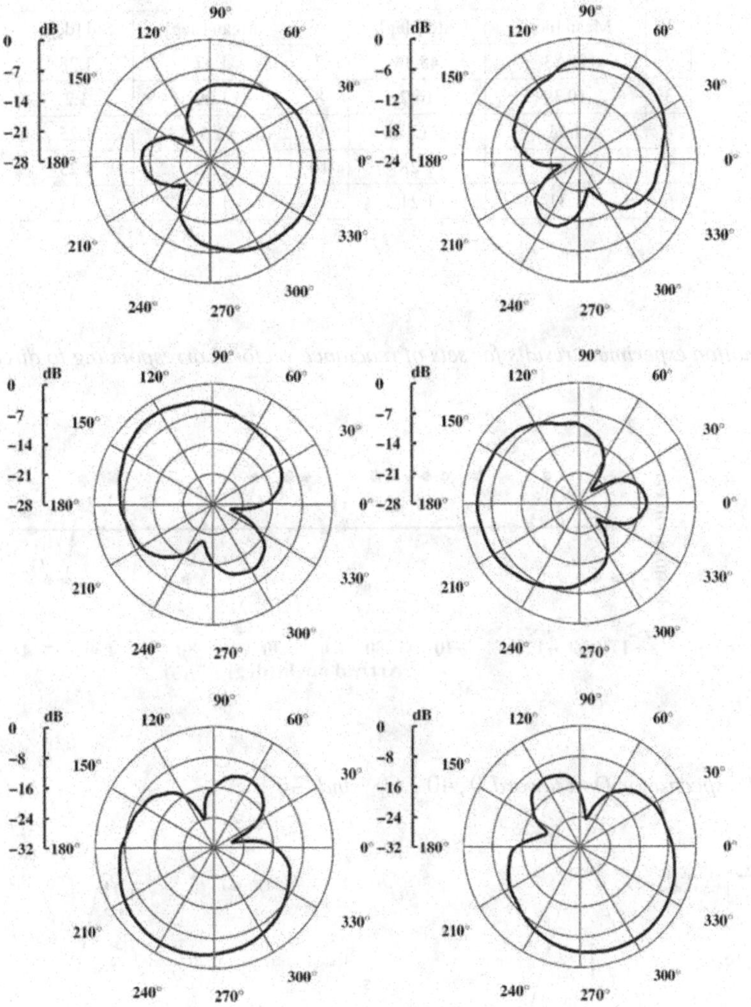


The experimental verification with the PPCC algorithm involved DoA estimation of one signal for angles regularly spread over the full azimuth in steps of $10^\circ$. The main purpose of the experiments was to investigate the PPCC method's performance according to both the number of reactance vectors $N$ and the case of directive beam patterns. The results in Table 1 show the estimation error mean and standard deviation (Std) against the number of reactance sets $N$. For $N >$ 4, we can verify that the estimation performance is better, since the estimation error mean and standard deviation are lower than $1.58^\circ$ and $1.28^\circ$, respectively. In these experiments, the reactances were arbitrarily chosen. In Figure 4, the PPCC method performance was evaluated for $N = 6$sets. In this case, the estimation error mean and standard deviation were $0.67^\circ$ and $0.59^\circ$, respectively. The RD correlation matrix in (14) was obtained by using $N \times K \times 360 = 21600$ measured snapshots with $K = 10$ and $N = 6$.

Similarly, the RD-MUSIC algorithm was verified with one impinging signal with results showed in Figure 5. The mutual impedance matrix **z** in (1) were obtained by measured data fitting with numerical electromagnetics code (NEC) (Iigusa, 2001) for a 2.484 GHz 7-element Espar antenna having $d = \lambda/4$inter-element spacing and the same element lengths. The experiment is done with the two methods explained in sections 3.2.3.1 and 3.2.3.2, and then the results obtained for each method are compared. It can be clearly seen that the estimated angle accuracy is better in the DM method than in the EWV method. Also, for the DM method, the relative angle estimation error is constant; the DoA spectrum is sharper and has a higher maximum peak value than in the EWV method. The RD correlation matrix in (14) was obtained by using $N \times K \times 360 = 432000$ measured snapshots with $K = 200$ and $N = 6$.

*Table 1. PPCC estimation precision versus number, N, of reactance sets*

| *N* | Mean [deg] | Std [deg] | *N* | Mean [deg] | Std [deg] |
|---|---|---|---|---|---|
| 2 | 28.83 | 45.38 | 7 | 1.47 | 1.28 |
| 3 | 10.39 | 16.2 | 8 | 1.39 | 1.2 |
| 4 | 1.53 | 1.1 | 9 | 1.5 | 1.25 |
| 5 | 1.5 | 1.28 | 10 | 1.58 | 1.25 |
| 6 | 1.31 | 1.21 | 11 | 1.36 | 1.12 |

*Figure 4. DoA estimation experiment results for sets of reactance vectors corresponding to directive beam patterns*

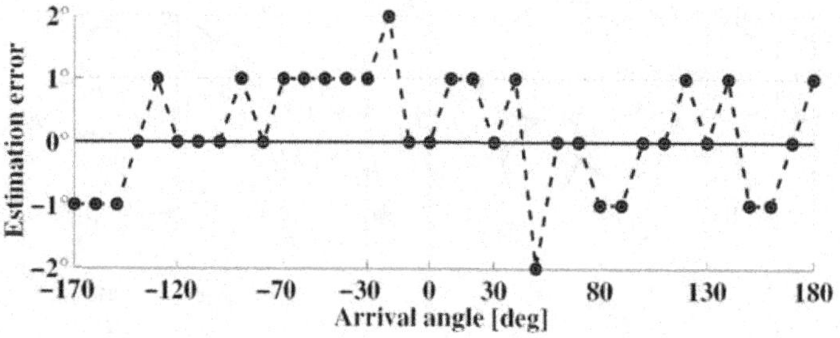


*Figure 5. RD-MUSIC spectra for DoA toward* $0^{\circ}, 40^{\circ}, 60^{\circ}$, *and* $140^{\circ}$

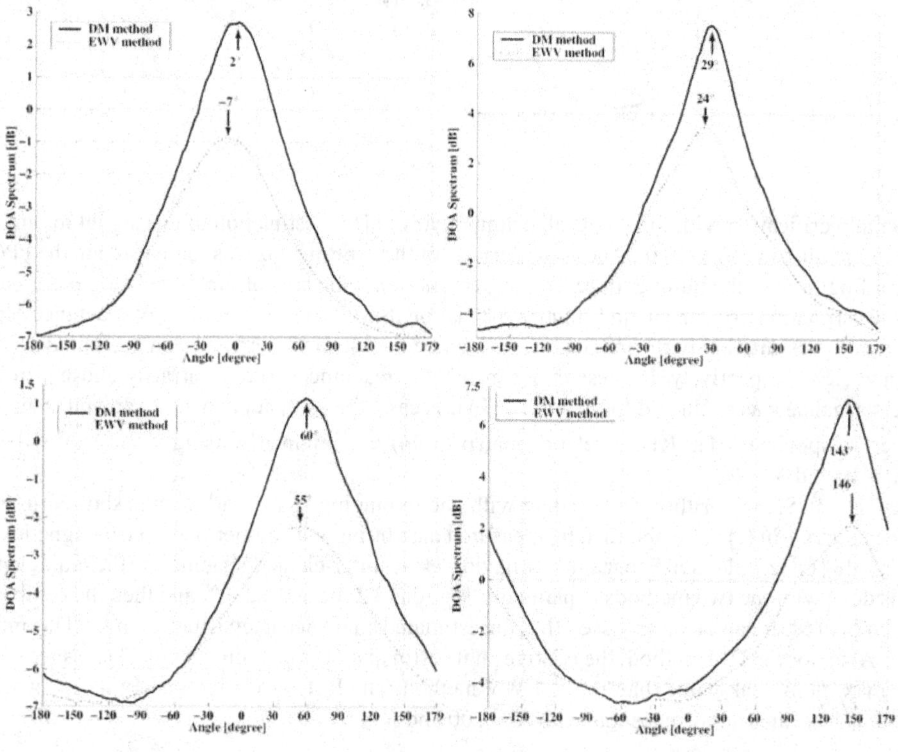

*Figure 6. RD-ESPRIT with a 7-element Espar antenna for one incoming signal*

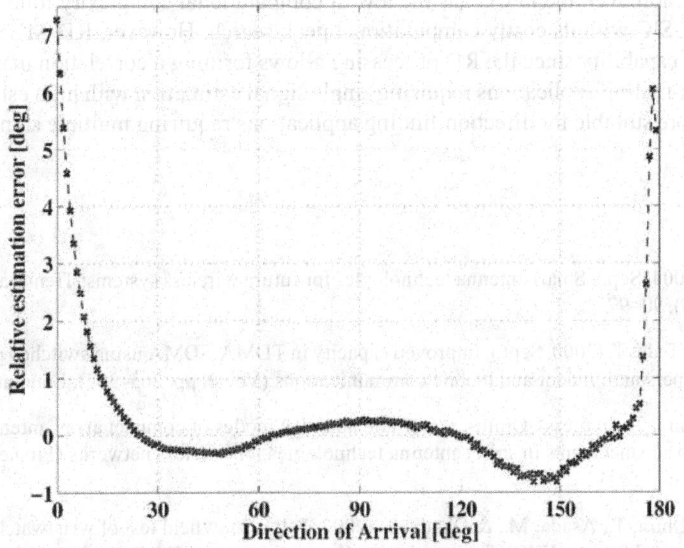


Finally, Figure 6 shows experiment for one signal with the RD-ESPRIT algorithm. The matrix $W$ was obtained with the previously explained calibration method by using $P$=36 regularly disposed and different angle values and $K$ = 100 snapshots for computing (14). Thus, a total of $(M+1)\times P\times K = 25200$ measured snapshots were used in the calibration procedure. It should be noted that whereas increasing $P$ increases performance, beyond the value of $P = 36$ no further increase in performance could be observed. The method exhibited the best DoA estimation precision when the signal DoA is toward the center area of the half azimuth plane. For the DoA in$\left[20^{\circ},160^{\circ}\right]$, the estimation precision taken as the distance between maximum and minimum estimation error is about 1.24°. However a distortion in the estimation error can be observed when the signal is toward an angle between two parasitic elements. This can be partially explained by the non-uniformity of the selected directive beam patterns due to varactor irregularities of the fabricated Espar antenna.

## CONCLUSION

Experimental verification results of DoA estimation in an anechoic room using three methods with an Espar antenna were shown. This experiments aim at investigating the DoA estimation ability of the low cost and low power consumption Espar antenna, in order to implement it in direction-finding applications. The three methods exploit the reactance diversity provided by an Espar antenna to correlate different antenna output signals measured at different times and for different reactance values. However, these methods required that $N$ pattern data to be measured and saved, which is costly in terms of measurement equipment.

The PPCC algorithm has the advantage of low computational cost and robustness against arrival signal data phase fluctuation, since its only make used of the antenna output power. With three pairs of opposite beampointing directive-beam-patterns, DoA estimation with less than 3 degrees of precision could be achieved in the full azimuth.

The RD-MUSIC algorithm was experimentally implemented through to methods, the Direct Measurement and Equivalent Weight Vector methods. The experiments show that the Direct Measurement method proposed in this paper did not only improve the estimated angle accuracy, but also solved the critical problem of calibrating the impedance parameters used in the Espar antenna output model.

The RD-ESPRIT algorithm was tested with a 7-element Espar antenna. The translational invariance configuration required by the ESPRIT algorithm could be found in the hexagonally shaped 7-element Espar antenna. Moreover, the practical application of the proposed algorithm requires an estimate of the RF equivalent weight matrix involved in the RD processing. However this parameter could be obtained during the calibration procedure, which is a required procedure even for the conventional array antenna. Experimental verification for one impinging signal showed that DoA estimation with high precision could also be achieved with RD-ESPRIT.

As a comparison, the three methods are equivalent in terms of measurement equipment required for the pre-measurement or calibration stage. Moreover, the PPCC has the lowest computational complexity followed by the RD-ESPRIT, and, ended by the RD-MUSIC with its costly computational peak search. However, RD-MUSIC and ESPRIT exhibit multiple signal estimation capability since the RD processing allows forming a correlation matrix. Therefore, PPCC is more suitable for direction-finding applications requiring single signal estimation with high estimation precision, while RD-based methods are more suitable for direction-finding applications requiring multiple signal estimation.

## REFERENCES

Alexiou, A., & Haardt, M. (2004, Sep.). Smart antenna technologies for future wireless systems: Trends and challenges. IEEE Communications Magazine, 42 (9), 90–97.

Almhdie, A., Kezys, V., & D. Todd, T. (2000, Sept.). Improved capacity in TDMA/SDMA using switched parasitic antennas. In IEEE international symposium on personal, indoor and mobile communications (Vol. 1, pp. 363–367). London.

Cheng, J., Taillefer, E., & Ohira, T. (in press). Omni-, sector and adaptive modes of compact array antenna. In C. Sun, J. Cheng, & T. Ohira (Eds.), Handbook on advancements in smart antenna technologies for wireless networks (First ed., pp. xxx–xxx). Hershey, USA: Idea Group Inc.

Furuhi, T., Hashiguchi, M., Ohira, T., Asada, M., & Okada, T. (2002, Feb.). Snowfield test of wristwatch-type microwave beacons and hand-held direction-of-arrival finders. IEICE Transaction on Communications, J86-B (2), 219–225. (in Japanese)

Harrington, R. F. (1978, May). Reactively controlled directive arrays. IEEE Transactions on Antennas and Propagation, 26 (3), 390–395.

Hirata, A., Yamada, H., & Ohira, T. (2003, Jun.). Reactance-domain MUSIC estimation using calibrated equivalent weight matrix of ESPAR antenna. In IEEE antennas and propagation society international symposium (Vol. 3, pp. 252–255). Columbus.

Iigusa, K., Hashiguchi, M., Hirata, A., & Ohira, T. (2001, Oct.). ESPAR antenna parameters fitting based on measured data. IEICE Technical Report, AP2001-104, RCS2001-143 , 93-100. (In Japanese)

Johnson, D. H., & Dudgeon, D. E. (1993). Array signal processing, concepts and techniques. PTR Prentice-Hall, Signal Processing Series. (ISBN 0-13-048513-6)

Krim, H., & Viberg, M. (1996, Jul.). Two decades of array signal processing research: The parametric approach. IEEE Signal Processing Magazine, 13 (4), 67–94.

Ohira, T., & Gyoda, K. (2000, May). Electronically steerable passive array radiator antennas for low-cost analog adaptive beamforming. In IEEE international conference on phased array systems and technology (pp. 101–104). Dana Point.

Ohira, T., & Gyoda, K. (2001, Dec.). Hand-held microwave direction-of-arrival finder based on varactor-tuned analog aerial beamforming. In IEEE asia-pacific microwave conference (Vol. 2, pp. 585–588). Taipei.

Ohira, T., & Iigusa, K. (2004, Sep.). Electronically steerable parasitic array radiator antenna. Wiley Periodicals, Electronics and Communications in Japan (Part II: Electronics), 87 (10), 25–45.

Ohira, T., & Cheng, J. (2004). Analog smart antennas. In S. Chandran (Ed.), Adaptive antenna arrays: Trends and applications (First ed., pp. 184–204). Berlin, Germany: Springer-Verlag.

Pierre, J., & Kaveh, M. (1991, Apr.). Experimental performance of calibration and direction-finding algorithms. In IEEE international conference on acoustics, speech, and signal processing (Vol. 2, pp. 1365–1368). Toronto.

Plapous, C., Cheng, J., Taillefer, E., Hirata, A., & Ohira, T. (2004, Dec.). Reactance domain MUSIC algorithm for electronically steerable parasitic array radiator antenna. IEEE Transactions on Antennas and Propagation, 52 (12), 3257–3264.

Roy, R., & Kailath, T. (1989, Jul.). ESPRIT- estimation of signal parameters via rotational invariance techniques. IEEE Transactions on Acoustics, Speech, and Signal Processing, 37 (7), 984–995.

Schlub, R., & Thiel, D. V. (2004, May). Switched parasitic antenna on a finite ground plane with conductive sleeve. IEEE Transactions on Antennas and Propagation, 52 (5), 1343–1347.

Schmidt, R. O. (1986, Mar.). Multiple emitter location and signal parameter estimation. IEEE Transactions on Antennas and Propagation, 34 (3), 276–280.

Scott, N. L., Leonard-Taylor, M. O., & Vaughan, R. G. (1999, Jun.). Diversity gain from a single-port adaptive antenna using switched parasitic elements illustrated with a wire and monopole prototype. IEEE Transactions on Antennas and Propagation, 47 (6), 1066–1070.

See, C. M. S. (1994, Mar.). Sensor array calibration in presence of mutual coupling and unknown sensor gains and phases. Electronics Letters, 30 (5), 373–374.

Shiskov, B., & Ohira, T. (2001, Dec.). Adaptive beamforming of ESPAR antenna based on stochastic approximation theory. In IEEE asia-pacific microwave conference (Vol. 2, pp. 597–600). Taipei.

Sun, C., Hirata, A., Ohira, T., & Karmakar, N. (2004, Jul.). Fast beamforming of electronically steerable parasitic array radiator antennas: Theory and experiment. IEEE Transactions on Antennas and Propagation, 52 (7), 1819–1832.

Svantesson, T., & Wennstr{m, M. (2001, Aug.). High-resolution direction finding using a o switched parasitic antenna. In Proc. of 11th ieee signal processing workshop on statistical signal processing (pp. 508–511). Singapore.

Taillefer, E., Chu, E., & Ohira, T. (2004, Oct.). ESPRIT algorithm for a seven-element regular-hexagonal shaped ESPAR antenna. In 7th european conference on wireless technology (pp. 153–156). Amsterdam.

Taillefer, E., Hirata, A., & Ohira, T. (2005a, Feb.). Direction-of-arrival estimation using radiation power pattern with an ESPAR antenna. IEEE Transactions on Antennas and Propagation, 53 (2), 678–684.

Taillefer, E., Hirata, A., & Ohira, T. (2005b, Nov.). Reactance-domain ESPRIT algorithm for a hexagonally shaped seven-element ESPAR antenna. IEEE Transactions on Antennas and Propagation, 53 (11), 3486–3495.

Taillefer, E., Plapous, C., Cheng, J., Iigusa, K., & Ohira, T. (2003, Mar.). Reactance-domain MUSIC for ESPAR antennas (experiment). In IEEE wireless communications and networking conference (Vol. 1, pp. 98–102). New Orleans.

Thiel, D. V. (2004, Jun.). Switched parasitic antennas and controlled reactance parasitic antennas: A systems comparison. In IEEE antennas and propagation society symposium (Vol. 3, pp. 3211–3214). Monterey.

Vaughan, R. (1999, Feb.). Switched parasitic elements for antenna diversity. IEEE Transactions on Antennas and Propagation, 47 (2), 399–405.

Wax, M., & Kailath, T. (1985, Apr.). Detection of signals by information theoretic criteria. IEEE Transactions on Acoustics, Speech, and Signal Processing, 33 (2), 387–392.

# Section II
# Performance Issues

# Chapter XI
# Physics of Multi-Antenna Communication Systems

**Santana Burintramart**
*Syracuse University, USA*

**Nuri Yilmazer**
*Syracuse University, USA*

**Tapan K. Sarkar**
*Syracuse University, USA*

**Magdalena Salazar-Palma**
*Universidad Carlos III de Madrid, Spain*

## ABSTRACT

*This chapter presents a concern regarding the nature of wireless communications using multiple antennas. Multi-antenna systems are mainly developed using array processing methodology mostly derived for a scalar rather than a vector problem. However, as wireless communication systems operate in microwave frequency region, the vector nature of electromagnetic waves cannot be neglected in any system design levels. Failure in doing so will lead to an erroneous interpretation of a system performance. The goal of this chapter is to show that when the vector nature of electromagnetic wave is taken into account, an expected system performance may not be realized. Therefore, the electromagnetic effects must be integrated into a system design process in order to achieve the best system design. Many researches are underway regarding this important issue.*

## INTRODUCTION

With an increasing demand in wireless devices, a system engineer is forced to increase quality of service, coverage, and bandwidth efficiency of a wireless system. As time and frequency domains are fully occupied, a space domain has been proposed to overcome these limitations and achieve the design criteria. To utilize the space domain, multiple antennas are required in a wireless system. Smart antenna utilizes multiple antennas and adaptive signal processing to direct energy to an intended receiver while rejecting all other interference signals in the same frequency band. Thus, it provides spatial dimension in wireless communications. However, most of the array theories used in multiple antenna system is based on an ideal antenna assumption, i.e. point sources, rather than real antennas. In practice, when real antennas are deployed, many assumptions for the ideal point sources are not valid any more. Thus, the processing algorithm needs to be modified to compensate for the non-ideal system. Such an effect introduced by real antenna comes from the vector nature of the electromagnetic waves.

In this chapter, the electromagnetic effects on multiple antenna systems will be discussed. We start from an antenna array and then move toward a multiple-input-multiple-output (MIMO) system. Next subsection, the mutual coupling effects on antenna arrays as well as some important performance parameters that are usually misinterpreted are discussed. Then, a compensation technique for the mutual coupling effects is introduced. Next, multiple antenna communications will be discussed followed by their performance metric known as channel capacity. The influence of electromagnetic effects on multiple antenna systems under the view of the channel capacity is discussed followed by conclusion.

## ANTENNA ARRAYS

An array of antennas has proved its significant role in both radar and communication systems. It has been shown that with multiple antennas at a receiver, the improvement in the signal-to-noise ratio (SNR) of the designed signal is feasible while any other interference is suppressed from the receiver. Similarly, by reciprocity property, multiple antennas can be used at the transmitter to concentrate energy to a particular receiver and to reduce its interference with other receivers. In an adaptive array processing context, it is customary to assume a receiver being an ideal point source situated in free space. This is generally valid for sonar and acoustic signals. However, when applied to electromagnetic waves, this assumption has never been met since there is always mutual coupling between antennas themselves. Mutual coupling destroys the linear wavefront assumption for the received signal, which is the first assumption in array processing. Moreover, when an antenna is mounted on a tower or any kind of platform, the coupling between the antenna and platform is also important as pointed out in (Sarkar, 2006).

In this section, the effect of mutual coupling on adaptive processing is demonstrated. A method for compensation of the effect of mutual coupling will be also discussed. The mutual coupling compensation, in general, is too tedious to be derived analytically. A computer code may be used for this purpose. However, to provide a physical picture to the readers, the work of (Adve & Sarkar, 2000) will be discussed here to show how the mutual coupling affects the array performance and how the mutual coupling can be integrated into the adaptive processing.

Let us consider an array of $N$ thin wire dipole antennas. The dipoles are assumed to be $z$-directed of length $L$ and radius $a$ and are placed along the $x$-axis, separated by a distance $\Delta x$. The port of each antenna is center loaded with an impedance of $Z_L$ Ohms primarily to make them resonant and thereby increase its efficiency. Figure 1 shows the model of the antenna arrays. To simplify the problem, the following assumptions are made (Djordjevic, 1995; Strait 1973):

1. The current flows only along the direction of the wire axes ($z$-axis in this case) and there is no circumferential variation of the current.
2. The current and charge densities on the wire are approximated by filaments of current and the charge distribution on the wire axes.
3. Surface boundary conditions can be applied to the relevant axial component on the wire axes.

Consider an incident electric field $E^{inc}$ impinging on the array, the relationship between the incident field and current on the wires that describes the behavior of the array can be expressed by the following integral equation (Adve, 2000; Sarkar, 2003):

$$E^{inc}(z) = \frac{1}{j\omega 4\pi\varepsilon_0}\left[\left(k^2 + \nabla\cdot\nabla\right)\int_{\ell} I(z')G(z,z')dz'\right]$$

$$= -\mu_0 \int_{axes} I(z')\frac{G(z,z')}{4\pi}dz' + \frac{1}{\varepsilon_0}\frac{\partial}{\partial z}\int_{axes}\frac{\partial I(z')}{\partial z'}\frac{G(z,z')}{4\pi}dz' \qquad (1)$$

*Figure 1. Linear array of thin wire dipole model*

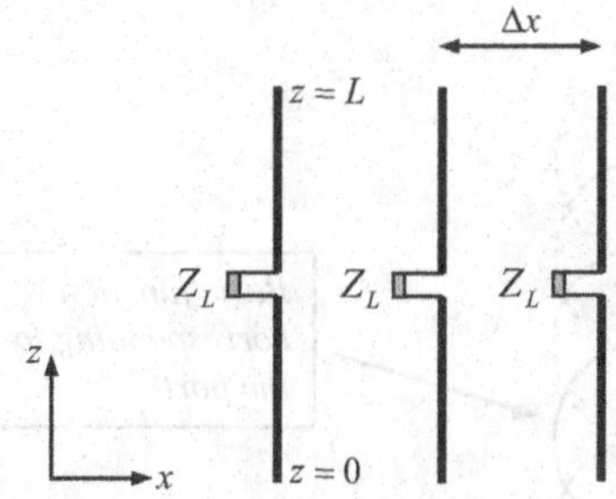


where the integration is carried out along the axis of the thin wires. $G(z,z') = e^{-jkR}/R$ denotes the Green's function where $k$ is the wave number in free space, and $R = \sqrt{(z-z')^2 + a^2}$ represents the displacement from a current filament $z'$ to a point of observation $z$. $\nabla$ is the divergence/gradient operator. In addition, $\mu_0$ and $\varepsilon_0$ are the permeability and permittivity of free space, respectively. To determine the current distribution along the wires, we expand the unknown currents in terms of some known basis functions with unknown constants:

$$I(z) = \sum_{n=1}^{N} \sum_{p=1}^{P} I_{p,n} f_{p,n}(z) \tag{2}$$

where $f_{p,n}$ is the $p$ th basis function on the $n$ th antenna element. We solve for the unknown constants by using the method of moment (MM) where the above equation can be arranged in matrix form as:

$$[V] = [\mathbf{Z}][I] \tag{3}$$

where $[I]$ is the MM current vector containing the coefficients of the expansion in (2). $[\mathbf{Z}]$ is the MM impedance matrix and is related to the array manifold only. The matrix $[V]$ is related to only the incident field, $E^{inc}$. If we assume the basis functions in our case are piecewise sinusoids as shown in Figure 2. $f_{p,n}$ can be written as (Sarkar, 2003):

$$f_{p,n}(z) = \begin{cases} \sin\left[k(z - z_{p-1,n})\right] & \text{for } z_{p-1,n} < z < z_{p,n} \\ \sin\left[k(z_{p+1,n} - z)\right] & \text{for } z_{p,n} < z < z_{p+1,n} \\ 0 & \text{elsewhere} \end{cases} \tag{4}$$

$$\Delta z \;\; = \;\; L/(P+1) \tag{5}$$

$$z_{p,n} \;\; = \;\; z_{0,n} + p\Delta z \tag{6}$$

$$V_i \;\; = \;\; \int_{z_{q-1,m}}^{z_{q+1,m}} f_{q,m}(z) E_z^{inc}(z) dz \tag{7}$$

Here $V_i$ denotes the $i$th component of $[V]$.

The element of the impedance matrix can be expressed as follows:

*Figure 2. Basis functions used in the method of moments*

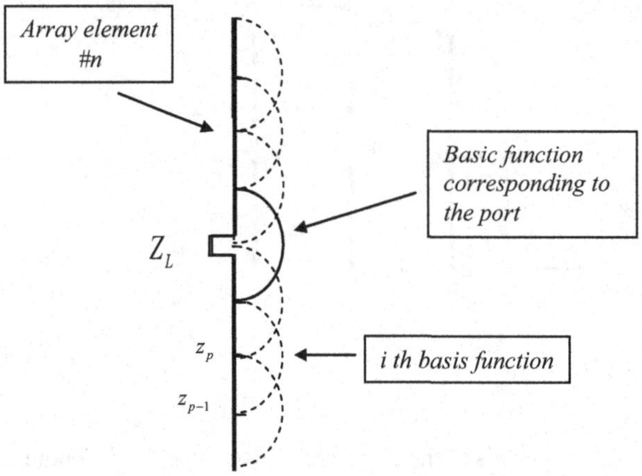


$$Z_{i,\ell} = \int_{z_{q-1,m}}^{z_{q+1,m}} f_{q,m}(z) \left\{ \begin{aligned} & j\omega\mu_0 \int_{z_{p-1,n}}^{z_{p+1,n}} f_{p,n}(z') \frac{e^{-jkR}}{4\pi R} dz' \\ & -\frac{1}{j\omega\varepsilon_0} \frac{\partial}{\partial z} \int_{z_{p-1,n}}^{z_{p+1,n}} \frac{d f_{p,n}(z')}{dz} \frac{e^{-jkR}}{4\pi R} dz' \end{aligned} \right\} dz' \tag{8}$$

where

$$i = [(m-1)p + q] \tag{9}$$

$$\ell = [(n-1)p + q] \tag{10}$$

Here $z_{0,n}$ is the $z$-coordinate of the bottom of the $n$ th antenna element. Note that other types of basis functions are also possible. In (3), the left-hand side of the equation is dependent only on the incident field. The [**Z**] matrix is of dimension $NP \times NP$ and [$V$] is $NP \times 1$. Here $NP$ is the total number of coefficients representing currents on the array.

Assuming an incident electric field is linearly polarized along the $z$-direction and its component is in the form of

$$E_z = E_0 e^{-j\mathbf{k}\cdot\mathbf{r}} \tag{11}$$

where the wave number $\mathbf{k} = -k[\hat{x}\cos\varphi \sin\theta + \hat{y}\sin\varphi\sin\theta + \hat{z}\cos\theta]$ represents the direction of arrival of the incident field $(\varphi, \theta)$. Substituting (11) in the above equations, we get the $i$ th component of [$V$] as

$$V_i = \frac{E_0 e^{jkx_m \cos\varphi \sin\theta}}{k\sin(k\Delta z)\sin^2\theta} 2e^{jkz_{q,m}\cos\theta} \left[\cos(k\Delta z\cos\theta) - \cos(k\Delta z)\right] \tag{12}$$

where $x_m$ represents the $x$-coordinate of the axis of the $m$ th antenna. Substituting (2) and (4) in (8), we obtain the entries of the impedance matrix [**Z**] by

$$Z_{i,\ell} = \frac{j30}{\sin^2(k\Delta z)} \int_{z_{q-1,m}}^{z_{q,m}} \sin\left[k(z - z_{q-1,m})\right] \times \left\{ \frac{e^{-jkR_1}}{R_1} - 2\cos(k\Delta z)\frac{e^{-jkR_2}}{R_2} + \frac{e^{-jkR_3}}{R_3} \right\} dz$$
$$+ \frac{j30}{\sin^2(k\Delta z)} \int_{z_{q,m}}^{z_{q+1,m}} \sin\left[k(z_{q+1,m} - z)\right] \times \left\{ \frac{e^{-jkR_1}}{R_1} - 2\cos(k\Delta z)\frac{e^{-jkR_2}}{R_2} + \frac{e^{-jkR_3}}{R_3} \right\} dz \quad (13)$$

where

$$R_1 = \sqrt{(x_m - x_n)^2 + (z - z_{p-1,n})^2} \quad (14)$$

$$R_2 = \sqrt{(x_m - x_n)^2 + (z - z_{p,n})^2} \quad (15)$$

When $m = n$, the term $(x_m - x_n)$ is set equal to the radius of the wire, $a$. Given the incident field, the currents on the array can be found by solving the inversed equation of (3). The inverse of the MM impedance matrix is called the MM admittance matrix, i.e., $[\mathbf{Y}] = [\mathbf{Z}]^{-1}$. However, the impedance matrix above does not take the load impedance $Z_L$ into account. To do so, we simply add the load impedance to the diagonal entries of [**Z**] that corresponds to the port of the antennas. We now define a new impedance matrix as follows:

$$[\mathbf{Z}_N]_{(i,i)} = Z(i,i) + Z_L \quad (17)$$

$$[\mathbf{Z}_N]_{(i,j)} = Z(i,j) \quad (18)$$

where $i$ corresponds to a port, and the parenthesis denotes the entry of the matrix. The new admittance matrix is then

$$[\mathbf{Y}_N] = [\mathbf{Z}_N]^{-1} \quad (19)$$

Since there is only one basis function representing the current distribution at the port, we thus can determine the voltage measured at the ports of the array from the following formulation

$$\begin{aligned} [V_{meas}] &= [\mathbf{Z}_L][I_{port}] = [\mathbf{Z}_L][\mathbf{Y}_{N,port}][V] \\ [V_{meas}] &= [\mathbf{C}][V] \end{aligned} \quad (20)$$

Here $[\mathbf{Z}_L]$ is the matrix of load impedance that is $[\mathbf{Z}_L] = diag(Z_L, Z_L, \ldots, Z_L)$. $[V_{meas}]$ denotes the measured voltage across the load impedance. We note that $[I_{port}]$ represents only currents at the $N$ ports of the array; thus, the dimension of $[\mathbf{Y}_{N,port}]$ is a rectangular matrix of dimension $N \times NP$ corresponding to the $N_w$ row of $[\mathbf{Y}_N]$ that corresponds to the $N_w$ port. The matrix [**C**] determines the relationship between an incident field and the measured voltages at the ports of the array including mutual coupling between antennas in the array. Note that [**C**] has dimension of $N \times NP$. In the following subsection, the mutual coupling compensation will be discussed.

One important issue in antenna array design is the difference between gain and directivity. The gain of an antenna array is the actual information that represents the performance of the array and not the directivity as the gain contains information about Ohmic losses in the array (Kraus, 1988). These two parameters are usually misinterpreted and falsely used interchangeably. The directivity only shows the radiation performance of the array when there is no loss in the system, which means that the antenna efficiency is 100%. However, when mutual coupling is presented, it creates mismatch in the array. As a result, some portions of the power are not transmitted, but is lost in the array structure causing reduction in the array efficiency. The following example gives an idea of mutual coupling effects on the array efficiency.

*Figure 3. Array setup – Transmitting array has 3 dipoles and the receiving antenna is a single dipole. The receiving antenna moves along the circumference of a 100 m. circle from* $\varphi = 0^{\circ}$ *to* $\varphi = 180^{\circ}$.

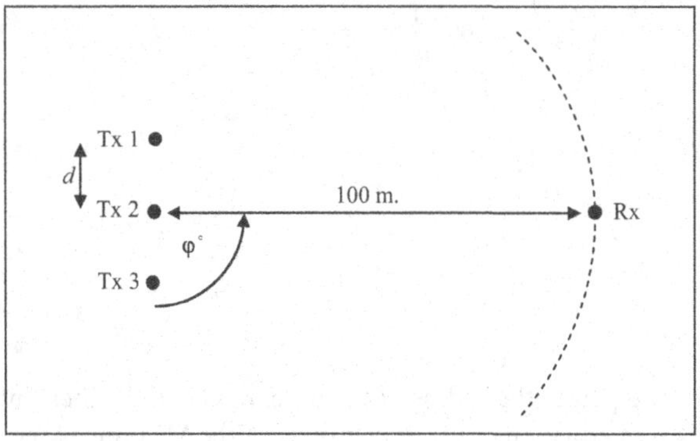


## Example I

In this example, we demonstrate the array efficiency through numerical simulations. On transmitters, two arrays of three half-wavelength dipoles are considered – one with 0.5λ spacing and the other with 0.25λ spacing. The transmitted power is constrained to be 1 *W.* for both the cases. The receiver is a dipole placed 100 m. away from the transmitting array. The receiver is placed along directions $\varphi = 0^{\circ}$ to $\varphi = 180^{\circ}$relative to the enfire of the transmitting array as shown in Figure 3 where *d* represents spacing between antenna elements.

Each of the dipoles is individually matched so that it radiates maximum power in free space. Noting that when put together as an array, the mutual coupling changes the input impedance of each antenna, thus, introducing mismatch in the array. Therefore, the array efficiency is reduced. For the three transmitting antennas, the directivity along the broadside of the array ($90^{\circ}$) is expected to be 3 time larger than that of the single transmitting antenna case. Figure 4 shows the received power at the receiving antenna when the antenna spacing at the transmitter are 0.5λ and 0.25λ. The received power is normalized to the received power obtained when there is 1 transmitting antenna.

*Figure 4. Received power normalized to the single transmitter case when the total transmitted power is 1 W. (Solid line: 0.5λ spacing, Dotted line: 0.25λ spacing.)*

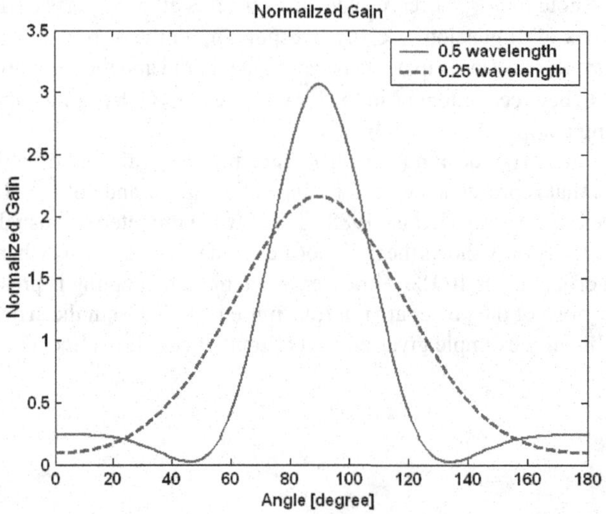

It is seen that when the spacing is reduced, the mutual coupling between the antenna elements has significant effects. The loss in the transmission is due to the loss in the mismatch at the transmitter terminals introduced by mutual coupling. Thus, the overall gain of the antenna array is degraded. The gain, not the directivity, of an array is an important parameter that needs to be concerned when designing an antenna array in the presence of strong mutual coupling.

## MUTUAL COUPLING COMPENSATION

It is seen that when multiple antennas are used as a transmitter and/or receiver, the mutual coupling destroys the received signal wavefront from its ideal condition. The effects of mutual coupling on array antennas have long been reported in the literature. (Gupta & Ksienski, 1983) proposed a method to compensate mutual coupling using the open-circuit voltages based on an *N* port network model. In the network, it is assumed that when the ports are opened, there will be no mutual coupling. Then, the relationship between the open-circuit and the measured voltages can be determined. However, this model is true only when the antennas are half-wavelength spaced (Adve & Sarkar, 2000). For other cases, the open-circuit voltages only reduce the effects of mutual coupling. In this section, the method of mutual coupling compensation will be discussed. This method has shown its effectiveness in eliminating the mutual coupling effects.

As seen from the previous section, equation (20) represents the relationship between incident fields and voltages measured at the ports of the array. One solution in eliminating mutual coupling is *a minimum norm technique* where one tries to find incident fields from a given set of measured voltages using matrix inversion in (20). Since [**C**] is a rectangular matrix, this approach tends to find the best solution of the following equation

$$[\tilde{V}] = [\mathbf{C}]^H \left\{ [\mathbf{C}][\mathbf{C}]^H \right\}^{-1} [V_{meas}] \tag{21}$$

where the superscript $H$ denotes the conjugate transpose of a matrix and $[\tilde{V}]$ represents the estimates of the incident fields. In fact, this compensation technique may be interpreted as finding the signal with minimum energy that results in the measured voltages. However, the matrix [**C**] needs to be calculated by the MM analysis *a priori*. A simpler mutual coupling compensation method based only on measured voltages will be discussed in the following context (Sarkar, 2003). This method is similar in concept to the procedure described in (Friedlander, 1993; Lee & Lin, 1997; Friedlander & Weiss, 1991, 1992) except that all the electromagnetic effects including mutual coupling are included in an accurate fashion through Maxwell's equations.

The goal is to select the best-fit transformation, $[\Im]$, between a real (measured) array manifold, $[\mathbf{A}(\varphi)]$, and the ideal array manifold, $[\mathbf{A}_v(\varphi)]$ called virtual array, corresponding to a uniform linear virtual array (ULVA) for all angles $\varphi$ within a predefined sector.

$$[\Im][\mathbf{A}(\varphi)] = [\mathbf{A}_v(\varphi)] \tag{22}$$

The procedure of finding the transformation $[\Im]$ independent of the angle $\varphi$ can be explained step-by-step as follows:

1. The field of view of the array is divided into $Q$ sectors.
2. In each sector, angles of arrival is uniformly defined to cover each sector:

$$[\mathbf{\Phi}_q] = [\varphi_q,\ \varphi_q + \Delta\phi,\ \varphi_q + 2\Delta\phi,\ \ldots,\ \varphi_{q+1}] \tag{23}$$

   where $\Delta\phi$ is the angular step size for the $q$ th sector.

3. The steering vectors (real array manifold) are measured/computed associated with the set $[\mathbf{\Phi}_q]$. This is done by placing a signal in a far-field along the angles defined in (23) and measuring the voltages at the ports. This can either be measured or simulated. Note that all of the undesired electromagnetic effects including that of near-field scatterers must be included in the process. The measured array is

$$[\mathbf{A}(\mathbf{\Phi}_q)] = [a(\varphi_q),\ a(\varphi_q + \Delta\phi),\ a(\varphi_q + 2\Delta\phi),\ \ldots,\ a(\varphi_{q+1})] \tag{24}$$

*Table 1. Interference angles-of-arrival and field intensities*

| | Azimuth angle (j ) | $E^{inc}$ [V/m] |
|---|---|---|
| Jammer 1 | 40° | 100 |
| Jammer 2 | 75° | 1000 |
| Jammer 3 | 120° | 100 |

*Figure 5. Output beam patterns of the array of 10 dipoles with the spacing of* $0.5\lambda$

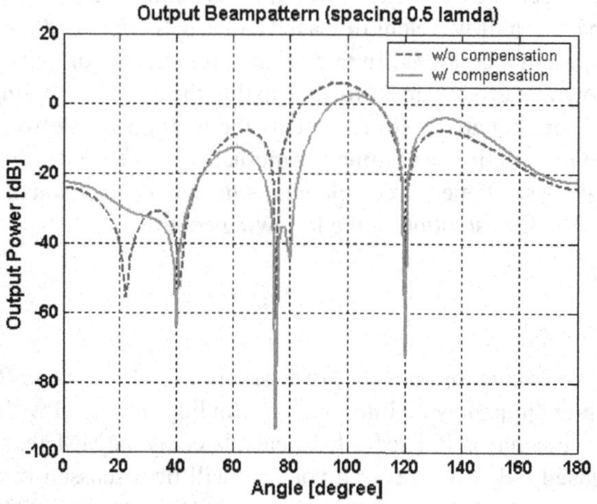


4. Next, we fix the virtual elements of the interpolated array corresponding to the ULVA consisting of point sources and calculate a set of theoretical steering vectors corresponding to the predefined angles in the previous steps. The theoretical array manifold is defined as follows

$$[\mathbf{A}_v(\mathbf{\Phi}_q)] = [a_v(\varphi_q),\ a_v(\varphi_q + \Delta\phi),\ a_v(\varphi_q + 2\Delta\phi),\ \ldots,\ a_v(\varphi_{q+1})] \tag{25}$$

5. We compute the transformation matrix $[\Im]$ for each of the sector using the least squares method such that $[\Im_q][\mathbf{A}(\mathbf{\Phi}_q)] = [\mathbf{A}_v(\mathbf{\Phi}_q)]$. The least squares solution of this equation is given by

$$[\Im_q] = [\mathbf{A}_v(\mathbf{\Phi}_q)][\mathbf{A}(\mathbf{\Phi}_q)]^H \left\{[\mathbf{A}(\mathbf{\Phi}_q)][\mathbf{A}(\mathbf{\Phi}_q)]^H\right\}^{-1} \tag{26}$$

6. Finally, we can use the transformation in (26) to transform the received voltages $[\mathbf{x}(m)]$ with mutual coupling effects into voltages that would be obtained from the ideal point sources

$$[\mathbf{x}_c(m)] = [\Im_q][\mathbf{x}(m)] \tag{27}$$

where $[\mathbf{x}(m)]$ is the measured voltage vector and $[\mathbf{x}_c(m)]$ is the compensated voltage vector. These corrected voltages can be applied to any signal processing algorithm

*Table 2. Amplitude estimation of the D3LS algorithm with the antenna spacing of* $0.5\lambda$

| | Exact Amplitude | Estimated Amplitude |
|---|---|---|
| w/o compensation | $1+j0$ | $0.006-j0.001$ |
| w/ compensation | $1+j0$ | $0.988+j0.011$ |

*Figure 6. Output beam patterns of the array of 10 dipoles with the spacing of* $0.25\lambda$

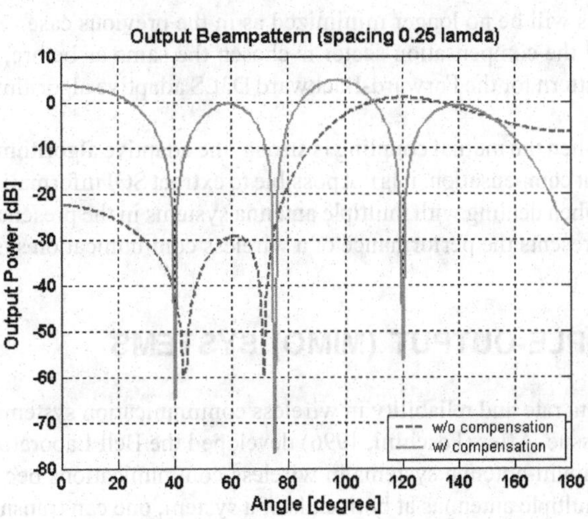


## Example II

In this example, the mutual coupling compensation method explained above will be demonstrated. A uniform linear array of half-wavelength dipoles is used as a receiver for a smart antenna system. The goal is to extract information about the signal of interest (SOI) embedded in a strong interference. The mutual coupling between antennas is calculated by using the MM code (Kolundzija, 1995). The receiving array is composed of 10 half-wavelength dipoles. The SOI electric field of 1 [V/m] incidents on the array at $\varphi = 110^{\circ}$ from the end-fire direction of the array. The directions and electric field intensities of three strong interferences are shown in Table 1. The Forward-Backward Direct Data-Domain Least Squares (D3LS) algorithm (Sarkar, 2003) is used to calculate the adaptive weights for this example.

Thermal noise of 20 *dB* SNR is added to the received signals. The weights are calculated for both with and without compensation. For the compensated case, the electric field of 1 [V/m] is placed along an observation sector, $[\ddot{O}_q] = [30^{\circ},\ 32^{\circ},\ \ldots,\ 150^{\circ}]$, then the transformation matrix is evaluated for the ULVA placed at the same positions of the real antennas. The output beam patterns for both the cases are shown in Figure 5 and the corresponding estimated amplitudes are shown in Table 2.

The output beam pattern of the compensated array provides deeper nulls at the interference directions. It is noted that the beam pattern is constrained to be 0 *dB* along the SOI direction. Obviously, the adaptive algorithm cannot extract the useful information about the SOI when the mutual coupling is not compensated. However, as we know that for the array of half-wavelength spacing, the mutual coupling is minimized. Thus, it is still possible to obtain a good beam pattern.

*Table 3. Amplitude estimation of the D3LS algorithm with the antenna spacing of* $0.25\lambda$

| | Exact Amplitude | Estimated Amplitude |
|---|---|---|
| w/o compensation | $1+j0$ | $0.056+j0.012$ |
| w/ compensation | $1+j0$ | $1.189+j0.028$ |

## Example III

In this example, the same setup as in the Example II will be used except that the antenna spacing is reduced to $0.25\lambda$. The mutual coupling between antennas will be no longer minimized as in the previous case. Again, the array is composed of 10 half-wavelength dipoles and the compensation sector is chosen the same as before, $[\mathbf{\Phi}_q]=[30^\circ,\ 32^\circ,\ \ldots,\ 150^\circ]$. Figure 6 shows the output beam pattern for the Forward-Backward D3LS adaptive algorithm. Their amplitude estimates are shown in Table 3.

In this case, it is obvious that when the mutual coupling is strong, the adaptive algorithm without compensation does not give satisfactory result. Without compensation, it is not possible to extract SOI information under strong interference. Therefore, one should be careful when dealing with multiple antenna systems in the presence of strong mutual coupling. In the next section, the metric represents the performance of a wireless communication system will be discussed.

## MULTIPLE-INPUT-MULTIPLE-OUTPUT (MIMO) SYSTEMS

With the increasing demand on data rate and reliability in wireless communication systems, multiple antenna concepts have been proposed to solve this issue. After (Foschini, 1996) developed the Bell Laboratories layered space-time system (BLAST), research on using multi-antenna systems in wireless communications becomes one of the most active research arenas. By introducing multiple antennas at both ends of a system, one can transmit data through a number of independent paths in a multipath fading environment. This results in a *multiplexing gain*, which is an increase in data rate of a system without affecting the system bandwidth. In this section, we will discuss an idea behind MIMO and how multiple independent paths can be utilized.

For a narrowband MIMO systems composed of $N_T$ transmit and $N_R$ receive antennas, a system model can be represented as

$$\mathbf{y} = \mathbf{H}\mathbf{x}+\mathbf{n} \tag{28}$$

where $\mathbf{y}=[y_1, y_2, \ldots, y_{N_R}]^T$ is the vector of received signals and $\mathbf{x}=[x_1, x_2, \ldots, x_{N_T}]^T$ is the vector of transmitted signals. **H** is an $N_R \times N_T$ channel matrix, whose $h_{ij}$ component represents the channel gain from the $j$ th transmit antenna to the $i$ th receive antenna. $\mathbf{n}=[n_1, n_2, \ldots, n_{N_R}]^T$ is the noise vector where its components are assumed to be additive white Gaussian noise with zero mean and unit variance. We note that the channel matrix **H** here is a deterministic model where the channel is assumed to be known. Figure 7 shows the MIMO system model in (28).

With the knowledge of the channel, applying the Singular Value Decomposition (SVD) to the channel matrix **H**, we get

$$\mathbf{H} = \mathbf{U}\mathbf{\Sigma}\mathbf{V}^H \tag{29}$$

where $\mathbf{U}=[\mathbf{u}_1, \mathbf{u}_2, \ldots, \mathbf{u}_{N_R}]\in C^{N_R\times N_R}$ and $\mathbf{V}=[\mathbf{v}_1, \mathbf{v}_2, \ldots, \mathbf{v}_{N_T}]\in C^{N_T\times N_T}$ are unitary matrices, and $\mathbf{\Sigma}$ is a diagonal matrix whose entries are singular values, $\sigma_i$, of **H**. Rank of the channel matrix is the number of singular values that are greater than zero. The rank of the channel matrix also gives information on how many sub-channels or independent paths can be utilized with a given channel. The channel is called a rich scattering environment when it has full rank or $rank(\mathbf{H})=\min(N_R, N_T)$. With this channel condition, we can utilize the full diversity of a MIMO system.

*Figure 7. MIMO system model*

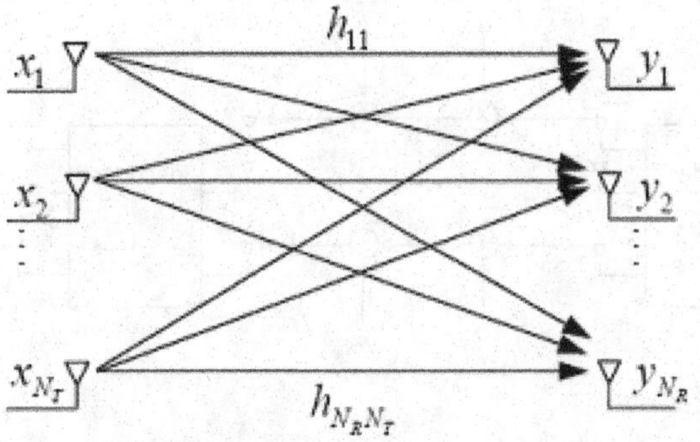


To demonstrate how one can transmit signals through multiple independent paths in a MIMO system, we assume that both transmitter and receiver have knowledge about the channel. To be more specific, the channel matrix **H** is known to the transmitter and receiver. The channel knowledge at the receiver is easy to obtain via training and tracking (Paulraj, 2003); thus, it is assumed throughout our discussion. The channel information at the transmitter requires some kinds of feedbacks to send the information back to the transmitter or can be obtained through estimation using reciprocity (Hwang, 2005). To transmit the signals, we first encode our input signals $\tilde{\mathbf{x}}$ (or messages we want to convey) by using the unitary matrix **V** so that the transmitted signals fed to the antenna are expressed by $\mathbf{x} = \mathbf{V}\tilde{\mathbf{x}}$. At the receiver end, we use the unitary matrix **U** to decode the received signals as $\tilde{\mathbf{y}} = \mathbf{U}^H\mathbf{y}$. We emphasize that the signal vectors with 'tilde' sign are pre- and post-processing signals and they are not transmitted directly through the MIMO channel. The signal vectors without 'tilde' are the actual signals that go through the MIMO channel. By using these pre- and post-processing, the messages can be conveyed through multiple independent paths as shown mathematically by

$$\tilde{\mathbf{y}} = \mathbf{U}^H\mathbf{y} = \mathbf{U}^H(\mathbf{H}\mathbf{x}+\mathbf{n})$$

$$\tilde{\mathbf{y}} = \mathbf{U}^H\mathbf{y} = \mathbf{U}^H\left(\mathbf{U}\boldsymbol{\Sigma}\mathbf{V}^H\mathbf{x}+\mathbf{n}\right)$$

$$\tilde{\mathbf{y}} = \boldsymbol{\Sigma}\mathbf{U}^H\mathbf{y} = \left(\mathbf{U}^H\mathbf{U}\right)\mathbf{V}\left(\mathbf{V}^H\mathbf{U}\right)\tilde{\mathbf{x}}+\mathbf{n}^H$$

$$\tilde{\mathbf{y}} = \boldsymbol{\Sigma}\,\tilde{\mathbf{x}}+\tilde{\mathbf{n}} \tag{30}$$

The equation (30) follows from the fact that **U** and **V** are unitary matrix so that $\mathbf{U}^H\mathbf{U}=\mathbf{I}$ and $\mathbf{V}^H\mathbf{V}=\mathbf{I}$. The noise vector $\tilde{\mathbf{n}} = \mathbf{U}^H\mathbf{n}$ is the transformed noise and since $\mathbf{U}^H\mathbf{n}$ is a linear combination of noise component at each receiving antenna, the noise distribution does not change. From (30), the transmitted messages $\tilde{\mathbf{x}}$ are transmitted in parallel and are attenuated by the factors represented by the singular values of the channel matrix. For large singular values, the messages are conveyed with less distortion compared to the one transmitted via small singular values. This parallel decomposition of the MIMO channel is shown in Figure 8.

When the channel is not full rank, $K = rank(\mathbf{H}) \leq \min(N_R, N_T)$, the maximum number of independent paths that can transmit the signals is *K*. This yields the multiplexing gain of *K* for this particular MIMO channel. In a rich scattering condition, the singular values of the channel matrix are expected to be comparable to each other and the channel matrix has full rank. Therefore, it is expected to increase the transmission rate by the factor of number of antennas in the system. However, in practice when real antennas are deployed, the mutual coupling effects also influence the rank of the channel matrix, which in turn alter the ability of parallel transmission. In the following examples, the channel parallel decomposition and the mutual coupling effects on the decomposition are demonstrated using real antenna system. In this discussion, the channel knowledge is assumed to be known both at the transmitter and receiver sides.

*Figure 8. Parallel decomposition of MIMO channel*

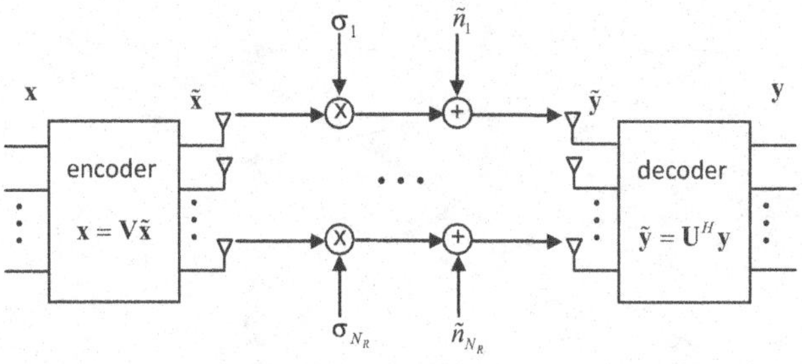


## Example VI

This example shows how the multiplexing gain of the MIMO systems can be achieved through parallel decomposition. Let us consider a narrowband $(2\times 2)$ MIMO system operating at 1 GHz. Each antenna is centrally loaded such that it is resonant in free space, which means that it provides maximum power transfer in free space. The two transmitters are separated by a wavelength (λ) and the receivers are separated similarly. The antennas are located inside a room of dimension $2m.\times 2m.$ The room has a metallic wall of height 0.5 m. and has no ceiling and floor. There is one metallic sphere with diameter of 0.4 m. and one metallic cube with dimension of $0.4m.\times 0.4m.\times 0.4m.$ situated inside the room. If we let the room to be centered at the origin, where the wall starts from $z=0$ m. as shown in Figure 9, the sphere is centered at (-0.2, 0.3, 0.25) m. and the cube is centered at (0.4, -0.1, 0.25) m.. Two transmit antennas are located at (0.15, -0.75, 0.175) m. and (-0.15, -0.75, 0.175) m. Similarly, the receive antennas are placed at (0.15, 0.75, 0.175) m. and (-0.15, 0.75, 0.175) m., respectively. The simulation is carried out through an accurate electromagnetic analysis (Kolundzija, 1995).

To apply channel decomposition, we first measure the channel matrix **H** as voltages induced at the receiving antennas. The load impedance at each antenna is $Z_L = 90.57 - j42.51$ [Ω]. This knowledge about the channel needs to be available at the both ends for the pre- and post-processing operations. In this example, the channel is

$$\mathbf{H} = \begin{bmatrix} 0.007 + j0.011 & -0.003 + j0.006 \\ -0.020 + j0.003 & 0.047 + j0.043 \end{bmatrix} \tag{30}$$

*Figure 9. MIMO simulation setup*

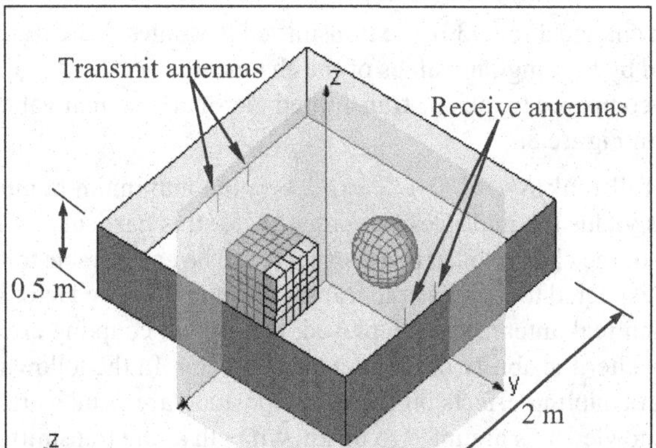

Applying the SVD to **H**, we get

$$\mathbf{U} = \begin{bmatrix} -0.002 - j0.045 & 0.539 + j0.840 \\ -0.992 + j0.118 & -0.036 + j0.027 \end{bmatrix}, \Sigma = \begin{bmatrix} 0.0668 & 0 \\ 0 & 0.0148 \end{bmatrix},$$

$$\mathbf{V} = \begin{bmatrix} 0.298 & 0.955 \\ -0.627 + j0.720 & 0.196 - j0.224 \end{bmatrix}.$$

By looking at the singular values, it is expected that the received signal after post-processing will be attenuated by these values as shown in (30). The signals to be conveyed are $\tilde{\mathbf{x}} = [\tilde{x}_1\ \tilde{x}_2]^T$ where $\tilde{x}_1 = 10.128$ and $\tilde{x}_2 = j10.128$. The amplitudes of these signals are selected such that the total transmitted power is 1 W. Thus, the input voltages fed to the MIMO system are:

$$\mathbf{x}_{1W} = \mathbf{V}\tilde{\mathbf{x}} = \begin{bmatrix} 3.014 + j9.669 \\ -4.082 + j9.269 \end{bmatrix}.$$

The received voltages across the load at the receiving antennas are

$$\mathbf{y} = \begin{bmatrix} -0.127 + j0.050 \\ -0.678 + j0.074 \end{bmatrix}.$$

Since there is no noise applied to the system, the output signals after post-processing are

$$\tilde{\mathbf{y}} = \begin{bmatrix} 0.677 \\ j0.150 \end{bmatrix},$$

which is equal to $\tilde{\mathbf{y}} = \Sigma\tilde{\mathbf{x}}$ as expected. It means that with this $(2 \times 2)$ MIMO system, one can transmit 2 data symbols at a time. Thus, the data rate is increased by 2. However, the received powers of the two signals are different according to the singular values of the channel. In this case, the received powers are $P_{\tilde{y}1} = 4.153$ $[mW]$ and $P_{\tilde{y}2} = 0.204$ $[mW]$, respectively.

## Example V

In this example, we want to show the effect of mutual coupling at the receiver compared to Example IV. To remove mutual coupling between receiving antennas, we will consider electric field at the ports of the receiving antennas instead. This is equivalent as changing the receiving antennas to be ideal point sources. The channel matrix corresponding to the electric field polarized along $z$-direction is

$$\mathbf{H} = \begin{bmatrix} -0.002 + j0.203 & -0.223 + j0.032 \\ -0.359 - j0.162 & 0.294 + j1.135 \end{bmatrix}$$

The SVD of channel matrix **H** is

$$\mathbf{U} = \begin{bmatrix} 0.095 - j0.095 & 0.152 - j0.979 \\ -0.882 - j0.451 & 0.133 - j0.021 \end{bmatrix}, \Sigma = \begin{bmatrix} 1.248 & 0 \\ 0 & 0.255 \end{bmatrix},$$

$$\mathbf{V} = \begin{bmatrix} 0.296 & -0.955 \\ -0.637 + j0.711 & -0.198 + j0.221 \end{bmatrix}.$$

The input voltages are

$$\tilde{\mathbf{x}} = [\tilde{x}_1 \ \tilde{x}_2]^T = \begin{bmatrix} 9.872 \\ j9.872 \end{bmatrix}$$

for the total transmitted power of 1 W. Thus, the input voltages to the transmitter are

$$\mathbf{x}_{1W} = \mathbf{V}\tilde{\mathbf{x}} = \begin{bmatrix} 2.925 - j9.428 \\ -8.473 + j5.065 \end{bmatrix}.$$

When the signals are transmitted through this channel, the decoded received signals (E-fields) are as expected

$$\tilde{\mathbf{y}} = \begin{bmatrix} 12.320 \\ j2.519 \end{bmatrix}.$$

An interesting point is that the ratio of the two received signals in this ideal receiving antenna case is larger than that of the real antenna case in the previous example. The ratio of the magnitudes of the received signals is 4.891 as opposed to the ratio of the previous example is 4.513. This indicates that the received signals in this ideal receiving antenna system will be more sensitive to perturbations than the received signal in the Example IV. Next example compares the sensitivities of these two systems.

## Example VI

In this example, we use the same simulation setups as in the previous two examples including channel matrices and the input signals. However, we introduce a small perturbation in to the systems by varying the two objects, i.e. cube and sphere, randomly. Their variations in the locations are considered as zero-mean Gaussian random variables with standard deviations $\underline{\sigma}_x = \underline{\sigma}_y = \underline{\sigma}_z = 0.01m.$, where $\underline{\sigma}_x$ denotes the standard deviation of the change in the location along $x$ direction, similarly for $y$ and $z$ directions. Now, we compare the performances of the two MIMO systems in the previous two examples, i.e. real and ideal receiving antennas, in the presence of this randomness. The simulations are run for 20 independent runs for both cases. Assuming that QPSK modulation is considered in this example, the message will be conveyed correctly if the phases of the received signals fall in the right sector. Thus, we only consider the phases of the received signals. Figure 10 shows the constellation diagram of the normalized received signals for the real antenna case. It is obvious that the two messages can be conveyed successfully.

On the other hand, for the ideal antennas case, the second message cannot be received successfully as shown in Figure 11. That is because the second signal in the ideal antenna case is more sensitive to perturbation than that of the real antenna one as observed from the singular values. In order to have two large singular values in this case, the channel matrix needs to be linearly independent which means that $rank(\mathbf{H}) = 2$. The columns of $\mathbf{H}$, representing channel responses from one transmitting antenna to the receiving antennas, have to be independent to each others. Mutual coupling destroys the dependency between these columns since it adds multiple reflections into the system. Thus, the mutual coupling in this case does increase number of useful singular values of the system, resulting in an improvement of the MIMO system performance. However, due to the complicated nature of a MIMO system, one cannot conclude that mutual coupling will always improve the system performance.

*Figure 10. QPSK constellation diagram of the real receiving antennas for $\sigma_x = \sigma_y = \sigma = 1cm$.*

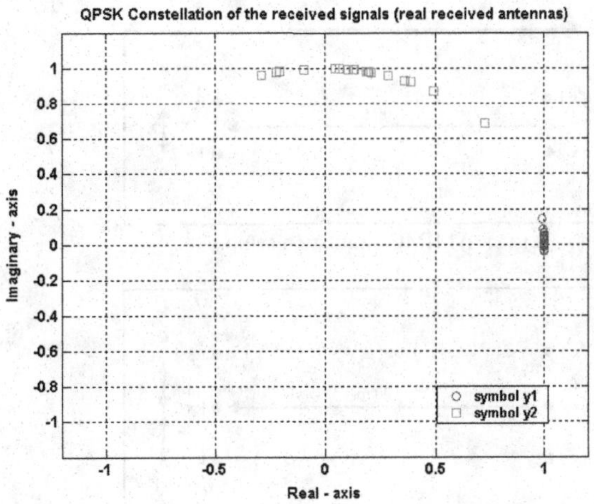


*Figure 11. QPSK constellation diagram of the ideal receiving antennas for $\sigma_x = \sigma_y = \sigma = 1cm$.*

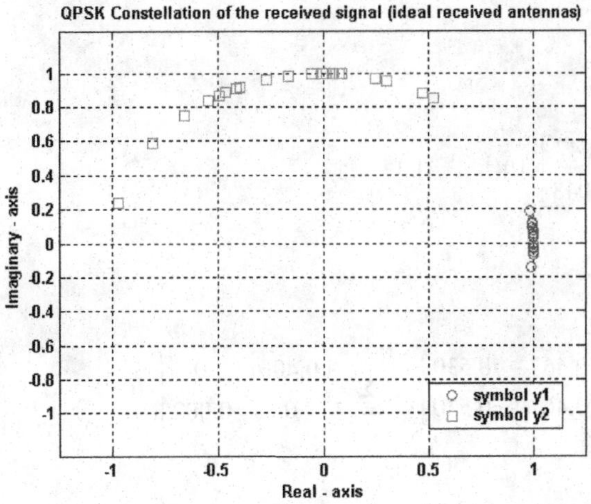


## Example VII

In this example, we will demonstrate that the maximum diversity may not be achieved due to the mutual coupling between antennas. It comes from the fact that the mutual coupling reduces the array efficiency resulting in a loss in the radiated power. Let us consider a $(2\times2)$ and a $(3\times3)$ MIMO systems. The input signal vector corresponding to the maximum eigenvalue will be transmitted to obtain the maximum SNR improvement or diversity gain. According to (Zheng and Tse, 2003), the maximum diversities for these two MIMO systems are 4 and 9, respectively, which are equal to $N_T N_R$. For both cases, the antennas are vertically polarized half-wave dipoles with a wavelength separation between antennas on each end. The transmitters and receivers are located 100 *m.* apart in free-space as shown in Figure 12. All the antennas are matched with load impedance $Z_L = 90.57 - j42.51$ [Ω].

For the $(2\times2)$ MIMO case, the measured voltage channel matrix is as follows

*Figure 12. MIMO simulation setups. All antennas are vertically polarized half-wave dipoles.*

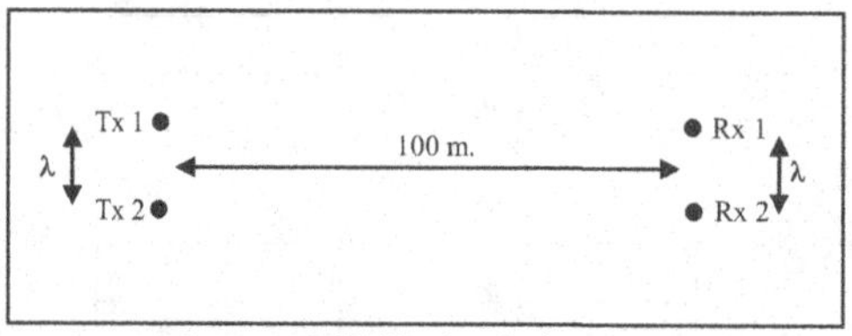


a.) $(2\times 2)$ MIMO simulation setup.

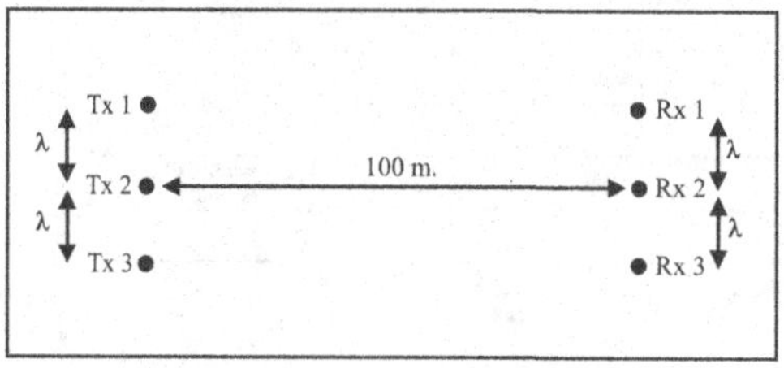


b.) $(3\times 3)$ MIMO simulation setup.

$$\mathbf{H}=\begin{bmatrix} 0.197+j0.048 & 0.198+j0.047 \\ 0.198+j0.047 & 0.197+j0.048 \end{bmatrix}\cdot 10^{-3} \quad [V]$$

and its SVD is

$$\mathbf{U}=[\mathbf{u}_1,\ \mathbf{u}_2]=\begin{bmatrix} -0.687-j0.167 & 0.467-j0.530 \\ -0.687-j0.167 & -0.467+j0.530 \end{bmatrix}, \quad \mathbf{\Sigma}=\begin{bmatrix} 0.406 & 0 \\ 0 & 0.002 \end{bmatrix}\cdot 10^{-3}$$

and

$$\mathbf{V}=[\mathbf{v}_1,\ \mathbf{v}_2]=\begin{bmatrix} -0.707 & -0.707 \\ -0.707 & 0.707 \end{bmatrix}.$$

To obtain the maximum diversity gain, $\mathbf{v}_1$ is used for the transmission. With 1 Watt transmitted power constraint, the received signals at the two receiving antennas are

$$\mathbf{y}=\begin{bmatrix} -3.880-j0.943 \\ -3.880-j0.943 \end{bmatrix} \ [mV].$$

To get the output signal, the received signals are weighted and summed by using the left singular vector $\mathbf{u}_1$, in the same way as shown in (30). Thus, the output signal is $\tilde{y}_{mimo}=\mathbf{u}_1^H\mathbf{y}=5.646\quad [mV]$. The corresponding received power is $P_{r,mimo}$ = 0.288 [μ$W$]. When compared with the SISO system, where only one transmitter and receiver are placed at the middle of the arrays, for 1 Watt transmitted power, the received power is $P_{r,mimo}$ = 0.078 [μ$W$]. The diversity gain of this

$(2\times2)$ MIMO system is 3.69, which is less than the theoretical diversity of 4. Similarly, for the $(3\times3)$ MIMO system, the channel matrix is

$$\mathbf{H}=\begin{bmatrix}0.191+j0.025 & 0.192+j0.112 & 0.196+j0.018\\ 0.192+j0.012 & 0.189+j0.002 & 0.192+j0.012\\ 0.196+j0.018 & 0.192+j0.112 & 0.191+j0.025\end{bmatrix}\cdot 10^{-3}\quad [V]$$

and its decomposition is

$$\mathbf{U}=\begin{bmatrix}-0.576-j0.065 & 0.393-j0.588 & -0.273-j0.298\\ -0.570-j0.035 & 0 & 0.584+j0.576\\ -0.576-j0.065 & -0.393+j0.588 & -0.273-j0.298\end{bmatrix}$$

$$\mathbf{\Sigma}=\begin{bmatrix}0.578 & 0 & 0\\ 0 & 0.008 & 0\\ 0 & 0 & 0.00003\end{bmatrix}\cdot 10^{-3},\ \mathbf{V}=\begin{bmatrix}-0.580 & -0.707 & 0.404\\ -0.571-j0.029 & 0 & -0.819-j0.042\\ -0.580 & 0.707 & 0.404\end{bmatrix}.$$

Using $\mathbf{v}_1$ for 1 Watt transmitted power constraint, the received signals are

$$\mathbf{y}=\begin{bmatrix}-4.753-j0.535\\ -4.704-j0.285\\ -4.753-j0.535\end{bmatrix}\ [mV].$$

Thus, the output signal is $\mathrm{y}_{mimo}=\mathbf{u}_1^H\mathbf{y}=8.245\ \ [mV]$ and its corresponding received power is $P_{r,mimo}=0.615\ [\mu W]$. This yields the diversity gain of 7.88 when compared with the SISO case. Even though these are simple systems, we could not obtain the maximum diversity gain as predicted by the theory since the mutual coupling between antennas reduces the radiation efficiency of the antenna array. This performance reduction is important and needs to be concerned when designing a multiple antenna system in practice especially for a compact array as on a mobile handset when antennas are very close together.

## CHANNEL CAPACITY

In this section, a fundamental limit on the spectral efficiency that one can reliably transmit on a multiple antenna system will be presented. *Channel capacity* represents a maximum error-free transmission rate in bit per second per unit bandwidth (bits/s/Hz). The channel capacity using a statistical model was first derived by Claude Shannon (Shannon, 1948). Let us consider a single-input-single-output (SISO) system first. Then, we will discuss the channel capacity for a MIMO system. For a SISO system, the relationship between a received signal $r(t)$ and a transmitted signal $s(t)$ is given by

$$r(t)=\sqrt{E_s}\alpha e^{-j\phi}s(t)+n(t). \qquad (31)$$

where $\alpha e^{-j\phi}$ is a channel gain and $E_s$ is a constrained total transmitted power. $n(t)$ denotes the zero-mean additive white Gaussian noise with variance $N_0$. The channel capacity of this SISO system is:

$$C_{SISO} = \log_2\left(1 + \frac{E_S}{N_0}\cdot\alpha^2\right) \quad \text{bits/s/Hz} \tag{32}$$

Noting that the second term in the parenthesis, $E_s\alpha^2 / N_0$, is the received signal-to-noise ratio. This channel capacity is valid only for an average radiated power constraint. Later, we will discuss an alternative way of representing the channel capacity based on signal and noise amplitudes instead of power. Let us consider a narrowband frequency flat model of $N_T$ transmit and $N_R$ receive antennas MIMO system with a constraint on total transmitted power of $E_s$ can be written as follows:

$$\mathbf{y} = \sqrt{\frac{E_s}{N_T}}\mathbf{Hx} + \mathbf{n} \tag{33}$$

The channel capacity of a MIMO system is defined as the maximum mutual information between the input signal $\mathbf{x}$ and the received signal $\mathbf{y}$ as $C = \max_{f(\mathbf{x})} I(\mathbf{x};\mathbf{y})$ where $f(\mathbf{x})$ is the probability distribution of the input vector $\mathbf{x}$ (Foschini, 1996) and the capacity of the MIMO system is given by (Paulraj, 2003)

$$C = \max_{Tr(R_{xx})=N_T} \log_2\left|\mathbf{I}_{N_R} + \frac{E_S}{N_T N_0}\mathbf{HR}_{xx}\mathbf{H}^H\right| \tag{34}$$

wherein $\mathbf{I}_{N_R}$ is an identity matrix of dimension $N_R \times N_R$, $\mathbf{R}_{xx} = E\{\mathbf{xx}^H\}$ is the covariance matrix of the transmitted signal $\mathbf{x}$. $E\{\cdot\}$ is the expected value operator. $|\cdot|$ and $Tr(\cdot)$ denote the determinant and the trace of a matrix, respectively. $Tr(\mathbf{R}_{xx}) = N_T$ is the total average power constraint at the transmitter. For a bandwidth of $B$ Hz, the MIMO channel capacity is $CB$ bits per second.

Another channel capacity formula based on voltage level is, in fact, the basic form of channel capacity derived by Shannon in (32). This basic form of the capacity states that (Shannon, 1998; Hartley_Theorem [online]; Sarkar, 2006) *to distinguish between M different signal functions of duration T on a channel, we can say that the channel can transmit $\log_2(M)$ bits in time T. And the rate of transmission is then equal to $\log_2(M)/T$.* Thus, this capacity can be defined as:

$$C' = \lim_{T\to\infty} \frac{\log_2 M}{T}. \tag{35}$$

To simplify the capacity without dealing with the concept of power, we let the received signal can be separated into $L = 2^Q$ distinct levels. This is equivalent as the number of quantization levels that can reasonably represent the received signal without distortion. Thus, $Q$ represents the number of bits used in the quantization. If the quantization level is $\Delta_q$, the noise floor in the receiver is then limited by the quantization noise $\Delta q/2$. Furthermore, as pointed out by Shannon that in the time interval $T$ there are $2BT$ independent signal amplitudes that can be transmitted, the total number of distinct signals is $M = (2^Q)^{2BT}$ where $B$ is the one-sided bandwidth of the signal. Therefore, the total number of information that can be transmitted is

$$C' = \lim_{T\to\infty} \frac{\log_2 M}{T} = 2BQ. \tag{36}$$

To link between these two capacities, we note that to obtain an error-free transmission, the noise floor in the capacity in (36) is limited by the quantization level $\Delta_q$. The average quantization noise power across the load impedance $Z_L$ can be expressed as

$$N_0 = \sigma_N^2 = \text{Re}\left\{\frac{1}{Z_L}\cdot\int_{-\Delta q/2}^{+\Delta q/2} e^2 p(e)de\right\}$$

$$= \text{Re}\left\{\frac{1}{Z_L}\cdot\int_{-\Delta q/2}^{+\Delta q/2} e^2 \frac{1}{\Delta q}de\right\}$$

$$= \text{Re}\left\{\frac{1}{Z_L}\right\}\cdot\frac{(\Delta q)^2}{12} \tag{37}$$

where $p(e) = 1/\Delta q$ is the probability density function of the quantization noise. This shows the relationship between the quantization noise level and its power (Sklar, 2002). $\text{Re}\{\cdot\}$ denotes the real part of a complex quantity. Assuming the received signal voltage (or current) level is $|V|$, then the received signal power, $P_R$, can be represented as $P_R = E_s\alpha^2 = |V|^2 \cdot \text{Re}(1/Z_L)$. Substituting the signal and noise power in (32), we get

$$C = \log_2\left(1 + \frac{12|V|^2}{(\Delta q)^2}\right) \tag{38}$$

For high SNR, substituting the number of distinct levels $L = 2^Q = \frac{|V|}{\Delta q}$ into (38), we get the channel capacity per unit bandwidth as follows

$$C = \log_2(12) + 2Q \tag{39}$$

For the signal of bandwidth $B$, the channel capacity in (39) becomes

$$C \approx 3.585B + 2BQ \tag{40}$$

Comparing (36) and (39), the relationship between the capacities derived from signal power and voltage level is $C' = C - 3.585B$. Even though the two channel capacities are closely related, the capacity related with voltage levels gives more pertinent meaning when dealing with near-field operations, which are usually the case for wireless communications (Sarkar, 2006).

## Example VIII

A $(1\times 3)$ SIMO system composed of one transmitting antenna and three receiving half-wave dipoles are placed in a metallic room without ceiling and floor. The receiving antennas are spaced half-wavelength apart. The room has a dimension of $2m.\times 2m.$ and it has a metallic wall of height 0.5 m. as shown in Figure 13. Inside the room, a metallic box of dimension $1m.\times 1m.\times 0.5m.$ is placed at the origin. The antenna is operated at $f = 1\,GHz$ and the system is assumed to be a narrowband system or $B = 1\ Hz$.

All of the antennas are conjugately matched in free space so that they deliver maximum power. The load impedance at each antenna is $Z_L = 90.57 - j42.51\ [\Omega]$. At the receiver, the received voltages from the receiving antennas are weighted

*Figure 13. SIMO simulation*

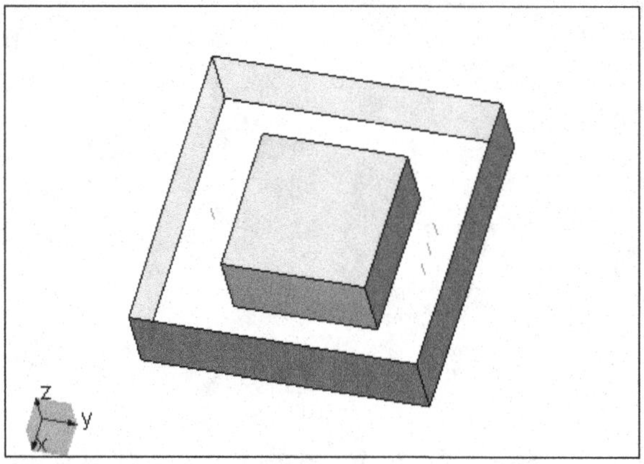


*Figure 14. SIMO receiver block diagram*

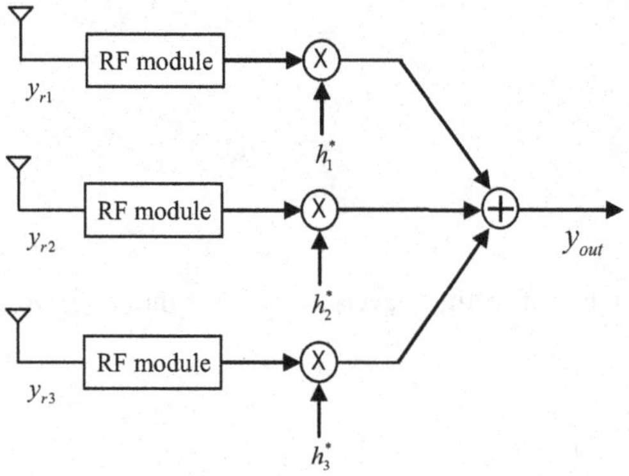


and summed to yield an output signal as shown in Figure 14. The weight, $h_i$, is selected such that the received signal from each antenna has unit gain and co-phased to each other. This can be achieved with the knowledge of the channels at the receiver through the use of pilot signals.

For 1 Watt transmitted power, the voltage at the generator is $V_{in} = 14.07$ [V] and the received voltages across the loads are $y_{r1} = 0.07 + j0.97$ [V], $y_{r2} = 0.89 - j0.01$ [V], and $y_{r3} = 0.07 + j0.97$ [V], respectively. According to the channel response, the weights are $h_1 = 0.07 + j0.99$, $h_2 = 0.89 - j0.01$, and $h_3 = 0.07 + j0.99$. Thus, the output signal from the receiver is $y_{out} = 2.856$ [V]. Assuming that the quantization level in this system is $\Delta q = 1\, mV$, the channel capacity $C' = 2\log_2\left(|y_{out}|/\Delta q\right) = 22.96$ [b/s/Hz].

When compared with a SISO system with only one receiving antenna placed at the center of the receiving array, for 1 Watt transmitted the SISO system receives $P_{siso} = 0.026\ [W]$ while the received power for the SIMO case is:

$$P_{simo} = \mathrm{Re}\left(\frac{|y_{out}|^2}{Z_L}\right) = 0.069\ [W].$$

The SIMO system gains 2.65 times better power reception over the SISO system. However, this benefit comes with the cost of the increasing complexity at the receiver chain. By theory, this SIMO system is expected to have 3 times as much power as in the SISO system if the array composes of ideal point sources and operates in free space. As shown in the previous subsection, the efficiency of the receiving array is reduced due to the mutual coupling between the array and near-field scatterers (walls). Thus, the improvement in the channel capacity when using multiple antennas is not always achieved as expected and it highly depends on the operating environment.

## MUTUAL COUPLING EFFECTS ON CHANNEL CAPACITY

As mentioned in the previous section, when the channel $\mathbf{H}$ is known at the transmitter, the channel decomposition is possible. Thus, the transmitter can efficiently transmit signals to the receiver. We repeat here the channel capacity in (34) for a MIMO system as

$$C = \max_{Tr(R_{xx})=N_T} \log_2 \left| \mathbf{I}_{N_R} + \frac{E_S}{N_T N_0} \mathbf{H}\mathbf{R}_{xx}\mathbf{H}^H \right| \tag{41}$$

Using Eigen decomposition we can represent $\mathbf{HR}_{xx}\mathbf{H}^H = \mathbf{Q\Lambda Q}^H$ and the capacity can be written as

$$C = \log_2 \left| \mathbf{I}_{N_R} + \frac{E_S}{N_0 N_T} \mathbf{Q\Lambda Q}^H \right| \tag{42}$$

$$C = \sum_{i=1}^{K} \log_2 \left( 1 + \frac{E_S}{N_0 N_T} \lambda_i \right) \tag{43}$$

where $\lambda_i$ is the eigenvalue and $K$ is the rank of the channel matrix $\mathbf{H}$. It means that total capacity is the sum of capacity of each independent path having a gain $\lambda_i$. In practice, the eigenvalues are not the same for every channel; thus, the best strategy is to transmit energy according to $\lambda_i$. This technique is well-known in the literature as a *water-filling* algorithm (Pauraj, 2003; Barbarossa, 2005). Note that mutual coupling is embedded in the channel matrix and it implicitly influences the channel capacity. Many attempts have been made to analytically separate channel matrix from the mutual coupling to understand the influence of the mutual coupling to the channel capacity (Janaswamy, 2002; Waldschmidt, 2002; Özdemir, 2003; Fletcher, 2003; Wallace, 2004). Most of the work done so far is mainly based on open-circuit assumption. We will briefly discuss the work by (Janaswamy, 2002) that shows the capacity of a fixed length linear array is reduced in the case of stronger mutual coupling. The open-circuit model of a $(N \times N)$ MIMO system with mutual coupling can be represented as:

$$\mathbf{v}_r = \frac{\mathbf{Z}_R \mathbf{H} \mathbf{Z}_T \mathbf{v}_s}{C_T C_R \sqrt{L_0}} \tag{44}$$

where $\mathbf{v}_r$ and $\mathbf{v}_s$ are received and transmitted signal vectors. $\mathbf{Z}_T = \mathbf{Z}^T(\mathbf{Z}^T + \mathbf{Z}_s)^{-1}$ is the mutual coupling at the transmitter end and $\mathbf{Z}_R = \mathbf{Z}_L(\mathbf{Z}^R + \mathbf{Z}_L)^{-1}$ is the mutual coupling at the receiver end. $\mathbf{Z}_s$ and $\mathbf{Z}_L$ represent source and load impedances, respectively. $\mathbf{Z}^T$ and $\mathbf{Z}^R$ are impedance matrices at the transmitter and receiver, respectively. $C_T$ and $C_R$ are defined as $C_T = Z_{11}^T / (Z_{11}^T + Z_{11}^{T*})$ and $C_R = Z_{11}^{R*} / (Z_{11}^R + Z_{11}^{R*})$. And, $L_0$ denotes the mean path loss of the channel $\mathbf{H}$. For detailed analysis, it can be found in (Janaswamy, 2002). The channel capacity of this MIMO system is represented by

$$C = \log_2\left(\det\left[I_N + \frac{\rho}{N}\times\frac{\mathbf{Z}_R\mathbf{\Psi}^R\mathbf{Z}_R^\dagger\mathbf{H}^u\mathbf{Z}_T^\dagger\mathbf{\Psi}^T\mathbf{Z}_T\mathbf{H}^{u\dagger}}{\left|C_T C_R\right|^2}\right]\right) \tag{45}$$

Here $H^u_{jk}$ is an uncorrelated complex Gaussian process with zero mean and unit variance and $\mathbf{\Psi}^T$ and $\mathbf{\Psi}^R$ are spatial correlation matrices at the transmitter and receiver. In this case, even though the open-circuit assumption is made for simplicity, the channel capacity is still a complicated function of channel and mutual coupling that cannot be completely separated. Moreover, in wireless communications, most of the communication occurs in near-field environment as mentioned in (Sarkar, 2006). The coupling does not occur only between antennas themselves, but also occurs between antennas and nearby objects including antenna towers. This makes the mutual coupling evaluation much more complicated and may not be separated from the channel. Mutual coupling effect on MIMO systems is still an opened research area to investigate. With understanding the mutual coupling effect, one could possibly design a wireless system more efficiently.

## CONCLUSION

This chapter presents a look at multi-antenna communication systems from electromagnetic point of view ranging from adaptive array antennas to MIMO systems. These systems are the promising tool in the nearby future generation of wireless communications. Multi-antenna systems offer an additional spatial dimension to the systems. Thus, an improvement in capacity is possible. However, when introducing multiple antennas into a system, the effect of electromagnetic needs to be considered. Otherwise, instead of improving a system performance, it might end up at degrading the overall system performance. Even though the mutual coupling seems to degrade the performance of an adaptive system by destroying the wavefront of the signals, it does improve the performance of other systems in a different way. As we have shown in this chapter, the mutual coupling increases the order of singular values in the channel decomposition for a MIMO system yielding a more reliable multiplexing gain. However, this cannot be guaranteed since it is very dependent on the operating environment that the system is deployed. This complicated interaction still remains opened for researchers and engineers to explore. Further development of a wireless communication system would be possible if the electromagnetic effect is properly integrated in to the system design.

## REFERENCES

Adve, R. S., & Sarkar, T. K. (2000, January). Compensation for the Effects of Mutual Coupling on Direct Data Domain Ddaptive Algorithms, *IEEE Transactions on Antennas and Propagation*, Vol. 48, No. 1 (pp. 86-94).

Barbarossa, S. (2005). *Multiantenna Wireless Communication Systems.* Massachusetts: Artech House.

Djordjevic, A. R., Bazdar, M. B., Sarkar, T. K., & Harrington, R. F. (1995). *Analysis of Wire Antennas and Scatterers: Software and User's Manual*, Norwood, MA: Artech House.

Fletcher, P. N., Dean, M., & Nix, A. R. (2003, February). Mutual Coupling in Multi-element Array Antennas and Its Influence on MIMO Channel Capaity, *Electronics Letters*, Vol. 39, No. 4 (pp. 342-344).

Foschini, G. J. (1996). Layered space-time architecture for wireless communication in a fading environment when using multi-element antennas, *Bell Labs Tech. J.*, Vol. 1, No. 2, (pp. 41-59).

Friedlander, B., & Weiss, A. J. (1991, March). Direction Finding in the Presence of Mutual Coupling, *IEEE Transactions on Aerospace and Electronics Systems*, Vol. 39, No. 3.

Friedlander, B., & Weiss, A. J. (1992, April). Direction Finding Using Spatial Smoothing with Interpolated Arrays, *IEEE Transactions on Aerospace and Electronics Systems*, Vol. 28, No. 2.

Friedlander, B. (1993). The Root-MUSIC Algorithm for Direction Finding with Interpolated Arrays, *Signal Processing*, Vol. 30, (pp. 15-29).

Gupta, I. J., & Ksienski, A. A. (1983, September). Effect of Mutual Coupling on the Performance of Adaptive Arrays, *IEEE Transactions on Antennas and Propagation*, Vol. AP-31, No. 5 (pp. 785-791).

Hartley_Theorem [online] Available: http://en.wikipedia.org/wiki/Shannon-Hartley_Theorem.

Hwang, S., Medouri, A., & Sarkar, T. K. (2005, February). Signal Enhancement in a Near-Field MIMO Environment through Adaptivity on Transmit, *IEEE Trans. on Antenna and Propagation*, Vol. 53, No. 2.

Janaswamy, R. (2002). Effect of Element Mutual Coupling on the Capacity of Fixed Length Linear Arrays, *IEEE Antenna and Wireless Propagation Letters*, Vol. 1 (pp.157-160).

Kolundzija, B. M., Ognjanovic, J. S., & Sarkar, T. K. (1995). *WIPL-D: Electromagnetic Modeling of Composite Metallic and Dielectric Structures*, Norwood, MA: Artech House.

Kraus, J. D. (1988). *Antennas*, 2nd ed., USA: McGraw-Hill, Inc.

Lee, T. S., & Lin, T. T. (1997). Adaptive Beamforming with Interpolation Arrays for Multiple Coherent Interferes, *Signal Processing*, Vol. 7 (pp. 177-194).

Özdemir, M. K., Arslan, H., & Arvas, E. (2003, December). Mutual Coupling Effect in Multi-antenna Wireless Communication Systems, *Global Telecommunications Conference*, Vol. 2 (pp. 829-833).

Paulraj, A., Nabar, R., & Gore, D. (2003). *Introduction to Space-Time Wireless Communications*, United Kingdom: Cambridge University Press.

Sarkar, T. K., Burintramart, S., Yilmazer, N., Hwang, S., Zhang, Y., De, A., & Salazar-Palma, M. (2006, December). A Discussion About Some of the Principles/Practices of Wireless Communication under a Maxwellian Framework, *IEEE Transactions on Antennas and Propagation*, Vol. 54, No. 12 (pp. 3727-3745).

Sarkar, T. K., Wicks, M. C., Salazar-Palma, M., & Bonneau, J. (2003). *Smart Antennas*, Hoboken, NJ: John Wiley & Sons.

Shannon, C. E. (1948). A Mathematical Theory of Communication, *Bell System Technical Journal*, Vol. 27., (pp. 379-423, July), (pp. 623-656, October).

Shannon, C. E. (1998). Communication in the presence of noise, *Proceedings of the IEEE*, Vol. 86, No. 2 (pp. 447-457).

Sklar, B. (2002). *Digital Communications*, 2nd Ed., India: Pearson Education.

Strait, B. J., Sarkar, T. K., & Kuo, D. C. (1973, June). *Special Programs for Analysis of Radiation by Wire Antennas* (Tech. Rep. AFCRL-TR-73-0399): Syracuse Univeristy.

Waldschmidt, C., Hagen, J. V., & Wiesbeck, W. (2002, June). Influence and Modelling of Mutual Coupling in MIMO and Diversity Systems, *IEEE Antenna and Propagation Society International Symposium*, Vol, 3 (pp. 190-193).

Wallace, J. W., & Jensen, M. A. (2004, July). Mutual Coupling in MIMO Wireless Systems: A Rigorous Network Theory Analysis, *IEEE Transactions on Wireless Communications*, Vol. 3, No. 4 (pp. 1317-1325).

Zheng, L., & Tse, D. N. C. (2003, May). Diversity and multiplexing: A fundamental tradeoff in multiple-antenna channels, *IEEE Transactions on Information Theory*, vol. 49, no. 5 (pp. 1073-1096).

# Chapter XII
# MIMO Beamforming

**Qinghua Li**
*Intel Corporation, Santa Clara, USA*

**Xintian Eddie Lin**
*Intel Corporation, Santa Clara, USA*

**Jianzhong (Charlie) Zhang**
*Samsung, Richardson, USA*

## ABSTRACT

*Transmit beamforming improves the performance of multiple-input multiple-output antenna system (MIMO) by exploiting channel state information (CSI) at the transmitter. Numerous MIMO beamforming schemes are proposed in open literature and standard bodies such as 3GPP, IEEE 802.11n and 802.16d/e. This chapter describes the underlying principle, evolving techniques, and corresponding industrial applications of MIMO beamforming. The main limiting factor is the cumbersome overhead to acquire CSI at the transmitter. The solutions are categorized into FDD (Frequency Division Duplex) and TDD (Time Division Duplex) approaches. For all FDD channels and radio calibration absent TDD channels, channel reciprocity is not available and explicit feedback is required. Codebook-based feedback techniques with various quantization complexities and feedback overheads are depicted in this chapter. Furthermore, we discuss transmit/receive (Tx/Rx) radio chain calibration and channel sounding techniques for TDD channels, and show how to achieve channel reciprocity by overcoming the Tx/Rx asymmetry of the RF components.*

## INTRODUCTION

It is well known that antenna phase array can form one directional radiation pattern (i.e., beam) to enhance transmit (or receive) signal energy at a desired direction. The directivity is obtained by constructive interference among multiple antenna signals in the desired direction. This is called transmit (or receive) beamforming. Beamforming can be applied to MIMO system by exploiting the multiple antennas at the transmitter and receiver. For example, $3\times 2$ a MIMO beamforming forms two beams and sends two data streams as shown in Figure 1. The received power can be increased by about 1.8 dB over that of $2\times 2$ MIMO. We call beamforming techniques in MIMO system MIMO beamforming. The principle of MIMO beamforming and the ideal beamforming algorithm, i.e. SVD beamforming (Telatar, 1995) are illustrated in Figure 2 and Figure 1. The transformation of input and output signal spaces is illustrated in Figure 2 for a $3\times 2$ example. In general, the singular value decomposition of an $M\times N$ channel matrix **H** is

$$\mathbf{H}=\underbrace{\left[\mathbf{u}_1\ \cdots\ \mathbf{u}_M\right]}_{\mathbf{U}}\begin{bmatrix}\sigma_1 & & & & & \\ & \ddots & & & & \\ & & \sigma_M & 0 & \cdots & 0\end{bmatrix}\underbrace{\left[\mathbf{v}_1\cdots\mathbf{v}_N\right]^*}_{\mathbf{V}} \tag{1}$$

where **U** and **V** are unitary matrix[b]; $\sigma_i$ is the *i*-th singular value with $\sigma_1\geq\cdots\geq\sigma_M$. The transformation of **H** rotates input vector $\mathbf{v}_i$ to output vector $\mathbf{u}_i$ and amplifies the length by $\sigma_i$ for $i=1,\cdots,M$. Since the transformation is linear, the input space $S\{\mathbf{v}_1,\cdots,\mathbf{v}_K\}_c$ is transformed to the output space $S\{\mathbf{u}_1,\cdots,\mathbf{u}_K\}$ for $K\leq M$, where *K* is the number of active data streams. As a result, the signal space $S\{\mathbf{v}_{M+1},\cdots,\mathbf{v}_N\}$ is converted into zero vector, i.e. null space. This is illustrated in Figure 2.

The capacity-achieving SVD beamforming algorithm (Telatar, 1995) exploits these properties and converts the matrix channel into *M* parallel scalar channels. The data symbol $x_i$ is sent by beamforming vector $\mathbf{v}_i$ as shown in Figure 1. The data symbol vector **x** is weighted by a beamforming matrix **Q** and beamformed signal $\mathbf{z}=\mathbf{Q}\mathbf{x}$ is sent by the antennas. There can be many choices of **Q**. In the SVD beamforming algorithm, **Q** – **V**. The signal model is

$$\mathbf{y}=\mathbf{H}\mathbf{Q}\mathbf{x}+\mathbf{n}, \tag{2}$$

where **n** is independent, identically distributed (i.i.d.) additive, white, Gaussian noise (AWGN) vector; the entry of **H** is modeled as i.i.d. complex, circularly symmetric, Gaussian random variables for non-light-of-sight (NLOS) channels. At the receiver end, the received signal **y** is spatially decoupled by $\mathbf{U}^*$ as

$$\hat{\mathbf{x}}=\mathbf{U}^*\mathbf{y}=\mathbf{U}^*\left(\mathbf{U}\Sigma\mathbf{V}^*\mathbf{V}\mathbf{x}+\mathbf{n}\right)=\begin{bmatrix}\sigma_1 x_1\\ \vdots\\ \sigma_M x_M\end{bmatrix}+\tilde{\mathbf{n}}, \tag{3}$$

where $\tilde{\mathbf{n}}$ has the same distribution as that of **n** because $\mathbf{U}^*$ is unitary. Since *M* parallel scalar channels are established in (3), transmission power can be optimally loaded across the channels using the water filling technique (Cover, 2006) to maximize the link capacity. For example, it is desirable to send less than *M* streams and put all power in the one or two strongest scalar channels in low signal to noise ratio (SNR) region. Therefore, only the first *K* columns of **V**, denoted $\mathbf{V}_B$, is required to be sent back. Quantization of $\mathbf{V}_B$ is one of the major topics in this chapter.

One of the main challenges of applying MIMO beamforming techniques to practical systems is the acquisition of CSI or beamforming matrix at the transmitter. In FDD and radio calibration absent TDD, channel reciprocity is not valid and CSI feedback from receiver to transmitter is essential. Quantization of CSI is also required, except in some analog modulation (Thomas, 2005; Marzetta, 2006). The transmitter computes the beamforming matrix based on the feedback and conducts the beamforming.

The simplest solution is to quantize the channel matrix element by element (i.e. scalar quantization) and feed the quantization indexes back. This is employed in fixed and slow fading wireless systems such as 802.16d[d] and 802.11n[e]. However, the feedback of the quantization indexes consumes significant amount of system bandwidth and reduces (and even cancels) the throughput gain delivered by MIMO beamforming. In practical systems, the feedback data is robustly modulated and heavily coded to prevent feedback error and the consequent loss. Therefore, the feedback channel has a low efficiency. The problem becomes more severe for mobile wireless systems due to time variations in mobile channels and consequently the need for frequent CSI updates. It is essential to design efficient feedback schemes for mobile wireless systems. The second challenge is the complexity in computing the feedback. The complexity in performing quantization and quantization codebook storage can be an issue for mobile device when high resolution feedback is needed. The third challenge is feedback latency. The latency is between 1-20 ms for cellular system like 3GPP and 802.16e[f]. Channel coherency in the order of this scale makes feedback stale and limits the application of MIMO beamforming to high mobility channel.

## LITERATURE SURVERY

Notable journal publications are overviewed in this section and the other literature related to industrial application will be introduced in the subsequent sections. MIMO transmit and receive beamforming was proposed in (Winters, 1987),

*Figure 1. A 3×2 MIMO beamforming example with system configuration and baseband signal model. The channel between the j-th transmit and i-th receive antenna is denoted by $h_{ij}$. The beamforming weight on the j-th transmit antenna for k-th data stream is denoted by $v_{jk}$.*

$$\begin{bmatrix} y_1 \\ y_2 \end{bmatrix} = \begin{bmatrix} h_{11} & h_{12} & h_{13} \\ h_{21} & h_{22} & h_{23} \end{bmatrix} \underbrace{\begin{bmatrix} v_{11} & v_{12} \\ v_{21} & v_{22} \\ v_{31} & v_{32} \end{bmatrix} \begin{bmatrix} x_1 \\ x_2 \end{bmatrix}}_{\begin{bmatrix} z_1 \\ z_2 \\ z_3 \end{bmatrix}} + \begin{bmatrix} n_1 \\ n_2 \end{bmatrix}$$

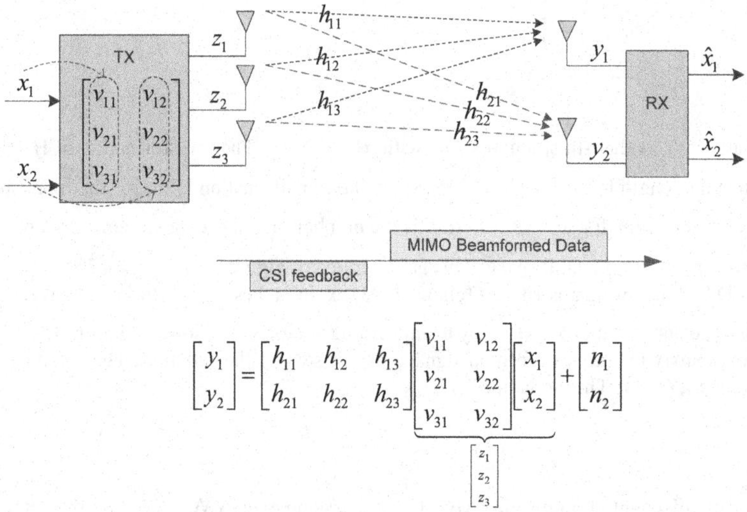

*Figure 2. Transformation of signal space over MIMO channel*

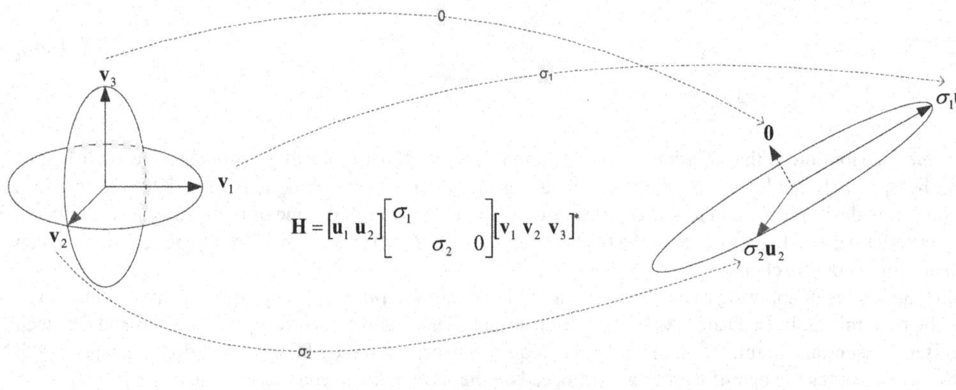

which is earlier than the well-known analytic work (Telatar, 1995). The performance of MIMO transmit beamforming depends on the amount of channel station information (CSI) at the transmitter[g]. Two types of assumptions are carried through, i.e. availability of complete CSI or partial CSI. With complete CSI, the MIMO beamforming gain can be quantified by the largest singular values of the channel matrix. The joint distribution of the ordered singular values is derived in (Edelman, 1989; M. Kang, 2006; M. Kang, 2003) for uncorrelated real Gaussian matrix, correlated complex Rayleigh channel, correlated complex Rician channel respectively. The analysis of performance metrics such as signal-to-noise ratio (SNR), symbol error rate, and outage probability is reported in (Maaref, 2005; H. Kang, 2006) for ideal beamforming of a single data stream. The analytic results for ideal beamforming are expressed in terms of generalized hypergeometric function, zonal polynomial, generalized Marcum Q-function, and modified Bessel function, which are difficult to evaluate.

The analysis with partial CSI is of practical interest and attracts more attention than that with the complete CSI. Research efforts focus on both the design of quantization codebooks and the analysis of performance loss due to quantization. It is shown in (Skoglund, 2003) that the channel capacity can be achieved by a message encoder and a beamformer, where the message encoder generates transmit symbols independently to the beamformer and only the beamformer is a function of the CSI. This removes the need to encode message jointly with the CSI and reduces the complexity of message encoding. Two types of partial CSI are considered, i.e. long term statistics and instantaneous information. The long term statistics are employed in fast fading channel, where the channel variation rate is greater than the feedback sampling rate. The feedback quantities include channel mean and channel covariance. The transmit beamforming and adaptation of modulation and power are optimized and characterized in (Visotsky, 2001; Zhou, 2004; Jorswieck, 2004; M. Kang, 2006). The quantities of instantaneous information feedback include quantized channel matrix and quantized beamforming matrix, both of which are sent over an error-free channel with a low data rate. Quantizing channel matrix, denoted by **H**, is employed in practice for its simplicity (IEEE P802.16d, 2004; IEEE P802.16e, 2006). Further research focuses on quantizing the beam forming matrix because only the columns $V_B$ with the largest singular values in the left unitary matrix of **H**'s SVD, **v** in (1), is needed at the transmitter. Note that $V_B$ is an orthogonal matrix and the set of $V_B$ forms a manifold, i.e. Grassmannian (or Stiefel) manifold. There is a probability distribution associated with $V_B$, which is determined by the joint distribution of **H**'s entries, e.g., correlated Gaussian complex. The source coding theory, i.e. rate distortion theory (Cover, 2006), is applied to the problem. It suggests that vector quantization is better than scalar quantization of $V_B$. In addition, the rate distortion theory is employed to derive the lower bounds for the feedback information rate, i.e., the number of bits per feedback. The distortion measure is the loss of performance metric, i.e. the received SNR, or the system mutual information, or the symbol error rate, or outage rate. Besides rate distortion theory, the Lloyd algorithm (Gersho, 1992) along with vector quantization for source coding is widely used in optimizing the quantization codebook of $V_B$.

Since multiple-transmit single-receive antenna system (MISO) is mathematically more tractable than MIMO, analytic results of MISO are usually obtained before that of MIMO. The connection between rate distortion theory and beamforming feedback is established for MISO in (Narula, 1998), where SNR and mutual information are used as performance metric. Lloyd algorithm is used for codebook search. Lower bound of outage rate is derived using the geometrical structure of MISO channel in (Mukkavilli, 2003). The lower and upper bounds of feedback information rate for a given distortion, i.e. the bounds for the rate distortion function, is derived for MISO channel in (Xia, 2006), where the distance measure is SNR loss and long Gaussian codeword is assumed. Close bounds of capacity loss and outage rate for MISO are derived in (Roh, 2006a) by assuming that the quantization regions, i.e. Voronoi regions, can be approximated by hyper-spheres with the same volume. Grassmannian line packing (Conway, 1996) is proposed in (Love, 2003; Love, 2005), where the performance metrics are SNR, SER, and mutual information. The distortion measure of the rate distortion function is approximated by chordal, projection two-norm, and Fubini-Study distances on Grassmannian manifold, which are used in the Grassmannian line packing targeted for maximum-likelihood and linear MIMO receivers. The criterion for codebook search is to maximize the minimum distance between any two distinct codewords, which is widely used in the literature afterward. The analysis of capacity loss for high SNR and high resolution codebook is reported in (June, 2006b) for MIMO with equal power transmission. A unified framework is proposed in (Zheng, 2007) for the performance analysis of high resolution codebook, where uniform input and congruent quantization region are assumed and high resolution quantization theory applies. Optimal codebook does not have a regular structure and results in high storage complexity for implementation. Structured codebook can be dynamically generated from limited parameters. A structured codebook is proposed in (Hochwald, 2000) for quantizing the Grassmannian manifold, where each dimension of the codeword has the same modulus. Structured codebooks that enable sequential column by column quantization of Stiefel manifold are proposed in (Roh, 2007; Li, 2007b), where Givens and Householder transformation are employed.

## TECHNIQUES FOR FDD

The feedback of beamforming matrix or channel matrix is essential in FDD because downlink/uplink[h] uses different frequency bands and channel reciprocity is not valid. The feedback information can be quantized by the receiver using a codebook. The codebook is known by both the transmitter and receiver beforehand and only the quantization index is fed back. Scalar quantization is usually applied to slow fading channel while vector quantization is applied to fast fading channel (IEEE P802.16d, 2004; IEEE P802.11n/D2.0, 2007). The former has a lower computational complexity and greater feedback overhead than the latter. The feedback information may also be sent using analog modulation without quantization.

In SVD beamforming (Telatar, 1995) shown in (3), only the right unitary matrix **v** of the singular value decomposition of the MIMO channel matrix needs to be known at the transmitter. It reduces the feedback amount by about a half compared to feeding back the channel matrix as in 802.16d (IEEE P802.16d, 2004). Various schemes (IEEE P802.16e, 2006; IEEE P802.11n/D2.0,2007; Hochwald, 2000; Onggosanusi, 2002; Love, 2003; Roh, 2004a; Roh, 2004b; Li, 2004b; Li, 2005a; Li, 2005b; Ikram, 2004; Mondal, 2005) are devised to quantize the orthogonal beamforming matrix and the analysis on the quantization loss is recently reported in (Roh, 2006b; Xia, 2006; Zheng, 2007). Two manifolds are defined for the set of beamforming matrix. The set of all $N \times K$ orthogonal matrixes is called Stiefel manifold and denoted by $\Omega_{N,K}$. The set of all $K$-dimensional subspace embedded in $N$-dimensional complex space is called Grassmann manifold and denoted by $\Phi_{N,K}$ (Muirhead, 1982). All the orthogonal matrixes in $\Omega_{N,K}$ whose columns span the same $K$-dimensional subspace are considered as one element in $\Phi_{N,K}$.

## Scalar Quantization

For an $N \times M$ MIMO channel, the scalar quantization quantizes each matrix element and will need to feedback $MNb$ bits of information for each channel matrix, where $b$ is the number of bits per complex element. The fixed wireless standard (IEEE P802.16d, 2004) uses an amplitude and phase format to represent each complex channel matrix element. The amplitude is encoded by one bit and the phase is encoded by 3 bits as shown in Figure 3. The advantage of the element-wise scalar quantization is low complexity. Only normalization, amplitude quantization and phase quantization are needed. On the other hand, the feedback payload is $4NM$ bits per matrix.

As we pointed out earlier, only $\mathbf{V}_B$ needs to be quantized and fed back for beamforming. Its orthogonal structure can be exploited to reduce the feedback load. We start by illustrating how a unit norm vector is represented. Taking a $4 \times 1$ unit norm vector for example, we may represent it as:

$$\mathbf{v}_1 = \begin{bmatrix} \cos\Psi_{11} \\ \sin\Psi_{11}\cos\Psi_{12}e^{j\phi_{11}} \\ \sin\Psi_{11}\sin\Psi_{12}\cos\Psi_{13}e^{j\phi_{12}} \\ \sin\Psi_{11}\sin\Psi_{12}\sin\Psi_{13}e^{j\phi_{13}} \end{bmatrix}. \tag{4}$$

The phase angle of the first element is redundant because it doesn't affect the signal power transformation shown in Figure 2 and can be factored into the singular value matrix in (1). Without loss of generality, we assume that the first element is real. Therefore, only 6 real parameters $\Psi_{1i} \in [0,\pi/2]$ and $\phi_{1i} \in [0,2\pi]$, $i = 1,2,3$ need to be quantized. For an $N \times 1$ vector, there are $2(N-1)$ real parameters. In the standard of wireless local area network (WLAN) (IEEE P802.11n/D2.0, 2007), the angles $\Psi_j$ and $\phi_{ij}$ are uniformly quantized by $b_\Psi$ and $b_\phi$ bits respectively. To quantize an $N \times K$ orthogonal matrix, we need the following unitary operation

$$\mathbf{Q}_1\mathbf{V}_B = \mathbf{Q}_1\begin{bmatrix}\mathbf{v}_1 & \mathbf{V}_2\end{bmatrix} = \begin{bmatrix} 1 & \mathbf{0} \\ \mathbf{0} & \tilde{\mathbf{V}}_2 \end{bmatrix}, \tag{5}$$

where $\mathbf{Q}_1$ is unitary; $\mathbf{v}_1$ and $\mathbf{V}_2$ are $N \times 1$ and $N \times (K-1)$ respectively. By the orthogonal condition, $\tilde{\mathbf{V}}_2$ is an $(N-1)\times(K-1)$ orthogonal matrix. The unitary matrix $\mathbf{Q}_1$ is not unique and there are many choices: Householder reflection or a series of Givens rotation (Golub, 1996), for example. The former will be discussed in subsection II.B.2 and the later is

$$\mathbf{Q}_1 = \mathrm{diag}\left(1, e^{-j\phi_{11}}, \cdots, e^{-j\phi_{1(N-1)}}\right)\mathbf{G}_{(N-1)N}\left(\Psi_{1(N-1)}\right)\mathbf{G}_{(N-2)(N-1)}\left(\Psi_{1(N-2)}\right)\cdots\mathbf{G}_{23}\left(\Psi_{12}\right)\mathbf{G}_{12}\left(\Psi_{11}\right), \quad (6)$$

where $\mathrm{diag}(a_1, \cdots, a_N)$ denotes a diagonal matrix with the diagonal elements $a_1, \cdots, a_N$. The Givens rotation $\mathbf{G}_{il}(\Psi)$ represents rotation of the $i$-th and $l$-th rows:

$$\mathbf{G}_{il}(\Psi) = \begin{bmatrix} \mathbf{I}_{i-1} & & & & \\ & \cos\Psi & & -\sin\Psi & \\ & & \mathbf{I}_{l-i-1} & & \\ & \sin\Psi & & \cos\Psi & \\ & & & & \mathbf{I}_{N-l} \end{bmatrix}, \quad (7)$$

where each $\mathbf{I}_m$ is the $m \times m$ identity matrix. We can then recursively quantize the first column of the matrix $\tilde{\mathbf{V}}_2$ in (5) as

$$\begin{bmatrix} 1 & & & \\ & \ddots & & \\ & & 1 & \\ & & & \mathbf{Q}_K \end{bmatrix} \cdots \begin{bmatrix} 1 & \\ & \mathbf{Q}_2 \end{bmatrix} \mathbf{Q}_1 \mathbf{V}_B = \begin{bmatrix} 1 & & \\ & \ddots & \\ & & 1 \end{bmatrix}. \quad (8)$$

Using (8), $\mathbf{V}_B$ can be written as

$$\mathbf{V}_B = \mathbf{Q}_1^* \begin{bmatrix} 1 & \mathbf{0} \\ \mathbf{0} & \tilde{\mathbf{V}}_2 \end{bmatrix} = \mathbf{Q}_1^* \begin{bmatrix} 1 & \mathbf{0} \\ \mathbf{0} & \mathbf{Q}_2^* \begin{bmatrix} 1 & \mathbf{0} \\ \mathbf{0} & \mathbf{Q}_3^* \begin{bmatrix} \ddots & \cdots \\ \cdots & \mathbf{Q}_K^* \begin{bmatrix} 1 \\ 0 \end{bmatrix} \end{bmatrix} \end{bmatrix} \end{bmatrix}, \quad (9)$$

*Figure 3. Feedback format of IEEE 802.16d channel matrix element*

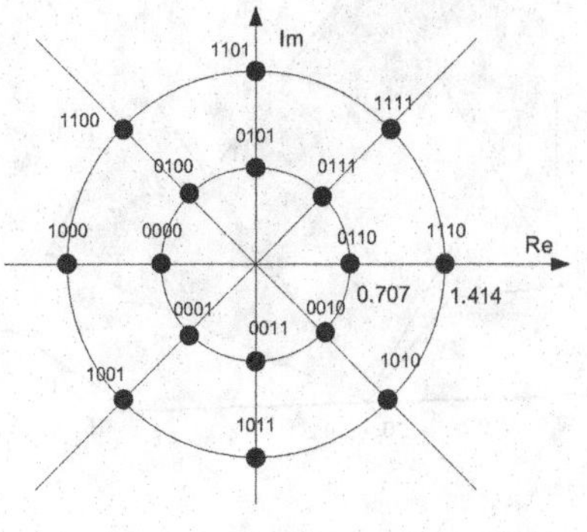

where each $\mathbf{Q}_l = \mathrm{diag}(1, e^{-j\phi_{l1}}, \cdots e^{-j\phi_{l(N-l)}}) \prod_{k=l}^{N-1} \mathbf{G}_{(N-k)(N-k+1)}(\Psi_{l(N-k)})$ is a $(N-l+1)\times(N-l+1)$ unitary matrix. Finally, the parameters $\phi_{li}$ and $\Psi_i$ for $l = 1, ..., N$ and $i = 1, ..., N-1$ are quantized individually and sent back to the transmitter. The transmitter computes $\mathbf{V}_B$ using (9). In the Givens angle quantization, the total number of real parameters is $2KN-(K+1)K$. In comparison, the scalar quantization of channel matrix entry effectively quantizes $2MN$ real parameters. Since $M \geq K$, i.e. number of the receive antennas can be greater than the number of data streams, the overhead of scalar quantization of matrix entry is much bigger than that of the Givens angle quantization. Even for $M = K$, the difference is still significant. For $2\times2$ and $3\times3$ matrix, the differences of quantization parameter numbers are 4 and 3 times respectively.

## Vector Quantization

The feedback overhead may be further reduced. For example, there are 10 Givens angles for a $4\times2$ beamforming matrix $\mathbf{V}_B$. Scalar quantization of each angle in 802.11n (IEEE P802.11n/D2.0, 2007) takes 2-5 bits on average to achieve significant beamforming gain. It consumes 20-50 bits per matrix feedback. Even though the scalar quantization of $\mathbf{V}_B$ in 802.11n reduces the feedback overhead by half as compared to the scalar quantization of channel matrix entry in 802.16d (IEEE P802.16d, 2004), there is room for further improvement. For instance, the overhead may be further reduced to 6 bits using vector quantization, i.e. quantizing the whole matrix jointly using a matrix codebook. In general, the scalar quantization needs to allocate integer, not fractional bit to each parameters. However, the differential entropy of Givens angle distribution of NLOS channels (Roh, 2004a; Roh, 2007; Muirhead, 1982) are not necessarily matched to the allocated integer bits. Therefore the angle by angle quantization is not efficient from the viewpoint of feedback overhead.

The orthogonal matrix $\mathbf{V}_B$ can be quantized as a whole by matrix codebooks. An example of matrix codebook is shown in Figure 4 and remarks about the codeword distribution are in the caption. The codebooks can be classified into two categories, structured and unstructured. The structured codebook (IEEE P802.16e, 2006; Hochwald, 2000; Roh, 2004a; Li, 2004b; Ikram, 2004) has a structure that parameterizes its generation with a few matrix operations (e.g. multiplication, Givens rotation, etc.), and thus only the parameters need to be stored. This property is desirable for mobile devices with limited storage. In addition, some structured codebooks with hierarchical structures enable low complexity quantization (Roh, 2004b; Li, 2004b; Li, January 2005). In contrast, the unstructured ones (Love, 2003) optimize each entry of the codebook over a distribution of the beamforming matrix set. Common optimization metrics are chordal distance (Hochwald, 2000; Love, 2003) and mutual information (Mondal, 2005). The performance of unstructured codebook is better than that of the structured due to relaxed optimization constraints. However, since the whole codebook for each antenna configuration needs to be stored, it is not desirable for codebooks with a large number of codewords. In addition, brute force search is usually needed in the quantization with unstructured codebook and the resultant complexity is prohibitive for large codebook.

*Figure 4. Illustration of the vector quantization codebook of the beamforming matrix*

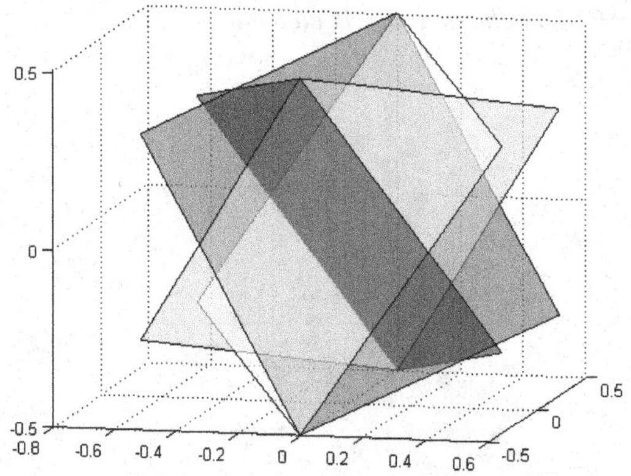

Four 3×2 orthorgonal beamforming matrixes are plotted, where the MIMO channel are real (not complex). Each matrix is represented by the plane spanned by the two matrix columns. The four matrixes are

$$\begin{bmatrix} -0.5774 & -0.5774 \\ 0.7887 & -0.2113 \\ -0.2113 & 0.7887 \end{bmatrix}, \begin{bmatrix} 0.5774 & -0.5774 \\ 0.7887 & 0.2113 \\ 0.2113 & 0.7887 \end{bmatrix}, \begin{bmatrix} 0.5774 & 0.5774 \\ 0.7887 & -0.2113 \\ -0.2113 & 0.7887 \end{bmatrix}, \text{and} \begin{bmatrix} -0.5774 & 0.5774 \\ 0.7887 & 0.2113 \\ 0.2113 & 0.7887 \end{bmatrix}.$$

They are uniformly distributed over 3-dimensional space. For channels with NLOS and large antenna spacing, the ideal beamforming matrix is uniformly distributed over the set of all 2-dimensional subspace in 3-dimensional space. The set forms Grassmann manifold $\Phi_{2,3}$ (Muirhead, 1982). This implies the quantization codeword should also be uniformly distributed over Grassmann manifold.

## Unstructured Codebook

Unstructured codebook consists of codewords on $\Phi_{N,K}$ (or $\Omega_{N,K}$), and there is no explicit parameterization (i.e. structures) among the codewords. The codewords need to be stored rather than be dynamically computed due to the lack of parameterization. The codewords are usually obtained by computer search to achieve a certain performance metric e.g. minimum codeword error rate and minimum channel capacity loss. The seminal work (Love, 2003) shows that the optimal codebook should spread the codewords as uniformly as possible over the whole $\Phi_{N,K}$. This can be solved by algorithms of Grassmann subspace packing (Conway, 1996). For example, one can search for codebook $\tilde{C}_V$ that maximizes the minimum chordal distance between any two codewords as

$$\tilde{C}_V = \arg\max_{C_V} \min_{V_i, V_j \in C_V, i<j} d(V_i, V_j), \tag{10}$$

where $d(V_i, V_j) = N - \|V_i^* V_j\|_F^2$ is the chordal distance between codewords $V_i$ and $V_j$ in $\Phi_{N,K}$, and $\|\|_F$ denotes Frobenius norm.

One application example of unstructured codebook is the small vector codebooks in 802.16e (IEEE P802.16e, 2006), which have eight codewords each. Another example of the unstructured codebook is the D-TxAA (Double-Transmit Adaptive Array) MIMO communication method adopted in the Release-7 of the 3GPP HSPA system. The codebook includes a set of four unitary matrix codewords:

$$V = \left\{ \begin{bmatrix} \frac{1}{\sqrt{2}} & \frac{1}{\sqrt{2}} \\ \frac{1+j}{2} & \frac{-1-j}{2} \end{bmatrix}, \begin{bmatrix} \frac{1}{\sqrt{2}} & \frac{1}{\sqrt{2}} \\ \frac{1-j}{2} & \frac{-1+j}{2} \end{bmatrix}, \begin{bmatrix} \frac{1}{\sqrt{2}} & \frac{1}{\sqrt{2}} \\ \frac{-1+j}{2} & \frac{1-j}{2} \end{bmatrix}, \begin{bmatrix} \frac{1}{\sqrt{2}} & \frac{1}{\sqrt{2}} \\ \frac{-1-j}{2} & \frac{1+j}{2} \end{bmatrix} \right\}.$$

To maximize the scheduling flexibility, BS have the option of transmitting one or two spatial streams, and the notion of 'primary' and 'secondary' precoding vectors is introduced. The first and second column of each matrix codeword is called the primary and secondary precoding vector respectively. SS chooses only the primary precoding vector that is the best for it among the four possible choices. BS receives the feedback of the choice, and sends one transport block using the primary precoding vector to the SS. In addition, BS can optionally choose to send another transport block using a secondary precoding vector subject to the constraint that both the primary and secondary are the two columns of a matrix codeword. For example, if SS chooses

$$\begin{bmatrix} \frac{1}{\sqrt{2}} & \frac{1+j}{2} \end{bmatrix}^T$$

to be the primary precoding vector, BS will send one transport block using this weight vector, and then optionally BS can send a second transport block using the weight vector

$$\begin{bmatrix} \frac{1}{\sqrt{2}} & \frac{-1-j}{2} \end{bmatrix}^T.$$

## Structured Codebook

Two types of structured codebooks are presented, which are Hochwald codebook and recursive codebook. The seminal work of structured matrix codebook (Hochwald, 2000) has several variants. It employs block-circulant construction. The method provides near-optimal codebooks with low storage since the parameterization of the codebook is very limited. The codebook is fully specified by the first codeword $\mathbf{V}_1$ and a diagonal rotation matrix **G**. The other codewords are given by

$$\mathbf{V}_l = \mathbf{G}^{l-1}\mathbf{V}_1,\ \text{for } l = 2,3,\cdots 2^L, \tag{11}$$

where $2^L$ is the codebooks size, each codeword is $N \times K$;

$$\mathbf{G} = \text{diag}\left( e^{j\frac{2\pi}{2^L}u_1} \ \cdots \ e^{j\frac{2\pi}{2^L}u_N} \right)$$

is specified by the integer vector $\mathbf{u} = [u_1, \ \cdots, \ u_N]^T$. Furthermore, $\mathbf{V}_1$ is chosen to be an $N \times K$ submatrix of the $N \times N$ DFT matrix **F**, whose entry on the *i*-th row and *j*-th column is

$$e^{j\frac{2\pi}{N}(i-1)(j-1)}.$$

Therefore, $\mathbf{V}_1 = [\mathbf{F}(:,k_1) \ \cdots \ \mathbf{F}(:,k_N)]_i$, where the column index $\mathbf{k} = [k_1, \ \cdots, \ k_K]^T$ uniquely specifies the choice.

One variant of the original Hochwald codebook (Hochwald, 2000) is adopted by 802.16e (IEEE P802.16e, 2006; Li, 2007b). It is observed that the magnitude of each entry in any codeword equals to the corresponding entry in the first codeword $\mathbf{V}_1$ because $\mathbf{G}^{l-1}$ only change the phase of $\mathbf{V}_1$'s entry to obtain $\mathbf{V}_l$ in (11). The magnitude remains $N^{-1/2}$. The constant modulus is an unnecessary constraint that can be removed to achieve better chordal distance properties. The variant (IEEE P802.16e, 2006; Li, 2007b) takes advantage of the energy-shifting property of eigen-coordinate transformation while preserving some of the nice features of the Hochwald construction, such as easy parameterization, block-circulant distance distribution (i.e. $d(\mathbf{V}_i,\mathbf{V}_j) = d(\mathbf{V}_{\text{mod}(i+k,2^L)},\mathbf{V}_{\text{mod}(j+k,2^L)})$, where $d(,)$ is the chordal distance (Conway, 1996)), etc. The minimum chordal distance between any pair of codewords can then be systematically maximized by removing the constant modulus constraint.

Since **G** is diagonal, it can be viewed as a part of an eigen-decomposition of some Hermitian matrix **S**:

$$\mathbf{S} = \mathbf{MGM}^*, \tag{12}$$

where **M** is a $N \times N$ unitary matrix. We replace **G** by **S** in (11) as

*Table 1. Codebooks based on improved Hochwald construction*

| $N$ | $K$ | $L$ | Column index for $\mathbf{V}_1$ | **u** | **b** | Chordal distance ratio: (Li, 2007b) / (Hochwald, 2000) |
|---|---|---|---|---|---|---|
| 3 1 | 6 | | $[1]$ | $[1 \quad 26 \quad 57]$ | [0.2518 – 0.6409*i*, -0.4570 - 0.4974*i*, 0.1177 + 0.2360*i*] | 0.1263/0.1166 |
| 4 | 1 | 3 | $[1]$ | $[1 \quad 2 \quad 7 \quad 6]$ | [0.2895 + 0.3635*i*, 0.5287 - 0.2752*i*, -0.2352 - 0.4247*i*, -0.4040+ 0.1729*i*] | 0.8282/ 0.7500 |
| 4 | 1 | 6 | $[1]$ | $[1 \quad 45 \quad 22 \quad 49]$ | [0.3954 - 0.0738*i*, 0.0206 + 0.4326*i*, -0.1658 - 0.5445*i*, 0.5487 - 0.1599*i*] | 0.3935/ 0.3643 |

$$\mathbf{V}_l = \mathbf{S}^{l-1}\mathbf{V}_1 = \left(\mathbf{MGM}^*\right)^{l-1}\mathbf{V}_1 = \mathbf{MG}^{l-1}\mathbf{M}^*\mathbf{V}_1, \text{ for } l = 2,3,\cdots 2^L. \tag{13}$$

The eigen-matrix **M** can be parameterized by imposing the Householder structure (Golub, 1996) on **M** as

$$\mathbf{M} = \mathbf{I} - 2\mathbf{bb}^*, \tag{14}$$

where **b** is an $N \times 1$ unit vector[j]. **Table 1** demonstrates the effectiveness of the modified method for several vector codebooks. Since the minimum chordal distance of the modified Hochwald construction is greater than that of the original Hochwald construction, the improved construction results in a more uniform codeword distribution and thus a smaller average quantization error. The first codeword $\mathbf{v}_1$ in **Table 1** is the same as that in (Hochwald, 2000). The only additional parameter is the vector **b**. The modified method also allows joint optimization over vector **b** and $\mathbf{V}_1$ that can be any unitary matrix.

The recursive codebook was proposed in (Roh, 2004b; Li, 2004b). A variant (MIMO, 2007) was adopted by 3GPP, which employs Householder reflection and has a constant modulus for each matrix entry. The matrix codebooks are constructed using the vector codebooks as building blocks. We first consider the distribution of the optimal beamforming matrix and then design the codebook accordingly. For NLOS channel, the distribution of **V**, called Harr distribution, is uniform on $\Omega_{N,N}$ (Muirhead, 1982). The Harr distribution has a recursive property (Roh, 2004b; Roh, 2007; Li, 2007b).

*Proposition 1*: Let **V** be uniformly distributed over $\Omega_{N,N}$ and $\mathbf{v}_1$ be the first column of **V**. Furthermore, let $\mathbf{Q}(\mathbf{v}_1)$ be an $N \times N$ unitary matrix whose first row is $\mathbf{v}_1^*$ and rest of rows are independent of **V** except $\mathbf{v}_1$. Then, $\mathbf{v}_1$ is uniformly distribution over $\Omega_{N,1}$ and

$$\mathbf{Q}(\mathbf{v}_1)\mathbf{V} = \begin{bmatrix} 1 & \\ & \mathbf{V}_2 \end{bmatrix}, \tag{15}$$

where $\mathbf{V}_2$ is uniformly distributed over $\Omega_{N-1,N-1}$.

Since both **V** and size-reduced $\mathbf{V}_2$ are uniformly distribution, this property enables a recursive quantization of **V** column by column with decreasing sizes. One can quantize the first column of **V** using a uniform vector codebook[k], reduce the problem size by one dimension, and recursively continue the quantization using another vector codebook with reduced dimension. Since only vector codebooks with uniform distribution are needed, the scheme is desirable for storage reduction and supports all antenna configurations. There are multiple choices for the unitary matrix $\mathbf{Q}(\mathbf{v}_1)$ satisfying Proposition 1. Household reflection matrix is chosen for low complexity implementation and consistency with vector codebook construction using the modified Hochwald scheme and (14) in (IEEE P802.16e, 2006; Li, 2007b). Householder reflection matrix (Golub, 1996) is defined as a unitary $N \times N$ matrix $\mathbf{Q}_H(\mathbf{v})$ that is a function of an $N \times 1$ unit vector $\mathbf{v}$:

$$\mathbf{Q}_H(\mathbf{v}) = \begin{cases} \mathbf{I}, & \mathbf{v} = \mathbf{e}_1 \\ \mathbf{I} - \rho\,\mathbf{w}\mathbf{w}^*, & \text{otherwise} \end{cases} \tag{16}$$

where $\mathbf{v}$'s first entry is real; $\mathbf{w} = \mathbf{v} - \mathbf{e}_1$ and

$\mathbf{e}_1 = [1 \quad 0 \quad \cdots \quad 0]^T$; $\rho = \frac{2}{\|\mathbf{w}^H\mathbf{w}\|}$; $\mathbf{I}$

is the $N \times N$ identity matrix and the first column and row of $\mathbf{Q}_H(\mathbf{v})$ are $\mathbf{v}$ and $\mathbf{v}^*$ respectively. Since $\rho$ is a real number that only depends on the input vector $\mathbf{v}$ and $\mathbf{v}$ is from a vector codebook, $\rho$ can be pre-computed, stored and no division operation is needed (16). As a result, the computation of (16) is of low complexity.

A recursive quantization scheme is illustrated in Figure 5 by a $4 \times 2$ example, where only the uniform vector codebooks and Householder reflection are employed. There are four and two transmit and receive antennas respectively. SS feeds back the first two columns of $\mathbf{V}$ to receive two data streams. The SS quantizes the first column of $\mathbf{V}$, $\mathbf{v}_1$, as

$$\hat{\mathbf{v}}_1 = \arg\max_{\mathbf{u} \in C_{4\times 1}} \left\|\mathbf{u}^H \mathbf{v}_1\right\|, \tag{17}$$

where $C_{4\times 1}$ is a uniform $4 \times 1$ vector codebook. We select $\hat{\mathbf{v}}_1$ with the maximum inner product among all unit vectors in the codebook. A Householder reflection matrix is constructed using (16) as

$$\mathbf{F}_1 = \mathbf{Q}_H(\hat{\mathbf{v}}_1). \tag{18}$$

If $\hat{\mathbf{v}}_1 = \mathbf{v}_1$, Householder reflection converts the first column and row of $\mathbf{V}$ into $[e^{j\phi_1} \quad 0 \quad 0 \quad 0]^T$ and $e^{j\phi_1}[1 \quad 0]$ as shown in (19), where $\phi_1$ is the phase of $v_1$. The Householder reflection results in

$$\mathbf{F}_1\mathbf{V} = \begin{bmatrix} e^{j\phi_1} & 0.0 \\ 0.0 & \begin{bmatrix} \hat{v}_{11} \\ \hat{v}_{21} \\ \hat{v}_{31} \end{bmatrix} \\ 0.0 & \\ 0.0 & \underbrace{\phantom{x}}_{\mathbf{v}_2} \end{bmatrix}, \tag{19}$$

where two properties are employed, i.e. $\hat{v}_1$ is real and $\mathbf{V}$ is orthogonal. Furthermore, $\mathbf{V}_2$ is unit norm. Since $\hat{\mathbf{v}}_1 \approx \mathbf{v}_1$, there will be nonzero residuals, noted by $0.0$, in the first column and row.

By Proposition 1, $\mathbf{V}_2$ is uniformly distributed over $\Omega_{N_t-1,1}$. From (19), the size of $\mathbf{V}_2$ is reduced from that of $\mathbf{V}$ by one row and one column. Recursively apply steps (18) and (19), one quantizes $\mathbf{V}_2$ using a $3 \times 1$ vector codebook. These steps are graphically illustrated in Figure 5. The quantization indexes $q_1$ and $q_2$ of $\mathbf{v}_1$ and $\mathbf{v}_2$ are fed back to BS. The BS computes the precoding matrix as shown in Figure 6, where the indexes are converted to vectors by codeword lookup and the vectors are concatenated by Householder reflection with low complexity. It is worth noting that the phases $\phi_i$ are not needed for beamforming. The $4 \times 2$ matrix codebook in the example is effectively formed by the concatenation of $4 \times 1$ and $3 \times 1$ vector codebooks as shown in Figure 6 by letting the $q_1$ and $q_2$ step through all the indexes.

*Figure 5. Recursive quantization of beamforming matrix using vector codebooks*

$$\mathbf{H} = \begin{bmatrix} h_{11} & h_{12} & h_{13} & h_{14} \\ h_{21} & h_{22} & h_{23} & h_{24} \end{bmatrix} \qquad \mathbf{V} = \begin{bmatrix} v_{11} & v_{12} \\ v_{21} & v_{22} \\ v_{31} & v_{32} \\ v_{41} & v_{42} \end{bmatrix} \qquad \mathbf{F}_1 \mathbf{V} = \begin{bmatrix} e^{j\theta_1} & 0.0 \\ 0.0 & \hat{v}_{11} \\ 0.0 & \hat{v}_{21} \\ 0.0 & \hat{v}_{31} \end{bmatrix}$$

$$q_1 \qquad\qquad\qquad q_2$$

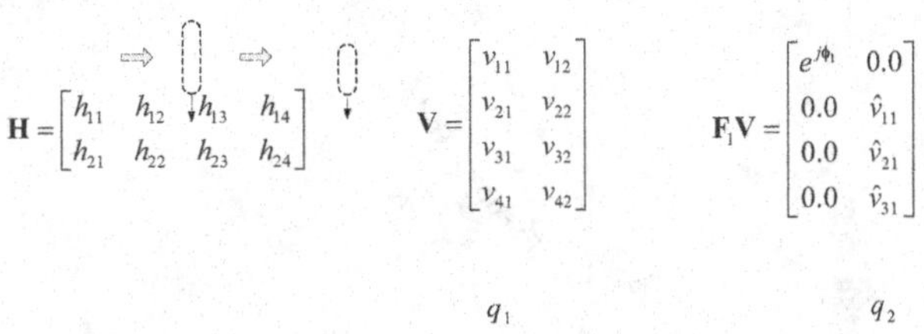

*Figure 6. Construction of beamforming matrix at transmitter according to feedback indexes*

$$\hat{\mathbf{V}} = \mathbf{F}_1 \begin{bmatrix} 1 \\ & \hat{\mathbf{v}}_2 \end{bmatrix}$$

$$q_1 \quad q_2$$

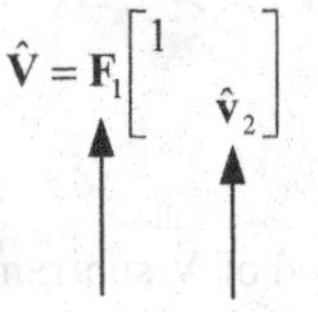

The feedback for $N \times (N-1)$ beamforming matrixes can be greatly reduced if SS feeds back the subspace spanned by the $N-1$ columns instead of the exact beamforming matrix. It is noted that the subspace is uniquely determined by its complementary, orthogonal space, which is a $N \times 1$ column vector as shown in Figure 7. Namely, for each $N \times (N-1)$ orthogonal matrix $\mathbf{B}$, the complementary vector $\mathbf{v}$ is the norm vector of the subspace spanned by $\mathbf{B}$'s columns. Therefore, only one, instead of $N-1$, vector is quantized and fed back. It is shown that the norm vector is uniformly distributed in $\Omega_{N,1}$ (Li, 2007b). Therefore, the uniform vector codebook for matrix codebook construction is again used for the complementary feedback. The matrix codeword of the complementary scheme can be easily computed from that of the uniform vector codebook using Householder reflection. Let $\mathbf{v}$ be the vector codeword and $\mathbf{Q}_H(\mathbf{v})$ be the Householder matrix computed by (16). It is noticed that the first column of $\mathbf{Q}_H(\mathbf{v})$ is $\mathbf{v}$ and $\mathbf{Q}_H(\mathbf{v})$ is a $N \times N$ unitary matrix. Then, column $2, \cdots, N$ forms a $N \times (N-1)$ matrix codeword whose norm vector is $\mathbf{v}$. When the strongest $N-1$ singular values of the channel matrix are identical, there is no performance degradation for the complementary feedback. On the other hand, when the singular values of $\mathbf{H}$, are substantially different, the vector feedback causes interference between beamformed channels and also prevents efficient adaptive bit loading and power loading at the transmitter. It results in a performance loss especially for linear receivers. The degradation of complementary feedback is shown to be within 1 dB for $4 \times 4$ and $4 \times 3$ channels (Intel, August 2006 I).

The recursive codebook, complementary scheme, and other techniques are jointly employed in 802.16e (IEEE P802.16e, 2006). The packet error rate (PER) performance for a $4 \times 2$ 802.16e downlink is shown in Figure 8. The $4 \times 2$ beamforming matrix is quantized by 6 bits and one 6-bit index per 36 OFDM subcarriers per 5 ms is sent. The MIMO beamforming outperforms space-time coding and transmit antenna selection by 3 and 1 dB respectively. Transmit antenna selection can be viewed as a special case of MIMO beamforming, which employs a coarse codebook and whose codewords are permutation matrixes of 0s and 1s. In this result, SS mobile speed of 3 km/h and a 10 ms feedback delay are assumed. The delay compensation is depicted in subsection II.E. The performance loss due to the codebook quantization can be conducted by following the frame work in (Roh, 2006; Xia, 2006; Zheng, 2007). The implementation complexity and feedback overhead are reported in (Intel, May 2006 II) for 3GPP long time evolution (LTE), which is similar to 802.16e. The complexity of codebook quantization is shown to be less than 1% of that of MMSE MIMO decoding.

## Differential Feedback

A beamforming scenario is depicted in Figure 9. BS specifies the feedback period of SS to be once every $N_f$ subframes. SS quantizes the beamforming matrix and feeds back the quantization indexes every period. BS reconstructs the beam-

*Figure 7. Subspace spanned by the beamforming vectors, i.e., $\mathbf{v}_1,\cdots,\mathbf{v}_{N-1}$, and the complementary feedback vector that is orthogonal to the subspace*

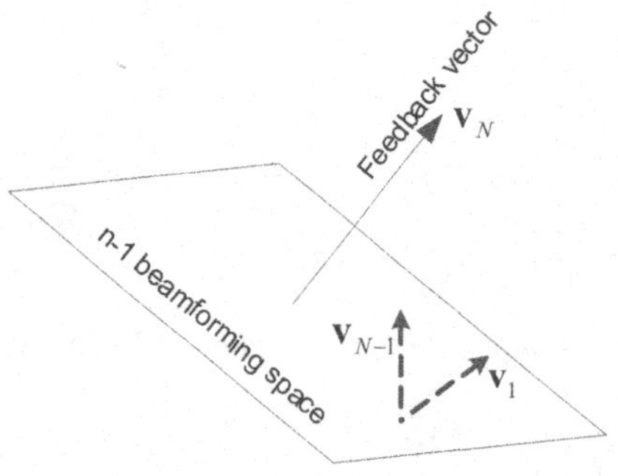


forming matrix and employs it for one transmission period of $N_f$ subframes. It should be noticed that the beamforming gain decreases with time within the period due to channel variation. It is desirable to have beamforming period less than the channel coherence time. Therefore, there is correlation between the two adjacent feedback matrixes.

Reducing feedback overhead is essential from a system design viewpoint, and various approaches (Intel, February 2006; Intel, May 2006 II; Nortel, February 2006; Nortel, March 2006; Li, 2007a) exploit the time/frequency correlation for feedback reduction, i.e. only feed back the difference in beamforming matrix. We focus on two recent approaches: spherical cap (Intel, February 2006; Intel, May 2006 II; Li, 2007a) and subspace tracking (Zhang, 2004; Nortel, February 2006; Nortel, March 2006). In subspace tracking (Nortel, February 2006; Nortel, March 2006), a beamforming codebook with codewords distributed on $\Phi_{N,K}$ is formed with a neighbor list the each codeword. The differential feedback specifies a move from the previous codeword to one of its neighbors in the list. The feedback overhead is reduced if the list is short. However, the scheme suffers from error propagation. If feedback error occurs, SS doesn't know which codeword BS is exactly using and thus which neighbor list should be employed for the next feedback. If SS assumes the feedback is always correct, beamforming error propagates in the subsequent beamforming after one feedback error occurs.

The spherical cap scheme (Intel, February 2006; Li, 2007a) addresses this error propagation problem. Both spherical cap and subspace tracking are gradient descent algorithms. The work (InterDigital, 2007) adopts a stochastic gradient approach, which is employed by the well-known LMS algorithm. A random direction is generated synchronously by

*Figure 8. PER performances of space-time code, antenna selection, codebook based precoding, and ideal SVD precoding over $4\times 2$ ITU Pedestrian B channels (ITU, 1997) with 3 km/h mobile speed. Two spatial streams are transmitted, each with QPSK modulation and rate ½ convolutional code.*

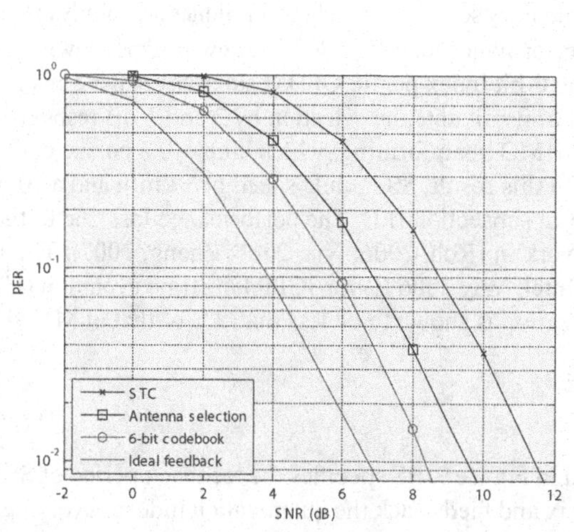

*Figure 9. Illustration of differential feedback and downlink transmit beamforming. BS sends beamformed signal* $\hat{\mathbf{V}}(j)\mathbf{d}(t)$ *, where j and t are transmit subframe index and feedback period index respectively;* $\hat{\mathbf{V}}(j)$ *is the employed beamforming matrix at BS; K is the number of data streams.*

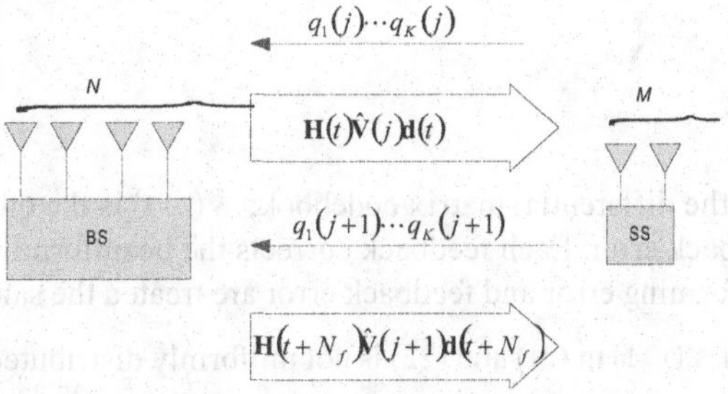


*Table 2. Overhead comparison between differential and non-differential feedbacks*

| Feedback matrix dimension | $2\times 1$ and $2\times 2$ | | $4\times 2$ | |
|---|---|---|---|---|
| | Differential | Non-differential | Differential | Non-differential |
| Overhead per matrix (bit) | 2 | 4 | 7 | 11 |
| Overhead reduction | 50% | | 36% | |

the receiver and transmitter for each update feedback. The gradient minimizing the beamforming error is projected onto the direction. The receiver decides to move forward or backward along the direction with a small step and one bit is fed back for the choice.

## Spherical Cap

SS sends back orthogonal matrixes depicting the changes of beamforming. BS multiplies the fed back matrixes cumulatively to track channel variation between two adjacent feedbacks. Namely, the current beamforming matrix is

$$\hat{\mathbf{V}}(j)=\hat{\mathbf{V}}(j-1)\tilde{\mathbf{V}}(j), \tag{20}$$

where $\hat{\mathbf{V}}(j-1)$ is the previous beamforming matrix and $\tilde{\mathbf{V}}(j)$ is the latest fed back matrix. If $\tilde{\mathbf{V}}(j)$ is $N\times K$ and $N>K$, it is expanded to $N\times N$ by adding $N-K$ orthogonal columns, where the added columns don't need to be unique. In the next period, SS observes the latest channel matrix from common pilots, the beamforming matrix $\hat{\mathbf{V}}(j)$ from control link or dedicated pilot, and computes the differential beamforming matrix $\tilde{\mathbf{V}}(j+1)$. It should be noticed that $\hat{\mathbf{V}}(j)$ may be far deviated from the ideal due to feedback error. It first computes singular value decomposition of $\hat{\mathbf{H}}$, which is the predicted channel matrix in middle of the next transmission period, as $\hat{\mathbf{H}}=\hat{\mathbf{U}}\hat{\Sigma}\hat{\mathbf{V}}^*$, where $\hat{\mathbf{V}}$ is $N\times N$. The differential update is computed (by subtracting the observed beamforming matrix $\hat{\mathbf{V}}(j)$ from $\hat{\mathbf{V}}$ in angle domain) as

$$\mathbf{V}(j+1)=\hat{\mathbf{V}}^*(j)\hat{\mathbf{V}}. \tag{21}$$

SS quantizes $\mathbf{V}(j+1)$ and sends the quantization indexes to BS for the next period. If rank adaptation is applied, only the first $K$ columns of $\mathbf{V}(j+1)$ need to be quantized and sent back, which correspond to the strongest $K$ singular modes of $\hat{\mathbf{H}}$. $\mathbf{V}(j+1)$ can be quantized by differential codebooks in $\Omega_{N,K}$ (or $\Phi_{N,K}$). The one in (Li, 2007a) employs $\Omega_{N,K}$ and is computed as

$$\tilde{\mathbf{V}}(j+1) = \arg\max_{\mathbf{V}_i} \left\| \operatorname{diag}\left(\mathbf{V}^*(j+1)\mathbf{V}_i\right) \right\|, \tag{22}$$

where $\mathbf{V}_i$ is the $i$-th codeword in the differential matrix codebook; $\tilde{\mathbf{V}}(j+1)$ is the quantized matrix.

This scheme is robust to feedback error. Each feedback corrects the beamforming error in the previous beamforming as in (20) and (21). The beamforming error and feedback error are treated the same and are corrected by subsequent feedback. It should be noticed that $\mathbf{V}(j+1)$ in (21) and (22) is not uniformly distributed over $\Omega_{N,K}$ or $\Phi_{N,K}$. Rather, $\mathbf{V}(j+1)$ concentrates around the identity matrix due to the similarity (or correlation) between $\hat{\mathbf{V}}(j)$ and $\hat{\mathbf{V}}$. The narrow distribution of $\mathbf{V}(j+1)$ implies little information and requires a fewer number of quantization bits. Extending the recursive structure depicted in subsection II.B.2 and (IEEE P802.16e, 2006; Li, 2004b; Li, 2007b), one can construct vector codebook (Intel, February 2006; Intel, May 2006 II; Li, 2007a) to quantize $\mathbf{V}(j+1)$ (recursively in a column by column fashion). In contrast to the uniform codebook in (IEEE P802.16e, 2006; Li, January 2005; Li, May 2005), the distribution of the codeword vectors concentrate about $[1,0,\cdots,0]^T$. Under the same PER performance, the feedback overheads of non-differential (Li, January 2005) and differential (Li, 2007a) are compared in **Table 2**, where the overhead is reduced by more than 35% for differential feedback.

## Subspace Tracking

The subspace tracking scheme is initially proposed to exploit frequency domain correlation (Zhang, 2004). The idea is simple: the channels of adjacent subcarriers are statistically correlated in an OFDM system, so should the precoding matrices for these subcarriers. More specifically, it is highly likely that for the two neighboring subcarriers, the two optimal precoding matrices reside within a small neighborhood in the high-dimensional Grassmann (or Stiefel) manifold. Consequently, one can select precoding matrices using the successive "neighborhood subspace search" across different subcarriers rather than an independent "global search" at each subcarrier.

The subspace tracking algorithm starts with the first subcarrier. The best precoding matrix $\mathbf{v}(1)$ is selected out of a Grassmann codebook with $2^L$ matrix codewords. Since the best precoding matrix for the second subcarrier, $\mathbf{v}(2)$, is in a small neighborhood of $\mathbf{v}(1)$ with high probability, one can narrow the search within this neighborhood. Even though the best codeword of the codebook may fall outside the neighborhood, the best codeword of the neighborhood still provide beamforming gain. Let the number of codewords in the neighborhood to be not greater than $2^D$ ($D < L$). Then, the number of bits to index the best codeword in the neighborhood is $D$, which is the number of feedback bits for the second subcarrier. Sequentially repeating this process to cover all $N_c$ subcarriers involved, and we end up with a total requirement of $L + (N_c - 1)D$ feedback bits, which can be much less than the $N_c L$ bits necessary for the non-differential scheme. The search for $\mathbf{v}(2)$ within the neighborhood of $\mathbf{v}(1)$ is illustrated in Figure 10.

The feedback overhead can be further reduced by down sampling, i.e. feeding back for every $N_p$ subcarriers. Since the correlation of the fed back subcarrier reduces as the down sample factor $N_p$ increases, the size of neighborhood needs to increase in order to capture the best codeword of the codebook. Namely, the number of bits in each differential feedback, $D$, needs to be increased. There exists an optimal combination of $N_p$ and $D$ to achieve the minimum feedback subject to a constraint of quantization loss. The neighborhood of the codeword $\mathbf{v}(j)$ for $j$-th subcarrier is computed as

$$C_{\mathbf{V}(j)} = \left\{\mathbf{V}_i : d\left(\mathbf{V}_i, \mathbf{V}(j)\right) \le \delta_{\mathbf{V}(j)}\right\}, \tag{23}$$

where $d(\mathbf{V}_i, \mathbf{V}(j))$ is the chordal distance between $\mathbf{V}_i$ and $\mathbf{V}(j)$. The radius $\delta_{\mathbf{V}(j)}$ is selected such that the size of the set is not greater than $2^D$.

The total number of bits required for $N_c$ subcarriers is

*Figure 10. Illustration of subspace tracking in Grassmann Manifold*

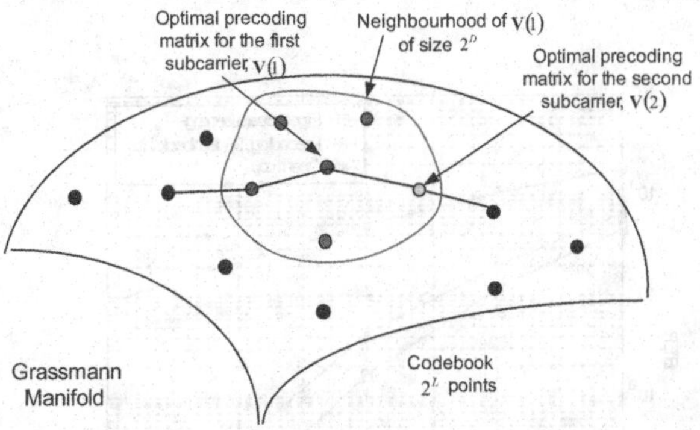


$$L+\frac{N_c}{N_p}D$$

bits. In Figure 11, we demonstrate the efficacy of subspace tracking scheme using a $4\times 2$ MIMO OFDMA system with used 96 subcarriers over a frequency flat channel. The number of subcarriers is $N_c = 128-32 = 96$ and the codebook size is $2^L = 64$, where $L = 6$ bits. Meanwhile, we choose the parameters $N_p = 4$ and $D = 4$. If we were to independently feed back a precoding matrix for each of the $N_c$ subcarriers, the feedback overhead would be $N_cL = 384$ bits. In comparison, the total number of feedback bits for subspace tracking is

$$L+\frac{N_c}{N_p}D=98$$

bits, which is much less than that of independent feedback. It is observed that the quantization loss is 1dB with only about 1 bit per subcarrier. Meanwhile, the precoding solution is 10 dB better than the open loop baseline solution.

## Analog Feedback

Analog modulation is proposed in (IEEE P802.16e, 2006; Thomas, 2005; Marzetta, 2006) to send CSI and skip the quantization. The scheme in (IEEE P802.16e, 2006; Thomas, 2005) is illustrated in Figure 12. SS estimates the channel response between each BS transmit antenna and each SS receive antenna in frequency domain using the training pilots in frequency domain as shown on the left of Figure 12. The frequency responses are converted into (delay) time domain as shown in the middle of Figure 12. Since the power of delay tap tends to exponentially decrease with the delay, the number of delay taps with significant power is limited. The response is truncated in time delay dimension. The complex response of each significant tap is fed back by one OFDM subcarrier using analog modulation. For overhead reduction, multiple truncated responses are placed side by side on the subcarriers of one OFDM symbol as shown on the right of Figure 12.

## Delay Compensation

The CSI from feedback can become outdated due to the relatively long feedback delay. Two examples of 802.16e are shown in Figure 13, where the real part of the fading process is plotted for 3 and 10 km/h in Figure 13(a) and (b), respectively. ITU Pedestrian B model (ITU, 1997) and a 2.6 GHz carrier frequency are assumed. This channel model is commonly used in 802.16 standard group. The sampling rate of the process is 200 Hz, which corresponds to the frame duration of 5 ms. The feedback delay in 802.16e is two frames, i.e. 10 ms. The correlation coefficients between two samples 10 ms apart are 0.95 and 0.56 for 3 and 10 km/h respectively. If a precoding matrix is select for $\mathbf{H}(t)_1$, it is very likely that the matrix is not optimal for the actual channel $\mathbf{H}(t+2)$ that carries the precoded signal.

*Figure 11. Performance of feedback-based MIMO precoding for a $4 \times 2$ MIMO OFDM link with 96 subcarriers. The other parameters are $L = 6$, $D = 4$, and $N_p = 4$*

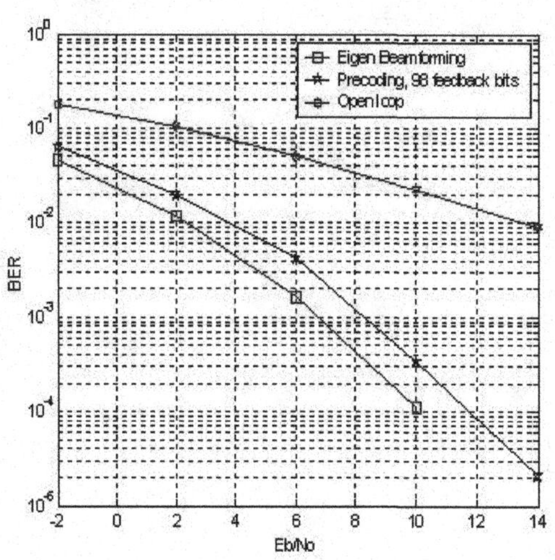


*Figure 12. Illustration of direct channel feedback*

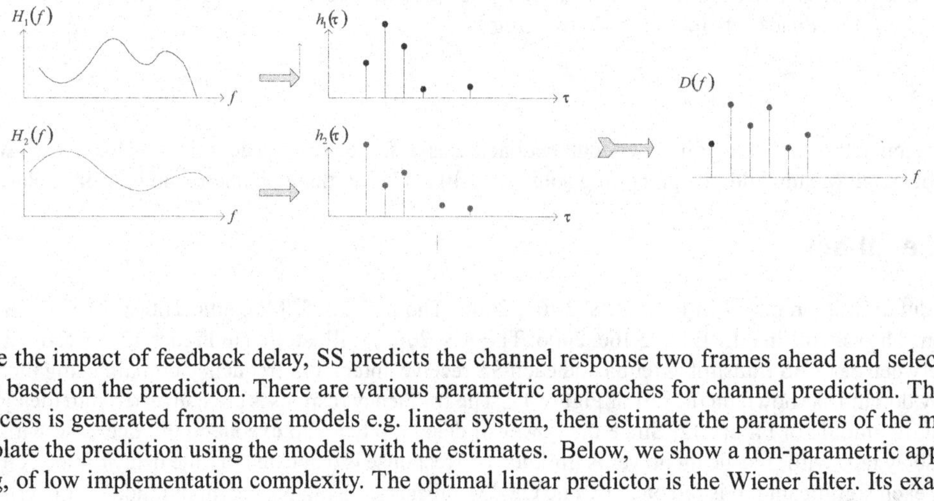


To mitigate the impact of feedback delay, SS predicts the channel response two frames ahead and selects the pre-coding matrix based on the prediction. There are various parametric approaches for channel prediction. They assume the fading process is generated from some models e.g. linear system, then estimate the parameters of the models, and finally extrapolate the prediction using the models with the estimates. Below, we show a non-parametric approach, i.e. linear filtering, of low implementation complexity. The optimal linear predictor is the Wiener filter. Its exact solution is written as

$$\mathbf{w} = \left(\mathbf{R}_h + \sigma^2 \mathbf{I}\right)^{-1} \mathbf{p}, \tag{24}$$

where

$$\mathbf{R}_h = \begin{bmatrix} r(0) & r(1) & \cdots & r(N_l - 1) \\ r^*(1) & r(0) & \cdots & r(N_l - 2) \\ \vdots & & \ddots & \vdots \\ r^*(N_l - 1) & r^*(N_l - 2) & & r(0) \end{bmatrix}; \mathbf{p} = \left[r^*(2) \;\; r^*(3) \;\; \cdots \;\; r^*(1 + N_l)\right]^T; N_l \text{ is the}$$

*Figure 13. Prediction of time varying fading channels for mobile speeds 3 and 10 km/h*

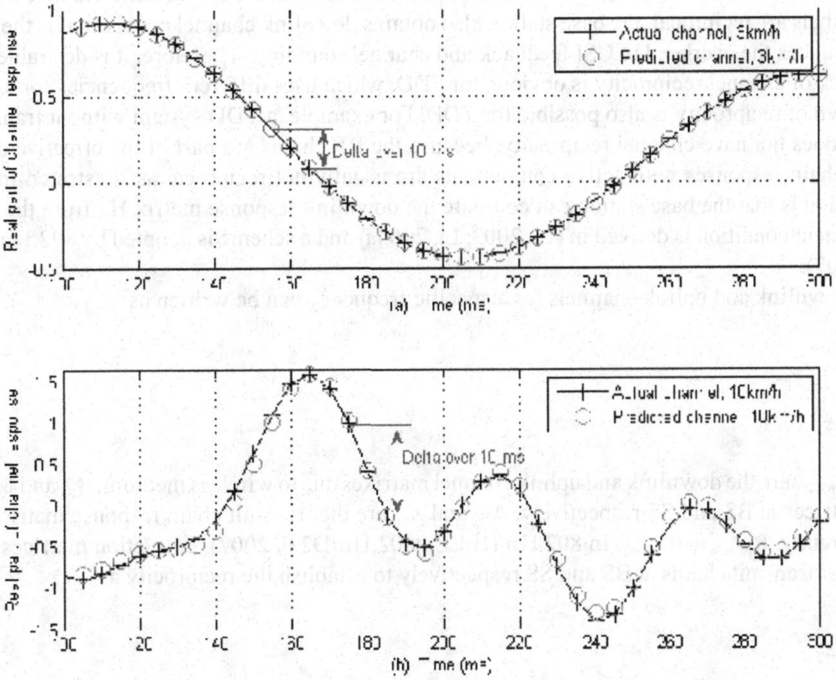

number of filter taps. The autocorrelation is computed as $r(\tau) = \underset{i,j}{\mathrm{E}}\left[\mathbf{H}_{i,j}(t)\mathbf{H}^*_{i,j}(t+\tau)\right]$, where the expectation is over antenna indexes $i$ and $j$ and may be further refined by averaging over $t$. The predicted channel is computed as

$$\hat{\mathbf{H}}_{i,j}(t+2) = \mathbf{w}^*\mathbf{h}, \tag{25}$$

where $\mathbf{h} = \left[\mathbf{H}_{i,j}(t) \;\; \cdots \;\; \mathbf{H}_{i,j}(t-N_t+1)\right]^T$. The inversion in (24) is not necessary. Adaptive implementations are available e.g. LMS and RLS, which have complexities of $O(N_t)$ and $O(N_t^2)$ respectively and also track the variation in the autocorrelations by incorporating windowing and forgetting factor. The predicted channels are also plotted in Figure 13, where the number of filter taps is five and the predicted channel matches well with the actual channel. The packet error rate (PER) performance of the channel prediction is shown in Figure 14 and the gain is explained in the caption.

## Miscellaneous

MIMO beamforming is integrated with other techniques in the real systems. For example, when reliable CSI is not available in fast fading, antenna group grouping (Chae, 2004) combines MIMO beamforming and space-time coding to balance the beamforming gain and the robustness of the link. In this case, the resolution of the beamforming codebook is low and the codewords are permutation unitary matrix with only zeros and ones. Similarly, transmit antennas selection can be viewed as a special case of MIMO beamforming with a coarse codebook. In another example, adaptive bit and power loading (Li, 2004c; Intel, May 2006 I) are applied over the beamformed spatial channels with distinct channel qualities in order to increase throughput. As the third example, in contrast to removing crosstalk between beamformed spatial channels in (Telatar, 1995), controlled crosstalk (Jiang, 2005; Intel, August 2006 II) is introduced across the channels to reduce link adaptation feedback by exploiting the power of successive interference cancellation of MIMO decoding.

## TECHNIQUES FOR TDD

The feedback overhead can be reduced in TDD by exploiting the channel reciprocity in a bidirectional link. The BS obtains the uplink channel matrix while it demodulates the uplink traffic from the SS for coherent detection. If the uplink and downlink channels are reciprocal, the base station also obtains downlink channel matrix that is the transpose of the uplink one. This reduces the overhead of CSI feedback and channel sounding. Therefore, it is desirable to maintain the reciprocity. The lack of channel reciprocity is obvious for FDD, which uses different frequencies for uplink and downlink. The breakdown of reciprocity is also possible for TDD. For example, a TDD system without transmit and receive chain calibrations does not have channel reciprocity, because the RF chains are part of the effective downlink/uplink channels and the chain responses with active components are usually not reciprocal as illustrated in Figure 15. The goal of the calibration is that the base station can compute the downlink response matrix $\mathbf{H}_D$ from the uplink response matrix $\mathbf{H}_U$. A sufficient condition is derived in (Li, 2003; Li, 2004a) and a scheme is adopted by 802.11n standard (IEEE P802.11n/D2.0, 2007).

The effective downlink and uplink channels for a specific frequency can be written as

$$\mathbf{H}_D = \mathbf{B}_{SS}\mathbf{H}_{BS\to SS}\mathbf{A}_{BS}, \tag{26}$$

$$\mathbf{H}_U = \mathbf{B}_{BS}\mathbf{H}_{SS\to BS}\mathbf{A}_{SS}, \tag{27}$$

where $\mathbf{H}_{BS\to SS}$ and $\mathbf{H}_{BS\to SS}$ are the downlink and uplink channel matrixes due to wireless medium; $\mathbf{B}_{BS}$ and $\mathbf{B}_{SS}$ are the receive chain response matrices at BS and SS respectively; $\mathbf{A}_{BS}$ and $\mathbf{A}_{SS}$ are the transmit chain response matrices at BS and SS respectively. Apparently, $\mathbf{H}_{BS\to SS} = \mathbf{H}^T_{BS\to SS}$. In 802.11n (IEEE P802.11n/D2.0, 2007), correction matrices $\mathbf{K}_{BS}$ and $\mathbf{K}_{SS}$ are added at the input of transmit chains at BS and SS respectively to establish the reciprocity as

$$\mathbf{H}_D\mathbf{K}_{BS} = \rho\left(\mathbf{H}_U\mathbf{K}_{SS}\right)^T, \tag{28}$$

*Figure 14. PER performance of MIMO beamforming at speeds of 3, 10, 15 and km/h. Space time block code (STBC) performance is plotted in black for reference. The performance of MIMO beamforming (B.F.) at 3 km/h without channel prediction (C.P.) is plotted in blue and it is about the same as the performance at 15 km/h with channel prediction. The modulation and coding schemes are 16QAM and 802.16e convolutional code with code rate 2/3 respectively. The antenna configuration is .*

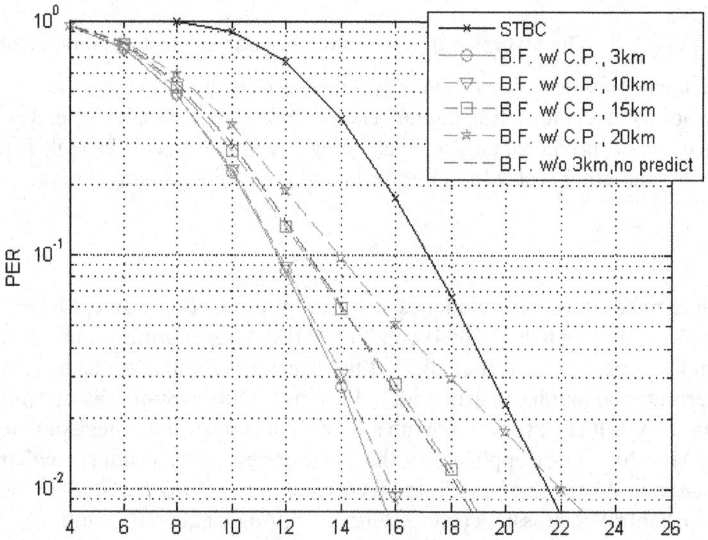

*Figure 15. Illustration of link response including components of transmit chain, wireless medium, and receive chain. It should be noticed that the uplink and downlink employ two distinct sets of transit/receive chains.*

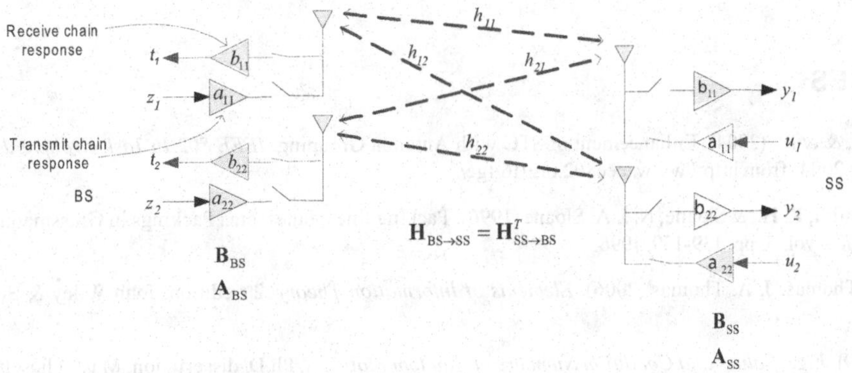


where $\rho$ is a scalar. There are multiple solutions for (28). One of them is

$$\mathbf{K}_{BS} = \rho_{BS}\mathbf{A}_{BS}^{-1}\mathbf{B}_{BS}, \tag{29}$$

$$\mathbf{K}_{SS} = \rho_{SS}\mathbf{A}_{SS}^{-1}\mathbf{B}_{SS}, \tag{30}$$

where $\rho_{BS}$ and $\rho_{SS}$ are scalars. The transmit chain at the SS usually has low efficiency and its transmission power is usually below the licensed level due to peak-to-average-power-ratio (PAPR) problem of OFDM. Since the correction matrix is added to the transmitter, it may further reduce the transmission power. It is desirable to add the correction at the receiver as in (Li, 2004a).

Since the cross coupling among transmit (receive) chains is usually less than 30 dB, which is below the AWGN level in practical SNR region, i.e. -6–30 dB. If the cross coupling is ignored, the matrixes $\mathbf{A}_{BS}$, $\mathbf{A}_{SS}$, $\mathbf{B}_{BS}$, and $\mathbf{B}_{SS}$d are diagonal. In (Li, 2004a), the reciprocity is established as

$$\mathbf{C}_{SS}\mathbf{H}_{D} = \rho\left(\mathbf{C}_{BS}\mathbf{H}_{U}\right)^{T}, \tag{31}$$

where $\mathbf{C}_{SS}$ and $\mathbf{C}_{BS}$ are diagonal correction matrices at the output of receive chains at SS and BS respectively. One simple solution to (31) is

$$\left(\mathbf{C}_{BS}\right)_{ii} = \rho_{BS}\frac{\left(\mathbf{A}_{BS}\right)_{ii}}{\left(\mathbf{B}_{BS}\right)_{ii}}, \text{ and } \left(\mathbf{C}_{SS}\right)_{ii} = \rho_{SS}\frac{\left(\mathbf{A}_{SS}\right)_{ii}}{\left(\mathbf{B}_{SS}\right)_{ii}}, \text{ for all } i. \tag{32}$$

Namely, the ratio of transmit response to receive response remains constant for all antennas at each device. The correction quantities $\mathbf{C}_{SS}$, $\mathbf{C}_{BS}$, $\mathbf{K}_{BS}$, and $\mathbf{K}_{SS}$ can be obtained by sending sounding signals over the bidirectional link and feeding back the received and quantized $\mathbf{H}_D$ and $\mathbf{H}_U$ to the other side.

## CONCLUSION AND FUTURE TREND

This chapter describes the issues, underlying principle, evolving techniques, and corresponding industrial applications of MIMO beamforming. Techniques in FDD focus on efficient quantization to reduce feedback overhead and robust channel prediction to compensate feedback delay. On the other hand, techniques in TDD centers on transceiver calibration to establish channel reciprocity. Examples in 3GPP, IEEE 802.11n and 80216d/e are depicted. MIMO beamforming delivers more than 2 dB gain for most practical antenna configurations.

Single user MIMO link and channel matrix with i.i.d. Gaussian entries are assumed in the chapter. Recently, there is research interest in beamforming feedback for multiuser MIMO (or MIMO broadcast channel) (Jindal, 2006) and the design of adaptive quantization codebook for channel matrix with correlated entries (Mondal, 2006).

## REFERENCES


Chae, C., W. Roh, & *et al.* (2004). Enhancement of STC with Antenna Grouping. *IEEE 802.16 Task Group e document* C802.16e-04/554r4, January 2004, from http://www.ieee802.org/16/tge/.

Conway, J. H., Hardin, R. H., & Sloane, N. J. A. Sloane (1996). Packing line, planes, etc.: Packings in Grassmannian spaces. *Experimental Mathematics*, vol. 5, pp. 139-159, 1996.

Cover, T. M., & Thomas, J. A. Thomas (2006). *Elements of Information Theory.* 2nd edition, John Wiley & Sons, Inc., Hoboken, New Jersey, 2006.

Edelman, A. (1989). *Eigenvalues and Condition Numbers of Random Matrices.* Ph.D. dissertation, Massachusetts Institute of Technology, May 1989.

Gersho, A., & Gray, R. M. (1992). *Vector Quantization and Signal Compression.* Boston, MA: Kluwer Academic, 1992.

Golub, G., & Van Loan, C. Van Loan (1996). *Matrix Computations.* 3rd edition, Johns Hopkins University Press, 1996.

Hochwald, B.M., Marzetta, T.L., Richardson, T.J., Sweldens, W., & Urbanke, R. (2000), Systematic design of unitary space-time constellations. *IEEE Transactions on Infomation Theory*, vol. 46, pp. 1962-1973, September 2000.

IEEE P802.16d (2004). *IEEE Standard for local and metropolitan area networks, Part 16: Air interface for fixed broadband wireless access systems.* October 2004.

IEEE P802.16e (2006). *IEEE standard for local and metropolitan area networks, Part 16: Air Interface for Fixed and Mobile Broadband Wireless Access Systems, Amendment 2: Physical and Medium Access Control Layers for Combined Fixed and Mobile Operation in Licensed Beands and Corrigendum 1.* February 2006.

IEEE P802.11n/D2.0 (2007). *Draft standard for Information Technology-Telecommunications and information exchange between systemes-local and metropolitan networks-Specific requirements-Part 11: Wireless LAN Medium Access Control (MAC) and Physical Layer (PHY) specifications: Enhancements for Higher Throughput.* February 2007.

Ikram, M., Onggosanusi, E. N. , Raghavan, V. , & *et al.* (2004, August). An enhanced closed-loop MIMO design for OFDM/OFDMA-PHY. *IEEE 802.16 Task Group e document* C802.16e-04/267, August 2004, from http://www.ieee802.org/16/tge/

Intel (2006, February). Codebook Design for Precoded MIMO. *3GPP Technical Specification Group of Radio Access Network, Working Group 1, meeting #44*, R1-060672, February 2006.

Intel (2006, May I). Text proposal for modulation and power adaptation of MIMO systems. *3GPP Technical Specification Group of Radio Access Network, Working Group 1, meeting #45*, document R1-061125, May 2006.

Intel (2006, May II). Scaleable Precoding and Implementation Complexities. *3GPP Technical Specification Group of Radio Access Network, Working Group 1, meeting #45*, R1-061126, Shanghai, China, May 2006.

Intel (2006, August I). A Low Feedback Scheme for Precoding. *3GPP Technical Specification Group of Radio Access Network, Working Group 1, meeting #46*, document T1-061991, August-September 2006.

Intel (2006, August II). Performance benchmark for a new unitary precoding scheme with uniform MCS allocation. *3GPP Technical Specification Group of Radio Access Network, Working Group 1, meeting #46*, document T1-061967, August-September 2006.

InterDigital (2007). Binary Differential Feedback Scheme for Downlink MIMO Pre-coding for E-UTRA. *3GPP Technical Specification Group of Radio Access Network, Working Group 1, meeting #49bis*, document R1-072782, June 2007.

ITU (1997). Recommendation ITU-R M.1225. Guidelines for evaluation of radio transmission technologies for IMT-2000. 1997.

Jiang, Y., Li, J., & Hager W. W. (2005). Uniform Channel Decomposition for MIMO Communications. *IEEE Transactions on Signal Processing*, vol. 53, no. 11, pp. 4283-4294, Nov. 2005.

Jindal, N. (2006). MIMO Broadcast Channels with Finite-Rate Feedback. *IEEE Transactions on Information Theory*, vol. 52, no. 11, pp. 5045-5060, November 2006.

Jorswieck, E. A., & Boche, H. (2004). Optimal Transmission Strategies and Impact of Correlation in Multiantenna Systems with Different Types of Channel State Information. *IEEE Transactions on Signal Processing*, vol. 52, no. 12, pp. 3440-3453, Dec. 2004.

Kang, H. (2006). *Multiple Antenna Systems in a Mobile-to-Mobile Environment*. Ph.D. dissertation, Georgia Institute of Technology, Dec. 2006.

Kang, M., & Alouini, M.-S. (2003). Largest Eigenvalue of Complex Wishart Matrices and Performance Analysis of MIMO MRC Systems. *IEEE Journal on Selected Areas in Communications*, vol. 21, no. 3, pp. 418-426, April 2003.

Kang, M., & Alouini, M.-S. (2006). Capacity of Correlated MIMO Rayleigh Channels. *IEEE Transactions on Wireless Communications*, vol. 5, no. 1, pp. 143-155, Jan. 2006.

Li, Q., & Lin, X. E. (2003). Antenna subsystem calibration apparatus and methods in spatial-division multiple-access systems. *Intel pending patent*, May 2003.

Li, Q., & Lin, X. E. (2004a). Calibration in MIMO systems. *Intel pending patent*, September 2004.

Li, Q., Lin, X. E., Ho, M., & *et al.* (2004b). Improved feedback for MIMO precoding. *IEEE 802.16 Task Group e document* C802.16e-04/527r4, November 2004, from http://www.ieee802.org/16/tge/.

Li, Q., Lin, X. E., Poon, A., & *et al.* (2004c). Adaptive bit loading for vertically encoded MIMO. *IEEE 802.16 Task Group e document* C802.16e-04/529r5, December 2004, from http://www.ieee802.org/16/tge/

Li, Q., Lin, X. E., & J. C. Zhang (2005, January, 2005a). Compact codebooks for transmit beamforming in closed-loop MIMO. *IEEE 802.16 Task Group e document* C802.16e-04/50r6, Jan. 2005, from http://www.ieee802.org/16/tge/.

Li, Q., & Lin, X. E. (2005, May, 2005b). Compact feedback for MIMO-OFDM systems over frequency selective channels. In proceedings of *the 61st IEEE Vehicular Technology Conference 2005-Spring*, vol. 1, pp. 187-191, May-June. 2005.

Li, Q., Li, G., Lin, X. E., & Zheng, S. (2007a). A Low Feedback Scheme for WMAN MIMO Beamforming. In proceedings of *IEEE Radio and Wireless Symposium 2007*, pp. 349-352, January 2007.

Li, Q., Lin, X. E., & J. C. Zhang (2007b). MIMO precoding in 802.16e WiMAX. *Journal of Communications and Networks*, in press, June 2007.

Love, D. J., & Heath R. W. Jr. (2003). Grassmannian Beamforming for Multiple-Input Multiple-Output Wireless Systems. IEEE Transactions on Information Theory, vol. 49, pp. 2735-2747, October 2003.

Love, D. J., & Heath R. W. Jr. (2005). Limited Feedback Unitary Precoding for Spatial Multiplexing Systems. IEEE Transactions on Information Theory, vol. 51, no. 8, pp. 2967-2976, August 2005.

Maaref, A., & Aissa, S. (2005). Closed-Form Expressions for Outage and Ergodic Shannon Capacity of MIMO MRC Systems. *IEEE Transactions on Communications*, vol. 53, no. 7, pp. 1092-1095, July 2005.

Marzetta, T. L., & Hochwald, B. M. (2006). Fast Transfer of Channel State Information in Wireless Systems. *IEEE Transactions on Signal Processing*, vol. 54, no. 4, pp. 1268-1278, April 2006.

MIMO ad-hoc session. (2007). Text Proposal for TS36.211 for 4-Tx Antenna SU-MIMO Codebook. *3GPP Technical Specification Group of Radio Access Network, Working Group 1, meeting #49bist,* document R1-073206, June 2007.

Mondal, B., Samanta, B. R., & Heath, R. W. Jr. (2005). Frame theoretic quantization for limited feedback MIMO beamforming systems. In proceedings of *International Conference on Wireless Networks, Communications and Mobile Computing 2005,* vol. 2, pp. 1065–1070, June 2005.

Mondal, B., & Heath, R. W. Jr. (2006). Channel Adaptive Quantization for Limited Feedback MIMO Beamforming Systems. *IEEE Transactions on Signal Processing,* vol. 54, no. 12, pp. 4717-4729, December 2006.

Muirhead, R J. (1982). *Aspects of multivariate statistical theory.* John Wiley & Sons, Inc., 1982.

Mukkavilli, K. K., Sabharwal, A., Erkip, E., & Aazhang, B. (2003). On Beamforming with Finite Rate Feedback in Multiple-Antenna Systems. *IEEE Transactions on Information Theory*, vol. 49, no. 10, pp. 2562-2579, October 2003.

Narula, A., Lopez, M. J., Trott, M. D., & Wornell, G. W. (1998). Efficient Use of Side Information in Mutliple-Antenna Data Transmission over Fading Channels. *IEEE Transactions on Communications*, vol. 16, no. 8, pp. 1423-1436, October 1998.

Nortel (2006, February). Differential Encoding/Decoding for an Arbitrary Codebook. *3GPP Technical Specification Group of Radio Access Network, Working Group 1, meeting #44*, R1-060660, Denver, Colorado, February 2006.

Nortel (2006, March). On MIMO-OFDM Downlink Pilots and Precoding Index Feedback. *3GPP Technical Specification Group of Radio Access Network, Working Group 1, meeting* #44bis, R1-060899, Athens, Greece, March 2006.

Onggosanusi, E. N., & Dabak, A. G. (2002). A feedback-based adaptive multi-input multi-output signaling scheme. In proceedings of *the 36th Asilomar Conference on Signals, Systems and Computers,* vol. 2, pp. 1694-1698, November 2002.

Roh, J. C., & Rao, B. D. (2004a). An efficient feedback method for MIMO systems with slowly time-varing channels. In proceedings of *Wireless Communications and Networking Conference 2004*, vol. 2, pp. 760–764, March 2004.

Roh, J. C., & Rao, B. D. (2004b). Channel feedback quantization methods for MISO and MIMO systems. In proceedings of *the 15th IEEE International Symposium on Personal, Indoor and Mobile Radio Communications, 2004.*, vol. 2, pp. 805-809, September 2004.

Roh, J. C., & Rao, B. D. (2006a). Transmit Beamforming in Multiple-Antenna Systems with Finite Rate Feedback: A VQ-Based Approach. *IEEE Transactions on Information Theory*, vol. 52, no. 3, pp. 1101-1112, March 2006.

Roh, J. C., & Rao, B. D. (2006b). Design and analysis of MIMO spatial multiplexing systems with quantized feedback. *IEEE Transactions on Signal Processing*, vol. 54, no. 8, pp. 2874-2886, August 2006.

Roh, J. C., & Rao, B. D. (2007). Efficient Feedback Methods for MIMO Channels Based on Parameterization. *IEEE Transactions Wireless Communications*, vol. 6, no. 1, pp. 282-292, January 2007.

Shannon, C. E. (1958). Channels with side information at the transmitter. *IBM J. Res. Develop.*, vol. 2, pp. 289-293, 1958.

Telatar, I. E. (1995). Capacity of multi-antenna Gaussian channels. *AT&T Bell Laboratories Technical Memo.*

Thomas, T. A., Baum, K. L., & Sartori, P. (2005). Obtaining channel knowledge for closed-loop multipstream broadband MIMO-OFDM communications using direct channel feedback. In proceedings of *Global Telecommunications Conference 2005*, vol. 6, November-December 2005.

Visotsky, E., & Madhow, U. (2001). Space-Time Transmit Precoding with Imperfect Feedback. *IEEE Transactions on Information Theory*, vol. 47, no. 6, pp. 2632-2639, September 2001.

Winters, J. (1987). On the Capacity of Radio Communication Systems with Diversity in a Rayleigh Fading Environment. *IEEE Journal on Selected Areas in Communications*, vol. 5, no. 5, pp. 871-878, June 1987.

Xia, P., & Giannakis, G. B. (2006). Design and Analysis of Transmit-Beamforming Based on Limited-Rate Feedback. *IEEE Transactions on Signal Processing*, vol. 54, no. 5, pp. 1853-1863, May 2006.

Zhang, J., Zhang, H., Waes, N. V., & *et al.* (2004). Closed-loop MIMO precoding to improve MIMO link performance with limited feedback. *IEEE 802.16 Task Group e document* C802.16e-04/262rl, August 2004, from http://www.ieee802.org/16/tge/

Zheng, J., & Rao, B. D. (2007). Analysis of Multiple-Antenna Systems with Finite-Rate Feedback Using High-Resolution Quantization Theory. *IEEE Transactions on Signal Processing*, vol. 55, no. 4, pp. 1461-1476, April 2006.

Zhou, S., & Giannakis, G. B. (2004). Adaptive Modulation for Multiantenna Transmissions with Channel Mean Feedback. *IEEE Transactions on Wireless Communications*, vol. 3, no. 5, pp. 1626-1636, September 2004.

## ENDNOTES

[a] The notation $N \times M$ in the context of MIMO, e.g. $N \times M$ MIMO (or channel), means there are $N$ and $M$ antennas at the transmitter and receiver respectively. It is worth noticing that the channel matrix from the transmitter to the receiver is $M$ by $N$, i.e., $M \times N$, not $N \times M$. We follow this conventional notation. For simplification, we assume $M \leq N$ without loss of generality in the presentation, and the extension to $M > N$ is straightforward.

[b] An $N \times K$ orthogonal matrix **O** has orthogonal columns with unit norm, i.e. $\mathbf{O}^*\mathbf{O} = \mathbf{I}_K$, where $\mathbf{I}_K$ is the identity matrix. A unitary matrix **U** is an $N \times N$ orthogonal matrix and satisfies $\mathbf{I}_N = \mathbf{U}^*\mathbf{U} = \mathbf{U}\ ^*$.

[c] $S\{\mathbf{x}_1, \cdots, \mathbf{x}_n\}$ denotes the space spanned by linear combination of vectors $\mathbf{x}_i$, for $i = 1, ..., n$.

[d] IEEE 802.16d standard is designed for cellular system experiencing fixed or slowly fading channels.

[e] IEEE 802.11n standard is designed for high throughput wireless local area network, which experiences slow fading.

[f] IEEE 802.16e and 3GPP long term evolution standards are designed for mobile cellular systems.

[g] Shannon fist studied the channel capacity with CSI in (Shannon, 1958).

[h] Uplink channel is from subscriber station to base station while downlink channel is the opposite. MIMO beamforming is usually conducted in the downlink for most of the systems.

i The notation $\mathbf{A}_{:,i}$ denotes the *i*-th column of matrix **A**.

j A unit vector is a vector that has the norm of unity.

k The *n*-dimensional uniform vector codebook consists of *n*-dimensional vectors that have unit norm and uniformly distribute over the *n*-dimensional sphere.

l $\mathbf{H}(t)$ denotes the channel matrix for a specific subcarrier of OFDM at time *t*.

m $(\mathbf{A})_{ii}$ denote the i-th diagonal entry of matrix **A**.

# Chapter XIII
# Joint Beamforming and Space-Time Coding for MIMO Channels

**Biljana Badic**
*Swansea University, UK*

**Jinho Choi**
*Swansea University, UK*

## ABSTRACT

*This chapter introduces joint beamforming (or precoding) and space-time coding for multiple input multiple output (MIMO) channels. First, we explain key ideas of beamforming and space- time coding and then we discuss the joint design of beamformer and space-time codes and its benefits. Beamforming techniques play a key role in smart antenna systems as they can provide various features, including spatially selective transmissions to increase the capacity and coverage increase. STC techniques can offer both coding gain and diversity gain over MIMO channels. Thus, joint beamforming and STC is a more practical approach to exploit both spatial correlation and diversity gain of MIMO channels. We believe that joint design will be actively employed in future standards for wireless communications.*

## INTRODUCTION

The last decade has seen an increased interest in the study of smart antenna and multiple input multiple output (MIMO) channels. In particular, after Foschini (1998) and Telatar (1999) showed that multiple antenna systems are capable of providing large capacity increase in wireless transmissions, there have been an increased number of research studies for MIMO channels in various aspects, including propagations, space-time transmission schemes, and receiver design. Smart antenna systems are some of the particular examples to exploit the potential of MIMO channels. Beamforming techniques are crucial for smart antenna systems. Conventional beamforming techniques have been well developed for the signal reception to exploit diversity gain and/or suppress interfering signals (Winters, 1998). Provided that the channel state information (CSI) is known at the transmitter equipped with multiple antennas, beamforming is available for the signal transmission to provide spatial selectivity. In multiuser communications, this spatial selectivity leads to the spatial multiplexing gain and space division multiple access (SDMA).

Transmit beamforming can provide not only spatial selectivity, but also transmit diversity gain (Lo, 1999; Choi, 2002a). Transmit diversity can be achieved if multiple (independent) fading channels available by the transmitter. Generally, transmit diversity gain can be exploited by beamforming if instantaneous CSI is available at the transmitter. On the other hand, if statistical properties of CSI (e.g., the spatial correlation of MIMO channels) are only available, beamforming gain can be achieved, but not diversity gain. Thus, the CSI at the transmitter is often critical to exploit the transmit diversity gain. However, one general drawback of the methods relying on instantaneous CSI at the transmitter is feasibility and the need of feedback of CSI from the receiver to the transmitter.

Limited feedback issue becomes important for closed-loop transmit diversity including the beamforming methods relying on instantaneous CSI. In (Choi, 2002a; Love, 2003; Mukkavilli, 2003), beamforming methods are considered with limited feedback that provides partial CSI to the transmitter.

If CSI is not available at the transmitter, space-time coding (STC) can be applied. A number of STC methods have been investigated to improve the system performance (Alamouti, 1998; Tarokh, 1998; Tarokh, 1999a; Tarokh, 1999b). So called Orthogonal Space-Time Block Codes (OSTBCs) became quite popular in the context of space-time transmit diversity since they can provide a full diversity with low complexity, however mostly with the drawback of reduced data rates. It is known that the performance can be degraded if channels are spatially correlated (Bolcskei, 2000). The code designs criteria in (Alamouti, 1998) and (Tarokh, 1998) assume that transmit and receive antennas are uncorrelated and each element of the MIMO channel matrix fades independently. However, this is not necessary to be true in practice. For example, in outdoor wireless systems, the base-station (BS) antennas are placed high above the ground and close to each other. In such a scenario, the BS antennas are unobstructed and see no local scatterers leading to high correlation between the BS antennas (Salz, 1994).

In general, the availability of CSI at the transmitter is important to the MIMO channel capacity. Provided that the CSI is known at the transmitter, multiple parallel channels (or eigenmodes) can be built using singular value decomposition (SVD). Then, an optimal power allocation algorithm called as *water-filling principle* (Telatar, 1999) across multiple parallel channels can be applied. The water-filling transmission scheme pours power on the eigenmodes of the MIMO channel in such a way that more power is delivered to stronger eigenmodes and less or no power to the weaker eigenmodes. Another strategy with full CSI at the transmitter is *beamforming* (Mukkavilli, 2003; Zhou, 2004) where only the strongest eigenmode is used.

In general, a transmission using a STC performs worse than a system using a beamforming technique (Mukkavilli, 2001; Larsson, 2002). This stems from the fact, that STBC systems spreads the available power uniformly in all directions in space, while beamforming uses information about the channel to steer energy in the direction of the receiver. The gap in the performance between the two methods can be quite significant, especially in highly correlated channels.

Recently, it has been attempted to jointly design beamformer and space-time encoder to compensate each other when MIMO channels are correlated and the instantaneous CSI is not available (Bahrami, 2006; Choi, 2004). In this chapter, we explain key ideas of beamforming and STC. In addition, more importantly, we discuss the joint design of beamformer and space-time codes and its benefits as mentioned above.

## BEAMFORMING DESIGN

In this section, we focus on beamformer design. Beamforming is possible with/without instantaneous CSI and the resulting performance varies (Choi, 2002a). Assuming that the receiver knows the perfect CSI, we show different beamforming methods depending on the availability of instantaneous CSI with diversity gain in this section.

### Beamforming over MIMO Channels

Suppose that there are $n_t$ transmit antennas and $n_r$ receive antennas. Denote by $h_{p,q}$ the channel coefficient from the $q$th transmit antenna to the $p$th receive antenna. Then, the received signal from the $p$th antenna over flat fading channels is written as

$$r_p = \sum_{q=1}^{n_t} h_{p,q} s_q + v_p, \quad p = 1,2,\ldots,n_r,$$

where $s_q$ denotes the signal transmitted by the $q$th transmit antenna and $v_p$ denotes the background noise of the $p$th antenna. We assume that the noise, $v_p$, is an independent circular-complex zero-mean Gaussian random variable with $E[|v_p|^2]=\sigma^2$. Using matrix-vector notation, we show that

$$\mathbf{r} = [r_1, r_2, \ldots, r_{n_r}]^T$$
$$= \mathbf{Hs} + \mathbf{v},$$

where $[\mathbf{H}]_{p,q} = h_{p,q}$, $\mathbf{s} = [s_1, s_2, \ldots, s_{n_t}]^T$, and $\mathbf{v} = [v_1, v_2, \ldots, v_{n_r}]^T$. We can see that **v** is a circular complex white Gaussian random vector with $E[\mathbf{v}] = \mathbf{0}$ and $E[\mathbf{v}\mathbf{v}^H] = \sigma^2\mathbf{I}$. For beamforming, the signal vector, **s**, is written as

$$\mathbf{s} = \mathbf{w}b,$$

where **w** and $b$ denote the beamforming vector and data symbol, respectively. Assuming that the instantaneous channel matrix, **H**, is available at the transmitter and the maximal ratio combining (MRC) is employed at the receiver, the maximum signal-to-noise ratio (MSNR) beamforming vector can be obtained as follows:

$$\mathbf{w}_{\text{MSNR}} = \arg\max_{\|\mathbf{w}\|\le 1} \frac{\|\mathbf{Hw}\,b\|^2}{E[\|\mathbf{v}\|^2]} = \arg\max_{\|\mathbf{w}\|\le 1} \mathbf{w}^H\mathbf{H}^H\mathbf{Hw}, \tag{2.1}$$

where $\|\cdot\|$ denotes 2-norm. For normalization purpose, the beamforming vector is normalized to have unit power. From (2.1), we can see that the MSNR beamforming vector is the eigenvector corresponding to the maximum eigenvalue of $\mathbf{H}^H\mathbf{H}$. The MSNR beamforming can achieve a full diversity gain as well as beamforming gain.

In time division duplex (TDD) mode, the instantaneous channel matrix, **H**, would be available at the transmitter using the channel reciprocity (Morgan, 2003) without any explicit feedback from the receiver. On the other hand, in frequency division duplex (FDD) mode, explicit feedback of CSI is required. As the amount of feedback is limited, various approaches are proposed to feed back the CSI. In (Love, 2003), the use of codebook is considered. A codebook consisting of $M$ beamforming vectors can be written as

$$W = \{\mathbf{w}_1, \mathbf{w}_2, \ldots, \mathbf{w}_M\}, \quad \|\mathbf{w}_m\| = 1 \text{ for all } m.$$

At the receiver, the best beamforming vector can be found from the codebook as follows and its index can be sent to the transmitter for beamforming:

$$\mathbf{w}_{\text{MSNR-CB}} = \arg\min_{\mathbf{w}\in W} \mathbf{w}^H\mathbf{H}^H\mathbf{Hw}.$$

Thus, the required number of bits for feedback is $\log_2 M$. This beamforming can be called the MSNR with codebook (MSNR-CB) beamforming. There is a trade-off between the amount of feedback and performance. As $M$ increases, more beamforming vectors are available and the performance can approach that of the MSNR beamforming. However, the number of the required bits for feedback increases with $M$. Note that the best beamforming vector varies as the channel matrix varies over time. Thus, the beamforming vector should be updated if the channel varies. The feedback rate depends on the channel variation, i.e., coherence time. For a mobile user of high mobility, the beamforming may require an excessive overhead for the feedback of CSI.

If any instantaneous CSI is not available, statistical beamforming can be considered to achieve beamforming gain without transmit diversity gain. Consider the beamforming that maximizes the *average* SNR:

$$\mathbf{w}_{MASNR} = \arg\max_{\|\mathbf{w}\|\le 1} \frac{E[\|\mathbf{Hw}\,b\|^2]}{E[\|\mathbf{v}\|^2]} = \arg\max_{\|\mathbf{w}\|\le 1} \mathbf{w}^H E[\mathbf{H}^H\mathbf{H}]\mathbf{w}. \quad (2.2)$$

If the spatial covariance matrix of MIMO channels, $\mathbf{R_H} = E[\mathbf{H}^H\mathbf{H}]$, is available at the transmitter, the maximum average SNR (MASNR) beamforming vector, $\mathbf{w}_{MASNR}$, in (2.2) can be found. The MASNR beamforming vector is the eigenvector corresponding to the maximum eigenvalue of $\mathbf{R_H} = E[\mathbf{H}^H\mathbf{H}]$. Since the MASNR beamforming vector is dependent on the spatial covariance matrix, not dependent on the instantaneous channel matrix, $\mathbf{H}$, the MASNR beamforming cannot exploit transmit diversity gain. As mentioned earlier, the MSNR beamforming can fully exploit transmit diversity gain. Thus, there is a significant difference between the MSNR and MASNR beamforming in exploiting transmit diversity gain. To see this difference clearly, we need to find the diversity gain for each beamforming.

Let $\lambda_1 \ge \lambda_2 \ge \ldots \ge \lambda_{n_t}$ denote the eigenvalues (which are random variables) of $\mathbf{H}^H\mathbf{H}$, and $\bar{\lambda}_1 \ge \bar{\lambda}_2 \ge \ldots \ge \bar{\lambda}_{n_t}$ denote the eigenvalues of $\mathbf{R_H}$. Denote by $\mathbf{e}_l$ and $\bar{\mathbf{e}}_l$ the eigenvectors corresponding to $\lambda_l$ and $\bar{\lambda}_l$, respectively. In general, $\bar{\lambda}_l \ne E[\lambda_l]$ and $\bar{\mathbf{e}}_l \ne E[\mathbf{e}_l]$. The MSNR and MASNR beamforming vectors are

$$\mathbf{w}_{MSNR} = \mathbf{e}_1; \quad \mathbf{w}_{MASNR} = \bar{\mathbf{e}}_1.$$

With the MSNR beamforming vector, the received signal becomes

$$\mathbf{r} = \mathbf{Hs} + \mathbf{v} = \underbrace{\mathbf{He}_1}_{=\mathbf{u}} b + \mathbf{v}.$$

After the MRC, we have

$$y = \mathbf{u}^H\mathbf{r} = \|\mathbf{u}\|^2 b + \mathbf{u}^H\mathbf{v}.$$

Since $E[\mathbf{u}^H\mathbf{vv}^H\mathbf{u}] = \mathbf{u}^H E[\mathbf{vv}^H]\mathbf{u} = \sigma^2\|\mathbf{u}\|^2$ and $\|\mathbf{u}\|^2 = \mathbf{e}_1^H\mathbf{H}^H\mathbf{He}_1 = \lambda_1$, the bit error rate (BER) for binary phase shift keying (BPSK) ($b \in \{-\sqrt{E_b}, +\sqrt{E_b}\}$), where $E_b$ represents the bit energy) is given as

$$P_{MSNR}(\mathbf{H}) = Q\left(\sqrt{\frac{\lambda_1 E_b}{\sigma^2}}\right) \le \exp\left(-\frac{\lambda_1}{2}\gamma\right), \quad (2.3)$$

where $\gamma = E_b/\sigma^2$ denotes the input SNR. The upper bound in (2.3) is obtained using the Chernoff bound. The average BER can be obtained by taking the expectation with respect to the maximum eigenvalue of $\mathbf{H}^H\mathbf{H}$, $\lambda_1$.

For spatially uncorrelated MIMO Rayleigh-fading channels, where each element of $\mathbf{H}$ is independent circular complex Gaussian random variable with zero mean and variance 1 (i.e., $h_{p,q} \sim CN(0,1)$), from (Kang, 2003; Lo, 1999; Dighe, 2001), the average BER at high SNR can be written as

$$\bar{P}_{MSNR} = E[P_{MSNR}(\mathbf{H})] \le E[\exp\left(-\frac{\lambda_1}{2}\gamma\right)] \approx c\gamma^{-n_t n_r}, \quad \gamma \gg 1, \quad (2.4)$$

where $c$ is a positive constant. Thus, we can see that the MSNR beamforming can achieve a diversity order of $n_t n_r$, which is the full diversity gain.

With the MASNR beamforming vector, $\mathbf{w}_{\text{MASNR}} = \bar{\mathbf{e}}_1$, the received signal becomes

$$\mathbf{r} = \mathbf{Hs} + \mathbf{v} = \underbrace{\mathbf{H}\bar{\mathbf{e}}_1}_{=\bar{\mathbf{u}}} b + \mathbf{v}.$$

After the MRC, we have

$$y = \bar{\mathbf{u}}^H \mathbf{r} = \|\bar{\mathbf{u}}\|^2 b + \bar{\mathbf{u}}^H \mathbf{v}.$$

Since $E[\bar{\mathbf{u}}^H \mathbf{v}\mathbf{v}^H \bar{\mathbf{u}}] = \bar{\mathbf{u}}^H E[\mathbf{v}\mathbf{v}^H]\bar{\mathbf{u}} = \sigma^2 \|\bar{\mathbf{u}}\|^2$, the BER becomes

$$P_{\text{MASNR}}(\mathbf{H}) = Q\left(\sqrt{\frac{\|\bar{\mathbf{u}}\|^2 E_b}{\sigma^2}}\right) \le \exp\left(-\frac{\|\bar{\mathbf{u}}\|^2}{2}\gamma\right). \tag{2.5}$$

Generally, the MASNR beamforming attempts to transmit signals through the statistically favourite eigenmode. Unfortunately, this statistically favourite eigenmode does not need to be the favourite eigenmode corresponding to the maximum eigenvalue of the instantaneous channel matrix, $\mathbf{H}$, as $\bar{\mathbf{e}}_1 \ne \mathbf{e}_1$. For spatially uncorrelated MIMO Rayleigh-fading channels, we can see that $\|\bar{\mathbf{u}}\|^2$ is a chi-square random variable with $2n_r$ degree of freedom, because $\|\bar{\mathbf{u}}\|^2$ is the square sum of $n_r$ circular-complex Gaussian random variables:

$$\|\bar{\mathbf{u}}\|^2 = \|\mathbf{H}\bar{\mathbf{e}}_1\|^2 = \sum_{p=1}^{n_r} |\mathbf{g}_p^H \bar{\mathbf{e}}_1|^2,$$

where $\mathbf{g}_p^H$ denotes the $p$th row vector of $\mathbf{H}$ and $\mathbf{g}_p^H \bar{\mathbf{e}}_1$ is a circular-complex Gaussian random variable and independent of $\mathbf{g}_p^H \bar{\mathbf{e}}_1$ for $q \ne p$. The pdf of $\|\bar{\mathbf{u}}\|^2$ can be expressed as

$$f(x = |\bar{\mathbf{u}}|^2) = \frac{1}{(n_r - 1)} x^{n_r - 1} e^{-x}.$$

The resulting average BER is

$$P_{\text{MASNR}} = E[P_{\text{MASNR}}(\mathbf{H})] \le E[\exp\left(-\frac{\|\bar{\mathbf{u}}\|^2}{2}\gamma\right)] \approx c'\gamma^{-n_r}, \quad \gamma >> 1, \tag{2.6}$$

where $c'$ is a positive constant. This shows that the receive antenna diversity can be achieved by MRC, but the transmit antenna diversity is not exploited. However, the MASNR can be effective if the channel is highly spatially correlated, where the diversity gain decreases.

## Statistical Eigenbeamforming with Selection Diversity

With limited feedback, we can exploit transmit antenna diversity with high beamforming gain for spatially correlated MIMO channels. Suppose that the spatial covariance matrix, $\mathbf{R}_\mathbf{H}$, is known at the transmitter. Consider the statistical eigenbeamforming with selection diversity:

$$\mathbf{w}_{\text{MSNR-SDE}} = \arg\min_{\mathbf{w} \in S} \mathbf{w}^H \mathbf{H}^H \mathbf{H} \mathbf{w},$$

*Table 2.1. Comparison of the three beamforming methods*

| Beamforming methods | MASNR | MSNR | MSNR-SDE |
|---|---|---|---|
| Diversity order | $n_r$ | $n_t n_r$ | $M_r$ |
| Type of CSI at transmitter | Spatial covariance matrix (no instantaneous CSI) | Instantaneous CSI | Spatial covariance matrix + best eigenmode |
| Amount of feedback of instantaneous CSI | no | High (proportional to $n_t$) | $\log_2 M$ |

where $S = \{\bar{\mathbf{e}}_1, \bar{\mathbf{e}}_2, \ldots, \bar{\mathbf{e}}_M\}$ and $M \le n_t$. This beamforming approach can be considered as a special case of the MSNR-CB beamforming, where the codebook is *S* which consists of the eigenmodes of the spatial covariance matrix, and called the MSNR with selection diversity and eigenmodes (MSNR-SDE) beamforming. In the codebook, *S*, there are the *M* most statistically favourite eigenmodes. Among those statistical eigenmodes, we can choose the best eigenmode that is the most favourite one for the instantaneous MIMO channel, **H**. In addition, since the number of the beamforming vector in *S* is $M \le n_t$, the number of bits for feedback is $\log_2 M \le \log_2 n_t$.

In (Choi, 2007), it is shown that if $\mathbf{g}_p^H$ (which is the *p*th row vector of **H**) is independent of $\mathbf{g}_q^H$, $q \ne p$ and a circular-complex Gaussian random vector with $E[\mathbf{g}_p] = \mathbf{0}$ and $E[\mathbf{g}_p \mathbf{g}_p^H] = \mathbf{R}$, the average BER is given by

$$P_{\text{MSNR-SD}} \approx c'' \left( \prod_{m=1}^{M} \lambda_m^{n_r} \right)^{-1} \gamma^{-Mn_r}, \quad \gamma >> 1, \tag{2.7}$$

where $c''$ is a positive constant. Note that

$$\mathbf{R_H} = E[\mathbf{H}^H \mathbf{H}] = \sum_{p=1}^{n_r} E[\mathbf{g}_p \mathbf{g}_p^H] = n_r \mathbf{R}.$$

As shown in (2.7), the average BER has two important terms: beamforming gain and diversity gain. The beamforming gain is

$$\prod_{m=1}^{M} \lambda_m^{n_r}.$$

Thus, we can see that the beamforming gain is maximized by choosing the eigenmodes that are statistically favourable. In addition, the diversity gain is $n_r M$, which can be maximized if $M = n_t$. The advantage of the MSNR-SDE beamforming is that *M* strong eigenmodes can be pre-selected to maximize the beamforming gain. Thus, it is very effective when the spatial correlation is high and $n_t$ is large. We can compare the three beamforming methods as in Table 2.1.

## Numerical Results

To see the performance difference depending on the beamforming methods, we consider an example with four transmit and two receive antennas. Spatially correlated Rayleigh fading MIMO channels have the spatial covariance matrix as follows:

$$[\mathbf{R_H}]_{p,q} = \rho^{|p-q|}, \quad p,q = 1,2,3,4, \tag{2.8}$$

where ρ denotes the spatial correlation, which is set to 0.6 in simulations. The resulting eigenvalues are

$$\{\bar{\lambda}_1, \bar{\lambda}_2, \bar{\lambda}_3, \bar{\lambda}_4\} = \{2.3870,\ \ 0.8993,\ \ 0.4290,\ \ 0.2847\}.$$

Fig. 2.1 shows the simulation results. As expected, the MSNR beamformer performs better than the MASNR beamformer as the instantaneous CSI is available at the transmitter for beamforming. For the MSNR-SDE beamforming, two bits are allocated for feedback of the partial CSI to choose one of four possible eigenvectors of $\mathbf{R}_\mathbf{H}$ (i.e., $M = n_t = 4$). The performance of the MSNR-SDE beamformer is better than that of the MASNR, but worse than that of the MSNR. However, as we can see from Fig. 2.1, the diversity gain of the MSNR-SDE is the same as that of the MSNR. Thus, the performance gap between the MSNR and MSNR-SDE beamformers would not be large as the SNR increases.

## Multiuser Beamforming and Other Issues

In multiuser systems, the beamforming can be generalized to support multiple users simultaneously. In (Schubert, 2004), an approach for multiuser beamforming is derived with singnal to interference-plus-noise ratio (SINR) constraints for individual users. With a total power constraint, the optimal beamforming for multiuser communications is addressed in (Gerlach, 1996). Multiuser beamforming can be combined with other multiple access methods in multiuser environments. In (Choi, 2004c; Choi, 2002b; Vojcic, 1998), multiuser beamforming for code division multiple access (CDMA) systems is investigated. For cellular systems, transmit beamforming with power control is studied in (Rashid-Farrokhi, 1998).

Beamforming can be used with channel coding to generate intentional fading effect. This beamforming can be seen as a means to convert the spatial diversity into the temporal diversity so that channel coding can exploit the diversity gain under slow fading environments. In (Hiroike, 1992), this idea was first introduced, and a further extension with the notion of eigenbeamforming is made in (Choi, 2004a). A generalization of this approach is studied for multiuser diversity in (Viswanath, 2002).

## SPACE-TIME CODE DESIGN

Space-time coding finds its application in cellular communications as well as in wireless local area networks. The design of STC amounts to find code matrices that satisfy certain optimality criteria. In particular, STC schemes optimize a trade-off between the three conflicting goals of maintaining a simple decoding algorithm, obtaining low error probability, and maximizing the information rate. There are different coding methods, namely, space-time trellis codes

*Figure 2.1. BER performance of various beamformers*

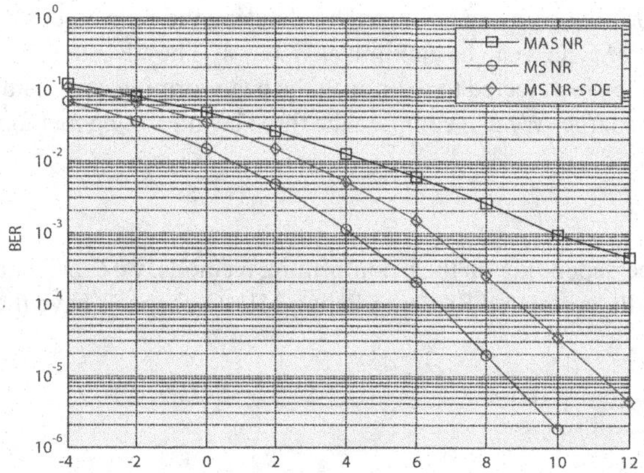

*Figure 3.1 Space-time coding scheme*

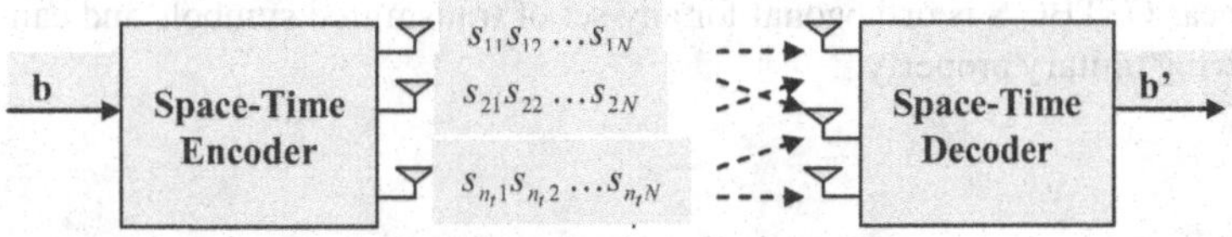


(STTC), space-time block codes (STBC), space-time turbo trellis codes and layered space-time (LST) codes. A main issue in all these schemes is the exploitation of redundancy to achieve high reliability, high spectral efficiency and high performance gain.

In this section, we focus on linear STBC. STBCs have been proposed by the pioneering work of Alamouti (1998). The Alamouti code achieves full diversity and full data rate in case of two transmit antennas. The key feature of this scheme is the *orthogonality* between the signal vectors transmitted over the two transmit antennas. This scheme can be applied to an arbitrary number of receive antennas by applying the theory of *orthogonal design* (Tarokh, 1998). The generalized schemes are referred to as *space-time block codes* (Tarokh, 1998; Tarokh, 1999a; Tarokh, 1999b). However, for more than two transmit antennas, no complex valued STBCs with full diversity and full data rate exist. Thus, many different code design methods have been proposed providing either full diversity or full data rate (Jafarkhani, 2001; Tirkkonen, 2002; Rupp, 2002; Badic, 2005b).

We describe and analyze different design principles of STBC that achieve full diversity gain and/or maximum rate gain for various MIMO scenarios in this section. In addition, we emphasize recent advances such as STBC with partial CSI at the transmitter.

## Space-Time Block Code Design

A general scheme of an STC system equipped $n_t$ transmit and $n_r$ receive antennas is depicted in Fig. 3.1.

Each block of incoming bits $\mathbf{b} = \{b_1, b_2, \ldots b_l\}$ passes through the space-time encoder, which maps the set of bits into an $(n_t \times N)$ space-time coded matrix **S**, where $N$ denotes the number of time instants composing the code. Different rows of **S** refer to different transmit antennas (space) and different columns refer to different time instants.

STBCs are designed to achieve maximum diversity order for a given number of transmit and receive antennas with the simple decoding algorithm. In a general form, an STBC can be seen as a mapping of $n_N$ complex symbols $\{s_1, s_2, \ldots, s_N\}$ onto a matrix **S** of dimension $n_t \times N$:

$$\{s_1, s_2, \ldots, s_N\} \rightarrow \mathbf{S}$$

An STBC code matrix **S** taking on the following form

$$\mathbf{S} = \sum_{n=1}^{n_N} \left(\bar{s}_n \mathbf{A}_n + j\tilde{s}_n \mathbf{B}_n\right)$$

where $\{s_1, s_2, \ldots, s_N\}$ is a set of symbols to be transmitted with $\bar{s}_n = \mathrm{Re}\{s_n\}$ and $\tilde{s}_n = \mathrm{Im}\{s_n\}$, and with fixed code matrices $\{\mathbf{A}_n, \mathbf{B}_n\}$ of dimension $n_t \times N$, is called linear STBCs.

The choice of specific STBC depends on performance parameter we want to improve for a given set of constraints. As a performance parameter we can use, for instance, average error probability, information rate or receiver complexity. In a choice of the specific coding designs three fundamental aspects have to be taken into account: simple decoding, minimum error probability, and maximum information rate. The essential question is: *How can we maximize the transmitted date rate using a simple coding and decoding algorithm at the same time as the bit error probability is minimized?*

## Orthogonal Space-Time Code (OSTBC)

A general structure of a linear OSTBC **S** is orthogonal for any set of transmitted symbols and can be represented in a matrix form with the following unitary property:

$$\mathbf{S}^{H} = \sum_{n=1}^{N} |s_n|^2 \, \mathbf{I}_{n_t} \qquad (3.1)$$

The *i*-th row of **S** corresponds to the symbols transmitted from the *i*-th transmit antenna in *N* transmission periods, while the *j*-th column of **S** represents the symbols transmitted simultaneously through $n_t$ transmit antennas at time *j*. $\mathbf{I}_n$ denotes an $(n \times n)$ identity matrix.

The columns of the transmission matrix **S** are orthogonal to each other. This means, in each block, the signal sequences from any two transmit antennas are orthogonal. The *orthogonality* enables to achieve full transmit diversity, and at the same time, it allows the receiver by means of simple MRC to decouple the signals transmitted from different antennas and consequently, it makes the maximum likelihood (ML) decoding simple.

## Alamouti Space-Time Block Code

Alamouti code (Alamouti, 1998) is valid for two transmit antennas and any number of receiver antennas. It is given in a matrix form as

$$\mathbf{S} = \begin{pmatrix} s_1 & s_2 \\ -s_2^* & s_1^* \end{pmatrix}. \qquad (3.2)$$

The first row represents the first transmission period and the second row the second transmission period. During the first transmission, the symbols $s_1$ and $s_2$ are transmitted simultaneously from antenna 1 and the antenna 2 respectively. In the second transmission period, the symbol $-s_2^*$ is transmitted from antenna 1 and the symbol $s_1^*$ from transmit antenna 2. It is clear that the encoding is performed in both time (two transmission intervals) and space domain (across two transmit antennas).

*Figure 3.2 Performance of the Alamouti Scheme in a Rayleigh fading channel, (QPSK)*

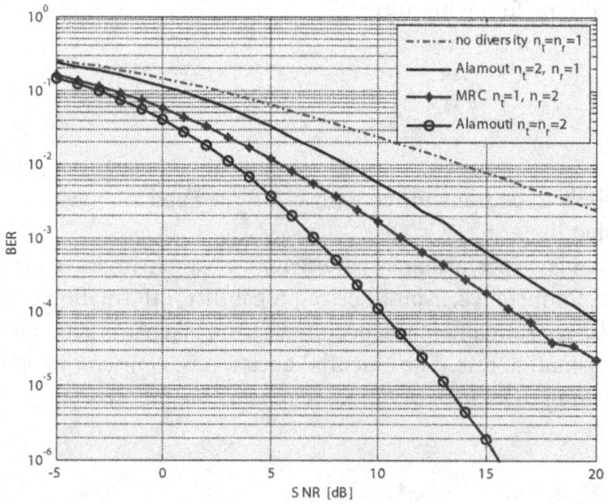

Suppose that there is one receive antenna. Assuming flat-fading channels with transmission coefficients $h_1$ and $h_2$, the received vector **r** is formed by stacking two consecutive received data samples $\mathbf{r} = [r_1, r_2]^T$ in time, resulting in

$$\mathbf{r} = \mathbf{S}\mathbf{h} + \mathbf{v}, \tag{3.3}$$

where $\mathbf{h} = [h_1, h_2]^T$ is the complex channel vector and **v** is the noise vector at the receiver. After some manipulations, from (3.3), it follows

$$\mathbf{y} = [r_1 \quad r_2^*]^T = \mathbf{H}_v\mathbf{s} + \mathbf{v}. \tag{3.4}$$

The resulting *equivalent virtual channel matrix* $\mathbf{H}_v$ has the similar structure as the Alamouti code:

$$\mathbf{H}_v = \begin{pmatrix} h_1 & h_2 \\ h_2^* & -h_1^* \end{pmatrix}. \tag{3.5}$$

Thus, by considering of the elements of **y** in Eqn. (3.4) as originating from two virtual receive antennas (instead of received samples at one antenna at two time slots) one could interpret the Alamouti STBC for $n_t = 2$ and $n_r = 1$ as a $(2 \times 2)$ spatial multiplexing transmission using one time slot. That can be also seen in Fig. 3.2. The Alamouti scheme for two transmit antennas and one receive antennas achieves the same diversity as the scheme with one single transmit and two receive antennas using MRC. However, the performance of Alamouti scheme is 3 dB worse due to the fact that the power radiated from each transmit antenna in the Alamouti scheme is half of that radiated from the single antenna and sent to two receive antennas and using MRC. In this way, the two schemes have the same total transmit power. In general, the Alamouti scheme with two transmit and $n_r$ receive antennas has the same diversity gain as an MRC receive diversity scheme with one transmit and $2n_r$ receive antennas. The key difference between the Alamouti scheme and a true $(2 \times 2)$ multiplexing system lies in the specific structure of $\mathbf{H}_v$. Unlike to a general i.i.d. MIMO channel matrix, the rows and columns of the virtual channel matrix are orthogonal:

$$\mathbf{H}_v\mathbf{H}_v^H = \mathbf{H}_v^H\mathbf{H}_v = \left(|h_1|^2 + |h_2|^2\right)\mathbf{I}_2 = |h|^2\mathbf{I}_2, \tag{3.6}$$

where $|h|^2$ is the power gain of the channel. Due to this *orthogonality*, the Alamouti scheme decouples the MISO channel into two virtually independent channels with channel gain $|h|^2$ and diversity $d = 2$.

## OSTBCs for More Than Two Transmit Antennas

Alamouti STBC has been a basis to create OSTBCs for more than two transmit antennas. Tarokh et al. (1999b) studied the error performance associated with unitary signal matrices and proved that full diversity, full data rate orthogonal STBC only exists for two transmit antennas. In the same work, some OSTBCs for more than two transmit antennas have been proposed achieving full-diversity but not full-rate. Ganesan at al. (2001) streamlined the derivations of many of the results associated with OSTBC and established an important link to the theory of the orthogonal and amicable orthogonal designs. OSTBCs are an important subclass of linear STBCs that guarantee that the ML detection of different symbols $\{s_n\}$ is *decoupled* and at the same time the transmission scheme achieves a diversity order equal to $n_t n_r$. The main disadvantage of OSTBCs is the fact that for more than two transmit antennas and complex-valued signals, OSTBCs only exist for code rates smaller than one symbol per time slot.

## Examples

For any arbitrary complex signal constellation, there are OSTBCs that can achieve a rate of 1/2 for any given number of transmit antennas. For example, the code matrices $\mathbf{S}_{n3}$ and $\mathbf{S}_{n4}$ in Eqn. (3.7) are OSTBCs for three and four transmit antennas, respectively and they have the rate 1/2 (Tarokh, 1998).

$$\mathbf{S}_{n_3} = \begin{pmatrix} s_1 & s_2 & s_3 \\ -s_2 & s_1 & -s_4 \\ -s_3 & s_4 & s_1 \\ -s_4 & -s_3 & s_2 \\ s_1^* & s_2^* & s_3^* \\ -s_2^* & s_1^* & -s_4^* \\ -s_3^* & s_4^* & s_1^* \\ -s_4^* & -s_3^* & s_2^* \end{pmatrix} \qquad \mathbf{S}_{n_4} = \begin{pmatrix} s_1 & s_2 & s_3 & s_4 \\ -s_2 & s_1 & -s_4 & s_3 \\ -s_3 & s_4 & s_1 & -s_2 \\ -s_4 & -s_3 & s_2 & s_1 \\ s_1^* & s_2^* & s_3^* & s_4^* \\ -s_2^* & s_1^* & -s_4^* & s_3^* \\ -s_3^* & s_4^* & s_1^* & -s_2^* \\ -s_4^* & -s_3^* & s_2^* & s_1^* \end{pmatrix} \tag{3.7}$$

With the code matrix $\mathbf{S}_{n_3}$, four complex symbols are taken at a time and transmitted via three transmit antennas in eight time slots. Thus, the symbol rate is 1/2. With the code matrix $\mathbf{S}_{n_4}$, four symbols are taken at a time and transmitted via four transmit antennas in eight time slots, resulting in a transmission rate of 1/2 as well.

The following code matrices $\mathbf{S}'_{n_3}$ and $\mathbf{S}'_{n_4}$ in Eqn. (3.8) are complex generalized designs for OSTBC with rate 3/4 for three and four transmit antennas, respectively (Tarokh, 1998):

$$\mathbf{S}'_{n_3} = \begin{pmatrix} s_1 & s_2 & \frac{s_3}{\sqrt{2}} \\ -s_2^* & s_1^* & \frac{s_3}{\sqrt{2}} \\ \frac{s_3^*}{\sqrt{2}} & \frac{s_3^*}{\sqrt{2}} & \frac{-s_1 - s_1^* + s_2 - s_2^*}{2} \\ \frac{s_3^*}{\sqrt{2}} & -\frac{s_3^*}{\sqrt{2}} & \frac{s_1 - s_1^* + s_2 + s_2^*}{2} \end{pmatrix}, \quad \mathbf{S}'_{n_4} = \begin{pmatrix} s_1 & s_2 & \frac{s_3}{\sqrt{2}} & \frac{s_3}{\sqrt{2}} \\ -s_2^* & s_1^* & \frac{s_3}{\sqrt{2}} & -\frac{s_3}{\sqrt{2}} \\ \frac{s_3^*}{\sqrt{2}} & \frac{s_3^*}{\sqrt{2}} & \frac{-s_1 - s_1^* + s_2 - s_2^*}{2} & \frac{s_1 - s_1^* - s_2 - s_2^*}{2} \\ \frac{s_3^*}{\sqrt{2}} & -\frac{s_3^*}{\sqrt{2}} & \frac{s_1 - s_1^* + s_2 + s_2^*}{2} & -\frac{s_1 + s_1^* + s_2 - s_2^*}{2} \end{pmatrix} \tag{3.8}$$

*Figure 3.3. OSTBC performance*

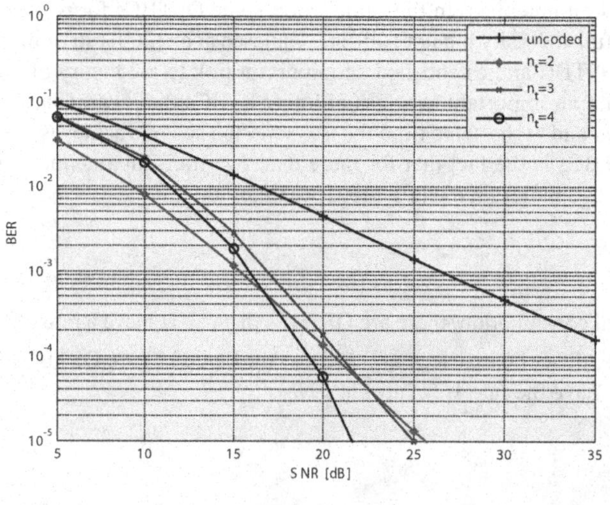

Fig. 3.3 depicts BER for the transmission of 2 bits/channel use using two, three, and four transmit antennas together with an uncoded 4-PSK transmission. The transmission using two transmit antennas employs the 4-PSK constellation and the Alamouti code from Eqn. (3.2). For three and four transmit antennas, the 16-QAM constellation and the codes $\mathbf{S}_{n_3}$ and $\mathbf{S}_{n_4}$ from Eqn. (3.7) are used. It can be seen that at a BER of $10^{-3}$ the rate 1/2 16-QAM code $\mathbf{S}_{n_4}$ gives about 8 dB gain over the use of an uncoded 8-PSK data transmission and at BER=$10^{-4}$ about 2 dB over the codes with two and three transmit antennas.

## Quasi-Orthogonal Space-Time Code (QSTBC)

The main characteristic of the code design methods explained in previous sections is the *orthogonality* of the codes. The codes are designed using such *orthogonal designs* using transmission matrices with orthogonal columns. It has been shown how simple decoding which can separately recover transmit symbols, is possible using an orthogonal design. Tarokh (1999) proved that a complex orthogonal design of STBCs which provides full diversity and full transmission rate is not possible for more than two transmit antennas. Jafarkhani (2001) introduced a new subclass of OSTBC so called Quasi Orthogonal Space-Time Block Codes (QSTBC). These codes achieve full data rate at the expense of a slightly reduced diversity.

In the proposed *quasi-orthogonal* code designs, the columns of the transmission matrix are divided into groups. While the columns within each group are not orthogonal to each other, different groups are orthogonal to each other. As an example of QSTBC for four transmit antennas we give (Jafarkhani, 2001)

$$\mathbf{S}=\begin{pmatrix}\mathbf{S}_1 & \mathbf{S}_2\\ \mathbf{S}_2^* & -\mathbf{S}_1^*\end{pmatrix}=\begin{pmatrix}s_1 & s_2 & s_3 & s_4\\ -s_2^* & s_1^* & -s_4^* & s_3^*\\ s_3^* & s_4^* & -s_1^* & -s_2^*\\ -s_4 & s_3 & s_2 & -s_1\end{pmatrix}, \quad (3.9)$$

where $\mathbf{S}_1$ and $\mathbf{S}_2$ are two (2 × 2) Alamouti codes from Eqn. (3.2) built with symbols ($s_1$, $s_2$) and ($s_3$, $s_4$), respectively. Using this code design, pairs of transmitted symbols can be decoded independently guaranteeing transmission with symbol rate one. The price is the loss of diversity in QSTBC is due to the coupling term between the estimated symbols and *non-orthogonality* of the matrix **S**.

*Figure 3.4. Comparison of OSTBCs and QSTBC*

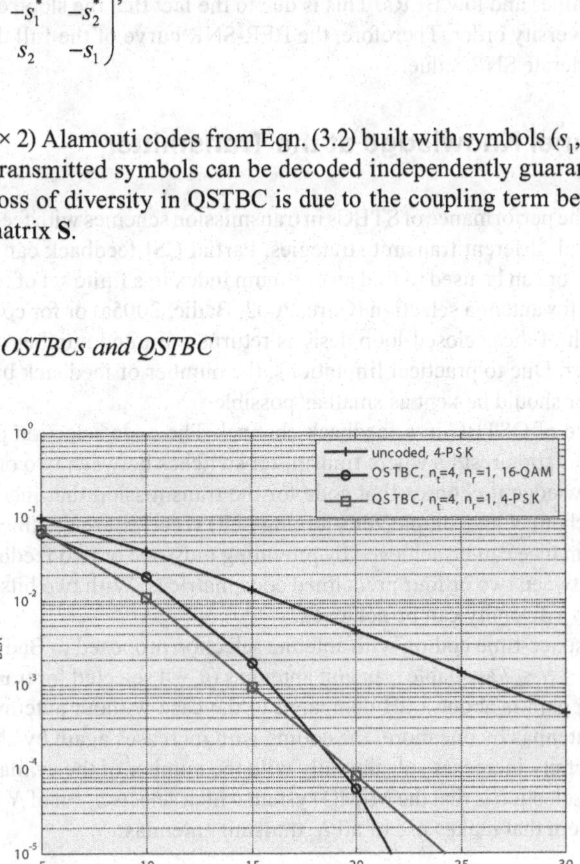

*Figure 3.5. Performance of STBC in a Feedback Channel, QPSK*

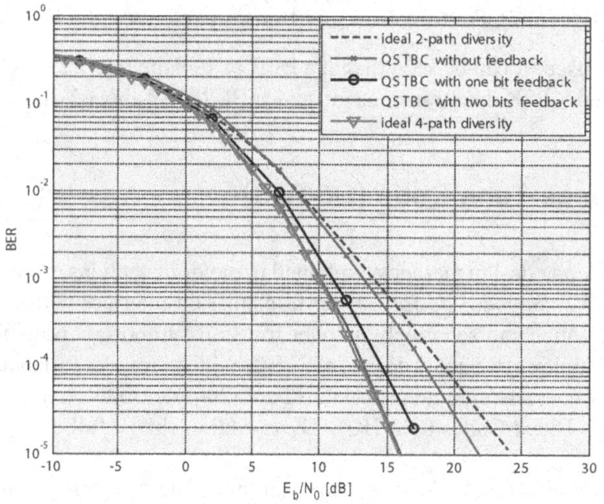


Fig. 3.5 compares rate one QSTBC from Eqn. (3.9) with rate-1/2 full diversity OSTBC from Eqn.(3.7). Simulation results show that full transmission rate is more important at very low SNR values and high BERs, whereas full diversity is the right choice for high SNR values and low BERs. This is due to the fact that the slope of the performance curve at high SNR is determined by the diversity order. Therefore, the BER-SNR curve of the full diversity scheme passes the curve for the QSTBC at some moderate SNR value.

## STBCs with Partial Channel Knowledge at the Transmitter

Designing STBCs and evaluating the performance of STBCs in transmission schemes with feedback has been an intensive area of research resulting in several different transmit strategies. Partial CSI feedback can correspond to a quantized channel estimate (Joengren, 2002), or can be used to find an optimum index in a finite set of precoder matrices (Larsson, 2002; Love, 2003), or can be used for antenna selection (Gore, 2002; Badic, 2005a) or for code selection (Akhtar, 2003; Badic, 2004a; Badic, 2004b). Each of these closed-loop designs returns a limited number of channel information bits from the receiver to the transmitter. Due to practical limitations, the number of feedback bits per code block returned from the receiver to the transmitter should be kept as small as possible.

Fig. 3.5 shows the performance of QSTBC in a feedback channel. The code selection published in Badic (2004b; 2004c) has been applied in the simulation results where transmitter switches between two or more predefined QSTBCs and based on partial channel knowledge it chooses that code for the transmission that increases the system diversity. An (4×1) MISO Rayleigh fading channel and simple zero-forcing (ZF) receiver are assumed. Obviously, a substantial improvement of the system performance can be achieved by providing only one or two feedback bits per code block enabling the transmitter to switch between two or four predefined code matrices. With two bits returned form the receiver to the transmitter, ideal 4-path ($n_t n_r$) diversity can be achieved.

In Fig. 3.6 the performance of space-time coding with antenna selection proposed in Badic (2005a) is shown. Uncorrelated Rayleigh fading channel, $4 \leq N_t \leq 7$ available transmit antennas ($n_t = 4$ selected for transmission), and one receive antenna are assumed. The coding gain is about 3 dB compared to the case without antenna selection. Increasing the number $N_t$ of available transmit antennas by one more, the coding gain increases again by about 1 dB. Most important, Fig. 3.6 shows that the system diversity increases substantially with the number of the available transmit antennas. The big advantage of transmit antenna selection is that the MIMO system which selects $n_t$ out of $N_t$ transmit antennas achieves the same diversity gain as the system that makes use of all $N_t$ transmit antennas.

*Figure 3.6. QSTBCs with Antenna Selection, QPSK*

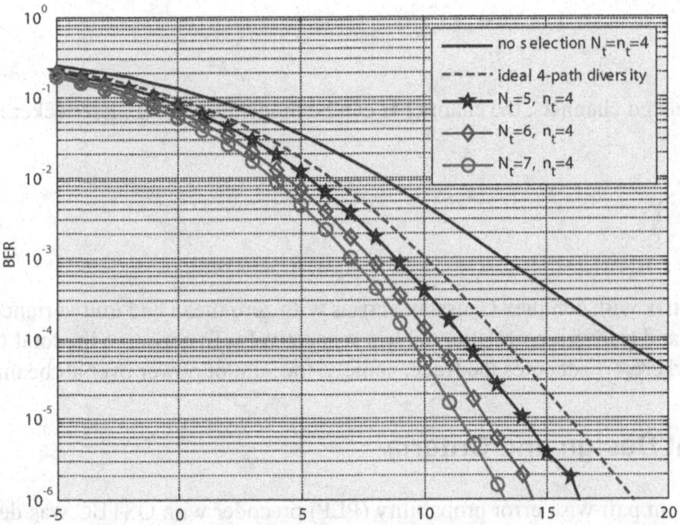


## STBC in Multiuser Scenario

STBC in a multiuser scenario have been studied in (Diggavi, 2001; Barbarossa, 2004; Naguib, 1998; Stamoulis, 2001). With simple linear combiner STBC using Alamouti coding strategy are capable of nulling the interference but price for this is loss in diversity ($d = 2$ instead of $d = 2n_r$). The loss of diversity is only in case of linear interference suppression. In case of non-linear interference suppression (joint optimal multiuser detection) the same performance as a single-user system at the cost of higher receiver complexity can be obtained. The proposed interference cancellers in (Naguib, 1998; Stamoulis, 2001) require only N antennas to suppress (N – 1) interference. These two approaches are conceptually very simple and computationally efficient. However, they require the explicit channel state information to decode the transmitted symbols and suppress interference.

# JOINT BEAMFORMING AND SPACE-TIME CODING

In general, beamforming or linear precoding (LP) exploits spatial correlation, while STC exploits transmit diversity effectively. Thus, it is natural to combine them together to improve the performance over spatially correlated fading channels. In this section we outline a general framework for the joint design of LP and STC in closed-loop MIMO systems. We review some recent works in this area and discuss design of optimal precoder and its application issues. Numerical results illustrate the benefits of joint LP and STC scheme.

## Joint Beamforming and Space-Time Block Coding

STC is usually designed to exploit spatial diversity, but spatial correlation can deteriorate most of their advantages. How to combat fading correlation is an open problem in multiple antenna based wireless communications. Combining the concept of LP with STC can overcome the spatial correlation utilizing multi-beam correlation characteristics. This joint LP and STC approach is illustrated in Fig. 4.1.

The transmitter consists of a space-time (ST) encoder and an LP with weighting matrix **W**. The input data sequence, **C** is mapped to a codeword **S** by the ST encoder, and then linear transformed by precoder **W**. The resulting signal, **WS**, is transmitted over the MIMO channel **H**. The ST codeword is generated independent of the channel, while the precoder adapts the transmitted signal to current channel conditions. Thus, the transmitter must have some knowledge of the channel when designing **W**. With the background noise, **V**, the received signal can be written as

$$\mathbf{Y} = \mathbf{HWS} + \mathbf{V}. \tag{4.1}$$

Assuming spatially correlated channels, the channel **H** can be modelled by (e.g., Kronecker model (Shiu, 2000))

$$\mathbf{H} = \frac{1}{\sqrt{Tr(\mathbf{R}_r)}} \mathbf{R}_r^{1/2} \mathbf{G} (\mathbf{R}_t^{1/2})^T,$$

where **G** an i.i.d. random matrix with complex Gaussian entries with zero mean and unit variance and $\mathbf{R}_r = E[\mathbf{HH}^H]$ and $\mathbf{R}_t = E[(\mathbf{H}^H\mathbf{H})^T]$ are receive and transmit correlation matrix respectively. To conserve the total transmit power, the LP must satisfy $\mathrm{Tr}(\mathbf{WW}^*) = p_0$ where $\mathrm{Tr}(\cdot)$ denotes the trace, namely, the sum of power over all beams must be a constant.

## Approaches for Joint Design and Criteria

In (Sampath, 2002) a minimum pair-wise error probability (PEP) precoder with OSTBC was designed assuming only knowledge of transmit correlation matrix. Joengren (2002) designed a precoder for a predetermined OSTBC based on knowledge of both transmit as well as receive correlation at the transmitter. Based on channel mean feedback Zhou (2002) derived an optimal precoder which turns out to be a generalized beamformer with multi-beams pointing to orthogonal directions along the eigenvectors of the correlation matrix. Coupled with OSTBCs, two-directional eigenbeamformer was designed outperforming conventional eigenbeamformer without rate reduction and without essential increase in complexity. In a similar way in (Liu, 2005) a QSTBC beamformer was presented as a four-directional or eight-directional eigenbeamformer for MIMO systems with four or eight transmit antennas, respectively. Vu (2006) provided a good tutorial of LP for single user MIMO wireless systems. It discusses principles of closed-loop MIMO transmission schemes considering their theoretical as well as practical issues.

In this subsection, we present approaches described in (Bahrami, 2006), where three different criteria are considered to design LP with STBC.

## Minimum Pair-Wise Error Probability Criterion

PEP is the probability that transmitted codeword **c** is erroneously received as different codeword **e** and can be upper-bounded as

$$PEP \le \exp\left(\frac{-E_s}{4N_0}\mathrm{Tr}\left(\mathbf{HW}(\mathbf{c}-\mathbf{e})(\mathbf{c}-\mathbf{e})^H\mathbf{W}^H\mathbf{H}^H\right)\right), \tag{4.2}$$

where $N_0/2$ is the noise variance per dimension, $E_s$ is the average constellation energy. To limit the total transmitted power, the power constraint $\mathrm{Tr}(\mathbf{WW}^H) = n_t$ should hold.

With eigen-decomposion of the nonnegative definite matrices, $\mathbf{R}_t^{1/2}\mathbf{W}(\mathbf{c}-\mathbf{e})(\mathbf{c}-\mathbf{e})^H\mathbf{W}^H(\mathbf{R}_t^{1/2})^H = \mathbf{U}\boldsymbol{\Lambda}\mathbf{U}^H$ and $\mathbf{R}_r^{1/2}(\mathbf{R}_r^{1/2})^H = \mathbf{VDV}^H$, (4.2) can be rewritten as

$$PEP \le \exp\left(\frac{-E_s}{4N_0}\mathrm{Tr}\left(\mathbf{VDV}^H\mathbf{GU}\boldsymbol{\Lambda}\mathbf{U}^H\mathbf{G}^H\right)\right).$$

Suppose that an OSTBC is used. With eigen-decomposition of the nonnegative-definite Hermitian matrices $\mathbf{W}(\mathbf{c}-\mathbf{e})(\mathbf{c}-\mathbf{e})^H\mathbf{W}^H = \mathbf{\Psi}\mathbf{P}\mathbf{\Psi}^H$ and $\mathbf{R}_t = \mathbf{\Phi}\mathbf{\Delta}\mathbf{\Phi}^H$ and due to special structure of OSTBCs, namely, $(\mathbf{c}-\mathbf{e})(\mathbf{c}-\mathbf{e})^H = \alpha\mathbf{I}$, where $\alpha$ is a positive scalar, further simplifications are possible. With the modified power constraint, $\mathrm{Tr}(\mathrm{P}) = \alpha n_t$, we can have $\mathbf{\Psi}\mathbf{P}\mathbf{\Psi}^H = \mathbf{W}\mathbf{W}^H$. Thus, the optimal precoder that minimizes the average PEP can be obtained by solving the following optimization problem:

$$\max_P \det\left(\mathbf{I} + \frac{E_s}{4N_0}(\mathbf{D}\otimes\mathbf{\Lambda})\tilde{\mathbf{U}}(\mathbf{I}\otimes\mathbf{P})\tilde{\mathbf{U}}^H\right)$$

$$\text{s.t.} \quad \mathrm{Tr}(\mathbf{P}) = \alpha n_t. \tag{4.3}$$

Since $\tilde{\mathbf{U}} = \mathbf{I}\otimes(\mathbf{\Phi}^H\mathbf{\Psi})$ is a unitary matrix it can be set as $\tilde{\mathrm{U}} = \mathbf{I}$ and $\mathbf{\Psi} = \mathbf{\Phi}$. The resulting optimal precoding matrix can be interpreted as an eigenbeamformer with orthogonal beams pointing to the eigenvectors of the transmit correlation matrix, called transmit eigenbeams, (which are the column vectors of $\mathbf{\Phi}$) and the solution of the optimization problem in Eqn. (4.3) is

$$W_{opt} = (\alpha^{1/2})\mathbf{\Phi}\mathbf{P}^{1/2}\mathbf{\Gamma}, \tag{4.4}$$

where $\mathbf{\Gamma}$ is an arbitrary unitary matrix which has no effect on the system performance and therefore can be set to identity.

The optimal power allocation across the beams varies for each design and it depends on SNR and requires complex numerical methods in the case of partial feedback. For the optimal PEP LP, it can be obtained by the maximization problem,

$$\max_P \det\left(\mathbf{I} + \frac{E_s}{2N_0}(\mathbf{D}\otimes\mathbf{\Lambda})(\mathbf{I}\otimes\mathbf{P})\right),$$

with the constraint, $\mathrm{Tr}(\mathbf{P}) = \alpha n_t$ The approximated solution is

$$p_i = \left(\nu - \left(\frac{E_s}{4N_0}\frac{1}{n_t}\mathrm{Tr}(\mathbf{D})\delta_i\right)^{-1}\right)^+, \tag{4.5}$$

where $\delta_i$ are the diagonal entries of $\mathbf{\Delta}$, $\nu$ is a constant determined by the power constraint, and $(x)^+ = \max(0,x)$.

## Minimum MSE Criterion

The mean-square error (MSE) is defined as $MSE = E[(C'-C)(C'-C)^H]$ where $C$ is transmitted codeword and $C'$ is detected at the receiver and for the system in Fig. 4.1 can be expressed as

$$MSE = \mathrm{Tr}\left(E_s\left(\frac{E_s}{N_0}\mathbf{W}^H\mathbf{H}^H\mathbf{H}\mathbf{W}\right)^{-1}\right).$$

Similar to the derivation of PEP, the MSE can be found as

$$\overline{MSE} = E_s \mathrm{Tr}\left(\mathbf{I} - \frac{E_s}{2N_0}(\mathbf{D} \otimes \mathbf{\Lambda})\right). \tag{4.6}$$

To obtain the MMSE LP, we need to minimize

$$\mathrm{Tr}\left(\mathbf{I} - \frac{E_s}{2N_0}(\mathbf{D} \otimes \mathbf{\Lambda})\right)$$

or to maximize

$$\mathrm{Tr}\left(\frac{E_s}{2N_0}(\mathbf{D} \otimes \mathbf{\Lambda})\right).$$

The optimal LP is the same as that for PEP as follows:

$$W_{opt} = \left(\alpha^{1/2}\right)\mathbf{\Phi}\mathbf{P}^{1/2}\mathbf{\Gamma}. \tag{4.7}$$

The solution for the optimal power across the beams can be obtained by solving

$$\max_{P} \mathrm{Tr}\left(\frac{E_s}{2N_0}(\mathbf{D} \otimes \mathbf{\Lambda})(\mathbf{I} \otimes \mathbf{P})\right)$$

subject to $\mathrm{Tr}(\mathbf{P}) = \alpha n_t$ and its solution is

$$p_i = \left(\nu \delta_i^{-1/2} - \left(\frac{E_s}{4N_0}\frac{1}{n_t}\mathrm{Tr}(\mathbf{D})\delta_i\right)^{-1}\right)^{+}. \tag{4.8}$$

Both, MMSE and PEP criteria lead to the same optimal LP designs expect for the power loadings. However, the MMSE criterion does not assume orthogonality and thus is applicable to both general coded and uncoded MIMO systems.

## Maximum Ergodic Channel Capacity Criterion

The capacity optimal precoder is designed to achieve maximum transmission rate and it is a solution of the following optimization problem:

$$\max_{P} E\left\{\log_2\left(\det\left(\mathbf{I} + \frac{1}{N_0}\mathbf{W}^H\mathbf{H}^H\mathbf{H}\mathbf{W}\right)\right)\right\} \tag{4.9}$$

with the constraint, $\mathrm{Tr}(\mathbf{P}) = \alpha n_t$. After some manipulations, it turns out that the optimal LP matrix is an eigenbeamformer with the transmit eigenbeams and it is given as

*Table 4.1. Summary of precoding criteria*

| Criterion | Optimization Problem | Precoder | Power Allocation |
|---|---|---|---|
| PEP | $\max_{tr(\mathbf{P})=\sigma n_t} \det\left(\mathbf{I}+\frac{E_s}{4N_0}(\mathbf{D}\otimes\Lambda)\tilde{\mathbf{U}}(\mathbf{I}\otimes\mathbf{P})\tilde{\mathbf{U}}^H\right)$ | $W_{opt}=\left(\alpha^{1/2}\right)\otimes\Phi\mathbf{P}^{1/2}\Gamma$ | $p_i=\left(\nu-\left(\frac{E_s}{4N_0}\frac{1}{n_t}\mathrm{Tr}(\mathbf{D})\delta_i\right)^{-1}\right)^+$ |
| MMSE | $\min_{tr(\mathbf{P})=\sigma n_t} E_s\mathrm{Tr}\left(\mathbf{I}-\frac{E_s}{2N_0}(\mathbf{D}\otimes\Lambda)\right)$ | $W_{opt}=\left(\alpha^{1/2}\right)\otimes\Phi\mathbf{P}^{1/2}\Gamma$ | $p_i=\left(\nu\alpha_i^{-1/2}-\left(\frac{E_s}{4N_0}\frac{1}{n_t}\mathrm{Tr}(\mathbf{D})\delta_i\right)^{-1}\right)^+$ |
| Channel Capacity | $\max_{tr(\mathbf{P})=\sigma n_t} E\left\{\log_2\left(\det\left(\mathbf{I}+\frac{1}{N_0}\mathbf{W}^H\mathbf{H}^H\mathbf{H}\mathbf{W}\right)\right)\right\}$ | $W_{opt}=\Phi\mathbf{P}^{1/2}\Gamma$ | $p_i=\left(\nu-\left(N_0\frac{\det(\mathbf{D})}{\mathrm{Tr}(\mathbf{D})\delta_i}\right)\right)^+$ |

*Figure 4.2. Performance of joint LP and OSTBC with rate-3/4 OSTBCs and ρ = 0.9*

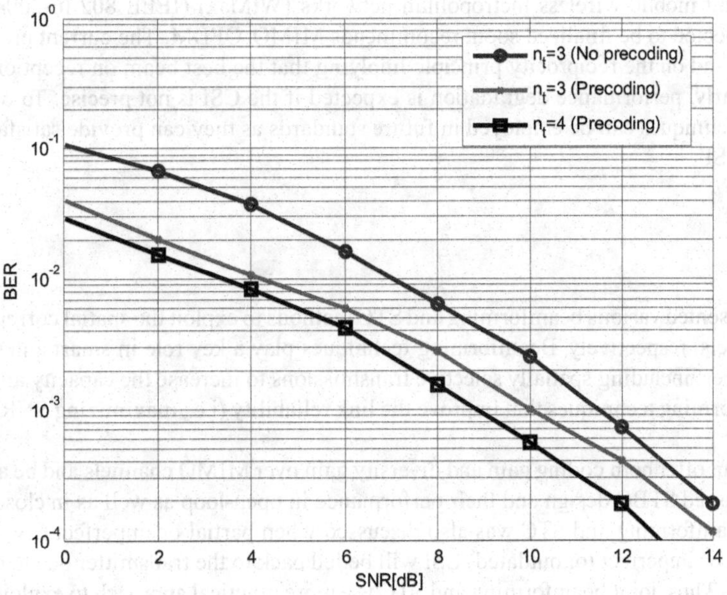


$$W_{opt} = \mathbf{\Phi P}^{1/2}\mathbf{\Gamma} \tag{4.10}$$

with the optimal power across the beams expressed as

$$p_i = \left(\nu - \left(N_0 \frac{\det(\mathbf{D})}{\mathrm{Tr}(\mathbf{D})\delta_i}\right)\right)^+. \tag{4.11}$$

As can be seen from above, the optimal LP has two key roles:

1) *decoupling the signals into orthogonal spatial directions and*
2) *allocating power to these directions.*

Thus, the optimal LPs can be seen as the eigenbeamformers which transmit the signal along the eigenvectors of the transmit correlation matrix. It turned out that the optimal LPs for all three criteria are the same in utilizing transmit eigenbeams, while the main difference among the LP solutions is the power allocation. Table 4.1 summarizes LP designs and power allocation across the eigenbeams.

The performance of the PEP LP is shown in Fig. 4.2 for QPSK modulated symbols. A spatially correlated Rayleigh fading MIMO channels (of $\rho = 0.9$ in (2.8)) with $n_t = 3, 4$ transmit antennas, $n_r = 2$ receive antennas and rate-3/4 OSTBC is considered. It can be seen from the simulation results that the PEP precoder has up to 3dB gain compared to case without precoding. Furthermore, the precoding gain increases with the number of transmit antennas up to 1dB.

## Application Issues

Limited feedback techniques have already been integrated in 3G cellular standards. These techniques are available for use by the transmit adaptive array (TXAA) mode in the closed-loop portion of the 3G Partnership Project (3GPP) standard providing both beamforming as well as diversity gain (Derryberry, 2002). However, the gain reduces in mobile scenarios due to inaccurate/outdated CSI. Some precoding techniques have been integrated into the IEEE 802.16e standard for broadband mobile wireless metropolitan networks (WiMax) (IEEE 802.16-2004, 2006). IEEE 802.11.n standard which is expected to be finalized soon, recommends MIMO-OFDM. The current precoding proposals use an open-loop method, based on the reciprocity principle implying that the best beam on reception must be the best beam for transmission. Clearly, performance degradation is expected if the CSI is not precise. To overcome imperfect CSI, joint LP and STBC techniques can be employed in future standards as they can provide satisfactory performance with partial or imperfect CSI.

## CONCLUSION

In this chapter, we presented various beamforming and STC methods to exploit the spatial correlation as well as diversity gain of MIMO channels, respectively. Beamforming techniques play a key role in smart antenna systems as they can provide various features, including spatially selective transmissions to increase the capacity and coverage increase. We focused on the beamforming techniques that improve the link reliability (i.e., maximizing SNR) with partial or full CSI of MIMO channels.

STC techniques can offer both coding gain and diversity gain over MIMO channels and be applied without knowing CSI. We mainly discussed STBC design and their performance in open-loop as well as in closed-loop MIMO systems. Furthermore, joint beamforming and STC was also discussed when partial or imperfect CSI is available. In practical situations, partial and/or imperfect (or outdated) CSI will be fed back to the transmitter due to channel estimation error and delay in feedback. Thus, joint beamforming and STC is a more practical approach to exploit both spatial correlation and diversity gain of MIMO channels. We believe that joint design will be actively employed in future standards for wireless communications.

## REFERENCES

Alamouti S.M. (1998). A simple diversity technique for wireless communications, *IEEE Journal on Selected Areas in Communications*, Vol. 16, No. 8, pp.1451-1458.

Akhtar J. & Gesbert D. (2003). Partial Feedback Based Orthogonal Block Coding, *The 57th IEEE Semiannual IEEE Vehicular Technology Conference, VTC 2003-Spring*, Vol. 1, pp. 287-291, Jeju, Korea.

Badic B., Herdin M., Gritsch G., Rupp M. & Weinrichter H. (2004a). Performance of various data transmission methods on measured MIMO channels, *The 59th IEEE Vehicular Technology Conference VTC 2004- Spring*, Vol. 2, pp. 651 – 655, Milan, Italy.

Badic B., Rupp M. & Weinrichter H. (2004b). Extended Alamouti codes in correlated channels using partial feedback, *IEEE International Conference on Communication ICC'04*, Vol. 2, pp. 896 - 900, Paris, France.

Badic B., Rupp M. & Weinrichter H. (2004c). Adaptive Channel-Matched Extended Alamouti Space-Time Code Exploiting Partial Feedback, *ETRI-Electronics and Telecommunication Research Institute Journ*al, Vol. 26, No. 5, pp. 443 - 451.

Badic B., Fuxjäger P. & Weinrichter H. (2005a). Optimization of Coded MIMO-Transmission with Antenna Selection, *The 61st IEEE Vehicular Technology Conference, VTC Spring 2005*, Vol. 2, pp. 905-909, Stockholm, Schweden.

Badic B., Rupp M. & Weinrichter H. (2005b). Quasi-Orthogonal Space-Time Block Codes: Approaching Optimality, *Thirteen European Signal Processing Conference*, Antalya, Turkey, September (invited paper).

Bahrami H. R. & Le-Ngoc T. (2006). Precoder design based on correlation matrices for MIMO systems, *IEEE Transactions on Wireless Communi*cation, Vol. 5, pp. 3579-3587.

Barbarossa S. (2004). Trace-orthogonal design of MIMO systems with simple scalar detectors, full diversity and (almost) full rate, *V IEEE Signal Processing Workshop on Signal Processing Advances in Wireless Communication, SPAWC 2004*, pp. 308- 312, Lisbon, Portugal.

Bolcskei H. & Paulraj A. J. (2000). Performance of space-time codes in the presence of spatial fading correlation, *Thirty-Fourth Asilomar Conference on Signals, Systems and Computers*, Vol. 1, pp. 687-693, Pacific Grove, CA.

Choi, J. (2002a). Performance analysis for transmit antenna diversity with/without channel information, *IEEE Transactions on Vehicular Technology*, Vol. 51, pp. 101-113.

Choi, J. (2002b). A semiblind method for transmit antenna arrays for CDMA Systems," *IEEE Transactions on Vehicular Technology*, vol. 51, pp. 624-635.

Choi, J., Yuan, J., Kim, S.R., Choi, I.-K., & Kwon, D.-S. (2004a) Coded downlink eigenbeamforming without feedback, *The 59th IEEE Vehicular Technology Conference VTC 2004- Spring*, Vol. 2, pp. 794-798, Milan, Italy.

Choi, J. & Yuan, J. (2004b). On the beamformer design for coded signals with known channel, *IEEE International Conferene on Communication ICC'04*, Vol. 5, pp. 2777-2781 Paris, France.

Choi, J. & Perreau, S. (2004c). MMSE multiuser downlink multiple antenna transmission for CDMA systems, *IEEE Transaction on Signal Processing*, Vol.52, pp.1564-1573.

Choi, J., Kim, S. R., & Choi, I.-K., (2007). Statistical eigenbeamforming with selection diversity for spatially correlated OFDM downlink, *IEEE Transactions on Vehicular Technology*, (to be published).

Derryberry R.T (2002). Transmit Diversity in 3G CDMA Systems, *IEEE Communications Magazine*, Vol. 40, pp. 1423-1436.

Diggavi. S. N. Al-Dhahir N. & Calderbank A. R. (2001). Algebraic properties of space-time block codes in intersymbol interference multiple-access channels, *IEEE Transactions on Information Theory*, Vol. 49, Issue 10, pp. 2403-2414.

Dighe, P. A., Mallik, R. K., & Jamuar, S. R. (2001), Analysis of transmit-receive diversity in Rayleigh fading, in *Proc. IEEE Global Telecommunication Conference, GLOBECOM*, San Antonio, TX, pp. 1132-1136.

Foschini G.J & Gans M.J. (1998). On limits of wireless communications in a fading environment when using multiple antennas, *Wireless Personal Communications*, Vol. 6, No. 3, pp. 311-335.

Ganesan, G. & Stoica, P. (2001). Space-time diversity using orthogonal and amicable orthogonal design, *Wireless Personal Communications,* Vol. 18, pp. 165-178, 2001.

Gerlach, D. & Paulraj, A. (1996). Base station transmitting antenna arrays for multipath environments, *Signal Processing archive*, Vol. 54, pp. 59-73.

Gore D.A. & Paulraj A.J. (2002). Mimo Antenna Subset Selection with Space-Time Coding, *IEEE Transactions on Signal Processing*, Vol. 50, Issue: 10, pp: 2580 – 2588.

Hiroike, A., Adachi, F., & Nakajima, N. (1992) Combined effects of phase sweeping transmitter diversity and channel coding, *IEEE Transactions on Vehicular Technology*, Vol. 41, pp. 170-176.

Jafarkhani, H. (2001). A Quasi-Orthogonal Space-Time Block Code, *IEEE Transactions on Communications*, Vol. 49, Issue: 1, pp: 1 – 4.

Joengren G., Skoglund M. & Ottersten B., (2002). Combining Beamforming and Orthogonal Space-Time Block Coding, *IEEE Transactions on Information Theory*, **Vol. 48, No. 3, pp. 611-627.**

Kang, M. & Alouini, M.-S. (2003). Largest Eigenvalue of Complex Wishart Matrices and Performance Analysis of MIMO MRC Systems, *IEEE Journal on Selected Areas in Communications*, Vol. 21, No. 3, pp: 418 – 426.

Larsson, G., Ganesan, G. & Stoica, P. (2002). On the Performance of Orthogonal Space-Time Block Coding with Quantized Feedback, *IEEE Communications Letters.*, Vol. 6, No. 11, pp. 487-489.

Liu L & Jafarkhani H. (2005). Application of Quasi-Orthogonal Space-Time Block Codes in Beamforming, *IEEE Transactions on Signal Processing,* Vol. 53, No. 1, pp. 54- 63.

Lo, T. K. Y. (1999). Maximum Ratio Transmission, *IEEE Transactions on Communications*, Vol. 47, No. 10, pp: 1458 – 1461.

Love D. & Heath R. (2003). Grassmannian beamforming for multiple-input multiple-output wireless systems, *IEEE Transactions on Information Theory*, Vol.49, pp. 2735-2747.

Morgan, D. R. (2003). Downlink adaptive array algorithms for cellular mobile communications, *IEEE Transactions on Communications*, Vol. 51, Issue: 3, pp: 476 – 488.

Mukkavilli K., Sabharwal A., Erkip E. & Aazhang B. (2003). On beamforming with finite rate feedback in multiple antenna systems, *IEEE Transactions on Information Theory*, Vol. 49, No. 10, pp. 2562-2579.

Mukkavilli K., Sabharwal A. & Aazhang B, (2001). Design of Multiple Antenna Coding Schemes with Channel Feedback", The *Asilomar Conference on Signals, Systems, and Computers*, pp. 1009-1013, Pacific Grove, CA.

Naguib A.F., Seshadri N. & Calderbank A.R. (1998). Applications of space-time block codes and interference suppression for high capacity and high data rate wireless systems," *The Asilomar Conference on Signals, Systems and Computers*, Pacific Grove, CA. November.

Rashid-Farrokhi, F. Liu, K. J. R., & Tassiulas, L. (1998). Transmit beamforming and power control for cellular wireless systems, *IEEE Journal on Selected Areas in Communications*, Vol. 16, No. 8, pp.1437-1450.

Rupp M. & Mecklenbrauker C.F. (2002). *On Extended Alamouti Schemes for Space-Time Coding, The 5th International Symposium on Wireless Personal Multimedia Communications*, Vol. 1, pp. 115 – 119, Sheraton Waikiki, Honolulu, Hawaii.

Sampath H. & Paulraj A. (2002). Linear precoding for space-time coded systems with known fading correlations, *IEEE Communication Letters,* Vol. 6, pp. 239-241.

Salz, J. & Winters, J. (1994). Effect of fading correlation on adaptive arrays in digital mobile radio, IEEE *Transactions Vehicular Technology*, Vol. 43, pp. 1049-1057.

Schubert, M. & Boche, H. (2004). Solution of the Multiuser Downlink Beamforming Problem with Individual SINR Constraints, *IEEE Transactions on Vehicular Technology*, Vol. 53, pp. 18-28.

Shiu D., Foschini G. J., Gans M. J., & Kahn J. M. (2000). Fading Correlation and Its Effect on the Capacity of Multielement Antenna Systems", *IEEE Transactions on Communications*, Vol. 48, No. 3, pp. 502-513.

Stamoulis A., Al-Dhahir N. & A. R. Calderbank (2001) Further results on interference cancellation and space-time block codes", *The Asilomar Conference on Signals, Systems and Computers*, Pacific Grove, CA., 2001.

Tarokh V., Seshadri N. & Calderbank A. R. (1998). Space-time codes for high data rate wireless communication: performance criterion and code construction, *IEEE Transactions on Information Theory,* Vol. 44, No. 2, pp. 744-765.

Tarokh, V., Jafarkhani, H. & Calderbank, A.R. (1999a). Space-time block coding for wireless communications: Performance results, *IEEE Journal Selected Areas in Communication*, Vol.17, No. 3, pp. 451-460.

Tarokh V., Jafarkhani H. & Calderbank A.R. (1999b). Space-time block codes from orthogonal designs, *IEEE Transactions on Information Theory*, Vol. 45, pp. 1456-1467.

Telatar I.E. (1999). Capacity of multi-antenna Gaussian channels", *European Transactions on Telecommunications*, Vol. 10, No. 6, pp. 585-595, 1999.

Tirkkonen O. & Hottinen A.( 2002 ). Complex Space-Time Block Codes for four TX antennas, *IEEE Global Telecommunication Conference 2002*, Vol. 2, pp. 1005-1009, San Francisco, CA, USA.

Viswanath, P., Tse, D., & Laroia, R. (2002) Opportunistic beamforming using dumb antennas, *IEEE Transactions on Information Theory*, Vol. 48, No. 6, pp.1277-1294.

Winters, J. H., (1998). Smart Antennas for Wireless Systems, *IEEE Personal Communications*, vol. 5, pp. 23-27.

Vojcic, B. R. & Jang, W. M. (1998). Transmitter Precoding in Synchronous multiuser communications, *IEEE Transactions on Communications*, Vol. 46, No. 10, pp: 1346 – 1355.

Vu M. & Paulraj A. (2006) MIMO Wireless Linear Precoding, accepted to *IEEE Signal Processing Magazine*, December 2006.

Zhou S., Wang Z. & Giannakis G.B. (2004). Performance analysis for transmit-beamforming with finite rate feedback, *Conference on Information Sciences and Systems,* Princeton University, Princeton, NJ.

Zhou S. & Giannakis G.B.( 2002). Optimal Transmitter Eigen-Beamforming and Space-Time Block Coding Based on Channel Mean Feedback, *IEEE Transaction on Signal Processing,* Vol. 50. No. 10., pp. 2599-2612.

IEEE 802.16-2004 (2006). IEEE Standard for Local and metropolitan area networks Part 16: Air Interface for Fixed and Mobile Broadband Wireless Access Systems Amendment 2: Physical and Medium Access Control Layers for Combined Fixed and Mobile Operation in Licensed Bands and Corrigendum 1, 2006.

# Chapter XIV
# Adaptive MIMO Systems with High Spectral Efficiency

**Zhendong Zhou**
*University of Sydney, Australia*

**Branka Vucetic**
*University of Sydney, Australia*

## ABSTRACT

*This chapter introduces the adaptive modulation and coding (AMC) as a practical means of approaching the high spectral efficiency theoretically promised by multiple-input multiple-output (MIMO) systems. It investigates the AMC MIMO systems in a generic framework and gives a quantitative analysis of the multiplexing gain of these systems. The effects of imperfect channel state information (CSI) on the AMC MIMO systems are pointed out. In the context of imperfect CSI, a design of robust near-capacity AMC MIMO system is proposed and its good performance is verified by simulation results. The proposed adaptive system is compared with the non-adaptive MIMO system, which shows the adaptive system approaches the channel capacity closer.*

## INTRODUCTION

High data rate communications have been one of the focuses in the telecommunication field because of the fast increase in demand for transmitting and exchanging information intense contents, such as multimedia, interactive and real-time materials. At the same time, the radio frequency spectrum is becoming a rare resource, as more and more services are brought into the world and the bandwidth demands for most services are increasing. Research in wired/wireless communications with high spectral efficiency has been gaining increasingly intense efforts since the capacity findings of Claude E. Shannon in 1948.

In the past decade, the pioneering work by Foschini (1998) and Telatar (1999) has proven that multiple-input multiple-output (MIMO) systems have significant higher capacity than conventional single-input single-output (SISO) systems. However, how to realize the high spectral efficiency promise of MIMO systems in a realistic application is still an open question.

In this chapter, the adaptive modulation and coding (AMC) technique is chosen as the candidate to approach the high spectral efficiency promise of MIMO systems. AMC technique, although in a simplified form, has gained its position in real-world applications, such as WiMAX. AMC was originally proposed in SISO systems to combat detrimental fading conditions. It adapts some of the system parameters according to the channel fading conditions. We give a systematic view of the integration of AMC into MIMO systems and present the possibility of achieving near capacity of MIMO systems with AMC by an example, the adaptive turbo-coded MIMO system.

# BACKGROUND

## MIMO Channel Model

Consider a point-to-point wireless communication channel equipped with $n_T$ transmit and $n_R$ receive antennas. This channel can be modeled by a multiple-input multiple-output (MIMO) system, which can be represented as a baseband channel matrix **H** of size $n_R \times n_T$. The entry of **H** at the *j*-th row and *i*-th column, $n_{j,i}$, represents the complex channel coefficient relating to the *i*-th transmit antenna and the *j*-th receive antenna. We assume that the antennas are placed at an enough distance between each other, such that the channel link between each transmit-receive antenna pair experiences independent fading. The fading is modeled as flat (non-frequency selective), since frequency selective fading can be combated by the orthogonal frequency division multiplexing (OFDM) technique.

The fading distribution is assumed to be Rayleigh, or equivalently, complex Gaussian, which represents the worst case fading situation. Proper normalization is assumed such that each entry $H_{j,i}$ can be modeled as a complex Gaussian random process with zero mean and unity variance (variance 0.5 for both real and imaginary parts). We denote this as $H_{j,i} \sim \mathcal{CN}(0,1)$, where $\mathcal{CN}(\mu,\sigma^2)$ denotes complex Gaussian distribution with mean $\mu$ and variance $\sigma^2$.

The input to the MIMO channel is represented by an $n_T \times 1$ vector $\mathbf{x} = [x_1, x_2, \cdots, x_{n_T}]^T$, where $x_i$ denotes the transmitted (modulated) symbol from the *i*-th transmit antenna. The output from the MIMO channel is represented by an $n_R \times 1$ vector $\mathbf{y} = [y_1, y_2, \cdots, y_{n_R}]^T$, where $y_j$ denotes the received symbol from the *j*-th receive antenna. The additive white Gaussian noise (AWGN) is represented by an $n_R \times 1$ vector $\mathbf{n} = [n_1, n_2, \cdots, n_{n_R}]^T$ where $n_j$ denotes the equivalent complex noise at the *j*-th receive antenna. The channel equation can be written as:

$$\mathbf{y} = \mathbf{H}\mathbf{x} + \mathbf{n}. \tag{1}$$

Graphically, this can be illustrated by Figure 1.
For convenience, we define

$$n = \max(n_T, n_R)$$
$$m = \min(n_T, n_R)$$
$$d = n - m.$$

We apply a singular value decomposition (SVD) to the matrix **H** and get

$$\mathbf{H} = \mathbf{U}\mathbf{D}\mathbf{V}^H \tag{2}$$

where **D** is an $n_R \times n_T$ diagonal matrix. Its diagonal entries are the singular values of **H**, that is, the non-negative square roots of the eigenvalues of $\mathbf{HH}^{\mathrm{H}}$. We denote the eigenvalues of $\mathbf{HH}^{\mathrm{H}}$ as $\lambda_1, \lambda_2, \cdots, \lambda_m$ thus

*Figure 1. MIMO channel model*

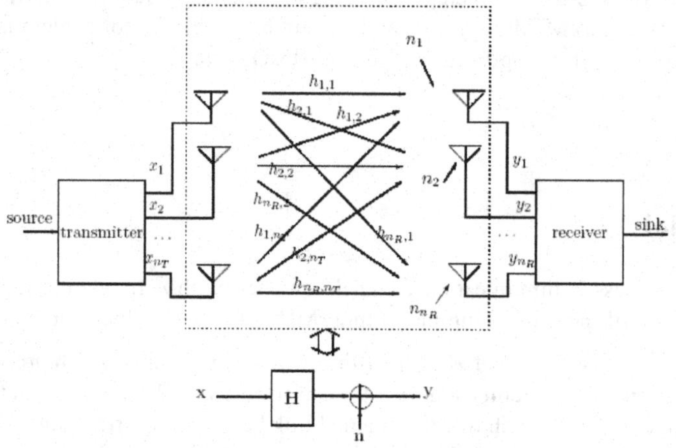


$$\mathbf{D}=\begin{cases}\begin{bmatrix}\sqrt{\lambda_1} & 0 & \cdots & 0 & 0 & \cdots & 0\\ 0 & \sqrt{\lambda_2} & \cdots & 0 & 0 & \cdots & 0\\ \cdots & & & & & & \\ 0 & 0 & \cdots & \sqrt{\lambda_m} & 0 & \cdots & 0\end{bmatrix}, & n_T \geq n_R\\ \begin{bmatrix}\sqrt{\lambda_1} & 0 & \cdots & 0\\ 0 & \sqrt{\lambda_2} & \cdots & 0\\ \cdots & & & \\ 0 & 0 & \cdots & \sqrt{\lambda_m}\\ 0 & 0 & \cdots & 0\\ \cdots & & & \\ 0 & 0 & \cdots & 0\end{bmatrix}, & n_T < n_R\end{cases}$$

$\mathbf{U}=[\mathbf{u}_1,\cdots,\mathbf{u}_{n_R}]$ and $\mathbf{V}=[\mathbf{v}_1,\cdots,\mathbf{v}_{n_T}]$ are $n_R \times n_R$ and $n_T \times n_T$ unitary matrices with left and right singular vectors of **H** as their columns, respectively. $\mathbf{u}_i$ and $\mathbf{v}_i$ are also called the eigenvectors of $\mathbf{HH}^{\mathrm{H}}$ and $\mathbf{H}^{\mathrm{H}}\mathbf{H}$ associated with $\lambda_i$, respectively. The eigenvalues and eigenvectors are related as

$$\mathbf{HH}^H\mathbf{u}_i=\lambda_i\mathbf{u}_i$$

$$\mathbf{H}^H\mathbf{H}\mathbf{v}_i=\lambda_i\mathbf{v}_i,\ i=1,\cdots,m.$$

Substituting Eq. (2) into (1), we obtain

$$\mathbf{y}=\mathbf{UDV}^H\mathbf{x}+\mathbf{n}$$
$$\mathbf{U}^H\mathbf{y}=\mathbf{DV}^H\mathbf{x}+\mathbf{U}^H\mathbf{n}$$
$$\mathbf{y'}=\mathbf{Dx'}+\mathbf{n'} \quad (3)$$

where

$$\mathbf{y'}=\mathbf{U}^H\mathbf{y}$$
$$\mathbf{x'}=\mathbf{V}^H\mathbf{x}$$
$$\mathbf{n'}=\mathbf{U}^H\mathbf{n}.$$

Eq. (3) readily shows a parallel channel model. To put it more explicitly, we can rewrite Eq. (3) in a component-wise form

$$y_i' = \sqrt{\lambda_i}\, x_i' + n_i', \ i = 1, \cdots, m \tag{4}$$

$$y_i' = n_i', \ n_T < n_R \text{ and } i = m+1, \cdots, n_R \tag{5}$$

Eq. (4) and (5) are further illustrated in Figure 2.

This clearly shows that the MIMO channel has been converted into *m* parallel subchannels. The equivalent channel input and output are x' and y', respectively. The subchannel power gains are $\{\lambda_i\}_{i=1}^{m}$.

## MIMO Channel Capacity

Telatar (1999) derived an elegant capacity expression for the MIMO system described in Eq. (1) when the channel state information (CSI) is known to the receiver but not to the transmitter.

$$C_{\text{no-CSI}} = mE_{\lambda_1}\left[\log_2\left(1 + \frac{P_t}{n_T \sigma^2}\lambda_1\right)\right] \tag{6}$$

where the expectation is carried over the unordered eigenvalue $\lambda_1$ and $P_t$ presents the average transmit power limit. Here the concept of ordered and unordered eigenvalue distribution is worth of some attention. Generally, the SVD algorithm produces sorted eigenvalues. Hence, each eigenvalue $\lambda_i$ should be treated as a random variable with a distribution different from others. This is called an ordered eigenvalue distribution. In contrast, if we do not sort the eigenvalues, or equivalently, if we randomly assign the order to them, each eigenvalue will have identical distribution. This is called an unordered eigenvalue distribution.

It has been shown that in high SNRs, Eq. (6) can be approximated by

$$C_{\text{no-CSI}} \approx m\log_2\left(\frac{P_t}{n_T \sigma^2}\right) + B = m\log_2\left(\frac{\gamma_s}{n_T}\right) + B \tag{7}$$

where *B* is a constant, and we define the signal-to-noise ratio (SNR) as

$$\gamma_s = \frac{P_t}{\sigma^2}.$$

From Eq. (7) we can see the relationship between the SNR $\gamma_s$ and the channel capacity $C_{\text{no-CSI}}$: for every 3 dB increase in the SNR, the capacity increases by *m* bits/s/Hz.

As we know that for a SISO system, the corresponding figure of merit is only 1 bit/s/Hz. The factor *m* in the MIMO system is hence defined as the *multiplexing gain*.

Note that the multiplexing gain is one of the measures that determine the theoretical spectral efficiency upper bound (the channel capacity) of a MIMO system in high SNRs. It represents the *incremental* relationship between SNR and capacity. If we plot the capacity versus SNR curve in a figure, the multiplexing gain reflects the slope of the curve.

Eq. (7) is for a MIMO system without transmitter side CSI. Following a similar approach, we can derive an asymptotic capacity formula for a MIMO system with transmitter side CSI.

$$C_{\text{CSI}} \approx m\log_2\left(\frac{\gamma_s}{m}\right) + B \tag{8}$$

Comparing Eq. (7) with (8), we can see the only difference between the systems with and without transmitter CSI is the factors dividing $\gamma_s$. In the system without transmitter CSI, $\gamma_s$ is divided by $n_T$. Physically, this means that the total transmit power $P_t$ has to be equally distributed across $n_T$ transmit antennas. In contrast, in the system with transmitter CSI, $\gamma_s$ is divided by *m*, which means that the total transmit power $P_t$ is equally distributed across *m* subchannels.

In a MIMO system with $n_T \le n_R$, $m = \min(n_T, n_R) = n_T$, and hence Eq. (7) and (8) are identical. On the contrary, if $n_T > n_R$, $m = \min(n_T, n_R) = n_R < n_T$. Thus the transmitter CSI brings an SNR gain of $n_T/m$ folds, or $10\lg(n_T / m)$ dB. In other words, to achieve the same capacity, a MIMO system with transmitter CSI needs an SNR of $10\lg(n_T / m)$ dB lower than the corresponding system without transmitter CSI.

This is particularly interesting in wireless downlink (base station to mobile station). In that case, the cost of multiple transmit antennas is much lower than that of multiple receive antennas. For instance, we can feasibly install 8 antennas in a base station, while 2 antennas in a mobile station. In this scenario, the SNR benefit brought by transmitter CSI is $10\lg(8/2) \approx 6$ dB, which means in the same noise and fading situation, three quarters of transmit power can be saved.

On the other hand, the above discussion does not indicate that the transmitter CSI is of no value when $n_T \le n_R$. Although the capacity expressions Eq. (7) and (8) are identical, we must keep in mind that the channel capacity is only a theoretical upper bound on the achievable spectral efficiency of a practical communication system. As we will show that even if they have identical capacity, how to approach this upper limit is a totally different story. The transmitter CSI can play an important role in approaching the capacity.

## Adaptive Modulation and Coding (AMC)

In a fading channel, the system performance is dominated by deep fades, where the channel gain drops significantly due to detrimental superposition of multipaths. Two categories of techniques are popular to combat the fading effect. One is interleaved coding technique. Basically, this technique uses interleaver to spread a deep fade across many codewords, and hence only a small fraction of a codeword is affected by the deep fade and is correctable by the decoder. Bit-interleaved coded modulation (BICM) is a typical example in this category, which shows good performance in fading channels.

Another technique to combat multipath fading is adaptive modulation and coding (AMC). Simply speaking, it adapts its transmission parameters, such as coding, modulation, transmit power, etc., according to the channel condition. For instance, when the channel gain is high, a high order modulation, say 64-QAM, and/or a high rate code, say rate-7/8 convolutional code, is used to achieve a high spectral efficiency. When the channel gain is low, a low order modulation, say BPSK, and/or a low rate code, say rate-1/3 convolutional code, is used to achieve an acceptable error probability performance. In an extremely deep fade, the transmitter might suspend the transmission temporarily to save the power.

AMC in SISO systems has been widely investigated in the open literature (Cavers, 1972; Vucetic, 1991; Goldsmith, 1997; Goldsmith, 1998; Goeckel, 1999; Hole, 2000; Chung, 2001; Örmeci, 2001; Vishwanath, 2003).

Obviously, to enable AMC, the CSI has to be known by the transmitter and coordination between the transmitter and receiver must be maintained. However, in most situations, the channel conditions vary relatively slow compared to the data rate (or the data block length), and hence within each block of transmission, the channel can be assumed as constant. In this case, the CSI can be fed back from the receiver to the transmitter reliably via a low rate feedback channel. Alternatively, in a time division duplex (TDD) system, the uplink channel estimate can readily be used as the downlink channel gain because the uplink and downlink share the same channel frequency band.

Compared to the interleaved coding technique, AMC has multiple advantages.

1. AMC does not need a long interleaver and hence the end-to-end delay caused by the interleaver can be avoided.
2. AMC is generally more efficient in power usage because the transmit power level is adapted according to the channel conditions.
3. AMC can be built upon existing coding and modulation techniques developed in AWGN channels.
4. Cross-layer design is readily applicable due to the inherent adaptive nature of AMC.
5. With one uniform design, AMC can achieve the highest spectral efficiency (or throughput) in various channel environments.

The major issues of AMC SISO systems include

1. Maximize the average spectral efficiency while maintaining a target error probability performance.
2. Minimize the average transmit power while maintaining a target error probability performance.
3. Identify the key adaptable parameters to achieve the above optimization.
4. Performance analysis in realistic wireless communication environments.
5. Design robust AMC systems in imperfect CSI environments.
6. Design AMC systems with reduced feedback requirement.

*Figure 2. Parallel equivalent subchannel model*

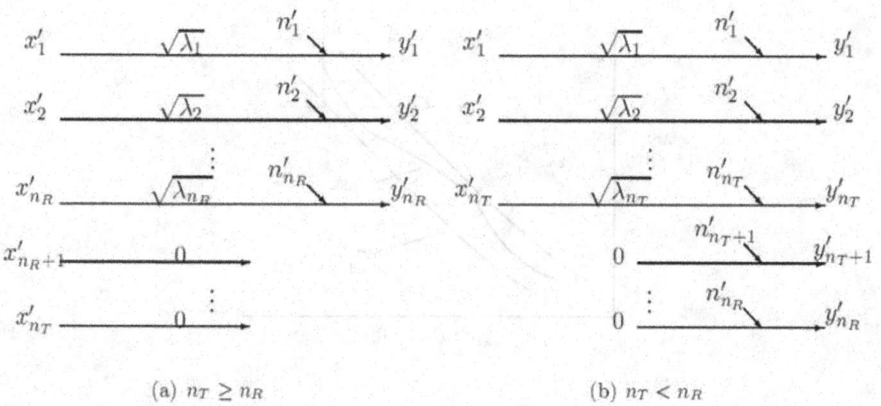

Although it has been proven that in SISO systems, utilizing the CSI at the transmitter does not improve the channel capacity significantly, AMC technique has been shown as a good candidate to approach the channel capacity in a practical application.

## AMC MIMO SYSTEMS

For the moment, we assume that both the transmitter and receiver have perfect knowledge about the channel matrix **H**. In this case, as shown in Eq. (3) and Figure 2, the MIMO channel can be seen as *m* parallel SISO subchannels. We can apply the AMC algorithms developed for SISO systems to each subchannel individually. However, AMC MIMO systems are not that simple, even in the perfect CSI condition.

In a SISO system, the parameters are adapted in time domain according to the channel condition. In contrast, a MIMO system provides both time and space domain by introducing multiple antennas (and hence subchannels). Although subchannels are parallel, they are not independent. Therefore, all the adaptable parameters need to be jointly adapted temporally and spatially, which is a complicated 2-D optimization problem.

In the event of imperfect CSI, where subchannels are no longer parallel, the situations will become even more complicated.

### Perfect CSI

The adaptable parameters of AMC system include modulation format, transmit power, coding scheme, block length, symbol duration, and so on. Among all these, the adaptation of the transmit power and the spectral efficiency (combined effect of modulation and coding scheme) have been shown to be most effective and feasible (Goldsmith, 1997; Chung, 2001).

The spectral efficiency is defined to be the information rate per unit bandwidth usage, having the unit of bit/s/Hz. It is determined by the modulation format and the coding rate. For instance, a 16-QAM with a rate-1/2 convolutional code will result in a spectral efficiency of

$\frac{1}{2}\log_2 16 = 2$ bits/s/Hz, assuming Nyquist spectrum shaping.

An AMC system may use different combinations of modulation and coding schemes according to the channel conditions. We refer to each combination of modulation and coding scheme as one *component mode* in the AMC system, and the set of all combinations as the *component mode set*, or mode set for short. Thus, adapting the spectral efficiency is equivalent to switching between component modes. For this reason, we interchangeably use the terms of mode adaptation and spectral efficiency adaptation. Later, we will show that the component mode set largely determines the average spectral efficiency performance of the overall AMC MIMO system.

*Figure 3. General definition of multiplexing gain*

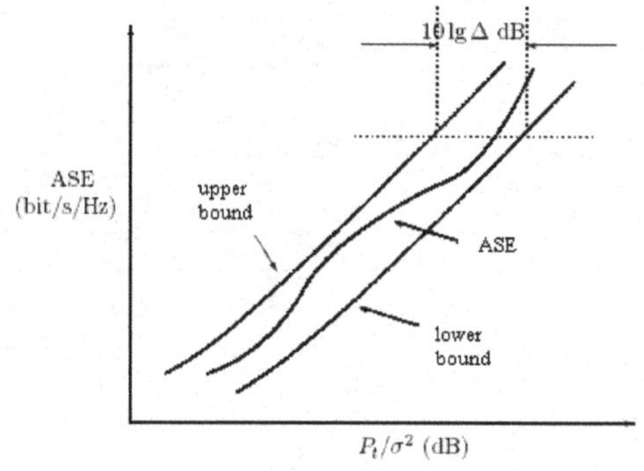


An AMC system that can adapt both the component mode and the transmit power is referred to as a variable-rate variable-power (VRVP) system. Here for historical reason, 'rate' is used as a synonym of 'spectral efficiency'.

We will confine ourselves to the VRVP framework since it has been proven to be an efficient combination of adaptable parameters.

Among various optimization objectives, we choose the maximization of the average spectral efficiency (ASE) as the goal function, while keeping the average transmit power and bit error probability at target levels. This will result in an adaptive system that has a variable ASE according to the channel conditions, such as the SNR. Thus, when we plot the ASE versus SNR curve, we will possibly get a similar pattern as the channel capacity expressed in (7) and (8). Therefore, we have the following definition.

If the ASE of an AMC MIMO system in high SNRs can be written as

$$b\log_2\gamma_s + c_1 \le \text{ASE} \le b\log_2\gamma_s + c_2 \tag{9}$$

where $b$, $c_1$ and $c_2$ are constants independent of $\gamma_s$, the AMC MIMO system is said to achieve a multiplexing gain of $b$. In particularly, when $b = m$, the system is said to achieve a full multiplexing gain.

This definition is illustrated in Figure 3. The inclusion of two constants $c_1$ and $c_2$ covers the situations where the ASE does not converge when the SNR approaches infinity, which is a common phenomenon in AMC MIMO systems.

Assume that the component mode set, denoted by $\mathcal{K}$, includes $N$+1 modes with spectral efficiencies of {$\text{SE}_0$, $\text{SE}_1$, …, $\text{SE}_N$}, where $\text{SE}_0$=0 represents transmission suspension. For each mode $j$, we denote its BER function in AWGN SISO channels as $\text{BER}_j(\gamma_s)$, where $\gamma_s$ represents the received SNR per modulation symbol. And for a given target BER, $\text{BER}_t$, we define the SNR threshold

$$\alpha_j = \text{BER}_j^{-1}(\text{BER}_t)$$

which is the received SNR value at which the BER is exactly equal to $\text{BER}_t$ in an AWGN SISO channel. In the parallel subchannel model, each subchannel at any time can choose one mode out of the mode set and determine its transmit power according to the adaptation algorithm. Assume that the spectral efficiency adopted in subchannel $i$ is $k_i$, the transmit power is $P_i$ and the instantaneous BER is $\text{BER}_i$. The question is how to optimally adapt the powers and modes in order to achieve the maximization of the goal function while keeping all the constraints met.

A generic framework of AMC MIMO systems can be represented as the following optimization problem.

$$\underset{k_i, P_i}{\text{Maximize}} \qquad \text{ASE} = E_{\mathbf{H}}\left[\sum_{i=1}^{m} k_i\right] \tag{10}$$

$$\text{Subject to} \qquad E_{\mathbf{H}}\left[\sum_{i=1}^{m} P_i\right] \le P_t \tag{11}$$

$$\text{BER}_i \leq \text{BER}_t \tag{12}$$

$$k_i \in \mathcal{K} \tag{13}$$

$$P_i \geq 0 \tag{14}$$

$$i = 1, 2, \cdots, m$$

where $E_x[\cdot]$ denotes expectation with respect to random variable $x$. Eq. (10) describes the goal function, to maximize the average spectral efficiency (ASE) by adapting mode $k_i$ and power $P_i$ in each subchannel; Eq. (11) represents the average transmit power constraint, where $P_t$ is the transmit power limit; Eq. (12) puts an instantaneous BER constraint on each subchannel; Eq. (13) expresses the spectral efficiency of component modes; and Eq. (14) simply states that the transmit power in each subchannel cannot be negative.

The formulation specified in Eq. (10) to (14) implies a 2-D optimization problem. The adaptable parameters, $P_i$ and $k_i$ can be adapted both along time axis as in conventional AMC SISO systems (temporal adaptation), and among multiple subchannels (spatial adaptation). As proven by Zhou (2005), this 2-D optimization problem can be decoupled into $m$ independent SISO 1-D optimization problems.

The basic idea about the decoupling is based on the unordered eigenvalue distribution. As mentioned in the previous section, the unordered eigenvalue distributions for all subchannels are identical. Hence, there is no reason to allocate more average power in one subchannel than the other. Therefore, this results in the equal power allocation policy.

$$E_{\mathbf{H}}[P_i] = \frac{P_t}{m}, \; i=1,2,\ldots,m. \tag{15}$$

The similar reasoning holds for the mode adaptation.

$$E_{\mathbf{H}}[k_1] = E_{\mathbf{H}}[k_2] = \cdots = E_{\mathbf{H}}[k_m]. \tag{16}$$

Note that in Eq. (15) and (16), only their expectation values, not their instantaneous values, are equal. Based on Eq. (15) and (16), the optimization problem in Eq. (10) to (14) can be reduced to

$$\underset{k_{1}, P_1}{\text{Maximize}} \qquad \text{ASE} = mE_{\lambda_1}[k_1] \tag{17}$$

$$\text{Subject to } E_{\lambda_1}[P_1] = \frac{P_t}{m} \tag{18}$$

$$\text{BER}_1 \leq \text{BER}_t \tag{19}$$

$$k_i \in \mathcal{K} \tag{20}$$

$$P_1 \geq 0 \tag{21}$$

Note that all the adaptations and constraints are on the first unordered subchannel now, since all subchannels have the same distribution and hence share the same adaptation policy. The strict proof of this decoupling approach can be found in the aforementioned publication.

The complete solution to this optimization problem is derived by Zhou (2005). Here a brief summary is recited. The optimum rate and power adaptation is

$$k_i = \mathrm{SE}_j, \nu_j \le \lambda_i < \nu_{j+1} \tag{22}$$

$$P_i = \alpha_j \frac{\sigma^2}{\lambda_i}, \nu_j \le \lambda_i < \nu_{j+1} \tag{23}$$

The adaptation in each subchannel is based on the subchannel gain $\lambda_i$. The subchannel gain range $[0,\infty)$ is divided into $N+1$ regions with the region boundaries $0 = \nu_0 < \nu_1 < \cdots < \nu_N < \nu_{N+1} = \infty$. Eq. (22) means that if the subchannel gain $\lambda_i$ lies in the *j*-th region, $[\nu_j, \nu_{j+1})$, the *j*-th mode is used in that subchannel. Note that if the feedback CSI is perfect, the received SNR in the *i*-th subchannel is $P_i\lambda_i / \sigma^2$. Thus, the power adaptation policy in Eq. (23) keeps the received SNR at the SNR threshold $\alpha_j$ if the *j*-th mode is used, which is equivalent to the BER constraint in Eq. (12) with equality.

The optimum subchannel gain boundaries are

$$\nu_j = \mu\delta_j, \delta_j = \frac{\alpha_j - \alpha_{j-1}}{\mathrm{SE}_j - \mathrm{SE}_{j-1}} \tag{24}$$

where the Lagrangian multiplier $\mu$ is detetermined by

$$\sum_{j=1}^{N} \alpha_j \left[\Theta_1(\mu\delta_j) - \Theta_1(\mu\delta_{j+1})\right] = \frac{P_t}{m\sigma^2} \tag{25}$$

The achieved maximum ASE can be evaluated as

$$\mathrm{ASE} = m\sum_{j=1}^{N} \mathrm{SE}_j \left[\Psi_1(1,\nu_j) - \Psi_1(1,\nu_{j+1})\right] \tag{26}$$

In Eq. (25) and (26) we use two functions $\Theta_1(\cdot)$ and $\Psi_1(\cdot)$ related to the probability density function (p.d.f.) of the unordered eigenvalue $p_{\lambda_1}(\lambda_1)$, defined as

$$\Theta_1(x) = \int_x^\infty p_{\lambda_1}(\lambda_1)/\lambda_1 d\lambda_1$$

$$\Psi_1(c,x) = \int_x^\infty e^{-(c-1)\lambda_1} p_{\lambda_1}(\lambda_1) d\lambda_1$$

Their closed-form expressions have been derived by Zhou (2005).

## Multiplexing Gain

So far, the problem specification and solution do not depend on any specific component modes. The only parameters related to the component modes are the SNR threshold $\alpha_j$ for each mode. In this section, we will establish the relationship between component modes and the multiplexing gain of the AMC MIMO system using those component modes.

We start from a well studied case: the adaptive uncoded QAM MIMO system.

If uncoded square QAMs, including 4-QAM, 16-QAM, 64-QAM, 256-QAM, etc., are adopted as the component mode set, Zhou (2005) shows that the maximum ASE can be expressed as

$$\mathrm{ASE} \approx m \log_2 \gamma_s + C_u \tag{27}$$

where

$$\gamma_s = \frac{P_t}{\sigma^2}$$

is the average SNR, and $C_u$ is a constant independent of $\gamma_s$.

According to the definition in Eq. (9), the AMC MIMO system based on the mode set of uncoded square QAMs achieves a full multiplexing gain of *m*.

The attainment of the full multiplexing gain is highly desired since it represents a consistent ASE performance in the wide range (from moderate to high) of SNRs. In other words, the achieved ASE is at a constant distance from the channel capacity regardless of what SNR level the system is operating at. It is known that most capacity approaching coding and modulation schemes, such as turbo codes and low density parity check (LDPC) codes, only work closely to the capacity in a small range of SNRs, usually at low SNRs. In contrast, an AMC MIMO system with a full multiplexing gain will work consistently in the wide range of moderate to high SNRs.

Now we study the property of the set of uncoded QAMs in the attempt to find the reason it results in a full multiplexing gain.

A simple calculation reveals that for an M-QAM, the symbol power $P_s$ can be expressed as

$$P_s = \frac{d_0^2}{6}(M-1)$$

where $d_0$ denotes the minimum Euclidean distance between constellation points. Then the spectral efficiency of the mode is

$$k = \log_2 M = \log_2\left(1+\frac{6P_s}{d_0^2}\right) = \log_2\left(1+\frac{6\sigma^2\gamma_s}{d_0^2}\right) \tag{28}$$

where $\gamma_s = P_s/\sigma^2$ is the symbol SNR. At high SNRs, we can obtain the following approximation,

$$k \approx \log_2\left(\frac{6\sigma^2\gamma_s}{d_0^2}\right) = \log_2\gamma_s + \log_2\left(\frac{6\sigma^2}{d_0^2}\right) \tag{29}$$

Since the noise normalized Euclidean distance $d_0/\sigma$ approximately determines the BER of M-QAM in AWGN SISO channels, given a target BER, $d_0/\sigma$ is fixed. Therefore, we can rewrite (29) as

$$k \approx \log_2\gamma_s + c_u \tag{30}$$

where $c_u$ is a constant determined by the target BER. Eq. (30) indicates an important property of uncoded QAMs as the component modes: given a target BER, the set of component modes exhibits 1 bit/s/Hz increase in spectral efficiency per 3 dB increase in SNR. Informally, we refer to a set of component modes satisfying this property as a *full-rate* mode set.

Now, we prove that a full rate mode set is a sufficient condition for an AMC MIMO system to achieve a full multiplexing gain.

More than that, we are to establish a general relationship between the component mode set and the multiplexing gain of the AMC MIMO system.

*Proposition 1*

If a component mode set satisfy the following condition in SISO AWGN channels for a given $BER_t$,

$$k \approx g\log_2\gamma_s + c \tag{31}$$

where *k* represents the spectral efficiency of component modes, $\gamma_s$ denotes the received symbol SNR and *g* and *c* are constants independent of $\gamma_s$ and *k*, then the AMC MIMO system using this mode set can achieve a multiplexing gain of *mg*.

**Proof.** See Appendix.

A direct consequence of this proposition is that a full-rate mode set, whose $g$=1, results in a full multiplexing gain $m$. Therefore, in designing an AMC MIMO system, a full-rate mode set is preferred.

In practice, it is not always easy to explicitly relate the spectral efficiency $k$ to the SNR threshold $\gamma_s$ as that in Eq. (31). However, we can derive a more practical method to estimate the multiplexing gain. Applying Eq. (31) to two consecutive component modes $SE_{j-1}$ and $SE_j$, we have

$$\mathrm{SE}_j - \mathrm{SE}_{j-1} = g\left(\log_2 \alpha_j - \log_2 \alpha_{j-1}\right)$$

Therefore,

$$g = \frac{\mathrm{SE}_j - \mathrm{SE}_{j-1}}{\log_2 \alpha_j - \log_2 \alpha_{j-1}} = \frac{10\lg 2 \times \left(SE_j - SE_{j-1}\right)}{10\lg\left(\mathrm{SE}_j / \mathrm{SE}_{j-1}\right) + \Delta\gamma_{b,j}[\mathrm{dB}]} \quad (32)$$

where the relationship between SNR per modulation symbol $\alpha_j$ and SNR per information bit $\gamma_{b,j}$

$$\alpha_j = \mathrm{SE}_j \cdot \gamma_{b,j}$$

is used. Eq. (32) shows a way to estimate the multiplexing gain based on the BER curves of the component modes in SISO AWGN channels. All one needs is to read the $\gamma_b$ value at the desired $BER_t$ for each component mode and substitute into Eq. (32). Note that it is usual that the values $g$ calculated from Eq. (32) for different $j$'s are different. However, the multiplexing gain resides in the high SNR range, thus, the $g$ value calculated for the largest $j$ will dominate the multiplexing gain of an AMC system, while those $g$ values for smaller $j$'s provide some estimate on moderate SNRs.

As an example to test the accuracy of Eq. (32), let us revisit the uncoded QAM mode set. In Figure 4, we plot the BER versus SNR curves of 4-QAM, 16-QAM, 64-QAM and 256-QAM in SISO AWGN channels. Assume that $BER_t = 10^{-4}$ is chosen. The SNR thresholds can be read from the figure as $\gamma_b$ =8.4, 12.2, 16.5 and 21.2 dB, respectively. By applying Eq. (32), we obtain $g$=0.9, 1 and 1. This verifies that a full multiplexing gain of $m$ can be achieved by an AMC MIMO system based on the component mode set of uncoded square QAMs.

*Figure 4. BER of uncoded square QAMs in SISO AWGN channels*

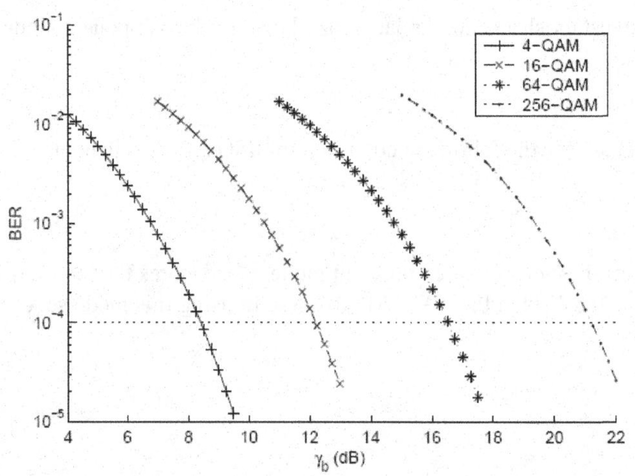

## Imperfect CSI

So far, we assume both the transmitter and receiver have perfect CSI. However, CSI imperfection is usually unavoidable in reality. Basically, we can categorize CSI imperfection in two types. The first type is caused by channel estimation errors, and the second caused by the feedback channel delay (in the case where a CSI feedback channel is adopted) or the channel variation between uplink and downlink (in TDD systems). The first type of CSI imperfection can be improved by advanced channel estimation techniques, training sequence/pilots insertion, etc. Particularly, in a low to moderate mobile environment, the state-of-the-art channel estimate algorithms will results in a negligible channel estimation error. In contrast, the second type of CSI imperfection is inherent to the CSI acquirement method. One possible technique to deal with the delay problem is channel prediction. However, the channel prediction in MIMO channels is still a challenging task, whose performance depends on the channel characteristics. Thus, the performance of an AMC MIMO system based on channel prediction will also be sensitive to the channel characteristics. Since we are interested in a robust AMC MIMO system, the channel prediction approach will not be considered.

Assume that the channel estimation error is negligible, and the total CSI delay is $\tau$. The channel variation can be modelled by a first order Markov chain,

$$\mathbf{H}(t) = \rho\mathbf{H}(t-\tau) + \sqrt{1-\rho^2}\,\mathbf{\Xi}(t) \tag{33}$$

where $\rho$ is the channel time variation correlation coefficient; $\mathbf{\Xi}(t)$ is the innovation term whose entries are assumed to be i.i.d. complex Gaussian random variables with zero mean and unity variance. Due to the CSI delay, at time $t$, the true channel matrix is $\mathbf{H}(t)$, while the transmitter is notified with $\mathbf{H}(t-\tau)$ For simplicity of notation, we drop the time variable by representing the outdated CSI as $\hat{\mathbf{H}}$ and the true CSI as $\mathbf{H}$. Thus, Eq. (33) can be rewritten as

$$\mathbf{H} = \rho\hat{\mathbf{H}} + \sqrt{1-\rho^2}\,\mathbf{\Xi} \tag{34}$$

Similar to Eq. (2), $\hat{\mathbf{H}}$ can be decomposed as

$$\hat{\mathbf{H}} = \hat{\mathbf{U}}\hat{\mathbf{D}}\hat{\mathbf{V}}^H \tag{35}$$

where $\hat{\mathbf{U}}$, $\hat{\mathbf{D}}$ and $\hat{\mathbf{V}}$ have similar meaning as their non-hatted versions in Eq. (2).

Due to the outdated CSI, the equivalent subchannel model expressed in Eq. (3) will not hold any more. If the transmitter is not aware of the delay and hence treats the outdated CSI $\hat{\mathbf{H}}$ as the true value, it will steer the modulated vector $\mathbf{x}'$ with the eigenvectors in $\hat{\mathbf{V}}$. This results in the following equivalent system equation.

$$\begin{aligned} \mathbf{y} &= \mathbf{Hx} + \mathbf{n} = \left(\rho\hat{\mathbf{H}} + \sqrt{1-\rho^2}\,\mathbf{\Xi}\right)\hat{\mathbf{V}}\mathbf{x}' + \mathbf{n} \\ \mathbf{y}' &= \hat{\mathbf{U}}^H\mathbf{y} = \rho\hat{\mathbf{U}}^H\hat{\mathbf{H}}\hat{\mathbf{V}}\mathbf{x}' + \sqrt{1-\rho^2}\,\hat{\mathbf{U}}^H\mathbf{\Xi}\hat{\mathbf{V}}\mathbf{x}' + \hat{\mathbf{U}}^H\mathbf{n} \\ \mathbf{y}' &= \rho\hat{\mathbf{D}}\mathbf{x}' + \sqrt{1-\rho^2}\,\hat{\mathbf{U}}^H\mathbf{\Xi}\hat{\mathbf{V}}\mathbf{x}' + \mathbf{n}' \end{aligned} \tag{36}$$

where $\mathbf{n}' = \hat{\mathbf{U}}^H\mathbf{n}$ has the same distribution as $\mathbf{n}$ since $\hat{\mathbf{U}}$ is a unitary matrix.

The second term of Eq. (36) has made it clear that the cross subchannel interference arises due to the CSI imperfection. Note that in Eq. (36), the outdated CSI $\hat{\mathbf{U}}$ is used. Because the the orthonormalization processing at the receiver is reversible, no information is lost. Actually we can use either $\hat{\mathbf{U}}$ or $\mathbf{U}$ to orthonormalize the received vector $\mathbf{y}$. The selection of $\hat{\mathbf{U}}$ is only for the ease of analysis.

Besides the cross subchannel interference, another effect of outdated CSI can be seen from the factor $\rho$ in the first term of Eq. (36). The effective subchannel gains have been reduced by this factor because $\hat{\mathbf{D}}$ and $\mathbf{D}$ have identical distribution.

In order to eliminate or reduce the effect of outdated CSI, two approaches can be taken. First, since the receiver has the perfect CSI, the cross subchannel interference as shown in Eq. (36) can be mitigated by interference cancellation algorithms. Second, the transmitter can modify its adaptation algorithm by taking into account some statistics of the outdated CSI. The second approach needs additional knowledge about the time variation characteristics of the channel, e.g. the correlation coefficient $\rho$. This will increase the feedback rate and more importantly, it will make the system more sensitive to the estimate of the channel statistics. This is against our goal to design a robust AMC MIMO system. For this reason, we opt for the interference cancellation at the receiver. Although this will result in higher complexity in the receiver, the benefit of this approach is multi-folds. Because it does not depend on the estimate of channel characteristics, its performance is more robust. The transmitter adaptation algorithm is completely the same as the perfect CSI case, thus, a simpler transmitter design is expected. Many state-of-the-art interference cancellation algorithms can be used straight away. And higher ASE is promising due to the transmitter adopts an ASE maximization algorithm.

## Design of Robust Near-Capacity AMC MIMO Systems

As indicated by Proposition 1, to design an AMC MIMO system with full multiplexing gain, we need a full-rate mode set. However, a full multiplexing gain is not the whole story. A full multiplexing gain only ensures a maximum slope of the ASE versus SNR curve (parallel to the capacity curve). It does not indicate how closely the ASE approaches the channel capacity. In order to attain a near-capacity ASE performance, a powerful component mode set is to be designed. Furthermore, to obtain a robust system in outdated CSI environment, an interference cancellation needs to be performed at the receiver. All these requirements add up to the following task: to design a powerful full-rate mode set and a receiver interference cancellation algorithm.

To the end of a near-capacity AMC MIMO system, the component modes must be capacity-approaching themselves. Both turbo codes and LDPC codes are good candidates in the sense of approaching the channel capacity. Here we choose turbo codes as an example. Similar design and result can be obtained by using LDPC codes. Having determined the coding scheme, we need to decide on the combination of coding and modulation to arrive at the component modes. If we only consider the spectral efficiency, the turbo coded modulation might be the best choice. However, turbo coded modulation has some disadvantages to be used as component mode. The adaptation between different modes is inflexible, restricted by block length. Even worse, the interference cancellation will become a big challenge for interfered turbo coded modulation modes. We realize that the equivalent channel model expressed in Eq. (36) can be rewritten as

$$\mathbf{y}' = \hat{\mathbf{U}}^H \mathbf{y} = \hat{\mathbf{U}}^H \mathbf{H}\hat{\mathbf{V}}\mathbf{x}' + \hat{\mathbf{U}}^H \mathbf{n} = \mathbf{G}\mathbf{x}' + \mathbf{n}' \tag{37}$$

where $\mathbf{G} = \hat{\mathbf{U}}^H\mathbf{H}\hat{\mathbf{V}}$ can be seen as the equivalent channel matrix relating **x'** to **y'**. This equivalent channel matrix **G** of size $m \times m$ is known to the receiver but not to the transmitter. This is a typical MIMO system without transmitter CSI and with the same number of transmit and receive antennas. A good technique coping with the cross antenna interference in this scenario is the turbo-BLAST architecture. For instance, Hochwald (2003) proposed a near-capacity turbo-BLAST scheme with an efficient iterative MIMO detection and channel decoding algorithm. Basically, the turbo-BLAST architecture is simply the concatenation of a turbo code and the V-BLAST structure (Foschini, 1996; Golden, 1999; Foschini, 1999). By applying the turbo-BLAST architecture into the equivalent system represented in Eq. (37), we arrive at the adaptive turbo coded MIMO system block diagram as shown in Figure 5.

The information bit stream is first encoded by a turbo encoder (ENC); the codeword bits are interleaved by the bit-interleaver (*I*) and then demultiplexed and mapped into *m* streams of QAM symbols according to the adaptation algorithm; after passing through the steering matrix $\hat{\mathbf{V}}$, the QAM symbol streams are transmitted from $n_T$ antennas. At the receiver, the received vector **y** is first multiplied by the orthonormalization matrix $\hat{\mathbf{U}}^H$ and then processed iteratively between the inner MIMO detector (MIMO DET) and the outer turbo decoder (DEC). Note that all the components at both the transmitter and receiver are controlled by the CSI estimation and adaptation control module, in which the CSI is estimated, the optimal coding and modulation parameters are determined and all necessary coordination signals are generated.

Up until now, we have not addressed how the turbo code rates and modulation formats are adapted according to the CSI. In the turbo-BLAST architecture, which is a non-adaptive system, the turbo code is fixed, usually rate-1/2 code. A straightforward and simple way to go for is that we fix the turbo code component and only adapt the modulation formats. However, this way will result in a significant loss of multiplexing gain. The resultant component mode set is not a full-rate mode set, because the fix turbo code reduces the spectral efficiency of the uncoded QAMs by a constant factor of *R*, where *R* is the turbo code rate.

Aiming at a full-rate mode set, we need to use different rate turbo codes with different modulation formats. For example, we can combine a rate-1/2 code with 4-QAM, rate-3/4 code with 16-QAM, rate-5/6 code with 64-QAM and

*Figure 5. Adaptive turbo coded MIMO system*

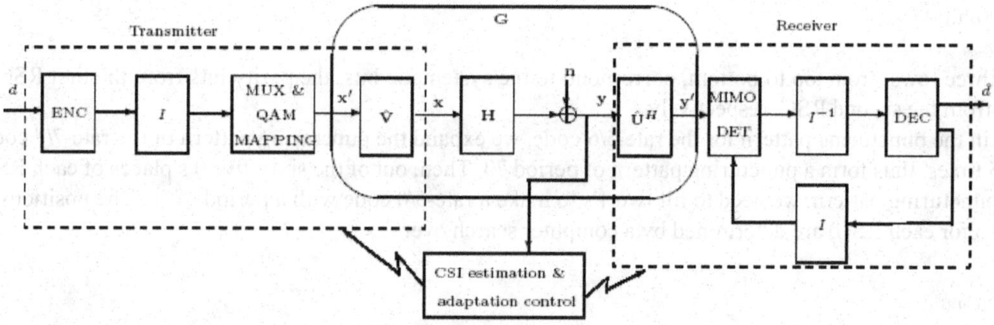


*Table 1. Multilevel punctured turbo codes*

| mode | code rate | QAM | SE |
|---|---|---|---|
| 0 | - | - | 0 |
| 1 | 1/2 | 4-QAM | 1 |
| 2 | 3/4 | 16-QAM | 3 |
| 3 | 5/6 | 64-QAM | 5 |
| 4 | 7/8 | 256-QAM | 7 |

rate-7/8 code with 256-QAM, etc. In such a way, the resultant spectral efficiencies for these component modes are 1, 3, 5 and 7 bits/s/Hz, respectively. However, these different rate turbo codes cannot be independently designed. Otherwise, the switching between component modes has to be restricted to be between turbo code frames. As we know, a good performance turbo code usually has a large frame length. Thus, frame-wise adaptation is not satisfactory due to channel variation within one frame. Even if short length turbo codes are used such that the channel does not change significantly within one frame, the frame-wise adaptation confines the bit-interleaving to within one frame, so the time diversity utilized by a conventional turbo-BLAST system by interleaving across independently faded channel blocks is not available. This will definitely impair the performance robustness in outdated CSI environments.

To this end, we adopt rate compatible punctured turbo codes (RCPTCs) as the code components. A special multilevel puncturing and interleaving technique is used. This technique is originally developed by Lau (2002) in order to implement a symbol-by-symbol adaptive bit-interleaved coded modulation system. Basically, this technique uses the rate-compatible punctured codes (RCPCs) as the component codes. It marks the puncturable bit positions without actually puncturing them before being bit-interleaved. A bank of bit-interleavers are used, each corresponding to one puncturing level. Then at the output of the interleavers, the marked positions are actually punctured. By this way, the bit-interleaving across different punctured codes is implemented.

We consider a rate-1/3 turbo code as the mother code to generate the set of RCPTCs. The code rates, modulation formats and spectral efficiencies of all component modes to be used by the adaptive turbo coded MIMO system are listed in Table 1.

Açıkel (1999) designed a set of high-rate punctured turbo codes and proposed a pragmatic way to generate RCPTCs. However, the generated RCPTCs do not satisfy our code rate requirement as specified in Table 1. So we do a code search for RCPTCs in accordance with Table 1. We choose the 16-state rate-7/8 punctured turbo code found by Açıkel (1999) as the starting point, whose generator sequences are $(g_0, g_1) = (23,31)$ in octets with $g_0$ as the feedback generator. The frame length (i.e. the length of the turbo code interleaver) is set as $L_F = 10500$ information bits. The puncturing period of the rate-7/8 code is $2 \times 7 = 14$. Only the parity bits from the constituent recursive systematic codes (RSCs) are punctured while the systematic bits are always transmitted. The puncturing pattern (0 for puncture and 1 for transmit) is

11111111111111
01000000000000
01000000000000

where the three rows, from top to bottom, correspond to the systematic bits, the parity bits from the first RSC and the parity bits from the second RSC, respectively.

To obtain the puncturing pattern for the rate-5/6 code, we expand the puncturing pattern of the rate-7/8 code by repeating it 5 times, thus form a puncturing pattern of period 70. Then, out of the sixty-five 0's places of each RSC in the expanded puncturing pattern, we need to fill two 1's to make a rate-5/6 code with a period of 70. The positions of these four 1's (two for each RSC) are determined by a computer search over

$$\binom{65}{2}^2 = 4{,}326{,}400$$

*Figure 6. BER curves of RCPTCs in SISO AWGN channels*

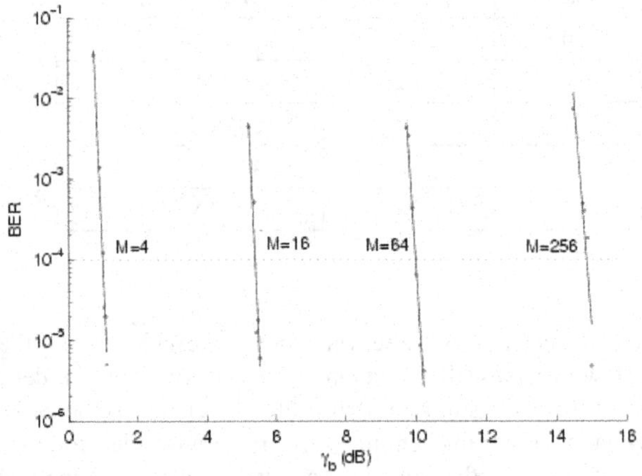


*Figure 7. ASE of adaptive uncoded and turbo coded MIMO systems*

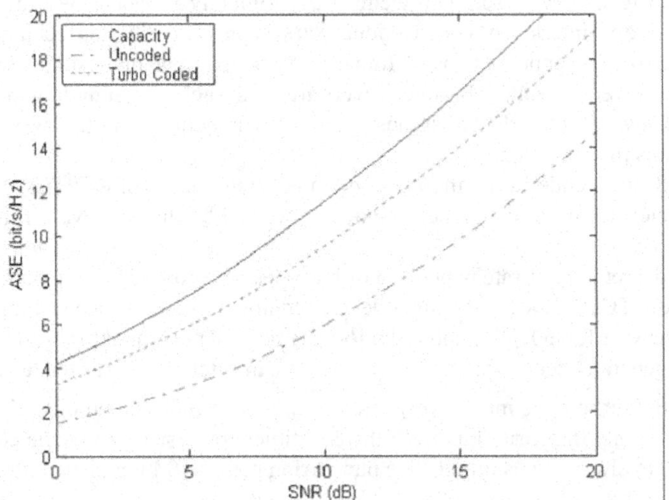

possibilities. We adopt the maximization of the minimum turbo codeword weight for weight-2 and weight-3 inputs as the optimum criterion for code search.

The similar procedure applies to the extension to the rate-3/4 code and then the rate-1/2 code. The final puncturing period is $14 \times 5 \times 3 = 210$. The resultant puncturing patterns and distance spectrum are not shown here due to space limitation and can be provided at request.

We simulate the obtained RCPTCs in SISO AWGN channels and show their BER versus SNR per information bit ( $\gamma_b$) curves in Figure 6. As an example, we set the target BER at $\text{BER}_t = 10^{-4}$ and the corresponding SNR thresholds for 4-QAM, 16-QAM, 64-QAM and 256-QAM are $\gamma_b$ =1.0, 5.4, 10.0 and 14.9 dB, respectively. Substituting the SNR thresholds into Eq. (32), we get $g$ =0.65, 0.89 and 0.95. Therefore, according to Proposition 1, the AMC MIMO system using the RCPTC mode set will achieve a near-full multiplexing gain of 0.95*m*.

We simulate the adaptive turbo coded MIMO system based on the obtained RCPTCs with $n_T = n_R = 4$, which we abbreviate as a (4,4) system, and show the ASE and BER performance in Figure 7 and Figure 8, respectively. The target BER is set at $\text{BER}_t = 10^{-4}$ for illustration purpose. Other BER values will present similar results. We use the iterative tree search algorithm (de Jong, 2005) for the MIMO detector and a log-MAP algorithm for the turbo decoder. The number of iterations in the detector-decoder loop is three, and that in the turbo decoder is eight. A block fading channel model is assumed with the channel fading block length $L_B = 105$ QAM symbols. For comparison, we also include the ASE and BER curves for the adaptive uncoded MIMO system and the channel capacity is also shown in Figure 7.

From Figure 7, we see that both curves for uncoded and turbo coded systems are parallel (or almost parallel) to the channel capacity curve, which indicates that a full (or near-full) multiplexing gain is achieved by both systems. In addition to the multiplexing gain, in the whole range of observed SNRs, the turbo coded system is within 3 dB of the capacity, exhibiting a 4 dB gain over the uncoded system.

In Figure 8, the actual BER is plotted versus the normalized CSI delay $f_D\tau$ with an average SNR of 20 dB, where $f_D$ is the maximum Doppler frequency shift. As seen from the figure, for both uncoded and turbo coded systems the actual BER does not meet the target BER requirement when the CSI delay increases. For uncoded system, when $f_D\tau$ goes beyond $3\times 10^{-3}$, the actually BER degrades significantly. In contrast, the turbo coded system keeps the target BER until $f_D\tau = 2\times 10^{-2}$, showing a seven-fold robustness against the CSI delay.

Compared to a non-adaptive turbo-BLAST system, the adaptive turbo coded MIMO system shows clear advantages. Besides the obvious adaptive nature, where the system can cope with different SNR situations and always maintain an optimum spectral efficiency, the adaptive turbo coded MIMO system approaches closer to the channel capacity than the corresponding turbo-BLAST system. To illustrate this, we reproduce the BER curves of a (4,4) turbo-BLAST system (Hochwald, 2003; de Jong, 2005) with different spectral efficiencies, which claim to be "achieving near-capacity" of MIMO channels. In Figure 9, we plot the BER curves for the cases when 4-QAM, 16-QAM, 64-QAM and 256-QAM

*Figure 8. BER of adaptive uncoded and turbo coded MIMO systems*

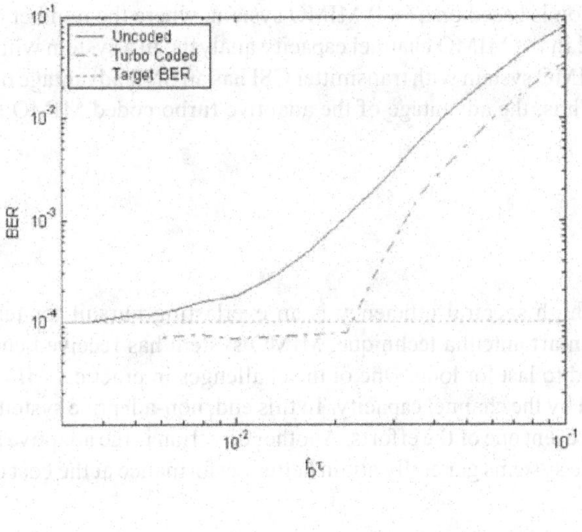

*Figure 9. BER of (4,4) turbo-BLAST MIMO system*

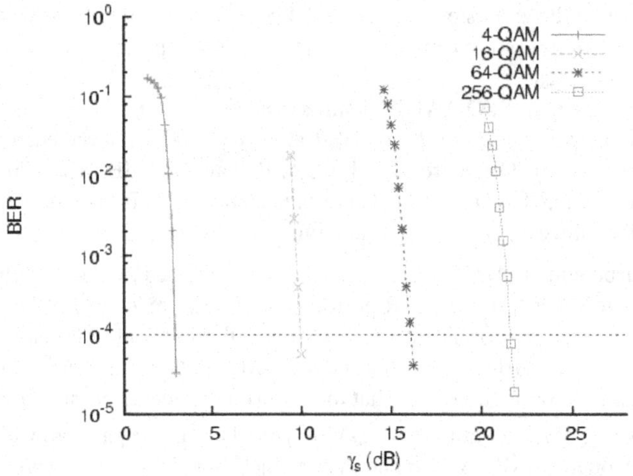


are used as the modulation formats. The turbo code has the 16-state generator sequences $(g_0, g_1) = (37,21)$ so as to do a fair comparison. A block fading channel with channel fading block length $L_B = 192$ information bits is assumed. And the turbo code interleaver length (frame length) $L_F = 9216$ information bits is used.

From Figure 9, we can read the required SNR values $\gamma_s$ to reach the target $\mathrm{BER}_t = 10^{-4}$ for 4-, 16-, 64- and 256-QAM as 2.9, 9.9, 16.0 and 21.5 dB, respectively. The corresponding spectral efficiencies of the turbo-BLAST system can be calculated as

$$\text{code rate} \times \log_2 M \times n_T$$

which results in 4, 8, 12 and 16 bits/s/Hz, respectively. From Figure 7, we can read the SNR values for the adaptive turbo coded MIMO system to reach these spectral efficiencies as $\gamma_s$=1.5, 8.2, 12.9 and 17.0 dB, respectively. Therefore, the SNR gains of the adaptive turbo coded MIMO system compared to the turbo-BLAST system are 1.4, 1.7, 3.1 and 4.5 dB, respectively.

This clearly shows that the adaptive system approaches the channel capacity closer. In high SNR and hence high spectral efficiency range, the gain is even larger. This seems to contradict the capacity result that indicates at high SNRs, the channel capacities of MIMO systems with and without transmitter CSI are asymptotically equal. However, as we emphasize earlier, even the transmitter CSI does not bring benefit when the channel capacity is the only concern, it does contribute to the superior near-capacity performance of a practical adaptive system. The transmitter CSI makes the task of approaching the channel capacity easier in a practical system.

Note that the above comparison is based on a (4,4) MIMO system, where the number of transmit and receive antennas are equal. As aforementioned in the MIMO channel capacity analysis, in a system with more transmit antennas than receive antennas, the adaptive MIMO system with transmitter CSI has an SNR advantage of $10\log_{10}(n_T / n_R)$ dB compared to its non-adaptive couterpart. Thus, the advantage of the adaptive turbo coded MIMO system over the non-adaptive turbo-BLAST will be bigger when $n_T > n_R$.

## FUTURE TRENDS

Wireless communications with high spectral efficiency is an everlasting pursuit for telecommunication researchers and engineers. As an extended smart antenna technique, MIMO system has received enormous attentions in the past decade, and this trend is expected to last for long. One of the challenges in practical MIMO systems is to approach the high spectral efficiency promised by the channel capacity. To this end, non-adaptive systems, including turbo and LDPC coded space-time architecture, present one of the efforts. Another direction is the adaptive system. Compared to the non-adaptive counterpart, the adaptive systems generally attain better performance at the cost of a CSI feedback channel and the extra complexity in the adaptation control. For low to moderate mobile environments (e.g. at pedestrian speeds), the CSI availability at the transmitter is not a big problem, and the adaptation overhead is not high since the adaptation only needs to be done when the channel condition changes. However, for a highly mobile wireless channel (e.g. at vehicular speeds), the channel condition varies quickly. In a system using feedback channel, the feedback channel needs to be operated at a high rate in accordance with the channel variation rate. The availability of this high capacity feedback channel is in problem. In a TDD system, the downlink channel may change significantly from the uplink channel, and hence the uplink channel estimation does not reflect the downlink channel any longer. When taking into account the challenges in synchronization, channel estimation and equalization, high spectral efficiency wireless communications in highly mobile environments is particularly far from resolved. For instance, theoretically a WiMAX system can support a throughput of up to 70 Mbits/s. However, in a mobile environment, the throughput might be reduced to only a few Mbits/s.

The proposed adaptive turbo coded MIMO system is suitable for a low to moderate mobile wireless environment, since its transmitter depends on the full CSI, **H**. It is crucial to design a good channel estimation algorithm such that the channel estimation error is negligible. Otherwise, the channel estimation error will introduce additional cross subchannel interference, which cannot be coped with by the iterative detection-decoding algorithm, since the channel estimation error is not known by the receiver.

In a system with feedback channel, a possible improvement is to reduce the feedback rate. Instead of feeding back the full CSI, it might be possible to only feedback the eigenspace spanned by the right eigenvectors contained in **V**. A further step in reducing the feedback rate is to design a quantized codebook such that only the indices of the codeword in the codebook need to be fed back. If the size of the codebook can be kept small while maintaining acceptable performance, the feedback rate can be reduced significantly. We notice that even with a powerful turbo code, the proposed system still exhibits a 3 dB distance from the channel capacity. This gap is partly attributed to the adoption of square QAMs as the modulation formats. In order to reduce the gap, the constellation shaping technique (Forney, 1992; Eyuboglu, 1992) might be effective.

## CONCLUSION

In this chapter, the MIMO channel capacity is analyzed with the focus on the multiplexing gain. Adaptive modulation and coding (AMC) is investigated as a technique that approaches the high spectral efficiency promise of the MIMO channel in a practical system. The general framework of AMC MIMO system is presented and the quantitative relationship between its multiplexing gain and its component mode set is established. The imperfect CSI effect on the AMC MIMO system is analyzed. All the above results along with other desired properties of AMC MIMO systems leads to the design criterion for a robust AMC MIMO system. Thereafter, the adaptive turbo coded MIMO system is proposed. Its performance is analyzed and simulated. It is shown that the proposed system achieves a near-capacity performance and is robust against the CSI imperfection. The adaptive system is compared with the non-adaptive system, which shows the adaptive system approaches the channel capacity closer.

## REFERENCES

Cavers, J. (1972). Variable-rate transmission for Rayleigh fading channels. *IEEE Transactions on Communications*, 20(1), 15-22.

Chung, S. T., & Goldsmith, A. J. (2001). Degrees of freedom in adaptive modulation: a unified view. *IEEE Transactions on Communications*, 49(9), 1561-1571.

de Jong, Y. L. C. & Willink, T. J. (2005). Iterative tree search detection for MIMO wireless systems. *IEEE Transactions on Communications*, 53(6), 930-935.

Eyuboglu, M. V. & Forney, G. D., Jr. (1992). Trellis precoding: combined coding, precoding and shaping for intersymbol interference channels. *IEEE Transactions on Information Theory*, 38(2), 301-314.

Forney, G. D., Jr. (1992). Trellis shaping. *IEEE Transactions on Information Theory*, 38(2), 281-300.

Foschini, G. J. (1998). On limits of wireless communication in a fading environment when using multiple antennas. *Wireless Personal Communications*, 6(3), 311-335.

Foschini, G. J. (1996). Layered space-time architecture for wireless communication in a fading environment when using multiple antennas. *Bell Laboratories Technical Journal*, 1(2), 41-59.

Foschini, G. J., Golden, G. D., Valenzuela, R. A. & Wolniansky, P. W. (1999). Simplified processing for high spectral efficiency wireless communication employing multi-element arrays. *IEEE Journal on Selected Areas in Communications*, 17(11), 1841-1852.

Goeckel, D. L. (1999). Adaptive coding for time-varying channels using outdated fading estimates. *IEEE Transactions on Communications*, 47(6), 844-855.

Golden, G. D., Foschini, G. J., Valenzuela, R. A. & Wolniansky, P. W. (1999). Detection algorithm and initial laboratory results using the V-BLAST space-time communication architecture. *IEE Electronics Letters*, 35(1), 14-15.

Goldsmith, A. J., & Chua, S.-G. (1997). Variable-rate variable-power MQAM for fading channels. *IEEE Transactions on Communications*, 45(10), 1218-1230.

Goldsmith, A. J., & Chua, S.-G. (1998). Adaptive coded modulation for fading channels. *IEEE Transactions on Communications*, 46(5), 595-602.

Hochwald, B. M. & ten Brink, S. (2003). Achieving near-capacity on a multiple-antenna channel. *IEEE Transactions on Communications*, 51(3), 389-399.

Hole, K. J., Holm, H., & Øien, G. E. (2000). Adaptive multidimensional coded modulation over flat fading channels. *IEEE Journal on Selected Areas in Communications*, 18(7), 1153-1158.

Lau, V. K. N. (2002). Performance analysis of variable rate: symbol-by symbol adaptive bit interleaved coded modulation for Rayleigh fading channels. *IEEE Transactions on Vehicular Technology*, 51(3), 537-550.

Örmeci, P., Liu, X., Goeckel, D. L., & Wesel, R. D. (2001). Adaptive bit-interleaved coded modulation. *IEEE Transactions on Communications*, 49(9), 1572-1581.

Telatar, I. E. (1999). Capacity of multi-antenna Gaussian channels. *European Transactions on Telecommunications*, 10, 585-595.

Vishwanath, S. & Goldsmith, A. J. (2003). Adaptive turbo-coded modulation for flat-fading channels. *IEEE Transactions on Communications*, 51(6), 964-972.

Vucetic, B. (1991). An adaptive coding scheme for time-varying channels. *IEEE Transactions on Communications*, 39(5), 653-663.

Zhou, Z. (2005). MIMO systems with adaptive modulation. *IEEE Transactions on Vehicular Technology*, 54(5), 1828-1842.

Zhou, Z. (2006). *MIMO systems with adaptive modulation and coding.* Unpublished doctoral dissertation, University of Sydney, Australia.

## APPENDIX

Now we give a formal proof to Proposition 1.

We define the incremental SNR between two consecutive component modes as

$$\Delta_j = \alpha_j / \alpha_{j-1},\ j=1, 2, \ldots \quad (38)$$

Thus, $10\log_{10}\Delta_j$ is the extra amount of SNR (in the unit of dB) needed to use the j-th mode instead of the (j − 1)-th mode without sacrificing the BER performance.

Now we have the following lemma.

## Lemma 1

A discrete-rate AMC MIMO system has the same multiplexing gain as its continuous-rate counterpart, if the incremental SNRs are finite, i.e.

$$\Delta_j < \Delta,\ j=1, 2, \ldots \quad (39)$$

where $\Delta$ is a finite constant.

A continuous-rate system is described the same as in Eq. (10)-(14) except that the discrete-rate constraint Eq. (13) is replaced by a non-negative constraint

$k_i \geq 0$, $i$=1,2,…,$m$

which means that the spectral efficiency can be varied on a continuous basis.

**Proof.**

At any instant $t$ and for any subchannel $i$, assume the power and rate allocation for the continuous-rate system is $P_i(t)$ and $k_i(t)$, respectively. Assuming $\mathrm{SE}_{j-1} \leq k_i < \mathrm{SE}_j$, we use $\Delta \cdot P_i(t)$ and $\mathrm{SE}_j$ as the power and rate in the discrete-rate system. Due to the definition of $\Delta_j$ and Eq.(39),($\Delta \cdot P_i(t), \mathrm{SE}_j$) must result in a BER below $\mathrm{BER}_t$. According to this modified power and rate allocation scheme, at any time instant and for each subchannel, the instantaneous spectral efficiency is not lower than that used in the continuous-rate counterpart and the instantaneous transmit power is at maximum $10\log_{10}\Delta$ dB higher than that used in the continuous-rate counterpart. Therefore, we have a discrete-rate system which has an ASE not lower than the corresponding continuous-rate system and an average power usage of up to $10\log_{10}\Delta$ dB higher. In other words, the SNR penalty for the discrete-rate AMC MIMO system, compared to its continuous-rate counterpart, is upper bounded by $10\log_{10}\Delta$ dB. Please see Figure 3 for an illustration. According to the definition of multiplexing gain, this readily concludes that the discrete-rate system has the same multiplexing gain as its continuous-rate counterpart. ■

We notice that the condition in Eq. (39) is generally satisfied in practical systems, so we will only consider continuous-rate systems hereafter.

Using Eq. (31) in the BER constraint Eq. (19), we get

$$k_1 \approx g\log_2(P_1\lambda_1/\sigma^2)+c$$

Then the optimization problem for the continuous-rate AMC MIMO system can be described as

$$\underset{P_1(\lambda_1)}{\text{Maximize}} \quad \mathrm{ASE} = m\left[\int\left(g\log_2\frac{P_1(\lambda_1)\lambda_1}{\sigma^2}+c\right)p_{\lambda_1}(\lambda_1)d\lambda_1\right] \quad (40)$$

$$\text{Subject to} \quad \int P_1(\lambda_1)p_{\lambda_1}(\lambda_1)d\lambda_1 = \frac{P_t}{m} \quad (41)$$

By applying a standard Lagrangian method, we obtain the optimal power allocation as

$$P_1 = \frac{P_t}{m}$$

which is simply the constant equal power distribution. Note that for an AMC MIMO system, there is always an SNR threshold, below which the transmission will be suspended, and this threshold is a constant independent of the current SNR. Assume this threshold is $\alpha_1$, Eq. (40) becomes

$$ASE \approx mg\int_{m\alpha_1/\gamma_s}^{\infty}\log_2\left(\frac{\gamma_s}{m\alpha_1}\lambda_1\right)p_{\lambda_1}(\lambda_1)d\lambda_1 + mg\log_2\alpha_1 + mc \quad (42)$$

Note that the first term in Eq. (42) is the same as Eq. (74) in the paper (Zhou, 2005) if we replace $K$ with $\alpha_1$. Thus, by using Eq. (26) of that reference, we can rewrite Eq. (42) at high SNRs as

$$ASE \approx mg\left[\log_2\left(\frac{\gamma_s}{m\alpha_1}\right)+\frac{\mathcal{B}}{m}\right]+mg\log_2\alpha_1+mc$$
$$= mg\log_2\gamma_s + c'$$

where $c' = g\mathcal{B} - mg\log_2 m + mc$ is a constant independent of $\gamma_s$. This has clearly shown that a full multiplexing gain of *mg* is achieved, which finishes the proof.

## REFERENCES


Cavers, J. (1972). Variable-rate transmission for Rayleigh fading channels. IEEE Transactions on Communications, 20(1), 15-22.

Chung, S. T., & Goldsmith, A. J. (2001). Degrees of freedom in adaptive modulation: a unified view. IEEE Transactions on Communications, 49(9), 1561-1571.

de Jong, Y. L. C. & Willink, T. J. (2005). Iterative tree search detection for MIMO wireless systems. IEEE Transactions on Communications, 53(6), 930-935.

Eyuboglu, M. V. & Forney, G. D., Jr. (1992). Trellis precoding: combined coding, precoding and shaping for intersymbol interference channels. IEEE Transactions on Information Theory, 38(2), 301-314.

Forney, G. D., Jr. (1992). Trellis shaping. IEEE Transactions on Information Theory, 38(2), 281-300.

Foschini, G. J. (1998). On limits of wireless communication in a fading environment when using multiple antennas. Wireless Personal Communications, 6(3), 311-335.

Foschini, G. J. (1996). Layered space-time architecture for wireless communication in a fading environment when using multiple antennas. Bell Laboratories Technical Journal, 1(2), 41-59.

Foschini, G. J., Golden, G. D., Valenzuela, R. A. & Wolniansky, P. W. (1999). Sim-plified processing for high spectral efficiency wireless communication employing multi-element arrays. IEEE Journal on Selected Areas in Communications, 17(11), 1841-1852.

Goeckel, D. L. (1999). Adaptive coding for time-varying channels using outdated fading estimates. IEEE Transactions on Communications, 47(6), 844-855.

Golden, G. D., Foschini, G. J., Valenzuela, R. A. & Wolniansky, P. W. (1999). Detec-tion algorithm and initial laboratory results using the V-BLAST space-time communi-cation architecture. IEE Electronics Letters, 35(1), 14-15.

Goldsmith, A. J., & Chua, S.-G. (1997). Variable-rate variable-power MQAM for fad-ing channels. IEEE Transactions on Communications, 45(10), 1218-1230.

Goldsmith, A. J., & Chua, S.-G. (1998). Adaptive coded modulation for fading chan-nels. IEEE Transactions on Communications, 46(5), 595-602.

Hochwald, B. M. & ten Brink, S. (2003). Achieving near-capacity on a multiple-antenna channel. IEEE Transactions on Communications, 51(3), 389-399.

Hole, K. J., Holm, H., & Øien, G. E. (2000). Adaptive multidimensional coded mod-ulation over flat fading channels. IEEE Journal on Selected Areas in Communications, 18(7), 1153-1158.

Lau, V. K. N. (2002). Performance analysis of variable rate: symbol-by symbol adap-tive bit interleaved coded modulation for Rayleigh fading channels. IEEE Transac-tions on Vehicular Technology, 51(3), 537-550.

Örmeci, P., Liu, X., Goeckel, D. L., & Wesel, R. D. (2001). Adaptive bit-interleaved coded modulation. IEEE Transactions on Communications, 49(9), 1572-1581.

Telatar, I. E. (1999). Capacity of multi-antenna Gaussian channels. European Trans-actions on Telecommunications, 10, 585-595.

Vishwanath, S. & Goldsmith, A. J. (2003). Adaptive turbo-coded modulation for flat-fading channels. IEEE Transactions on Communications, 51(6), 964-972.

Vucetic, B. (1991). An adaptive coding scheme for time-varying channels. IEEE Transactions on Communications, 39(5), 653-663.

Zhou, Z. (2005). MIMO systems with adaptive modulation. IEEE Transactions on Vehicular Technology, 54(5), 1828-1842.

Zhou, Z. (2006). MIMO systems with adaptive modulation and coding. Unpublished doctoral dissertation, University of Sydney, Australia.

# Chapter XV
# Detection Based on Relaxation in MIMO Systems

**Joakim Jaldén**
*Royal Institute of Technology, Sweden*

**Björn Ottersten**
*Royal Institute of Technology, Sweden*

## ABSTRACT

*This chapter takes a closer look at a class of MIMO detention methods, collectively referred to as relaxation detectors. These detectors provide computationally advantageous alternatives to the optimal maximum likelihood detector. Previous analysis of relaxation detectors have mainly focused on the implementation aspects, while resorting to Monte Carlo simulations when it comes to investigating their performance in terms of error probability. The objective of this chapter is to illustrate how the performance of any detector in this class can be readily quantified thought its diversity gain when applied to an i.i.d. Rayleigh fading channel, and to show that the diversity gain is often surprisingly simple to derive based on the geometrical properties of the detector.*

## INTRODUCTION

A central component of wireless multiple input-multiple output (MIMO) systems is the symbol detector or demodulator where the receiver produces estimates of the symbols (or bits) transmitted over the MIMO channel given a set of received signals and an estimate of the channel state. However, unlike their single input-single output (SISO) equivalents, naive implementations of optimal detectors for MIMO channels often tend to be prohibitively computationally complex. This is partially due to the fact that in MIMO channels, each output signal tends to be influenced by *all* input signals. The nature of this influence is determined by the channel fading which is not a priori known.

The MIMO channel is frequently modeled in vector/matrix form according to

$$\mathbf{y} = \mathbf{H}\mathbf{s} + \mathbf{v} \tag{1}$$

where $\mathbf{y}$ is the vector of *received signals* (after matched filtering and sampling), where $\mathbf{s}$ is the vector of *transmitted symbols* drawn from some constellation alphabet $\mathcal{S}$, where $\mathbf{v}$ is additive *noise*, and where $\mathbf{H}$ is the *channel matrix* modeling the input-output relation of the MIMO channel. In the particular case of the narrow-band multiple antenna channel with spatial multiplexing across antennas, the elements of $\mathbf{H}$ would have the physical interpretation of baseband equivalent complex gains between the transmitting and receiving antennas (Tse & Viswanath, 2005). The model in itself is however more general and essentially applicable to any scenario where a group of symbols are linearly modulated and transmitted over a linear channel (Barbosa, 1989). In some cases, the channel matrix will have special structure that is exploitable in the transmission and detection process. However, in the multiple antenna MIMO scenario the matrix, **H**, is often unstructured and this calls for general detection methods.

Under the assumption of uncorrelated Gaussian noise and that the receiver has access to both **H** as well as **y** the maximum likelihood (ML) detector of **s** can be expressed according to

$$\hat{\mathbf{s}}_{\mathrm{ML}} = \arg\min_{\hat{\mathbf{s}}\in\mathcal{S}^m} \| \mathbf{y} - \mathbf{H}\hat{\mathbf{s}} \|^2. \tag{2}$$

The ML detector thus amounts to, among all possible noise-free hypotheses $\mathbf{H}\mathbf{s}$, finding the one which most closely matches the vector of received signals and provides the smallest possible probability of detection error. Unfortunately, implementing the ML detector requires solving an constrained minimization problem which is computationally difficult. In fact, it can be shown that (2) is NP-hard for general **H** and **y** (Verdu, 1989) which implies that there are no known polynomial time algorithms for its solution. This has generated a large body of research into optimal and sub-optimal implementations of the ML detector. Part of this research effort has been devoted to the development of a class of detectors which may collectively bereferred to as relaxation detectors (Yener et. al., 2002; Thoen et. al., 2003; Cui et. al. 2005; Cui & Tellembura, 2006). The literature on these detectors has previously focused mainly on implementation aspects. Specifically, in the case of the *convex* relaxations studied in (Yener et. al., 2002; Thoen et. al., 2003), there are well known efficient implementations based on convex optimization (Boyd & Vandenberghe, 2004). This makes the relaxation detectors appealing from a computational complexity point of view. However, the error probability performance of these detectors is traditionally investigated through Monte Carlo simulations while the general detection performance is still not well understood.

The purpose of this chapter is to introduce some analytic tools to study the detection performance of these detectors without having to resort to simulations. In particular, we will show how ideas which are very similar to the pairwise error probability (PEP) analysis of traditional detectors may also be successfully applied to the study of the relaxation detectors. To this end, we shall throughout the chapter adopt some rather simplistic modeling assumptions regarding the quantities in (1) to enable a uniform and comprehensive treatment of the performance of the relaxation detectors while attempting to avoid a range of technical difficulties. We shall assume that $\mathbf{y}\in\mathbb{R}^n$, $\mathbf{H}\in\mathbb{R}^{n\times m}$, $\mathbf{s}\in\mathcal{S}^m\subset\mathbb{R}^m$ and $\mathbf{v}\in\mathbb{R}^n$ are real valued quantities. This is in contrast most literature on MIMO communications where a complex valued model is typically used but since most relaxation detectors are more easily treated under such an assumption we find it preferable. It should also be noted that in many cases (although with a few exceptions) the extension to the complex case is straightforward. In fact, whenever the complex symbol constellation is separable (i.e., *M*-QAM) the complex model can be written on an equivalent real valued form. We shall further assume that the components of **v** and **H** are zero-mean, Gaussian distributed, with variance $\sigma^2$ and $n^{-1}$ respectively, i.e., $\mathbf{v} \sim N(0,\sigma^2\mathbf{I}_n)$ and $\mathrm{vec}(\mathbf{H}) \sim N(0,n^{-1}\mathbf{I}_{nm})$. The performance of the detectors will be addressed by averaging over the realization of **H** and **v**. We will also for simplicity only consider the case of a binary constellation, i.e., $\mathcal{S}=\{\pm1\}$.With these assumptions the signal to noise ratio (SNR), herein denoted $\rho$, is defined according to

$$\rho \equiv \frac{E\{\| \mathbf{Hs} \|^2\}}{m\sigma^2} = \frac{1}{\sigma^2}. \tag{3}$$

We will assume that $n \geq m$ which implies that $\mathbf{H}^T\mathbf{H}$ is invertible with probability one and, based on (3), for the most part use $\rho^{-1}$ in place of $\sigma^2$ in the analysis.

## Optimal and Sub-Optimal Detection

As stated in the introduction, the ML detector is given by

$$\hat{\mathbf{s}}_{\mathrm{ML}} = \arg\min_{\hat{\mathbf{s}}\in\mathcal{S}^m} \| \mathbf{y} - \mathbf{H}\hat{\mathbf{s}} \|^2 \tag{4}$$

and its straightforward implementation requires a search over $2^m$ hypotheses corresponding to the $2^m$ different vectors in $\mathcal{S}^m$. A computationally less cumbersome, heuristic, approach to the detection problem is provided by the zero forcing (ZF) detector. The ZF detector attempts to remove the impact of the channel matrix **H** by first multiplying **y** by the pseudo inverse of **H** and then finding the closest candidate vector. The ZF estimate is given by

$$\hat{\mathbf{s}}_{\mathrm{ZF}} = \arg\min_{\hat{\mathbf{s}} \in \mathcal{S}^m} \| \mathbf{H}^{\dagger}\mathbf{y} - \hat{\mathbf{s}} \|^2 \quad (5)$$

where $\mathbf{H}^{\dagger} \equiv (\mathbf{H}^T\mathbf{H})^{-1}\mathbf{H}^T\mathbf{y}$ denotes the pseudo-inverse of **H**. The minimization in (5) can be accomplished by componentwise rounding and it follows that $\hat{\mathbf{s}}_{\mathrm{ZF}} = \mathrm{sgn}(\mathbf{H}^{\dagger}\mathbf{y})$ under the binary constellation assumption. Solving (5) in order to find an estimate of **s** is however not optimal from a probability of error point of view.

There is a large body of works that propose and studies MIMO detectors that range from the simplest cases, to which the ZF detector belong, all the way to exact ML performance. In the high performance range we find the so called sphere decoders (Damen et. al., 2003; Murugan et. al. 2006) which represent exact implementations of the ML detector at lower computational cost than the brute force implementation. The sphere decoder has however been shown to have an average complexity which is exponential in the number of transmit antennas (Jaldén & Ottersten, 2005) which limits the range of problem sizes to which it may be applied. Additionally, the use of lattice reductions techniques have been suggested in order to design detectors which have the same diversity as the optimal ML detector at a significantly reduced computational cost (Yao & Wornell, 2002; Taherzadeh et al., 2007). As stated in the introduction, we will study a class of detectors collectively referred to as relaxation detectors.

To arrive at analytical performance results, we will study these detectors in the high SNR regime by means of their diversity order, *d*, defined according to

$$d \equiv -\lim_{\rho \to \infty} \frac{\ln P(\hat{\mathbf{s}} \neq \mathbf{s})}{\ln \rho}. \quad (6)$$

Thus, the diversity order specifies the rate (on a log-log scale) at which the error probability tends to zero with increasing SNR. A higher diversity order indicates better high SNR performance. It is previously known that the diversity order of the ML detector, for the model considered herein is

$$d_{\mathrm{ML}} = \frac{n}{2} \quad (7)$$

while the diversity order of the ZF detector is given by

$$d_{\mathrm{ZF}} = \frac{n-m+1}{2}. \quad (8)$$

It is also worth noting that the factor $\frac{1}{2}$ which appears in both (7) and (8) is due to the fact that we are explicitly considering a real valued model. For the complex model the corresponding results would have been $n$ and $n - m + 1$. The proofs of (7) and (8) are analogous to their complex valued counterparts (Tse & Viswanath, 2005).

Although the diversity order does not fully characterize the behavior of a detector, especially at low to moderate SNR, it is a commonly used measure of a detector's robustness toward fading when applied to the wireless MIMO channel. Also, as it results in a single scalar measure of performance, it is very useful as a first comparison of two different detectors. In particular, it can be seen that the ZF detector yields the lowest diversity order for the detectors we consider and that the ML detector yields the highest diversity order.

It should also be pointed out that we explicitly study the case of uncoded systems. This can be motivated in several ways. For one, the uncoded system yields the shortest possible delay in communications. Further, also in the (arguably more realistic) coded case the insight gained from the structurally simpler uncoded case is useful when understanding the performance difference of various detection strategies. In fact, almost all soft output MIMO demodulators considered in practice either exploit a hard output detector as a component of the demodulator or is similar to a related hard decision detector. This former case is exemplified in (Steingrimsson et. al. 2003) where the hard output semidefinite relaxation detector is used together with the log-max approximation to yield soft information in a coded system. The latter case is exemplified in (Hochwald & Brink, 2003) by the use of a list sphere decoder which may be seen as an extension of the related, original, hard decision sphere decoder.

## Exponential Equality

Following (Zheng & Tse, 2003), we make use of the $\doteq$ notation to denote *exponential equality*, defined according to

$$f(\rho) \doteq \rho^a \Leftrightarrow \lim_{\rho\to\infty} \frac{\ln f(\rho)}{\ln \rho} = a \tag{9}$$

where $f(\rho)$ is any function of $\rho$. Similar definitions will apply to $\dot{\leq}$ and $\dot{\geq}$. The value of introducing this notation is that it gives a relatively compact and suggestive way of expressing results regarding diversity. For example, the results of (7) and (8) reduce to

$$P(\hat{\mathbf{s}}_{\mathrm{ML}} \neq \mathbf{s}) \doteq \rho^{-\frac{n}{2}} \quad \text{and} \quad P(\hat{\mathbf{s}}_{\mathrm{ZF}} \neq \mathbf{s}) \doteq \rho^{-\frac{n-m+1}{2}}.$$

A few consequences of the definition in (9) which we will make frequent use of are:

- An invariance to shifts in SNR, i.e.

$$f(\rho) \doteq f(\alpha\,\rho) \tag{10}$$

for any $\alpha \in \mathbb{R}$.

- A scaling invariance, i.e.

$$\alpha f(\rho) \doteq f(\rho) \tag{11}$$

for any $\alpha \in \mathbb{R}$.

- An additive property, i.e.

$$f(\rho) \doteq \rho^a, g(\rho) \doteq \rho^b \Rightarrow f(\rho) + g(\rho) \doteq \rho^{\max(a,b)}. \tag{12}$$

- A multiplicative property, i.e.

$$f(\rho) \doteq \rho^a, g(\rho) \doteq \rho^b \Rightarrow f(\rho)g(\rho) \doteq \rho^{a+b}. \tag{13}$$

As these properties are straightforwardly derived from (9) and we omit their explicit proofs.

## The Class of Relaxation Detectors

In order to introduce the relaxation detectors, consider the two step approach of obtaining an estimate, $\hat{\mathbf{s}}$, of $\mathbf{s}$ given by first solving

$$\mathbf{z}^{\star} \equiv \arg\min_{\mathbf{z}\in\mathcal{R}} \| \mathbf{y} - \mathbf{H}\mathbf{z} \|^2, \tag{14}$$

where $\mathcal{R}$ is some given set for which $\mathcal{S}^m \subset \mathcal{R} \subset \mathbb{R}^m$, and then solving

$$\hat{\mathbf{s}} \equiv \arg\min_{\hat{\mathbf{s}} \in \mathcal{S}^m} \| \mathbf{z}^\star - \hat{\mathbf{s}} \|^2 . \tag{15}$$

The ML and ZF detectors considered above may both be viewed as special instances of this approach and are identified by the choice of the relaxation set, $\mathcal{R}$. In the case of the ML detector $\mathcal{R}$ corresponds to the original constellations, i.e., $\mathcal{R} = \mathcal{S}^m$, and in the case of the ZF detector $\mathcal{R}$ corresponds to the set of real numbers, i.e., $\mathcal{R} = \mathbb{R}^m$. Naturally, for the ML detector the second step in (15) becomes somewhat of a formality. For the ZF detector it follows that $\mathbf{z}^\star = \mathbf{H}^\dagger \mathbf{y}$ which is implicit in (5). However, making the two step procedure explicit opens the possibility of selecting $\mathcal{R}$ somewhere between the two extremes given by $\mathcal{S}^m$ and $\mathbb{R}^m$. This also leads to a generic description of the class of *relaxation detectors*.

In this context of the relaxation detectors, the first optimization in (14) may be viewed as a relaxation of the optimization problem in (2) while the second optimization step may be viewed as a rounding procedure which is required since $\mathbf{z}^\star \in \mathcal{S}^m$ can not be guaranteed for general $\mathcal{R}$. Naturally, for a relaxation detector to be useful in the first place, there must be an efficient way of solving (14) and this is typically the main topic of the papers written on the subject (Yener et al., 2002; Thoen et al., 2003; Cui et. al. 2005; Cui & Tellembura, 2006). We shall however not discuss this issue in detail but instead consider a generic relaxation detector defined by $\mathcal{R}$ and simply assume that there is an algorithm available that will solve (14). This assumption is made in order to focus on the discussion of the "probability of error" performance of the generic detector and to make the analysis generally applicable. The implementation details of the specific relaxation detectors discussed in this work are treated in the references cited above.

In order to provide some concrete examples, note that two straightforward relaxations are given by

$$\mathcal{R} = \mathcal{R}_1 \equiv \{\mathbf{s} \mid \| \mathbf{s} \|_\infty \leq 1\} \tag{16}$$

and

$$\mathcal{R} = \mathcal{R}_2 \equiv \{\mathbf{s} \mid \| \mathbf{s} \|^2 \leq m\}. \tag{17}$$

These relaxations were suggested in (Yener et al., 2002) in the context of CDMA but are equally applicable to the generic detection problem considered herein. In short, $\mathcal{R}_1$ constrains the transmitted vector to an $m$-dimensional hypercube while $\mathcal{R}_2$ constrains it to a hypersphere. As $\mathcal{R}_1$ and $\mathcal{R}_2$ are convex sets we refer to the corresponding relaxation detectors as convex relaxation detectors. Also, well known convex optimization techniques (Boyd & Vandenberghe, 2004) yields efficient solutions to (14) in these cases.

The relaxation proposed in (Thoen et al., 2003) under the name constrained least squares (CLS), is given by

$$\mathcal{R} = \mathcal{R}_3 \equiv \{\mathbf{s} \mid \| \mathbf{s} \|^2 = m\} \tag{18}$$

and thus constrains the feasible points in (14) to belong to the *surface* of a hyper-sphere. Although this does not lead to a convex optimization problem in (14) it can be shown that strong duality will nonetheless hold and the optimization problem in (14) can still be efficiently solved using convex optimization techniques involving the Lagrange dual function (Thoen et al., 2003; Boyd & Vandenberghe, 2004).

Although the relaxation detectors obtained by choosing $\mathcal{R}$ according to the above three strategies all improve upon the ZF detector in terms of error provability we shall see that none of them are capable of providing a diversity order greater than the one experienced by the ZF detector. The performance of the CLS detector (based on (18)) is illustrated in Fig. 1 where also the performance of the ML and ZF detectors are included for reference. In order to clarify why no diversity increase is observed for the CLS detector we shall in what follows take a deeper look at the error probability of the generic relaxation detector. We shall specifically address the question of how the set $\mathcal{R}$ would have to be structured in order to provide an increase in diversity and explain why the relaxation detectors discussed in (Cui et. al. 2005; Cui & Tellembura, 2006) does see such an increase in diversity.

Before proceeding, it should also be made clear that there are relaxation detectors that are not directly expressed in the form of (14) and (15). The now well known semidefinite relaxation (SDR) detector proposed in (Tan & Rasmussen, 2001; Ma et al., 2002) and polynomial moment relaxation detector (Cui et al., 2006) are examples of such. These detectors are interesting since they have through numerical simulations been shown to provide high diversity while having computationally attractive implementations. In the case of the SDR detector it has also recently been rigorously proven that it achieves the maximal diversity order when applied to the model considered herein (Jaldén & Ottersten, 2008). We shall however not consider these detectors further in this work but remain focused on the basic structure offered by (14), the idea being that it is more useful for the understanding of the general relaxation detector to clarify *why* these basic relaxation detectors have a specific diversity order and also show how this relates to the geometry of $\mathcal{R}$.

# DIVERSITY OF THE ML AND ZF DETECTORS

In order to obtain insight into the performance of the class of relaxation detectors, especially in terms of diversity, it is convenient to first review the "classical" pairwise error probability (PEP) analysis of the ML and ZF detectors and then extrapolate the insight gained to the relaxation detectors. We will in the following briefly review the PEP analysis and then in Section 3 discuss how these concepts relate to the performance of the relaxation detectors. However, since the aim is only to provide some intuition before addressing the error probability analysis of the relaxation detectors (which is the main topic of this chapter) we will make no attempt to provide any stringent proofs of the statements made. Instead, the reader previously unfamiliar with the error probability analysis of the ML and ZF detector is referred to other sources, see e.g. (Tse and P. Viswanath, 2005, Appendix A), for a comprehensive treatment of the topic.

## Pairwise Error Probability

### ML Detection

For the ML detector, the pairwise error probability is defined as

$$P(\mathbf{s} \to \hat{\mathbf{s}}) \equiv P\left(\|\mathbf{y}-\mathbf{H}\hat{\mathbf{s}}\|^2 < \|\mathbf{y}-\mathbf{H}\mathbf{s}\|^2\right). \tag{19}$$

Specifically, the pairwise error probability is the probability that $\hat{\mathbf{s}}$ is (according to the ML metric in (2)) considered more likely than the vector $\mathbf{s}$ which was actually transmitted. The occurrence of such an event is guaranteed to yield a detection error in the sense that $\hat{\mathbf{s}} \neq \mathbf{s}$. For the ML detector the conditional PEP may be explicitly computed according to (Paulraj, 2003)

$$P(\mathbf{s} \to \hat{\mathbf{s}} \mid \mathbf{H}) = Q\left(\frac{\|\mathbf{H}\boldsymbol{\delta}\|}{2\rho^{-\frac{1}{2}}}\right) \tag{20}$$

where $\boldsymbol{\delta} \equiv \mathbf{s} - \hat{\mathbf{s}}$. By taking the fading of $\mathbf{H}$ into account, one obtains

$$P(\mathbf{s} \to \hat{\mathbf{s}}) \equiv P\left(\|\mathbf{y}-\mathbf{H}\hat{\mathbf{s}}\|^2 < \|\mathbf{y}-\mathbf{H}\mathbf{s}\|^2\right) = E\left[Q\left(\frac{\|\mathbf{H}\boldsymbol{\delta}\|}{2\rho^{-\frac{1}{2}}}\right)\right], \tag{21}$$

where the expectation is taken with respect to the distribution of $\mathbf{H}$. By using the union bound (Durret, 1996), it can be shown that the overall error probability of the ML detector is bounded according to

$$p_{e\mathrm{ML}} \equiv P(\hat{\mathbf{s}}_{\mathrm{ML}} \neq \mathbf{s}) \leq \sum_{\hat{\mathbf{s}} \neq \mathbf{s}} P(\mathbf{s} \to \hat{\mathbf{s}}), \tag{22}$$

a bound which is known to be tight in the sense that it gives the correct diversity order. The diversity order of the ML detector (for the setup considered herein) is given by (7). What determines the diversity order of the ML detector is the probability that the argument of the $Q$-function in (20) is small, i.e. that

$$\frac{\|\mathbf{H}\boldsymbol{\delta}\|}{2\rho^{-\frac{1}{2}}} \leq \gamma \quad \Leftrightarrow \quad \|\mathbf{H}\boldsymbol{\delta}\|^2 \leq 4\gamma^2\rho^{-1}$$

for some $\gamma > 0$. It can be shown that (Tse & Viswanath, 2005)

$$P\left(\|\mathbf{H}\boldsymbol{\delta}\|^2 \leq 4\gamma^2\rho^{-1}\right) \doteq \rho^{-\frac{n}{2}}$$

from which it follows that

$$P(\mathbf{s} \to \hat{\mathbf{s}}) \doteq \rho^{-\frac{n}{2}}.$$

Since all terms appearing in the union bound in (22) decay at the same rate with increasing SNR it follows by property (12) that

$$p_{e\mathrm{ML}} \equiv P(\hat{\mathbf{s}}_{\mathrm{ML}} \neq \mathbf{s}) \doteq \rho^{-\frac{n}{2}}$$

which is equivalent to the statement in (7).

## ZF Detection

Similar to the ML detector it is also relevant to treat the ZF detector performance through the use of pairwise error events. In the ZF case the pairwise error event is naturally defined as the event that the ZF detector, using the metric in (5), decides in favor of $\hat{\mathbf{s}} \neq \mathbf{s}$ when in fact $\mathbf{s}$ was transmitted. Specifically, the ZF-PEP may be defined accordion to

$$P(\mathbf{s} \to \hat{\mathbf{s}}) \equiv P\left(\|\mathbf{H}^{\dagger}\mathbf{y} - \hat{\mathbf{s}}\|^2 \leq \|\mathbf{H}^{\dagger}\mathbf{y} - \mathbf{s}\|^2\right).$$

The ZF-PEP simplifies to (Biglieri et. al., 2002)

$$P(\mathbf{s} \to \hat{\mathbf{s}}) = E\left[Q\left(\frac{\|\boldsymbol{\delta}\|^2}{2\rho^{-\frac{1}{2}}\sqrt{\boldsymbol{\delta}^T(\mathbf{H}^T\mathbf{H})^{-1}\boldsymbol{\delta}}}\right)\right], \quad (23)$$

where $\boldsymbol{\delta} \equiv \mathbf{s} - \hat{\mathbf{s}}$ denotes the codeword difference vector and where the expectation is again taken over the fading distribution of $\mathbf{H}$. Based on (23) it is fairly straightforward to prove that

$$p_{e\mathrm{ZF}} \equiv P(\hat{\mathbf{s}}_{\mathrm{ZF}} \neq \mathbf{s}) \doteq \rho^{-\frac{n-m+1}{2}}$$

which is equivalent to (8), see e.g. (Biglieri et. al., 2002).

## The Geometry of Error Events

In the case of ML detection we can argue that for an error to occur in the high SNR regime where $\|\mathbf{v}\|^2 \approx m\rho^{-1}$ the channel matrix realization must also satisfy

$$\|\mathbf{H}(\mathbf{s} - \hat{\mathbf{s}})\|^2 \approx m\rho^{-1}$$

for some $\hat{\mathbf{s}} \neq \mathbf{s}$. This event can be given a rather natural and intuitive geometrical interpretation when the model in (1) is viewed as a (random) linear map of the input symbol vector followed by a perturbation by $\mathbf{v}$. More concretely, consider the conditional PEP in the example where

$$\mathbf{H} = \begin{bmatrix} 3 & 2 \\ 4 & 1 \end{bmatrix}, \quad \mathbf{s} = \begin{bmatrix} 1 \\ 1 \end{bmatrix}, \quad \text{and} \quad \hat{\mathbf{s}} = \begin{bmatrix} -1 \\ 1 \end{bmatrix}.$$

The situation is illustrated in Fig. 2 where on the left hand side the four possible symbol vectors (or points) in $\mathcal{S}^2$ are illustrated by circles. The specific points corresponding to $\mathbf{s}$ and $\hat{\mathbf{s}}$ are shown in black and the codeword difference vector given by $\boldsymbol{\delta} = \hat{\mathbf{s}} - \mathbf{s}$ is also indicated. The right hand side of Fig. 2 illustrates the same points under the linear mapping induced by $\mathbf{H}$. We will in what follows refer to the left hand side of Fig. 2 as *input space* and the right hand size as *output space*. The vector of received signals, $\mathbf{y}$, may thus be regarded as a point, $\mathbf{Hs}$, in the output space perturbed by the noise vector $\mathbf{v}$.

Since the ML detector in (2) selects the feasible point which is closest in the output space it follows that in order for a pairwise error event to occur the noise perturbation must move the received point, $\mathbf{y}$, closer to $\mathbf{H}\hat{\mathbf{s}}$ than $\mathbf{Hs}$. Equivalently, the projection of the noise vector $\mathbf{v}$ onto the difference vector $\mathbf{H}(\hat{\mathbf{s}} - \mathbf{s})$ mush have a length of at least $\mu_{\mathrm{ML}} \equiv \frac{1}{2} \| \mathbf{H}\boldsymbol{\delta} \|$. The probability of this event is what is given by (20). Thus, intuitively the overall error probability of the ML detector is determined by the probability that $\mu_{\mathrm{ML}}$ is small in comparison with the noise component. Specifically, it can be shown that

$$P\left(\mu_{\mathrm{ML}}^2 \leq \rho^{-1}\right) \doteq \rho^{-\frac{n}{2}} \tag{24}$$

which is to be compared with the diversity order of the ML detector, i.e. the error probability of the ML detector is proportional to the probability of obtaining a value of $\mu_{\mathrm{ML}}^2$ which is on the order of $\rho^{-1}$.

A similar geometric interpretation of the ZF pairwise error is also possible for the ZF detector. Although it is certainly possible to consider the metric (5) directly we will instead consider the relaxation formulation of the ZF detector offered by (14) and (15) where $\mathcal{R} = \mathbb{R}^m$. To this end, let $\mathcal{A}$ denote the set of all possible realizations of the intermediate quantity $\mathbf{z}^*$ in (14) for which the ZF detector decides in favor of $\hat{\mathbf{s}}$ as opposed to $\mathbf{s}$. In short, $\mathcal{A}$ consists simply of all points closer to $\hat{\mathbf{s}}$ than $\mathbf{s}$. For the particular example considered previously this set is illustrated by the shaded region on the left hand side of Fig. 3 The pairwise error event of the ZF detector is equivalent to $\mathbf{y} \in \mathbf{H}\mathcal{A}$ where $\mathbf{H}\mathcal{A}$ denotes the image of $\mathcal{A}$ in the output space. In the example illustrated in Fig. 3 a pairwise error occurs if and only if the vector of received signals falls within the shaded region on the right. Analogously with the ML detector analysis it follows that for an error to occur the noise must perturb the vector $\mathbf{Hs}$ by a distance of at least $\mu_{\mathrm{ZF}}$, where $\mu_{\mathrm{ZF}}$ is indicated in Fig. 3 and formally defined as

$$\mu_{\mathrm{ZF}} \equiv \min_{\mathbf{y} \in \mathbf{H}\mathcal{A}} \| \mathbf{y} - \mathbf{Hs} \| = \min_{\mathbf{z} \in \mathcal{A}} \| \mathbf{H}(\mathbf{z} - \mathbf{s}) \|. \tag{25}$$

The value of $\mu_{\mathrm{ZF}}$ may be computed explicitly (it amount to solving a quadratic program subject to linear constrains (Boyd & Vandenberghe, 2004)) and is given by

$$\mu_{\mathrm{ZF}} = \frac{\| \boldsymbol{\delta} \|^2}{2\sqrt{\boldsymbol{\delta}^T (\mathbf{H}^T \mathbf{H})^{-1} \boldsymbol{\delta}}}. \tag{26}$$

This is (maybe not surprisingly) the same expression as the expression which appears in the argument of the $Q$-function in (23). Similar to the ML case it is also possible to examine the distribution of $\mu_{\mathrm{ZF}}$ in order to conclude that

$$P\left(\mu_{\mathrm{ZF}}^2 \leq \rho^{-1}\right) \doteq \rho^{-\frac{n-m+1}{2}}, \tag{27}$$

an observation which should be compared to the ZF detector diversity, i.e. the error probability of the ZF detector is proportional to the probability of obtaining a value of $\mu_{\mathrm{ZF}}^2$ which is on the order of $\rho^{-1}$.

## RELAXATION DETECTOR DIVERSITY

We will now tackle the problem of determining $p_e = P(\hat{s} \neq s \mid s)$ for the generic relaxation detector outlined by (14) and (15). In short, we will be interested in determining the probability that (given that $s$ was transmitted) the detector decides in favor of $\hat{s} \neq s$. Specifically, we shall by considering the geometry of the relaxation detector derive bounds on the diversity order, given the geometric properties of the set $\mathcal{R}$. Our main result states that the diversity order of the relaxation detector is mainly determined by the topological dimension of $\mathcal{R}$.

### The Geometry of the Relaxation Detector

In order to determine the diversity order of the generic relaxation, consider again the generic relaxation detector introduced in Section 1.3 where the estimate, $\hat{s}$, is obtained by first solving

$$\mathbf{z}^{\star} \equiv \arg\min_{\mathbf{z} \in \mathcal{R}} \| \mathbf{y} - \mathbf{H}\mathbf{z} \|^2$$

and then enforcing the constellation constraint by mapping $z^{\star}$ to $\mathcal{S}^m$ according to

$$\hat{\mathbf{s}} \equiv \arg\min_{\hat{\mathbf{s}} \in \mathcal{S}^m} \| \mathbf{z}^{\star} - \hat{\mathbf{s}} \|^2 .$$

To understand what events may cause a detection error it is useful to again consider the geometric model discussed in Section 2.2. To make the discussion a little less abstract, we will in the examples consider the CLS relaxation of (Thoen et. al. 2003) obtained by choosing $\mathcal{R} = \mathcal{R}_3$ for $\mathcal{R}_3$ given by (18). This relaxation is obtained by noting that $\| s \|^2 = m$ for all $\mathbf{s} \in \mathcal{S}^m$. The detection process of the CLS relaxation detector is illustrated in Fig. 4 where $\mathcal{R}$ and its image, $\mathbf{H}\mathcal{R}$, in the output space are illustrated. The transmitter selects a message, **s**, to be transmitted and the detector obtains **y** which lies a distance of $\| v \|$ away from $\mathbf{Hs}$ in the output space. The relaxation detector solves (14) and finds the point $\mathbf{Hz}^{\star}$ for $\mathbf{z}^{\star} \in \mathcal{R}$ which is closest to **y** in the *output space* and provides an estimate of **s** by finding the symbol vector $\hat{s}$ which in closest to $z^{\star}$ in the *input space*. The quantities of interest are illustrated in Fig. 4. In this case the relaxation detector would also obtain the correct estimate, namely $\hat{s} = s$.

It is apparent that if an error is to occur this would require that $z^{\star}$ falls far from **s**. However, in the high SNR range of interest, where **v** would correspond to a small perturbation of $\mathbf{Hs}$ this would require that some point in $\mathcal{R}$, say $\mathbf{z} \in \mathcal{R}$, which is far from $\mathbf{s}$ has an image (under transformation by $\mathbf{H}$) which is close to $\mathbf{Hs}$. Specifically, it would require that

$$\| \mathbf{Hz} - \mathbf{Hs} \| = \| \mathbf{H}(\mathbf{z} - \mathbf{s}) \|$$

is small compared to $\| v \|$. In order to make these statements precise consider the following lemma where $\mathcal{R}_e$ is used to denote the feasible points in the input space which are mapped to some $\hat{\mathbf{s}} \neq \mathbf{s}$.

**Lemma 1.** Let $\mathcal{R}_e$ be defined by

$$\mathcal{R}_e \equiv \mathcal{R} \cap \{\mathbf{z} \in \mathbb{R}^m \mid \exists \hat{\mathbf{s}} \neq \mathbf{s}, \| \mathbf{z} - \hat{\mathbf{s}} \| \leq \| \mathbf{z} - \mathbf{s} \| \} \tag{28}$$

and let

$$\mu \equiv \min_{\mathbf{z} \in R_e} \| \mathbf{H}(\mathbf{z} - \mathbf{s}) \| . \tag{29}$$

Then

$$\mu > 2 \| \mathbf{v} \| \Rightarrow \hat{\mathbf{s}} = \mathbf{s} . \tag{30}$$

*Proof:* Let $\mathcal{R}_e$ be given by (28) and assume that $\mathbf{z} \in \mathcal{R}_e$. Then it follows that

$$\| \mathbf{y} - \mathbf{Hz} \| = \| \mathbf{H}(\mathbf{s}-\mathbf{z}) + \mathbf{v} \| \geq \| \mathbf{H}(\mathbf{s}-\mathbf{z}) \| - \| \mathbf{v} \| > \| \mathbf{v} \|$$

where the last inequality follows by the definition of $\mu$ and the assertion on the left hand side of (29). At the same time we have

$$\| \mathbf{y} - \mathbf{Hs} \| = \| \mathbf{v} \|$$

which implies that **z** can not be optimal in (14) (it does not have the smallest objective value). Thus, it follows that $\mathbf{z}^\star \notin \mathcal{R}_e$, or equivalently that $\|\mathbf{z}^\star - \mathbf{s}\| < \|\mathbf{z}^\star - \hat{\mathbf{s}}\|$, for all $\hat{\mathbf{s}} \neq \mathbf{s}$ which implies $\hat{\mathbf{s}} = \mathbf{s}$.

The geometric interpretation of Lemma 1 is given in Fig. 5. Again, analogous to the PEP analysis of the ML and ZF detectors, it states that unless **H** is such that any point in $\mathcal{R}_e$ is close to **Hs** in the output space, the detector will make a correct decision. A natural question is thus if the diversity order of the relaxation detector may be obtained by studying the probability that $\mu$ is small in a similar manner to (24) and (27). Unfortunately it turns out that it is difficult to give a generally valid answer to this question without making *any* further assumptions regarding $\mathcal{R}$. However, as Lemma 1 only specifies a sufficient condition for correct decision one may expect to obtain a lower bound on the diversity order of the detector based solely on the statistics of $\mu$. This is also correct and given a mathematically precise meaning by Proposition 2 below.

**Proposition 2.** Let $\mu$ be defined according to (29), then

$$P\left(\mu^2 \leq \rho^{-1}\right) \doteq \rho^{-d} \quad \Rightarrow \quad P(\hat{\mathbf{s}} \neq \mathbf{s} \mid \mathbf{s}) \dot{\leq} \rho^{-d}. \tag{31}$$

*Proof:* By assumption it follows that given $\delta > 0$ it holds that

$$-\frac{\ln P\left(\mu^2 \leq \rho^{-1}\right)}{\ln \rho} \geq d - \delta \Leftrightarrow P\left(\mu^2 \leq \rho^{-1}\right) \leq \rho^{-d+\delta}$$

provided $\rho$ is sufficiently large. Further, since $P\left(\mu^2 \leq \rho^{-1}\right) \leq 1$ for any value of $\rho$ it follows that there is a constant, $\alpha$, for which

$$P\left(\mu^2 \leq \rho^{-1}\right) \leq \alpha \rho^{-d+\delta}$$

for all $\rho \geq 0$. Let $\xi \equiv \rho \|\mathbf{v}\|^2$ and note that by Lemma 1 it is known that $\mu^2 > 4\rho^{-1}\xi$ represents a sufficient condition for $\hat{\mathbf{s}} = \mathbf{s}$. Note also that $\xi$ is $\chi_n^2$ distributed and let $f_\xi(x)$ denote the probability density function (PDF) of $\xi$. It follows that

$$P(\hat{\mathbf{s}} \neq \mathbf{s} \mid \mathbf{s}) \leq P\left(\mu^2 \leq 4\rho^{-1}\xi\right) = \int P\left(\mu \leq 4\rho^{-1}x\right) f_\xi(x)dx$$

$$\leq \int \alpha (4\rho^{-1}x)^{d-\delta} f_\xi(x)dx = \alpha (4\rho^{-1})^{d-\delta} E[\gamma^{d-\delta}]$$

Since the moments of $\gamma$ are finite and independent of $\rho$ it follows that

$$P(\hat{\mathbf{s}} \neq \mathbf{s} \mid \mathbf{s}) \dot{\leq} \rho^{-d+\delta}.$$

However, since $\delta > 0$ was arbitrary the assertion made in (31) follows.

Proposition 2 effectively states that it is sufficient to study the probability that $\mu$ is small in order to provide a lower bound on the diversity order of the relaxation detector. In order to fully characterize the diversity order one would of course also need to prove that the inequality which appears in (31) is tight. Intuitively, this follows from the statement that if $\mu < \rho^{-1}$ the detector will make an error with high probability which is certainly plausible based on the geometric insight. It should also be noted that although the right hand side of (31) provides a bound, conditioned on **s**, most relaxation detectors of interest are symmetric in the sense that $P(\hat{\mathbf{s}} \neq \mathbf{s} \mid \mathbf{s})$ could be replaced by $P(\hat{\mathbf{s}} \neq \mathbf{s})$.

## Typical Errors and Diversity Order

Analogous to with the information theoretic notion of channels in outage it is useful to define the notion of *typical errors* (Tse & Viswanath, 2005). In light of Proposition 2 a typical error would be characterized by an event for which $\mu < \rho^{-1}$ and the diversity order or the relaxation detector is given by

$$-\lim_{\rho\to\infty} \frac{\ln P\left(\mu^2 \leq \rho^{-1}\right)}{\ln \rho} \equiv d. \tag{32}$$

Determining $d$ is however still a non-trivial issue which must be addressed at some point. Thus, in what follows, we will provide a general methodology for computing $d$ based on the geometry or $R$.

In order to solve this problem it is convenient to first address it in a slightly more abstract setting. To this end, let $\tau(\mathcal{A})$ be given by

$$\tau(\mathcal{A}) \equiv \min_{\mathbf{a}\in\mathcal{A}} \| \mathbf{Ha} \|^2 \tag{33}$$

and $\mathcal{A} \subset \mathbb{R}^n$ is an arbitrary, non-empty, closed set which does not include the origin (if $\mathcal{A}$ would include the origin we could make the trivial observation that $\tau(\mathcal{A}) = 0$). Further, let $d(\mathcal{A})$ be defined according to

$$d(\mathcal{A}) \equiv -\lim_{\rho\to\infty} \frac{\ln P\left(\tau(\mathcal{A}) \leq \rho^{-1}\right)}{\ln \rho}.$$

The question at hand is now how $d(\mathcal{A})$ depends on the geometry or structure of $\mathcal{A}$. Note also that this is closely related to (29), where $\mu^2 = \tau(\mathcal{A})$ and $\mathcal{A}$ is equal to $\mathcal{R}_e - \mathbf{s}$.

It is illustrative to first consider the simplest case where $\mathcal{A}$ contains only one element, $\mathcal{A} = \{\mathbf{a}_1\}$. This case has actually already been encountered in the study of the ML-PEP. Specifically, in the one element case $\tau(\mathcal{A})$ is simply given by $\tau(\mathcal{A}) = \| \mathbf{Ha}_1 \|^2$ and

$$P\left(\tau(\mathcal{A}) \leq \rho^{-1}\right) \doteq \rho^{-\frac{n}{2}} \tag{34}$$

which follows by (24). By applying the union bound, the same result follows in a straightforward fashion for any set, $\mathcal{A} = \{\mathbf{a}_k\}_{k=1}^{K}$, composed of a finite number of points, i.e.

$$P\left(\tau \leq \rho^{-1}\right) \leq \sum_{k=1}^{K} P\left(\| \mathbf{Ha}_k \|^2 \leq \rho^{-1}\right) \doteq \rho^{-\frac{n}{2}}.$$

For the general case where $\mathcal{A}$ may contain an infinite number of elements it is naturally not possible to naively rely on the union bound in the same way as was done above. Specifically, a straightforward use of the union bound would only yield the obvious statement that $P\left(\tau(\mathcal{A}) \leq \rho^{-1}\right) \leq \infty$ which is not very useful. However, the events that $\| \mathbf{Ha}_1 \|^2$ and $\| \mathbf{Ha}_2 \|^2$ are small are highly dependent when $\| \mathbf{a}_1 - \mathbf{a}_2 \|$ is small. Therefore, we argue that itis sufficient to sample the set $\mathcal{A}$ with a sample spacing of $\rho^{-\frac{1}{2}}$ in order to obtain a sufficiently tight bound on $P\left(\tau(\mathcal{A}) \leq \rho^{-1}\right)$. To make this statement precise consider the following lemma.

**Lemma 3.** Let $\mathcal{A}_\rho(\mathbf{a}_0)$ be given by

$$\mathcal{A}_\rho(\mathbf{a}_0) \equiv \{\mathbf{a} \mid \| \mathbf{a} - \mathbf{a}_0 \| \leq \rho^{-\frac{1}{2}}\} \tag{35}$$

and

$$\tau_\rho(\mathbf{a}_0) \equiv \min_{\mathbf{a}\in \mathcal{A}_\rho(\mathbf{a}_0)} \| \mathbf{Ha} \|^2 .$$

Then

$$P\big(\tau_\rho(\mathbf{a}_0)\big) \dot{\leq} \rho^{-\frac{1}{2}}$$

*Proof:* The proof is similar to the proof of (Jaldén & Ottersten, 2008, Lemma 3) and omitted.

The interpretation of the lemma is that the single point argument leading up to (34) may also be extended to a spherical neighborhood of radius $\rho^{-\frac{1}{2}}$. The union bound can thus be applied to a covering of such $\rho^{-\frac{1}{2}}$-spheres instead of every single point in $\mathcal{A}$. Specifically, let $\{\mathbf{a}_k\}_{k=1}^{K_\rho}$ be a set of point for which for which

$$\mathcal{A} \subset \cup_{k=1}^{K_\rho} \mathcal{A}_\rho(\mathbf{a}_k).$$

Then it follows by the union bound that

$$P\big(\tau(\mathcal{A}) \leq \rho^{-1}\big) \leq \sum_{k=1}^{K_\rho} P\big(\tau_\rho(\mathbf{a}_k) \leq \rho^{-1}\big) \tag{36}$$

where $\tau$ and $\tau_\rho(\mathbf{a}_k)$ are defined by (33) and (35) respectively. It is also useful to note that as the sample spacing (i.e., $\rho^{-\frac{1}{2}}$) is decreased, the number of spheres in the covering, $K_\rho$, must grow in order to provide a complete covering of $\mathcal{A}$. Nevertheless, we may conclude that

$$P\big(\tau_\rho(\mathbf{a}_k) \leq \rho^{-\frac{1}{2}}\big) \dot{\leq} K_\rho \, \rho^{-\frac{n}{2}} \tag{37}$$

where $K_\rho$ is the smallest number of $\rho^{-\frac{1}{2}}$-spheres required to cover the set $\mathcal{A}$. Thus, all we need to do in order to complete our argument is to determine the growth rate of $K_\rho$. Specifically, we need to ompute

$$\lim_{\rho\to\infty} \frac{\ln K_\rho}{\ln \rho}. \tag{38}$$

Interestingly enough, the limit (38) is strongly related to a quantity which in fractal geometry is referred to as the box-counting dimension. Specifically, let $N_\varepsilon(\mathcal{A})$ be the minimum number of $\varepsilon$-spheres required to cover a set $\mathcal{A}$. Then, the *box-counting dimension* of $\mathcal{A}$, $d_B(\mathcal{A})$, is defined as

$$d_B(\mathcal{A}) = -\lim_{\varepsilon\to 0} \frac{\ln N_\varepsilon(\mathcal{A})}{\ln \varepsilon} \tag{39}$$

(assuming that the limit exists) (Falconer, 1990, Definition 3.1; Massopust 1995). For compact smooth manifolds the box-counting dimension coincides with the standard topological dimension (Boothby, 1986), i.e. a point has dimension 0, a line dimension 1, a smooth surface dimension 2 etc.

Combining (36), (37) and (39) now yields

$$P\big(\tau(\mathcal{A}) \leq \rho^{-1}\big) \dot{\leq} \rho^{-\frac{n-d_B(\mathcal{A})}{2}}. \tag{40}$$

It is however in certain situations possible to further improve on the bound given in (40). To see this, let $\gamma > 0$ be a constant such that $\| \mathbf{a} \| \geq \gamma$ for $\mathbf{a} \in \mathcal{A}$. That such a $\gamma$ exists follow by the assumption that $\mathcal{A}$ is closed and excludes the origin. Let $h_\gamma(\mathcal{A})$ denote the *radial projection* of $\mathcal{A}$ onto the $\gamma$-sphere centered around the origin, i.e., $h_\gamma(\mathcal{A})$ is given by

$$h_\gamma(\mathcal{A}) \equiv \{\mathbf{x} \mid \mathbf{x} = \frac{\gamma}{\|\mathbf{a}\|}\mathbf{a}, \mathbf{a} \in \mathcal{A}\}.$$

The concept is illustrated in Fig. 6 for 3 different subsets of $\mathbb{R}^2$ with different topological dimensions. By construction it now follows for $\mathbf{a} \in \mathcal{A}$ that

$$\|\mathbf{Ha}\| \geq \frac{\gamma}{\|\mathbf{a}\|}\|\mathbf{Ha}\| = \|\mathbf{Hx}\|, \quad \text{where} \quad \mathbf{x} = \frac{\gamma}{\|\mathbf{a}\|}\mathbf{a}$$

since $\|\mathbf{a}\| \geq \gamma$. This however imply that

$$\tau(A) \geq \tau(h_\gamma(\mathcal{A})).$$

Thus, by applying the result of (40) to $h_\gamma(\mathcal{A})$, rather than to $\mathcal{A}$ directly, we obtain the bound given in

$$P\left(\tau(\mathcal{A}) \leq \rho^{-1}\right) \leq P\left(\tau(h_\gamma(\mathcal{A})) \leq \rho^{-1}\right) \dot{\leq} \rho^{-\frac{n-d_B(h_\gamma(\mathcal{A}))}{2}}. \quad (41)$$

Further, since the dimension of $h_\gamma(\mathcal{A})$ is always smaller than or equal to the dimension of $\mathcal{A}$ (a projection can never increase the dimension) it also follows that (41) represents an improvement over (40) in the general case. This is illustrated in Fig. 6 (a) where $d_B(\mathcal{A}) = 2$ but where $d_B(h_\gamma(\mathcal{A})) = 1$. Note also that $d_B(\mathcal{A}) = d_B(h_\gamma(\mathcal{A}))$ in both Fig. 6 (b) and (c) in which case the bound in (41) yields no improvement over (40). It is also illustrative to note that $d_B(h_\gamma(\mathcal{A}))$ does not actually depend on the value of $\gamma$ and therefore $\gamma = 1$ could be chosen without loss of generality when determining $d(h_\gamma(\mathcal{A}))$.

In order to illustrate the usefulness of the bound in (41), consider its use in determining the probability that the minimum eigenvalue of $\mathbf{Q} = \mathbf{H}^T\mathbf{H}$ is small in comparison to $\rho^{-1}$. To this end let $\lambda_1(\mathbf{Q})$ denote the minimum eigenvalue of $\mathbf{Q}$ and note that (Horn & Johnsson, 1985)

$$\lambda_1(\mathbf{Q}) = \min_{\|\mathbf{a}\|=1} \|\mathbf{Ha}\|^2.$$

Since $\mathcal{A} = \{\mathbf{a} \mid \|\mathbf{a}\| = 1\}$ is simply the surface of a sphere in $\mathbb{R}^m$ it has a topological (and box-counting dimension) of $m-1$ and the bound in (40) directly yields

$$P\left(\lambda_1(\mathbf{Q}) \leq \rho^{-1}\right) \dot{\leq} \rho^{-\frac{n-(m-1)}{2}} = \rho^{-\frac{n-m+1}{2}}, \quad (42)$$

a bound which in retrospect is known to be tight. Note also that in this case no improvement is offered by (41). It is also illustrative do compare (42) with the expression for the ZF diversity and note that this supports the argument that the diversity order of the ZF detector is determined by the probability that the channel matrix is nearly singular.

By applying the statement in (41) to the set $\mathcal{R}_e$ encountered in the analysis of the relaxation detector we may now give a precise statement regarding the diversity order of the basic relaxation detector. We summarize the result in the following proposition.

**Proposition 4.** Let $\hat{\mathbf{s}}$ be the output of the basic relaxation detector based on $\mathcal{R}$. Let $\mathcal{R}_e$ be defined according to (28) and let $h(\mathcal{R}_e, \mathbf{s})$ be the radial projection of $\mathcal{R}_e$ onto the unit sphere centered at $\mathbf{s}$. Then,

$$-\lim_{\rho\to\infty} \frac{\ln P(\hat{\mathbf{s}} \neq \mathbf{s} \mid \mathbf{s})}{\ln \rho} \geq \frac{n - d_B(h(\mathcal{R}_e, \mathbf{s}))}{2},$$

where $d_B(h(\mathcal{R}_e,\mathbf{s}))$ denotes the box-counting dimension of $h(\mathcal{R}_e,\mathbf{s})$.

*Proof:* By Proposition 2 it follows that

$$P(\hat{\mathbf{s}} \neq \mathbf{s} \mid \mathbf{s}) \dot{\leq} P(\mu^2 \leq \rho^{-1})$$

where $\mu$ is defines according to (29). Further, by appealing to the statement made in (41) for $\mathcal{A} = \mathcal{R}_e - \mathbf{s}$ the proposition follows.

The statement of Proposition 4 yields a universally valid bound on the diversity of any relaxation detector in terms of the geometry of $\mathcal{R}$. We conjecture that the bound is also the tightest possible in the sense that it actually provides the exact diversity for any relaxation detector and note that it is easily verified to be tight in the cases of the ML and ZF detectors, where the exact diversity order is known. It should also be noted that it is numerically confirmed for the CLS detector based on Fig. 1. We proceed under the assumption that the conjecture is true.

Naturally, it may still be difficult to determine the quantity $d_B(h(\mathcal{R}_e,\mathbf{s}))$ which appears in Proposition 4. However, the following corollary provides a lower bound on the diversity which is typically much easier to verify and compute and which may often by obtained simply by inspection. Also, we note also that this bound is tight for many cases of interest but not always.

**Corollary 5.** Let $\hat{\mathbf{s}}$ be the output of the basic relaxation detector based on $\mathcal{R}$. Assume further that $\mathcal{R}$ is a smooth manifold of dimension $d_M$. Then,

$$\frac{n-d_M}{2} \leq -\lim_{\rho\to\infty} \frac{\ln P(\hat{\mathbf{s}} \neq \mathbf{s})}{\ln \rho}. \tag{43}$$

*Proof:* Note first that since $\mathcal{R}$ is a smooth manifold it follows that $d_B(\mathcal{R}) = d_M$ (Falconer, 1990). Further, since $\mathcal{R}_e \subset \mathcal{R}$ it follows that $d_B(\mathcal{R}_e) \leq d_B(\mathcal{R}) = d_M$, regardless of which point $\mathbf{s}$ was transmitted. Since a projection can not lead to an increase in dimension it follows that $d_B(h(\mathcal{R}_e,\mathbf{s})) \leq d_B(\mathcal{R}_e) \leq d_M$ which establishes (43).

## Discussion

Based on Proposition 4 and Corollary 5 it is relatively straightforward to find the diversity order of the relaxation detectors considered in Section 1.3, namely those given by $\mathcal{R}_1$, $\mathcal{R}_2$ and $\mathcal{R}_3$. We can find the diversity order of these detectors simply by inspection. The particular case where $m = 2$ is illustrated by Fig. 7. In both the case of $\mathcal{R}_1$ and $\mathcal{R}_2$ for $m = 2$ we see that the projection of $\mathcal{R}_e$ onto the unit sphere located at the transmitted vector $\mathbf{s}$ has a topological dimension of 1 (it is a line segment). In other words, $d_B(h(\mathcal{R}_e,\mathbf{s})) = 1$. By Proposition 4 this implies that the diversity order or the detector is equal to $(n - 1)/2$ in both cases. In the general case where $m > 2$ the dimension of $h(\mathcal{R}_e,\mathbf{s})$ will be equal to $m - 1$. Thus, the diversity of the relaxation detectors employing $\mathcal{R}_1$ or $\mathcal{R}_2$ will be equal to $(n - m + 1)/2$ which is equal to the diversity order of the ZF detector.

Intuitively, and based on the analysis of the previous section, the difference in diversity order of the ZF, ML and relaxation detectors may be explained as follows. For all of the detectors it is *necessary* that the minimum eigenvalue of $\mathbf{H}^T\mathbf{H}$ is small compared to $\rho^{-1}$ for an error to be likely in the high SNR regime. For the ZF detector this is also *sufficient* and the error probability is proportional to

$$P(\lambda_1 \leq \rho^{-1}) \doteq \rho^{-\frac{n-m+1}{2}}.$$

In the case of the ML detector an additional requirement for high error probability is that the eigenspace corresponding to the smallest eigenvalue, that is the set of point on the form $\alpha\mathbf{u}_1$ for $\alpha \in \mathbb{R}$, is nearly parallel to one of the vectors $\boldsymbol{\delta} = \hat{\mathbf{s}} - \mathbf{s}$. Otherwise, $\mathbf{H}\hat{\mathbf{s}}$ will still not be close to $\mathbf{Hs}$ in the output space for any $\hat{\mathbf{s}} \in \mathcal{S}^m$. The probability of this event can be shown to be on the order of $\rho^{-\frac{m-1}{2}}$ and the overall probability of ML detection error decays as $\rho^{-\frac{n-m+1}{2}}\rho^{-\frac{m-1}{2}} = \rho^{-\frac{n}{2}}$ where $\rho^{-\frac{n-m+1}{2}}$ is the probability that $\lambda_1$ is small. For the general relaxation detector it is (apart from one eigenvalue being small) also necessary that the eigenspace passes within a distance of $\rho^{-\frac{1}{2}}$ from the set $\mathcal{R}_e$. This probability decays as

$$\rho^{-\frac{m-d_B(h_\gamma(\mathcal{R}_e))}{2}}.$$

The probability of the event thus depend on the *dimension* of $\mathcal{R}_e$ or more precisely, on the dimension of its projection onto the unit sphere around **s**. The concept is illustrated in Fig. 8.

With this in mind, it is fairly straightforward to see that choosing $\mathcal{R}_3$ instead over $\mathcal{R}_2$ will not increase the diversity order of the relaxation detector. The reason is that $h(\mathcal{R}_e, \mathbf{s})$ will be the same in both cases. However, by partitioning **s** according to (as suggested in (Cui et. al., 2005))

$$\mathbf{s} = \begin{bmatrix} \mathbf{s}_1 \\ \mathbf{s}_2 \end{bmatrix}$$

where $\mathbf{s}_1 \in \mathbb{R}^k, \mathbf{s}_2 \in \mathbb{R}^{m-k}$ for some $k \in [1, m]$, while placing individual constraints on $\mathbf{s}_1$ and $\mathbf{s}_2$ according to $\|\mathbf{s}_1\|^2 = g$ and $\|\mathbf{s}_2\|^2 = m - k$, would yield a detector of diversity order $(n - m + 2)/2$. This follows by applying Corollary 5 and noting that the solution set of vectors **s** where the two sub-vectors satisfy individual norm constraint forms an $m - 2$ dimensional smooth manifold. (It is the solution set to a system of equations with $m$ variables and 2 constrains.) By further partitioning **s** into $g$ sub-vectors, constraining the norm of each sub-vector would yield a diversity order or

$$\frac{n-m+g}{2}.$$

In the extreme case where, $g - m$ we would obtain

$$\mathcal{R} = \{\mathbf{s} \mid \mathbf{s} = (s_1, \dots, s_m)^T, s_i^2 = 1\, i = 1, \dots, m\} \tag{44}$$

and a diversity order of $\frac{n}{2}$ which equals the diversity order of the ML detector. This should however come as no surprise since $\mathcal{R} = \mathcal{S}^m$ in this case and the relaxation detector becomes equivalent to the ML detector.

A problem is of course that whenever $g > 1$ the problem in (14) is difficult to solve exactly. However, as explained, when $g - 1$ the relaxation detector will experience no diversity gain. This is of course disappointing as it seems to suggest that that the only way to improve upon the diversity order of the ZF detector would rely on a computationally intractable detector structure. One way to circumvent this would be to only solve (14) approximately which is also what was done in (Cui et. al., 2005) where an increase in diversity order was observed thorough numerical simulations. One could also argue that this is a problem only because it is hard to within $\mathbb{R}^m$ find a suitable relaxation set, $\mathcal{R}$, which provides high diversity as well as a solution to the relaxed problem that is easily computed. A possible solution may thus be to lift the detection problem into a higher dimensional space prior to the relaxation. This is also the strategy offered by *the semidefinite relaxation* detector proposed in (Tan & Rasmussen, 2001; Ma et al. 2002), and the *polynomial moment relaxation detector* in (Cui et al. 2006). In these cases the ML detection problem is recast as an equivalent problem in a higher dimensional space where there are more degrees of freedom for choosing a suitable relaxation. In fact, it can be shown that the particular relaxation offered by the semidefinite relaxation detector is strong enough for achieving maximal diversity (Jaldén & Ottesten, 2008) while still having a polynomial time algorithm for solving the relaxed optimization problem (Boyd & Vandenberghe, 2004; Tan & Rasmussen, 2001; Ma et. al. 2002).

## CONCLUSION

In this chapter we have considered a class of relaxation detectors that are approximations to the optimal ML detector. This was done by studying the diversity order of the detectors, the motivation behind such a study being that is provides a first order characterization of the detectors robustness towards fading in the wireless MIMO system. The conclusion is that the diversity, or equivalently high SNR behavior, is determined by the topological dimension of the relaxation set. This attaches a qualitative value to the intuition that by selecting a "looser" relaxation in order to obtain a computationally less complex detector, the error probability will also be worse. In essence, the diversity of a relaxation detector may be obtained by simply considering the number of receive antennas, herein denotes by $m$, and then simply subtracting the number of free dimensions in the relaxation set, $\mathcal{R}$. To exemplify this statement, we note that in the case of the ML detector, which may be seen as a special case of a relaxation detector, the number of free dimensions in $\mathcal{R} = \mathcal{S}^m$ is zero (since $\mathcal{S}^m$ is a discrete set) and the diversity is $(m-0)/2 = m/2$. In the case of the CLS detector the number of free dimensions of $\mathcal{R}$ is $n-1$ (here $\mathcal{R}$ is the surface of a sphere in $n$-dimensions) and the diversity becomes $(m-(n-1))/2 = (m-n+1)/2$. This technique of establishng the diversity order applies generally and can be used to establish the diversity order of a wide range of relaxation detectors. Specifically, we obtain the "rule of thumb" that every real valued dimension of the input space which is constrained yields an increase in diversity by $\frac{1}{2}$. This is also in a sense why the relaxations which simply bound the input variables to some range, such as the detectors of (Yener et. al., 2002), do not see an increase in diversity when applied to the task of detecting spatially multiplexed symbols which have been transmitted over the wireless, multiple antenna, MIMO channel.

Finally, it is also useful to note that although we have not touched upon more potent relaxations such as the one used in the SDR detector, we expect similar tools to be useful also when establishing the performance of these detectors.

## REFERENCES

Barbosa, L. (1989). Maximum likelihood sequence estimators: a geometric view. *IEEE Transactions on Information Theory, 35*(2), 419-427.

Biglieri, E., Taricco, G., & Tulino, A. (2002). Performance of spacetime codes for a large number of antennas. *IEEE Transactions on Information Theory, 48*(9), 1794-1803.

Boothby, W. M. (1986). *An Introduction to Differential Manifolds and Riemannian Geometry,* 2nd ed. New York: Academic Press.

Boyd, S., & Vandenberghe, L. (2004). *Convex Optimization.* Cambridge, UK: Cambridge University Press.

Cui, T., Ho. T., & Tellambura, C. (2006). A motivational approach to self: Integration in personality. In *Proceedings of the IEEE International Conference on Communications, ICC.*

Cui, T., & Tellambura, C. (2006). Polynomial-constrained detection using a penalty function and a differential-equation algorithm for MIMO systems. *IEEE Signal Processing Letters, 13*(3), 133-136.

Cui, T., Tellambura, C., & Wu, Y. (2005). A motivational approach to self: Integration in personality. In, *Proceedings of the IEEE Global Telecommunications Conference, GLOBECOM.*

Damen, M. O., Gamal, H. E., & Caire, G., (2003). On maximum-likelihood detection and the search for the closest lattice point. *IEEE Transactions on Information Theory, 49*(10), 2389-2401.

Durett, R. (1996). *Probability: Theory and Examples,* 2nd ed. Duxbury Press.

Falconer, K. (1990). *Fractal Geometry: Mathematical Foundations and Applications.* Wiley.

Hochwald, B. M., & Brink, S., (2003). Achieving near-capacity on a multiple-antenna channels. *IEEE Transactions on Communications, 51*(3), 389-399.

Horn, R. A., & Johnson, C. R. (1985). *Matrix Analysis.* Cambridge, UK: Cambridge University Press.

Jaldén, J., & Ottersten, B. (2005). On the complexity of sphere decoding in digital communications. *IEEE Transactions on Signal Processing, 53*(4), 1474-1484.

Jaldén, J., & Ottersten, B. (2008). The diversity order of the semidefinite relaxation detector. *IEEE Transactions on Information Theory, 54*(4), 1406-1422

Ma, W.-K., Davidson, T. N., Wong, K., Luo, Z.-Q., & Ching, P.-C., (2002). Quasi-maximum-likelihood multiuser detection using semi-definite relaxation with application to synchronous CDMA. *IEEE Transactions on Signal Processing, 50*(4), 912-922.

Massopust, P. R. (1995). *Fractal Fractal Functions, Fractal Surfaces, and Wavelets.* Academic Press.

Murugan, A. D., Gamal, H. E., Damen, M. O., & Caire, G., (2006). A unified framework for tree search decoding: rediscovering the sequential decoder. *IEEE Transactions on Information Theory, 52*(3), 933-953.

Paulraj, A., Nabar, R. U. & Gore, D. (2003). *Introduction to Space Time Wireless Communications.* Cambridge, UK: Cambridge University Press.

Steingrimsson, B., Luo, Z.-Q. & Wong, K., (2003). Soft quasi-maximum-likelihood detection for multiple-antenna wireless channels. *IEEE Transactions on Signal Processing, 51*(11), 2710-2719.

Taherzadeh, M., Mobasher, A., &Khandani, A. K. (2007).LLL Reduction achieves the receive diversity in MIMO decoding. *IEEE Transactions on Information Theory, 53*(12), 4801-4805.

Tan, P., & Rasmussen, L. (2001). The application of semidefinite programming for detection in CDMA. *IEEE Journal on Selected Areas in Communications, 19*(8), 1442-1449.

Thoen, S., Deniere, L., Van der Perre, L., Engels, M., & De Man, H. (2003). Constrained least squares detector for OFDM/SDMA-based wireless networks. *Transactions on Wireless Communications, 50*(1), 129-140.

Tse, D., & Viswanath, P (2005). *Fundamentals of Wireless Communication.* Cambridge, UK: Cambridge University Press.

Verdú, S. (1989). Maximum likelihood sequence estimators: a geometric view. *Algoritmica, 4*, 303-312.

Yao, H., & Wornell, G. W. (2002). Lattice reduction aided detectors for MIMO communication systems. In, *Proceedings of the IEEE Global Telecommunications Conference, GLOBECOM.*

Yener, A., Yates, R. D., & Ulukus, S., (2002). CDMA multiuser detection: a nonlinear programming approach. *IEEE Transactions on Communications, 50*(6), 1016-1024.

Zheng, L., & Tse, D. (2003). Diversity and multiplexing: A fundamental tradeoff in multiple-antenna channels. *IEEE Transactions on Information Theory, 49*(5), 1073-1096.

## APPENDIX

*Figure 1. Probability of error defined according to* $p_e \equiv P(\hat{\mathbf{s}} \neq \mathbf{s})$ *for various detectors in the m = n = 4 case*

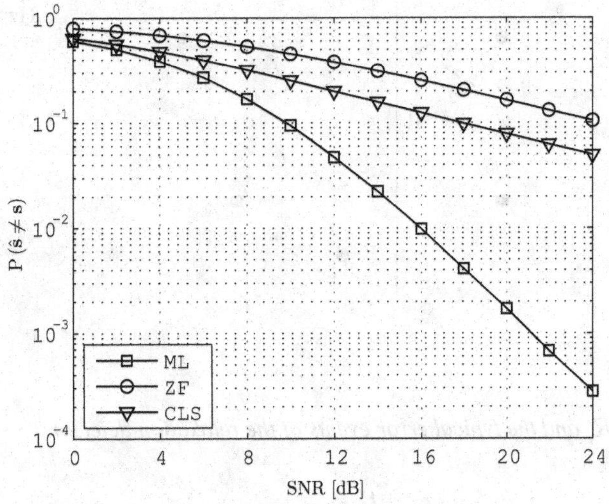


*Figure 2. The pairwise error events of the ML detector*

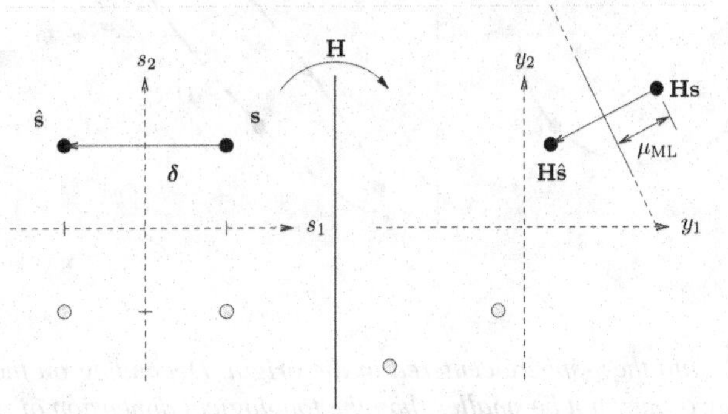


*Figure 3. The pairwise error events of the ZF detector*

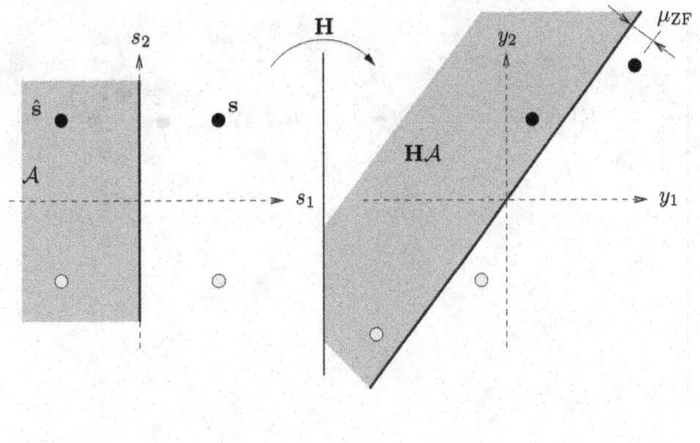

*Figure 4. The geometry of the relaxation detector*

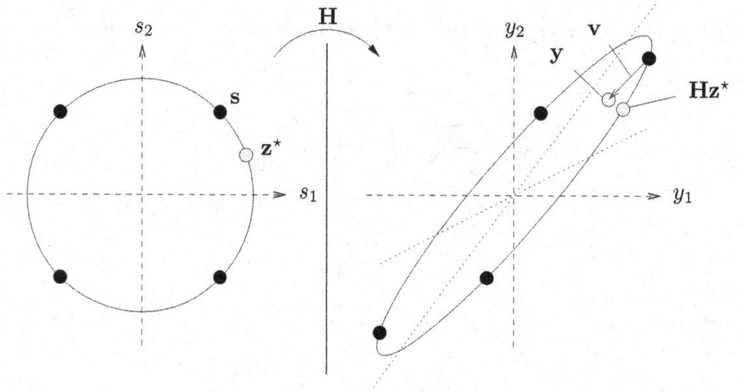


*Figure 5. The deformation of $\mathcal{R}_e$ and the typical error events of the relaxation detector*

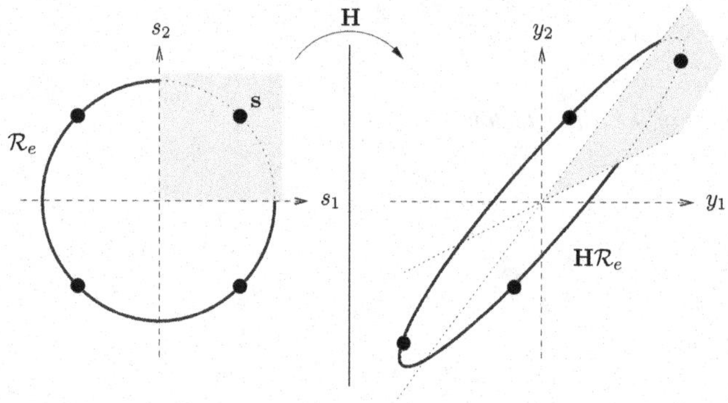


*Figure 6. The projection of $\mathcal{A}$ onto the γ-sphere centered in the origin. Depending on the shape of $\mathcal{A}$ the topological dimension of the projection may or may not be smaller than the topological dimension of $\mathcal{A}$.*

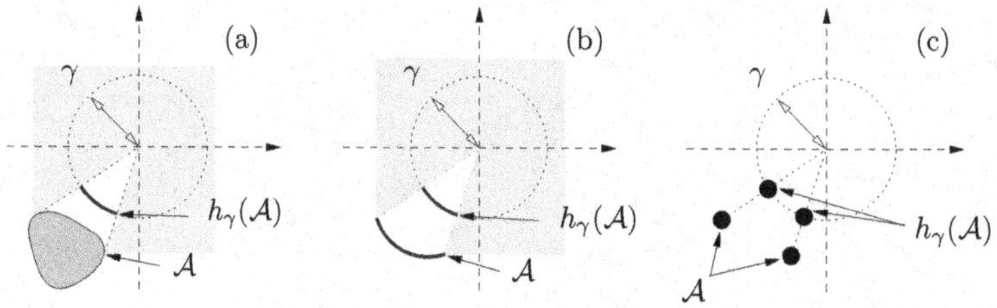

*Figure 7. The projection of $\mathcal{R}_e$ onto the unit sphere centered in $\mathbf{s}$ is illustrated for the case of $\mathcal{R}_1$ and $\mathcal{R}_2$ defined in (16) and (17) respectively*

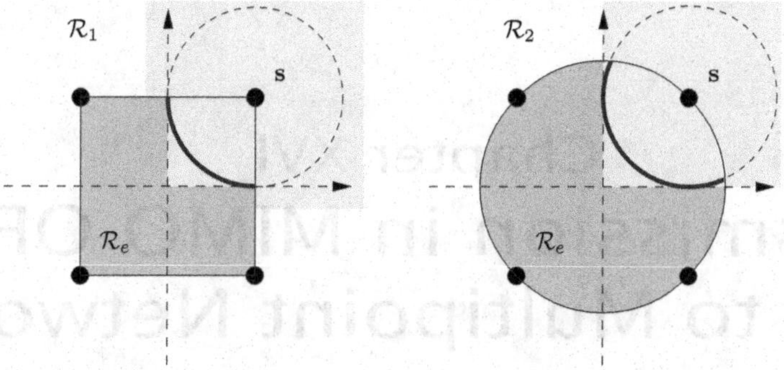


*Figure 8. Error events in the relaxation and ML detector. When the smallest eigenvalue of $\mathbf{H}^T\mathbf{H}$ is small in relation to $\rho^{-1}$ the detector will be likely to make an error when one of the points in $\mathcal{R}_e$ falls close to the corresponding eigenspace.*

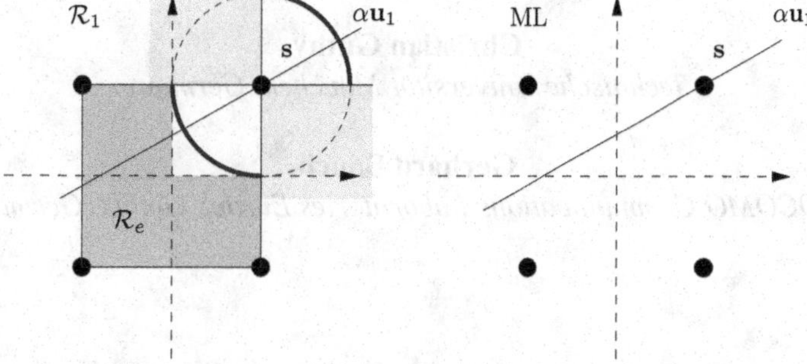

# Chapter XVI
# Transmission in MIMO OFDM Point to Multipoint Networks

**Wolfgang Utschick**
*Technische Universität München, Germany*

**Pedro Tejera**
*Technische Universität München, Germany*

**Christian Guthy**
*Technische Universität München, Germany*

**Gerhard Bauch**
*DOCOMO Communications Laboratories Europe GmbH, Germany*

## ABSTRACT

*This chapter discusses four different optimization problems of practical importance for transmission in point to multipoint networks with a multiple input transmitter and multiple output receivers. Existing solutions to each of the problems are adapted to a multi-carrier transmission scheme by considering the special structure of the resulting space-frequency channels. Furthermore, for each of the problems, suboptimum approaches are presented that almost achieve optimum performance and, at the same time, do not have the iterative character of optimum algorithms, i.e., they deliver a solution in a fixed number of steps. The purpose of this chapter is to give an overview on optimum design of point to multipoint networks from an information theoretic perspective and to introduce non-iterative algorithms that are a good practical alternative to the sometimes costly iterative algorithms that achieve optimality.*

## INTRODUCTION

Increasing demand for broadband services calls for higher data rates in future communication systems. For instance, in fourth generation wireless communication systems data rates of several Mb/s in high mobility and 1 Gb/s in low mobility scenarios are expected (Tachikawa, 2003). Particularly challenging is the accomplishment of this goals in point to multipoint networks. In such networks, an access point transmits independent information to a number of users that compete for the available system resources, i.e., transmission time, transmit power, spectrum and space. Thus, the challenge consists in providing a satisfactory service to all users by making an adequate allocation of resources. Prominent examples of point to multipoint networks are the downlink of wireless local area networks, the downlink of mobile networks and the downstream direction of wired or wireless last mile access networks.

Spectral limitations, due to scarcity of spectrum in wireless systems and narrow bandwidths of customary copper wire in the last mile, together with the general frequency selectivity of wideband channels are two major barriers to be overcome in the way to data rates beyond those of current communication systems. Sending information over multiple inputs at the transmitter and retrieving information from multiple outputs at the receiver has the potential to increase the amount of information reliably transmitted per time and frequency unit, i.e., it allows a more efficient use of the spectrum (Foschini, 1998).[a] On the other hand, Orthogonal Frequency Division Multiplexing (OFDM) is able to transform the frequency selective channel into a set of non-interfering frequency flat channels, which enormously simplifies equalization at the receiver (Raleigh, 1998). Hence, combination of MIMO and OFDM seems to be key for implementation of future high rate communication systems (Sampath, 2002).

In this chapter we describe MIMO OFDM transmission schemes for point to multipoint networks that achieve optimum rates, that is, rate vectors at the boundary of the capacity region. Computation of optimum transmit parameters is performed by means of iterative algorithms involving a complexity that strongly depends on the a priori unknown number of iterations required to reach convergence. In addition, the optimum solution allocates interfering spatial dimensions to users, which makes it necessary to inform each user about the statistics of the interference that it receives. For each transmission scheme suboptimum allocation algorithms are presented that are able to closely approach performance of optimum approaches and exhibit two crucial advantages. Computation of optimum transmit parameters requires a complexity similar to that of only one iteration of the optimum approaches and users are assigned decoupled spatial dimensions, which makes possible the reduction of the required signaling overhead.

## System Model

We consider a typical multiuser MIMO OFDM system model where the signal $\mathbf{y}_{k,n} \in \mathbb{C}^{r_k \times 1}$ received by user $k \in \{1, 2, \ldots, K\}$ on subcarrier $n \in \{1, 2, \ldots, N\}$ is given by

$$\boldsymbol{y}_{k,n} = \boldsymbol{H}_{k,n} \boldsymbol{x}_n + \boldsymbol{n}_{k,n}, \tag{1}$$

where $\mathbf{H}_{k,n} \in \mathbb{C}^{r_k \times t}$ is a matrix modeling the channel between the transmitter and user $k$ on subcarrier $n$, $\mathbf{x}_n \in \mathbb{C}^{t \times 1}$ is the vector of transmit signals on subcarrier $n$ and $\mathbf{n}_{k,n} \in \mathbb{C}^{r_k \times 1}$ is a vector representing additive white Gaussian noise with covariance matrix $\mathrm{E}\{\boldsymbol{n}_{k,n}\boldsymbol{n}_{k,n}^{\mathrm{H}}\} = \boldsymbol{I}$, where $\mathrm{E}\{\bullet\}$ denotes the expectation operator. The variables $t$ and $r_k$ denote the number of transmit and receive antennas, respectively. Perfect channel knowledge is assumed at both the transmitter and the receivers. This is a valid assumption if the channel is invariant over time as, for instance, in wired networks. In these cases, enough time is available to learn the channel at both ends of the communication link by employing an adequate signaling scheme. Also in wireless networks there exist scenarios where this assumption can be considered reasonable. These are typically indoor scenarios such as offices or hotspots in airports or other public areas. In such scenarios mobility is low and long coherence times allow to learn the channel with a moderate signaling overhead. The transmit signal is formed by superposition of the signals intended for each of the users as

$$\boldsymbol{x}_n = \sum_{k=1}^{K} \boldsymbol{x}_{k,n}.$$

In terms of capacity, the optimum choice for the statistics of the transmit signals $x_{k,n}$ is a zero mean Gaussian distribution (Weingarten, 2006). The covariance matrices $\boldsymbol{Q}_{k,n} = \mathrm{E}\{\boldsymbol{x}_{k,n}\boldsymbol{x}_{k,n}^{\mathrm{H}}\}$ are required to satisfy the following transmit power constraint

$$\sum_{n=1}^{N}\sum_{k=1}^{K} \mathrm{Trace}\{\boldsymbol{Q}_{k,n}\} \leq P \tag{2}$$

In the sequel we introduce four of the most significant problem statements related to the optimization of transmit parameters in the broadcast channel. For each problem we first briefly describe the existing optimum solution before we introduce the corresponding suboptimum approach.

## SUM RATE MAXIMIZATION

The first design problem that we address is maximization of the sum rate, i.e., maximization of the total amount of information that an access point can transmit to the users in the cell. Mathematically, the problem can be stated as follows,

$$\max_{\{Q_{k,n}\}} \sum_{n=1}^{N} \sum_{k=1}^{K} R_{k,n} \tag{3}$$

where the maximization is subject to the power constraint (2) and the achievable rates $R_{k,n}$ are given by (Vishwanath, 2003)

$$R_{k,n} = \log \frac{\left| I + H_{k,n} \sum_{i \geq k} Q_{i,n} H_{k,n}^{H} \right|}{\left| I + H_{k,n} \sum_{i > k} Q_{i,n} H_{k,n}^{H} \right|} \tag{4}$$

These rates are achieved by coding the information sent to different users successively, i.e., first information for user 1 is encoded, then, information for user 2 is encoded and so on until, finally, information for user $K$ is encoded. As a consequence, at the time information for user $k$ is encoded, signals intended for users $1,2,...,k-1$ are known. These signals constitute a source of interference for user $k$. However, it is theoretically possible to encode information for user $k$ taking into account this known interference so as to completely nullify its impact on the achievable rate (Costa, 1983). This is the reason why the terms corresponding to users $i < k$ do not appear in (4).

## Optimum Approach

The optimization problem (3) is not convex, which prevents from the direct application of convex optimization algorithms. However, this problem can be transformed into a convex optimization problem by applying an interesting duality result between point to multipoint transmission (broadcast channel) and multipoint to point transmission (multiple access channel) (Vishwanath, 2003). In Vishwanath (2003), this result was presented for general memoryless MIMO channels. In the following description this result is straightforwardly applied to the MIMO OFDM channel. The broadcast channel with system model given by (1). has a dual multiple access channel with model given by

$$r_n = \sum_{k=1}^{K} H_{k,n}^{\mathrm{H}} w_{k,n} + n_n ,$$

where $\mathbf{r}_n \in C^{t \times 1}$ is the vector of received signals on subcarrier $n$, $\mathbf{H}_{k,n}^{\mathrm{H}} \in C^{t \times r_k}$ is the conjugate transpose of $\boldsymbol{R}_{k,n}$ and models the channel between user $k$ and the receiver on subcarrier $n$, $\mathbf{w}_{k,n} \in C^{r_k \times 1}$ is the signal transmitted by user $k$ and $\mathbf{n}_n \in C^{t \times 1}$ is a vector representing additive white Gaussian noise with covariance matrix equal to the identity matrix. As we can see, in the multiple access channel the roles of receivers and transmitters are exchanged. The transmitter in the broadcast channel receives in the dual multiple access channel the superposition of signals transmitted by the users, which act as receivers in the broadcast channel. For the multiple access channel successive decoding is optimum in terms of capacity (Cover, 1993). Let us assume that user $K$ is decoded first, then user $K-1$ is decoded and so on until, finally, user 1 is decoded. The achievable rates for user $k$ in the multiple access channel are then given by

$$\overline{R}_{k,n} = \log \frac{\left| \mathbf{I} + \sum_{i \leq k} \mathbf{H}_{i,n}^{H} \mathbf{\Gamma}_{i,n} \mathbf{H}_{i,n} \right|}{\left| \mathbf{I} + \sum_{i < k} \mathbf{H}_{i,n}^{H} \mathbf{\Gamma}_{i,n} \mathbf{H}_{i,n} \right|} \tag{5}$$

where $\Gamma_{k,n} = \mathrm{E}\{w_{k,n} w_{k,n}^{\mathrm{H}}\}$ is the transmit covariance matrix of user $k$ on subcarrier $n$. Observe that interference terms corresponding to users decoded before user $k$ do not appear in the above expression as at the time of decoding user $k$ this interference terms are known and can be subtracted from the received signal. The duality result in Vishwanath (2003) establishes that under the assumption of an overall power constraint in the multiple access channel, i.e.,

$$\sum_{n=1}^{N}\sum_{k=1}^{K}\mathrm{Trace}\{\boldsymbol{\Gamma}_{k,n}\}\leq P \tag{6}$$

if certain rates are achievable in the broadcast channel, they are also achievable in its dual multiple access channel. The inverse is also valid, i.e., if some rates are achievable in the dual multiple access channel, the same rates are also achievable in its corresponding broadcast channel. In Vishwanath (2003) an algorithm is given in order to transform covariance matrices achieving certain rates in the multiple access channel into covariance matrices that achieve the same rates in the broadcast channel. An important feature of this algorithm is that the encoding order that must be used in the broadcast channel to achieve the same rates as in the dual multiple access channel is the reverse of the decoding order applied in the multiple access channel.

A consequence of this duality result is that the solution to (3) can be found by solving

$$\max_{\{\boldsymbol{\Gamma}_{k,n}\}}\sum_{n=1}^{N}\sum_{k=1}^{K}\overline{R}_{k,n} \tag{7}$$

in the multiple access channel and converting the resulting optimum covariance matrices into optimum covariance matrices in the broadcast channel. Substituting (5) into (7), we obtain

$$\max_{\{\boldsymbol{\Gamma}_{k,n}\}}\sum_{n=1}^{N}\log\left|\boldsymbol{I}+\sum_{k}\boldsymbol{H}_{k,n}^{\mathrm{H}}\boldsymbol{\Gamma}_{k,n}\boldsymbol{H}_{k,n}\right| \tag{8}$$

Note that the objective function is now a concave function of the covariance matrices $\boldsymbol{\Gamma}_{k,n}$, i.e., the problem is convex and can be solved applying standard gradient based iterative optimization methods. This is a major benefit of the duality between broadcast and multiple access channels. Another feature of the objective function worth noting is its independence of the decoding order, i.e., the maximum sum rate can be achieved with any decoding order in the multiple access channel and, as a consequence, it can also be achieved with any encoding order in the broadcast channel.

More specialized algorithms than standard gradient methods have been proposed in (Vishwanath, 2003; Jindal, 2005; Yu, 2006) for general memoryless MIMO channels, i.e., for N=1. Among them, the iterative waterfilling algorithm presented in Vishwanath (2003) has become especially popular. This algorithm has been applied to a MIMO OFDM setting in Tejera (2006). For the MIMO OFDM channel this algorithm consists of the following steps.

1. Initialization: $l=1$, $\boldsymbol{\Gamma}_{k,n}^{(0)}=\boldsymbol{0},\forall k,n$
2. Compute effective channels for all users and subcarriers, i.e.,

   $$\boldsymbol{H}_{k,n}^{\mathrm{eff},(l)}=\boldsymbol{H}_{k,n}\left(\boldsymbol{I}+\sum_{k'\neq k}\boldsymbol{H}_{k,n}^{\mathrm{H}}\boldsymbol{\Gamma}_{k,n}^{(l)}\boldsymbol{H}_{k,n}\right)^{-1/2},\ \forall n,k$$

3. Compute

   $$\{\boldsymbol{M}_{k,n}^{(l)}\}=\arg\max_{\{\boldsymbol{A}_{k,n}\}}\sum_{n=1}^{N}\sum_{k=1}^{K}\log\left|\boldsymbol{I}+\sum_{k}\left(\boldsymbol{H}_{k,n}^{\mathrm{eff},(l)}\right)^{\mathrm{H}}\boldsymbol{A}_{k,n}\boldsymbol{H}_{k,n}^{\mathrm{eff},(l)}\right|,$$

   subject to the constraints

   $$\sum_{k,n}\mathrm{Trace}\{\boldsymbol{A}_{k,n}\}\leq P \text{ and } \boldsymbol{A}_{k,n}\geq\boldsymbol{0},\forall k,n.^{\mathrm{b}}$$

   The optimum matrices $\boldsymbol{M}_{k,n}^{(l)}$ have the eigenvectors of $\boldsymbol{H}_{k,n}^{\mathrm{eff},(l)}\left(\boldsymbol{H}_{k,n}^{\mathrm{eff},(l)}\right)^{\mathrm{H}}$ and the eigenvalues are obtained by performing waterfilling over the eigenmodes of the effective channels.
4. Update $\boldsymbol{\Gamma}_{k,n}^{(l)}=\left((K-1)\boldsymbol{\Gamma}_{k,n}^{(l-1)}+\boldsymbol{M}_{k,n}^{(l)}\right)/K$, $\forall k,n$, and $l=l+1$.

Steps 2, 3 and 4 should be repeated until convergence is reached, i.e., until the difference between $\boldsymbol{\Gamma}_{k,n}^{(l)}$ and $\boldsymbol{\Gamma}_{k,n}^{(l-1)}$ is negligible. The main merit of the algorithm presented in the following section is that it almost achieves optimum performance and it has a complexity which is essentially the same as that involved in just one iteration of any of the existing optimum iterative algorithms as the one described above.

## Suboptimum Approach: The Successive Encoding Successive Allocation Method (SESAM)

One way to simplify the optimum solution to problem (3) is to completely suppress multi-user interference. This is part of the Successive Encoding Successive Allocation Method or shortly SESAM (Tejera, 2006). The main part of the algorithm consists of the determination of an allocation strategy

$$A_{\max}(i) = \{\pi_n(i), n\},$$

wherein $n$ denotes the subcarrier index and $\pi_n(i) \in \{1,\ldots,K\}$ is an encoding order, i.e., user $\pi_n(1)$ is encoded first on carrier $n$, user $\pi_n(2)$ second and so on. This encoding order is determined, as the name SESAM already implies, successively as follows. In each step, the user to be encoded next and the corresponding transmit beamforming vector are found by solving

$$\left(\pi_n(i), \boldsymbol{v}_{\pi_n(i)}\right) = \arg\max_k\left(\max_{\boldsymbol{v}}\left\|\boldsymbol{H}_{k,n}\boldsymbol{v}\right\|_2\right) \tag{9}$$

subject to $\|\boldsymbol{v}\|_2 = 1$ and $\boldsymbol{v} \in \operatorname{null}\{\boldsymbol{v}_{\pi_n(1)}^{\mathrm{H}},\ldots,\boldsymbol{v}_{\pi_n(i-1)}^{\mathrm{H}}\}$. The norm of the transmit beamforming vector is restricted to 1, as we will perform power allocation in a later step. Problem (9) can also be written as

$$\left(\pi_n(i), \boldsymbol{v}_{\pi_n(i)}\right) = \arg\max_k\left(\max_{\boldsymbol{v}}\left\|\boldsymbol{H}_{k,n}\boldsymbol{P}_i\boldsymbol{v}\right\|_2\right) \tag{10}$$

subject to $\|\boldsymbol{v}\|_2 = 1$. Here, $\boldsymbol{P}_i$ is a matrix projecting the users' channels into the nullspace of the vectors $\mathbf{v}_{\pi_n(1)}^{\mathrm{H}},\ldots,\mathbf{v}_{\pi_n(i-1)}^{\mathrm{H}}$, i.e.,

$$\boldsymbol{P}_i = \boldsymbol{I} - \sum_{j=1}^{i-1}\boldsymbol{v}_{\pi_n(j)}\boldsymbol{v}_{\pi_n(j)}^{\mathrm{H}}.$$

That is, $\pi_n(i)$ is chosen to be the user with maximum principal singular value in the subspace given by $\boldsymbol{P}_i$. The transmit beamforming vector is then given by the corresponding right singular vector. At the receivers each user applies a matched filter

$$\mathbf{g}_{\pi_n(i)} = \frac{\mathbf{H}_{\pi_n(i),n}\mathbf{v}_{\pi_n(i)}}{\left\|\mathbf{H}_{\pi_n(i),n}\mathbf{v}_{\pi_n(i)}\right\|_2} \tag{11}$$

to each subchannel assigned to it. These filters can be shown to be capacity preserving in this case. That leads to the following expression for the received signal $y_{\pi_n(i)}$:

$$y_{\pi_n(i)} = \boldsymbol{g}_{\pi_n(i)}^{\mathrm{H}}\boldsymbol{H}_{\pi_n(i),n}\boldsymbol{v}_{\pi_n(i)}\sqrt{p_{\pi_n(i)}}s_{\pi_n(i)} + \sum_{j\neq i}\boldsymbol{g}_{\pi_n(i)}^{\mathrm{H}}\boldsymbol{H}_{\pi_n(i),n}\boldsymbol{v}_{\pi_n(j)}\sqrt{p_{\pi_n(j)}}s_{\pi_n(j)} + \boldsymbol{g}_{\pi_n(i)}^{\mathrm{H}}\boldsymbol{n}_{\pi_n(i)} \tag{12}$$

where $p_{\pi_n(i)}$ is the power allocated to user $\pi_n(i)$ on the $i$th spatial dimension on subcarrier $n$, $s_{\pi_n(i)}$ with $\mathrm{E}\{s_{\pi_n(i)}s_{\pi_n(i)}^*\} = 1$ is the desired signal for that user in that dimension and $\boldsymbol{n}_{\pi_n(i)}$ is the corresponding noise vector. The second term in (12) contains the multi-user interference. As $\mathbf{v}_{\pi_n(i)}$ is the right singular value of $\mathbf{H}_{\pi_n(i),n}\mathbf{P}_i$ (cf. (10)) and beamforming vectors are chosen to be mutually orthogonal, the following relationship holds,

$$\boldsymbol{v}^{\mathrm{H}}_{\pi_n(i)} \boldsymbol{P}_i \boldsymbol{H}^{\mathrm{H}}_{\pi_n(i),n} \boldsymbol{H}_{\pi_n(i),n} \boldsymbol{P}_i \boldsymbol{v}_{\pi_n(j)} = \left\| \boldsymbol{H}_{\pi_n(i),n} \boldsymbol{P}_i \right\|_2^2 \boldsymbol{v}^{\mathrm{H}}_{\pi_n(i)} \boldsymbol{v}_{\pi_n(j)} = 0, \quad \forall j \neq i,$$

where the 2-norm of a matrix $\|\bullet\|_2$ denotes its principal singular value. Furthermore, we can trivially write

$$\boldsymbol{P}_i \boldsymbol{v}_{\pi_n(j)} = \left( \boldsymbol{I} - \sum_{l=1}^{i-1} \boldsymbol{v}_{\pi_n(l)} \boldsymbol{v}^{\mathrm{H}}_{\pi_n(l)} \right) \boldsymbol{v}_{\pi_n(j)} = \boldsymbol{v}_{\pi_n(j)} \quad \forall j \geq i. \tag{13}$$

Using these two results and (10) we obtain

$$\boldsymbol{v}^{\mathrm{H}}_{\pi_n(i)} \boldsymbol{P}_i \boldsymbol{H}^{\mathrm{H}}_{\pi_n(i),n} \boldsymbol{H}_{\pi_n(i),n} \boldsymbol{P}_i \boldsymbol{v}_{\pi_n(j)} = \boldsymbol{g}^{\mathrm{H}}_{\pi_n(i)} \boldsymbol{H}_{\pi_n(i),n} \boldsymbol{v}_{\pi_n(j)} \left\| \boldsymbol{H}_{\pi_n(i),n} \boldsymbol{v}_{\pi_n(i)} \right\|_2 = 0, \quad \forall j > i.$$

Therefore, we observe that by means of the nullspace constraint in (9) multi-user interference from users $\pi_n(j > i)$ on users $\pi_n(i)$ is eliminated and the received signal can be written as

$$y_{\pi_n(i)} = \boldsymbol{g}^{\mathrm{H}}_{\pi_n(i)} \boldsymbol{H}_{\pi_n(i),n} \boldsymbol{v}_{\pi_n(i)} \sqrt{p_{\pi_n(i)}}\, s_{\pi_n(i)} + \sum_{j<i} \boldsymbol{g}^{\mathrm{H}}_{\pi_n(i)} \boldsymbol{H}_{\pi_n(i),n} \boldsymbol{v}_{\pi_n(j)} \sqrt{p_{\pi_n(j)}}\, s_{\pi_n(j)} + \boldsymbol{g}^{\mathrm{H}}_{\pi_n(i)} \boldsymbol{n}_{\pi_n(i)}.$$

The remaining multi-user interference is in contrast to linear zero-forcing approaches (Spencer, 2004) generally not suppressed by beamforming. This is due to the fact that (13) does not hold for $j < i$. As the signals of users $\pi(j < i)$ are known when the signal of user $\pi(i)$ is encoded, this interference cancellation can be efficiently neutralized (Costa, 1983). Consequently, user $\pi(i)$ experiences a scalar subchannel without any multi-user interference, whose signal-to-noise-ratio (SNR) $\lambda^2_{\pi_n(i)}$ is given by

$$\lambda^2_{\pi_n(i)} = \frac{\left| \boldsymbol{g}^{\mathrm{H}}_{\pi_n(i)} \boldsymbol{H}_{\pi_n(i),n} \boldsymbol{v}_{\pi_n(i)} \right|^2}{\boldsymbol{g}^{\mathrm{H}}_{\pi_n(i)} \boldsymbol{g}_{\pi_n(i)}} = \left| \boldsymbol{g}^{\mathrm{H}}_{\pi_n(i)} \boldsymbol{H}_{\pi_n(i),n} \boldsymbol{v}_{\pi_n(i)} \right|^2.$$

Note that one user can be assigned to several subchannels on one subcarrier if it has more than one antenna. In turn, on the same carrier up to $i$ data streams can be simultaneously transmitted, if the total number of receive antennas is greater than the number of transmit antennas. That is a direct consequence of the nullspace constraint in (9). Furthermore it is worth mentioning that this encoding and allocation process can be run independently on each subcarrier. After the allocation process the available transmit power $P$ is finally distributed over all resulting scalar subchannels according to the waterfilling principle (Cover, 1991).

Performance of this scheme is illustrated in Fig. 1 for a broadcast channel with $K = 10$ users, $t = 4$ transmit antennas and $r_k = 2$ receive antennas. For averaging purposes, the entries of the channel matrices have all been independently drawn from a zero mean circularly symmetric complex Gaussian distribution of unit variance, i.e., no frequency correlation and no spatial correlation has been considered. As we see the performance gap with respect to the optimum solution is almost negligible. The narrow gap between optimum performance and the performance delivered by SESAM seems to be independent of system settings and statistical assumptions (Tejera, 2006).

## WEIGHTED SUM RATE MAXIMIZATION

Sum rate maximization results in a distribution of rates that might not be desired. For instance, it might happen that some users with weak channels are not served at all while other users obtain far too high rates for the service that they requested. In order to prevent such situations, it is advisable to choose transmission parameters according to criteria that include some mechanism to control the final distribution of resources among users. A first approach that allows some control on the quality of service finally obtained by the users in the network is weighted sum rate maximization. In this case, the rates of the users are weighted with so called priorities, which, as the name indicates, establish a ranking

*Figure 1. Average optimum and SESAM sum rate for a BC with $K = 10$, $t = 4$, $r_k = 2$ and $n = 16$*

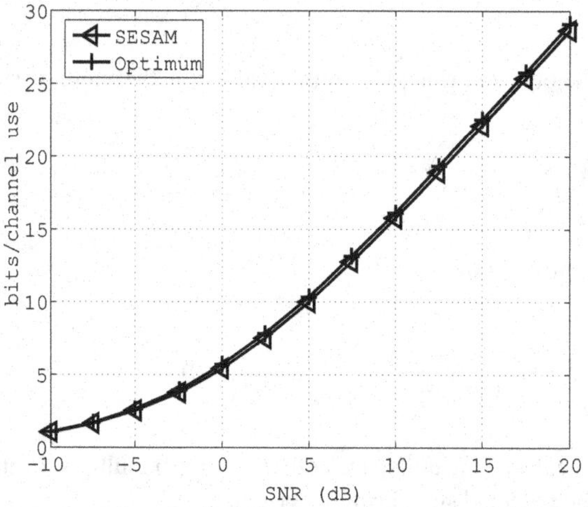


among users according to the quality of service that they should be given. Let $\mu_k$ be the priority assigned to user $k$. The corresponding optimization problem reads

$$\max_{\{Q_{k,n}\}} \sum_{k=1}^{K} \mu_k \sum_{n=1}^{N} R_{k,n} \tag{14}$$

which is subject to (2).

## Optimum Approach

Problem (14) is non-convex due to the fact that the objective function is neither a convex, nor a concave function of the covariance matrices. Fortunately, the duality result that we briefly described in the previous section can also be here applied in order to transform this non-convex problem into a convex one. The dual problem is given by

$$\max_{\{\Gamma_{k,n}\}} \sum_{k=1}^{K} \mu_k \sum_{n=1}^{N} \overline{R}_{k,n} \tag{15}$$

subject to (6). Let us assume, without loss of generality, that the priorities are such that $\mu_1 \geq \mu_2 \geq \cdots \geq \mu_K$ In the dual multiple access channel the optimum decoding order is obtained by decoding users with low priority first (Tse, 1998). According to this result and the priority order that has been assumed, optimally, user $K$ is decoded first and user 1 is decoded last. As a consequence, substituting (5) in (15) the weighted sum rate optimization problem can be written in the dual multiple access channel as follows,

$$\max_{\{\Gamma_{k,n}\}} \sum_{k=1}^{K} \eta_k \sum_{n=1}^{N} \log\left|\boldsymbol{I} + \sum\nolimits_{i\leq k} \boldsymbol{H}_{i,n}^{\mathrm{H}} \boldsymbol{\Gamma}_{i,n} \boldsymbol{H}_{i,n}\right|, \tag{16}$$

where $\eta_k = \mu_k - \mu_{k+1}$, for $1 \leq k \leq K-1$, and $\eta_K = \mu_K$. Note that the objective function is now an addition of concave functions of the covariance matrices $\boldsymbol{\Gamma}_{k,n}$ and, therefore, a concave function itself. In the context of memoryless MIMO channels, an iterative waterfilling algorithm similar to that used for solving the sum rate problem has been presented for $r_1 = r_2 = \cdots = r_K = 1$ in Kobayashi (2006). For the general setting of users with multiple antennas a steepest ascent algorithm was presented in Vishwanathan (2003). A straightforward extension of the latter algorithm has been applied to a MIMO OFDM setting in Tejera (2006 b). In each iteration, this algorithm computes the gradient of the objective function in (16). The computation of the gradient with respect to any covariance matrix $\boldsymbol{\Gamma}_{k,n}$ yields

$$\boldsymbol{G}_{k,n} = \sum_{j=k}^{K} \eta_j \boldsymbol{H}_{k,n} \left( \boldsymbol{I} + \sum_{i \le j} \boldsymbol{H}_{i,n}^{\mathrm{H}} \boldsymbol{\Gamma}_{i,n} \boldsymbol{H}_{i,n} \right)^{-1} \boldsymbol{H}_{k,n}^{\mathrm{H}}.$$

Let $\lambda_{k,n}^{l}$ denote the principal eigenvalue of the gradient matrix $\boldsymbol{G}_{k,n}^{l}$ in the $l$th iteration. Then, in order to search for an improved set of transmit covariance matrices, the one-dimensional subspace is considered that is defined by the unit norm eigenvector $\boldsymbol{v}_{k',n'}^{l}$ associated with the maximum principal eigenvalue,

$$\lambda_{k',n'}^{l} = \max\left\{\lambda_{k,n}^{l} \mid k = 1,\ldots,K,\ n = 1,\ldots,N\right\}.$$

Correspondingly, the new covariance matrices are computed as

$$\boldsymbol{\Gamma}_{k,n}^{l+1} = \xi \boldsymbol{\Gamma}_{k,n}^{l} + (1-\xi) P \boldsymbol{v}_{k',n'}^{l} \boldsymbol{v}_{k',n'}^{l,\mathrm{H}} \delta_{n,n'} \delta_{k,k'},$$

where $0 \le \xi \le 1$, and $\delta_{s,s'} = 1$ if $s = s'$ and $\delta_{s,s'} = 0$ otherwise. As indicated in Vishwanathan (2003), the optimum value of $\xi$ can be found applying bisection. Henceforth, this algorithm will be referred to as line search (LS) algorithm. Although, theoretically the LS algorithm converges to the optimum, in practice, the number of iterations required to achieve convergence appears to increase linearly with the number of subcarriers. This is related to the fact that the step size per iteration diminishes as the number of subcarriers increases, i.e., $\xi$ approaches 1. This is, in turn, a consequence of the fact that at each iteration only the structure of the covariance matrix of one user at one subcarrier is updated. As a result, the algorithm becomes very inefficient and eventually impracticable if applied to systems with a large number of subcarriers.

## Divide and Conquer

In order to speed up computation of covariance matrices, problem (14) can be divided into a number of smaller problems. To this end, for each subcarrier, we factorize $\boldsymbol{\Gamma}_{k,n} = p_n \overline{\boldsymbol{\Gamma}}_{k,n}$ such that

$$\sum_{k=1}^{K} \mathrm{Trace}\{\overline{\boldsymbol{\Gamma}}_{k,n}\} \le 1 \tag{17}$$

and $\sum_{n=1}^{N} p_n \le P$. Taking this factorization into account, optimum covariance matrices are found iterating the following two steps.

First, for given $\boldsymbol{p} = [p_1, p_2, \ldots, p_N]^{\mathrm{T}}$, solve

$$\max_{\{\overline{\boldsymbol{\Gamma}}_{k,n}\}} \sum_{k=1}^{K} \eta_k \log\left| \boldsymbol{I} + p_n \sum_{i \le k} \boldsymbol{H}_{i,n}^{\mathrm{H}} \overline{\boldsymbol{\Gamma}}_{i,n} \boldsymbol{H}_{i,n} \right|, \tag{18}$$

subject to (17) and $\overline{\boldsymbol{\Gamma}}_{k,n} \ge 0$, for every $n$ = 1, ..., $N$.

Second, for a given matrices $\overline{\boldsymbol{\Gamma}}_{k,n}$, solve

$$\max_{\boldsymbol{p}} \sum_{k=1}^{K} \eta_k \sum_{n=1}^{N} \log\left| \boldsymbol{I} + p_n \sum_{i \le k} \boldsymbol{H}_{i,n}^{\mathrm{H}} \overline{\boldsymbol{\Gamma}}_{i,n} \boldsymbol{H}_{i,n} \right|, \tag{19}$$

subject to $\sum_{n=1}^{N} p_n \leq P$ and $p_n \geq 0$.

Both problems are convex. In the second step, an optimum power allocation over subcarriers $\boldsymbol{p}$ is found for a given set of normalized covariance matrices. In the first, given the optimum power allocation $\boldsymbol{p}$ obtained in the previous iteration, an optimum set of normalized covariance matrices is found for every subcarrier. It is clear that each step improves the value of the objective function in (16) and hence convergence is guaranteed.

In the first step, optimization of the normalized covariance matrices can be done applying the algorithm presented in Vishwanathan (2003). In the second step, the Karush-Kuhn-Tucker (KKT) conditions of the optimization problem (Boyd, 2006) yield the following set of equations,

$$\sum_{k=1}^{K} \eta_k \operatorname{Trace}\left\{ \left( \boldsymbol{I} + p_n \boldsymbol{A}_{k,n} \right)^{-1} \boldsymbol{A}_{k,n} \right\} - \upsilon + \zeta_n = 0, \quad \forall n \tag{20}$$

$$P - \sum_{n=1}^{N} p_n \geq 0, \quad \upsilon \geq 0, \quad p_n \geq 0, \quad \zeta_n \geq 0, \quad \forall n$$

$$\upsilon \left( P - \sum_{n=1}^{N} p_n \right) = 0, \quad p_n \zeta_n = 0, \quad \forall n$$

where $A_{k,n} = \sum_{i=1}^{k} \boldsymbol{H}_{i,n}^{\mathrm{H}} \overline{\boldsymbol{\Gamma}}_{i,n} \boldsymbol{H}_{i,n}$. Considering the eigenvalues $\lambda_{k,n}^{s}$, $s = 1, 2, \ldots, t$, of matrix $\boldsymbol{A}_{k,n}$, (20) can be rewritten as

$$\sum_{k=1}^{K} \sum_{s=1}^{t} \frac{\eta_k \lambda_{k,n}^{s}}{1 + p_n \lambda_{k,n}^{s}} - \upsilon + \zeta_n = 0 \quad \forall n.$$

An efficient algorithm can be implemented that computes the power allocation $\boldsymbol{p}$ satisfying these conditions based on the following two observations.

***Observation 1***: For a given $\upsilon$, $p_n \neq 0$ if and only if $\sum_{k=1}^{K} \sum_{s=1}^{t} \eta_k \lambda_{n,k}^{s} > \upsilon$. In that case, $\zeta_n = 0$ and

$$\sum_{k=1}^{K} \sum_{s=1}^{t} \frac{\eta_k \lambda_{k,n}^{s}}{1 + p_n \lambda_{k,n}^{s}} - \upsilon \tag{21}$$

is a monotonically decreasing function of the transmit power $p_n$.

*Observation 2*: The optimum $\upsilon$ is a monotonically decreasing function of the transmit power *P*. Moreover,

$$\max_{n} \left\{ \sum_{k=1}^{K} \sum_{s=1}^{t} \eta_k \lambda_{n,k}^{s} \right\} > \upsilon,$$

i.e., at least one subcarrier gets some power.

From *observation 1* it becomes clear that for a given $\upsilon$ there is a unique power allocation $\boldsymbol{p}$ which can be efficiently computed. On the other hand, according to *observation 2*, if this power allocation exceeds the available transmit power, $\upsilon$ should be increased, otherwise it should be decreased. In this way, bisection can be used in order to compute $\upsilon$ corresponding to the particular transmit power constraint.

The number of iterations per tone required by the divide and conquer (DC) algorithm to reach convergence is essentially independent of the number of subcarriers. Indeed, the number of subcarriers does not appear in the formulation of problem (18). Thus, for a given power allocation vector, computation of the optimum normalized covariance matrices is independent of this parameter. In turn, for fixed normalized covariance matrices the power allocation on a certain

*Table 1. Number of iterations needed by DC and LS algorithms for convergence with N=16 (N=64)*

| $\mu_1$ | 0 | 0.2 | 0.4 | 0.6 | 0.8 | 1 |
|---|---|---|---|---|---|---|
| Inner iterations (DC) | 38.4 (38.4) | 30.0 (29.7) | 15.1 (15.7) | 10.1 (10.0) | 23.6 (23.4) | 38.6 (37.6) |
| Outer iterations (DC) | 3 (3) | 9 (9) | 3 (3) | 2 (2) | 7 (7) | 3 (2) |
| Iterations (LS) | 895 (3665) | 965 (3852) | 607 (2986) | 415 (2755) | 604 (2525) | 897 (3602) |

subcarrier $n$ is computed by equating (21) to zero and solving for $p_n$. We also note that this computation does not essentially depend on the number of subcarriers. Figs. 2 and 3 show the boundary of the capacity regions of two randomly chosen broadcast channels (BC) with $N = 16$ and $N = 64$, respectively. Each circle represents a point on the boundary of the capacity region corresponding to a particular choice of priorities, $\mu_1 = 0.1m$, $m \in \{0,1,\ldots,10\}$, and $\mu_2 = 1 - \mu_1$. These points have been computed with the DC algorithm presented in this section. As initial point a uniform power allocation over the frequency and scaled identity covariance matrices have been chosen. Table I shows the number of iterations that the DC and the LS algorithms need in order to reach some of these points. For the DC algorithm, outer iterations means the number of times that the power allocation problem (19) needs to be solved. Inner iterations refers to the average number of iterations needed to solve (18) accumulated over the total number of outer iterations. This number is equivalent to the average number of gradient computations per user and subcarrier that is required to reach the final solution. It can be observed that these numbers are almost invariant with respect to the number of subcarriers. By contrast, the number of iterations needed by the LS algorithm in order to reach the same performance as the DC algorithm is observed to increase by approximately factor 4 when passing from $N = 16$ to $N = 64$ subcarriers. In the LS algorithm the number of iterations are equivalent to the number of gradient computations performed per user and subcarrier. Limiting this number to that required by the DC algorithm, the rate vectors achieved by the LS algorithm are

*Figure 2. Capacity region and rate balancing points for a BC with $K = 2$, $t = 4$, $r_k = 2$ and $N = 16$*

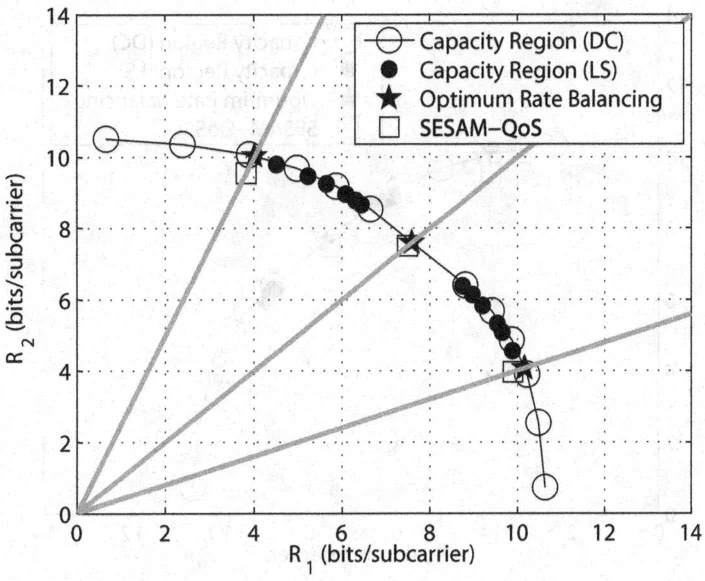

represented by the black dots in Figs. 2 and 3. For an approximately constant number of operations per subcarrier, we observe that points computed by the LS algorithm tend to accumulate around the starting point for increasing number of subcarriers.[c]

## Approximation of the Capacity Region with SESAM

SESAM can almost reach sum capacity. In order to also approximate points at the boundary of the capacity region different from those achieving sum capacity, SESAM must be modified at two steps (Brehmer, 2006). Firstly, the user allocation policy

$$A_{region}(i) = \{\pi_n(i), n\}$$

must be modified. For this purpose, the following problem can be solved in a successive manner as before,

$$\left(\pi_n(i), \boldsymbol{v}_{\pi_n(i)}\right) = \arg\max_k \left( \max_{\boldsymbol{v}} \mu_k \left\| \boldsymbol{H}_{k,n} \boldsymbol{v} \right\|_2 \right)$$

subject to $\|\boldsymbol{v}\|_2 = 1$ and $\boldsymbol{v} \in \text{null}\{\boldsymbol{v}^{\mathrm{H}}_{\pi_n(1)}, \ldots, \boldsymbol{v}^{\mathrm{H}}_{\pi_n(i-1)}\}$. The difference between this problem and (9) is the consideration of the priorities $\mu_k$ in the maximization. That is, users with high priority are more likely to get dimensions assigned than users with low priority. The rest of the encoding and allocation process is conducted as for the sum capacity maximizing algorithm described in the previous section.

Secondly, for a given vector of priorities, the optimum power allocation over the allocated subchannels is, in general, different from the pure waterfilling solution and is computed by solving

$$\max_{p_{\pi_n(i)}} \sum_{k=1}^{K} \mu_k R_k = \max_{p_{\pi_n(i)}} \sum_{k=1}^{K} \mu_k \left( \sum_{\pi_n(i)=k} \log_2\left(1 + p_{\pi_n(i)} \lambda^2_{\pi_n(i)}\right) \right)$$

*Figure 3. Capacity region and rate balancing points for a BC with $K = 2$, $t = 4$, $r_k = 2$, and $N = 64$*

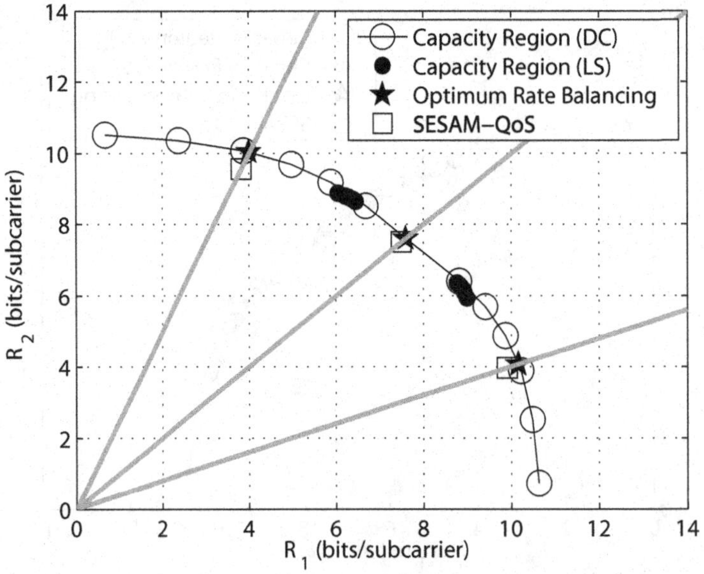

subject to $\sum_n \sum_i p_{\pi_n(i)} \leq P$ and $p_{\pi_n(i)} \geq 0,\ \forall n,i$. The solution to this problem is a generalized waterfilling power allocation of the form

$$p_{\pi_n(i)} = \max\{(\mu_{\pi_n(i)}\eta - \lambda_{\pi_n(i)}^{-2}), 0\},$$

where $\eta$ is chosen so that the transmit power constraint is satisfied. Fig. 4 shows the boundaries of the capacity region and the region achieved by the modified SESAM algorithm for a broadcast channel with $K = 2$, $t = 4$, $r_k = 2$ and $N = 16$. In general, for a given vector of priorities the optimum rate vector obtained from solving (14) is not well approximated by the rate vector obtained by SESAM. However, the point obtained by SESAM will be also close to the capacity region boundary. I.e., priorities have, in general, a different impact on the outcome of SESAM and the optimum approach.

## RATE BALANCING

Weighted sum rate maximization is a suitable policy in the context of communication systems with stationary random arrival of information and buffering capability. In such systems, if, at each time slot, the priorities are chosen to be proportional to the length of the queue corresponding to each user, the system can be stabilized, i.e., the average delay is bounded for all users (see Kobayashi (2006) and references therein). However, in the case of very stringent delay constraints and limited mobility, assigning priorities to users and optimizing weighted sum rate capacity does not guarantee that the final ranking of the users, as given by the rates they obtain, corresponds to the intended prioritization. For instance, it may happen that a high priority user obtains far a lower rate than a low priority user. This is generally the case if the channel of the latter is good enough as compared to the channel of the former. Sometimes it might be desirable to have a stronger control upon the relative performance achieved by the users in the network with respect to each other. Here is where the rate balancing problem formulation becomes relevant. Now the users are assigned relative rates $q_k$ rather than priorities. A relative rate expresses the share of rate each user should get out of the total transmitted rate. Mathematically, the optimization problem can be written as

*Figure 4. Capacity region and SESAM achievability region for a BC with $K = 2$, $t = 4$, $r_k = 2$ and $N = 16$*

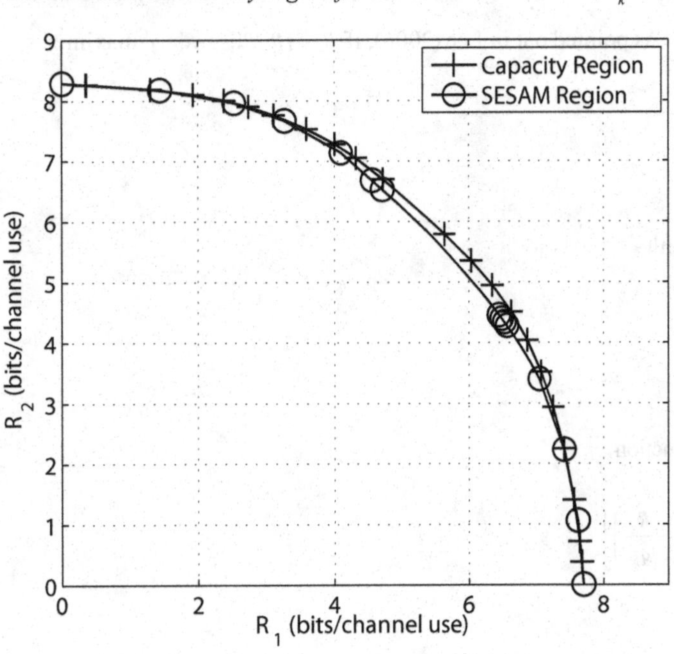

$$\max \gamma \quad \text{s.t.} \quad \gamma \leq \frac{R_k}{q_k}, \forall k \tag{22}$$

where the maximization is over the variable $\gamma$ and the rate vectors $\boldsymbol{R} = [R_1, R_2, \ldots, R_K]^T$ lying in the capacity region corresponding to the particular channel realization and the given transmit power constraint. It can be easily shown that the maximum in (22) is achieved with equality in the constraints. Thus, this problem is equivalent to finding the intersection between the straight line defined by the constraints $R_k = \gamma q_k$ and the boundary of the capacity region (cf. Figs. 2 and 3).

## Optimum Approach

As the capacity region is a convex set, the current optimization problem is also convex. Furthermore, the feasibility region has always a non-empty interior and, therefore, strong duality holds (Boyd, 2006). Accordingly, its solution can be found by solving the dual minimization problem. The dual function of problem (22) can be written as

$$g(\boldsymbol{\lambda}) = \max_{\gamma, \mathbf{R}} \left[ \gamma + \sum_{k=1}^{K} \mu_k \left( \frac{R_k}{q_k} - \gamma \right) \right],$$

where $\mu_k \geq 0 \; \forall k$. This function is equal to $\infty$ unless $\sum_{k=1}^{K} \mu_k = 1$. As a result, the dual problem reads

$$\min_{\boldsymbol{\mu}} \max_{\boldsymbol{R}} \sum_{k=1}^{K} \mu_k \frac{R_k}{q_k} \quad \text{s.t.} \quad \sum_{k=1}^{K} \mu_k = 1, \; \mu_k \geq 0 \forall k,$$

where $\boldsymbol{\mu} = [\mu_1, \mu_2, \ldots, \mu_K]^T$. Alternatively, the first constraint can be incorporated into the objective function in order to obtain (Lee, 2006)

$$\min_{\bar{\boldsymbol{\mu}}} \max_{\mathbf{R}} \left[ \frac{R_K}{q_K} + \sum_{k=1}^{K-1} \mu_k \left( \frac{R_k}{q_k} - \frac{R_K}{q_K} \right) \right] \quad \text{s.t.} \quad \sum_{k=1}^{K-1} \mu_k \leq 1 \lambda_k \geq 0 \, k = 1, \ldots, K-1 \tag{23}$$

where $\bar{\boldsymbol{\mu}} = [\mu_1, \mu_2, \ldots, \mu_{K-1}]^T$. As pointed out in Lee (2006), if $\boldsymbol{R}' = [R_1', R_2', \cdots, R_{K-1}']^T$ maximizes

$$\max_{\mathbf{R}} \left[ \frac{R_K}{q_K} + \sum_{k=1}^{K-1} \mu_k \left( \frac{R_k}{q_k} - \frac{R_K}{q_K} \right) \right],$$

$\widetilde{\boldsymbol{R}} = [\widetilde{R}_1, \widetilde{R}_2, \cdots, \widetilde{R}_{K-1}]^T$ with

$$\widetilde{R}_k = \left( \frac{R_k'}{q_k} - \frac{R_K'}{q_K} \right)$$

is a subgradient of the function

$$h(\bar{\boldsymbol{\mu}}) = \max_{\mathbf{R}} \left[ \frac{R_K}{q_K} + \sum_{k=1}^{K-1} \mu_k \left( \frac{R_k}{q_k} - \frac{R_K}{q_K} \right) \right]$$

in the feasible region given by the constraints $\sum_{k=1}^{K-1}\mu_k \leq 1,\ \mu_k \geq 0\ k=1,\ldots,K-1$. This fact and convexity of problem (23) enable a straightforward application of the ellipsoid algorithm (see references in Lee (2006)) in order to compute the optimum solution. Roughly speaking, the algorithm works as follows. First, consider an ellipsoid that comprises the whole feasibility region of problem (23). In this way it is guaranteed that this first ellipsoid contains the optimum solution. As a starting point consider the center of such ellipsoid, e.g., $\bar{\boldsymbol{\mu}}^{(0)} = \left[\mu_1^{(0)}, \mu_2^{(0)}, \ldots, \mu_{K-1}^{(0)}\right]^T$. Then, in order to compute the subgradient of $h(\bar{\boldsymbol{\mu}})$ at $\bar{\boldsymbol{\mu}}^{(0)}$ solve the following weighted sum rate maximization problem

$$\max_{\mathbf{R}} \sum_{k=1}^{K} \mu_k^{(0)} \frac{R_k}{q_k} \tag{24}$$

From the solution of this problem the subgradient $\tilde{\boldsymbol{R}}^{(0)}$ can be computed, which is used in order to construct a new ellipsoid of smaller volume but still containing the optimum solution. As starting point for the next iteration, the center of this new ellipsoid is chosen. It might be that the initial point at a certain iteration falls outside the feasibility region. In such case, a subgradient of the violated constraint must be computed and based on this subgradient an ellipsoid is computed which reduces the volume of the previous one and still comprises the optimum $\bar{\boldsymbol{\mu}}$. The ellipsoid algorithm provably converges to the optimum. However, the convergence of this algorithm is known to be very slow.

Summing up, in order to optimally solve the rate balancing problem (22) an iterative algorithm is needed that in each iteration requires computation of a weighted sum rate maximization problem, which is itself solved iteratively. This constitutes the main motivation for the introduction of an algorithm in the following section that, with essentially the same complexity of one inner iteration of the weighted sum rate maximizing DC algorithm discussed in the previous section, is able to closely approach the solution of (22). Table II shows the number of iterations that are required for the convergence of the optimum rate balancing algorithm described above for $t = 4$, $r_k = 2$, $q_k = 1$, $\forall k$. For the ellipsoid method two stop conditions have been considered. First,

$$\max_k \left\{ \left| \frac{R_k}{q_k} - \frac{1}{K}\sum_{j=1}^{K}\frac{R_j}{q_j} \right| \right\} < \varepsilon'$$

i.e., the vector of rates obtained at a certain iteration from (24) is almost parallel to the vector of constraints $\boldsymbol{q} = \left[q_1, q_2, \ldots, q_K\right]^T$, which means that the intersection of the straight line defined by vector $\boldsymbol{q}$ and the boundary of the capacity region has been almost reached. The second stop condition requires that the square of the largest radius of the ellipsoid is smaller than $\varepsilon$. This condition is needed as there exist points on the boundary of the capacity region that are not achievable by solving a weighted sum rate optimization problem. These are called time sharing points and can only be reached by switching between several transmit policies over time. The number of iterations required by the ellipsoid method in order to achieve convergence is given in the field "iterations" for $\varepsilon = 0.01$. As stop conditions for the iterative solution of problems (18) and (19) we require the increment in the value of the respective objective function at a certain iteration to be smaller than 0.1% and 1% of the value achieved in the previous iteration, respectively. The field "inner iterations (DC)" indicates the average number of gradient computations that are needed per user, subcarrier and iteration of the ellipsoid method in order to achieve convergence. Especially significant is the degradation in convergence speed of the ellipsoid method as the number of users increases.

*Table 2. Number of iterations needed by rate balancing optimum algorithms in order to reach convergence for a system with $q_1 = q_2 = \cdots = q_K = 1$, $r_1 = r_2 = \cdots = r_K = 2$, $t = 4$ and $K = 2/5/10$ users*

| SNR (dB) | 0 | 5 | 10 | 15 | 20 |
|---|---|---|---|---|---|
| Inner iterations (DC) | 11.3/17.8/19.9 | 9.9/19.2/21.7 | 7.0/17.8/20.3 | 4.0/15.3/18.0 | 2.5/13.2/16.0 |
| Iterations | 6.4/85.7/440.6 | 6.8/85.1/437.4 | 7.0/84.3/422.2 | 7.0/83.8/418.4 | 7.0/83.3/411.6 |

## Suboptimum Approach

SESAM can also be applied to approximate the solution to the rate balancing problem (Tejera, 2008). To this end, both the subchannel allocation policy and the subsequent power loading of the original algorithm must be adjusted to the new design criterion.

### Subchannel Allocation

In contrast to previous allocation policies, now, at each step $i$, the allocation of a subchannel on a certain subcarrier depends on the subchannel allocation on all other subcarriers. That is, allocation can not be conducted independently on every subcarrrier. The allocation $A_{\mathrm{RB}}(i)$ at step $i$ is performed in two substeps.

1. Determination of number of subcarriers:
   The numbers of subcarriers that each user will occupy in the current allocation step are computed. To this end, we subsume these numbers into the vector $N(i)$, $N(i) = \boldsymbol{\beta}(i)N$.
   As the number of subcarriers must sum up to $N$, $\boldsymbol{\beta}^{\mathrm{T}}(i)\boldsymbol{1} = 1$ must hold, where $\boldsymbol{1}$ denotes the all ones vector. $\boldsymbol{\beta}(i)$ is derived from the following set of equations,

$$\begin{aligned} \gamma \boldsymbol{q} &= \boldsymbol{R}_{\mathrm{su}}(i)\boldsymbol{\beta}(i), \\ \boldsymbol{\beta}^{\mathrm{T}}(i)\boldsymbol{1} &= 1, \end{aligned} \tag{25}$$

   where $\boldsymbol{R}_{\mathrm{su}}(i)$ denotes a $K \times K$ diagonal matrix containing the "single user rates " on its diagonal and $\boldsymbol{q}$ is the vector of rate balancing constraints. According to (25) the number of subcarriers obtained by a user at a given step should be proportional to the ratio between the QoS constraint and the rate obtained by this user should all subcarriers be assigned to it at this step. Thus, if a user $k$ has high QoS demands, i.e., $q_k$ is high compared to other users, the user will get more subcarriers allocated than users with lower demands, which makes fully sense. An exception to that rule occurs when the user has very strong channels, i.e., its single user capacity is high. In this case, the demands of that user may be met by assigning less subcarriers to it. Due to the diagonal structure of $\boldsymbol{R}_{\mathrm{su}}(i)$, (25) is numerically easy to solve. A few words have to be said about the computation of the single user rates. Theoretically, they can be obtained by performing waterfilling over the scalar subchannels that the corresponding user could get allocated at a certain step. However, the transmit power available for the allocated channels at a certain step is unknown a priori. We, therefore, assume that at the $i$-th step the available power is equal to $P/i$ and base the computation of single user rates on this assumption. The factor $1/i$ takes into account that the channel gains become smaller with each allocation step and, as with waterfilling, the applied power allocation method provides stronger subchannels with higher power than weaker ones. Obviously, $N(i)$ must only contain integer numbers. This is achieved by rounding the elements of $\boldsymbol{\beta}(i)N$ and making small readjustments, if the sum of the rounded elements is not equal to $N$.
2. Actual subchannel allocation:
   In this substep the actual allocation of users to subcarriers is performed. The goal is to preserve most of the maximum possible cumulated subchannel gain in step $i$ under consideration of the number of subcarriers computed in the previous substep. The maximum cumulated subchannel gain is given by the sum of subchannel SNRs obtained with the sum capacity maximizing allocation strategy $A_{\max}(i)$. Mathematically, this optimization problem can be written as

$$\min_{A_{\mathrm{RB}}(i)} \left( \sum_{\{k,n\} \in A_{\max}(i)} \lambda_{k,n}^{(i)2} - \sum_{\{k,n\} \in A_{\mathrm{RB}}(i)} \lambda_{k,n}^{(i)2} \right) \tag{26}$$

   subject to $N(i) = \boldsymbol{Z}(i)\boldsymbol{1}$, where $\lambda_{k,n}^{(i)2}$ is the SNR corresponding to user $k$, on subcarrier $n$, at step $i$, and

$$[\mathbf{Z}(i)]_{k,n} = \begin{cases} 1; & \{k,n\} \in A_{\mathrm{RB}}(i) \\ 0; & \text{otherwise.} \end{cases}$$

*Figure 5. Average rate balancing points for a balanced BC with $K = 2$, $t = 4$, $r_k = 2$ and $N = 16$*

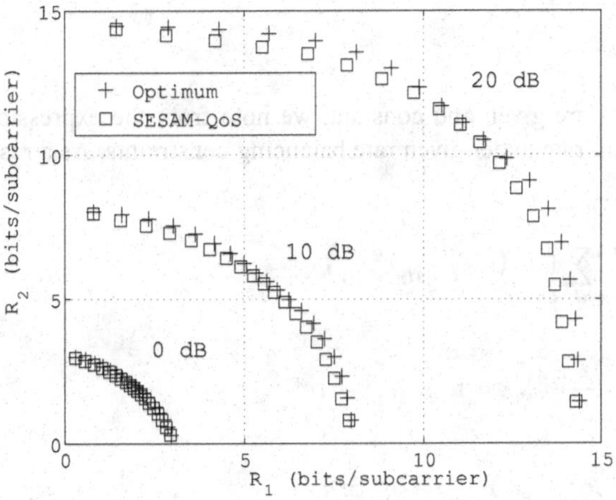


*Figure 6. Average optimum and suboptimum rates per user with a fairness constraint $t = 4$, $r_k = 2$ and $N = 16$*

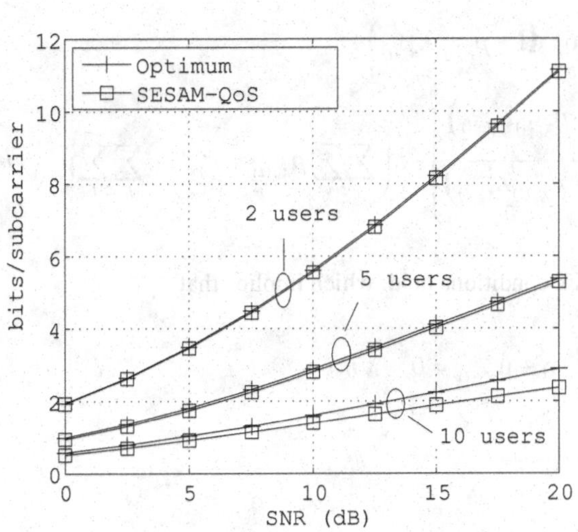


In order to determine $A_{RB}(i)$ from (26), we start by allocating each subcarrier to that user with the strongest SNR. This corresponds to determining the allocation strategy $A_{max}(i)$. Users are then divided into two groups: users in one group occupy more subcarriers than they should according to $\mathbf{N}(i)$, for users in the other group the opposite holds. Then a reallocation of subchannels is performed consecutively such that the minimum in (26) is achieved. In each step, that user from the first group is replaced by a user from the second group on a certain subcarrier such that the difference between the two users' SNRs is the smallest over all possible user pairs and over all subcarriers. Once a user has received the desired number of subcarriers, it is removed from the corresponding group.

## Power Allocation

Power must be allocated to the resulting subchannels such that the QoS constraints $\boldsymbol{q}$ are achieved. A suboptimum algorithm for that problem has been proposed previously in Shen (2005). The optimum solution is given in the following. Without loss of optimality, equality can be considered in the constraints of (22). Doing so, we can write,

$$\gamma \sum_{k=1}^{K} q_k = \sum_{k=1}^{K} R_k.$$

Since the QoS constraints $q_k$ are given and constant, we note from the expression above that maximizing $\gamma$ is equivalent to maximizing the sum rate under given rate balancing constraints. As a result, the rate balancing problem can be stated as,

$$\max_{p_{\pi_n(i)}} \sum_{k=1}^{K} R_k = \sum_{k=1}^{K} \sum_{\pi_n(i)=k} \log_2\left(1 + p_{\pi_n(i)} \lambda^2_{\pi_n(i)}\right)$$

$$\text{s.t.} \quad \frac{\sum_{\pi_n(i)=k} \log_2\left(1 + p_{\pi_n(i)} \lambda^2_{\pi_n(i)}\right)}{\sum_{\pi_n(i)=\text{ref}} \log_2\left(1 + p_{\pi_n(i)} \lambda^2_{\pi_n(i)}\right)} = \frac{q_k}{q_{\text{ref}}}, \ \forall k = 1, \ldots, K, k \neq \text{ref},$$

$$\sum_n \sum_i p_{\pi_n(i)} \leq P_{\text{Tx}}, \qquad p_{\pi_n(i)} \geq 0, \quad \forall n, i \tag{27}$$

where "ref" denotes an arbitrarily chosen reference user. The Lagrangian of this problem reads as

$$L\left(p_{\pi_n(i)}, \eta, \nu_k, \mu_{\pi_n(i)}\right) = \sum_{k=1}^{K} \sum_{\pi_n(i)=k} \log_2\left(1 + p_{\pi_n(i)} \lambda^2_{\pi_n(i)}\right) +$$

$$+ \sum_{\substack{k=1 \\ k \neq \text{refuser}}}^{K} \nu_k \left( \frac{q_k}{q_{\text{ref}}} - \frac{\sum_{\pi_n(i)=k} \log_2\left(1 + p_{\pi_n(i)} \lambda^2_{\pi_n(i)}\right)}{\sum_{\pi_n(i)=\text{ref}} \log_2\left(1 + p_{\pi_n(i)} \lambda^2_{\pi_n(i)}\right)} \right) - \eta \left( \sum_n \sum_i p_{\pi_n(i)} - P_{\text{Tx}} \right) + \sum_n \sum_i \mu_{\pi_n(i)} p_{\pi_n(i)}.$$

At the optimum point the KKT conditions hold, which implies that

$$\frac{\partial L}{\partial p_{\pi_n(i)}} = \frac{\lambda^2_{\pi_n(i)}}{\left(1 + p_{\pi_n(i)} \lambda^2_{\pi_n(i)}\right) \ln 2} a_k - \eta + \mu_{\pi_n(i)} = 0, \quad \forall n, i \tag{28}$$

with

$$a_k = \begin{cases} 1 + \nu_k \dfrac{1}{\sum_{\pi_n(i)=\text{ref}} \log_2\left(1 + p_{\pi_n(i)} \lambda^2_{\pi_n(i)}\right)}, & k \neq \text{ref}, \\[2ex] 1 - \sum_{\substack{k=1 \\ k \neq \text{ref}}}^{K} \nu_k \dfrac{\sum_{\pi_n(i)=k} \log_2\left(1 + p_{\pi_n(i)} \lambda^2_{\pi_n(i)}\right)}{\left(\sum_{\pi_n(i)=\text{ref}} \log_2\left(1 + p_{\pi_n(i)} \lambda^2_{\pi_n(i)}\right)\right)^2}, & k = \text{ref}. \end{cases}$$

Furthermore, $\mu_{\pi_n(i)} = 0$ if $p_{\pi_n(i)} \neq 0$. Considering this and the fact that $a_k$ does not depend on either $n$ or $i$, it can be directly concluded that, if greater than zero, $p_{\pi_n(i)}$ must be chosen such that the first factor in (28) is independent of $\lambda_{\pi_n(i)}$, i.e.,

$$p_{\pi_n(i)} = \max\left\{ \left( \xi_k - \frac{1}{\lambda^2_{\pi_n(i)}} \right), 0 \right\}$$

with $\xi_k = a_k/(\ln 2\eta)$. This result has the form of a waterfilling solution with a user dependent waterlevel $\xi_k$. The waterlevels $\xi_k$ need to be determined such that the relative rate constraints

$$\frac{q_k}{q_{\text{ref}}} = \frac{R_k}{R_{\text{ref}}} = \frac{\sum\limits_{\substack{\pi_n(i)=k \\ p_{\pi_n(i)}\neq 0}} \log_2(\xi_k \lambda^2_{\pi_n(i)})}{\sum\limits_{\substack{\pi_n(i)=\text{ref} \\ p_{\pi_n(i)}\neq 0}} \log_2(\xi_{\text{ref}} \lambda^2_{\pi_n(i)})}, \quad \forall k = 1,\ldots,K, k \neq \text{ref} \tag{29}$$

are fulfilled for all users and the power budget is fully exploited. Fixing $\xi_{\text{ref}}$ the waterlevels of all other users can be numerically computed from (29). $\xi_{\text{ref}}$ itself is a monotonically increasing function of the total transmit power. Therefore, a bisection method can be employed in order to find $\xi_{\text{ref}}$ that fulfils the power constraint with equality.

In Figs. 2 and 3, the rate balancing points have being plotted that are reached by the optimum approach and the approach described in this section, which we call SESAM-QoS, for three different rate balancing constraints, which are represented by the three straight lines departing from the origin. For the same settings and three different SNR values,[d] Fig. 5 shows average results under the assumption of independent and identically distributed complex Gaussian entries in the channel matrices. Finally, Fig. 6 shows performance curves for different numbers of users and a "maximum fairness" constraint, i.e., $q_k = 1$, $\forall k$. In all cases performance loss is observed to be certainly small and yet, the complexity of the suboptimum algorithm is similar to that of just one iteration of the optimum algorithm.

## SUM THROUGHPUT MAXIMIZATION UNDER MINIMUM RATE REQUIREMENTS

The rate balancing problem leads to a predefined point on the boundary of the capacity region but does not enable throughput maximization on the boundary itself. In particular, an unfavourable choice of rate ratios may lead to strong

*Figure 7. Comparison of the transmission rates achieved with the presented heuristic algorithm under minimum rate requirements with two points on the boundary of the capacity region, $t = 4$, $r_k = 2$ and $N = 16$*

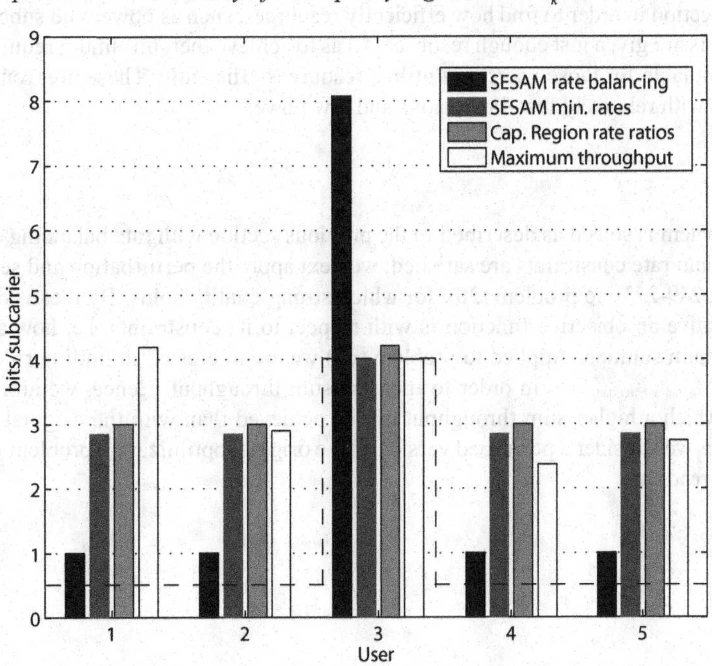

decreases in sum capacity as compared to the unconstrained sum capacity maximization. On the other hand, as it has already been discussed, sum capacity maximization without quality of service constraints may lead to an unfair distribution of resources. In order to solve this trade-off between fairness and maximum throughput, minimum transmission rates for each user can be introduced and the problem of maximizing sum capacity under these minimum rate requirements and a maximum power restraint can be stated, i.e.

$$\max \sum_{k=1}^{K} R_k \quad \text{s.t.} \quad R_k \geq R_{k,\min}, \quad \forall k = 1,\ldots,K \tag{30}$$

and subject to (2).

## Optimum Approach

An iterative algorithm that solves problem (30) can be found along the lines of the steps described in the previous section (Michel, 2007). First, we note that this is a convex optimization problem that has a feasibility region with a non-empty interior provided that $\boldsymbol{R}_{\min} = [R_{1,\min}, R_{2,\min}, \ldots, R_{K,\min}]^T$ belongs to the interior of the capacity region. Note that if no feasible points exist or if $\boldsymbol{R}_{\min}$ is on the boundary of the capacity region the problem is trivial.[e] Thus, in the relevant case, strong duality holds. As a result, in order to solve (30) we can minimize its dual function, which is given by

$$g(\boldsymbol{\lambda}) = \max_{\boldsymbol{R}} \sum_{k=1}^{K} (1 + \lambda_k) R_k - \sum_{k=1}^{K} \lambda_k R_{k,\min},$$

subject to $\lambda_k \geq 0$. The rate vector achieving the maximum in this expression is a subgradient of $g(\boldsymbol{\lambda})$. Based on this subgradient the ellipsoid method can be applied in order to minimize $g(\boldsymbol{\lambda})$. Again, the resulting algorithm is doubly iterative. That is, at each iteration of the ellipsoid algorithm, a weighted sum rate optimization problem must be iteratively solved.

## Sum Throughput Enhancements Under Minimum Rate Requirements

So far a non-iterative suboptimum approach to problem (30) has not been found yet. Instead we present a heuristic approach in this section which increases throughput as compared to the rate balancing solution when certain minimum rates constraints are given (Guthy, 2007). To this end, we rely on a sensitivity analysis of the optimum power allocation described in the previous section in order to find how efficiently resources, such as power and subchannels, are utilized by the users. Inefficient users are given just enough resources so as to achieve their minimum required rates. In this way, more resources become available for those users that utilize resources efficiently. These are, roughly speaking, users that can achieve high rates with relatively few dimensions and low power.

### User Classification

First the rate balancing problem is solved as described in the previous section with rate balancing constraints $q_k = R_{k,\min}$. Assuming that the minimum rate constraints are satisfied, we next apply the perturbation and sensitivity analysis described in Boyd (2006, pp. 249-253) to problem (26), for which strong duality holds. By means of sensitivity analysis one can find out how sensitive an objective function is with respect to its constraints, i.e. how a change of a certain constraint affects the optimum solution. Applied to problem (27) we are interested in finding rules how to change the ratio constraints $q_k / q_{\text{ref}} = R_{k,\min} / R_{\text{ref},\min} = \rho_k$ in order to increase sum throughput. Hence, we aim at finding new ratio constraints $\rho_k + u_k$, with which a higher sum throughput can be achieved than with the original problem. In order to determine the appropriate $u_k$ we consider a perturbed version of the original optimization problem with ratio constraints $\rho_k$. Its Lagrange function reads as

$$L\left(p_{\pi_n(i)},\eta,\nu_k,\mu_{\pi_n(i)},u_k\right)=\sum_{k=1}^{K}\sum_{\pi_n(i)=k}\log_2\left(1+p_{\pi_n(i)}\lambda^2_{\pi_n(i)}\right)+$$

$$+\sum_{\substack{k=1\\k\neq\mathrm{ref}}}^{K}\nu_k\left(\rho_k+u_k-\frac{\sum_{\pi_n(i)=k}\log_2\left(1+p_{\pi_n(i)}\lambda^2_{\pi_n(i)}\right)}{\sum_{\pi_n(i)=\mathrm{ref}}\log_2\left(1+p_{\pi_n(i)}\lambda^2_{\pi_n(i)}\right)}\right)-\eta\left(\sum_n\sum_i p_{\pi_n(i)}-P_{\mathrm{Tx}}\right)+\sum_n\sum_i\mu_{\pi_n(i)}p_{\pi_n(i)}.$$

Due to strong duality the optimum sum throughput $R_{\mathrm{sum,opt}}$ is equal to the minimum of the dual function $g\left(\eta,\upsilon_k,\mu_{\pi_n(i)},u_k\right)$, i.e.,

$$R_{\mathrm{sum,opt}}=\min_{\mu_{\pi_n(i)},\nu_k,\eta}g\left(\eta,\upsilon_k,\mu_{\pi_n(i)},u_k\right)=\min_{\mu_{\pi_n(i)},\nu_k,\eta}\left(\sup_{p_{\pi_n(i)}}L\left(p_{\pi_n(i)},\eta,\upsilon_k,\mu_{\pi_n(i)},u_k\right)\right) \tag{31}$$

With this relationship the partial derivative of $R_{\mathrm{sum,opt}}$ with respect to $u_k$ can be written as

$$\frac{\partial R_{\mathrm{sum,opt}}}{\partial u_k}=\frac{\partial L\left(p^*_{\pi_n(i)},\eta^*,\nu^*_k,\mu^*_{\pi_n(i)},u_k\right)}{\partial u_k}=\nu^*_k \tag{32}$$

where the superscript * denotes the optimum values leading to $R_{\mathrm{sum,opt}}$ according to (30). Once the corresponding primal optimization problem has been solved for $u_k=0,\forall k=1,\ldots,K,k\neq\mathrm{ref}$, the $K-1$ optimum Lagrange multipliers $\nu^*_k$ can be computed from the $K$ equations $\xi_k=a_k/(\ln 2\eta)$ (c.f. Equation (28)). The result is given by

$$\nu^*_k(u_k=0)=\left(1-\frac{\xi_k\sum_{i=1}^{K}q_i}{\sum_{i=1}^{K}q_i\xi_i}\right)R_{\mathrm{ref,opt}}=\tilde{\nu}_k R_{\mathrm{ref,opt}}. \tag{33}$$

If $\nu^*_k>0$, the maximum sum rate will increase for $u_k>0$, i.e., the relative rate requirement $\rho_k$ should be further increased to achieve the desired goal. On the other hand, a reduction of rate requirements for users with $\nu^*_k>0$ is necessary for an increase in sum rate. Unfortunately, expressing the feasible domain in terms of rate ratios is too complex and therefore it is difficult to determine how far the $\rho_k$ can be increased or decreased without violating the minimum rate constraints. As a result, we resort to a heuristic approach in order to find a new resource allocation that increases the throughput while fulfilling the minimum rate constraints. The following user classification is made:

- Users with $\tilde{\nu}_k<0$ are put into the so-called "looser" group, which will be referred to as **Group L** in the following. They will be only served with the corresponding minimum required rates.
- The remaining users are comprised in the "winner" **Group W**, i.e., during the next steps they will obtain more resources than in the current step of the algorithm. Consequently their individual rates will increase as well.

Note that there are only $K-1$ Lagrange multipliers, but $K$ users. Let us for the moment assume that the reference user belongs to Group W. From (33) it can be seen that $\tilde{\nu}_k$ is independent of the reference user. Thus, although we started with an arbitrarily chosen reference user, the later change of the reference user does not affect the sign of the optimum Lagrange multipliers. Consequently, the assumption that the reference user belongs to the winner group can be made without loss of generality. Based on this classification, next a method is described, that assigns users of Group L the resources that they just need in order to reach their minimum rates. With the remaining resources the rate balancing problem is solved again for the users of Group W, which are able to improve their rates due to the larger amount of resources available.

## Subchannel Allocation

Assume that a user $k$ belonging to Group L obtained $N_k^{(1)}(i)$ subcarriers assigned at step $i$ during the execution of the SESAM-QoS algorithm, and its total rate was $R_k^{(1)}$. Now, during the new execution of the SESAM-QoS algorithm this user will be given

$$N_k(i) = \frac{R_{k,\min}}{R_k^{(1)}} N_k^{(1)}(i)$$

subcarriers at step $i$. The remaining subcarriers

$$N - \sum_{k \in \text{Group L}} N_k(i)$$

are distributed amongst the users in Group W. To this end, the ratios of subcarriers are determined according to (25) considering the rate balancing constraints of these users. The actual allocation of subcarriers to the users is then performed according to (26).

## Power Allocation

Power allocation is conducted in two steps: First, power is distributed to the users in Group L such that the minimum rates can be achieved with as little transmit power as possible, i.e., the following optimization problem is solved:

$$P_{\text{tot,L}} = \min \sum_{k \in \text{Group L}} \sum_{\pi_n(i)=k} p_{\pi_n(i)}$$

$$\text{s.t.} \sum_{\pi_n(i)=k} \log_2(1 + p_{\pi_n(i)} \lambda^2_{\pi_n(i)}) = R_{k,\min}, \quad \forall k \in \text{Group L}, \quad p_{\pi_n(i)} \geq 0, \forall \text{n, i}.$$

The solution of this problem can again be computed via the KKT conditions. The remaining power is then distributed to the users in Group W such that sum rate is maximized and the rate ratios $\rho_k$ are fulfilled, i.e.,

$$\max \sum_{k \in \text{Group W}} \sum_{\pi_n(i)=k} \log_2(1 + p_{\pi_n(i)} \lambda^2_{\pi_n(i)})$$

$$\text{s.t.} \sum_{\pi_n(i)=k} \log_2(1 + p_{\pi_n(i)} \lambda^2_{\pi_n(i)}) = \rho_k \sum_{\pi_n(i)=\text{ref}} \log_2(1 + p_{\pi_n(i)} \lambda^2_{\pi_n(i)}), \quad \forall k \in \text{Group W},$$

$$\sum_{k \in \text{Group W}} \sum_{\pi_n(i)=k} p_{\pi_n(i)} = P_{\text{Tx}} - P_{\text{tot,L}}, \quad p_{\pi_n(i)} \geq 0, \forall \text{n, i}.$$

## Simulation Results

Fig. 7 shows performance of the described algorithm for a broadcast channel with $t = 4$, $r_k = 2$, $N = 16$ and SNR = 20 dB. In this figure, the dashed line represents the minimum rates requested by the different users. Compared to the solution of the rate balancing problem with minimum rates as rate balancing constraints, a gain of 27% in sum throughput is possible. This is achieved by forcing user 3 to be served with its minimum required rate. Two points on the the boundary of the capacity region are also represented in this figure. The one denoted as "rate ratios" lies on the same balancing constraint (straight line) as the rate vector resulting from the heuristic approach described in this section. The other one denoted as "maximum throughput" represents the solution to (30). It can be concluded that the presented method is able to almost achieve the boundary of the capacity region. In order to approach the point of maximum sum capacity knowledge of the rate ratios in this point would be required.

## CONCLUSION

In this chapter an overview of four of the most significant design problems in point to multipoint MIMO OFDM networks has been given. State-of-the-art optimum iterative algorithms have been discussed for each of the design tasks and the main shortcomings of these approaches have been identified. These are the iterative nature of the algorithms in general, and the slow convergence speed of subgradient approaches in particular. For each of the design problems discussed, suboptimum algorithms have been presented that are non-iterative, exhibit a complexity similar to that of one iteration of the optimum approaches and, at least for a broad class of common scenarios, show insignificant performance losses with respect to the optimum approaches. A further advantage of these suboptimum schemes is reduced signaling overhead, which is due to the zero forcing constraints enforced by the schemes, on the one hand, and to the non-consideration of time-sharing rate vectors, on the other.

Both optimum and suboptimum approaches considered here are based on the assumption of perfect channel state information at the transmitter. While this is key for the conceptual and mathematical tractability of the problems studied and might be deemed to be a realistic assumption in wired networks, full channel knowledge is, in general, difficult to obtain in wireless networks, where the channel is usually time varying. Realistic modeling of imperfections in the channel state information and solution of the discussed design problems under consideration of these imperfections constitute two of the most exciting and challenging open topics of research in the area of point to multipoint networks.

## REFERENCES

Boyd, S., & Vandenberghe, L. (2006). *Convex Optimization.* Cambridge University Press.

Brehmer, J., Molin, A., Tejera, P., & Utschick, W. (2006, June). *Low Complexity Approximation of the MIMO Broadcast Channel Capacity Region.* Paper presented at the IEEE International Conference on Communications (ICC), Istanbul, Turkey.

Costa, M. (1983). Writing on Dirty Paper. *IEEE Transactions on Information Theory, 29*(3), 439-441.

Cover, T. M., & Thomas, J. A. (1993). *Elements of Information Theory.* New York: John Wiley & Sons.

Foschini, G. J., & Gans, M. J. (1998). On Limits of Wireless Communications in a Fading Environment when Using Multiple Antennas. *Wireless Personal Communications, 6*(3), 311-335.

Guthy, C., Utschick, W., Bauch, G., & Nossek, J.A. (2007). *Sum Throughput Enhancements in Quality of Service Constrained Multiuser MIMO OFDM Systems.* Wireless personal communication. Retrieved July 2, 2008, from http://www.springerlink.com/content/81pqOn7552830w6k1

Jindal, N., Rhee, W., Vishwanath, S., Jafar, S.A., & Goldsmith, A.J. (2005). Sum Power Iterative Water-filling for Gaussian Broadcast Channels. *IEEE Transactions on Information Theory, 51(5),* 1570-1580.

Kobayashi, M. & Caire, G. (2006). An Iterative Water-Filling Algorithm for Maximum Weighted Sum-Rate of Gaussian MIMO-BC. *IEEE Journal on Selected Areas in Communications, 24(8)*, 1640-1646.

Lee, J. & Jindal, N. (2006). *Symmetric Capacity of MIMO Downlink Channels.* Paper presented at the IEEE International Symposium on Information Theory (ISIT), Seattle, USA.

Michel, T., & Wunder, G. (2007, February). *Achieving QoS and Efficiency in the MIMO Downlink with Limited Power.* Paper presented at the ITG/IEEE Workshop on Smart Antennas, Vienna, Austria.

Raleigh, G. G., & Cioffi, J. M. (1998). Spatio-Temporal Coding for Wireless Communication. *IEEE Transactions on Communications, 46(3)*, 357-366.

Sampath, H., Talwar, S., Tellado, J., Erceg, V., & Paulraj, A. (2002). A fourth-generation MIMO-OFDM broadband wireless system: Design, performance, and field trial results. *IEEE Communications Magazine, 40(9)*, 143-149.

Spencer, Q.H., Swindlehurst, A.L., & Haardt, M. (2004). Zero-forcing methods for downlink spatial multiplexing in multi-user MIMO channels. *IEEE Transactions on Signal Processing, 52*(2), 461-471.

Tachikawa, K. (2003). A perspective on the evolution of mobile communications. *IEEE Communications Magazine, 41(10)*, 66-73.

Tejera, P., Utschick, W., Bauch, G., & Nossek, J.A. (2006). Subchannel Allocation in Multiuser Multiple Input Multiple Output

Systems. *IEEE Transactions on Information Theory, 52(10)*, 4721-4733.

Tejera, P., Utschick, W., Bauch, G., & Nossek, J.A. (2006b). *Efficient Implementation of Successive Encoding Schemes for the MIMO OFDM Broadcast Channels.* Paper presented at the IEEE International Conference on Communications (ICC), Istanbul, Turkey.

Tejera, P., Utschick, W., Bauch, G., & Nossek, J.A. (2008). Rate balancing in multiuser MIMO OFDM systems. *IEEE Transactions on Communications*, accepted for publication.

Tse, D. & Hanly, S. (1998). Multi-Access Fading Channels: PartI: Polymatroid Structure, Optimal Resource Allocation and Throughput Capacities. IEEE Transactions on Information Theory, 44(7), 2796-2815.

Vishwanath, S., Jindal, N. & Goldsmith, A.J. (2003). Duality, Achievable Rates, and Sum-Rate Capacity of Gaussian MIMO Broadcast Channels. *IEEE Transactions on Information Theory, 49*(10), 2658-2668.

Vishwanath, S., Rhee, W., Jindal, N., Jafar, S.A., & Goldsmith, A.J. (2003). *Sum Power Iterative Waterfilling for Gaussian Vector Broadcast Channels.* Paper presented at the IEEE International Symposium on Information Theory (ISIT), Yokohama, Japan.

Vishwanathan, H., Venkatesan, S., & Huang, H. (2003). Downlink Capacity Evaluation of Cellular Networks With Known-Interference Cancellation. *IEEE Journal on Selected Areas in Communications, 21*(5), 802-811.

Weingarten, H., Steinberg, Y., & Shamai, S.S. (2006). The Capacity Reigion of the Gaussian Multiple-Input Multiple-Output Broadcast Channel. *IEEE Transactions on Information Theory, 52*(9), 3936-3964.

Yu, W. (2006). Sum-Capacity Computation for the Gaussian Vector Broadcast Channel via Dual Decomposition. *IEEE Transactions on Information Theory, 52(2)*, 754-759.

## ENDNOTES

[a] Input and output are general terms to denote antennas in wireless systems and wires in wired transmission systems. Systems employing multiple inputs and multiple outputs are called MIMO.

[b] The expression $A \geq 0$ means that matrix $A$ is positive semidefinite.

[c] Note that the initial covariance matrices in these examples yield a solution close to both ends of the time-sharing segment.

[d] SNR is here define as *P/N*.

[e] One can check whether the constraint can be fulfilled by first solving the rate balancing problem with constraint vector $R_{\min}$. If $\gamma > 1$ the problem is not feasible. If $\gamma = 1$, $R_{\min}$ is the only feasible point. If $\gamma < 1$ strong duality holds.

# Section III
# Applications of Smart Antennas

# Chapter XVII
# Smart Antennas for Code Division Multiple Access Systems

**Salman Durrani**
*The Australian National University, Australia*

**Marek E. Bialkowski**
*The University of Queensland, Australia*

## ABSTRACT

*This chapter discusses the use of smart antennas in Code Division Multiple Access (CDMA) systems. First, we give a brief overview of smart antenna classification and techniques and describe the issues that are important to consider when applying these techniques in CDMA systems. These include system architecture, array antennas, channel models, transmitter and receiver strategies, beamforming algorithms, and hybrid (beamforming and diversity) approach. Next, we discuss modeling of smart antennas systems. We present an analytical model providing rapid and accurate assessment of the performance of CDMA systems employing a smart antenna. Next, we discuss a simulation strategy for an adaptive beamforming system. A comparison between the analytical results and the simulation results is performed followed by a suitable discussion.*

## INTRODUCTION

Over the last two decades smart antenna technologies, referred to as Multiple Input Single Output (MISO) or Single Input Multiple Output (SIMO), have received much attention (Alexiou, 2004). Conventional cellular systems make use of omnidirectional or sectorized antenna systems. "The major drawback of such antenna systems is that electromagnetic energy is radiated unnecessarily in too many directions within the entire cell thus causing interference to other users in the system. One way to limit this interference and direct the energy to the desired user is to use smart antennas" (Osseiran, 2006). Smart antenna systems consist of an array of antenna elements whose signals are intelligently combined to exploit the spatial domain and improve the performance of a wireless system. For narrowband variety of these systems, the "basic principle is to multiply the signals at different antenna elements with complex weights before the signals are transmitted or when the received signals are summed up" (Pedersen, 2003). This allows a narrow beam (with gain) to be created towards the desired user and low side-lobes towards other users. Smart antennas offer a broad range of ways to improve the performance of wireless systems. For example, in the initial development of the system they offer enhanced range and in a longer term increased capacity of wireless systems. Due to the capability of spatial separation of signals, they allow different subscribers to share the same spectral resources (Space Division Multiple Access). The same capability of spatial separation of signals allows for mitigating adverse multi-path effects.

Smart antennas can be generally divided into three main groups depending upon the level of intelligence. These are (i) fixed beam systems (Osseiran, 2006) or fixed multi-beam (also called switched beam) systems (Allen, 2004), (ii) phased arrays (Janaswamy, 2001) and (iii) adaptive array antennas (Godara, 1997). Adaptive implementations have better performance than fixed or switched beam systems, at the expense of higher implementation cost and complexity. The advent of low cost Digital Signal Processor (DSP) chips, Application Specific Integrated Circuits (ASIC's), system-in-a-package or system-on-a-chip realizations have made adaptive antenna systems practical for commercial use (Kaiser, 2005).

According to (Alexiou, 2004), "the adoption of smart antennas in future wireless systems relies on two main investigation approaches to be performed:

- Consideration of the smart antenna features early in the design phase of future systems.
- Realistic performance evaluation of smart antennas according to the critical parameters associated with system requirements".

As far as first approach is concerned, smart antennas are already an integral part of Universal Mobile Telecommunications System (UMTS) and third generation Code Division Multiple Access (CDMA) standards and their evolutionary roadmaps (Boche, 2006). Furthermore, there is currently increasing interest in the incorporation of smart antenna techniques for IEEE wireless LAN/MAN (802.11n, 802.16 and 802.20) and smart antennas are expected to play a significant role in enabling broadband wireless communication (Alexiou, 2004). As far as second approach is concerned, performance evaluation based on accurate modeling with suitable simulation methodology is required.

This chapter addresses these two approaches and provides an overview of smart antenna technologies for Code Division Multiple Access system. The remainder of the chapter is organised as follows. In Section 2, we first summarize the background information and then present the state of the art in smart antenna systems from key system architecture and performance analysis perspectives. In Section 3, we discuss smart antenna system modeling for CDMA and present a performance analysis based on analytical expressions, including simulation results for verification. Finally in Section 4, we present conclusions and identify some avenues for future work.

## SMART ANTENNA TECHNOLOGIES

This section is devoted to the review of background information for smart antenna technologies. The emphasis is on summarizing the state of the art by focusing on well-chosen highlights, i.e. classification, system architecture and modeling and performance analysis aspects. The presented considerations are limited to the narrowband case. By the narrowband system, we mean the system whose operational frequency band is no more than a few percents.

### Classification

A smart antenna is defined as an array of antennas with a digital signal processing unit, which can change its pattern dynamically to adjust to noise, interference and multipaths. The conceptual block diagram of a smart antenna system is shown in Figure 1.

We can identify three main blocks: (i) array antenna (ii) complex weights and (iii) adaptive signal processor. The antenna elements are usually identical and have an omni-directional pattern in the azimuth plane. The array geometry usually comprises of a Uniform Linear Array (ULA) or Uniform Circular Array (UCA) of antenna elements. The signals received at the different antenna elements are multiplied with complex weights and then summed up. The complex weights are continuously adjusted by the Adaptive Signal Processor which uses all available information such as pilot or training sequences or knowledge of the properties of the signal to calculate the weights. This is done so that the main beam tracks the desired user and/or nulls are placed in the direction of interferers and/or side lobes towards other users are minimized. It should be noted that the term "smart" refers to the whole antenna system and not just the array antenna alone.

"The fundamental idea behind a smart antenna is not new but dates back to the early sixties when it was first proposed for electronic warfare as a counter measure to jamming" (Janaswamy, 2001). The smart antenna systems for cellular base stations can be divided into the following three main categories (Janaswamy, 2001), which are illustrated in Figure

2 (Lehne, 1999). These are (i) switched beam system (ii) phased arrays and (iii) adaptive arrays. It has to be noted that this division is not rigid and switched beam and phased array systems are simpler physical approaches to realising fully adaptive antennas. This step by step migration strategy has been used to lower the initial deployment costs to service providers. These categories are discussed in detail below:-

## Switched Beam Systems

A switched beam antenna system consists of several highly directive, fixed, pre-defined beams which can be formed by means of a beamforming network e.g., the Butler matrix (Butler, 1966; Pattan, 2000) which consists of power splitters and phase shifters. The system detects the signal strength and chooses from one of the several beams that gives the maximum received power.

A switched beam antenna can be thought of as an extension of the conventional sector beam antennas in that it divides a sector into several micro-sectors. It is the simplest technique and easiest to retro-fit to existing wireless technologies. However switched beam antenna systems are effective only in low to moderate co-channel interfering environments owing to their lack of ability to distinguish a desired user from an interferer, e.g. if a strong interfering signal is at the centre of the selected beam and the desired user is away from the centre of the selected beam, the interfering signal can be enhanced far more than the desired signal with poor quality of service to the intended user.

## Phased Arrays

Phased arrays make use of the Angle of Arrival (AOA) information from the desired user to steer the main beam towards the desired user (Janaswamy, 2001; Sun, 2004). The signals received by each antenna element are weighted and combined to create a beam in the direction of the mobile. Only the phases of the weights are varied and the amplitudes are held constant.

Phased arrays improve upon the capabilities of a switched beam antenna. They can be considered as a generalization of the switched lobe concept and have an increased number of possible beam directions (Lehne, 1999). This number is limited by the number of antenna elements in the array and the lowest phase bit, when binary phase shifters are employed. In general, the phased arrays cannot offer the capability of steering nulls towards undesired users. The limitations of phased array can be overcome using fully adaptive arrays.

## Adaptive Arrays

In an adaptive array, signals received by each antenna are weighted and combined using complex weights (magnitude and phase) in order to maximise a particular performance criterion e.g. the Signal to Interference plus Noise Ratio

*Figure 1. Block diagram of a smart antenna system*

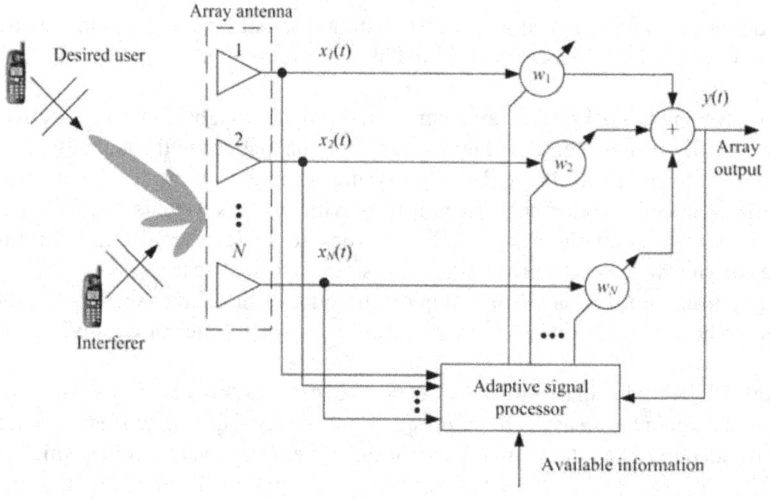

*Figure 2. Classifications of smart antenna systems (Lehne, 1999)*

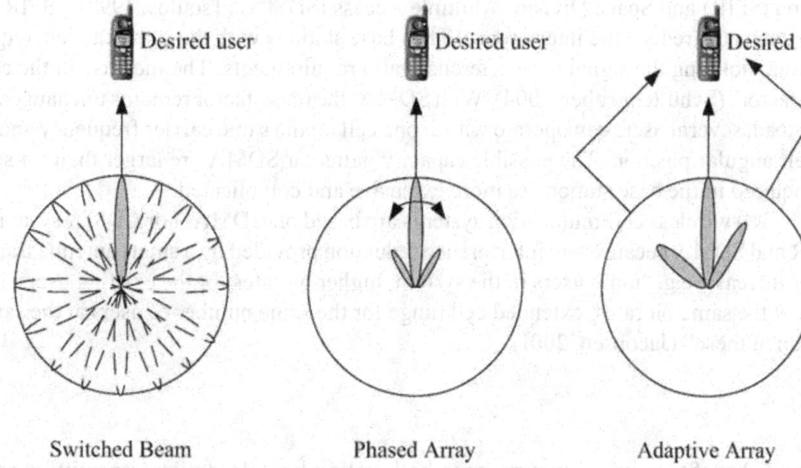


(SINR) or the Signal to Noise Ratio (SNR). Fully adaptive system use advanced signal processing algorithms to locate and track the desired and interfering signals to dynamically minimize interference and maximize intended signal reception (Stevanovic, 2003).

The main difference between a phased array and an adaptive array system is that the former uses beam steering only, while the latter uses beam steering and nulling. For a given number of antennas, adaptive arrays can provide greater range (received signal gain) or require fewer antennas to achieve a given range (Winters, 1999). However the receiver complexity and associated hardware increases the implementation costs.

## Smart Antenna vs. Diversity

With respect to wireless (indoor or outdoor) communication systems, antenna arrays are designed to operate in one of two distinct modes: smart antenna or diversity. In the two cases, the array is located at a base station, where there is enough space to accommodate it. Smart antenna techniques are used when signals arriving at antenna elements are coherent. This coherence is the basis for forming a narrow beam towards a desired user. "Coherence between antenna signals in a compact array with half wavelength spacing can be expected in environments such as typical urban macro cells, where the azimuth dispersion of the radio channel seen at the antenna array is typically small" (Pedersen, 2003). Diversity techniques are applied when signals arriving at antenna elements are independent. This independence usually takes place between two extreme parts of the array. Switching between the signals received by two sections of the array or suitable combining helps to increase signal to noise ratio and reduce the likelihood of deep fades. Traditionally, diversity arrays are considered separate from smart antenna systems and fall outside the scope of this chapter. The relationship of the three smart antenna categories (discussed above) to general space-time processing techniques is illustrated in Figure 3.

## System Aspects Influencing Smart Antenna Performance

The choice of a smart antenna receiver is highly dependent on the air interface parameters such as multiple access method, the type of duplexing and pilot availability (Boukalov, 2000). Besides the compatibility with the air interface, the number of antenna elements is also a very important consideration. The critical parameters are discussed in detail below:

## CDMA vs. TDMA

The different air interface techniques have significant impact on the design and optimum approach for smart antennas because of the different interference scenarios. In TDMA systems, the users are separated by orthogonal time slots. TDMA systems employ frequency reuse plan, which leads to a small number of strong interferers for both uplink and downlink (Tsoulos, 1999). By comparison, CDMA systems employ a total frequency reuse plan and the different users are multiplexed by distinct code waveforms. Thus in CDMA, each user's transmission is a source of interference for all other users.

The utilization of smart antennas in TDMA systems can be divided into two main stages. These are Spatial Filtering for Interference Reduction (SFIR) and Space Division Multiple Access (SDMA) (Tsoulos, 1999). "SFIR uses the beam directivity from smart antennas to reduce the interference. Thus base stations with the same carrier frequencies can be put closer together, without violating the signal to interference ratio requirements. The increase in the capacity is then the decrease in the reuse factor" (Schuttengruber, 2004). With SDMA, the reuse factor remains unchanged compared to a conventional system. Instead, several users can operate within one cell on the same carrier frequency and the same time slot distinguished by their angular position. The possible capacity gains for SDMA are larger than for spatial filtering. However, the changes required in the base station are more extensive and complicated.

The Third Generation (3G) wireless communication systems are based on CDMA. For CDMA systems, there is less difference between SFIR and SDMA because any interference reduction provided by a smart antenna translates directly into a capacity or quality increase, e.g. "more users in the system, higher bit rates for the existing users, improved quality for the existing users at the same bit rates, extended cell range for the same number of users at the same bit rates, or any arbitrary combination of these" (Jacobsen, 2001).

## Uplink vs. Downlink

Smart antennas can provide benefits to wireless systems in both uplink (mobile station transmitting and base station receiving) and downlink (base station transmitting and mobile station receiving). The uplink and downlink calibration (coherency requirement) is crucial e.g. to translate angles of arrival into angle of departure for adaptive array system.

Current 2G systems such as GSM and IS-95 CDMA and 3G systems such as cdma2000 and WCDMA are Frequency Division Duplex (FDD) systems. In FDD systems, the downlink channel characteristics are independent of the uplink characteristics due to the frequency difference. Thus the processing performed on the uplink cannot be exploited directly in the downlink without any additional processing (Tsoulos, 1999). By comparison in Time Division Duplex (TDD) systems (e.g. TD-CDMA systems), "the uplink and downlink can be considered reciprocal, provided that the channel conditions have not changed considerably between the receive and transmit time slots" (Tsoulos, 1999). Under these conditions the weights calculated by the smart antenna for the uplink can also be used for the downlink. Application of smart antennas to the downlink transmission for current FDD systems is therefore one of the major challenges related to smart antenna technology (Pedersen, 2003; Tsoulos, 1999). In this regard, retro-directive arrays for both receive and transmit applications have recently been proposed (Miyamoto, 2002). An excellent overview of opportunities and constraints for application of smart antenna techniques in downlink of 3G Universal Mobile Telecommunications Systems (UMTS) is given in (Pedersen, 2003).

## Pilot Availability

Pilot availability is a crucial element for downlink beamforming in smart antennas systems. A pilot signal (understood as the signal known to both sides of the communication link) transmitted by the Base Station (BS) to each user provides an accurate channel estimation to each Mobile Station (MS) during downlink beamforming. The dedicated pilot signal should have enough power such that the channel estimation is sufficiently good to ensure adequate signal to noise ratio (SNR) at the demodulator (Osseiran, 2006).

"In IS-95 CDMA forward link, a common pilot channel is broadcast throughout the sector to provide cell identification, phase reference and timing information to the mobile stations" (Giuliano, 2002). However this common pilot cannot be used for channel estimation in smart antenna applications because "the reference signal (pilot) used for channel estimation must go through exact same path as the data" (Giuliano, 2002). The IS-95 CDMA reverse link has no pilot signal to maintain a coherent reference. Hence non-coherent demodulation is used on the reverse link (Garg, 2000).

Recognizing the potential of smart antennas in improving the performance of CDMA systems, some additional channels are dedicated in 3G wireless communication systems for potential use by smart antenna receivers, e.g. cdma2000 has auxiliary carriers to help with downlink channel estimation in forward link beamforming (Rappaport, 2002) while WCDMA has connection dedicated pilot bits to assist in downlink beamforming. This includes the Primary Common PIlot Channel (P-CPICH) which is transmitted in the entire cell and Secondary Common Pilot Channel (S-CPICH) which is transmitted over a specific area of the cell (Garg, 2000). In (Osseiran, 2006), it was shown that the WCDMA system throughput gain, with fixed beam smart antennas, was higher when Secondary Common Pilot Channel (S-CPICH) is used as phase reference, as compared to Primary Common Pilot Channel (P-CPICH). Thus it is important to take into account the pilot availability when assessing the impact of smart antennas on the performance of wireless systems.

*Figure 3. Classification of smart antenna and diversity techniques for array antennas*

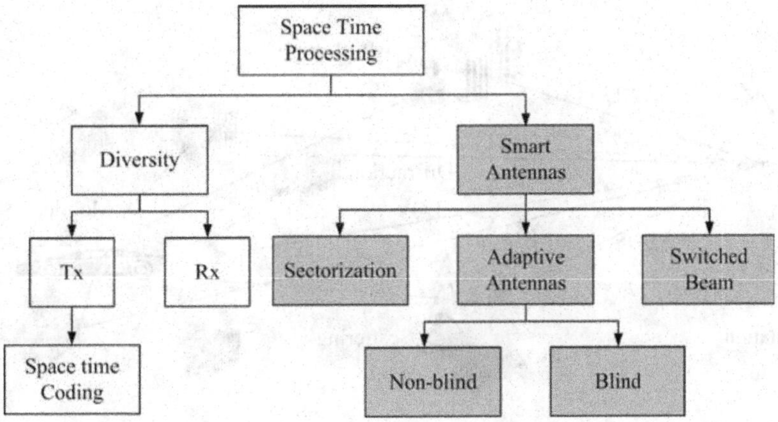


## Array Size

Smart antennas are usually physically located at the Base Station (BS) only. Due to power consumption and size limitations, it may not be practical to have multiple antennas at the Mobile Station (MS) in the downlink. The initial applications of smart antennas/MIMO are expected to be in "high speed data links for laptop terminals, where multiple antennas can be easily supported for the area and power considerations" (Kaiser, 2005).

"The number of elements in the array antenna is a fundamental design parameter, as it defines the number of interference sources the array can eliminate and/or reduce and the additional gain the array will provide. The achievable improvement in the system spectral efficiency increases with the number of elements in the array" (ITUR, 2003).

Because of practical considerations regarding costs, hardware implementation and installation, the number of horizontally separated antenna elements is usually in the range 4-12 (ITUR, 2003; Durrani, 2004c). Typical element spacing used is half wavelength in order to minimise mutual coupling and avoid grating lobes (Balanis, 2005). This corresponds to an array size of approximately 1.2 m at 900 MHz and 60 cm at 2 GHz for an 8 element uniform linear array antenna. Environmental issues may also have an impact on the array size, especially with recent growing public concern for reduced visible pollution and less visible base stations.

# Performance Analysis of Smart Antenna Technologies

## Channel Models for Smart Antennas

Channel modeling is one of the most important and fundamental research areas in wireless communications. It plays a crucial role in the design, analysis and implementation of smart antennas in wireless communication systems (Parsons, 1992; Simon, 2000; Cavers, 2000; Patzold, 2002). When a radio signal is transmitted in a wireless channel, the waves propagate through the physical medium (atmosphere or free space) and interact with the physical objects that constitute the propagation environment such as buildings, hills, trees and moving vehicles, giving rise to many subpaths of the transmitted signal. Multipath fading is caused by the constructive and destructive combination of these randomly delayed, reflected, scattered and diffracted subpath signal components. A typical wireless propagation environment is illustrated in Figure 4.

A successful adoption of smart antenna technology in a given wireless environment requires its extensive testing. In order to minimize the costs associated with this design and development stage, the use of realistic channel models that can accurately characterize spatial as well as temporal variations of the channel is a crucial requirement. In the past, classical channel models have focused mainly on the modeling of temporal aspects, such as fading signal envelopes, Doppler shifts of received signals and received power level distributions (Clark, 1968; Jakes, 1974; Sklar, 1997; Sklar, 1997b). The use of smart antennas introduces a new dimension, spatial dimension, in the channel models. The spatial properties of the channel, e.g. the angle of arrival and the distribution of arriving waves in azimuth, have an enormous impact on the performance of smart antenna systems and hence need to be accurately characterized (Ertel, 1998).

*Figure 4. Illustration of propagation in a typical wireless environment*

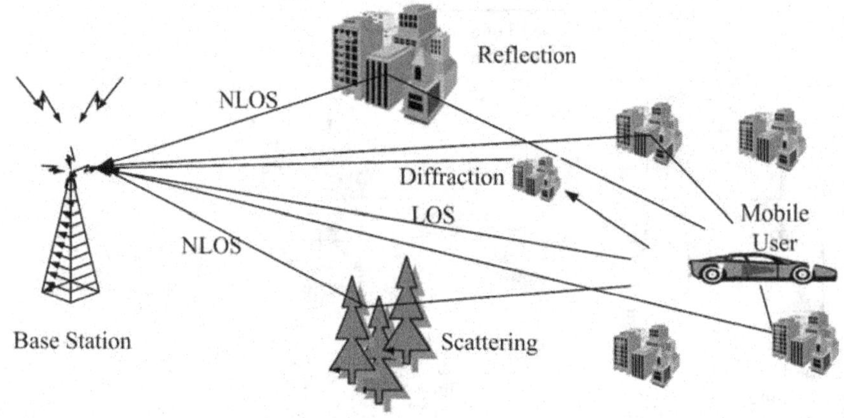


The spatial channel models for smart antennas have received much attention in literature (Almers, 2007; Yu, 2002; Ertel, 1998). It has to be noted that majority of the channel models are Two-Dimensional (2-D) in nature i.e. they assume that radio propagation takes place in the azimuth plane containing the transmitter and the receiver. Work has also been undertaken with regard to Three Dimensional (3-D) models (Aulin, 1979; Parsons, 1991; Mohasseb, 2002; Athanasiadou, 2000).

The channel models for smart antennas can be generally divided into four groups; empirical, deterministic, geometric and parametric. Empirical models are based on field measurements (Pedersen, 2000) while deterministic models use ray-tracing techniques to model the channel impulse response (Rizk, 1997). The advantage of these models is that they provide greater accuracy with site-specific results. However complexity becomes an issue for link-level simulations. Geometric models are defined by a particular distribution of scatterers and assume that the propagation between transmit and receive antennas takes place via scattering from intervening obstacles e.g. Circular Scattering Model (CSM) (Petrus, 1996) and Gaussian Scatter Density Model (GDSM) (Janaswamy, 2002). The main advantage of geometric models is that once the coordinates of the scatterers are drawn from a random process, all necessary spatial information can be easily derived. However the limitation is that resulting simulation time is large (especially when it is required to generate scatterer distributions for different channel environments).

By contrast, parametric channel models use important physical parameters such as phases, delays, Doppler frequency, angle of departure (AOD), angle of arrival (AOA) and angle spread to provide a description of MIMO channel characteristics (Yu, 2004). These parameters are described by "prescribing underlying probability distribution functions without assuming an underlying geometry for scatterers" (Almers, 2007) (as is done in geometric models). It is fundamentally

*Figure 5. Illustration of Parametric Channel Model angular parameters [Adapted from (Durrani, 2007)]*

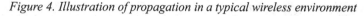

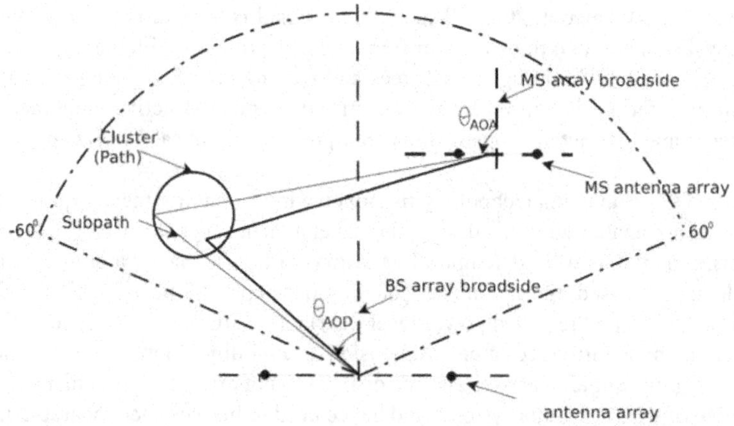

assumed that the transmitted signal is propagated through the environment via scattering from clusters. The total received signal can be modelled as time delayed replicas of the transmitted signal with each path further consisting of subpaths. The paths (corresponding to a cluster) are considered to be resolvable while the subpaths (corresponding to scatterers in a cluster) are considered to be unresolvable. When all the parameters are defined, they can be related using an analytical formulation. This can then be used to model different environments by choosing appropriate parameter values. A simplified sketch of a parametric model showing the angular parameters is given in Figure 5 (Durrani, 2007).

An example of standardized parametric channel model is the Spatial Channel Model (SCM) proposed by Third Generation Partnership Project Two (3GPP2) (SCM,2003). This detailed model is applicable for a variety of environments. However a limitation of the above model for smart antenna applications is that it does not take into account Mobile Station (MS) mobility. Modelling the movement of the MS is crucial as it influences both the spatial and temporal channel characteristics. Also for the case of adaptive beamforming, the smart antenna must track and steer its beam towards the desired user. Therefore, the performance of a smart antenna cannot be realistically evaluated without simulating MS mobility (Jelitto,1999; Ihara, 2001). An important extension to the SCM model for smart antennas was developed in (Durrani, 2006) which proposed a thorough framework for the incorporation of MS mobility. The proposed model in (Durrani, 2006) allows for efficient and accurate characterization of MS mobility in smart antenna channels while maintaining low complexity for link-level simulations.

## Adaptive Beamforming Algorithms

Smart antenna is essentially an optimal adaptive digital beamformer. Several adaptive algorithms have been developed in literature to determine the optimal weights based on optimality criteria such as minimum mean square error (MMSE), least square error (LSE), maximum signal-to-interference plus noise-ratio (SINR) (Litva, 1996). "In both the classical least mean square (LMS) and the recursive least square (RLS) approaches, the desired signal is supplied using either a training sequence or decision direction. The training sequence approach requires that the environment to be stationary from one training period to the next" (Garg, 2004). To overcome this limitation, other techniques have been developed, which do not require training sequences. These techniques are generally referred to as blind adaptive algorithms. They determine the beamforming weights by attempting to restore some known property of the received signal, e.g. constant modulus property of transmitted signals in CDMA system or information of spreading codes to distinguish different users occupying the same frequency band (Garg, 2004).

Several blind adaptive beamforming algorithms have been proposed in literature for 2G and 3G CDMA systems (SongTh, 2001; NaguibTh, 1996; Thompson, 1996; Grant, 1998; Thompson, 1996b; Tsoulos, 2000; ChoiJ, 2000; Choi, 2002; Li, 2003; ChoiJ, 2003). These algorithms are based on criteria of either (optimal) Maximum Signal to Interference plus Noise Ratio (SINR) or (sub-optimal) Maximum Signal to Noise Ratio (SNR). A technique to implement Maximum SINR beamforming, which utlilised the pre- and post-array correlation matrices, was first proposed in (NaguibTh, 1996). However the disadvantage of the above procedure was its heavy computational load. From point of view of practical implementation, it is desirable to develop smart antenna algorithms which have smaller computational load. Recently, more simplified Maximum SINR beamforming algorithms have been proposed in (Song, 2001). Maximum SNR beamforming is computationally even simpler as it utilises post-array correlation matrices only. Simple smart antennas utilising Maximum SNR beamforming have been proposed based on Modified Conjugate Gradient Method (MCGM) (Choi, 1997), Lagrange multipliers (Choi, 2000) and power method (Lee, 2003) respectively. It was shown in (Anna, 1999) that for moderate number of interferers and/or multipaths per user and low to moderate angle spreads, the performance of antenna arrays with Maximum SNR beamforming is close to the performance with Maximum SINR beamforming.

When adaptive beamforming algorithms are incorporated in CDMA systems, the most practical and widely used implementation is the two-dimensional (2-D) RAKE receiver, where beamforming for each path is followed by a conventional RAKE receiver. The detailed structure of such a receiver is discussed in detail in Section 3.2.

## Analytical Models

It is well known that array antennas with a suitable signal processing algorithm can improve the performance of Direct Sequence Code Division Multiple Access (DS-CDMA) systems by reducing the Multiple Access Interference (Godara, 1997). In this regard, it is important to analyse the mean Bit Error Rate (BER) performance of a DS-CDMA system, with $M$-ary orthogonal modulation and noncoherent detection, employing a smart antenna. This is because this type of modulation has been successfully used in the reverse link of IS-95 CDMA system and is also specified in radio configurations

1 and 2 of the reverse link in cdma2000 standard (Garg,2000). A major challenge in the analysis is to derive closed-form expressions for the BER, which are a very important tool in the planning and design of smart antenna systems.

The BER analysis of CDMA systems with noncoherent *M*-ary orthogonal modulation has been done by a number of researchers (Bi, 1992; Kim, 1992; Bi, 1992b; Jalloul, 1994; Patel, 1994; Aalo, 1998). In (Bi, 1992) and (Kim, 1992), the analysis was presented for an Additive White Gaussian Noise (AWGN) channel. Extensions to the case of a multipath fading channel for the Rayleigh distribution was presented in (Jalloul, 1994; Patel, 1994) and for the case of more general Nakagami fading in (Aalo, 1998) (the Nakagami distribution includes Rayleigh distribution as a special case and can also accurately approximate Rician fading). In both these papers, the mean Bit Error Rate (BER) was calculated by using the standard Gaussian Approximation (GA) (Holtzman, 1992) by first replacing the values of all the fading coefficients in the interference terms by their expectations and then using Stirling's formula (Jalloul, 1994) or averaging over a known fading distribution in order to reflect the effect of fading (Aalo, 1998). Recently, an analysis of multicode CDMA with noncoherent *M*-ary orthogonal modulation was published in (Iskander, 2004). It has to be noted that all the above considerations were restricted to the case of single antenna receivers.

An exact analysis of the BER of CDMA systems with array antennas is difficult. Thus different approximate analytical methods have been proposed to analyse the performance of CDMA smart antenna systems. Analytical results for a CDMA system with noncoherent M-ary orthogonal modulation and employing an array antenna operating in a Rayleigh fading environment were presented in (Naguib, 1996), which used the analysis procedure given in (Jalloul, 1994). No closed-form expression for the BER was given in (Naguib,1996). This analysis procedure given in (Jalloul, 1994; Naguib, 1996) was also used to analyse the performance of a W-CDMA based smart antenna system in (Anna, 1999). An alternative simplified technique utilizing the interference suppression coefficient was proposed in (Song, 1999; SongTh, 2001) and illustrated for the case of a cdma2000 based smart antenna system in (Song, 2001).

Recently in (Spagnolini, 2004; Spagnolini, 2001), a simple analytical method was described to analyse the performance of a DS-CDMA system employing an array antenna. The proposed method was shown to provide a more accurate assessment than the method of (Song, 1999; SongTh, 2001). However, the application of the proposed method was considered only for the simple case of coherent Binary Phase Shift Keying (BPSK) modulation. In (Durrani, 2005) the proposed technique was used to analyse the performance of CDMA, with noncoherent M-ary orthogonal modulation, employing a receiving adaptive antenna in a Nakagami fading environment. An expression of the SINR at the output of the 2-D RAKE receiver was derived which permits the BER to be readily evaluated using a closed form expression. The results in (Durrani, 2005; Durrani, 2004b), obtained using this analytical model, showed good agreement with the (computationally intensive) simulation results.

## Hybrid Smart Antenna Systems

A smart antenna can mitigate Multiple Access Interference (MAI) by beamforming (spatial filtering) operation and consequently improve the performance of a CDMA system. However smart antennas may not be effective in all circumstances. This has led to the creation of novel hybrid applications of smart antennas. Recently smart antennas have been considered in combination with multi-user detectors/interference cancellation (Nazar, 2002; Dahlhaus, 2000; Mohamed, 2002), PN code acquisition (Wang, 2003b; Wang, 2003) and power control (Knopp, 2002; Mercado, 2002). For application to CDMA systems, an important hybrid is the combination of diversity and smart antennas. This is because in addition to MAI, the performance of CDMA systems is limited by multipath fast fading. Therefore further improvement in performance can be expected if efforts are made to jointly combat MAI and fading.

Diversity is a very effective technique which has been traditionally employed to combat fading. It uses multiple antennas to provide the receiver with multiple uncorrelated replicas of the same signal. The signals received on the disparate diversity branches can then be combined using various combining techniques, e.g., Equal Gain Combining (EGC) (Proakis, 1995; Eng, 1996; Stuber, 2001). Transmit diversity using two antennas at the base station has been adopted for the W-CDMA standards being developed within the Third Generation Partnership Project (3GPP) (Bjerke, 2004; 3GPP, 2003; Alamouti,1998). According to (Osseiran, 2006), transmit diversity has a similar hardware complexity as a 3-sector site equipped with two fixed beams each. A comparison of performance of transmit diversity and fixed beam smart antenna systems is given in (Osseiran, 2004). However diversity arrays have limited interference rejection and fail to eliminate the error probability floor in CDMA (Godara, 1997; Rooyen, 2000). Diversity and beamforming also have conflicting requirements for optimum performance, e.g. diversity arrays employ widely spaced antenna elements ($5\lambda$ or $20\ \lambda$, where $\lambda$ denotes the wavelength) while conventional beamforming arrays employ closely spaced antenna elements, with typical inter-element spacing of half wavelength.

*Figure 6. Hierarchical beamforming array geometry*

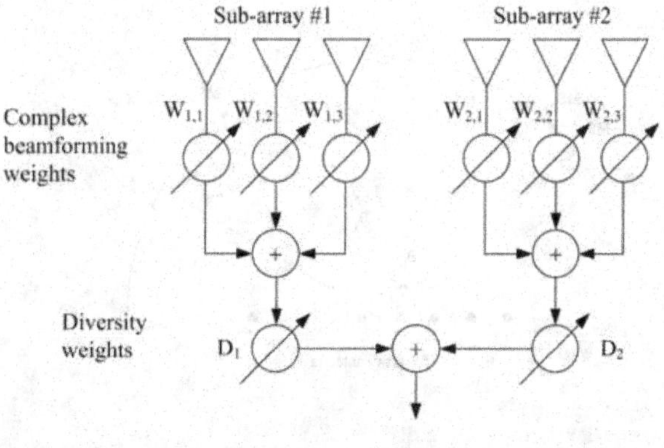


A hybrid scheme of diversity and beamforming called Hierarchical Beamforming (HBF), was proposed in (Zhou, 2003; Zhou, 2001). In HBF, the array elements are divided into groups to form several sub-beamforming arrays, as illustrated in Figure 6. The inter-element spacing within a sub-array is assumed half wavelength, while the distance between the adjacent sub-arrays is large (e.g. 5λ or 20 λ or more) to ensure independent fading between sub-arrays. At any particular time one of the subarrays with the best channel state is selected to form a beam for signal transmission. The performance of a generic DS-CDMA system employing such an array in the downlink was analysed in (Zhou, 2003). However the analysis assumed zero angle spread. This assumption is reasonable in suburban areas where the coverage is from elevated BS antennas as the multipath rays arrive at the BS with a small angle spread (e.g. a few degrees). However when the base stations are located within or near urban clutter, they can consequently experience a much larger angle spread than the elevated base stations, e.g. 5°-15° depending on the height of the BS (Pedersen, 2000). Thus it is important to consider the effect of angle spread.

In (Durrani, 2004), the work was extended by comparing the performance of conventional and Hierarchical Beamforming (HBF) while taking angle spread into account. It was shown that the consideration of angle spread affects the performance comparison between conventional and HBF. While assuming zero angle spread, the performance of HBF is superior to Conventional Beamforming (CBF) due to space diversity gain afforded by the well separated sub-arrays. The inclusion of angle spread produces spatial fading across the array, which results in additional diversity gain and improves the performance of both CBF and HBF. With small angle spread of 5°, the performance of HBF is still better than CBF. However for larger angle spreads of 10°-15°, when path diversity is exploitable and the system is heavily loaded, CBF yields better mean BER results than HBF. These findings are based on the assumption of perfect channel estimation and provide a lower bound on the actual system performance.

## PERFORMANCE OF ADAPTIVE BEAMFORMING FOR CDMA

In this section an analytical model for investigating the mean Bit Error Rate (BER) performance of a Direct Sequence Code Division Multiple Access (DS-CDMA) system employing adaptive beamforming is described. The analytical results are compared with simulation results to validate the proposed model.

### System Model

Consider a BS serving a single 120° angular sector. Without loss of generality, it is assumed that the BS employs a Uniform Linear Array (ULA) of $N$ identical omni-directional antenna elements, with inter-element spacing of $d=\lambda/2$, as shown in Figure 7.

Let $K$ denote the total number of active Mobile Stations (MS) in the system, which are randomly distributed in the azimuthal direction, along the arc boundary of the sector cell in the far field of the array. It is assumed that each MS uses $M$-ary orthogonal modulation. The block diagram of the MS transmitter is shown in Figure 8. The $k$=1 user is assumed to be the desired user. For simplicity, an uncoded system is considered i.e. the convolutional encoder and interleaver are ignored. This approach is widely used (see, for instance (Anna, 1999; Naguib, 1996; Song, 2001; Bjerke, 2004)) as it allows the analysis of wireless communication systems employing multiple antennas. It has to be noted that, in practice, channel coding is an essential component of DS-CDMA wireless communication systems.

Figure 7. Smart antenna BS serving a single 120° angular sector of CDMA system

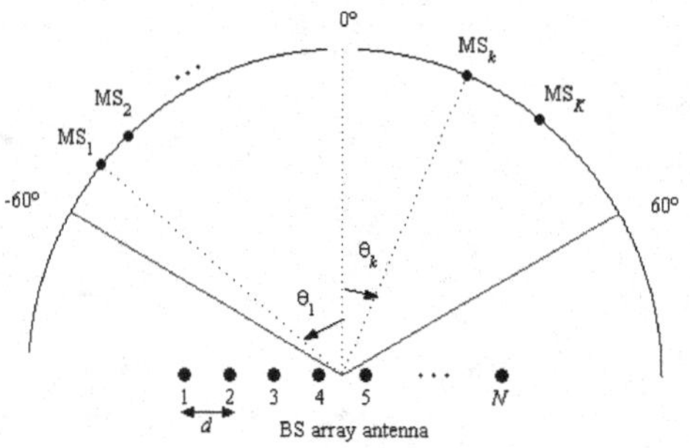


Figure 8. Block diagram of mobile station transmitter

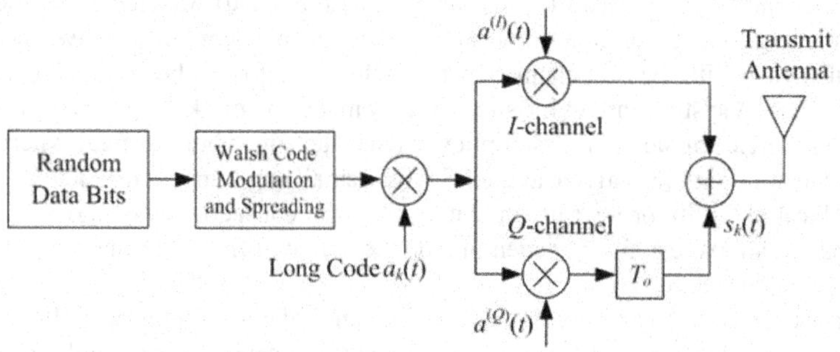


The transmitted signal $s_k(t)$ of the $k$th user can be written as

$$s_k(t) = W_k^{(p)}(t)[a_k^{(I)}(t)\cos(\omega_c t) + a_k^{(Q)}(t)\sin(\omega_c t)] \tag{1}$$

where $p$=1,2,...,$M$, $W_k^{(p)}(t)$ is a Hadamard-Walsh function of dimension $M$ which represents the $p$th orthogonal signal of the $k$th user, $a_k^{(I)}(t) = a^{(I)}(t)a_k(t)$, $a_k^{(Q)}(t) = a^{(Q)}(t)a_k(t)$, $a_k^{(I)}(t)$ is the $k$th users long code sequence, $a^{(I)}(t)$ and $a^{(Q)}(t)$ are the in-phase and quadrature channel Pseudo-Noise (PN) random sequences respectively, $\omega_c = 2\pi f_c$ and $f_c$ is the carrier frequency. The transmitted power of each user is assumed unity. This reflects the assumption of perfect power control.

The $k$th user's signal propagates through a multipath fading channel with Angle of Arrival (AOA) $\theta_k$. The channel impulse response between the $k$th transmitting user and the $n$th element of the array antenna is given by

$$h_{n,k} = \sum_{l=1}^{L} \beta_{k,l} \exp[-j(\phi_{k,l} + 2\pi \tfrac{d}{\lambda}(n-1)\sin\theta_k)]\delta(t-\tau_{k,l}) \tag{2}$$

where $\beta_{k,l}$, $\phi_{k,l}$ and $\tau_{k,l}$ are the path gain, phase and delay respectively and $\lambda$ is the wavelength. The path gains are assumed to follow the Nakagami-$m$ distribution with parameters ($m$, $\Omega_{k,l}$=E[($\beta_{k,l}{}^2$]), where $E[\cdot]$ denotes the expectation. The Nakagami-$m$ probability density function is given by

$$f_\alpha(\alpha) = \frac{2m^m \alpha^{2m-1}}{\Omega^m \Gamma(m)} \exp\left(-\frac{m\alpha^2}{\Omega}\right) \tag{3}$$

*Figure 9. Block diagram of smart antenna receiver*

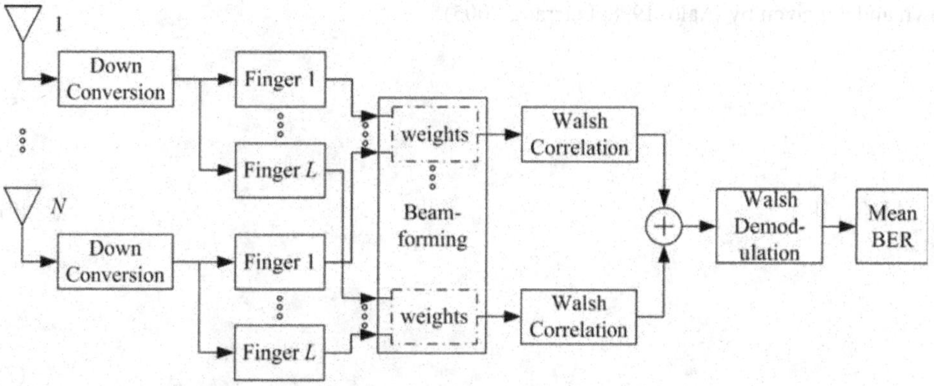


where $\Gamma(\cdot)$ is the gamma function and $m$ ranges from ½ to $\infty$. The Nakagami-$m$ distribution includes Rayleigh distribution as a special case for $m$=1 and can also accurately approximate Rician fading when $m$>1 (Simon, 2000).

Under the above assumptions, the total received signal at the $n$th antenna is given by

$$x_n(t) = \sum_{k=1}^{K}\sum_{l=1}^{L}[\beta_{k,l}W_k^{(p)}(t-\tau_{k,l})a_k^{(I)}(t-\tau_{k,l})\cos(\omega_c t+\varphi_{n,k,l})$$
$$+\beta_{k,l}W_k^{(p)}(t-\tau_{k,l})a_k^{(Q)}(t-\tau_{k,l})\sin(\omega_c t+\varphi_{n,k,l})]+\eta_n(t) \quad (4)$$

where $\varphi_{n,k,l}$ is the overall phase which includes the path phase and the difference in propagation delays between the antennas and $\eta(t)$ is the Additive White Gaussian Noise (AWGN).

## 2-D Rake Receiver

The functional block diagram of the smart antenna receiver is shown in Figure 9. The BS receiver consists of a two-dimensional (2-D) RAKE receiver (Naguib, 1996) with a Maximum Signal to Noise Ratio beamformer (Hudson, 1981) followed by an $L$ finger noncoherent RAKE combiner (Price, 1958). From the figure, the following main blocks can be identified (i) array antenna, (ii) PN despreading, (iii) beamforming and (iv) Walsh correlation and demodulation. The function and modeling of these blocks is briefly explained below.

The first step for the receiver is to obtain the quadrature components at each antenna. This is achieved by multiplying the received waveforms by $\cos(\omega_c t)$ and $\sin(\omega_c t)$ respectively and then lowpass filtering to remove the double frequency components that result from multiplication (Lee, 1998). After filtering, each resolvable path is then detected by one of the RAKE fingers immediately following the radio-frequency stages. The RAKE receiver is assumed to have perfect knowledge of the multipath delays. To detect the $l$th path, the signal is despread using the sequence of the respective mobile and synchronized to the delay of the $l$th path. Additionally, the cross-correlation of the $I$ and $Q$ channels is found in order to noncoherently combine the signals. Thus the in-phase signal is also correlated with the quadrature arm and an equivalent operation performed for the quadrature signal.

A beamformer is then constructed for each resolvable multipath and the signal after PN despreading is combined by the beamformer. In the beamforming operation, the signals received by the antenna elements are weighted by complex weights and then summed up. It is assumed that the weights are determined using the maximum Signal to Noise Ratio (SNR) beamforming criteria. We assume perfect channel estimation and use the ideal maximum SNR beamforming weight vectors in the analysis, i.e. the weight vector is set equal to the vector channel coefficients which are assumed to be perfectly known (Song, 2001; Anna, 1999; Thompson, 1996). This provides an estimate of the best case smart antenna system performance (when there are no errors in the beamforming weights).

The next step is the correlation of the smart antenna output with stored replicas of the Walsh functions. This is required to form the decision variable for demodulation. The overall decision variable is obtained by Equal Gain Combining (EGC) of all the decision variables from the $L$ multipaths. This output of the RAKE combiner, which contains the desired user, noise, Multiple Access Interference (MAI) and self interference components, is then used to estimate the transmitted data.

For the case of a single antenna receiver, the variances of the noise $\sigma_N^2$, MAI $\sigma_M^2$ and self interference components $\sigma_I^2$, are well known and are given by (Aalo, 1998; Durrani, 2005)

$$\sigma_N^2 = \frac{N_o}{2} \tag{5}$$

$$\sigma_M^2 = \frac{E_s}{3N_c}\sum_{k=1}^{K-1}\sum_{l=1}^{L} E[(\beta_{k,l})]^2 \tag{6}$$

$$\sigma_I^2 = \frac{E_s}{3N_c}\sum_{l=1}^{L-1} E[(\beta_{k,l})]^2 \tag{7}$$

where $N_c$ is the spreading gain, $E_s$ is the symbol energy and $N_o$ is the noise power spectral density.

## Ber Analysis Technique

We follow the analysis technique in (Spagnolini, 2004) and develop a simple model, with a closed form expression for the BER, which can be used to calculate the system performance with array antennas operating in a Nakagami-*m* fading environment. In the technique, the following assumptions are made (Spagnolini, 2004; Spagnolini, 2001):-

- The interferers are uniformly distributed in the coverage area comprising an angular sector of 120°.
- The interferers are partitioned into two spatial equivalence classes (Sweatman, 1999) - in-beam and out-of-beam - based on whether their Angles of Arrivals lie inside or outside the beam formed toward the desired user.
- A piece-wise beampattern is used to approximate the actual ULA beampattern. In the pass band (or in-beam) the beampattern is modeled by a triangular function while in the stop-band (or out-of-beam) the beampattern is approximated by a constant attenuation factor representing an average side-lobe level.
- Because of the piece-line approximation, the energy of each in-beam interferer is a random variable uniformly distributed within [1/2, 1] (half power beamwidth region). Hence a correction factor of 3/4 (average value) is applied for in-beam interferers.
- For the purpose of evaluation of the error probability, the in-beam interferers are counted as interference while the out-of-beam users increase the additive noise level.

The beampattern assumption and the partitioning of interferers is illustrated in Figure 10 (Durrani, 2005). Using the above assumptions, the procedure proposes to adapt the single antenna BER results to array antenna systems by manipulating the terms accounting for noise and interference in the error probability formulas for single antenna receivers (Spagnolini, 2004).

## Performance Analysis

Let the modified variances of the noise, self interference and MAI be denoted as $\bar{\sigma}_N^2$, $\bar{\sigma}_M^2$ and $\bar{\sigma}_I^2$, respectively. Let $\kappa$ denote the number of in-beam interferers. Then number of out-of-beam interferers is $K-\kappa-1$. Applying the partitioning of the interferers, we get (Durrani, 2005).

$$\bar{\sigma}_M^2 = \frac{E_s}{3N_c}\left[ f\sum_{k=1}^{\kappa}\sum_{l=1}^{L} E[(\beta_{k,l})]^2 + \alpha_o \sum_{k=\kappa+1}^{K-1}\sum_{l=1}^{L} E[(\beta_{k,l})]^2 \right] \tag{8}$$

where $\alpha_o$ is the attenuation factor for out-of-beam interferers and $f$=3/4 is a correction factor for in-beam interferers.

The self interference due to the desired user's multipaths cannot be reduced by beamforming as the spatial channel in Eq. (2) is based on one AOA $\theta_k$. Hence we have

*Figure 10. Illustration of the beampattern approximation and partitioning of interferers [Adapted from (Durrani, 2005)]*

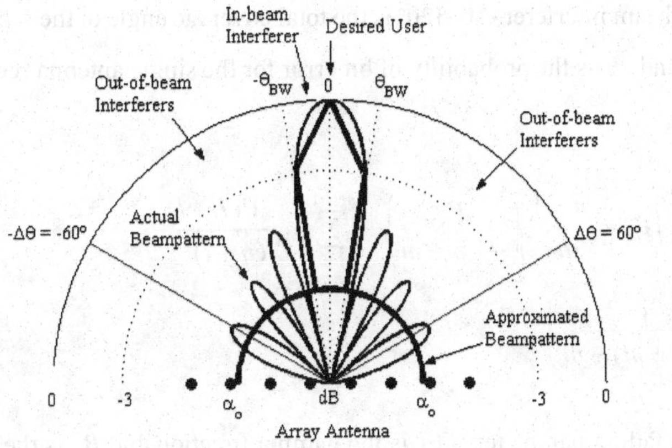


$$\bar{\sigma}_I^2 = \sigma_I^2 \tag{9}$$

In comparison with the power of the desired signal, the noise power at the output of the antenna array is reduced by $N$ times, where $N$ is the number of antenna elements (Choi, 1997). Hence, we get

$$\bar{\sigma}_N^2 = \frac{\sigma_N^2}{N} = \frac{N_o/2}{N} \tag{10}$$

The modified SINR at the output of the 2-D RAKE receiver is thus given by

$$\rho = \frac{E_s}{2\bar{\sigma}_T^2}$$

where $\bar{\sigma}_T^2 = \bar{\sigma}_M^2 + \bar{\sigma}_I^2 + \bar{\sigma}_N^2$ is the total variance. For simplicity, we consider a uniform power delay profile i.e. $\Omega_{k,1} = \Omega_{k,2} \cdots = \Omega_{k,L} = \Omega / L$. Substituting the values and simplifyng, we get

$$\rho = \frac{\gamma}{\frac{1}{N} + \frac{2\gamma}{3N_c}\left[(L-1) + \alpha_o(K-\kappa-1)L + fL\kappa\right]} \tag{11}$$

where

$$\gamma = \log_2(M)\frac{\Omega}{L}\frac{E_b}{N_o}.$$

Using Eq. (11) and assuming a uniform distribution of interferers in the angular sector, the average bit error probability of the 2-D RAKE receiver as (Spagnolini,2004)

$$P_b = \sum_{\kappa=0}^{K-1} \eta^{\kappa}(1-\eta)^{K-\kappa-1}\binom{K-1}{\kappa}P_b' \tag{12}$$

where ρ is given by Eq. (11) and

$$\eta = \frac{2\theta_{BW}}{\Delta\theta}$$

is the probability of an in-beam interferer, $\Delta\theta$=120° is the total coverage angle of the sector, 2 $\theta_{BW}$ is the total beamwidth towards the desired user and $P_b'$ is the probability of bit error for the single antenna receiver in Nakagami fading given by (Aalo, 1998)

$$P_b' = \frac{M/2}{M-1}\frac{1}{\Gamma(mL)}\sum_{n=1}^{M-1}(-1)^{n+1}\binom{M-1}{n}\left(\frac{m}{\frac{n}{n+1}\rho+m}\right)^{mL}\sum_{l=0}^{n(L-1)}\mathrm{B}_{ln}\frac{\Gamma(L+l)}{(n+1)^{L+l}}$$
$$\times\sum_{i=0}^{l}\binom{l}{i}\frac{\Gamma(mL+i)}{\Gamma(L+i)}\left(\frac{\rho}{n(\rho+m)+m}\right)^{i} \qquad (13)$$

where $m$ is the Nakagami fading parameter, $\Gamma(\cdot)$ is the gamma function and $B_{ln}$ is the set of coefficients which can be computed recursively (Aalo,1998). The optimized values of equivalent beamforming parameters 2 $\theta_{BW}$ and $\alpha_o$ used in this work are given in Table 1 (Spagnolini,2004).

## Results

The analytical model described in the last section can be readily evaluated using a mathematical software package such as Matlab or Mathematica. We also perform simulations to confirm the analytical results. The simulation results are obtained by using the simulation model developed for the purpose (DurraniTh, 2004). In order to generate simulation results that represent an average measure of the cellular system's performance, the following Monte Carlo simulation strategy is adopted:

- A simulation 'drop' (3GPP2, 2003) is defined as the transmission and reception of 125 frames, which corresponds to the time required by the desired user to traverse the entire azimuth range [-60°, 60°] with angle change 0.01 ° per snapshot.
- In each simulation run over the variable of interest (e.g. $E_b/N_o$ or number of users $K$ or number of antennas $N$), the number of bit errors is collected over a minimum of 50 drops. If the number of bit errors exceeds the specified minimum number of bit errors (i.e. 100 (Jeruchim, 1992)), then the simulation is terminated and the mean BER is calculated. Otherwise, the simulations are continued until at least 100 bit errors have been detected or the number of drops reaches a maximum of 100. This upper limit on the number of drops is imposed to limit the overall simulation time.
- At the beginning of each drop, the desired user's AOA is set to -60°, while the AOA's of the interferers are uniformly randomly distributed over the azimuth sector range.
- During a drop, the MS's AOA increases or decreases linearly with angle change 0.01° per snapshot. Also the channel undergoes fast fading according to the motion of the MS's.

This procedure ensures that the BER performance of the system is averaged over the ensemble of channel parameters and AOA of the multiple users and is not conditioned on a particular spatial or temporal state of the system. We consider the case of $m$=1 (Rayleigh fading), $M$=64 and spreading gain $N_c$=256 which is relevant to the uncoded IS-95 CDMA and cdma2000 systems.

*Table 1. Equivalent beamforming parameters*

| Number of antenna elements $N$ | 4 | 6 | 8 |
|---|---|---|---|
| Attenuation $\alpha_o$ (dB) | -12 | -14 | -16 |
| Beamwidth $2\theta_{BW}$ (degs) | 30 | 20 | 15 |

*Figure 11. Mean BER vs. Number of users K for Eb/No = 10 dB, assuming L = 1,2 Rayleigh fading paths/user and N = 1,4,6,8 antennas respectively (lines: analytical model, markers: simulations)*

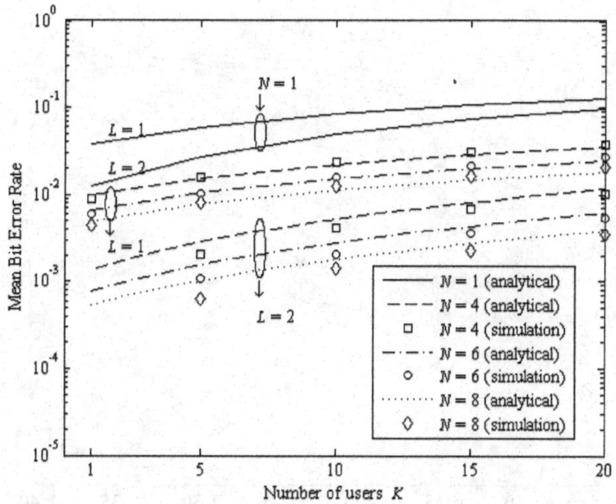


*Figure 12. Mean BER vs. Eb/No for N = 8 antennas, assuming K = 5,20 users and L = 2,3 Rayleigh fading paths/user respectively (lines: analytical model, markers: simulations) [Adapted from (Durrani, 2005)]*

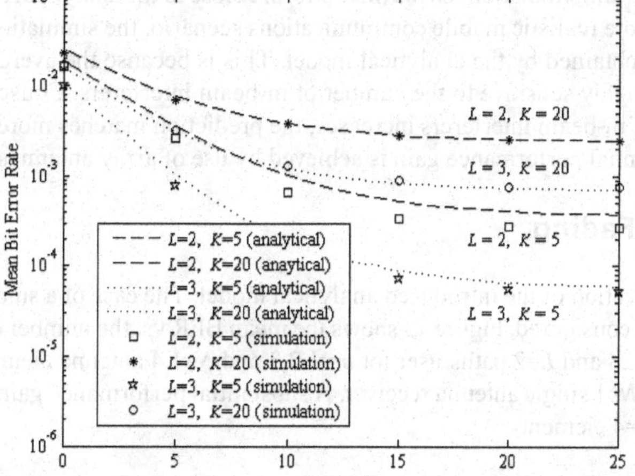


## Effect of Number of Users

Figure 11 shows the Mean BER vs. Number of users for $E_b/N_o$=10 dB, assuming $L$=1,2 Rayleigh fading multipaths and $N$=1,4,6,8 antennas respectively. Comparing with the reference curves for single antenna ($N$=1), it is clear that beamforming improves the performance of the system considerably, e.g. for $K$=20 users and $L$=2 paths/user, a $N$=4 element smart antenna system improves the BER by a factor of approximately 8 as compared to the performance of a conventional receiver. It can be seen from the figure that as the number of users increases, the analytical model provides a closer match with simulation results, especially for the case of $L$=2 paths/user.

*Figure 13. Mean BER vs. Number of users K for fixed Eb/No =10 dB, M=8, Nc =128, L=2 paths/user, assuming N=1,4 antennas and m=1,2,3 respectively (lines: analytical model) [Adapted from (Durrani, 2005)]*

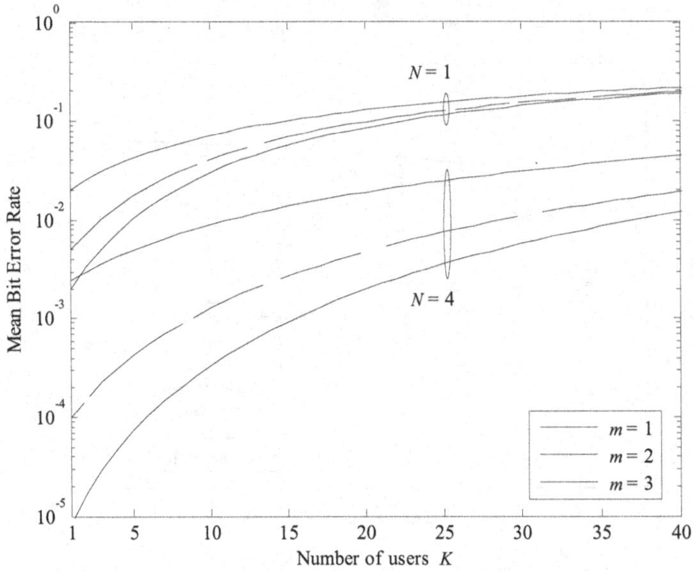


## Effect of Number of Antennas

Figure 12 shows the mean BER versus $E_b/N_o$ (dB) with $N$=8 antennas, $K$=5, 20 users and $L$=2,3 paths/user respectively. It can be seen that for $K$=5 users the simulation results (markers) are close to the analytical results (lines). However for $K$=20 users, which represents a more realistic mobile communications scenario, the simulation results provide an even better agreement with the results obtained by the analytical model. This is because the average BER performance predicted by the analytical model is highly sensitive to the number of in-beam interferers. Consequently, as the number of users and hence the probability of in-beam interferers increases, the prediction matches more closely with the simulations. It can also be seen that substantial performance gain is achieved by use of array antennas.

## Effect of Nakagami Fading

Here, we illustrate the application of the introduced analytical model. The case of a smart antenna operating in a Nakagami fading environment is considered. Figure 13 shows the mean BER vs. the number of users $K$ at fixed $E_b/N_o$=10 dB, $M$=8, processing gain $N_c$=128 and $L$=2 paths/user for $m$=1,2,3 and $N$=1,4 antenna elements respectively. The reference curves in this figure are for $N$=1 single antenna receiver. A substantial performance gain is achieved when the array size is increased from $N$=1 to $N$=4 elements.

## CONCLUSION

This chapter presented an overview of smart antenna technologies for Code Division Multiple Access Systems. It has summarized key system and modelling aspects influencing the performance of smart antenna systems. A system model providing rapid and accurate assessment of the performance of CDMA systems employing a smart antenna has been discussed. It has been shown that smart antennas can have a significant role in improving the performance of cellular CDMA systems. This chapter has focused on the impact of smart antennas on physical layer of CDMA systems. The impact of smart antennas on higher layers of the communication protocol stack, e.g. Medium Access Control (MAC) layer is an avenue for future exploration.

## ACKNOWLEDGMENT

The work presented in this chapter has been supported by the Australian Research Council Discovery grant DP0450118 and ANU Faculty Research Grants Scheme R66810RG73.

## REFERENCES

[Aalo, 1998] Aalo V., Ugweje O., & Sudhakar R. (1998). Performance analysis of a DS/CDMA system with noncoherent *M*-ary orthogonal modulation in nakagami fading. *IEEE Transactions on Vehicular Technology*, 47(1), 20–29.

[Alamouti, 1998] Alamouti S. M. (1998). A simple transmit diversity technique for wireless communication. *IEEE Journal on Selected Areas in Communications*, 16, 1451–1458.

[Alexiou, 2004] Alexiou A., & Haardt M. (2004). Smart antenna technologies for future wireless systems: trends and challenges. *IEEE Communications Magazine*, 42(9), 90–97.

[Allen, 2004] Allen B. & Beach M. (2004). On the analysis of switched-beam antennas for theW-CDMA downlink. *IEEE Transactions on Vehicular Technology*, 53(3), 569–578.

[Almers, 2007] Almers P., et. al., (2004). Survey of Channel and Radio propagation Models for wireless MIMO systems. *EURASIP Journal on Wireless Communications and Networking*, Volume 2007. Retrieved from http://www.hindawi.com/journals/wcn/volume-2007/si.stcm.html

[Anna, 1999] Anna M. D. & Aghvami A. H. (1999). Performance of optimum and suboptimum combining at the antenna array of a W-CDMA system. *IEEE Journal on Selected Areas in Communications*, 17(12), 2123–2137.

[Athanasiadou, 2000] Athanasiadou G. E. & Nix A. R. (2000). A novel 3-D indoor ray-tracing propagation model: The path generator and evaluation of narrow-band and wide-band predictions. *IEEE Transactions on Vehicular Technology*, 49(4), 1152–1168.

[Aulin, 1979] Aulin T. (1979). A modified model for the fading signal at a mobile radio channel. *IEEE Transactions on Vehicular Technology*, 29, 182–203.

[Balanis, 2005] Balanis C. A. (2005). *Antenna Theory: Analysis and Design*, 3rd ed. John Wiley.

[Bi, 1992] Bi Q. (1992). Performance analysis of a cellular CDMA system. In *IEEE Vehicular Technology Conference (VTC)*, Denver, May 10-13, 43–46.

[Bi, 1992b] Bi Q (1992b). Performance analysis of a CDMA system in the multipath fading environment. In *IEEE International Symposium on Personal, Indoor and Mobile Radio Communications (PIMRC)*, Boston, MA, 108–111.

[Bjerke, 2004] Bjerke B. A., Zvonar Z., and Proakis J. G. (2004). Antenna diversity combining aspects for WCDMA systems in fading multipath channels. *IEEE Transactions on Wireless Communications*, 3(1), 97–106.

[Boche, 2006] Boche H., Bourdoux A., Fonollosa J.R., Kaiser T., Molisch A. & Utschick W. (2006). Smart antennas: state of the art. *IEEE Vehicular Technology Magazine*, 1(1), 8–17.

[Boukalov, 2000] Boukalov A. O., & Haggman S. G. (2000). System aspects of smart-antenna technology in cellular wireless communications –an overview. *IEEE Transactions on Microwave Theory and Techniques*, 48(6), 919–929.

[Butler, 1966] Butler J. L. (1966). Digital, matrix and intermediate frequency scanning. In R. C. Hansen (Ed.), *Microwave Scanning Antennas*, Academic Press.

[Cavers, 2000] Cavers J. K. (2000). *Mobile Channel Characteristics*. Kluwer Academic Publishers.

[Choi, 2002] Choi S., Choi J., Im H.-J. &. Choi B. (2002). A novel adaptive beamforming algorithm for antenna array CDMA systems with strong interferers. *IEEE Transactions on Vehicular Technology*, 51(5), 808–816.

[Choi, 2000] Choi S. & Shim D. (2000). A novel adaptive beamforming algorithm for a smart antenna system in a CDMA mobile communication environment. *IEEE Transactions on Antennas and Propagation*, 49(5), 1793–1806.

[Choi, 1997] Choi S. & Yun D. (1997). Design of adaptive antenna array for tracking the source of maximum power and its application to CDMA mobile communications. *IEEE Transactions on Antennas and Propagation*, 45(9), 1393–1404.

[ChoiJ, 2003] Choi J. & Choi S. (2003). Diversity gain for CDMA systems equipped with antenna arrays. *IEEE Transactions on*

*Vehicular Technology*, 52(3), 720–725.

[ChoiJ, 2000] Choi J. (2000). Pilot channel-aided techniques to compute the beamforming vector in CDMA system with antenna array. *IEEE Transactions on Vehicular Technology*, 49(5), 1760–1775.

[Clark, 1968] Clark R. H. (1968). A statistical theory of mobile radio reception. *Bell Labs Syst. Tech. J.*, 47, 957–1000.

[Dahlhaus, 2000] Dahlhaus D. & Zhenlan C. (2000). Smart antenna concepts with interference cancellation for joint demodulation in the WCDMA UTRA uplink. In *IEEE International Symposium on Spread Spectrum Techniques and Applications ISSSTA*, 1, 244 – 248.

[Durrani, 2007] Durrani S., Bialkowski M.E. & Latif S. (2007). A Parametric Channel Model for Smart Antennas Incorporating Mobile Station Mobility. In *IEEE International Symposium on Personal, Indoor and Mobile Radio Communications* (PIMRC), Athens, Greece.

[Durrani, 2006] Durrani S. & Bialkowski M.E. (2006). A Parametric Channel Model for Smart Antennas Incorporating Mobile Station Mobility. In *IEEE Vehicular Technology Conference* (VTC), Melbourne, 2803 - 2807.

[Durrani, 2005] Durrani S. & Bialkowski M.E. (2005). Analysis of the error performance of adaptive array antennas for CDMA with non-coherent M-ary orthogonal modulation in Nakagami fading. *IEEE Communications Letters*, 9(2), 148–150.

[Durrani, 2004] Durrani S. & Bialkowski M.E. (2004). Performance of hierarchical beamforming in a Rayleigh fading environment with angle spread. In *International Symposium on Antennas* (ISAP), vol.2, Sendai, Japan,Aug.17-21, 937–940.

[Durrani, 2004b] Durrani S. & Bialkowski M.E. (2004). A simple model for performance evaluation of a smart antenna in a CDMA system. In *IEEE International Symposium on Spread Spectrum Techniques and Applications* (ISSSTA), 379-383.

[Durrani, 2004c] Durrani S. & Bialkowski M.E. (2004). Effect of mutual coupling on the interference rejection capabilities of linear and circular arrays in CDMA systems. *IEEE Transaction on Antennas and Propagation*, 52(4), 1130-1134.

[DurraniTh, 2004] Durrani S. (2004). Investigations into Smart Antennas for CDMA Wireless Systems. Ph.D. dissertation, The University of Queensland, Brisbane, Australia.

[Eng, 1996] Eng T., Kong N. & Milstein L. B. (1996). Comparison of diversity combining techniques for rayleigh fading channels. *IEEE Transactions on Communications*, 44(9), 1117–1129.

[Ertel, 1998] Ertel R., Cardieri P., Sowerby K. W., Rappaport T. S. & Reed J. H. (1998). Overview of spatial channel models for antenna array communication systems. *IEEE Personal Communications Magazine*, 5(1), 10–22.

[Garg, 2004] Garg V. K. Laxpati S. R. & Wang D. (2004). Use of Smart Antenna System in Universal Mobile Communications Systems (UMTS). *IEEE Antennas and Wireless Propagation Letters*, 3, 66–70.

[Garg, 2000] Garg V. K. (2000). *IS-95 CDMA and Cdma2000, Cellular/PCS System Implementation.* Prentice Hall PTR.

[Giuliano, 2002] Giuliano R., Mazzenga F. & Vatalaro F. (2002). Smart cell sectorization for third generation CDMA systems. *Wireless Communications and Mobile Computing*, 2, 253–267.

[Godara, 1997] Godara L.C. (1997). Application of antenna arrays to mobile communications, part II: Beam-forming and direction-of-arrival consideration. *Proceedings of the IEEE*, 85(8), 1195–1245.

[Grant, 1998] Grant P. M., Thompson J. S. & Mulgrew B. (1998). Adaptive arrays for narrowband CDMA base stations. *Electronics & Communication Engineering Journal*, 10(4), 156–166.

[Holtzman, 1992] Holtzman J. M. (1992). A simple accurate method to calculate spread spectrum multiple access probabilities. *IEEE Transactions on Communications*, 40(3), 461–464.

[Hudson, 1981] Hudson J. E. (1981). *Adaptive Array Principles.* Peter Peregrinus Ltd..

[Ihara, 2001] Ihara T., Tanaka S., Sawahashi M. & Adachi F. (2001). Fast two-step beam tracking algorithm of coherent adapative antenna array diversity receiver in W-CDMA reverse link. *IEICE Transactions on Communications* ,E84-B(7), 1835–1848.

[Iskander, 2004] Iskander C.-D. & Mathiopoulos P. T. (2004). Performance of multicode DS/CDMA with *M*-ary orthogonal modulation in multipath fading channels. *IEEE Transactions on Wireless Communications*, 3(1), 209–223.

[ITUR, 2003] International Telecommunications Union (2003). Adaptive antennas concepts and key technical aspects (Draft New Report) , ITU-R M [adapt] [Doc. 8/10].

[Jacobsen, 2001] Jacobsen (2001). *Smart antennas for dummies* (Tech. Rep.).Telenor.

[Jakes, 1974] Jakes W. C. (1974). *Microwave Mobile Communications*. John Wiley.

[Jalloul, 1994] Jalloul L. M. & Holtzman J. M. (1994). Performance analysis of DS/CDMA with noncoherent *M*-ary orthogonal modulation in multipath fading channels. *IEEE Journal on Selected Areas in Communications*, 12(5), 862–870.

[Janaswamy, 2002] Janaswamy R. (2002). Angle and time of arrival statistics for the gaussian scatter density model. *IEEE Transactions on Wireless Communications*, 1(3), 488 –497.

[Jelitto, 1999] Jelitto J., Stege M., Lohning M., Bronzel M. & Fettweis G. (1999). A vector channel model with stochastic fading simulation. In *IEEE International Symposium on Personal, Indoor and Mobile Radio Communications (PIMRC)*, Osaka, Japan.

[Kaiser, 2005] Kaiser T. (2005). When will smart antennas be ready for the market? Part I. *IEEE Signal Processing Magazine*, 22(2), 87–92.

[Kim, 1992] Kim K. I. (1992). On the error probability of a DS/SSMA system with a noncoherent Mary orthogonal modulation. In *IEEE Vehicular Technology Conference (VTC)*, vol. 1, Denver, May 10-13, 482–485.

[Knopp, 2002] Knopp R. & Caire G. (2002). Power control and beamforming for systems with multiple transmit and receive antennas. *IEEE Transactions on Wireless Communications*, 1(4), 638–648.

[Lee, 2003] Lee W.-C., Choi S., Choi J. &. Suk M (2003). An adaptive beamforming technique for smart antennas in WCDMA system. *IEICE Transaction on Communications*, E86-B(9), 2838–2843.

[Lee, 1998] Lee J. S. & Miller L. E. (1998), *CDMA Systems Engineering Handbook*. Artech House.

[Lehne, 1999] Lehne P. H., & Pettersen M. (1999). An overview of smart antenna technology for mobile communication systems. *IEEE Communications Survey*, 2(4).

[Li, 2003] Li H. J. & Liu T. Y. (2003). Comparison of beamforming techniques for W-CDMA communication systems. *IEEE Transactions on Vehicular Technology*, 52(4), 752–760.

[Litva, 1996] Litva J. & Lo T.K. (1996). *Digital Beamforming in Wireless Communications*. ArtechHouse.

[Mercado, 2002] Mercado A. & Liu K. J. R. (2002). Adaptive QoS for wireless multimedia networks using power control and smart antennas. *IEEE Transactions on Vehicular Technology*, 51(5), 1223–1233.

[Miyamoto, 2002] Miyamoto R. Y. & Itoh T. (2002). Retrodirective arrays for wireless communications. *IEEE Microwave Magazine*, 3(1), 71–79.

[Mohamed, 2002] Mohamed N. A. & Dunham J. G. (2002). A low-complexity combined antenna array and interference cancellation DS-CDMA receiver in multipath fading channels. *IEEE Journal on Selected Areas in Communications*, 20(2),248–256.

[Mohassed, 2002] Mohassed Y. Z. & Fritz M. P. (2002). A 3-D spatio-temporal simulation model for wireless channels. *IEEE Journal on Selected Areas in Communications*, 20(6),1193–1203.

[Naguib, 1996] Naguib A. F. & Paulraj A. (1996). Performance of wireless CDMA with *M*-ary orthogonal modulation and cell site antenna arrays. *IEEE Journal on Selected Areas in Communications*, 14(9), 1770–1783.

[NaguibTh, 1996] Naguib A. F. (1996). Adaptive antennas for CDMA wireless networks. Ph.D. dissertation, Stanford University.

[Nazar, 2002] Nazar S. N., Ahmad M., Swamy M. N. S. & Zhu W.-P. (2002). An adaptive parallel interference canceler using adaptive blind arrays. In *IEEE Vehicular Technology Conference (VTC)*, 1, 425 – 429.

[Osseiran, 2004] Osseiran A. & Logothetis A. (2004). System Performance of TX Diversity and 2 Fixed beams in W-CDMA. *Kluwer Journal of Wireless Personal Communication*, 31(1) , 33–50.

[Osseiran, 2006] Osseiran A., & Logothetis A. (2006). Smart antennas in a wcdma radio network system: Modeling and evaluations. *IEEE Transactions on Antennas and Propagation*, 54(11) , 3302–3316.

[Parsons, 1992] Parsons J. D. (1992). *The Mobile Radio Prpagation Channel*. Halsted.

[Parsons, 1991] Parsons J. D. & Turkmani M. D. (1991). Characterization of mobile radio signals: Model description. *Proc. Inst. Elect. Eng.*, 138, 459–556.

[Patel, 1994] Patel P. & Holtzman J. M. (1994). Analysis of simple successive interference cancellation scheme in a DS/CDMA system. *IEEE Journal on Selected Areas in Communications*, 12(5), 796–807.

[Patzold, 2002] Patzold M. (2002). *Mobile Fading Channels*. John Wiley.

[Pattan, 2000] Pattan B. (2000). *Robust Modulations Methods and Smart Antennas in Wireless Communications.* Prentice Hall PTR.

[Pedersen, 2000] Pedersen K. I. & Mogensen P. E. (2000). A stochastic model of the temporal and azimuth dispersion seen at the base station in outdoor propagation environments. *IEEE Transactions on Vehicular Technology*, 49(2), 437–447.

[Pedersen, 2003] Pedersen K. I., Mongensen P. E., & Moreno J. R. (2003). Application and Performance of Downlink Beamforming Techniques in UMTS. *IEEE Communications Magazine*,134–143.

[Petrus, 1996] Petrus P., Reed J. H., & Rappaport T. S. (1996). Geometrically based statistical channel model for macrocellular mobile environments. In *IEEE Global Telecommunications Conference (GLOBECOM)*, London, UK, 1197–1201.

[Price, 1958] Price R. & Green P. E. (1958). A communication technique for multipath channels. *Proceedings of Institute of Radio Engineers)*, 46, 555–570.

[Proakis, 1995] Proakis J. G. (1995). *Digital Communications*, 3rd ed. McGraw-Hill.

[Rappaport, 2002] Rappaport T. S. (2002). *Wireless Communications: Principles and Practice*, 2nd ed. Prentice Hall.

[Rizk, 2002] Rizk K. , Wagen J.-F. & Gardiol F. (2002). Two-dimensional ray-tracing modeling for propagation prediction in microcellular environments. *IEEE Transactions on Vehicular Technology*, 46(2), 508–518.

[Rooyen, 2000] Rooyen P. V., Lotter M. & Wyk D. V. (2000). *Space-Time processing for CDMA Mobile Communications.* Kluwer Academic Publishers.

[Schuttengruber, 2004] Schuttengruber W., Molisch A. F. & Bonek E. (2004). Smart antennas for mobile communications - tutorial. [Online]. Available: http://www.nt.tuwien.ac.at/mobile/research/smart antennas tutorial/

[Simon, 2000] Simon M. K. &. Alouini M.-S. *Digital Communication over Fading Channels.* John Wiley & Sons.

[Sklar, 1997] Sklar B. (1997). Rayleigh fading channels in mobile digital communications part I: Charaterization. *IEEE Communications Magazine*, 90–100.

[Sklar, 1997b] Sklar B. (1997). Rayleigh fading channels in mobile digital communications part II: Mitigation. *IEEE Communications Magazine*, 102–109.

[Song, 1999] Song Y. S. & Kwon H. M. (1999). Analysis of a simple smart antenna for CDMA wireless communications. In *IEEE Vehicular Technology Conference (VTC)*, San Antonio, Texas, 254–258.

[Song, 2001] Song Y. S., Kwon H. M. & Min B. J. (2001). Computationally efficient smart antennas for CDMA wireless communications. *IEEE Transactions on Vehicular Technology*, 50(6) 1613–1628.

[SongTh, 2001] Song Y. S. (2001). Smart antenna algorithms and PN code tracking. Ph.D. dissertation, Wichita State University.

[Spagnolini, 2004] Spagnolini U. (2004). A simplified model to evaluate the probability of error in DS-CDMA systems with adaptive antenna arrays. *IEEE Transactions on Wireless Communications*, 3(2), 578–587.

[Spagnolini, 2001] Spagnolini U. (2001). A simplified model for probability in error in DS-CDMA systems with adaptive antenna arrays. In *IEEE International Conference on Communications (ICC)*, 2271–2275.

[Stevanovic, 2003] Stevanovic I., Skrivervik A. & Mosig J. R. (2003). *Smart antenna systems for mobile communications* (Tech. Rep.). Ecole Polytechnique Federale De Lausanne.

[Stuber, 2001] Stuber G. L. (2001). *Principles of Mobile Communication*, 2nd ed. Kluwer Academic Publishers.

[Sweatman, 1999] Sweatman C. Z.W. H., Mulgrew B., Thompson J. S. & Grant P. M. (1999), Spatial equivalence classes for CDMA array processing. In *IEEE International Conference on Communications (ICC)*, Vancouver, Canada, 544–548.

[Sun, 2004] Sun C., Hirata A., Ohira T. & Karmakar N. C. (2004). Fast beamforming of electronically steerable parasitic array radiator antennas: Theory and experiment. *IEEE Trans. on Antennas and Propagation,* 52(7), 1819-1832.

[Thompson, 1996] Thompson J. S., Grant P. M., & Mulgrew B. (1996). Performance of antenna array receiver algorithms for CDMA. In *IEEE Global Telecommunications Conference (GLOBECOM)*, 1, 18–22.

[Thompson, 1996b] Thompson J. S., Grant P. M., & Mulgrew B. (1996). Smart antenna arrays for CDMA systems. *IEEE Personal Communications*, 3(5), 16–25.

[Tsoulos, 2000] Tsoulos G. V., Athanasiadou G. E. & Piechocki R. J. (2000). Low-complexity smart antenna methods for third-generation W-CDMA systems. *IEEE Transactions on Vehicular Technology*, 49(6), 2382–2396.

[Tsoulos, 1999] Tsoulos G. V. (1999). Smart antennas for mobile communication systems: Benefits and challenges. *Electronics and Communication Engineering Journal*, 11(2), 84–94.

[Wang, 2003] Wang B. & Kwon H. M. (2003). PN code acquisition using smart antenna for spread spectrum wireless communications - part I. *IEEE Transactions on Vehicular Technology*, 52(1), 142–149.

[Wang, 2003b] Wang B. & Kwon H. M. (2003). PN code acquisition for DS-CDMA systems employing smart antennas - part II. *IEEE Transactions on Wireless Communications*, 2(1), 108–117.

[Winters, 1999] Winters J. H., & Gans M.J. (1999). The range increase of adaptive versus phased arrays in mobile radio systems. *IEEE Transactions on Vehicular Technology*, 48(2), 353–362.

[Yu, 2002] Yu K. & Ottersten B. (2004). Models for MIMO propagation channels: A review. *Wireless Communications and Mobile Communications*, 2(7), 653–666.

[Zhou, 2003] Zhou Y., Chin F., Liang Y. C. & Ko C. C. (2003). Performance comparison of transmit diversity and beamforming for downlink of DS-CDMA system. *IEEE Transactions on Wireless Communications*, 2(2), 320–334.

[Zhou, 2001] Zhou Y., Chin F., Liang Y. C., & Ko C. C. (2001). A novel beam selection transmit diversity scheme for DS-CDMA system. *IEICE Trans. Commun.*, E84-B(8), 2178–2185.

[3GPP, 2003] Third Generation Partnership Project (3GPP) (2003). Technical specification group radio access network, physical layer procedures, 3GPP TS25.214 v3.7.0.

[3GPP2, 2003] Third Generation Partnership Project Two (3GPP2) (2003). Spatial Channel Model Text Description (SCM Text v2.3).

# Chapter XVIII
# Cross-Layer Performance of Scheduling and Power Control Schemes in Space-Time Block Coded Downlink Packet Systems

**Aimin Sang**
*NEC Laboratories America, USA*

**Guosen Yue**
*NEC Laboratories America, USA*

**Xiaodong Wang**
*Columbia University USA*

**Mohammad Madihian**
*NEC Corporation of America, USA*

## ABSTRACT

*In this chapter, we consider a cellular downlink packet data system employing the space-time block coded (STBC) multiple-input-multiple-output (MIMO) scheme. Taking the CDMA high data rate (HDR) system for example, we evaluate the cross-layer performance of typical scheduling algorithms and a point-to-point power control scheme over a time division multiplexing (TDM)-based shared MIMO channel. Our evaluation focuses on the role of those schemes in multi-user diversity gain, and their impacts on medium access control (MAC) and physical layer performance metrics for delay-tolerant data services, such as throughput, fairness, and bit or frame error rate. The cross-layer evaluation shows that the multi-user diversity gain, which comes from opportunistic scheduling schemes exploiting independent channel oscillations among multiple users, can increase the aggregate throughput and reduce the transmission error rate. It also shows that STBC/MIMO and one-bit and multi-bit power control can indeed help the physical and MAC layer performance but only at a risk of limiting the multiuser diversity gain or the potential throughput of schedulers for delay-tolerant bursty data services.*

## INTRODUCTION

Very high rate physical-layer transmission and scheduling schemes have recently drawn significant attentions for the design of the next-generation wireless cellular system. The downlink transmission and throughput are in particular considered to be a primary bottleneck in the current system design. In anticipation of the high demand for wireless data service, two solutions of downlink data systems come out by utilizing existing CDMA systems for high rate data transmission. Both support high-rate packet data services on a shared channel. One of them is High Data Rate (HDR) system which is based on the techniques of cdma2000 (3GPP2, 2002), while the other is High Speed Data Packet Access (HSDPA) which is based on the WCDMA systems (3GPP, 2001). In these two standards, multiple active data users access the downlink channel in a time-division multiple accessing (TDMA) manner with certain scheduling scheme (Bender, 2000). Based on the channel state information (CSI), the scheduler at the base station (BS) selects a user to transmit according to certain scheduling criterion that should use transmission capacity efficiently to achieve high throughput. In this chapter, we take multiple-input-multiple-output (MIMO) HDR systems for an example, but our study applies to MIMO HSDPA systems as well.

MIMO techniques have been studied extensively in the recent past. Various MIMO schemes could be distinguished by different design goals (Alamouti, 1998; Foschini, 1996; Telatar, 1999). Among them the orthogonal space-time block coding (STBC) aiming at full transmitter-diversity was recently adopted for implementation as one of the transmit diversity modes of 3G wireless networks. The STBC schemes, originally proposed by Alamouti (1998) and Tarokh et al. (1999), introduce a simple and elegant mechanism with spatial or antenna diversity that improves the spectral efficiency over wireless channels. In this chapter, the STBC/MIMO systems are considered to support delay-tolerant bursty data services over a TDM-based downlink shared channel.

Focusing on MAC-layer throughput and fairness, we consider several typical scheduling algorithms for the delay-tolerant data services. Among them, the "greedy" or maximum carrier-to-interference ratio (Max-C/I) scheduling (Knopp and Humblet, 1995) routes each transmission time slot to the user with the best instantaneous channel conditions. This scheduling scheme explores independent channel dynamics of multiple users, the so called "multiuser diversity" (Viswanath et al., 2002). Another scheduling algorithm, the Proportional Fair (PF) scheduling (Jalali et al., 2000), balances between instantaneous channel status and long-term throughput performance of different users, i.e., it balances between the transmission rate and the fairness among users. The third scheduling scheme we consider is the Round-Robin (RR) that picks the user for transmission purely randomly. Note that the HDR systems has 11 channel states, which thus renders invalid the following scheduling schemes: Wireless Fluid Fair Queueing (WFFQ) and its packet-level approximation Idealized Wireless Fair Queueing (IWFQ) (Lu et al., 1999), and Channel-condition Independent Packet Fair Queueing (CIF-Q) (Ng et al., 1998), both are variants of wired scheduling schemes based on the over-simplified On-Off channel model. We refer readers to the references (Fattah and Leung, 2002; Cao and Li, 2001) for further information.

Our study shows that the performance of the three scheduling algorithms over STBC/MIMO differs significantly in throughput and multiuser diversity gain, while the STBC/MIMO channel may have inherent statistical limitations to support high-rate data services. This reveals that the global spectral efficiency in a single cellular data system depends on efficient cross-layer collaborations or an integral design of transmission and scheduling algorithms.

To see the above point more clearly, we also study a simple power control scheme given multi-fold multi-user diversity by channel-dependent scheduling schemes. Given the feedback knowledge of channel statistics and diversity gain, we derive the 1-bit and multi-bit power control scheme to minimize the frame/bit error rate (FER/BER). A numerical approach is given to obtain the optimal power level. We compare performance with and without the power control scheme in the CDMA/HDR STBC/MIMO systems. The results shows that although the power control scheme provides certain FER gain at the physical layer, it degrades the multiuser diversity gain at the scheduling layer. Therefore, its application in STBC/MIMO systems has dual impacts that may offset each other. Again we see the necessity of a cross-layer design and evaluation in designing such systems.

## SYSTEM DESCRIPTIONS

We consider a downlink high-speed data service system with space-time block coding using $n_T$ transmit antennas and $n_R$ receive antennas, as shown in Fig. 1. The channel is time slotted. Infinite data from different users are buffered and will be selected by the scheduler. The output data bits are QPSK modulated, then time multiplexed with pilot signal. The data sequences are then STBC coded and transmitted. At the receiver, the signals are received from $n_R$ receive antennas. After matched filtering and symbol-rate sampling, the received discrete-time signal at the $k$th terminal is modeled by

*Figure 1. Downlink transmitter and receiver structure*

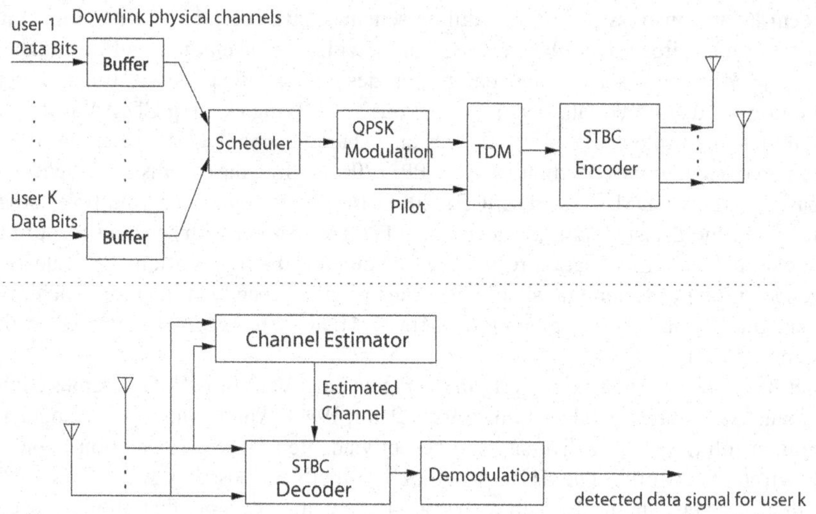


$$\boldsymbol{r}_k(t) = \sqrt{\frac{E_s}{n_T}}\boldsymbol{H}_k\boldsymbol{c}_k(t) + \boldsymbol{n}_k(t) \qquad t = 1, \cdots, T,$$

with

$$\mathbf{r}_k(t) = \left[r_{1,k}(t), \cdots, r_{n_R,k}(t)\right],$$

$$\mathbf{H}_k(t) = \left[\mathbf{h}_{1,k}^T(t), \cdots, \mathbf{h}_{n_R,k}^T(t)\right]^T,$$

$$\mathbf{h}_{j,k}(t) = \left[h_{1j,k}(t), \cdots, h_{n_T j,k}(t)\right],$$

$$\mathbf{c}_k(t) = \left[c_{1,k}(t), \cdots, c_{n_T,k}(t)\right],$$

$$\mathbf{n}_k(t) = \left[n_{1,k}(t), \cdots, n_{n_R,k}(t)\right],$$

where $E_s$ is the total energy in one time slot; $c_{i,k}(t)$ $i = 1, \cdots n_T$, is the symbol transmitted to the $k$th user from the $i$th transmit antenna at the $t$th time slot; $\boldsymbol{H}_k$ is an $n_R \times n_T$ complex matrix with each entry $h_{j,k}(t)$ representing the complex channel gain from the $i$th transmit antenna to the $j$th receive antenna of the $k$th user; $h_{ij,k}(t)$ is a complex Gaussian random variable with mean 0 and variance 0.5 per dimension; $\boldsymbol{n}_k$ is a complex Gaussian random vector with mean 0 and covariance matrix $\sigma^2\boldsymbol{I}$, i.e., $\boldsymbol{n}_k(t) \sim \mathcal{N}_c(\boldsymbol{0}, s^2\boldsymbol{I})$.

For orthogonal STBC systems, the instantaneous mutual information (Gozali et al., 2002; Hassibi and Hochwald, 2002) is given by:

$$r_k(t) = R\log(1 + \text{SNR}_{STBC}). \qquad (1)$$

where $R$ is the code rate. For $2 \times 2$ (2Tx-2Rx) and $2 \times 1$ STBC/MIMO systems, Alamouti code derives $R$ = 1. For $3 \times n_R$ STBC/MIMO systems, there are only $\frac{3}{4}$-rate codes available (Hassibi and Hochwald, 2002), i.e., $R = \frac{3}{4}$. $\text{SNR}_{STBC}$ is the instantaneous effective SNR at the receiver side. For the STBC decoder at the $k$th user terminal, its output is given by

$$z_k(t) = \sqrt{\frac{E_s}{n_T}\sum_{j=1}^{n_R}\sum_{i=1}^{n_T}|h_{ij,k}|^2}\, c_k(t) + v_k(t), \quad (2)$$

where

$$v_k(t) \sim N_c\left(0, \sum_{j=1}^{n_R}\sum_{i=1}^{n_T}|h_{ij,k}|^2 \sigma^2 I\right). \quad (3)$$

Therefore, the effective SNR at either the receiver or decoder can be written as

$$\text{SNR}_{STBC} = \frac{E_s}{n_T\sigma^2}\sum_{j=1}^{n_R}\sum_{i=1}^{n_T}|h_{ij,k}|^2.$$

## SCHEDULING SCHEMES AND CHANNEL DYNAMICS

To determine the performance of the above-described system, channel-dependent scheduling schemes play a crucial role. Given a large number of mobile users of independent channel statistics, chances are that at any moment, there is always someone whose instantaneous effective SNR is close to its peak. Based on the CSI feedbacks from all the users, BS can pick the user of the best instantaneous channel at any moment and thus increase the long-term downlink aggregate throughput. This is the concept of multiuser diversity (Knopp and Humblet, 1995).

### Greedy Scheduling

In this section, we describe the multiuser diversity gain in effective SNR given the "greedy" (or Max-C/I) scheduling scheme. Without loss of generality, we assume that the supportable rate of each user is in proportion to its effective SNR. Define $\gamma$ as the effective SNR at the output of the space-time decoder of the $k$th user when $k$ is the only user in the system:

$$\gamma \equiv \text{SNR}_{STBC} = \frac{\text{SNR}}{n_T}\sum_{j=1}^{n_R}\sum_{i=1}^{n_T}|h_{ij,k}|^2, \quad (4)$$

where with QPSK modulation, $E_s = 2E_b$ for 2bits/symbol, $\text{SNR} = \frac{E_S}{\sigma^2} = \frac{2E_b}{\sigma^2}$. We know $|h_{ij,k}|^2$ is a sum of square of two i.i.d. Gaussian random variables with distribution $\mathcal{N}(0,0.5)$. Thus, $\gamma$ follows the chi-square distribution with $2n_Tn_R$ degrees of freedom. Its probability density function (pdf) is given by

$$f_\gamma(\gamma) = \frac{1}{(n_Tn_R-1)!(\frac{E_s}{n_T\sigma^2})^{n_Tn_R}}\gamma^{n_Tn_R-1}e^{-\frac{\gamma}{E_s/(n_T\sigma^2)}}. \quad (5)$$

Its cumulative distribution function (cdf) is given by

$$F(y) = 1 - \mu^{n+1}e^{-\mu t}\sum_{i=0}^{n}\frac{y^i}{i!\mu^{n-i+1}}, \quad (6)$$

where $\mu = \frac{n_T \sigma^2}{E_s}$ and $n = n_T n_R - 1$.

Now suppose that there are $K$ users in the system. With the RR scheduling scheme, the base station routes the transmission of packets equally randomly across all users. Therefore, the pdf of the effective SNR after scheduling, i.e., the effective SNR of the aggregate channel, remain the same as in (5). Now consider the "greedy" or Max-C/I scheduling algorithm, by which the base station selects the user $k$ of the best instantaneous channel. That is,

$$k^* = \arg \max_{k \in \{1,2,\cdots,K\}} \sum_{j=1}^{n_R} \sum_{i=1}^{n_T} | h_{ij,k} |^2 . \tag{7}$$

Define $\tilde{\gamma}$ as the effective SNR at the output of the Max-C/I scheduler with $K$-fold multiuser diversity. Then

$$\tilde{\gamma} = \max_{k \in \{1,2,\cdots,K\}} \frac{E_s}{n_T \sigma^2} \sum_{j=1}^{n_R} \sum_{i=1}^{n_T} | h_{ij,k} |^2 . \tag{8}$$

By order statistics (David, 1981), the pdf of $\tilde{\gamma}$ is

$$g_{\tilde{\gamma}}(y) = K f_\gamma(y) F(y)^{K-1}. \tag{9}$$

Fig. 9 is the plot of this pdf (i.e., the STBC without one or two-bit power control), where the SNR increases with $K$ for more multiuser diversity gain.

## Fair Scheduling

In the above, we have considered multiuser diversity gain in effective SNR with the Max-C/I scheduling. Now we focus on fair scheduling algorithms with multiuser diversity gain measured by throughput increment (in kbps). A downlink scheduler has two basic tasks with packet data services over the high-speed packet data channel. One is the high aggregate throughput. Another is the fairness guarantees among different users. Being a subjective topic, fairness has different criteria. Here we consider a resource-based fairness criterion. Namely, time-slot allocation to different users should be equal regardless of user statistics. No doubt the RR scheduling provides the strictest fairness by the definition. However, the throughput is quite low for the lack of multiuser diversity gain. In contrast, by picking the best channel at any time slot, Max-C/I maximizes aggregate throughput but starves users with constantly poor channel, e.g., the users far away from the cell site.

To balance between the two extreme schemes, the channel-dependent Proportional Fair (PF) scheduling (Jalali et al., 2000; Viswanath et al., 2002) was proposed as follows:

$$k^*(t) = \arg \max_{k \in 1,2,\cdots,K} \frac{r_k(t)}{\tilde{r}_k(t)},$$

$$\tilde{r}_k(t+1) = (1 - \frac{1}{t_c}) \tilde{r}_k(t) + \frac{1}{t_c} r_k(t) \mathbf{1}_{(k^*(t)=k)}, \tag{10}$$

where $r_k(t)$ denotes the instantaneous "supportable channel rate" at time slot $t$ according to the feedbacks from the data rate control (DRC) channel of the $k$th user; $\tilde{r}_k(t)$ is its low-pass filtered mean throughput

$$\mathbf{1}_{(k^*(t)=k)} = \begin{cases} 1, & \text{scheduler picks user} \\ 0, & \text{otherwise} \end{cases};$$

and $t_c$ is the filtering time interval. In Jalali et al. (2000), $t_c$ is set to be 1000 slots (1.667s) to enable a fast adaptation of scheduling performance to channel dynamics. It also controls the transmission latency of each individual user. Note that $\mathbf{1}_{(k^*(t)=k)}$ is a necessary term to differentiate the mean throughput $\tilde{r}_k(t)$ from the mean supportable channel rate $\bar{r}_k(t+1)=(1-\frac{1}{t_c})\bar{r}_k(t)+\frac{1}{t_c}r_k(t)$. In contrary to claims by some papers, we will see in Section 6 that $\bar{r}_k(t)$ cannot replace $\tilde{r}_k(t)$ in the PF scheduling.

According to the definition, PF considers both instantaneous channel dynamics and long-term mean throughput of each individual user. Its fairness is provided in the following way: a user who did not get scheduled due to the poor channel status in the past has a low transmission rate, and thus a high priority for transmission by the next moment. This guarantees an asymptotically equal fraction of time among users (Viswanath et al., 2002). In a long run, it meets the following target:

$$U=\max_{r_k}\sum_{k=1}^{K}U_k(r_k)=\max_{r_k}\sum_{k=1}^{K}\log r_k, \tag{11}$$

where $r_k \geq 0$ is the stationary throughput expectation for user $k$; $U_k$ is a strictly concave and continuously increasing utility function, reflecting the network economy for elastic traffic of best-effort services (Shenker, 1995). To understand the background of the optimization target, look at an asymptotic format of the utility function

$$U(t)=\sum_{k=1}^{K}U_k(\tilde{r}_k(t))=\sum_{k=1}^{K}\log\tilde{r}_k(t), \tag{12}$$

$$U=\lim_{t\to+\infty}\max_{\tilde{r}_k(t)}U(t), \tag{13}$$

and its derivative in time domain

$$\frac{dU(t)}{dt}=\sum_{k=1}^{K}\frac{\partial U_k}{\partial\tilde{r}_k(t)}\frac{d\tilde{r}_k(t)}{dt}=\sum_{k=1}^{K}\frac{\tilde{r}_k(t)'}{\tilde{r}_k(t)}. \tag{14}$$

Asymptotically the concave utility function should adopt the fastest growth in time. Therefore, by the steepest ascend the scheduler should select the user of the maximum $\frac{\tilde{r}_k(t)'}{\tilde{r}_k(t)}$ if only one user is to be transmitted at any moment. In discrete time interval $\Delta t$ denotes one time slot. When the $k$th user is to be selected, $\mathbf{1}_{(k^*(t)=k)}=1$ and $\tilde{r}_k(t+1)=(1-\frac{1}{t_c})\tilde{r}_k(t)+\frac{1}{t_c}r_k(t)$. So $\tilde{r}_k(t)'$ can be approximated by

$$\frac{\tilde{r}_k(t+1)-\tilde{r}_k(t)}{\Delta t}=\frac{r_k(t)-\tilde{r}_k(t)}{t_c\Delta t}. \tag{15}$$

As a result, we get the following proportional relationship:

$$\frac{\tilde{r}_k(t)'}{\tilde{r}_k(t)}\sim\frac{1}{t_c\Delta t}\left(\frac{r_k(t)}{\tilde{r}_k(t)}-1\right)\sim\frac{r_k(t)}{\tilde{r}_k(t)}. \tag{16}$$

The above shows that at any time slot, the PF scheduler picks the user of the largest $\frac{r_k(t)}{\bar{r}_k(t)}$ to maximize the utility. Comparatively RR is purely random and does not have a channel-dependent and throughput-related utility target, while Max-C/I optimizes $\sum_{k=1}^{K} r_k$, i.e., the aggregate throughput. Therefore, PF's throughput would be somewhere in-between of Max-C/I and RR, as we will see in the section of results.

## A POWER CONTROL SCHEME

In the scenario of high-speed downlink data channel, what is the performance of the typical point-to-point power control scheme? As we know, a base station may compensate channel fading or fluctuations of each mobile user using a point-to-point power adaptation scheme. This is very important for time-critical voice services that require dedicated stable channels to support the traffic which is relatively smooth and of low rates. However, for delay-tolerant data services where traffic tends to be bursty and of high rates, dedicated channels are costly. In this case, by pooling channel resources at the system level, in other words, by using a shared high-speed downlink channel, the bursty and asynchronous fluctuations of many users will be better accommodated at the base station. This increases the spectral efficiency of the data services. Under this circumstance, one would wonder how the power control performs.

An idealistic assumption is that the transmission power for the data channel is not strictly upper bounded. For example, certain reservation has been made for the power adaptation purpose. With this assumption we will show that the power control has two aspects: although it enhances the reliability (in terms of FER and BER) at the physical layer, it also reduces multiuser diversity gain. Generally speaking, the typical point-to-point control mechanism has different impacts on scheduling and physical layers, which may offset each other. As a starting point of our analysis, let us describe the BER first.

Denote a new random variable $X$ as the channel condition. For the RR scheduling, $X$ is given by

$$X = \sum_{j=1}^{n_R} \sum_{i=1}^{n_T} | h_{ij} |^2, \tag{17}$$

with pdf $$f_X(x) = \frac{1}{(n_T n_R - 1)!} x^{n_T n_R - 1} e^{-x}; \tag{18}$$

while with the Max-C/I scheme,

$$X = \max_k \sum_{j=1}^{n_R} \sum_{i=1}^{n_T} | h_{ij,k} |^2 \tag{19}$$

with pdf

$$f_X(x) = \frac{K}{(n_T n_R - 1)!} x^{n_T n_R - 1} e^{-x} \left( 1 - e^{-x} \sum_{m=0}^{n_T n_R - 1} \frac{x^m}{m!} \right)^{K-1}. \tag{20}$$

We can write the effective SNR at the output of the space-time decoder as follows:

$$\gamma = \frac{\text{SNR}}{n_T} X = \frac{E_s}{n_T \sigma^2} X.$$

Thus, the bit error probability of the STBC decoder is

$$P_b = \int_0^{\infty} Q\left(\sqrt{\frac{E_s}{n_T\sigma^2}x}\right)\cdot f_X(x)dx$$

$$= \int_0^{\infty} Q\left(\sqrt{\frac{2E_b}{n_T\sigma^2}x}\right)\cdot f_X(x)dx, \qquad (21)$$

where $f_X(x)$ is the pdf of $X$ given by (18) or (20), depending on the scheduling scheme to be used.

## 1-Bit Power Control Scheme

Based on the system described above, we now introduce a simple power control scheme. First, consider the case of 1-bit feedback power control that works on the aggregate channel after scheduling and uniformly switches $E_s$ to two levels: $E_1$ and $E_2$. Denote $x_1$ as the median value of $X$. Thus we have

$$\int_0^{x_1} f_X(x)dx = \frac{1}{2}. \qquad (22)$$

The power feedback can be obtained by the rules

$$\begin{cases} X < x_1 & \rightarrow \quad \text{power } E_1 \\ X > x_1 & \rightarrow \quad \text{power } E_2 \end{cases} \qquad (23)$$

The total average power is $\frac{1}{2}E_1 + \frac{1}{2}E_2 = E_s$. $E_1$ and $E_2$ are chosen to minimize

$$P_b = \int_0^{x_1} Q\left(\sqrt{\frac{E_1}{n_T\sigma^2}x}\right) f_X(x)dx + \int_{x_1}^{\infty} Q\left(\sqrt{\frac{E_2}{n_T\sigma^2}x}\right) f_X(x)dx.$$

Set

$$\frac{\partial P_b}{\partial E_1} = g(E_1) = 0. \qquad (24)$$

The solution of the above equation is the optimized 1-bit power control value, which is

$$E_1^* = \arg\min_{E_1} P_b, \qquad (25)$$

and $E_2^* = 2E_s - E_1^*$. Define

$$F(E,x)=\frac{\partial}{\partial E}Q\left(\sqrt{\frac{E}{n_T\sigma^2}x}\right) \qquad =-\frac{1}{2\sqrt{2\pi}}\sqrt{\frac{x}{n_T\sigma^2 E}}\exp\left(-\frac{E}{2n_T\sigma^2}x\right), \tag{26}$$

$$F'(E,x)=\frac{\partial^2}{\partial E^2}Q\left(\sqrt{\frac{E}{n_T\sigma^2}x}\right)=\frac{1}{2\sqrt{2\pi}}\left(\sqrt{\frac{x}{n_T\sigma^2 E}}\frac{x}{2n_T\sigma^2}+\frac{1}{2}\sqrt{\frac{x}{n_T\sigma^2 E^3}}\right)\cdot\exp\left(-\frac{E}{2n_T\sigma^2}x\right). \tag{27}$$

Then, we have

$$g(E_1)=\int_0^{x_1}F(E_1,x)f_X(x)dx-\int_{x_1}^{\infty}F(2E_s-E_1,x)f_X(x)dx, \tag{28}$$

and the derivative of $g(E_1)$ given by

$$g'(E_1)=\int_0^{x_1}F'(E_1,x)f_X(x)dx+\int_{x_1}^{\infty}F'(2E_s-E_1,x)f_X(x)dx. \tag{29}$$

Then, we can solve $E_1^*$ iteratively by

$$E_1^{(n)}=E_1^{(n-1)}-\frac{g(E_1^{(n-1)})}{g'(E_1^{(n-1)})}. \tag{30}$$

Now we present the performance analysis in the form of the effective SNR with the power control technique applied. Denote $Y$ as the effective SNR with power control. According to the control rules in (23), we have

$$Y=\begin{cases}\dfrac{E_1^*}{n_T\sigma^2}X, & \text{if } X<x_1\\[2ex] \dfrac{E_2^*}{n_T\sigma^2}X, & \text{if } X>x_1\end{cases}. \tag{31}$$

Denote

$$a_1=\frac{E_1^*}{n_T\sigma^2},\quad a_2=\frac{E_2^*}{n_T\sigma^2}. \tag{32}$$

The pdf of the effective SNR $Y$ with 1-bit feedback power control is then given by

$$g_Y(y) = \begin{cases} \frac{1}{a_1} f_X\left(\frac{y}{a_1}\right), & y < a_2 x_1 \\ \frac{1}{a_1} f_X\left(\frac{y}{a_1}\right) + \frac{1}{a_2} f_X\left(\frac{y}{a_2}\right), & a_2 x_1 < y < a_1 x_1, \\ \frac{1}{a_2} f_X\left(\frac{y}{a_2}\right), & y > a_1 x_1 \end{cases}$$

where the pdf of random variable $X$ is given in (18) for the RR scheduling or (20) for the Max-C/I (greedy) scheduling scheme.

## Extension to Multi-Bit Power Control

Now we consider an $M$-bit power control scheme, which corresponds to $2^M$ power values, denoted as $E_p$, $p = 1, ..., 2^M$. Define $x_p$, $p = 1, ..., 2^M - 1$, as the critical channel condition for power selection. Let $x_0 = 0$ and $x_{2^M} = +\infty$. Thus, $x_p$ satisfies

$$\int_{x_{p-1}}^{x_p} f_X(x)dx = \frac{1}{2^M}, p = 1, \cdots, 2^M. \tag{33}$$

The rules of $M$-bit power feedback become

$$X \in \left[x_{p-1}, x_p\right) \longrightarrow \text{power } E_p, \quad p = 1, \cdots, 2^M. \tag{34}$$

Thus, similar to (24), the bit error probability for $M$-bit power control is given by

$$P_b(E_1, \cdots, E_{2^M}) = \sum_{p=1}^{2^M} \int_{x_{p-1}}^{x_p} Q\left(\sqrt{\frac{E_p}{n_T \sigma^2} x}\right) f_X(x)dx, \tag{35}$$

with the constraint of $E_p$ given by

$$\frac{1}{2^M} \sum_{p=1}^{2^M} E_p = E_s. \tag{36}$$

The optimal solution of $\left\{E_p\right\}_{p=1}^{2^M}$ is

$$\left\{E_p^*\right\}_{p=1}^{2^M} = \arg \min_{E_1, \cdots, E_{2^M}} P_b(E_1, \cdots, E_{2^M}). \tag{37}$$

Similar to the procedure of 1-bit power control, we can solve $\left\{E_p^*\right\}_{p=1}^{2^M}$ numerically. First, substitute the constraint of $E_p$ in (36) to (35). Thus, $P_b$ can be written as

$$P_b\left(E_1,\cdots,E_{2^M-1}\right)=\sum_{p=1}^{2^M-1}\int_{x_{p-1}}^{x_p}Q\left(\sqrt{\frac{E_p}{n_T\sigma^2}x}\right)f_X(x)dx+\int_{x_{2^M-1}}^{+\infty}Q\left(\sqrt{\frac{E_{2^M}}{n_T\sigma^2}x}\right)f_X(x)dx,$$

where $E_{2^M}=2^M E_s-\sum_{p=1}^{2^M-1}E_p.$

Define

$$\phi_p\left(E_1,\cdots,E_{2^M-1}\right)=\int_{x_{p-1}}^{x_p}F(E_p,x)f_X(x)dx-\int_{x_{2^M-1}}^{+\infty}F(E_{2^M},x)f_X(x)dx,$$

$$p=1,\cdots,2^M-1.$$

Thus, we have $2^M-1$ nonlinear equations with $2^M-1$ unknown variables given by

$$\phi_1\left(E_1,\cdots,E_{2^M-1}\right)=0,$$
$$\phi_2\left(E_1,\cdots,E_{2^M-1}\right)=0,$$
$$\vdots$$
$$\phi_{2^M-1}\left(E_1,\cdots,E_{2^M-1}\right)=0.$$

Denote $\boldsymbol{E}=\left[E_1,\cdots,E_{2^M-1}\right]^T$, $\phi_p(\boldsymbol{E})=\phi_p\left(E_1,\cdots,E_{2^M-1}\right)$, $\boldsymbol{\Phi}(\boldsymbol{E})=\left[\phi_1(E),\cdots,\phi_{2^M-1}(E)\right]^T$, and the Jacobian matrix $J(E)$ with

$$[\boldsymbol{J}(\boldsymbol{E})]_{pq}=\frac{\partial\phi_p(\boldsymbol{E})}{\partial E_q}=\delta_{pq}\int_{x_{p-1}}^{x_p}F'(E_p,x)f_X(x)dx+\int_{x_{2^M-1}}^{+\infty}F'(E_{2^M},x)f_X(x)dx,$$

where $\delta_{pq}=1$ for $p=q$ and $\delta_{pq}=0$ for $p\neq q$. Thus, the optimal solution of $M$-bit power feedback can be obtained by iteratively computing

$$\boldsymbol{E}^{(n)}=\boldsymbol{E}^{(n-1)}+\delta\,\boldsymbol{E}^{(n-1)},$$

and $E_{2^M-1}^{(n)}=2^M E_s-\sum_{p=1}^{2^M-1}E_p^{(n)},$

where $\boldsymbol{J}\left(\boldsymbol{E}^{(n-1)}\right)\cdot\delta\,\boldsymbol{E}^{(n-1)}=-\boldsymbol{\Phi}\left(\boldsymbol{E}^{(n-1)}\right).$

Then the effective SNR for $M$-bit power control is given by

$$\overline{Y}=\frac{E_p^*}{n_T\sigma^2}X\quad\text{when }x_{p-1}<X<x_p,\quad p=1,\cdots,2^M.$$

Define

$$a_p = \frac{E_p^*}{n_T \sigma^2}, p = 1, \cdots, 2^M.$$

The pdf of effective SNR is W

$$g_{\bar{Y}}(y) = \sum_{\{p: a_p x_{p-1} < y < a_p x_p\}} \frac{1}{a_p} f_X\left(\frac{y}{a_p}\right).$$

## RESULTS

In this section we provide numerical and simulation results for a CDMA/HDR system. The focus is to evaluate the cross-layer performance under multiuser and spatial diversities. That is, we intend to evaluate the combination of fair scheduling, STBC/MIMO, and power control in serving a high-rate data system like HDR, where multiuser diversity is an inherent characteristics to explore. The performance criteria includes the following aspects: the physical layer metrics including FER and mean effective SNR (per bit); the Media Access Control (MAC) layer metrics including aggregate downlink throughput, mean effective SNR (per slot), and fairness among heterogeneous users.

### Performance of The Scheduling Schemes over STBC/MIMO

*Simulation setup:* In the following simulation tests, we focus on the case of users who differ only in distance from the base station, and their STBC/MIMO channels have the same statistics as characterized by (4) with or without Doppler fading. To focus on the scheduling performance, we assume a very powerful coding scheme and continuous rate value by Shannon capacity. Following the settings in the works by Bender et al. (2000) and Jalali et al. (2000), mobile users are uniformly located within a cell. Their mean effective SNR for the case of one-transmitter-and-one-receive antenna (1Tx-1Rx), i.e., the SNR in (4), has the same typical cdf (see Fig. 2(a)) as the CDMA/HDR system in Bender et al. (2000). Sample cases of the normalized effective SNR are shown in Fig. 2(b), together with their mean and variance. Later we will see that the scheduling performance is largely determined by the mean and tail distributions of the effective SNR.

Compared to the real scenario where rich scattering, moving patterns, and interference differ for mobile users who have discrete supportable rates, this simplified simulation setup allows us to focus on scheduling performance under multiuser and antenna diversity. For each given number of users *K*, scheduling algorithm, and antenna diversity, 50 runs of simulations are executed to get a stable value. For each run, users' mean effective SNRs are selected according to Fig. 2(a), and each run lasts for 30 seconds (18000 time slots). For all the test cases we assume infinitely-backlogged traffic to each users, and unless explicitly mentioned the user speed is assumed to be zero.

*Issues with the scheduling and STBC schemes:* Among the three scheduling algorithms, RR is purely random. Its throughput and the average post-scheduling SNR are the worst due to the lack of channel exploitation. Fig. 3 shows its SNR gain represented by the SNR increment from $K = 1$. First, RR's performance remains largely flat with increasing user number *K*. Second, the three cases of (1Tx-1Rx, 2Tx-2Rx, 3Tx-3Rx) with RR totally merge. To a very large extent, its performance fails to improve with higher multiuser and antenna diversity. Obviously RR is not a good choice for scheduling high-rate shared data channel.

The other scheduling schemes perform better than RR for their constructive exploitations of channel fading. However, for a fixed number of users in Fig. 3, the SNR gain with PF and Max-C/I gets lower with an increasing order of antenna arrays. It is because that the STBC technique uses spatial antennas to compensate temporal channel fading, and implicitly decreases the multiuser diversity gain. This limits the increase of scheduling throughput at the MAC layer with higher spatial diversity. Therefore, the design of physical channel should consider scheduling requirements as well. To support the channel-dependent scheduling algorithms, BLAST/MIMO and opportunistic beamforming (Viswanath et al., 2002) are more suitable than STBC/MIMO.

Now let us focus on the combination of PF and STBC/MIMO. Fig. 4 reveals the factors determining the scheduling throughput. Firstly and obviously, the throughput increases for antenna diversity according to the comparisons between 1Tx-1Rx and 2Tx-2Rx or 3Tx-3Rx for any scheduling algorithm and any *K*. Secondly, in spite of its advantages in fair-

*Figure 2. (a) Typical (mean) SNR distribution of CDMA/HDR systems (b) pdf of the normalized effective SNR*

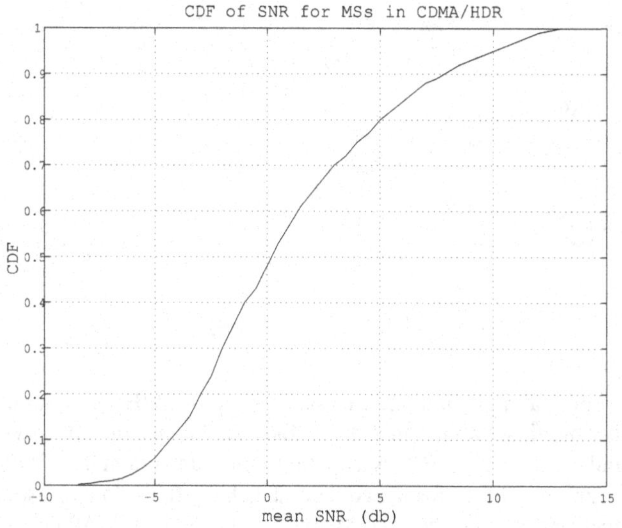


*(a)*

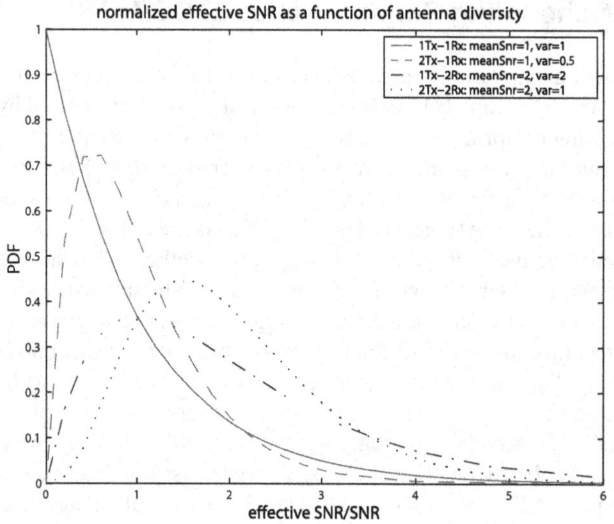


*(b)*

ness guarantee, the PF scheduling is lagged farther behind of Max-C/I in throughput with higher numbers of antennas or users. The increasing gap shows the conservativeness of PF in exploiting STBC capacity. Further research shows that we can trade off slight fairness for significant throughput improvement using a more flexible scheduling scheme. Finally, refer to (1), there are only $\frac{3}{4}$-rated codes available for 3Tx-3Rx STBC channel, i.e., $R=1$ in (1), while Alamouti code makes $R=1$. That is why the throughput actually decreases from 2Tx-2Rx to 3Tx-3Rx in spite of the increasing antenna diversity. This is just another point in support of cross-layer system design.

We also tested scenarios with Doppler fading of different speeds: 0km/h, 3km/h, 10km/h, and 40km/h. In those tests, chip rate is 1.2288Mbps with 2048 chips per slot, i.e. the slot frequency is 600, or slot interval is 1.667ms. The bandwidth is 1.25MHz. The carrier frequency is 1.88GHz. With mobility, the relationship between the mean throughput

*Figure 3. Performance of STBC and scheduling: K-fold multiuser diversity gain*

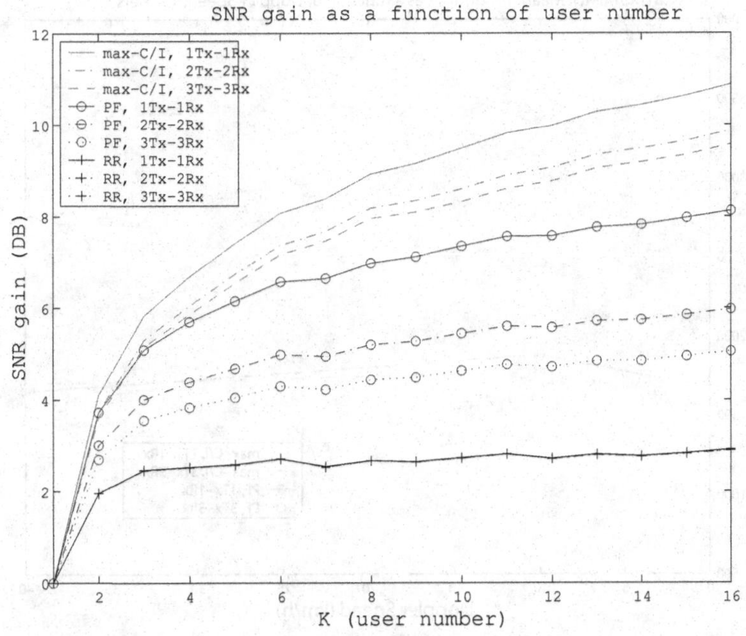


*Figure 4. Mean aggregate throughput as a function of user number*

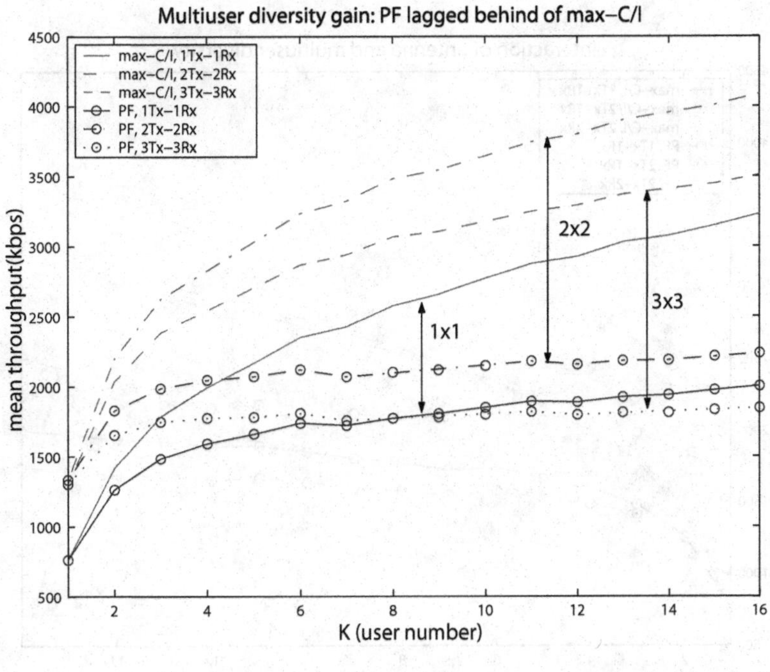

*Figure 5. Mean transmission rate as a function of Doppler speed*

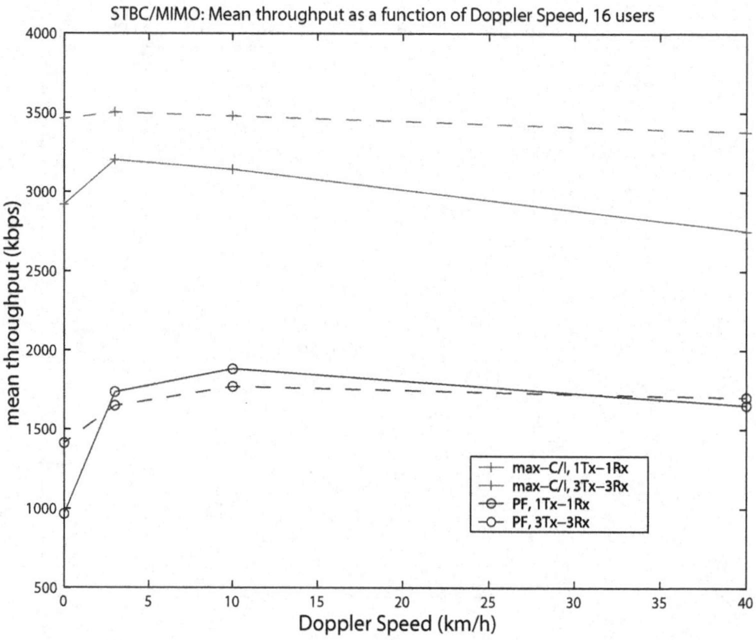


*Figure 6. STBC/MIMO: multiuser diversity gain as a function of the number of transmit antennas and users*

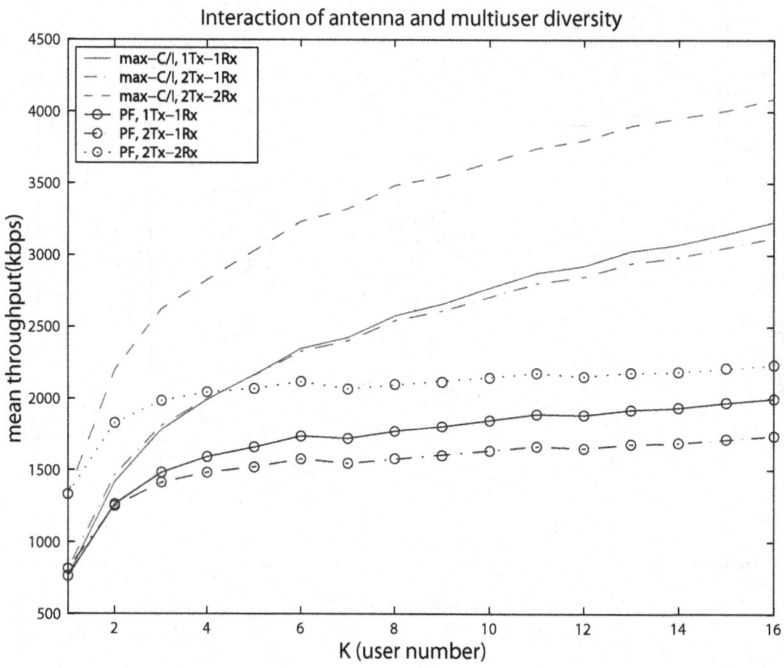

and multiuser or antenna diversity are generally unchanged. However, the throughput as a function of speeds increases first then decreases within an acceptable range, as shown in Fig. 5. That is because mobility increases channel dynamics, and thus the multiuser diversity, but decreases the accuracy of CSI feedbacks. Therefore it has double-faced impacts on the scheduling throughput. At the physical layer, as we will see later, mobility has only a single impact with degraded FER.

*Roles of diversity in designing scheduler and channel:* An efficient data system should have compatible layers. To design such a system, we need a better understanding of the multiuser and antenna diversity, especially their relevant impacts on scheduler and channel design. For this purpose we conduct more tests.

First is about the PF scheduler. As aforementioned, we note that by the definition of the PF scheme, i.e., by the (11), $\tilde{r}_k(t)$ is the mean throughput or mean transmission rate of the $k$th user, where $\mathbf{1}_{(k^*(t)=k)}$ is to differentiate it from mean supportable channel rate $\bar{r}_k(t)$. Intuitively $\tilde{r}_k(t)$ reflects the throughput performance of the scheduler in the eye of the $k$th user. Users who has not been transmitted for a while will have a higher scheduling priority by the PF. Now, suppose $\bar{r}_k(t)$ replaces $\tilde{r}_k(t)$ in the scheduling and do the similar test as before for the PF. Not shown here, the downlink throughput would actually start dropping when the number of users reached certain level, e.g., when $K = 3{\sim}5$. This simple test reveals the fundamental role of $\tilde{r}_k(t)$ in weighting dynamically the scheduling priority with multiuser diversity gain.

Now look at the antenna diversity in characterizing in STBC/MIMO channels and the scheduling throughput. Refer to (4). The effective SNR of the $k$th user has the expectation of $E[\gamma] = n_R\text{SNR}$. Its coefficient of variance is $\frac{1}{\sqrt{n_T n_R}}$. Therefore, for a fixed transmission power (or a fixed SNR), an increasing $n_T$ reduces the channel fluctuations and hence the multiuser diversity. Similarly, although $n_R$ increases the mean channel linearly, it reduces the multiuser diversity. Thus with the STBC scheme the multiuser diversity does not increase linearly with antenna diversity. The associated impacts on scheduling throughput can be observed in Fig. 6, where the 1Tx-1Rx STBC offers better throughput than the corresponding 2Tx-1Rx STBC in spite that the latter has higher spatial diversity. Refer to Fig. 2(b). 1Tx-1Rx has the same mean as but a larger coefficient of variance than 2Tx-1Tx. The longer-tailed 1Tx-1Rx distribution offers more opportunity to multiuser-diversity scheduling. Therefore, more transmit antennas in STBC do not necessarily guarantee higher throughput at the scheduling level, especially when we consider the lack of full-rate codes given three or more transmit antennas.

Compared to the STBC, other transmission code schemes for MIMO channels may offer more compatible antenna and multiuser diversity, and thus higher throughput. One scheme is the opportunistic beamforming that keeps instead of mitigating the fades in channel dynamics. Another scheme is the Selective Transmit Diversity (STD), which selects only one "good" antenna for transmission at each time slot. Those coding schemes provide more bursty channel fluctuations than STBC. Thus they have higher inherent multiuser diversity in favor or high-speed packet data schedulers. The service-awareness at physical channel constitutes another aspect in the cross-layer system design.

*Fairness of the scheduling over STBC/MIMO:* As we mentioned earlier, fair scheduling schemes trade off aggregate throughput for fairness among mobile users. Here we consider the fairness to be either resource-based or performance-based. The resource-based fairness metrics is the time-fraction allocated to individual users, while the performance-based fairness metrics refers to $\frac{E[\tilde{r}_k]}{E[r_k(t)]}$, i.e., the user-specific mean throughput normalized by its mean channel rate. A perfect fairness is denoted by equal fairness metrics among all users. To check the fairness performance, we randomly pick sixteen mobile users of heterogeneous SNRs in (4), following the cdf of CDMA/HDR in Fig. 2(a). Then we rank and label them in an increasing order of mean SNR: users of smaller IDs are farther away from the BS and has a constantly poorer channel. The mean of their supportable rate ranges from 153kbps to 3.767Mbps. For such a simulation setup, we test the three scheduling algorithms in the CDMA/HDR systems of STBC/MIMO scheme.

Our simulation results are shown in Fig. 7 and 8, respectively. Both figures reveal the lack of support for fairness with Max-C/I, where users of smaller IDs have lower fairness metrics. RR and PF have similar performance, both having almost equal metrics among all users for any setup (1Tx-1Rx, 2Tx-2Rx, and 3Tx-3Rx STBC/MIMO). In other words, they treat users equally fairly. Comparatively as shown by Fig. 8, PF is not only fair, but also more efficient than RR for its consistently higher throughput.

Furthermore, regardless of PF or Max-C/I in Fig. 8, $\frac{E[\tilde{r}_k]}{E[r_k(t)]}$ for the 1Tx-1Rx case is always the highest, then the 2Tx-2Rx case, and 3Tx-3Rx is the lowest. Consistent with our previous observations, this interesting phenomenon implies that higher spatial diversity in STBC/MIMO smoothes more channel variations and thus reduces the multiuser diversity gain. Recall the assumption that users differ only in their distance to the BS. In other words, in case that users have the

*Figure 7. STBC/MIMO: per-user resource allocation of scheduling schemes for different users*

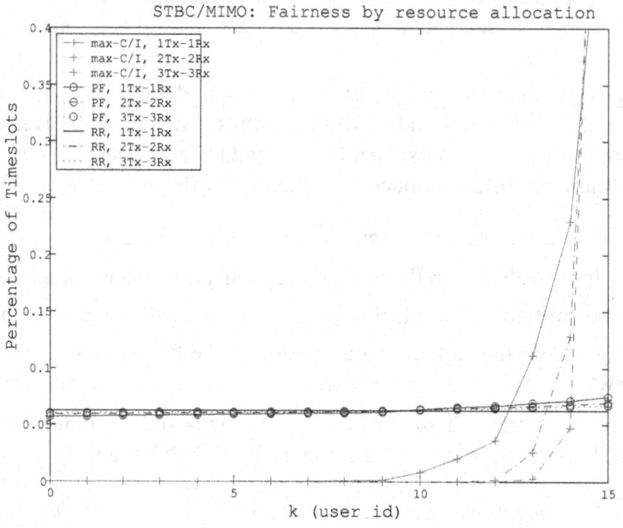


*Figure 8. STBC/MIMO: per-user throughput of scheduling schemes for different users*

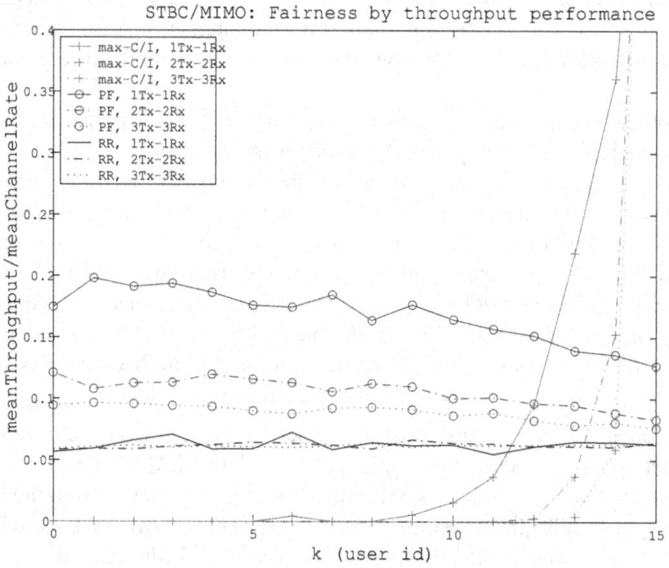

same mean but different channel fluctuations, those of more bursty channels may have higher $\frac{E[\tilde{r}_k]}{E[r_k(t)]}$ and time-fraction than those of smoother ones. Therefore, fairness is not only decided by scheduling schemes, but also user-specific channel characteristics. As a result, schedulers of more generic fairness guarantees than PF should take user-specific channel statistics into account. Again this reflects the necessity of a cross-layer design.

## Impacts of the Power Control Scheme on STBC and Scheduling

*Numerical Results:* Next we numerically present the impact of the power control scheme on STBC channel with different antenna arrays. As we already knew, the channel statistics directly affects the scheduling throughput at the MAC layer. So our focus here is the multiuser diversity gain in terms of BER and FER at physical layer.

First, the impacts of power control scheme on the cdf of effective SNR are plotted in Fig. 9 using Monte Carlo simulations. The analysis is for a 2Tx-1Rx STBC with 1, 5, 10 and 30-fold multiuser diversity. The result of 2-bit power control is also presented as an example of multi-bit power control. As seen from the figure, for a given cdf level of 1%, the effective SNR is enhanced by 9.1, 11.3, 12.6, and 13.1dB for $K=5$, 10, 20, and 30 users. The 1-bit power control scheme gives 0.5 to 1dB gain for each case.

Using the first moment of the effective SNR in (8), Fig. 10 illustrates the average effective SNR with multiuser diversity gain for various configurations, namely 1Tx-2Rx, 2Tx-2Rx, 4Tx-4Rx, and 8Tx-8Rx. The results with 1-bit and 2-bit power control are also plotted. Note that the SNR gain reaches a saturation point for all array sizes when the number of users increases to certain level. The average effective SNR gain with power control is slightly less than the one without power control in each case. It means that the power control scheme reduces multiuser diversity gain slightly if combined with the Max-C/I scheduling. Not shown here due to analytical difficulties, the multiuser diversity loss with PF scheduling can be more significant and may offset the control gain.

As we can see from Fig. 9, contracting the effective SNR towards its mean value, the power control scheme has two impacts. On the one hand, it compensates fading by raising the valley and reducing the peak. At the physical layer this smoothness implies better reliability and smaller FER, as will be shown shortly by detailed simulations. However, when spatial diversity is high, the STBC/MIMO channel is already smoothed by the temporal redundancy in transmission. Then the power control scheme would have smaller impacts. On the other hand, the smaller variations of effective SNR

*Figure 9. The impact of $K$-fold multiuser diversity on the cdf of mean effective SNR: with or without 1-/2-bit power control*

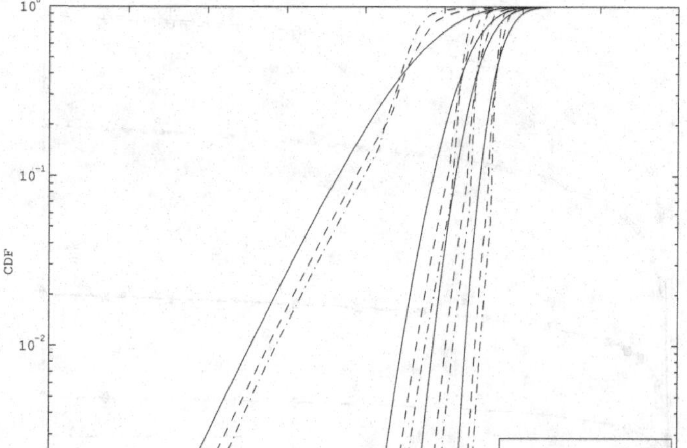

due to power control reduces scheduling throughput at the MAC layer for the less gain from fading exploitation by channel-dependent schedulers. That is the destructive role of power control on throughput performance, which may offset the FER improvement at the physical layer. Thus, such a cross-layer evaluation suggests a conservative advantage to employ the power control scheme, especially considering the associated increase in system complexity. Comparatively schemes like Waterfilling (Holtzman, 2000) may improve the channel reliability without sacrificing the multiuser diversity in the high-rate packet data systems.

*Simulation Scenario:* Now let us evaluate the power control scheme at the physical layer by simulations. We simulate the HDR systems with greedy (Max-C/I) scheduling only, employing two transmit antennas and one receive antenna. As a simple example, we use Alamouti code with the code matrix $\mathcal{G}$ defined by

$$\mathcal{G} = \begin{bmatrix} c_1 & c_2 \\ -c_2^* & c_1^* \end{bmatrix}. \tag{2}$$

The simulation environment mainly follows 3GPP2 (2002). The data rate of coded bits is 1843.2kbps. Each physical layer packet takes exactly one slot. Each active slot of forward channel contains 2048 chips with chip rate 1.2288Mbps. Each 2048-chip consists of 64 preamble, 192 pilot, 256 MAC, and 1536 QPSK data chips. In our simulation, the pilot chips are used for channel estimation, while the preamble and MAC chips are dummy inputs. The slot structure is shown in Fig. 11. The carrier frequency used here is 1.88GHz.

*Simulation Results:* The simulation results in fading channels with different Doppler speeds of 3km/hr, 10km/hr, 40km/hr are illustrated in Fig. 12–14. Fig. 12 illustrates performance curves of FER versus SNR for $K$-fold multiuser diversity, $K$ = 1, 5, and 20, respectively. The results with 1-bit and 2-bit control feedback are also plotted together with the one without power control. The user speed is $V$ = 3km/hr. It is seen that there is more than 5 dB gain for $K$ = 5 and 7dB for 20-fold multiuser diversity at FER=1%. The 1-bit power control gives further 0.5 dB gain for $K$ = 5 and 1dB gain for $K$ = 20. With 2-bit power control, there is 0.5 dB gain for $K$ = 5, while only slightly performance gain for $K$ = 20. This means when $K$ is large, 1-bit control is quite enough to achieve almost all of the performance gain. For $K$ = 1, where only high error region is plotted, the FER performance slightly gets worse using the power control. Fig. 15 illustrates BER performance for $K$–fold multiuser diversity in fading channels with Doppler speed of 3km/hr. From Fig. 15,

*Figure 10. Mean effective SNR gain as a function of multiuser diversity (i.e., user number K) and antenna diversity: with or without power control*

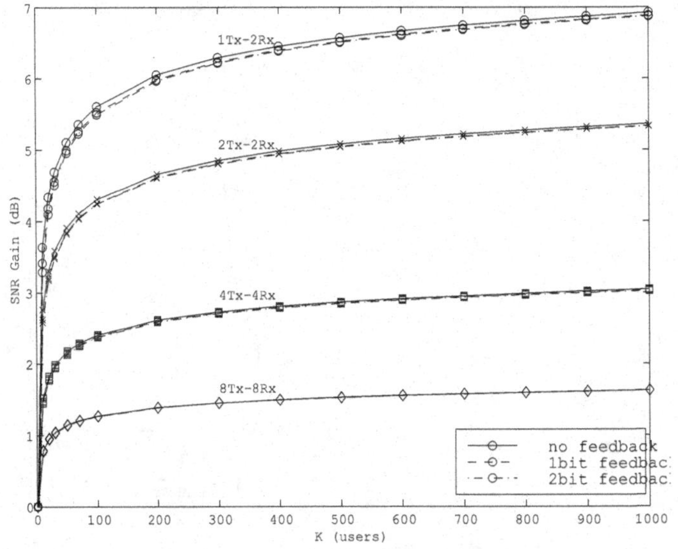

*Figure 11. Forward link active slot structure*

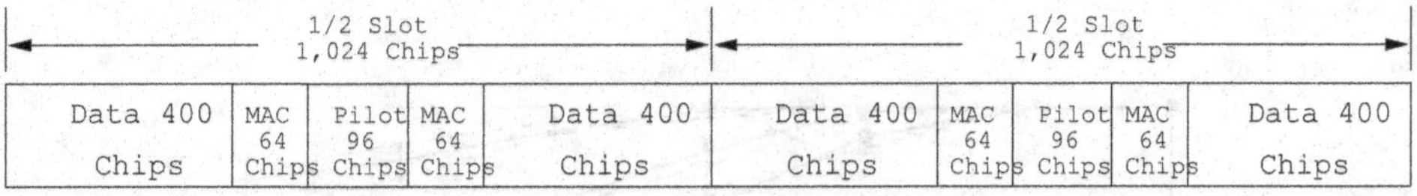


*Figure 12. FER performance of STBC as a function of bit-based SNR with K -fold multiuser diversity: 2Tx-1Rx, QPSK, V =3km/hr*

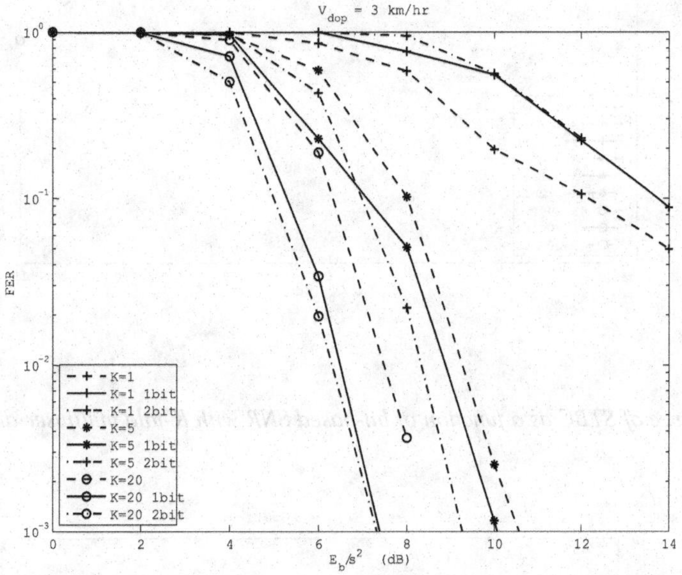


*Figure 13. FER performance of STBC as a function of bit-based SNR with K -fold multiuser diversity: 2Tx-1Rx, QPSK, V =10km/hr*

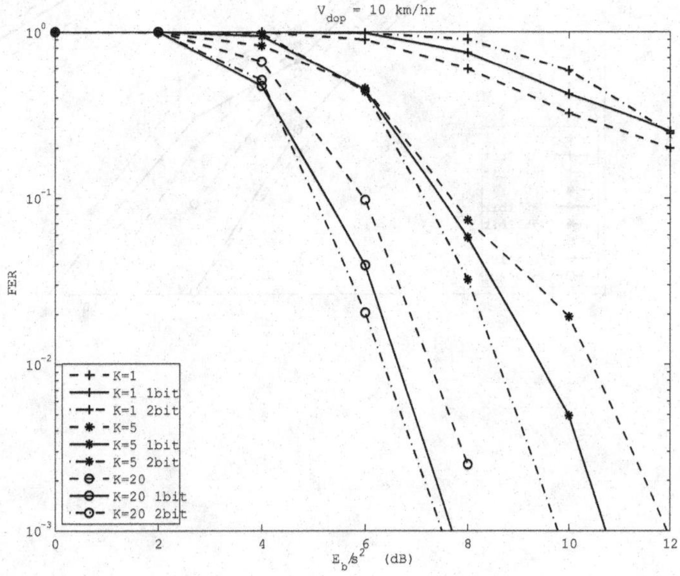

*Figure 14. FER performance of STBC as a function of bit-based SNR with K-fold multiuser diversity: 2Tx-1Rx, QPSK, V=40km/hr.*

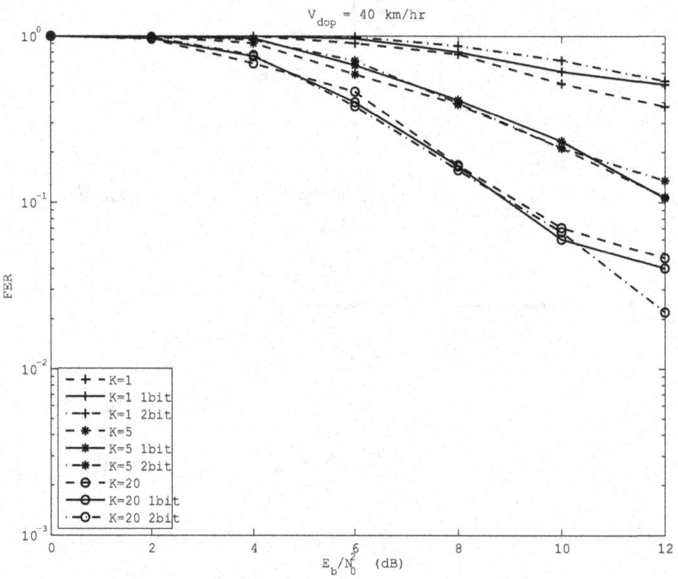


*Figure 15. BER performance of STBC as a function of bit-based SNR with K-fold multiuser diversity: 2Tx-1Rx, QPSK, V=3km/hr*

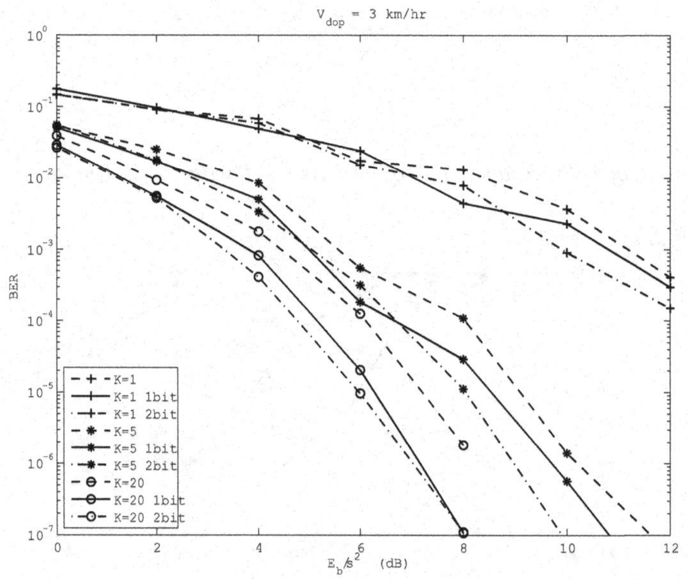

*Figure 16. FER performance of STBC as a function of bit-based SNR with K-fold multiuser diversity and configuration (2Tx-1Rx, QPSK, V = 3km/hr): No power control, off-optimal power control, and off-optimal power control with 5% error rate*

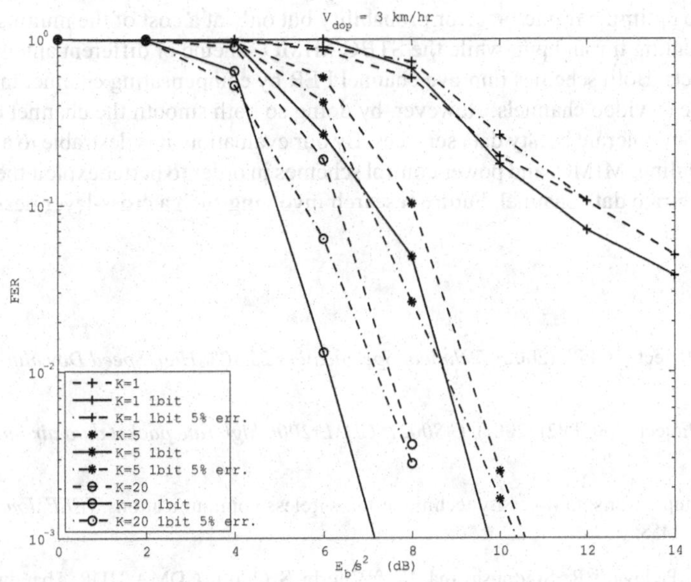


we see that the BER performance with power control is better for almost all the cases including $K = 1$ in the low error regions where performance is concerned. In addition, we note that the power control scheme targets minimum BER instead of FER.

Fig. 13 shows the FER performance results at $V = 10$ km/hr. In this case, the fading is still slow, thus the performance is close to the results in Fig. 12. The power control scheme has 1dB performance gain for 1-bit feedback and 2dB gain for 2-bit feedback when $K = 5$. Unfortunately for a normal BS of $K = 20$, this gain tends to be less.

When $V$ increases to 40km/hr, the fading varies more across consecutive slots. And the assumption of constant fading gain within one slot does not hold any more. The performance is getting worse, as shown in Fig. 14. At this time, the systems performance with the power control scheme is almost the same as the open-loop systems. In other words, the power control scheme has only marginal advantages under disastrous channel conditions.

In the end, we check the performance if the controlled power is fixed at a certain level for all SNR without using the optimal value. The results are shown in Fig. 16, where we fix $E_1 = 1.25E_s$ and $E_2 = 0.75E_s$. This tests the robustness of the power control scheme in three scenarios: no power control, off-optimal power control, off-optimal power control with 5% feedback error. We can see that the control scheme provides certain gains with off-optimal power and 5% control error-rate given $K =$ 5 and $K = 20$. With this advantages at the physical layer, but considering the loss of multiuser diversity gain at the MAC layer and the complexity for introducing such a control, the overall benefit is not significant. It reflects the limited usage of the point-to-point power control scheme in the high-rate downlink data channel.

## CONCLUSION

We evaluated three scheduling algorithms in high-speed packet data cellular systems, taking STBC/MIMO-based CDMA/HDR TDM-shared downlink channel for an example. We also evaluated the point-to-point power control scheme associated with such scheduling schemes with or without Doppler fading. The purpose is to study quantitatively the cross-layer performance of these techniques in terms of throughput, fairness, and FER from the point of view of multiuser diversity gain. The results reveal the performance limitations, and the fundamental reasons behind, for combining the existing scheduling algorithms with power control and STBC/MIMO.

On the one hand, it is shown that the channel-dependent scheduling schemes can enjoy a reduced physical-layer FER and SNR of the shared channel due to multiuser diversity gain. On the other hand, the 1-bit and multi-bit power control scheme is shown to optimize frame/bit error probability but only at a cost of the multiuser diversity gain or the potential MAC-layer scheduling throughput, while the STBC/MIMO scheme of different antenna setups may have the similar double-faced impacts. Both schemes improve channel FER by compensating channel fading, and thus are very suitable for dedicated voice or video channels. However, by doing so both smooth the channel oscillation and thus the potential throughput for delay-tolerant bursty data services. By our evaluation, it is desirable to adopt a more compatible cross-layer design of scheduling, MIMO, and power control schemes in order to better exploit the multiuser and antenna diversity over a shared high-rate data channel. Future research involving such a cross-layer design is underway.

## REFERENCES

3rd Generation Partnership Project (3GPP). (2002) *Technical Specification 25.308: High Speed Downlink Packet Access (HSDPA) (Version 5.2.0).*

3rd Generation Partnership Project 2 (3GPP2) (2001). *C.S0024: CDMA2000 high rate packet data air interface specification (Version 4.0).*

Alamouti, S. M. (1998). A simple transmit diversity technique for wireless communications. *IEEE Journal on Selected Areas in Communications*, 16(8), 1451–1458.

Bender, P., Black, P., Grob, M., Padovani, R., Sindhushyana, N., & Viterbi, S. (2000). CDMA/HDR: a bandwidth-efficient high-speed wireless data service for nomadic users. *IEEE Communication Magazine*. 38(7), 70–77.

Cao, Y., & Li, V. (2001). Scheduling algorithms in broad-band wireless networks. *IEEE Proceedings of the IEEE*. 89(1), 76–87.

David, H. A. (1981). *Order Statistics, (2nd Ed)*. New York: John Wiley & Sons Inc.

Fattah, H., & Leung, C. (2002). An overview of scheduling algorithms in wireless multimedia networks. *IEEE Wireless Communications*, 9(5), 76–83.

Foschini, G. J. (1996). Layered space-time architecture for wireless communication in a fading environment when using multi-element antennas. *Bell Labs Technical Journal*. 1(2), 41–59.

Gozali, R., Buehrer, R. M., & Woerner, B. D. (2002). The impact of multiuser diversity on space-time block coding. *Proceedings of IEEE Vehicular Technology Conference (VTC), fall*, 1, 420–424.

Gozali, R., & Woerner, B. D. (2002). Theoretic bounds on orthogonal design transmit diversity. *Proceedings of IEEE International Symposium on Information Theory (ISIT)*. 304–304.

Hassibi, B., & Hochwald, B. M. (2002). High-rate codes that are linear in space and time. *IEEE Transactions on Information Theory*. 48(7), 1804–1824.

Holtzman, J. M. (2000). CDMA forward link waterfilling power control. *Proceedings of IEEE Vehicular Technology Conference (VTC) spring*. 1663–1667.

Jalali, A., Padovani, R., & Pankaj, R. (2000). Data throughput of CDMA-HDR a high efficiency high data rate personal communication wireless system. *Proceedings of IEEE Vehicular Technology Conference (VTC) spring*. 1854–1858.

Knopp, R., & Humblet, P. A. (1995). Information capacity and power control in single cell multiuser communications. *Proceedings of IEEE International Conference on Communications*. 331–335.

Lu, S., Bharghavan, V., & Sirkant, R. (1999). Fair scheduling in wireless packet networks. *IEEE/ACM Transactions on Networking*. 7(4), 473–489.

Ng, T., Stoica, I., & Zhang, H. (1998). Packet fair queueing algorithms for wireless networks with location-dependent errors. *Proceedings of IEEE INFOCOM*. 3, 1103-1111.

Shenker, S. (1995). Fundamental design issues for the future internet. *IEEE Journal on Selected Areas in Communication*. 13(7), 1176–1188.

Tarokh, V., Jafarkhani, H., & Calderbank, A. R. (1999). Space-time block codes from orthogonal designs. *IEEE Transactions on Information Theory.* 45(5), 1456–1467.

Telatar, I. E. (1999). Capacity of multi-antenna gaussian channels. *European Transactions on Telecommunications*. 10, 585–595.

Viswanath, P., Tse, D. N. C., & Laroia, R. (2002). Opportunistic beamforming using dumb antennas. *IEEE Transactions on Information Theory.* 48(6), 1277–1294.

# Chapter XIX
# Mobile Ad Hoc Networks Exploiting Multi-Beam Antennas

**Yimin Zhang**
*Villanova University, USA*

**Xin Li**
*Villanova University, USA*

**Moeness G. Amin**
*Villanova University, USA*

## ABSTRACT

*This chapter introduces the concept of multi-beam antenna (MBA) in mobile ad hoc networks and the recent advances in the research relevant to this topic. MBAs have been proposed to achieve concurrent communications with multiple neighboring nodes while they inherit the advantages of directional antennas, such as the high directivity and antenna gain. MBAs can be implemented in the forms of multiple fixed-beam directional antennas (MFBAs) and multi-channel smart antennas (MCSAs). The former either uses multiple predefined beams or selects multiple directional antennas and thus is relatively simple; the latter uses smart antenna techniques to dynamically form multiple adaptive beams and thereby provides more robust communication links to the neighboring nodes. The emphases of this chapter lie in the offerings and implementation techniques of MBAs, random-access scheduling for the contention resolution, effect of multipath propagation, and node throughput evaluation.*

## I. INTRODUCTION

Traditional wireless networks require single-hop wireless connectivity to the wired network. Recently, mobile ad hoc networks have yielded considerable advances to support communications among a group of mobile hosts where no wired backbone infrastructure is available (Lal, 2004; Choudhury, 2006; Ramanathan, 2005). User nodes in ad hoc networks traditionally employ omnidirectional antennas, where a transmission on a given channel requires all other nodes in range keep silent or use alternative channels with a different time slot, frequency, or spreading code. As such, the use of omnidirectional antennas does not provide effective channel use and, subsequently, wastes a large portion of the network capacity (Huang, 2002a; Bandyopadhyay, 2006). Incorporation of directional antennas has been proposed to achieve improved network capacity and quality of service. Compared to omnidirectional antennas, directional antennas have higher directivity and antenna gain. Therefore, directional antennas not only significantly reduce the power necessary for the service coverage and packet transmission, but also mitigate the interference in the directions away from that of the desired users. As a result, the use of directional antennas provides a platform to serve increased number of nodes and network throughput. The antenna gain due to directional transmission and reception enables extended communication range of each hop, thereby reducing the number of hops between distant source and sink nodes, and increasing the efficiency and reliability of the network (Ko, 2000; Nasipuri, 2000; Wang, 2002; Zhang, 2005).

A directional antenna with a single beam, however, does not fully utilize the offering of multi-sensor systems. In addition, the deployment of directional antennas may result in new problems. For example, the deafness problem appears when a node is tuned to a specific direction and thus cannot hear a node in another direction, even they are closely located. The deafness problem not only impedes dynamic resource allocation, but also increases the possibility of network outage for certain services (Choudhury, 2004; Jain, 2006a). To mitigate the deafness problem and enhance the network capacity, multi-beam antennas (MBAs) have been proposed to achieve concurrent communications with multiple neighboring nodes while inheriting the advantages of directional antennas, such as the high directivity and antenna gain. MBAs can be implemented in the forms of multiple fixed-beam directional antennas (MFBAs) and multi-channel smart antennas (MCSAs). To form multiple fixed-beams, MFBAs and multiple radios (MRs) with a directional antenna equipped in each radio can be exploited (Bahl, 2004; Draves, 2004). As a result, high network throughput can be achieved. In a stationary environment, the antenna patterns can be optimized to further improve network performance. However, the performance of MFBAs and MRs degrades in a time-varying multipath propagation environment, which is typically experienced in indoor and low-altitude outdoor wireless networks (Winters, 2006).

Another approach to implement MBAs is to use MCSAs (Singh, 2005; Zhang, 2006; Li, 2007). By using smart antenna techniques, multiple beams can be adaptively and dynamically formed by a node so as to provide robust communication links with multiple users. At the expense of higher complexity, an MCSA-based approach takes the same advantages as the MFBA implementation, but its performance does not degrade in time-varying multipath environment (Zhang, 2006; Li, 2007).

The purpose of this chapter is to discuss the recent advances of MBA approaches for wireless ad hoc network applications. To bridge the gap between omnidirectional antennas and MBAs, the concept and offerings of ad hoc networks with directional antennas are first reviewed and a brief introduction of the medium access control (MAC) protocols and routing approaches developed for directional antennas is provided. Beamforming techniques and random-access scheduling (RAS) schemes in the contention resolution are then introduced. The respective node throughput performance and probability of concurrent communications are examined using a simplified ideal sector-based model as well as a precise output signal-to-interference-plus-noise ratio (SINR) based model.

This chapter is organized as follows. Section II reviews the concept of directional antennas as well as the associated MAC protocols and routing schemes for ad hoc networks. Section III discusses multi-channel beamforming techniques in detail, including adaptive multi-channel beamforming, fixed-beam antennas, and the analysis of output SINR performance. Section IV provides RAS schemes respectively based on the prioritized packet delivery and throughput maximization criteria, where two different models, respectively based on idealized sectors and the output SINR, are considered. The analysis and numerical evaluation of node throughput performance of the two RAS schemes in single-path and multipath environments are presented in Sections V and VI, respectively. Relevant issues to the MBAs are addressed in Section VII to broaden understanding of this topic. Finally, the conclusion of this chapter and some important remarks are provided in Section VIII.

## II. AD HOC NETWORKS WITH DIRECTIONAL ANTENNAS

### Concept and Offerings

A directional antenna is typically implemented using the switch beam scheme, where a set of predefined beams are formed and the one that best receives the signal from a particular desired user is selected. It is relatively simple in terms of hardware implementation and processing complexity. As such, it has become a conveniently accessible and adoptable technology for use in wireless LANs and ad hoc networks (Bandyopadhyay, 2006).

In an ad hoc network, co-channel interference is one of the key factors that limit the overall network capacity and quality. Refer to Fig. 1(a). When nodes S and D communicate using omnidirectional antennas, all other nodes depicted in this figure are within the respective ranges of S and D and, therefore, should remain silent to avoid co-channel interference. When the network is multihop, another key limiting factor is the forwarding burden of the intermediate nodes which increases with the number of hops.

Directional antennas can enhance the network capacity by reducing the above two limiting factors. As shown in Fig. 1(b), when directional antennas are used, multiple parallel links can be constructed without interfering to each other. Yi (2003) has shown that, due to the reduction of the interference area, the capacity gain can be increased with an improvement factor inversely proportional to the beamwidth of the transmit and receive antennas. Furthermore, due to the beamforming gain, the use of directional antennas yields a longer communication range and a higher receive signal level, leading to improved power efficiency, increased signal quality, and reduced number of hops. For example, nodes X and K may not have direct link when they are separated beyond the communication range corresponding to omnidirectional antennas, whereas they can be directly linked when the antennas are directional. Thus, the use of directional antennas provides improved routing performance as well as enhanced capacity (Choudhury, 2003; Ramanathan, 2005; Das, 2006).

## MAC Protocols Using Directional Antennas

The node capability enhancement using directional antennas can be effectively leveraged only through appropriate changes to higher layer network protocols. Below, we summarize MAC protocol schemes that take directional antennas into account.

So far, a variety of directional antenna-based MAC protocols have been developed (e.g., Dai, 2006; Korakis, 2003; Jurdak, 2004). These MAC protocols can be classified into two major categories: random access and scheduling access. Random access based protocols can be further classified into different collision avoidance approaches: 1) pure-RTS/CTS protocols; 2) tone-based protocols; and 3) other protocols using additional control packets (Dai, 2006). Refer to Fig. 1(a), RTS/CTS (Request-to-Send/Clear-to-Send) based protocols using an omnidirectional antenna reserve the wireless media over a large area. Thus, effective network capacity is not achieved. Several directional MAC (DMAC) schemes have been proposed to take the advantages of the directivity of antennas. The DMAC proposed in (Nasipuri, 2000) is based on omnidirectional RTS and omnidirectional CTS (oRTS/oCTS). As such, the neighboring nodes sense the communication links to avoid collision in a fashion similar to omnidirectional MAC algorithms. This protocol does not assume the *a prior* knowledge of each node's location information, and the respective direction of the senders can be estimated from the beam position corresponding to the strongest signal power of the oRTS and oCTS packets. The co-channel interference is reduced by directionally transmitting and receiving data packets. When there are ongoing communication links around the sender, however, collision may occur if the sender initiates an oRTS. In this situation, sending directional RTS (DRTS) enables the establishment of transmission links in the unblocked directions (Ko, 2000). Directional transmission can be used in both RTS and CTS between a pair of nodes (Bandyopadhyay, 2001). In this case, neighboring nodes may not be aware of the communication link between the pair, and thus the deafness problem

*Figure 1. Ad hoc networks using omnidirectional and directional antennas*

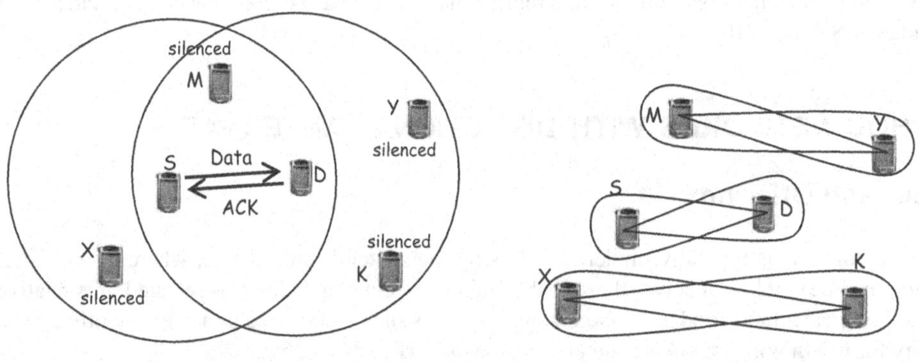


(a) Omnidirectional antennas (b) Directional antennas

may occur. In the directional virtual carrier sensing (DVCS) protocol (Takai, 2002), each node caches the estimated directions-of-arrival (DOAs) of all its neighboring nodes when it hears any signal from them. The RTS is transmitted directionally if the location information of a neighboring node is available. The directional network allocation vector (DNAV) is used to indicate the directions reserved by neighboring nodes. The multihop MAC (MMAC) proposed by Choudhury *et al* (2006) incorporates the DMAC and DVCS protocols, exploits the extended transmission range provided by directional antennas and, depending on the channel status, establishes directional-omni (DO) or directional-directional (DD) wireless links. In the circular-DMAC protocol (Korakis, 2003), a DRTS is transmitted consecutively at each switch beam and the location information of the neighboring nodes is recorded. Such information can be used to solve the hidden terminal and deafness problems.

The dual busy tone multiple access protocol with directional antennas (DBTMA/DA) (Huang, 2002b) is a tone-based directional MAC protocol that uses transmission and reception busy tones to notify neighboring nodes of the channel use. A node defers to transmit/receive when a busy tone is sensed. The toneDMAC (Choudhury, 2004) uses a time slot to transmit the tone signal. As such, it simplifies the system as transmission is needed only at a single frequency band. For the DOA-MAC protocol (Singh, 2004), each time frame contains three miniframes, respectively for tone transmission, packet transmission, and acknowledgement (ACK) transmission. In the tone transmission period, each transmitter sends a tone to its intended receivers. The receivers, with the help of DOA estimation algorithms, then point their respective beams towards the sender.

Practically, it is often not feasible for a node to receive signals from neighboring nodes when it makes transmission to other nodes. As such, for an MBA, synchronization across the active beams connecting to neighboring users is required, in addition to the use of directional transmission and reception. The explicit synchronization via intelligent feedback (ESIF) protocol (Jain, 2006a/b) is a random access based DMAC protocol for MBAs to achieve node synchronization with the use of embedded feedback information. By using a control packet (SCH), it is desirable to mitigate the deafness problem.

Another category of directional antenna based MAC protocols exploits scheduling. For example, the receiver-oriented multiple access (ROMA) protocol can simultaneously form multiple beams for the transmission or reception through proper scheduling (Bao, 2002). In each time slot, the active nodes are equally divided into transmitters and receivers, and they couple together in pairs to maximize the throughput. Either end of the transmission can use a directional mode. In the directional transmission and reception algorithm (DTRA) (Zhang, 2005), each time frame is divided into three subframes: neighbor discovery and handshaking period, connection confirmation and data reservation period, and data transmission period. The location information among neighboring nodes is exchanged via directional scanning.

## Routing

In an ad hoc network, each node has a limited transmission range. Two nodes cannot communicate directly when they are separated beyond their node transmission range. In this case, multihop communication becomes necessary with some intermediate nodes acting as routers (Rajaraman, 2002). In addition, an ad hoc network may experience rapid and unpredictable topology change, as the nodes move with an arbitrarily pattern. Consequently, proper network routing, i.e., the determination of the path of data delivery from one node to another, becomes an important issue in ad hoc networks. Routing protocols have to fulfill two major tasks: route discovery and route maintenance. While routing protocols in ad hoc networks resemble those developed for wired networks, they should consider the special limitations of ad hoc networks, such as limited bandwidth, highly dynamic topology, and limited range of links. Routing protocols available for ad hoc networks are either of reactive, proactive, or fixed nature (Bandyopadhyay, 2006). Traditional network routing protocols like the destination sequenced distance vector (DSDV) are proactive, where they maintain a route to all nodes within the network, including those to which no packets are to be sent. They also react to dynamic topology changes, even if these changes have no effect on the traffic. Reactive routing protocols, like the dynamic source routing (DSR) and the ad hoc on-demand distance vector (AODV), on the other hand, only react when a route is needed between the source and destination nodes. They do not maintain the routes to the nodes that they are not communicating with. Lang (2003) provides a comprehensive overview of the ad hoc network routing protocols when the nodes are equipped with omnidirectional antennas.

When directional antennas are used, directional routing algorithms have been developed to improve the spatial reuse. Most existing directional routing schemes either assume a complete network topology beforehand or use omnidirectional routing schemes to forward packets in a directional environment. For example, the directional routing protocol (DRP) (Gossain, 2006) is an on-demand directional routing protocol for single switch beam antennas. This protocol assumes a cross-layer interaction between the routing and MAC layer, and includes an efficient route discovery mechanism, establishment and maintenance of the directional routing table (DRT) and directional neighbor table (DNT), and novel directional route recovery mechanisms. Simulation results show that the DRP considerably improves the packet delivery ratio and decreases the end-to-end packet latency. When MBAs are employed, it is desirable that the directional routing protocols support concurrent data links at a node. Development of routing protocols for MBAs still remains an open issue.

## Security

With lack of infrastructural support, security in ad hoc networks becomes inherently vulnerable to susceptible wireless link attacks. Achieving security within an ad hoc network is a challenging problem due to several reasons (Zhou, 1999). 1) Dynamic topologies and membership, in which the network topology of ad hoc network may be dynamic because the mobility and the membership of nodes can be random and time-varying. 2) Vulnerable wireless link, in which passive/active link attacks like eavesdropping, spoofing, denial of service, masquerading, impersonation are possible. 3) Roaming in dangerous environments, characterized by any malicious node or misbehaving node that might create hostile attacks or deprive all other nodes from providing any service.

Secure communication among nodes requires secure communication links. Before establishing a secure communication link, a node should have the capability of identifying another node by virtue of the identity and the associated credential information, which needs to be authenticated and protected so that the authenticity and integrity cannot be questioned. Thus, it is essential to provide security architecture and a seamless privacy protection to harness the use of ad hoc networks. The deployment of directional antennas and MBAs provides another degree-of-freedom, i.e., spatial dimension, to strengthen the security of ad hoc networks (Hu, 2004; Caballero, 2006; Wu, 2007).

# III. MULTI-CHANNEL BEAMFORMING TECHNIQUES

In this section, we introduce multi-channel transmit and receive beamforming techniques. MBA implementations with MFBAs and MCSAs are provided. The performance of MCSAs and MFBAs, in terms of throughput gain and output SINR, is respectively examined.

## Multiple Fixed-Beam Antennas and Multi-Channel Smart Antennas

In MFBAs, multiple active beams can be selected from the predefined beams, whereas in MRs, each radio is equipped with its own predefined directional antenna (Bahl, 2004). Both directional structures achieve concurrent communications with multiple users in addition to inheriting the advantage of the switch beam antennas. When the propagation environment is stationary, beam design and allocation can be optimized to provide high network throughput. Beam optimization, however, becomes impractical in a mobile wireless network where the channels are time-varying. More importantly, in a multipath propagation environment, which is typically the case in indoor and low-altitude outdoor wireless networks (Winters, 2006), the paths originated from the same neighboring node are likely to occupy multiple beams, increasing the probability of collisions (see Fig. 2(a)) (Zhang, 2006). Neither MFBA nor MR is effective in accommodating a multipath propagation environment, even with sophisticated MAC and higher-layer controls.

MCSAs, on the other hand, are flexible in beam steering and thus support concurrent communications with multiple nodes with a substantially reduced probability of collision (see Fig. 2(b)). In contrast to MFBAs, MCSAs can be designed to provide multi-fold advantages (Zhang, 2006): (1) They can adaptively form nulls in the directions of interfering signals (co-channel interferers or jammers). (2) When a signal arrives with multiple paths, they achieve spatial diversity without exhausting additional degree-of-freedom. (3) They can tradeoff array gain, spatial multiplexing gain, and interference mitigation gain so as to obtain the optimum transmission performance. (4) They allow coexistence of dedicated channels to specified users as well as shared standby channels for other active and potential users, thus eliminating the deafness problem at no additional cost of overload/resource and dynamically optimizing the resource allocation among different data streams.

## Adaptive Multi-Channel Beamforming

A node equipped with an MCSA in the network can form $M_a$ beams to separate up to $M_S$ spatially independent signal streams through electronic steering of an array consisting of $M$ omnidirectinal antennas or the use of $M$ antennas with different directivities, where $M_s \le M_a \le M$. In the following, we consider multi-channel transmit beamforming and multi-channel receive beamforming using $M$ omnidirectional antennas with appropriate array weights.

## Multi-Channel Transmit Beamforming

Assume that a node has $M_S$ signal streams to be simultaneously sent to different neighboring nodes, and the signal vector corresponding to these streams at time $t$ is denoted as $\mathbf{d}(t) = [d_1(t) \cdots, d_{M_S}(t)]^T$, where $(\cdot)^T$ denotes transpose. The streams may have different modulations to provide appropriate data rates according to the required data size, priority class, and channel quality. We only consider flat-fading environment and, therefore, the time index $t$ is omitted from all subsequent expressions for notational simplicity. The transmitted signal vector can be expressed as

$$\mathbf{x}_s = \mathbf{A}\mathbf{Q}^{\frac{1}{2}}\mathbf{d}, \tag{1}$$

where $\mathbf{A} \in \mathbb{C}^{M \times M_S}$ is the beamformer matrix (to be determined later), $\mathbf{Q} = \mathrm{diag}(q_1, \cdots, q_{M_S})$ is the power loading matrix subject to the total power constraint $\mathrm{tr}(\mathbf{Q}) = \sum_{i=1}^{M_S} q_i = P$, with $P$ denoting the total transmit power.

The multi-channel transmit beamforming problem can be solved by either linear or nonlinear techniques (Costa, 1993; Peel, 2005). Assume that the transmit node has the channel state information (CSI), denoted as $\mathbf{H}_T \in \mathbb{C}^{M_S \times M}$. If a linear zero-forcing technique is used, the beamforming matrix can be obtained as (Peel, 2005)

$$\mathbf{A} = \frac{1}{\sqrt{\lambda}}\mathbf{H}_T^H\left(\mathbf{H}_T\mathbf{H}_T^H\right)^{-1}, \tag{2}$$

where $\lambda = \mathrm{tr}\left[\left(\mathbf{H}_T\mathbf{H}_T^H\right)^{-1}\right]$, $(\cdot)^H$ denotes conjugate transpose, and $\mathrm{tr}(\cdot)$ denotes matrix trace.

## Multi-Channel Receive Beamforming

Consider a target node (TN) with receivers designed to simultaneously receive up to $S_R \le M$ signals. Assume that there are $n \le S_R$ neighboring users communicating with the TN. Denote $s_i$ and $\mathbf{u}_i \in \mathbb{C}^{M \times 1}$ as, respectively, the signal stream transmitted from the $i$th source node and the corresponding equivalent channel vector, where $i$=1, ..., $n$. Then, the $M$×1 received signal vector at the TN is given by

$$\mathbf{x} = \sum_{i=1}^{n}\mathbf{u}_i s_i + \mathbf{n} = \mathbf{U}_R\mathbf{s} + \mathbf{n}, \tag{3}$$

where $\mathbf{U}_R = [\mathbf{u}_1, \cdots, \mathbf{u}_n] \in \mathbb{C}^{M \times n}$, $\mathbf{s} = [s_1, \cdots, s_n]^T \in \mathbb{C}^{n \times 1}$, and $\mathbf{n}$ is the noise vector.

We can design a set of weights for the TN to simultaneously receive and separate the $n$ signals. The well-known optimum weights in the minimum mean square error (MMSE) sense, assuming that the CSI can be estimated at the TN, is given by (Monzingo, 1980; Winters, 1984)

$$\mathbf{W}_0 = \left(\sigma^2\mathbf{I} + \mathbf{R}_\mathbf{x}\right)^{-1}\mathbf{U}_R \in \mathbb{C}^{M \times n}, \tag{4}$$

where $\sigma^2$ is the noise variance in each receive channel, $\mathbf{I}$ is the identity matrix with an appropriate dimension, and $\mathbf{R}_\mathbf{x} = E(\mathbf{x}\mathbf{x}^H) \in \mathbb{C}^{M \times M}$ is the covariance matrix of the received data. It is noted that an improved detection performance can be achieved using optimum or suboptimum multi-user detection methods (Verdu, 1998).

To avoid the deafness problem, a standby channel can be secured using a dedicated or shared spatial channel so that the TN can hear new communication requests (Zhang, 2006). The standby channel can either be omnidirectional or perform directional scanning.

*Figure 2. MFBA and MCSA in a multipath environment*

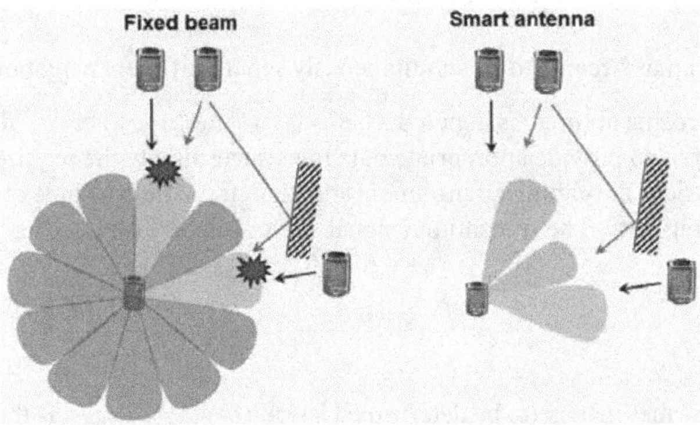


(a) MFBA (b) MCSA

## Fixed-Beam Antennas

MFBAs form multiple fixed beams in different directions. When there is no *a priori* information of the neighboring users to be communicated, as in a typical wireless ad hoc network, the beams are designed to cover the entire azimuth plane. Some wireless networks, such as fixed mesh networks, are designed to serve certain sectors or users located in specific directions and, thus, the beams are designed only to cover these directions.

In this chapter, we consider that wireless nodes are mobile and their connections are dynamic. Therefore, the beams should be equipped with the capability of serving all the azimuth directions. Such beams can be implemented using multiple directional antennas, which collectively serve the entire azimuth plane, or an array of omnidirectional antennas with multiple sets of array weights. We use the array model, where each set of array weights corresponds to a directional array pattern (Li, 2007). For example, an array of $M$ beams can be designed to have a half-power beamwidth (HPBW) of $2\pi/M$. Adjusting the array configurations may alter the array pattern and, as such, changes the tradeoff between the signal transmission/reception quality and the interference mitigation capability in different spatial locations.

## Output SINR Analysis

The output SINR of MFBAs and MCSAs is first analyzed in a single-path environment, and the results are then generalized to the case of multipath propagation.

### Single-Path Propagation Environment

A single or multiple active neighboring nodes (ANNs) make transmission to a TN equipped with multiple beams. In the absence of a proper contention resolution scheme, e.g., without the use of substantial data coding or spectrum spreading, it is practical to consider that, at each TN beam, only one ANN signal can achieve sufficiently high output SINR for successful data decoding. For a specific TN beam, therefore, one signal is considered as the signal-of-interest (SOI), and others become signals-not-of-interest (SNOIs). Generally, we assume that, at a certain time, there are $n$ ANNs, i.e., one SOI and $n$–1 SNOIs with respect to a node. Denote $s_d$ and $s_j$ as the SOI and the $j$th SNOI, respectively, with $E[|s_d|^2]=E[|s_j|^2]=1, j=1,\ldots, n-1$. Then, the received signal vector $\mathbf{x}$ in (3) can be rewritten as

$$\mathbf{x} = \mathbf{x}_d + \mathbf{x}_I + \mathbf{n} = \mathbf{u}_d s_d + \sum_{j=1}^{n-1} \mathbf{u}_j s_j + \mathbf{n}, \quad (5)$$

where $\mathbf{x}_d$, $\mathbf{x}_I$, and $\mathbf{n}$ are respectively the SOI, SNOI, and noise vectors, and $\mathbf{u}_d$ and $\mathbf{u}_j$ are the propagation vectors of the SOI and *j*th SNOI.

Assume that the SOI, the SNOIs, and the noise are statistically independent. In addition, the elements of the noise vector $\mathbf{n}$ are assumed to be independent and identically distributed (i.i.d.) complex Gaussian processes. Furthermore, the channels are assumed to be quasi-stationary, i.e., they do not change over the time interval of an array processing operation but impose random variation over a long time period. Then, for each SOI, we can properly define the covariance matrix of the received data as $\mathbf{R}_x = E[\mathbf{x}\mathbf{x}^H]$, and the covariance matrix of the interference-plus-noise as $\mathbf{R}_{IN} = E[(\mathbf{x}_I+\mathbf{n})(\mathbf{x}_I+\mathbf{n})^H] = \mathbf{R}_I + \sigma^2\mathbf{I}$, where $\mathbf{R}_I = E[\mathbf{x}_I\mathbf{x}_I^H]$ is the covariance matrix of the SNOIs.

For an MCSA, the beam for the SOI can be formed with the optimum weight vector in the MMSE sense, given by (Monzingo, 1980; Winters, 1984)

$$\mathbf{w}_o = \mathbf{R}_{IN}^{-1}\mathbf{u}_d, \tag{6}$$

and the corresponding output SINR is

$$\gamma = \mathbf{u}_d^H \mathbf{R}_{IN}^{-1}\mathbf{u}_d. \tag{7}$$

For an MFBA, its beams are predefined and a free beam can be selected to receive the SOI. Denote $\mathbf{w}_b$ as the corresponding weight vector of the beams corresponding to the SOI. Then, the output SINR can be written as

$$\gamma^{FBMA} = \frac{\mathbf{w}_b^H \mathbf{u}_d \mathbf{u}_d^H \mathbf{w}_b}{\mathbf{w}_b^H \mathbf{R}_{IN} \mathbf{w}_b}, \tag{8}$$

which, in general, is not optimum. Therefore, an MFBA may suffer from an SINR loss, as opposite to an MCSA which maximizes the output SINR.

As an example of an MFBA (Li, 2007), a simple coverage scheme is shown in Fig. 3(a), where four fixed beams are formed using a uniform circular array (UCA) consisting of four omnidirectional antennas with an array radius of 0.235 $\lambda$, where $\lambda$ is the carrier wavelength. This array pattern covers the entire azimuth plane, where each beam has a HPBW of 90° and a maximum sidelobe gain of −15.5 dB. The beam that receives the highest signal level from a desired ANN is selected as the receive beam. For example, the beam with direction of maximum gain toward 90° is selected if the incident wave falls into the angular range over [45°, 135°].

On the other hand, an MCSA consists of an array followed by *M*-channel adaptive processing circuitries to form up to *M* dynamic beams adapted to the incident waves. Hence, the angular width and position of each beam is reconfigured real-time. For a fair comparison, we consider an MCSA consisting of an identical UCA with *M*=4 omnidirectional antennas and an array radius of 0.235 $\lambda$.

Consider the case of two ANNs. The SOI arrives with a DOA of 30°, whereas that of the SNOI is 40°. Assume that the received power at the TN corresponding to each ANN is the same and the resulting input SNR is 20 dB. For the MFBA, the beam with the maximum gain towards 0° is selected to receive the SOI, shown as the solid line in Fig. 3(a). The output SINR is obtained as merely 1 dB. Obviously, a collision between the SOI and SNOI occurs in this beam. On the contrary, when the MCSA is used, a dynamic beam is formed as shown in Fig. 3(b). Although the two signals are closely spaced, a null is formed and directed toward the direction of the SNOI, resulting in an output SINR of 11.5 dB. As such, the advantage of an MCSA over an MFBA is evidently demonstrated.

For MFBA, the output SINR is 1 dB. (b) For MCSA, the output SINR is 11.5 dB.

## Multipath Propagation Environment

A wireless channel often experiences multipath propagation, i.e., the TN receives not only the direct path of the transmitted signal, but also its reflected signals that propagate over other paths. Multipath propagation is a typical problem in many wireless systems. Specifically, the multipath propagation phenomenon may become even richer (with a higher angular spread) in ad hoc networks, since the nodes are typically located in indoor or low-altitude outdoor environments (Winters, 2006; Babich, 2006).

It is well known that, in a frequency-nonselective multipath environment, the paths arriving from the same ANN form an equivalent path with a generalized steering vector, often referred to as the spatial signature (Lin, 1982). For example, consider one signal, whose waveform is represented as $s_1(t)$. When this signal arrives through $K$>1 paths, the received signal vector is expressed as $\sum_{i=1}^{K} \mathbf{u}_{1,i} s_1 = \tilde{\mathbf{u}}_1 s_1$, where $\mathbf{u}_{1,i}$ is the channel vector corresponding to its $i$th path. Therefore, it becomes obvious that the contribution of the $K$ paths is equivalently represented by a single spatial signature $\tilde{\mathbf{u}}_1$ which, in general, does not have an angular bearing.

Using the above equivalent model, the received signal vector in a multipath environment can be expressed as

$$\tilde{\mathbf{x}} = \tilde{\mathbf{u}}_d s_d + \sum_{j=1}^{n-1} \tilde{\mathbf{u}}_j s_j + \mathbf{n}, \tag{9}$$

where $\tilde{\mathbf{u}}_d = \sum_{i=1}^{N_d} \mathbf{u}_{di}$ and $\tilde{\mathbf{u}}_j = \sum_{k=1}^{N_{Ij}} \mathbf{u}_{j,k}$ are the spatial signatures of the desired and the $j$th SNOI propagation vectors, respectively, $N_d$ is the number of paths of SOI, and $N_{Ij}$ is the number of paths of the $j$th SNOI. Similar to the single-path case, we can properly define $\mathbf{R}_{\tilde{x}} = E\left[\tilde{\mathbf{x}}\tilde{\mathbf{x}}^H\right]$ as the covariance matrix of the received signal samples, and $\mathbf{R}_{\tilde{I}} = E\left[\tilde{\mathbf{x}}_I \tilde{\mathbf{x}}_I^H\right]$ and $\mathbf{R}_{\widetilde{IN}} = E\left[\left(\tilde{\mathbf{x}}_I + \mathbf{n}\right)\left(\tilde{\mathbf{x}}_I + \mathbf{n}\right)^H\right] = \mathbf{R}_{\tilde{I}} + \sigma^2\mathbf{I}$ as that of the SNOIs and interference-plus-noise, respectively. For the MCSA, the optimum beam in the MMSE sense can be formed as

$$\tilde{\mathbf{w}}_o = \mathbf{R}_{\widetilde{IN}}^{-1} \tilde{\mathbf{u}}_d, \tag{10}$$

and the corresponding output SINR can be expressed as

$$\tilde{\gamma} = \frac{\tilde{\mathbf{w}}_o^H \tilde{\mathbf{u}}_d \tilde{\mathbf{u}}_d^H \tilde{\mathbf{w}}_o}{\tilde{\mathbf{w}}_o^H \mathbf{R}_{\widetilde{IN}} \tilde{\mathbf{w}}_o} = \tilde{\mathbf{u}}_d^H \mathbf{R}_{\widetilde{IN}}^{-1} \tilde{\mathbf{u}}_d \tag{11}$$

In contrast, for the MFBA, a free beam is selected to receive the SOI. Due to the fixed-beam nature, the weights $\mathbf{w}_{b'}$ used in a multipath environment are the same as that used in a single-path propagation environment. Consequently, the output SINR of the MFBA is given by (Li, 2007)

$$\tilde{\gamma}^{FBMA} = \frac{\mathbf{w}_{b'}^H \tilde{\mathbf{u}}_d \tilde{\mathbf{u}}_d^H \mathbf{w}_{b'}}{\mathbf{w}_{b'}^H \mathbf{R}_{\widetilde{IN}} \mathbf{w}_{b'}}. \tag{12}$$

Again, the solution of an MFBA is not optimal and is expected to provide inferior performance to an MCSA counterpart in a multipath environment.

*Figure 3. Comparison between the array pattern of the MFBA and MCSA (SOI: 30°, SNOI: 40°). (a) For MFBA, the output SINR is 1 dB. (b) For MCSA, the output SINR is 11.5 dB.*

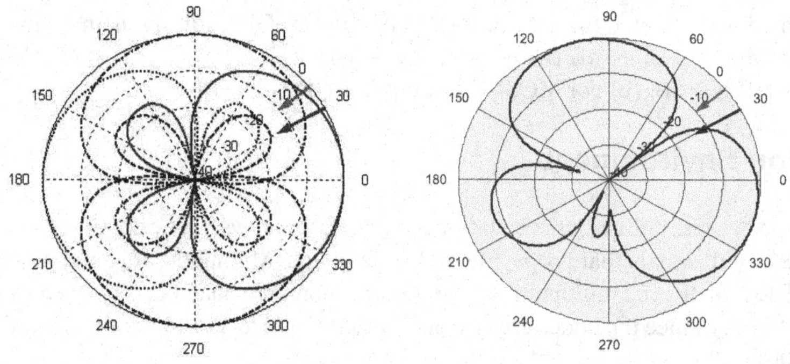


*(a)* *(b)*

As an example (Li, 2007), we present a performance comparison between an MFBA and an MCSA operating in a multipath propagation scenario as depicted in Fig. 4(a), where each of the four ANNs has two signal paths. Consider that the SOI arrives at –60° and 0°, whereas the three SNOIs arrive at 30° and 90°, 120° and 180°, and 210° and 270°, respectively. Assume that each path has the same receive power with an input SNR of 20 dB evaluated at the TN. The antenna array model depicted in Fig. 3 is used. For the MFBA, the beam with maximum gain toward 0° is selected to serve the SOI, and the resulting output SINR is very low (3.9 dB). A rather promising result is obtained when the MCSA is used, where a dynamic beam is formed as shown in Fig. 4(b) for the SOI, and an output SINR of 24.5 dB is yielded.

In this example, the four ANNs have the same output SINR due to the symmetric incidence. Therefore, the MCSA can provide high quality links to concurrently communicate with all four ANNs, whereas the MFBA fails.

## IV. RANDOM-ACCESS SCHEDULING TECHNIQUES

In an ad hoc network exploiting MBAs, it is crucial to use a proper RAS scheme in contention resolution to coordinate the node access and resolve the contention (Li, 2004). Although several control schemes have been proposed for wireless networks (Bao, 2002; Choudhury, 2002), few are feasible for MBAs, particularly when operating in a multipath propagation environment (Jain, 2006b). This section discusses RAS techniques that support MBAs in both single-path and multipath propagation environments.

Without loss of generality, consider a scenario where a TN has *N* neighboring nodes, which are independently and randomly located around the TN with a uniform angular distribution. In a single-path propagation environment, the signal transmitted from each ANN falls into only one beam of the TN. In a multipath environment, on the other hand, the received signals transmitted from an ANN may fall into multiple beams of the TN. Regardless of the propagation environment, when multiple signal arrivals originated from different ANNs fall into the same beam, collisions occur in the beam. As a result, this beam cannot successfully receive packets and thus does not contribute to the node throughput. Only those beams that successfully receive packets contribute to the node throughput. It is evident that, in the absence of a proper contention resolution scheme, collisions may occur as a result of random packet transmission. Furthermore, multipath propagation is likely to yield more frequent collisions. When the contention resolution scheme utilizes proper RAS, collisions can be avoided and, as a result, network performance can be significant improved.

To take advantages of concurrent link capability of MBAs achieved through the exploitation of multiple beams, we focus on RAS schemes which are incorporated into contention resolution to utilize the spatial dimensionality. Assume that all nodes in a region of interest share the same wireless channel (i.e., the same frequency, code, or time channel). Moreover, time is assumed to be slotted and all data packets have the same length *T*. Data packets, including both newly generated and retransmitted ones, arrive at each node according to a Poisson process with an arrival rate $\lambda$ packets/sec. Thus, the corresponding offered load of each node is $R=\lambda T$. Each arrived packet is intended for a single TN. In order to illustrate the performance of RAS schemes, we focus on the one-hop case.

*Figure 4. Comparison of the array pattern of the MFBA and MCSA operating in multipath environment in the presence of four ANNs (SOI: –60° and 0°; SNOI 1: 30° and 90°; SNOI 2:120° and 180°; SNOI 3: 210° and 270°). (a) For MFBA, the output SINR is 3.9 dB. (b) For MCSA, the output SINR is 24.5 dB.*

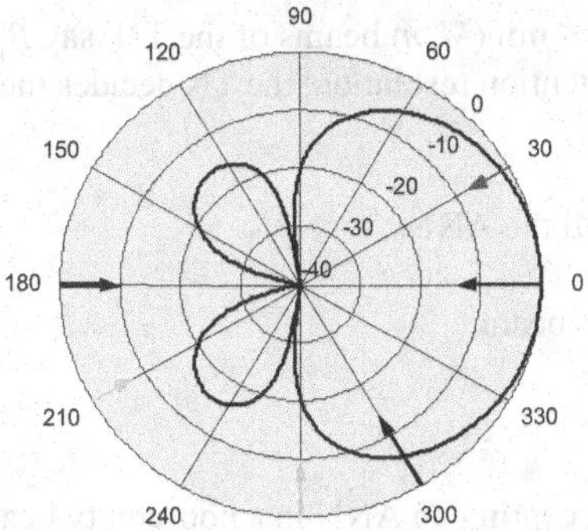


(a)

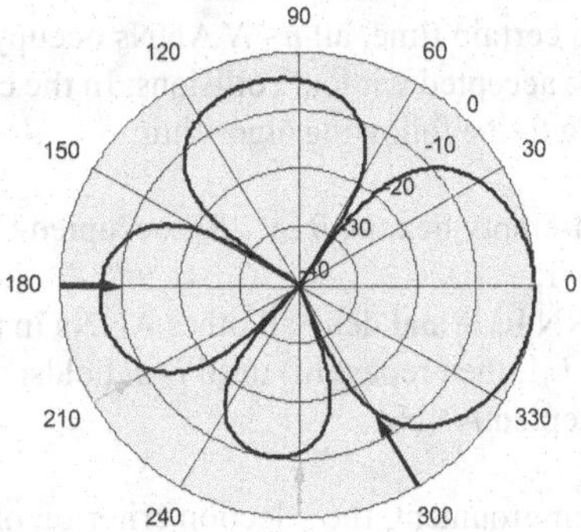


(b)

For convenience, denote $p$ as the probability that each neighboring user attempts to transmit a packet in a time slot. During a time slot, the probability that $n$ out of the $N$ neighboring nodes simultaneously attempt to transmit a packet is given by (Li, 2007)

$$P_a(N,n,p) = C_n^N p^n (1-p)^{N-n}, \tag{13}$$

where $C_n^N$ denotes the combination operation representing the number of different ways of selecting $n$ out of $N$ neighboring nodes. Note that, when $p$ is sufficiently small, the offered load $R$ can be approximated by $R=Np$.

To simplify the problem, we first introduce a relatively simple sector model. Under this model, two RAS techniques, respectively based on packet prioritization and throughput-maximization, are developed and compared. RAS techniques based on output SINR are then discussed.

## Sector-Based RAS

### Sector-Based Beam Model

For an MFBA, each node is equipped with $M$ fixed-beam antennas, each forming a conical beam that spans a sector of $\theta_{BW}=2\pi/M$ radians. Hence, all antennas of the TN collectively form $M$ non-overlapping beams that cover the entire azimuth plane. The antenna gain is considered constant within the beam region whereas it drops to zero outside this region. As such, interferers arriving from outside of the beamwidth are entirely filtered out and do not affect the signal received in the beam. This is known as the sector-based beam model or pie-type model. When multiple neighboring nodes simultaneously transmit packets which fall into the same beam at the TN, the MFBA cannot avoid collisions between these packets because the angular resolution of the beams, once they are predefined, cannot be adjusted.

An MCSA, on the other hand, independently forms multiple dynamically defined and real-time steered beams toward different ANNs. The angular resolution, or the beamwidth, of each beam is adaptively adjusted corresponding to the signal environment. To perform a sector-based analysis, an equivalent sector-based MCSA model can be used (Zhang, 2006). Similar to the sector-based beam model developed for an MFBA, an MCSA with $M$ antennas forms up to $M/\beta$ virtual beams, each with beamwidth $\alpha=\beta\theta_{BW}$, where $\beta \le 1$ is the beamwidth ratio that reflects the capability of the MCSA to form narrower beams. When the angular separation between any adjacent signal arrivals is larger than a threshold $\alpha/2$, the MCSA can filter out the interfering signals. Note that, for the MCSA, the maximum number of actual beams remains to be $M$, and the exact value of $\alpha$ depends on the array configuration, required SINR, and the number and type of interferers.

### Single-Path Propagation Environment

In a single-path environment, the signal transmitted from each ANN falls into only one beam of the TN. Collision occurs when two or multiple signals fall into a beam when on-demand protocols are used. In order to avoid such collisions, a proper RAS scheme incorporated into contention resolution can be exploited such that the TN only accepts one ANN in each non-empty beam and denies all others.

Assume that, at a certain time, all $n \le N$ ANNs occupy $m \le \min(M, n)$ beams of the TN, say $B_1, \ldots, B_m$. In this case, up to $m$ ANNs can be accepted without collisions. In the contention resolution, the TN decides the acceptance or denial of an ANN according to the following algorithm:

i) Find the $m$ non-empty beams, $B_1, \ldots, B_m$, occupied by all the ANNs;
ii) Initialize $i$ as $i=1$;
iii) Accept one ANN in $B_i$ and deny all other ANNs in this beam;
iv) Update $i$ as $i \leftarrow i+1$, then repeat iii) until $i>m$ holds;
v) Output the accepted ANNs.

In a single-path environment, the selection criterion of accepting an ANN in a non-empty beam does not affect the total number of accepted ANNs in the TN. Therefore, when the ANNs have different priority classes, proper prioritization can be performed without sacrificing the overall node throughput.

## Multipath Propagation Environment

Compared to the single-path propagation case, the RAS becomes more complicated in a multipath propagation environment. In this case, the signal transmitted from an ANN may arrive at the TN through multiple paths and fall into multiple beams.

Similar to the single-path case, we can exploit proper RAS to deny some ANNs such that collision-free communications from the accepted ANNs to the TN are guaranteed. Depending on the design criteria, different RAS schemes can be implemented. In the following, we first discuss the RAS scheme that is based on priority consideration, and that based on the throughput-maximization is then introduced.

## Priority-Based Algorithm

With the growth of real-time applications, supporting real-time flows with delay constraints is an important and challenging issue in ad hoc wireless networks. In order to provide differentiated services for real-time and non-real-time packets, the employed RAS schemes must adopt certain mechanisms to incorporate differentiated priority-based access, such that the packets with higher priority can be transmitted in preference to packets with lower priority (Deng, 1999; Barry, 2001; Veres, 2001; Aad, 2001; Kanodia, 2001; Yang, 2002; Yin, 2003; Xiao, 2003). Priority-based RAS is discussed below.

Assume that, at a certain time, there are $n \leq N$ ANNs, each transmitting its respective signal through $K$ quasi-stationary paths and each path signal has sufficiently high strength since all node are located in a similar multipath environment (Note that such assumption of equal number of paths is only for the convenience of analysis, and it does not affect the effectiveness of the RAS algorithm.). The $K$ paths are assumed to follow a uniform angular distribution and some or all of these paths may fall into the same beams. Denote $B_a$ as the number of beams occupied by the signal arrivals from all the $n$ ANNs. It is obvious that $B_a \leq \min(Kn, M)$. In the priority-based algorithm, first of all, the TN sorts the $n$ ANNs based on their priority classes to form a queue: $U_1, U_2, \ldots, U_n$, where $U_1$ has the highest priority and $U_n$ has the lowest priority. To solve the contention problem, the ANNs accepted by the TN are determined according to the following RAS algorithm:

i) Unconditionally accept $U_1$;
ii) Accept $U_2$ only if none of its paths falls into the beams occupied by $U_1$, otherwise deny it;
iii) For $i > 2$, accept $U_i$ only if none of its paths falls into the $B_{oc}$ beams, where $B_{oc}$ is the number of beams occupied by the previously accepted ANNs. Update $B_{oc}$ to include the beams occupied by the newly accepted user;
iv) Update $i$ as $i \leftarrow i+1$, then repeat iii) until either $B_{oc} = B_a$ or $i > n$ holds;
v) Deny all the remaining ANNs if there are any.

It can be seen that this RAS scheme protects the accepted ANNs to the maximum extent by denying the ANNs that have paths falling into any beams occupied by the previously accepted ANNs. That is, all accepted ANNs individually have mutually exclusive beam occupancies.

## Throughput-Maximization Algorithm

In some applications, the priority issue of the ANNS may have less importance compared with the overall network throughput. Rather, the RAS scheme should aim to maximize the node throughput (Li, 2001; Toumpis, 2003; Spyropoulos, 2003a/b). Below, a throughput-maximization (TM) RAS scheme is developed to maximize the node throughput gain (NTG) of an MFBA-equipped node in various propagation environments. The NTG is defined as the mean number of ANNs accepted by the TN of interest.

Assume that, at a certain time, there are $n \leq N$ ANNs, each transmitting its respective signal through $K$ quasi-stationary paths. Denote $B_o$ as the number of beams occupied by the signal arrivals from all the $n$ ANNs, and $S$ as the maximum number of the accepted ANNs. Towards the contention resolution, the accepted ANNs are determined according to the following RAS algorithm:

i) If each ANN occupies all the $M$ beams, let $S$=1, select one ANN and deny the other $n$–1 ANNs. Go to v);
ii) Deny any ANNs that individually occupy all the $M$ beams. Update the number of beams, $m$, occupied by the remaining ANNs;

iii) When redundant ANNs with overlapping occupancy (i.e., they exactly occupy the same number and indices of beams) exist, only one ANN is preserved and the others are denied;
iv) Update $n$, the number of remaining ANNs, let $S$=1 and determine the possible acceptance of ANNs according to the following procedure:
 1) Go to v) if $n$=1, otherwise let $N_s$=min($n,m$) and go to 2);
 2) Search in the entire search space composed of $C_{N_s}^n$ possible ways to determine whether $N_s$ out of the $n$ ANNs can be accepted simultaneously, i.e., at least one path of each of the $N_s$ ANNs does not collide. Once one possible way that $N_s$ out of the $n$ ANNs can be accepted is found, then stop searching, let $S$=$N_s$, and go to v). If no way for simultaneously accepting $N_s$ out of the $n$ ANNs can be found, go to 3);
 3) Update $N_s$ as $N_s \leftarrow (N_s-1)$ and repeat 2) until $N_s$=1.
v) Output the $S$ ANNs accepted by the TN.

Note that the complexity of the TM search process is high. A simplified search procedure can be developed based on the concept of *mutually exclusive set* (Li, 2008).

## SINR-Based RAS

The sector-based model provides a simple platform to analyze the collision problem and develop RAS approaches. However, as we discussed in Section III, this model is approximate and does not precisely represent actual array beams. The use of actual output SINR is an appropriate measure to accurately determine the reliability performance of communication links. In this following, we consider output SINR-based RAS schemes. We focus on two perspectives: the priority class of an ANN and the output SINR.

Two criteria should be satisfied for an ANN to be accepted at a TN. First, to decode the SOI, the ANN should yield sufficiently high output SINR in the presence of other ANNs. Second, it should not impose significant interference to the nodes that are already accepted. Similar to the sector-based model, the TN first sorts all the ANNs according to their priority classes to form a queue: $U_1$, $U_2$, ... , $U_n$. For convenience, we define the following parameters:

$\gamma_0$: output SINR threshold required to accept an ANN
$\gamma_i$ : ouput SINR of $U_i$
$U_a$: a set of ANNs that are accepted by the TN
$S$: number of elements of $U_a$ ( i.e., the number of accepted ANNs)

In the following, priority-based RAS algorithms are considered in the contention resolution and the output SINR is analyzed respectively for MCSAs and MFBAs.

## MCSA-Based RAS Algorithm

For a TN exploiting an MCSA, the following procedure is used to determine the acceptance of the ANNs:

i) Initialization: $U_a=\Phi$ (empty set), $S$=0, $i$=1;
ii) Calculate $\gamma_1$ in the absence of $U_2$, ..., $U_n$. If $\gamma_1 \geq \gamma_0$, accept $U_1$ by rendering both $U_a=U_1$ and $S$=1, otherwise deny $U_1$;
iii) Update $i$ as $i \leftarrow i+1$. In the presence of all $U_a$ and $U_i$, recalculate the output SINR of the $S$ already accepted ANNs and $\gamma_i$. If the output SINR for each of the $S$+1 users exceeds $\gamma_0$, accept $U_i$ by appending $U_i$ into $U_a$ and updating $S$ as $S \leftarrow S+1$, otherwise deny $U_i$;
iv) Repeat iii) until either $S$=$M$ or $i$=$n$ holds;
v) Deny all the remaining ANNs if there are any.

Clearly, this RAS scheme takes the priority class of ANNs into account and satisfies the abovementioned output SINR and interfering criteria.

## MFBA-Based RAS Algorithm

When considering the output SINR of MFBAs, we exploit actual antenna beams with actual antenna patterns and sidelobes, rather than idealized sectors. For example, in Fig. 3(a), four actual fixed beams are formed. In a multipath environment, the signal arrivals originated from an ANN may fall into multiple beams. The output SINR corresponding to an ANN is evaluated at the beam that provides the maximum output SINR. For convenience, we define the following parameters.

$N_i$: number of beams occupied by $U_i$
$U_a$: a set of ANNs that are accepted by the TN
$b_{Ui}$: the beam corresponding to maximum output SINR for $U_i$
$B_a$: a set of beams that respectively yield the highest output SINR for the accepted ANNs

For an MFBA, the following RAS procedure is used in the contention resolution to determine the accepted ANNs:

i) Initialization: $U_a=\Phi$, $B_a=\Phi$, $S=0$, $i=1$;
ii) Calculate the output SINR for each of the $N_1$ beams occupied by $U_1$ in the absence of $U_2$, ..., $U_n$. If the highest output SINR corresponding to beam $b_{U1}$ exceeds $\gamma_0$, accept $U_1$ by rendering $U_a=\{U_1\}$ and $B_a=\{b_{U1}\}$ as well as $S=1$, otherwise deny $U_1$;
iii) Update $i$ as $i \leftarrow i+1$. In the presence of all $U_a$ and $U_i$, recalculate the output SINR of all the $S$ accepted ANNs. If the output SINR for any of the $S$ ANNs is lower than $\gamma_0$, deny $U_i$ and go to iv); otherwise find the remaining $N_{ri}$ beams $B_{ri}$, by excluding the beams $B_a$ from the $N_i$ beams occupied by $U_i$, calculate the output SINR for each of the $B_{ri}$ beams, and obtain the maximum value $\gamma_{imax}$ corresponding to beam $b_{Ui}$. If $\gamma_{imax}$ exceeds $\gamma_0$, accept $U_i$ by appending $U_i$ into $U_a$, appending $b_{Ui}$ into $B_a$, and updating $S$ as $S \leftarrow S+1$, otherwise deny $U_i$;
iv) Repeat iii) until either $S=M$ or $i=n$ holds;
v) Deny all the remaining ANNs if there are any.

Similarly, this RAS scheme also considers the priority class of the ANNs and their mutual interference impact. In contrast to the MCSA-based RAS algorithm, the throughput performance may, however, degrade due to the fact that the beams are predefined, i.e., they cannot adapt to time-varying propagation channels.

The performance of the above two RAS algorithms is evaluated in the next two sections.

## V. SECTOR-BASED PERFORMANCE EVALUATION

In this section, we use the sector-based model to evaluate the performance, in terms of the probability of concurrent packet reception (CPR) and NTG, of MBAs with both MFBA and MCSA structures, operating in single-path and multipath environments.

## Probability of Concurrent Packet Reception

Consider a TN equipped an $M$-beam MBA with $N$ neighboring nodes. Each of these neighboring nodes attempts to transmit packets to the TN with a probability $p$. A TN beam is considered collided when signal arrivals from more than one ANNs fall into this beam when the on-demand protocols are used. The probability of CPR is defined as the probability that two or more ANNs are successfully received by the TN (Zhang, 2006; Jain, 2006b), which can be expressed as

$$P_{cpr} = \sum_{n=2}^{M} P_a(N,n,p)\cdot Q(M,M_a,n), \tag{14}$$

where $P_a(N,n,p)$ is defined in (13), and $Q(M,M_a,n)$ is the probability that all the packets transmitted from the $n$ nodes are successfully received without collision at some $M_a$ beams, where $n \le Ma \le M$.

For MFBA-based and MCSA-based TNs, respectively, $Q(M,M_a,n)$ is given by

*Figure 5. Comparison of the CPR probability (p = 0.1). Markers: simulations, lines: analytical*

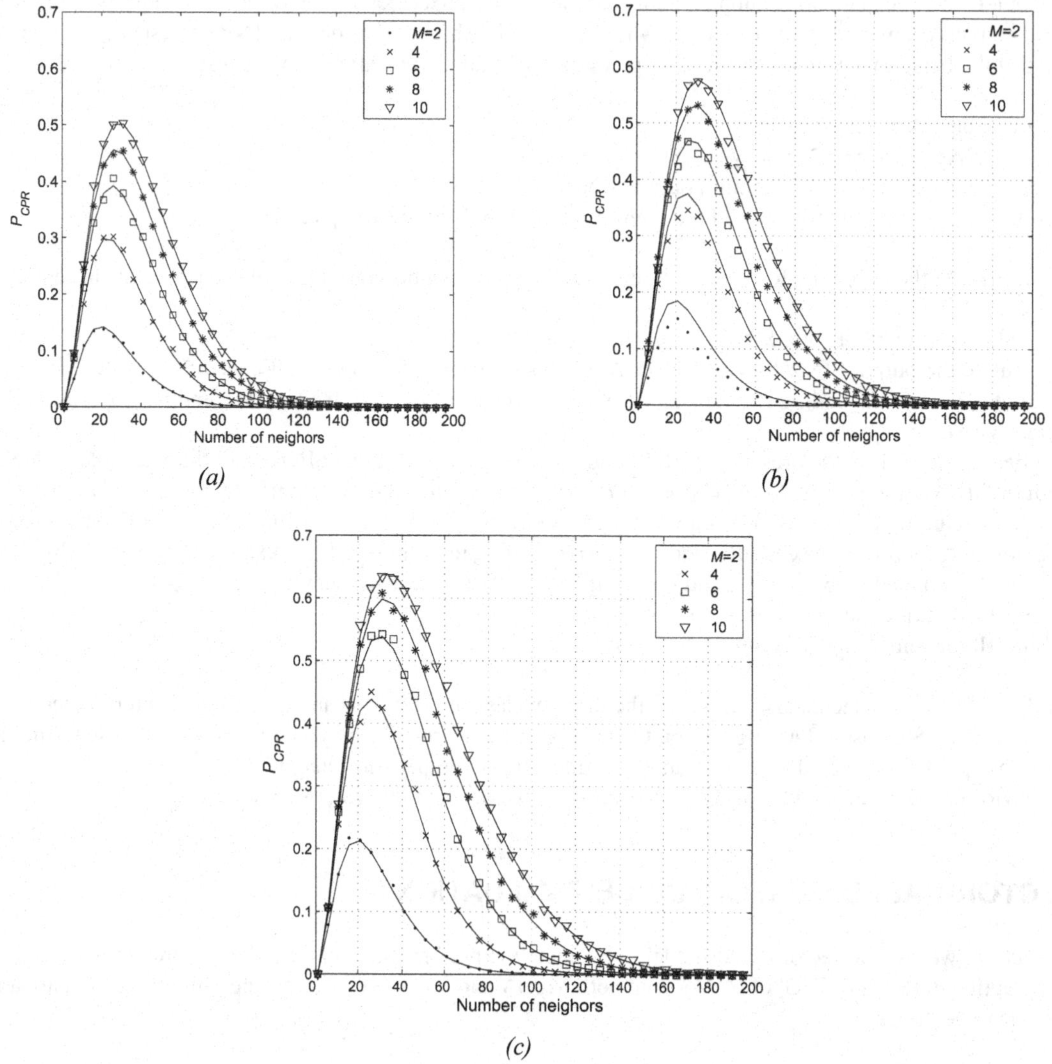


$$Q_{\mathrm{MFBA}}(M, M_a, n) = \prod_{i=0}^{n-1}\left(1 - \frac{i}{M}\right), \qquad Q_{\mathrm{MCSA}}(M, M_a, n) = \prod_{i=0}^{n-1}\left(1 - \frac{i\beta}{M}\right). \tag{15}$$

Figure 5 shows the probability of CPR for different values of $N$ and $M$, where $M$=$M_a$ and $p$=0.1 are assumed, and $\beta$ is set to 1 for the MFBA and to 0.7 and 0.5 for the MCSA, respectively. The improvement of the probability of CPR of the MCSA over the MFBA is evident.

## Node Throughput Gain

In addition to the probability of CPR, another important measure of the network performance is the NTG, denoted as $G$ (Chockalingam, 1998; Li, 2007). This measure highlights the effect of the RAS schemes used in the contention resolution and important MAC layer parameters, and diminishes some physical layer parameters. As such, it reflects not only the advantage of the MBAs over omnidirectional antennas and single-beam directional antennas, but also the efficiency of the employed MAC protocols. When the application is specified, the NTG can be mapped to the real data rates so as to directly indicate the channel utilization efficiency.

## Single-Path Propagation Environment

### On-Demand Protocol Case

For on-demand protocols, the NTG can be derived from the number of beams of a TN where signal arrivals are accepted. We first consider the probability that $m$ out of the $M$ beams accept packets transmitted from $n$ ANNs ($m \le \min(M,n)$), i.e., the probability that the signal arrivals from $m$ out of $n$ ANNs are collision-free, whereas those from the other $n-m$ ANNs are collided. This probability can be expressed as

$$P_m(M,n,m) = C_m^M \cdot {}^nP_m \cdot N^c(M-m,n-m)/M^n, \tag{16}$$

where ${}^nP_m$ denotes the permutation operation representing the number of different ways of selecting $m$ neighboring nodes from all the $n$ ANNs, and

$$N^c(M-m,n-m) = (M-m)^{n-m} \cdot \left[1 - \sum_{i=1}^{\min(M-m,n-m)} (-1)^{i+1} C_i^{M-m} \cdot {}^{n-m}P_i \cdot \frac{(M-m-i)^{n-m-i}}{(M-m)^{n-m}}\right] \tag{17}$$

is the number of different possible ways that the remaining $n-m$ ANNs fall into the other $M-m$ beams and collide (there may exist empty beams in the $M-m$ beams).

When $n$ neighboring nodes attempt to transmit, the mean number of beams at the TN that successfully receive collision-free signals from different ANNs can be written as

$$G_n^{\text{on}}(M,n) = \sum_{m=1}^{\min(M,n)} m \cdot P_m(M,n). \tag{18}$$

Therefore, the NTG of an MFBA, i.e., the mean number of collision-free non-empty beams, is expressed as

$$G_{\text{MFBA}}^{\text{on}}(M,N,p) = \sum_{n=1}^{N} P_a(N,n,p) \cdot G_n^{\text{on}}(M,n). \tag{19}$$

In the case of an MCSA with $M$ directly available beams, there are $M_e = \lfloor M/\beta \rfloor$ dynamic virtual beams due to the beamwidth reduction. Thus, similar to (16), the probability that $m$ out of the $M_e$ beams are accepted in the presence of $n$ ANNs ($m \le \min(M,n)$) can be written as $P_m(M_e,n,m)$. Consequently, for an MCSA, when $n$ neighboring users attempt to transmit, the mean number of beams at the TN that successfully receive collision-free signals from different ANNs can be expressed as

$$G_n^{\text{on}}(M_e,n,\beta) = \sum_{m=1}^{\min(M_e,n)} \min(m,M) \cdot P_m(M_e,n,m). \tag{20}$$

Finally, the NTG of an MCSA is given by

$$G_{\text{MCSA}}^{\text{on}}(M,N,p,\beta) = \sum_{n=1}^{N} P_a(N,n,p) \cdot G_n^{\text{on}}(M_e,n,\beta). \tag{21}$$

*Figure 6. Comparison of NTG in the presence of on-demand protocols (p = 0.1). Markers: simulations, lines: analytical*

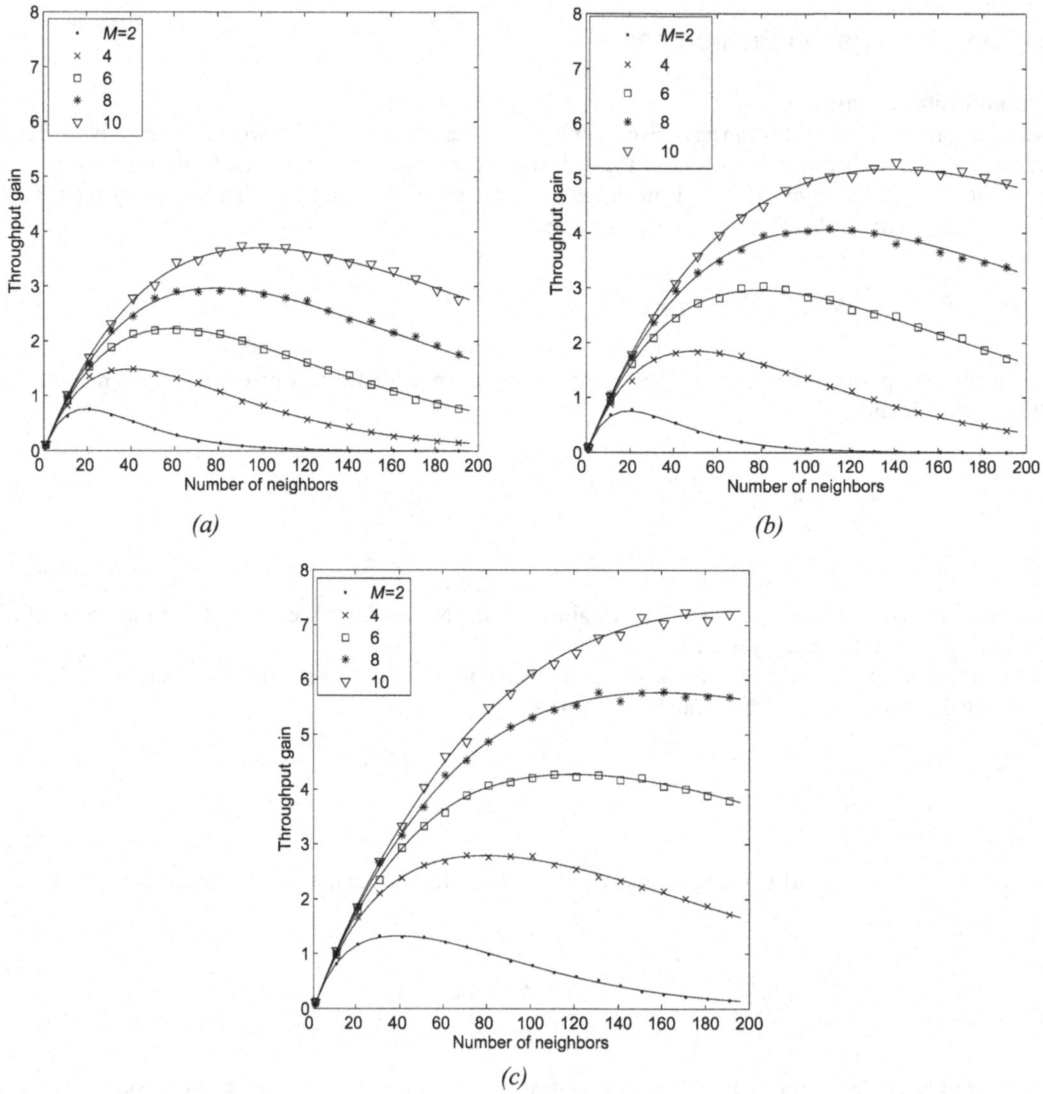


Figure 6 compares the NTG for different values of $N$ and $M$ under conditions of $M=M_a$ and $p=0.1$. Again, we set $\beta$ to 1 for the MFBA and to 0.7 and 0.5 for the MCSA, respectively. The improvement of the NTG of the MCSA over the MFBA is significant.

## RAS Case

When a certain kind of RAS is employed for the contention resolution, the TN may deny some attempted transmissions to avoid collisions. That is, each beam can accept only one ANN that has a path falling in that beam. In a single-path environment, both priority-based and TM RAS algorithms reach this goal and yield the same NTG. It is clear that the NTG becomes the mean number of non-empty beams and is expressed for the MFBA and MCSA as (Zhang, 2006)

*Figure 7. Comparison of NTG upper bounds in the presence of RAS schemes (p = 0.1). Markers: simulations, lines: analytical*

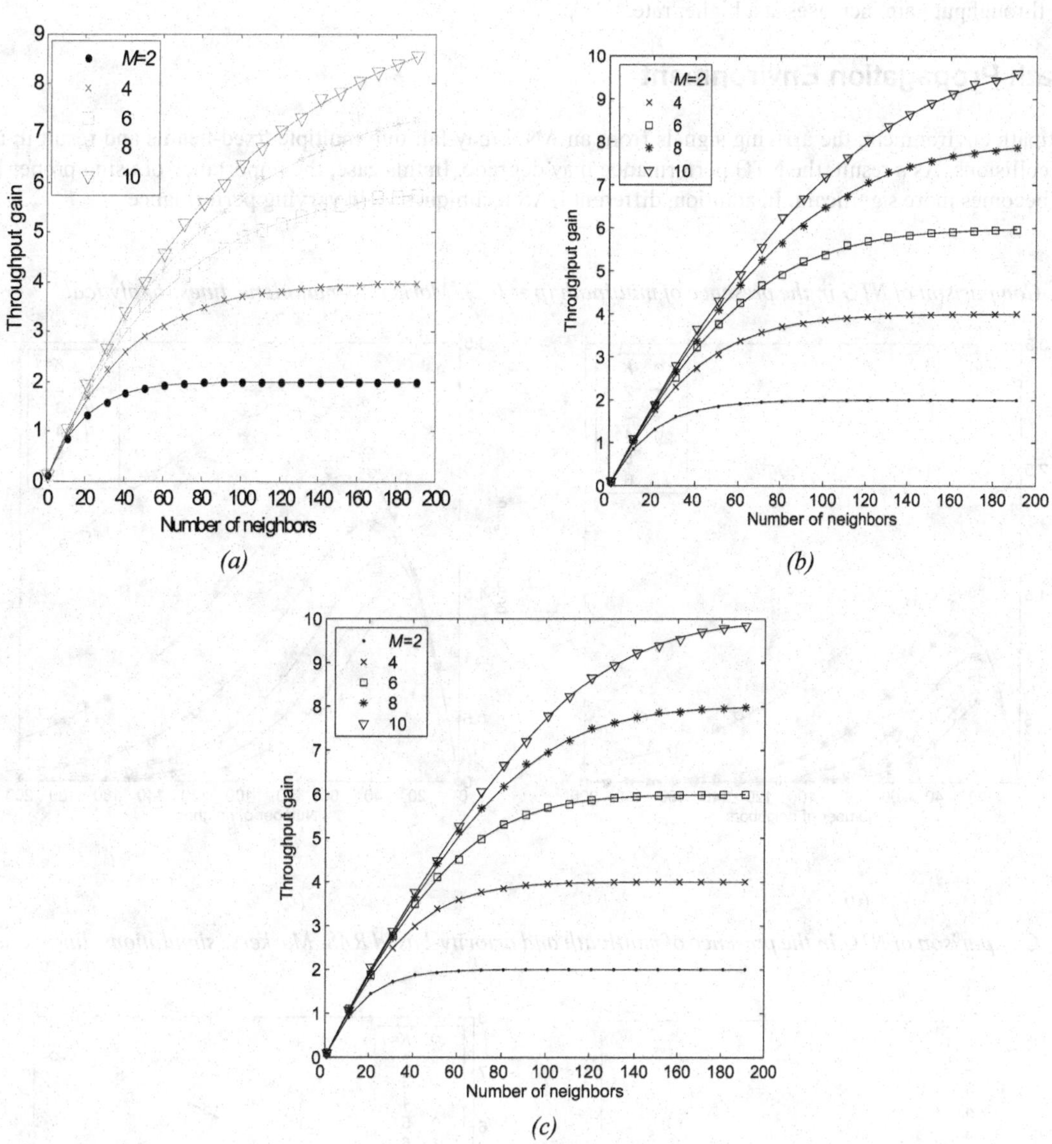


$$G_{\text{MFBA}}^{\text{RAS}} = \sum_{n=1}^{N} P_a(N,n,p)\cdot G_n^{\text{RAS}}(M,M_a), \qquad G_{\text{MCSA}}^{\text{RAS}} = \sum_{n=1}^{N} P_a(N,n,p)\cdot G_n^{\text{RAS}}(M_e,M_a), \tag{22}$$

respectively, where $P_a(N,n,p)$ is defined in (13), and

$$G_n^{\text{RAS}}(M,M_a) = \sum_{m=1}^{M} \min(n,M_a)\cdot C_m^M \cdot \frac{m!\text{S}(n,m)}{M^n}, \tag{23}$$

where $\text{S}(n,m) = \frac{1}{m!}\sum_{i=0}^{m-1}(-1)^i \cdot C_i^m \cdot (m-i)^n$ is the Stirling number of the second kind (Graham, 1994), representing the number of different ways that $n$ ANNs occupy $m$ beams without empty beams.

Figure 7 compares the NTG performance in the presence of RAS schemes (Zhang, 2006). Again $M$=$M_a$ and $p$=0.1 are assumed, and $\beta$ is chosen as 1 for the MFBA and as 0.7 and 0.5 for the MCSA. It shows that, when the MCSA is used, the throughput gain increases at a higher rate.

## Multipath Propagation Environment

In a multipath environment, the arriving signals from an ANN may fall into multiple fixed-beams and result in more frequent collisions. As a result, the NTG performance may degrade. In this case, the importance of using proper RAS methods becomes more significant. In addition, different RAS techniques yield varying performance.

*Figure 8. Comparison of NTG in the presence of multipath (p = 0.1). Markers: simulations, lines: analytical*

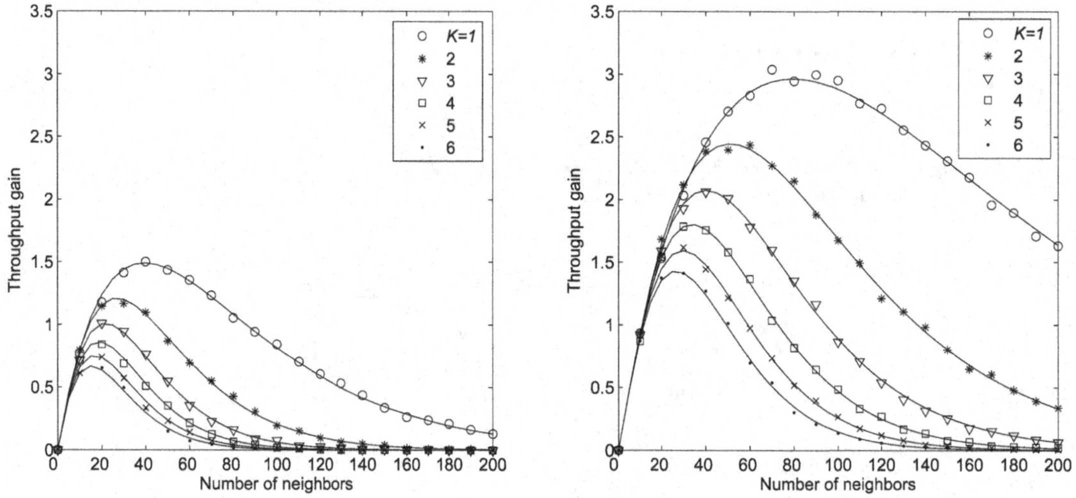


*(a)* *(b)*

*Figure 9. Comparison of NTG in the presence of multipath and priority-based RAS. Markers: simulations, lines: analytical*

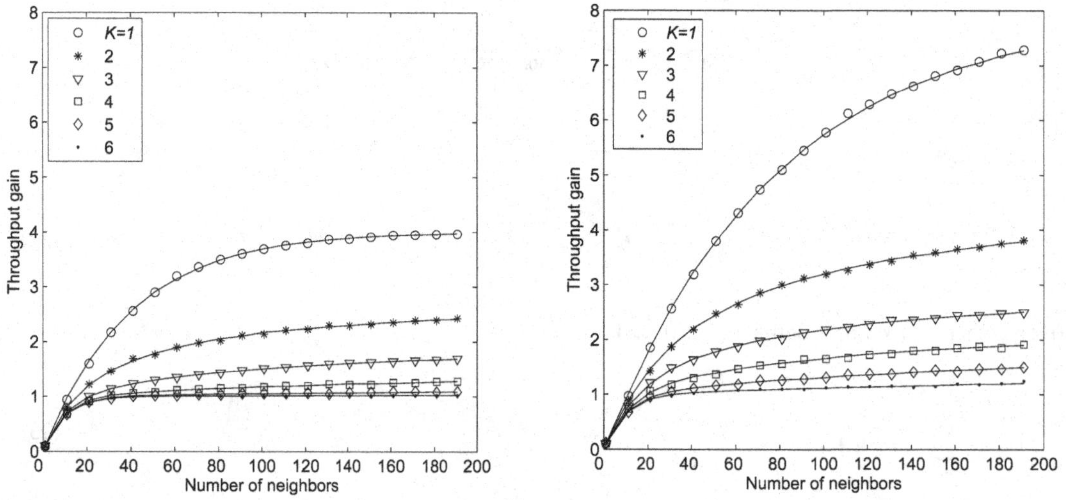


*(a)* *(b)*

## On-Demand Protocol Case

For on-demand protocols, the collision analysis developed for multi-beam operations in a single-path propagation environment can be extended to multipath propagation scenarios. Assume that at a certain time, there are $n \leq N$ ANNs, each transmitting its respective signal through $K$ quasi-stationary paths. As such, the signal arrivals from an ANN may fall into at most $B=\min(K,M)$ fixed beams. In this case, the $i$th ANN, denoted as $u_i$, is considered collided if all its $b$ beams collide with signal arrivals transmitted from the other nodes. The NTG in the multipath propagation environment can be expressed as

$$G_{\text{MFBA}}^{\text{on-mp}} = \sum_{n=1}^{N} n \cdot P_a(N,n,p) \cdot P_s(M,K,n)\, G_{\text{MFBA}}^{\text{on-mp}} = \sum_{n=1}^{N} n \cdot P_a(N,n,p) \cdot P_s(M,K,n), \tag{24}$$

where $P_s(M,K,n)$ is the probability that each ANN is accepted by the TN in the presence of $n$ ANNs, each with $K$-path. The probability is given by

$$P_s(M,K,n) = 1 - \sum_{b=1}^{B} \frac{S(K,b) \cdot {}^{M}P_b}{M^K} \cdot \left[ 1 - \sum_{i=1}^{b} (-1)^{i+1} \cdot C_i^b \cdot \left(1 - \frac{i}{M}\right)^{K(n-1)} \right]. \tag{25}$$

As an example, Fig. 8 shows the NTG with four and eight fixed beams, respectively, operated in a quasi-stationary multipath environment with different number of paths. It is evident that the performance of the MFBA degrades as the number of paths, $K$, increases due to the increased probability of collision among different ANN signals. Note that, as $K$ increases, the maximum NTG decreases and corresponds to a smaller number of neighboring nodes ($N$). The negative impact of multipath propagation on the NTG becomes insignificant in an under-saturated situation, i.e., the number of ANNs is sufficiently small and thus collision is not the primary limiting factor of the NTG.

## Priority-Based RAS Case

When the priority-based and throughput maximization RAS schemes are used, the analytical expressions of the NTG are rather complicated and thus are omitted due to the space limitation (Li, 2008). The NTG of an ad hoc network with four ($M$=4) and eight ($M$=8) fixed beams in a quasi-stationary environment using the priority-based RAS is shown in Fig. 9, where the number of paths ($K$) varies from 1 to 6, and $p$=0.1 is assumed. The results show that, due to the use of RAS strategy, the NTG becomes a non-decreasing function of the number of neighboring nodes ($N$), as opposed to the on-demand protocol case where the NTG is not monotonic with $N$. It is also evident that the NTG improves as the number of beams increases, since a higher number of available beams can concurrently support more users.

When the number of paths is large, each ANN is likely to occupy a high number of beams, thus yielding a low saturation value of the NTG. Note that the asymptotical NTG for $K$>>1 is one, i.e., only the ANN with the highest priority is accepted.

## Throughput-Maximization RAS Case

The NTG of an ad hoc network with four ($M$=4) and eight ($M$=8) fixed beams in a quasi-stationary environment using the TM is plotted in Figs. 10 and 11, where $p$ is set to 0.1. It is seen that, compared to the priority-based algorithm, the TM algorithm achieves a higher NTG. Similar to the priority-based RAS, the NTG improves as $M$ increases. The NTG remains a decreasing function with respect to $K$. For both priority- and TM-based RAS schemes, the asymptotic NTG for $K$>>1 is unity.

# VI. OUTPUT SINR-BASED PERFORMANCE EVALUATION

In this section, the NTG is considered from a practical output SINR perspective. That is, only the ANNs whose signals achieve sufficiently high output SINR at the TN are accepted and contribute to the NTG. The probability that an ANN

*Figure 10. Comparison of NTG in the presence of TM RAS in a multipath environment. Markers: simulations, lines: analytical*

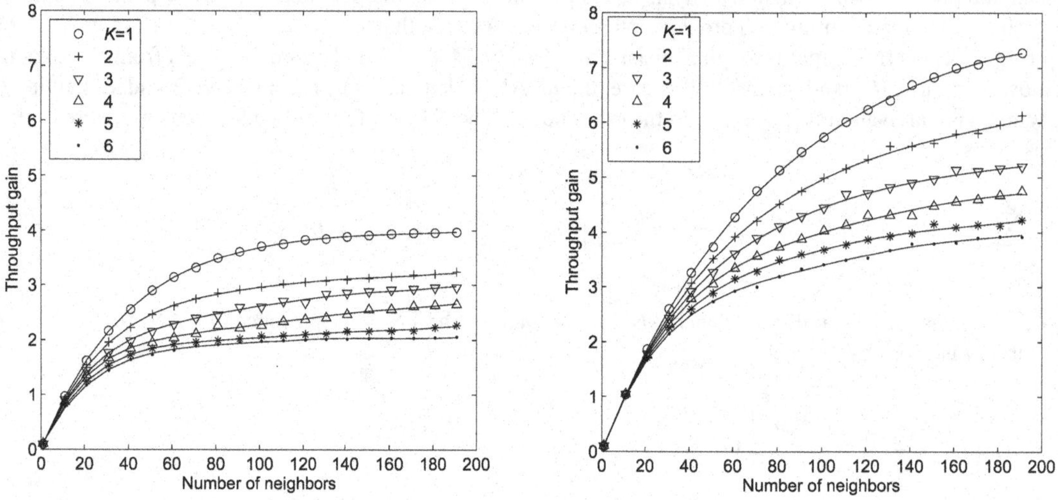


*(a)* *(b)*

*Figure 11. Comparison of NTG in the presence TM RAS in a multipath environment. Markers: simulations, lines: analytical*

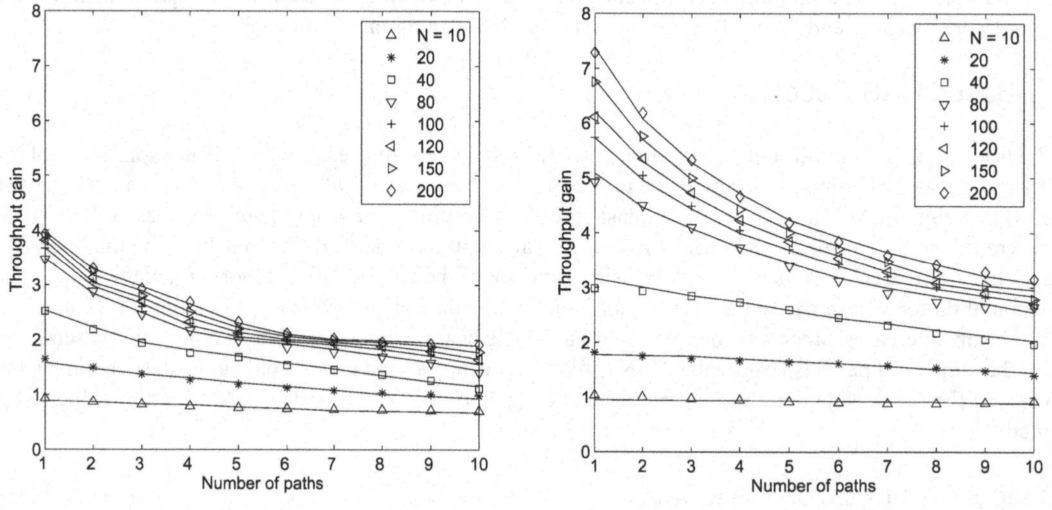


*(a)* *(b)*

is accepted can be defined as

$$P_{acc}(n,\gamma_0) = \Pr(\gamma_{out} \geq \gamma_0). \quad (26)$$

where $\gamma_{out}$ is the output SINR of this ANN if it is accepted, and $\gamma_0$ is the required output SINR threshold. Thus, the mean number of accepted ANNs in the presence of $n$ ANNs is obtained as

$$N_a\left(n, P_{acc}(n,\gamma_0)\right) = \sum_{i=1}^{n} i \cdot C_i^n \cdot P_{acc}(n,\gamma_0)^i \cdot \left(1 - P_{acc}(n\gamma_0)\right)^{n-i}. \tag{27}$$

Further, the NTG in the presence of $N$ neighboring users is given by (Li, 2007)

$$G(M,N,p) = \sum_{n=1}^{N} P_a(N,n,p) \cdot N_a\left(n, P_{acc}(n,\gamma_0)\right) = \sum_{n=1}^{N} n \cdot P_a(N,n,p) \cdot P_{acc}(n,\gamma_0). \tag{28}$$

This result shows that the NTG depends on the number of neighboring users ($N$), each neighboring user's transmission probability ($p$), and the probability of acceptance for each ANN ($P_{acc}$).

In the following, the probability of acceptance and the NTG are evaluated through numerical simulations. Each path is assumed to have the same propagation gain and yields an input SNR of 20 dB at the TN. The number of paths of each ANN varies from $K$=1 to 3, and $p$ is set to 0.1.

Figure 12 shows the probability of acceptance of each ANN for a four-beam MFBA and MCSA, respectively, where $\gamma_0$ is set to 16 dB, which implies a 10 dB SINR degradation tolerance to co-channel interference. The results demonstrate that, while the probability of acceptance decreases for both MFBA and MCSA as $n$, the number of ANNs, increases, the MFBA shows much sharp reduction. For the MFBA, the probability of acceptance reduces as $K$ increases, since more paths of each ANN are likely to yield higher number of collisions and reduced output SINR. On the contrary, for the MCSA, the probability of acceptance slightly increases with $K$ as a result of enhanced output SINR due to the combining gain of multiple coherent SOI paths. The MFBA has a low probability of acceptance even when there are only two ANNs, whereas the MCSA maintains a high probability of acceptance when $n$ ranges from 2 to 4.

Figure 13(a) compares the corresponding NTG, where $K$ is 1 and 2, and the required output SINR threshold is set to $\gamma_0$=23 and 16 dB, respectively, in the on-demand protocol case. This figure clearly shows that, as $K$ increases from one to two, the NTG of an MFBA-based network is reduced, whereas the MCSA achieves a higher throughput. For example, when $\gamma_0$=23 dB, the maximum NTG of the MCSA in the presence of two paths is 30% higher than that in the case of single path. Fig. 13(a) illustrates that more ANNs results in higher NTG loss once the NTG reaches the corresponding maximum value.

Figure 13(b) compares the corresponding NTG when the priority-based RAS scheme is used. $\gamma_0$ is set to 16 dB. While the NTG curves become a non-decreasing function of $N$ for both MFBA and MCSA, the advantage of MCSA

*Figure 12. Comparison of the probability of acceptance and NTG for four-beam MFBA and MCSA (p=0.1, $\gamma_0$=16 dB)*

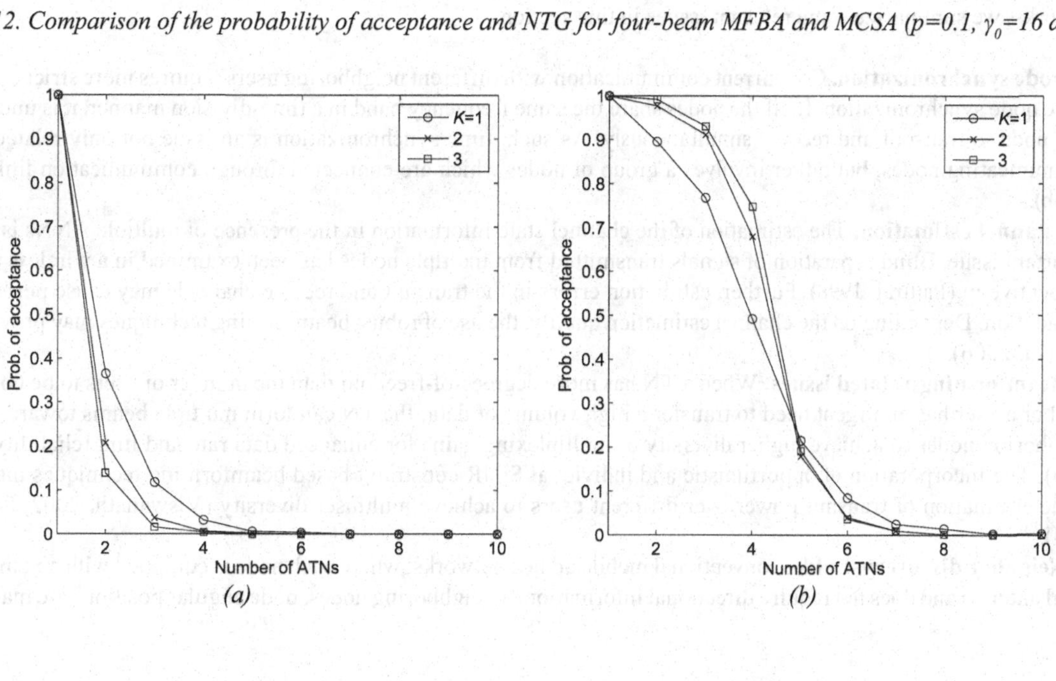

(a) (b)

*Figure 13. Comparison of the simulated NTG for the MFBA and MCSA operated in a multipath environment where the on-demand protocol and the priority-based RAS schemes are used, respectively*

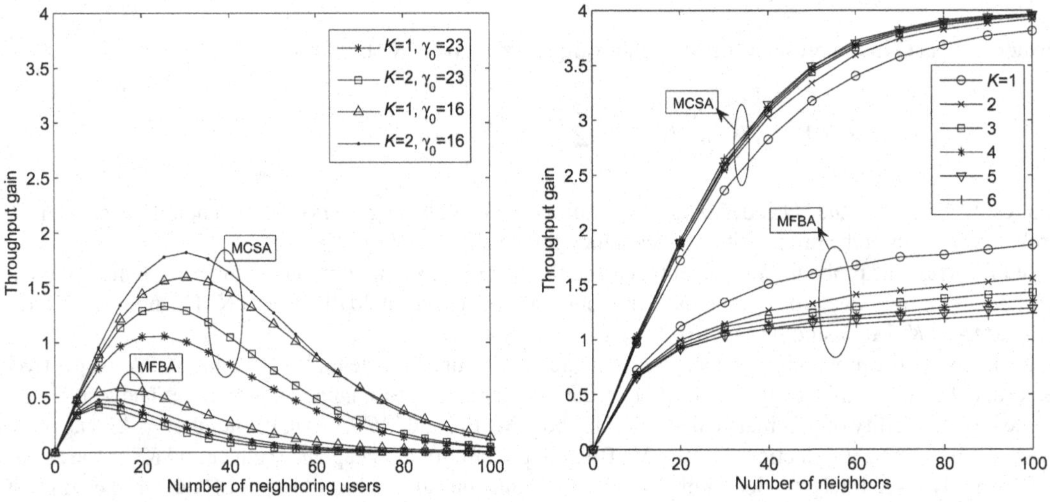


(a) (b)

over MFBA is evident. In the former, the NTG approaches 4 for a large value of $N$. The multipath effect reduces the NTG for the MFBA, whereas the MCSA benefits from the multipath propagation.

## VII. RELEVANT ISSUES

Having discussed the offerings and performance of MBAs and the impacts of using proper RAS schemes in ad hoc networks, we summarize some relevant issues in this section.

**Node synchronization.** Concurrent communication with different neighboring users requires more strict constraint on the node synchronization. If all the nodes share the same frequency band in a time-division manner, it is undesirable for a node to transmit and receive simultaneously. As such, time synchronization is an issue not only related to two communicating nodes, but rather involves a group of nodes which are connected through communication links (Jain, 2006b).

**Channel estimation.** The estimation of the channel state information in the presence of multiple ANNs is another important issue. Blind separation of signals transmitted from multiple nodes has been examined in a wireless network perspective in (Paulraj, 1998). Further, estimation errors in the transmit and receive channels may cause performance degradation. Depending on the channel estimation quality, the use of robust beamforming techniques may prove necessary (Li, 2006).

**Beamforming-related issues.** When a TN has more degrees-of-freedom than the number of users to be communicated or a user has an urgent need to transfer a high volume of data, the TN can form multiple beams towards certain neighboring nodes to achieve higher diversity or multiplexing gains for enhanced data rate and link reliability (Chen, 2006). The incorporation of opportunistic and individual SINR constraint based beamforming techniques may guide the determination of transmit power over different users to achieve multiuser diversity (Viswanath, 2002; Schubert, 2004).

**Neighbor discovery.** Unlike conventional mobile ad hoc networks, where each node is equipped with an omnidirectional antenna and does not require directional information of neighboring nodes, node angular positions are maintained in MBAs for effective beamforming and random-access scheduling. Several efforts have been made to support neighbor discovery to locate and track the TN's neighbors (Jakllari, 2005; Zorzi, 2006; Bandyopadhyay, 2006).

**Scheduling schemes.** When MBAs are employed, more sophisticated MAC and routing mechanisms including power control are necessary so as to exploit spatial reuse and control the amount of interference and collision. Our discussion in this chapter was focused on the space-domain approaches, whereas most current work is involved in the design of frame structures in the time domain (Bandyopadhyay, 2006). Scheduling schemes combining the spatial and time dimensionality may increase the network flexibility and efficiency. Furthermore, cross-layer design is desirable to yield joint physical layer, MAC, and routing optimization (Chen, 2002; Martinez, 2004; Zorzi, 2006).

## VIII. CONCLUSION

Multi-beam antenna (MBA) techniques for wireless ad hoc network applications were introduced in this chapter. Two implementations, namely, multiple fixed-beam antennas (MFBAs) and multi-channel smart antennas (MCSAs), were discussed. The performance in terms of the node throughput and the probability of concurrent communications was examined with the exploitation of two random-access scheduling (RAS) schemes incorporated into the contention resolution process, respectively, for the node priority consideration and throughput maximization. Two antenna models were used in the performance analysis. The sector-based model is relatively simple, whereas the output SINR based model provides accurate performance evaluation. In time-varying multipath propagation environments, MFBAs show significant throughput degradation, whereas MCSAs achieve path gain that enhances the output SINR. The impact of using proper RAS in the contention resolution schemes becomes more evident in a multipath propagation environment. Finally, some important issues relevant to the MBAs are addressed for broad understanding of the challenges in this area of research and development.

## REFERENCES


[Aad01] Aad, I., & Castelluccia, C. (2001). Differentiation mechanisms for IEEE 802.11. *IEEE Conference on Computer Communications*, 1, 209-218.

[Akyildiz05] Akyildiz, I. F., Wang, X., & Wang, W. (2005). Wireless mesh networks: a survey. *Computer Networks*, 47, 445-487.

[Babich06] Babich, F., Comisso, M., D'Orlando, M., & Mania, L. (2006). Interference mitigation on WLANs using smart antennas. *Wireless Personal Communications Journal*, 36(4), 387-401.

[Bahl04] Bahl, P., Adya, A., Padhye, J., & Walman, A. (2004). Reconsidering wireless systems with multiple radios." *ACM SIGCOMM Computer Communication Review*, 34(5), 39-46.

[Bandyopadhyay01] Bandyopadhyay, S., Hausike, K., Horisawa, S., & Tawara, S. (2001). An adaptive MAC and directional routing protocol for ad hoc wireless networks using ESPAR antenna. *ACM International Symposium on Mobile Ad Hoc Networking & Computing*, 243-246.

[Bandyopadhyay06] Bandyopadhyay, S., Roy, S., & Ueda, T. (2006). *Enhancing the Performance of Ad Hoc Wireless Networks with Smart Antennas*. Boca Raton, FL:Auerbach.

[Bao02] Bao, L., & Garcia-Luna-Aceves, J. J. (2002). Transmission scheduling in ad hoc networks with directional antennas. *ACM International Conference on Mobile Computing and Networking*, 48-58.

[Barry01] Barry, M., Campbell, A. T., & Veres, A. (2001). Distributed control algorithms for service differentiation in wireless packet networks. *IEEE Conference on Computer Communications*, 1, 582-590.

[Blake01] Li, J., Blake, C., Decouto, D. S. J., Lee, H. I., & Morris, R. (2001). Capacity of ad hoc wireless networks. *ACM International Conference on Mobile Computing and Networking*, 61-69.

[Caballero06] Caballero, E. J. (2006). Vulnerabilities of intrusion detection systems in mobile ad-hoc networks - The routing problem. *TKK T-110.5290 Seminar on Network Security, from http://www.tml.tkk.fi /Publications/C/22/papers/Jimenez_final.pdf.*

[Chen02] Chen, K., Shah, S. H., & Nahrstedt, K. (2002). Cross-layer design for data accessibility in mobile ad hoc networks. *Wireless Personal Communications*, 21, 49-76.

[Chen06] Chen, B., & Gans, M. J. (2006). MIMO communications in ad hoc networks. *IEEE Transactions on Signal Processing*, 54(7), 2773–2783.

[Chockalingam98] Chockalingam, A., & Rao, R. R. (1998). MAC layer performance with steerable multibeam antenna arrays. *IEEE International Symposium on Personal, Indoor and Mobile Radio Communications*, 2, 973-977.

[Choudhury02] Choudhury, R. R., Yang, X., Vaidya, N. H., & Ramanathan, R. (2002). Using directional antennas for medium access control in ad hoc networks. *ACM International Conference on Mobile Computing and Networking*, 59-70.

[Choudhury03] Choudhury, R. R., & Vaidya, N. H. (2003). Impact of directional antennas on ad hoc routing. *Personal and Wireless Communications*, 590-600.

[Choudhury04] Choudhury, R. R., & Vaidya, N. H. (2004). Deafness: A MAC problem in ad hoc networks when using directional antennas. *IEEE International Conference on Network Protocols*, 283-292.

[Choudhury06] Choudhury, R. R., Yang, X., Ramanathan, R., & Vaidya, N. H. (2006). On designing MAC protocols for wireless networks using directional antennas. *IEEE Transactions on Mobile Computing*, 5(5), 477-490.

[Dai06] Dai, H., Ng, K.-W., & Wu, M.-Y. (2006). An overview of MAC protocols with directional antennas in wireless ad hoc networks. *IEEE International Conference on Wireless and Mobile Communications*, 84-84.

[Das06] Das, S. M., Pucha, H., Koutsonikolas, D., Hu, Y. C., & Peroulis, D. (2006). DMesh: incorporating practical directional antennas in multi-channel wireless mesh networks." *IEEE Journal on Selected Areas in Communications*, 24(11), 2028-2039.

[Deng99] Deng, D.-J., & Chang, R.-S. (1999). A priority scheme for IEEE 802.11 DCF access method. *IEICE Transactions on Communications*, E82-B(1), 96-102.

[Draves04] Draves, R., Padhye, J., & Zill, B. (2004). Routing in multi-radio, multi-hop wireless mesh networks. *ACM International Conference on Mobile Computing and Networking*, 114-128.

[Elbatt03] Elbatt, E., Anderson, T., & Ryu, B. (2003). Performance evaluation of multiple access protocols for ad hoc networks using directional antennas. *IEEE Wireless Communications and Networking Conference*, 2, 982-987.

[Gossain06] Gossain, H., Joshi, T., Cordeiro, C., & Agrawal, D. P. (2006). DRP: an efficient directional routing protocol for mobile ad hoc networks. *IEEE Transactions on Parallel and Distributed Systems*, 17(12), 1438-1541.

[Graham94] Graham, R. L., Knuth, D. E., & Patashnik, O. (1994). *Concrete Mathematics: A Foundation for Computer Science*, 2nd ed., MA: Addison-Wesley.

[Gupta00] Gupta, P., & Kumar, P. R. (2000). The capacity of wireless networks. *IEEE Transactions on Information Theory*, 46(2), 388-404.

[Hu04] Hu, L., & Evans, D. (2004). Using directional antennas to prevent wormhole attacks. *Network and Distributed System Security Symposium.*

[Huang02a] Huang, Z., & Shen, C. C. (2002). A comparison study of omnidirectional and directional MAC protocols for ad hoc networks. *IEEE Global Telecommunications Conference*, 1, 57-61.

[Huang02b] Huang, Z., Shen, C., Srisathapornphat, C., & Jaikaeo, C. (2002). A busy-tone based directional MAC protocol for ad hoc networks. *Military Communications Conference*, 2, 1233-1238.

[Jain06a] Jain, V., & Agrawal, D. P. (2006). Mitigating deafness in multiple beamforming antennas. *IEEE Sarnoff Symposium.*

[Jain06b] Jain, V., & Agrawal, D. P. (2006). Concurrent packet reception bounds for on-demand MAC protocols for multiple beam antennas. *IEEE International Symposium on Personal, Indoor and Mobile Radio Communications*, 1-5.

[Jakllari05] Jakllari, G., Luo, W., & Krishnamurthy, S. V. (2005). An integrated neighbor discovery and MAC protocol for ad hoc networks using directional antennas. *IEEE International Symposium on a World of Wireless, Mobile and Multimedia Networks*, 11-21.

[Jurdak04] Jurdak, R., Lopes, C. V., & Baldi, P. (2004). A survey, classification and comparative analysis of medium access control protocols for ad hoc networks. *IEEE Communication Surveys & Tutorials*, 6(1), 2-16.

[Kanodia01] Kanodia, V., Li, C., Sabharwal, A., Sadeghi, B., & Knightly, E. (2001). Distributed multi-hop scheduling and medium access with delay and throughput constraints. *ACM International Conference on Mobile Computing and Networking*, 1, 200-209.

[Ko00] Ko, Y. B., Shankarkumar, V., & Vaidya, N. H. (2000). Medium access control protocols using directional antennas in ad hoc networks. *IEEE Conference on Computer Communications*, 1, 13-21.

[Korakis03] Korakis, T., Jakllari, G., & Tassiulas, L. (2003). A MAC protocol for full exploitation of directional antennas in ad hoc wireless networks. *ACM International Symposium on Mobile Ad Hoc Networking & Computing*, 98-107.

[Lal04] Lal, D., Jain, V., Zeng, Q., & Agrawal, D. P. (2004). Performance evaluation of medium access control for multiple-beam antenna nodes in a wireless LAN. *IEEE Transactions on Parallel and Distributed Systems*, 15(12), 1117-1129.

[Lang03] Lang, D. (2003). A comprehensive overview about selected ad hoc networking routing protocols. *Tech. Rep. TUM-I0311*, Technische Univ. at Munchen.

[Li04] Li, X. (2004). Contention resolution in random-access wireless networks based on orthorgonal complementary codes. *IEEE Transactions on Communications*, 52(1), 82-89.

[Li06] Li, J., & Stoica, P. (2006). *Robust Adaptive Beamforming*, New Yowk: Wiley.

[Li07] Li, X., Zhang, Y., & Amin, M. G. (2007). Performance of wireless networks exploiting multi-channel smart antennas in multipath environments. *International Waveform Diversity and Design Conference.*

[Li08] Li, X., Zhang, Y., & Amin, M. G. (2008). Multibeam antenna scheduling in ad hoc wireless networks. *SPIE Defense & Security Symposium.*

[Lin82] Lin, H. C. (1982). Spatial correlations in adaptive arrays. *IEEE Transactions on Antennas and Propagation*, 30(2), 212-223.

[Martinez04] Martinez, I., & Altuna, J. (2004). A cross-layer design for ad hoc wireless networks with smart antennas and QoS support. *IEEE International Symposium on Personal, Indoor and Mobile Radio Communications*, 589-593.

[Monzingo80] Monzingo, R. A., & Miller, T. W. (1980). *Introduction to Adaptive Arrays*. New York: John Wiley.

[Nasipuri00] Nasipuri, A., Ye, S., You, J., & Hiromoto, R. (2000). A MAC protocol for mobile ad hoc networks using directional antennas. *IEEE Wireless Communications and Networking Conference*, 3, 1214-1219.

[Paulraj98] Paulraj, A. J., Papadias, C. B., Reddy, V. U., & van der Veen, A. J. (1998). Space-Time Blind Signal Processing for Wireless Communication Systems, 179-210, chapter 4 in *Wireless Communications: Signal Processing Perspectives*, Poor, H. V., & Wornell, G. W., Prentice Hall.

[Peel05] Peel, C. B., Hochwald, B. M., & Swindlehurst, A. L. (2005). A vector-perturbation technique for near-capacity multi-antenna multi-user communication - Part I: channel inversion and regularization. *IEEE Transactions on Communications*, 53(1), 95-202.

[Rajaraman02] Rajaraman, R. (2002). Topology control and routing in ad hoc networks: a survey. *ACM SIGACT News*, 33(2), 60-73.

[Ramanathan05] Ramanathan, R., Redi, J., & Santivanez, C. (2005). Ad hoc networking with directional antennas: a complete system solution. *IEEE Journal on Selected Areas in Communications*, 23(3), 496-506.

[Schubert04] Schubert, M., & Boche, H. (2004). Solution of the multiuser downlink beamforming problem with individual SINR constraints. *IEEE Transactions on Vehicular Technology*, 53(1), 18-28.

[Singh04] Singh, H., & Singh, S. (2004). Tone based MAC protocol for use with adaptive array antennas. *IEEE Wireless Communications and Networking Conference.*

[Singh05] Singh, A., Ramanathan, P., & Van Veen, B. (2005). Spatial reuse through adaptive interference cancellation in multi-antenna wireless networks. *IEEE Global Telecommunications Conference*, 5, 3092-3096.

[Spyropoulos03a] Spyropoulos, A., & Raghavendra, C. S. (2003). Capacity bounds for ad-hoc networks using directional antennas. *IEEE International Conference on Communications*, 1, 348-352.

[Spyropoulos03b] Spyropoulos, A., & Raghavendra, C. S. (2003). Asymptotic capacity bounds for ad-hoc networks revisited: the directional and smart antenna cases. *IEEE Global Telecommunications Conference*, 3, 1216-1220.

[Sudaresan03] Sudaresan, K., & Sivakumar, R. (2003). On the medium access control problems in ad hoc networks with smart antennas." *ACM SIGMOBILE Mobile Computing and Communications Review*, 7(3), 25-26.

[Takai02] Takai, M., Martin, J., Ren, A., & Bagrodia, R. (2002). Directional virtual carrier sensing for directional antennas in mobile ad hoc networks. *ACM International Symposium on Mobile Ad Hoc Networking & Computing (MobiHoc)*, 183-193.

[Toumpis03] Toumpis, S., & Goldsmith, A. J. (2003). Capacity regions for wireless ad hoc networks. *IEEE Transactions on Wireless Communications*, 2(4), 736-748.

[Verdu98] Verdu, S. (1998). *Multiuser Detection*, Cambridge University Press.

[Veres01] Veres, A., Campbell, A. T., Barry, M., & Sun, L.-H. (2001). Supporting service differentiation in wireless packet networks using distributed control." *IEEE Journal on Selected Areas in Communications*, 19(10), 2081-2093.

[Viswanath02] Viswanath, P., Tse, D. N. C., & Laroia, R. (2002). Opportunistic beamforming using dumb antennas. *IEEE Transactions on Information Theory*, 48(6), 1277-1294.

[Wang02] Wang, Y., & Garcia-Luna-Aceves, J. J. (2002). Spatial reuse and collision avoidance in ad hoc networks using directional antennas. *IEEE Global Telecommunications Conference*, 1, 112-116.

[Winters84] Winters, J. H. (1984). Optimum combining in digital mobile radio with cochannel interference. *IEEE Transactions on Vehicular Technology,* 33(3), 144-155.

[Winters06] Winters, J. H. (2006). Smart antenna techniques and their application to wireless ad hoc networks. *IEEE Wireless Communications*, 77-83.

[Wu07] Wu, B., Chen, J., Wu, J., & Cardei, M. (2007). A survey on attacks and countermeasures in mobile ad hoc networks, Chapter 12 in *Wireless/Mobile Network Security*, Xiao, Y., Shen, X., & Du, D., Springer.

[Xiao03] Xiao, Y. (2003). A simple and effective priority scheme for IEEE 802.11. *IEEE Communication Letters*, 7(2), 70-72.

[Yang02] Yang, X., & Vaidya, N. H. (2002). Priority scheduling in wireless ad hoc networks. *ACM International Symposium on Mobile Ad Hoc Networking & Computing*, 71-79.

[Yi03] Yi, S., Pei, S., & Kalyanaraman, S. (2003). On the capacity improvement of ad hoc wireless networks using directional antennas. *ACM International Conference on Mobile Computing and Networking*, 108-116.

[Yin03] Yin, J., Zeng, Q.-A., & Agrawal, D. P. (2003). A novel priority based scheduling scheme for ad hoc networks." *IEEE Vehicular Technology Conference*, 3, 1637-1641.

[Zhang05] Zhang, Z. (2005). Pure directional transmission and reception algorithm in wireless ad hoc networks with directional antennas. *IEEE International Conference on Communications,* 5, 3386-3390.

[Zhang06] Zhang, Y., Li, X., & Amin, M. G. (2006). Multi-channel smart antennas in wireless networks. *Asilomar Conference on Signals, Systems, and Computers*, 305-309.

[Zhou99] Zhou, L., & Haas, Z. J. (1999). Securing ad hoc networks. *IEEE Networks Magazine*, 13(6), 24-30.

[Zorzi06] Zorzi, M., Zeidler, J., Anderson, A., Rao, B., Proakis, J., Lee, A., Jensen, M., & Krishnamurthy, S. (2006). Cross-layer issues in MAC protocol design fro MIMO ad hoc networks. *IEEE Wireless Communications*, 62-76.

# Chapter XX
# Key Generation System Using Smart Antenna

**Toru Hashimoto**
*ATR Wave Engineer Laboratories, Japan*

**Tomoyuki Aono**
*Mitsubishi Electric Corporation, Japan*

## ABSTRACT

*The technology of generating and sharing the key as the representative application of smart antennas is introduced. This scheme is based on the reciprocity theorem of radio wave propagation between the two communication parties. The random and intentional change of antenna directivity that is electrically changed by using such an ESPAR antenna as variable directional antenna is more effective for this scheme, because the propagation environment can be undulated intentionally and the reproducibility of the propagation environment can be decreased. In this chapter, experimental results carried out at many environments are described. From these results, this system has a potential to achieve the "unconditional security."*

## INTRODUCTION

The wireless communication has become more popular and convenient by progressing of the key technologies. PDC (Personal Digital Cellular) and WLAN (Wireless LAN) have become an indispensable for many persons. On the other hands, wireless communications have danger that it is hard to notice to tapping by eavesdroppers unlike the wired communication. The common key encryption scheme is the best way to protect the wireless communication data from the eavesdroppers.

The common encryption scheme consists of two parts: encrypt and decrypt part and key management part. The encrypt and decrypt part adopts the cryptographic algorithm like AES (Advanced Encryption Standard) standardized by NIST (National Institute of Standards and Technology) in recent systems. The key management part is more important part for the communication system, especially wireless communication system. Although it has many kinds of key management scheme, many schemes only have "computational security." If the eavesdroppers have much computational power like quantum computer, they can obtain the key in practical time.

Thus the information-theoretic (unconditional) security technology attracts attention recently in various communications. One of the technologies for key generation that achieve "information-theoretic (unconditional) security" is using fluctuation of the communication channel with such smart antenna as variable directional antenna. The "information-theoretic (unconditional) security" means that although the eavesdroppers have much computational power, it can not be unguessable from the information the eavesdroppers obtain.

The generated key is used to the common key encryption that is suitable for wireless communication data encryption between legitimate terminals. In general, the reciprocity theorem of radio wave propagation is established between legitimate terminals in wireless communications. The propagation environment of the third party listening in another place is different from that of legitimate terminals, so the eavesdropper is difficult to generate an identical key that generated between legitimate terminals. In addition, the random and intentional change of antenna directivity that is electrically changed by using such smart antenna as variable directional antenna is more effective for this system, because the propagation environment can be undulated intentionally and the reproducibility of the propagation environment can be decreased.

The technology of generating and sharing the key for common key encryption as the application of smart antenna is described in this chapter including the principle, component of this system, procedure of key generation and the experimental results at various environments performed by Aono et al(Aono, Higuchi, Ohira, Komiyama & Sasaoka, 2005) (Aono, Higuchi, Ohira, Komiyama & Sasaoka, 2006) (Aono, Higuchi, Taromaru, Ohira & Sasaoka, 2005).

## KEY GENERATION SYSTEM USING THE FLUCTUATION OF RADIO WAVES

In this section, to achieve "information-theoretic (unconditional) security" in encrypted wireless communication, adopting the key generation method that the fluctuation of the radio wave channel response is used is the most suitable.

Smart antenna like an ESPAR antenna (Ohira & Cheng, 2004) is effective to an intentional fluctuation of the radio waves. This fluctuation of radio waves and the reciprocity theorem of radio wave propagation are the key technology to achieve security.

### The Principle of the Key Generating and Sharing System

This system is based on the reciprocity theorem of radio wave propagation between the two terminals such as Access Point (AP) and User Terminal (UT). In addition, a smart antenna like an ESPAR antenna is used for intentional fluctuation. The key is made by this fluctuation of radio waves.

The principle of key generation and sharing is described. The beam-forming technique of the ESPAR antenna; that is, adjusting the DC voltage given to the varactors with reverse bias is used effectively. Furthermore thanks to the reciprocity theorem of radio wave propagation, the Received Signal Strength Indicator (RSSI) obtained by alternately transmitting short packets between the two terminals has the proportional relation. From this relation, RSSI profile obtained by AP and UT independently has also the proportional relation and the same encoded value is provided by making these profiles multilevel coding, for example, binary coding. As a result, these encoded values can treat as generated and shared keys.

The propagation characteristics between AP and eavesdropper (EV) is different from it between AP and UT so that an EV listening at another place has difficulty to obtain the same encoded value. Thus "unconditional (information-theoretic) Security" is achieved practically by using the fluctuation of radio waves.

### The Procedure of the Key Generating and Sharing System

#### Configuration

The configuration of this key generating and sharing system describes below. This system consists of two terminals, AP and UT.

AP's outline drawing is shown in Fig.1, and its function block diagram is shown in Fig. 2. It consists of four parts: "Communication device," "Microcontroller," "D/A converter," and "7-elements ESPAR antenna."

- The "Communication device" sends and receives such data as syndrome, initial value, etc. which are set by a microcontroller and communicate with packets through an ESPAR antenna. The received level of packets is measured and converted into an RSSI value in this device.
- "Microcontroller" carries out the "making RSSI profile" and "generating the key" steps in the "key generator" function. These steps are based on the RSSI value from the "Communication device." The "7-element ESPAR

antenna" needs DC voltage to control the beam pattern of the antenna. The seeds of DC voltage, called reactance vectors, are made in the "Pseudo random number generator" function in the "Microcontroller," which is set to the "D/A converter."

- The "D/A converter" changes the digital value of the reactance vectors into analog DC voltage and sets this voltage to the varactor in the "ESPAR antenna." The resolution of "D/A converter" has effects on the number of radiation patterns of the ESPAR antenna. For example, when the resolution of "D/A converter" is 4bit, the radiation patterns of 7-element ESPAR antenna result in $(2^4)^6 = 16^6 = 16777216$ patterns, because it has six parasitic elements.
- The "ESPAR antenna," which is a variable-directional array antenna with a single central active radiator surrounded by parasitic elements, is a component of AP. The six parasitic elements are located at equal intervals around a single central active radiator, and each is loaded with a varactor diode, which is a variable-capacitance diode. By adjusting the DC voltage given to the varactors with reverse bias, the antenna's beam can be formed.

UT's outline drawing is shown in Fig. 3, and its function block diagram is shown in Fig. 4.They consist of three parts: "an omni-directional patch antenna," the "communication device," and the "microcontroller." The role of each block is identical to AP except for the antenna.

## Procedure

The procedure of this key generating and sharing system is described in detail. The precondition of wireless communication system is following, AP and UT can communicate at the same frequency by a method such as Time Division Duplex (TDD). The DC voltage is set to each varactor diode that is connected to the parasitic element of the 7-element ESPAR antenna by using a pseudorandom number.

The procedure of this system is shown in Fig. 5. This procedure consists of four processes; RSSI profile generation process, data deletion process, binary coding process, and key disagreement correction process, and describes in detail as follows.

# RSSI PROFILE GENERATION PROCESS

To generate RSSI profile at both terminals independently, beam-steering technique of ESPAR antenna is used. AP decides a certain directional beam pattern by obtaining pseudorandom numbers and sends a short packet to UT, which receives a packet from AP and computes the RSSI value by measuring the received level. A computed RSSI value is recorded in UT's memory. UT returns an identical short packet to AP as soon as possible to maintain a propagation environment. AP receives a packet from UT and computes and records the RSSI value as well as UT. If a packet isn't received by UT or AP, a packet-send-error is recognized by timeout, and AP will change the beam pattern and retry transmission.

Next, AP changes the directional beam pattern and repeats the above procedure "N" times. Here, "N" consists of key length "K" and the length of redundant bit "$\alpha$", $N = K + \alpha$. AP and UT can independently compile a RSSI data profile.

*Figure 1. AP's outline drawing with 7-element ESPAR antenna*

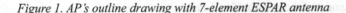

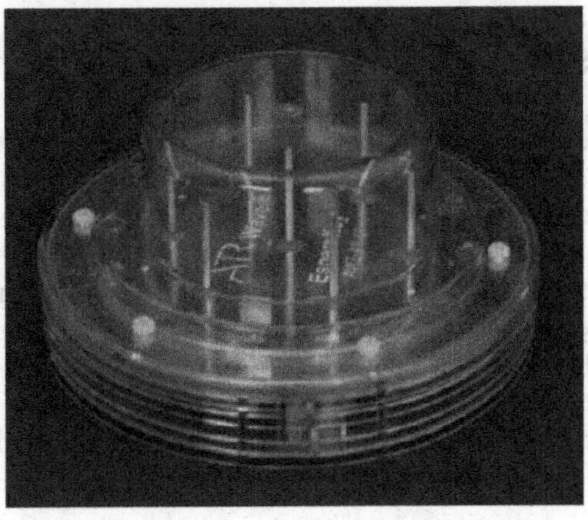

*Figure 2. AP's block diagram*

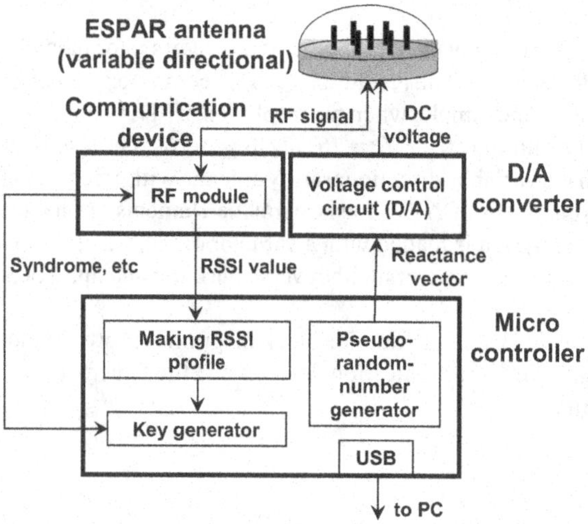


*Figure 3. UT's outline drawing with omni-directional antenna*

Thanks to the reciprocity theorem of radio wave propagation between AP and UT, this profile in the terminals has a proportional relation, except for the random noise, the differences in transmission power, the receiver's noise figures, and such antenna performance factors as sensitivity or directivity. From each profile, AP and UT independently define a threshold level that is considered the median value of each RSSI profile in this system. Now the value of the RSSI profile includes noise components that affect the rate of agreement of the generated and shared key.

## DATA DELETION PROCESS

To reduce the bad influence like noise for key generation, data deletion process of the threshold neighborhood is adopted, which is shown in Fig. 6.

AP and UT have an RSSI profile the size of "N". As shown above, "N" consists of key length "K" and the length of redundant bit "$\alpha$". In Fig. 6, "K" is defined as 6 and "$\alpha$"as 4. From this profile, the threshold level is decided by the median value of each RSSI profile sorted afterwards in each terminal by small order.

At AP, a subset of the RSSI value is chosen as the most susceptible place for noise by picking the largest "$K/2 + \beta$" and the smallest "$K/2 + \beta$" RSSI values, with "$\beta < \alpha/2$" (in Fig. 6, "$\beta$" is defined as 1). The positions of the unchosen RSSI values are transmitted to UT and deleted at AP and UT.

At UT, the RSSI values undeleted by transmitted information from AP are sorted again and the process is repeated, this time choosing the largest "K/2" and smallest "K/2" RSSI values from the remaining RSSI values. Again, the posi-

Figure 4. UT's block diagram

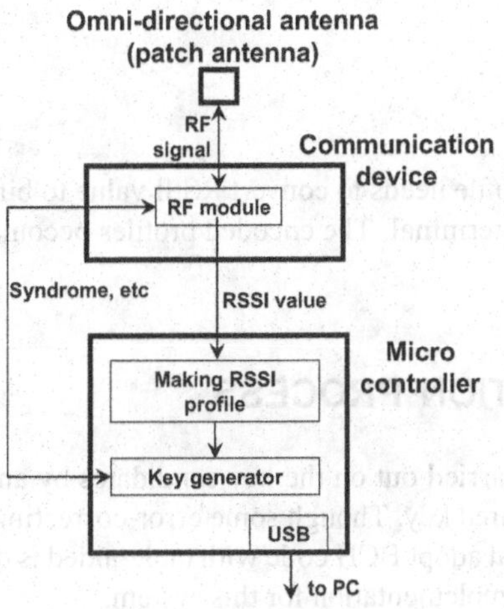


Figure 5. Key generating and sharing procedure

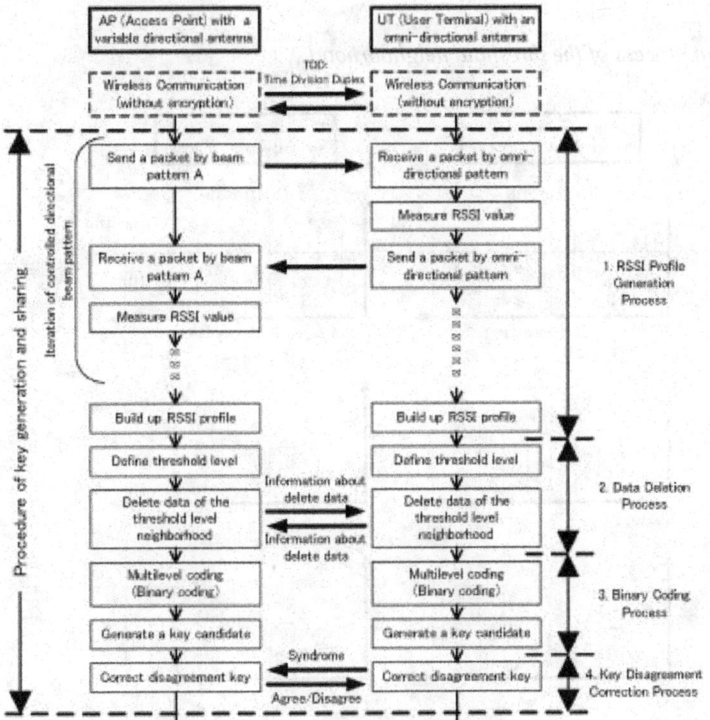

tions of the unchosen RSSI values are transmitted to AP, and unchosen RSSI values are deleted at both terminals. The remaining RSSI profiles whose size is "K" are sorted again by original number.

## BINARY CODING PROCESS

The key for encryption is binary data. The profile needs to convert RSSI value to binary data and independently is encoded binary by using threshold level at each terminal. The encoded profiles become candidates for the generated and shared key.

## KEY DISAGREEMENT CORRECTION PROCESS

The key disagreement correction process is carried out on the key candidates by an error-correcting code to improve the rate of agreement of the generated and shared key. Though some error-correcting codes such as BCH code, LDPC, etc. are considered for this system, the case that adopt BCH code with table-aided is described following as an example. The BCH code has an advantage in terms of implementation for this system.

In this process, the bit patterns of key candidates are basically considered as a series of block code. Furthermore each terminal has a table for error correction. In this condition, each terminal calculate syndrome, $S_{AP} = x_{AP}\mathbf{H}^{\mathrm{T}}$ and $S_{UT} = x_{UT}\mathbf{H}^{\mathrm{T}}$, where $x_{AP}$ and $x_{UT}$ are bit patterns in the key candidates of AP and UT, respectively, $\mathbf{H}^{\mathrm{T}}$ denotes a check matrix, and superscript $\mathbf{T}$ is the transpose of the matrix. Calculated syndrome $S^{UT}$ is transmitted from UT to AP. AP

*Figure 6. Data deletion process of the threshold neighborhood*

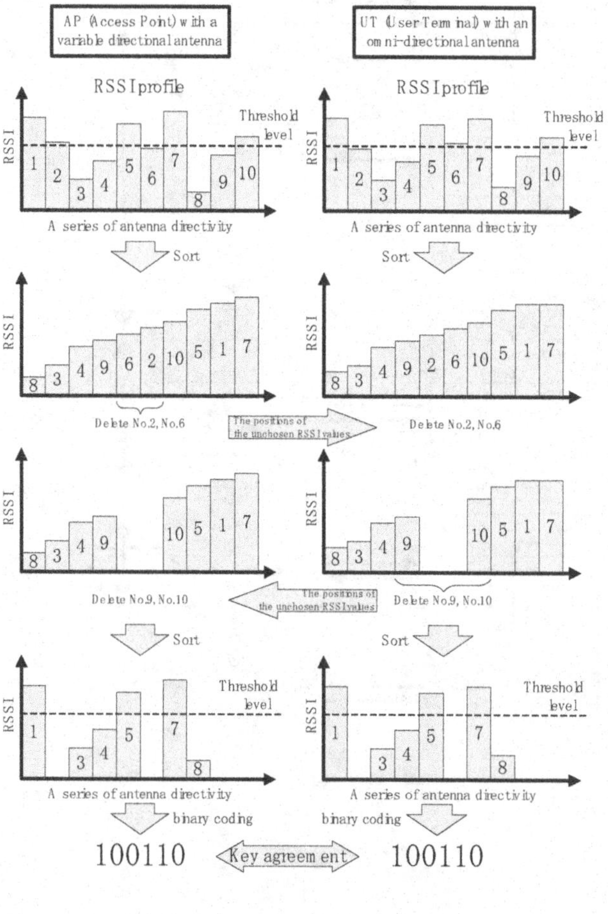

define the differences in the syndrome as $S = S_{AP} - S_{UT}$ and in the bit patterns as $e = x_{AP} - x_{UT}$. The relationship between these parameters is expressed as $S = e\mathbf{H}^{T}$. If $S = 0$ is true, then $e = 0$ is true, and both bit patterns of the key candidates are considered to be in agreement. If $S = 0$ is false, then AP will correctly estimate $e$ to minimize the number of disagreement bits by an error correction technique. If no agreement is obtained after the key disagreement correction process, the generated key is rejected, and the entire process is repeated.

## EXPERIMENTAL RESULT

In this section, the experimental results that are performed by Aono are introduced.

### System Configuration for Experimental Equipment

For feasibility experiment, experimental terminals based on IEEE 802.15.4 are established. AP's outline is shown in Fig. 1 and UT's outline is shown in Fig. 3. IEEE 802.15.4 is the international standard specification for wireless PAN (Personal Area Network) and regulates the physical layer and the MAC (Media Access Control) layer. The simple specifications of the IEEE802.15.4 standard are shown in Table. 1. In this experiment, each terminal is connected to PC by a USB interface.

### Precondition

The preconditions of the key generation method are described as follows. Key length "K" is 128 bits and the length of redundant bit "α", described in Subsection 2.2.2, is basically 256 bits. In other words, the number of RSSI value "N" is 384 at each terminal.

This experimental system has an EV, whose figure and function block diagram is the same as UT. The difference between UT and EV is only the positions in which the packet was received.

### Experiments

#### Experiment When One Wireless Channel is Used for Generating and Sharing the Key

In this feasibility experiment, a sketch of experimental room is shown in Fig. 7 and there are three evaluation indexes; distribution of disagreement bits in generated keys, generated key agreement ratio, and independence of generated key.

- Distribution of disagreement bits in generated keys

The distribution of disagreement bits for the generated secret keys is shown in Fig. 8 and Fig. 9. In this case, the eavesdropper was located at the position of "eavesdropper 4". In the case of Fig. 8, secret keys were generated without a data deletion process shown in Fig. 5, that is, "α"=0 in Subsection 2.2.2. The upper graph shows the disagreement bits

*Table 1. Specifications of IEEE802.15.4*

| | |
|---|---|
| Frequency | 2.4 GHz (ISM Band)<br>• Same as wireless LAN (IEEE 802.11b/g) |
| Channel number | 16 CH (CH 11-CH 26)<br>• use arbitrary 1 CH for control |
| Transmission power | 1 mW |
| Data rate | 250 kbps |

*Figure 7. Sketch of experimental room*

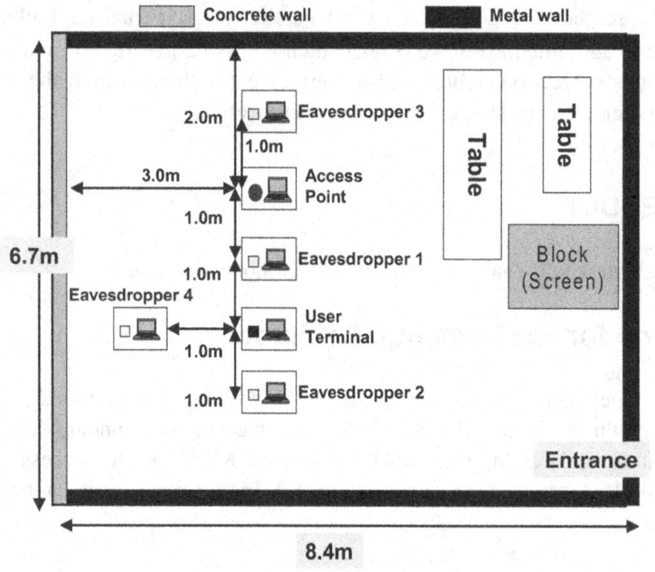


*Figure 8. Disagreement bits distribution (without data deletion process: $\alpha = 0$)*

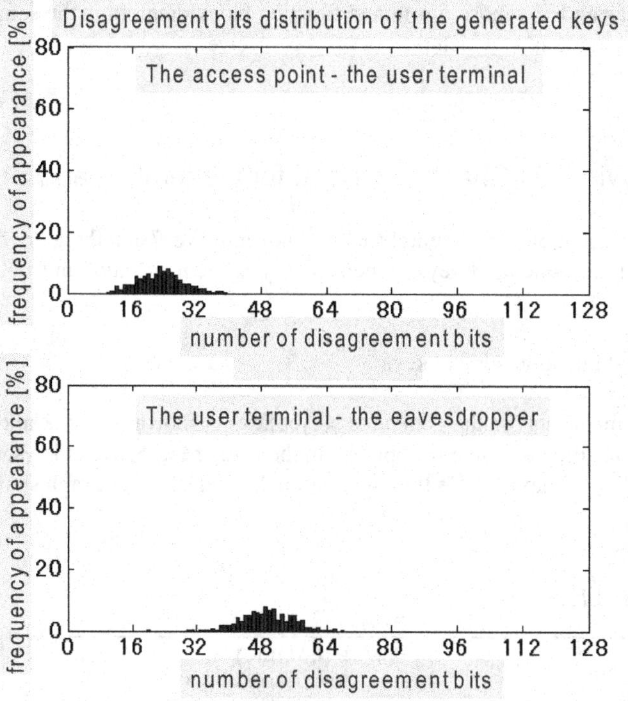

distribution of the generated secret keys between AP and UT, and the lower graph shows that between UT and EV. Both graphs show that many disagreements occur. Although the upper case involves regular communication parties, there are too many disagreement bits to correct the disagreement. On the other hand, in Fig. 9, the upper graph shows that the disagreement bits are not so serious, and thus the disagreement correcting process is effective.

- Generated key agreement ratio

Fig. 10 shows the generated key agreement ratio between AP and UT. The agreement ratio means the percent of generated keys for which complete agreement is obtained. This graph shows the capability of an error-correcting codes needed for disagreement correction in a 32-bit process, and, by using 8-bit disagreement correction (2-bit correction for a 32-bit situation), the agreement ratio improved to more than 90% regardless of where the eavesdropper was located as a barrier.

- Independence of generated key

The independence of generated key is evaluated as the correlation coefficient of two generated keys: one is generated by UT and the other is generated by EV. The former is expressed as $X = [x_1, \cdots, x_i, \cdots, x_K]$, and the latter is expressed as $Y = [y_1, \cdots, y_i, \cdots, y_K]$. "K" is the key length in bits. The correlation coefficient is defined as follows:

$$\rho = \frac{S_{xy}}{\sigma_x \sigma_y}$$

where $S_{xy}$ is the covariance between $X$ and $Y$ given by

*Figure 9. Disagreement bits distribution (with data deletion process: $\alpha = 256$)*

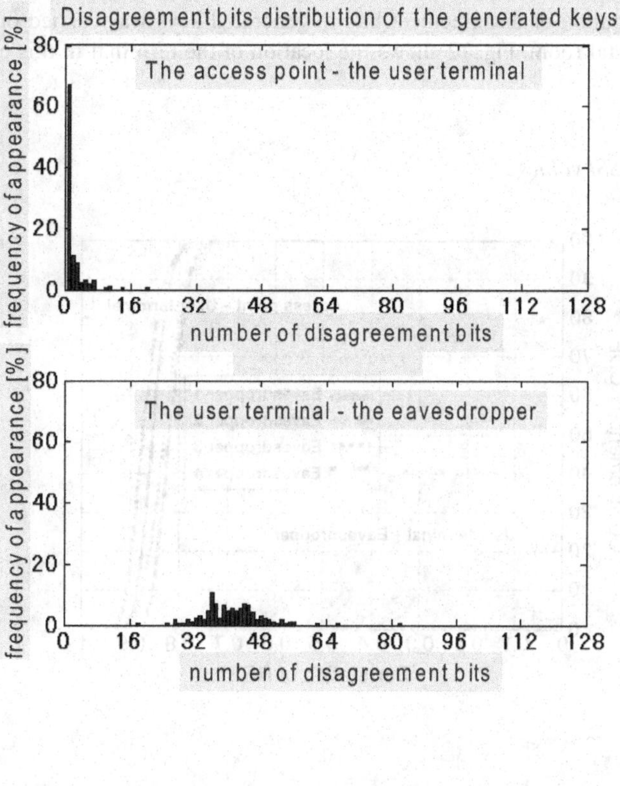

*Figure 10. Key agreement ratio*

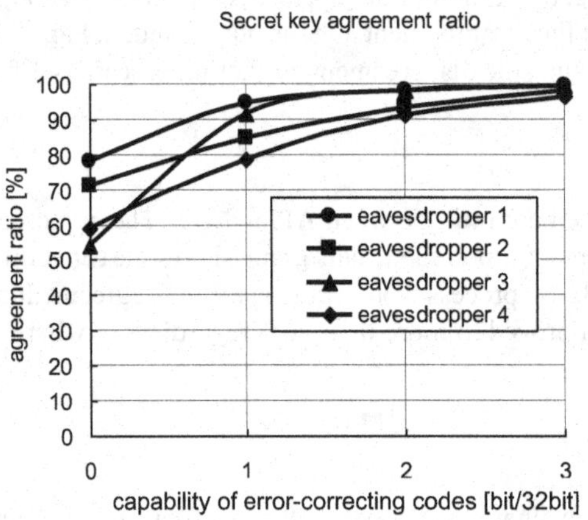


$$S_{xy} = \frac{1}{K}\sum_{i=1}^{K}(x_i - \bar{x})(y_i - \bar{y})$$

where $\sigma_x$ and $\sigma_y$ are the standard deviations of *X* and *Y*, respectively.

Fig. 11 shows the correlations of measured RSSI. The right-side lines are the correlations of AP and UT, and the left-side lines are those of UT and EV. The vertical scale of this graph is the complementary Cumulative Distribution Function (CDF). Although the lines of the graph are differently influenced by the eavesdropper's position, the correlation of UT and EV is nearly as low as 0.5. As a result, this scheme has the ability to generate and share keys with sufficient independence.

## Experiment When the Access Point is Located Near the Ceiling of the Experimental Room

The evaluation about distribution of disagreement bits in generated keys was carried out when AP was set up near the ceiling of the experimental room. Fig. 12 shows the location of the terminal in this experiment. Fig.13 shows the

*Figure 11. Correlations of RSSI values*

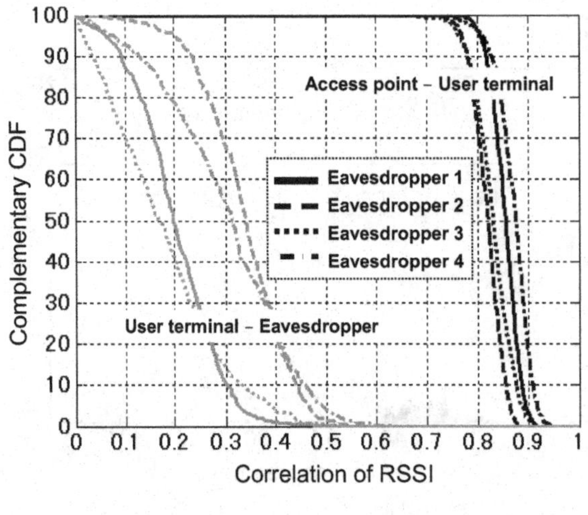

*Figure 12. Location of each terminal in the experiment at a different height*

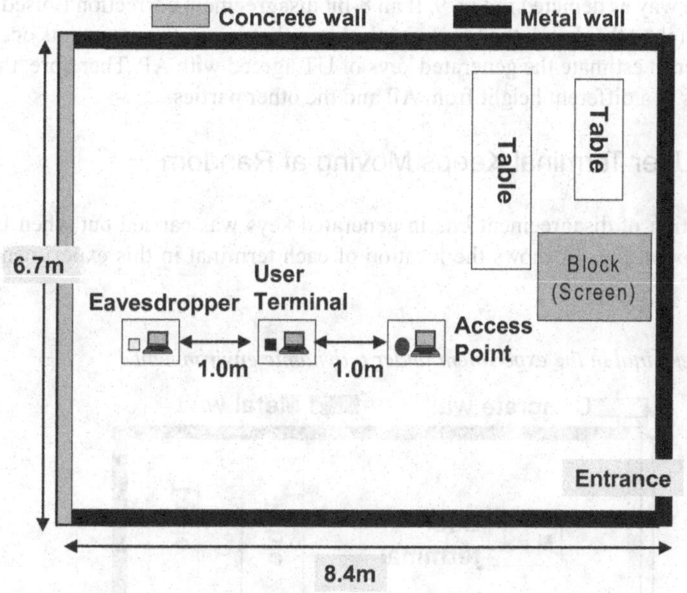


*Figure 13. Disagreement bits distribution at a different height from AP and other parties*

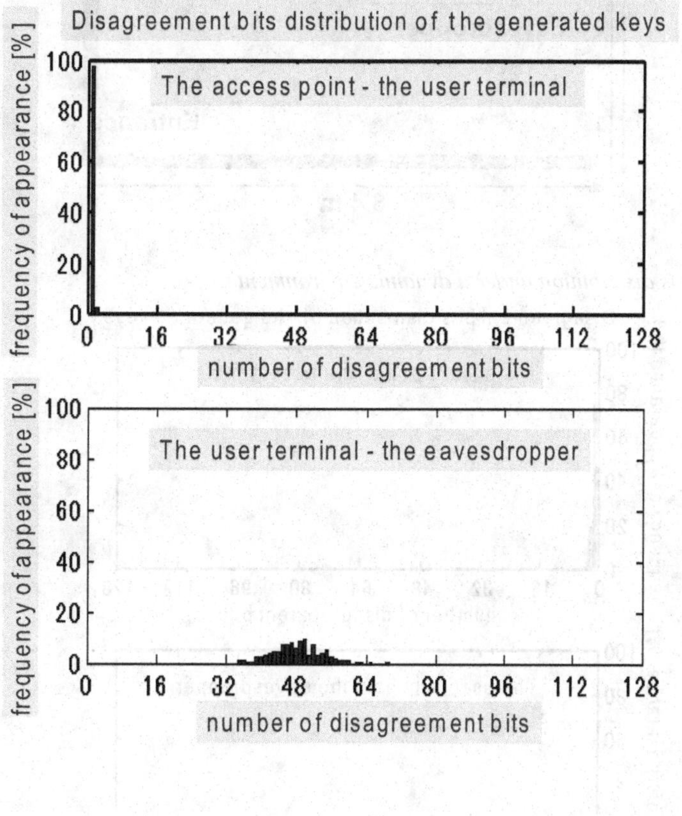

experimental results for the disagreement bits distribution. The upper graph shows that the disagreement bits are very few in many cases, in the same way as depicted in Fig. 9. If an 8-bit disagreement correction is used, key agreement can succeed in almost any case. On the other hand, the lower graph shows that many disagreements occur, indicating that it is difficult for the eavesdropper to estimate the generated keys of UT agreed with AP. Therefore, there was no problem in generating and sharing keys at a different height from AP and the other parties.

## Experiment When the User Terminal Keeps Moving at Random

The evaluation about distribution of disagreement bits in generated keys was carried out when UT keeps moving at random in the experimental room. Fig. 14 shows the location of each terminal in this experiment. Fig. 15 shows the

*Figure 14. Location of each terminal in the experiment under a dynamic environment*

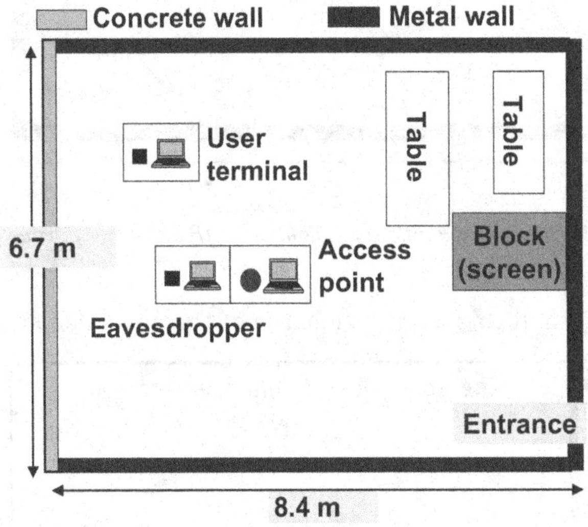


*Figure 15. Disagreement bits distribution under a dynamic environment*

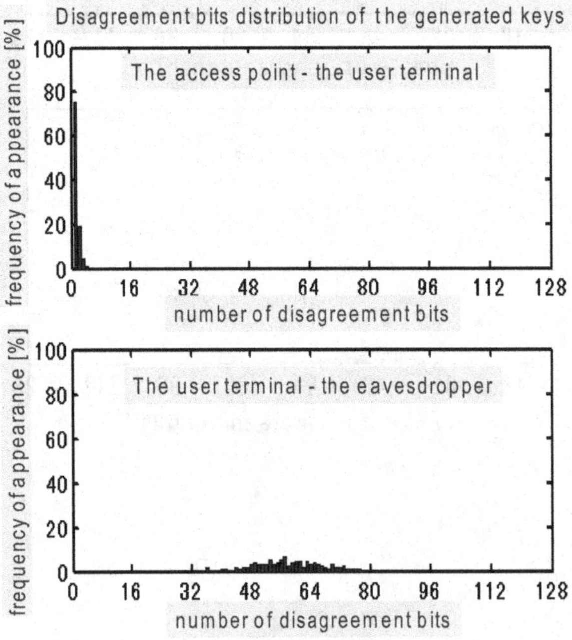

experimental results for the disagreement bits distribution. The upper graph shows the distribution of the generated keys between AP and UT, and the lower graph shows that between UT and EV. The upper graph shows that there are very few disagreement bits shown in another cases. On the other hand, the lower graph shows that many disagreements occur at the same as previous case.

Thinking about an experiment result from a different viewpoint, other results are found. An example of the experimental result is shown in Fig. 16. This graph shows normalized RSSI profiles of AP, UT and EV. Although these two lines of AP and UT are nearly proportional to each other, the key generated from these RSSI profiles is not applicable because this key is biased and weak with regard to security. Fig. 17 shows the bit distribution of the generated keys of AP and UT. One line of this graph indicates one generated key, and the white and the black parts of the line indicate "zero" and "one", respectively. Since there are many runs of zero and one, these are not available for use with secret keys.

To acquire the unguessable keys by observing the location of UT, a data interleaving technique was applied to this system. The interleave scheme applied in this system that the RSSI value is measured in time series is shown in Fig.18. Consequently, the generated key agreement ratio was 99.6% in the case of using 8-bit disagreement-correction, which is the same as the result obtained without interleaving technique. The RSSI profile shown in Fig.16 was, nevertheless, unguessable as shown in Fig.19 because of the randomized profile. The bit distribution of the generated keys is improved as shown in Fig.20 and runs of zero and one are decreased. By this improved scheme, the generated secret keys were made unguessable against the eavesdropper.

*Figure 16. An example of the experimental result*

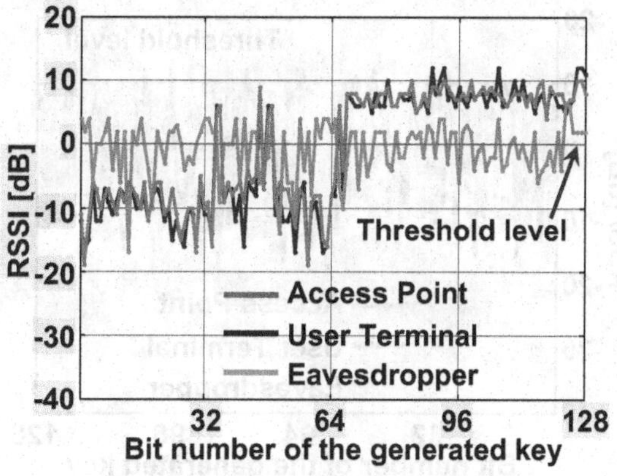


*Figure 17. The distribution of generated key*

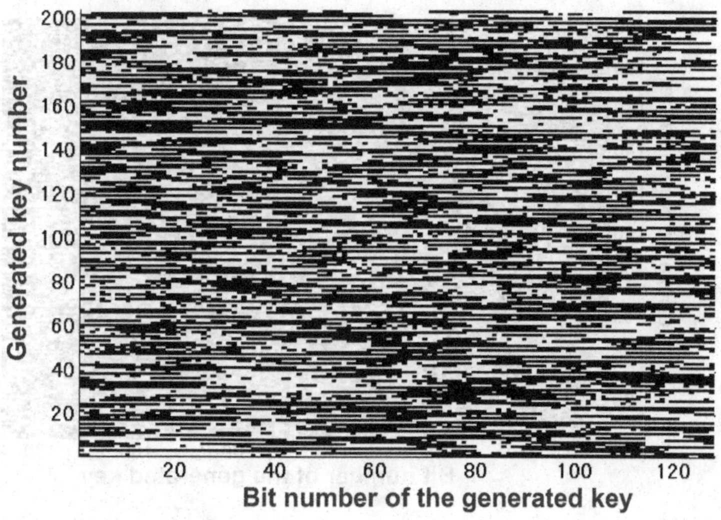

Figure 18. RSSI interleave scheme

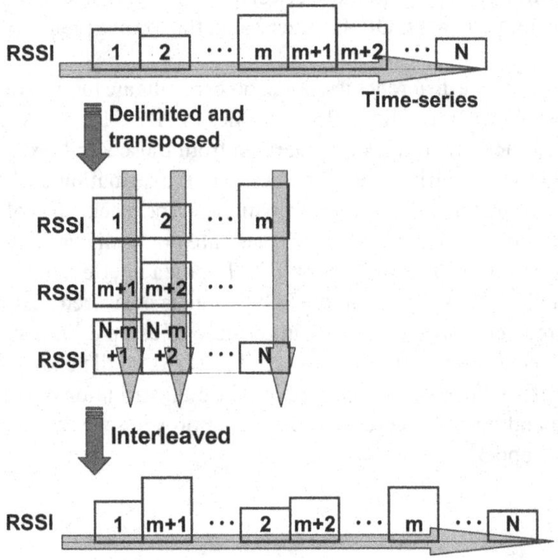


Figure 19. An example of the experimental result applying RSSI interleave scheme

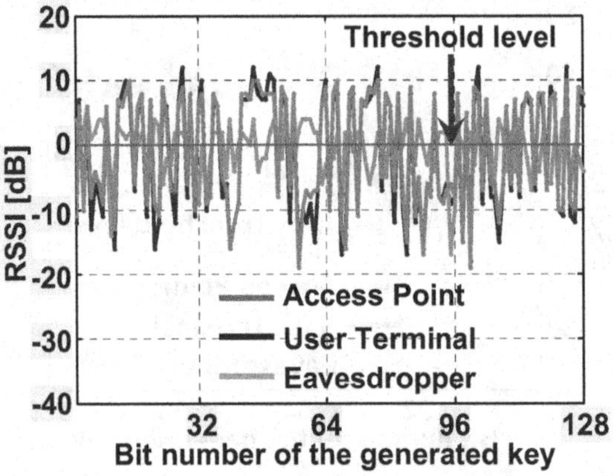


Figure 20. The distribution of generated key applying RSSI interleave scheme

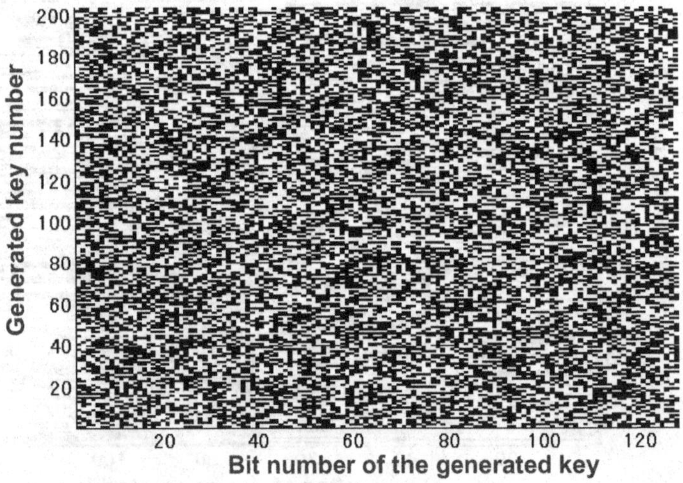

*Figure 21. Sketch of experimental room*

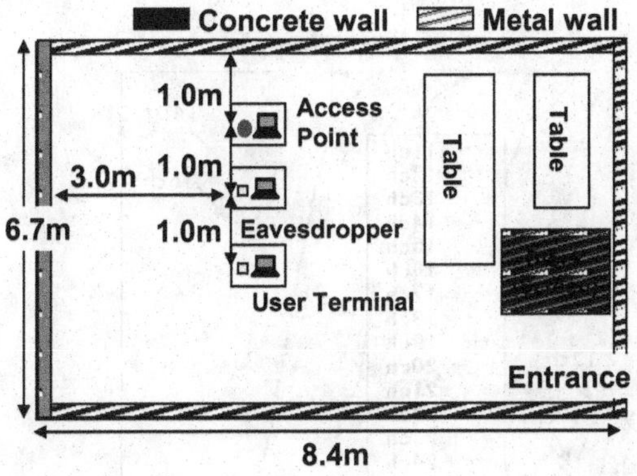


*Figure 22. RSSI correlation between Access Point and User Terminal*

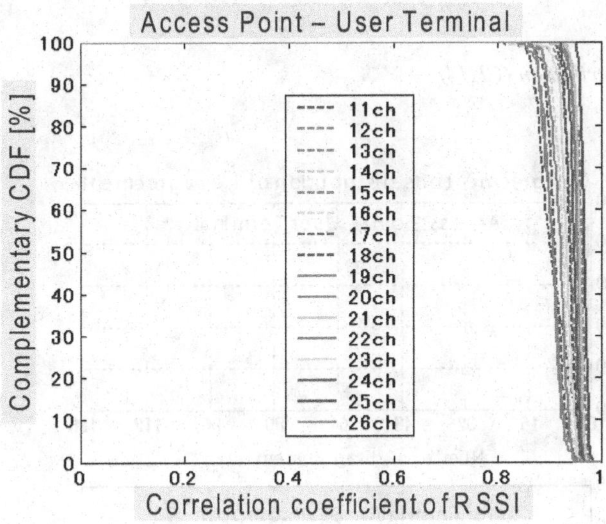


## Experiment in the Experimental Room with the Frequencies Switching Key Generation Scheme

The experiment that was performed in the experimental room is presented. The experimental room is shown in Fig. 21 that is same room to previous experiments. The experimental results about the correlations of the measured RSSI values are shown in Fig. 22 and Fig. 23. In these graphs, the RSSI correlation coefficient values are represented by the horizontal axis, and the complementary cumulative distribution function (complementary CDF) is represented by the vertical axis. In Fig. 22, it is found that the correlation between Access Point and User Terminal is sufficiently high that agreement of the generated key is successful. On the other hands, it is found that there are cases with high correlation between User Terminal and Eavesdropper in different channels. The disagreement bit distribution for CH14 as an example of the high correlation characteristic is presented in Fig. 24. The lower graph presents the disagreement bit distribution between User Terminal and Eavesdropper. The distribution is very close to that in the upper graph, indicating the disagreement bit distribution between the legitimate terminals. The tendency is similar for CH16, with a high correlation characteristic.

Since the RSSI correlation with eavesdropper is dependent on the locations of the terminals and other ambient conditions, it is difficult to estimate a channel with high correlation. In this measurement, the terminals are located on

*Figure 23. RSSI correlation between User Terminal and Eavesdropper*

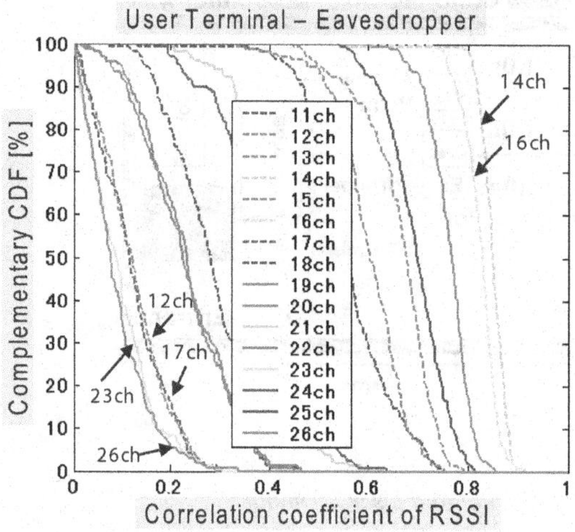


*Figure 24. Disagreement bits distribution (CH14)*

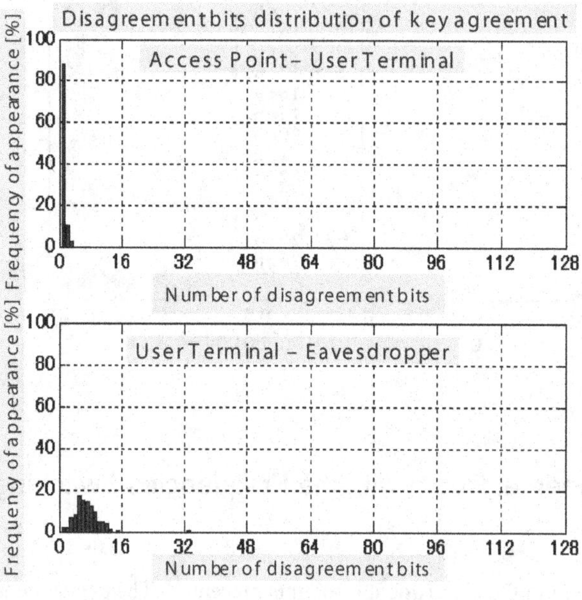


one straight line and are close to each other. Nevertheless, there exist channels whose correlation is not high. Therefore, it is considered possible to reduce the correlation with the eavesdropper by generating the key while switching among several carrier frequencies.

Fig. 25 shows the RSSI correlation characteristics when the keys are generated while switching among 16 channels. In order to generate a 128 bit secret key from 384 packets of measured RSSI values, the channels are switched every 24 packets. It is then found from each graph that the correlation with eavesdropper can be reduced. Also, Fig. 26 show the disagreement bit distributions for this measurement. This result shows the frequencies switching key generation scheme is effective for key generation.

*Figure 25. RSSI correlation between User Terminal and Eavesdropper with the frequencies switching key generation scheme*

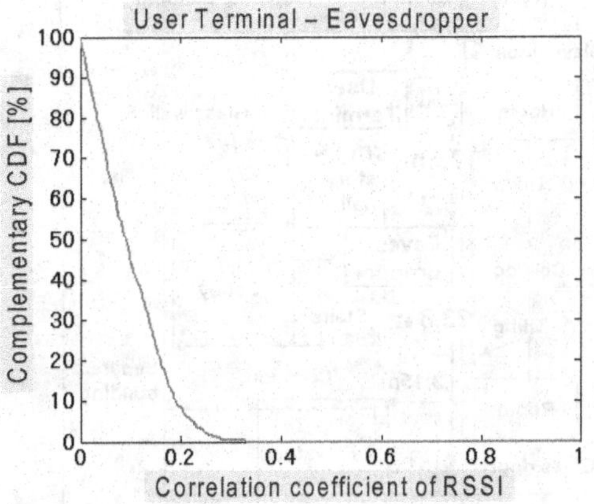


*Figure 26. Disagreement bit distribution with the frequencies switching key generation scheme*

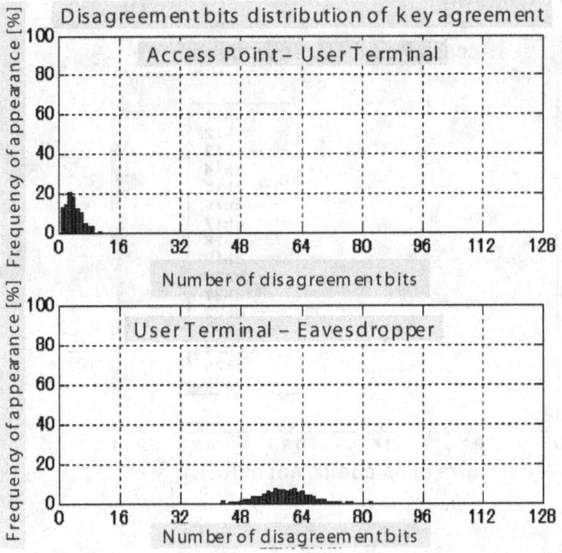


## Indoor Experiment with the Frequencies Switching Key Generation Scheme

In this subsection, the experiment that carried out another place and same condition at previous subsection is presented. Fig. 27 shows the location of the terminal in this experiment. The correlations of the measured RSSI values are shown in Fig. 28 and Fig. 29. Furthermore the RSSI correlation characteristics when the keys are generated while switching among 16 channels are shown in Fig.30 and its disagreement bits distributions is shown in Fig. 31. The tendency of all results is similar to the results shown in previous subsection.

*Figure 27. Sketch of the experimental environment*

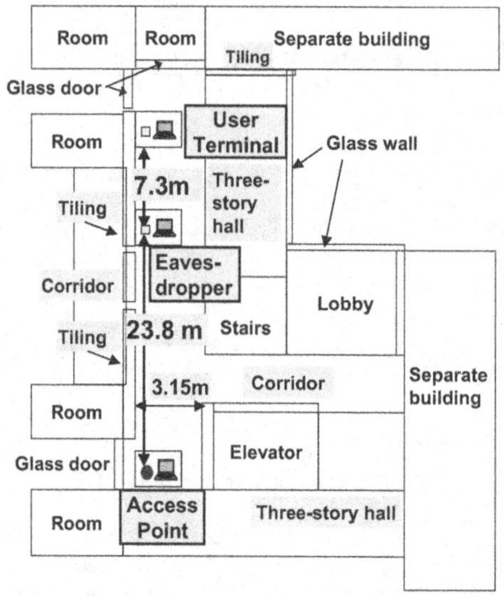


*Figure 28. RSSI correlation between Access Point and User Terminal*

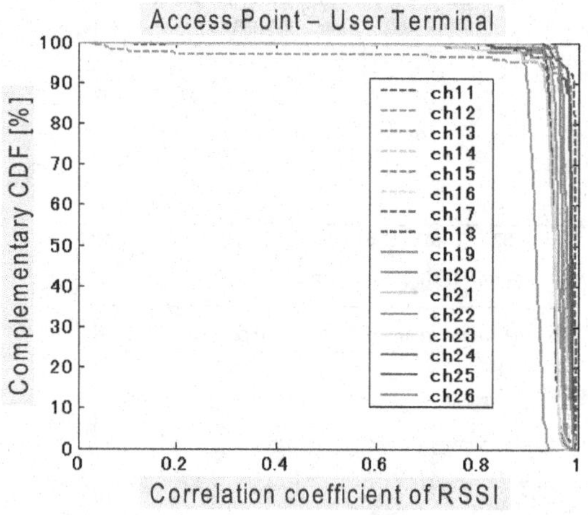


## Experiment When User Terminal and Eavesdropper are Adjacent

The experiment was performed at the same experimental room shown in Fig. 21. The distance between User Terminal and Eavesdropper is 10cm and the distance between Access Point and Eavesdropper is 1m. Access Point, User Terminal, and Eavesdropper are located along a straight line. Although the correlation between Access Point and User Terminal is high in all channels, that between User Terminal and Eavesdropper is lower or higher depending on the channel, as shown in Fig. 32.

Fig. 33 shows the RSSI correlation characteristics when the keys are generated while switching among 16 channels and its disagreement bits distribution is shown in Fig. 34. The tendency of these results is similar to the result written in subsection 3.3.5.

*Figure 29. RSSI correlation between User Terminal and Eavesdropper*

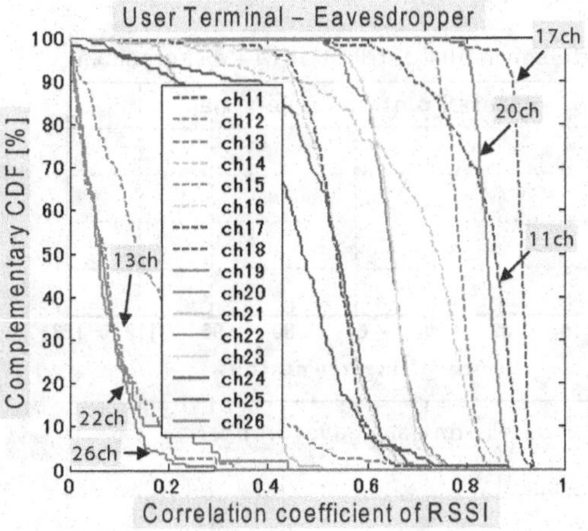


*Figure 30. RSSI correlation between User Terminal and Eavesdropper with the frequencies switching key generation scheme*

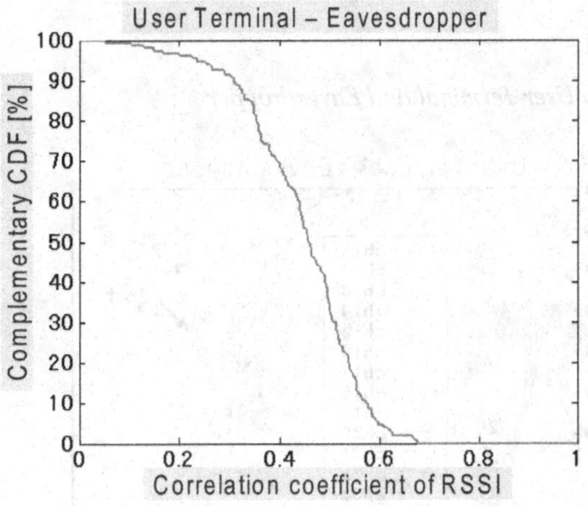


## Outdoor Experiment

As shown in Fig. 35, a key generation experiment is performed outdoors without reflecting objects such as walls. User Terminal is placed adjacent to Eavesdropper, about 10 cm away on Access Point side. The terminals are placed on plastic carts. The distance between Access Point and User Terminal is 3 m. Access Point, User Terminal and Eavesdropper is placed on a straight line. As the result, the correlation between User Terminal and Eavesdropper is lower or higher depending on the channel, as shown in Fig. 36. The results when the keys are generated while switching among 16 channels are shown in Fig. 37 and Fig. 38. The tendency of these results is also similar to the result written in previous subsection.

From these results, the frequencies switching key generation scheme is effective.

# FUTURE TRENDS

The progress of such wireless communication technologies as RF devices, signal processing, modulation methods, antennas, etc. have greatly contributed to the cost reduction of terminals and speed-up of wireless communication. As a result,

*Figure 31. Disagreement bit distribution with the frequencies switching key generation scheme*

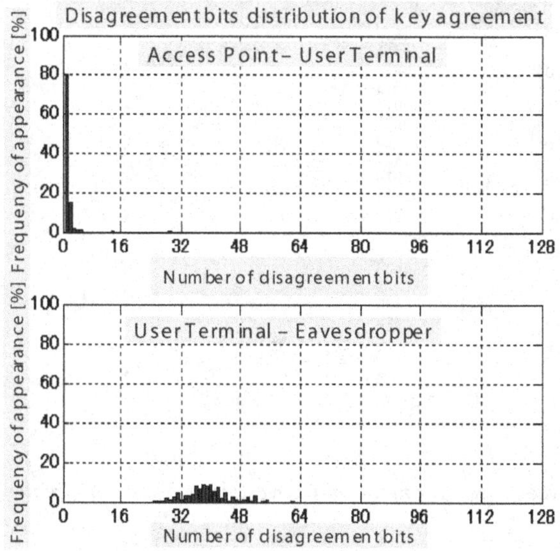


*Figure 32. RSSI correlation between User Terminal and Eavesdropper*

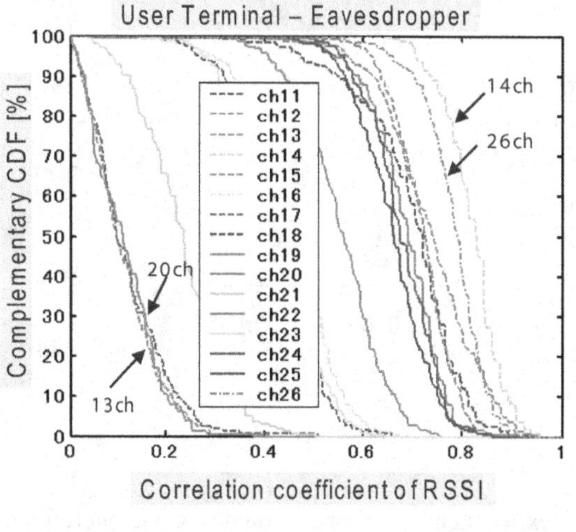


such wireless communication as Wireless LAN or Personal Digital Cellular has become more popular and convenient.

On the other hand, this rapid spreads of wireless communication causes threats, such as tapping by eavesdroppers (passive attacks) and impersonation from third party (active attacks). For active attacks, mutual authentication is the best solution in general, many kinds of which are incorporated in much application software. Thus, it is necessary to improve the security for wireless system. To improve the security, some schemes are considered. The easiest scheme to protect wireless communication data from passive attackers is to encrypt the data with either public key encryption or common key encryption.

The public key encryption system does not need an identical key for encryption and decryption, which is advantage from view of the key management that includes key distribution, generation, sharing, and so on. But it has a disadvantage of low speed. Therefore, it has mainly been used for key exchange protocol, authentication schemes, and digital signatures.

*Figure 33. RSSI correlation between User Terminal and Eavesdropper with the frequencies switching key generation scheme*

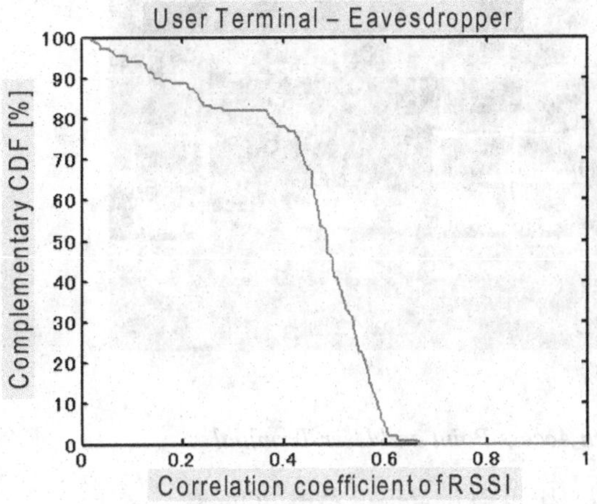


*Figure 34. Disagreement bit distribution with the frequencies switching key generation scheme*

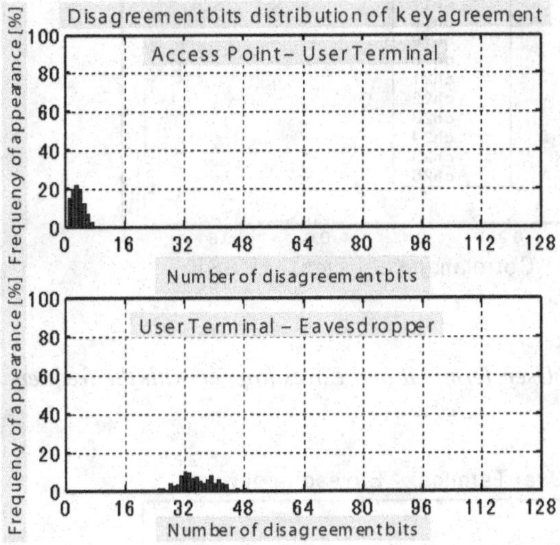


Though common key encryption system is suitable for protecting wireless communication data because it can encrypt and decrypt at high speed, it must manage common secret keys between players. For secure encryption using common key encryption scheme, key management between players is crucial. To manage the key with less effort without decreasing the safety of communication channel is the most important for secure wireless communication.

Many kinds of key management schemes, especially key generating and sharing methods, have been proposed and adopted at many established wireless networks. Representative examples include the Diffie-Hellman key exchange algorithm (Diffie & Hellman, 1976) and key distribution methods using public key encryption such as RSA cryptosystem (Rivest, Shamir & Adleman, 1978). Both methods can be decoded in practical time if eavesdroppers have high computational power, which continues to increase rapidly, because the security of both is based on the difficulty of calculation, such as the Discrete Logarithm Problem or prime factorization called "computational security."

*Figure 35. The location of each terminal at outdoor experiment*

*Figure 36. RSSI correlation between Access Point and User Terminal*

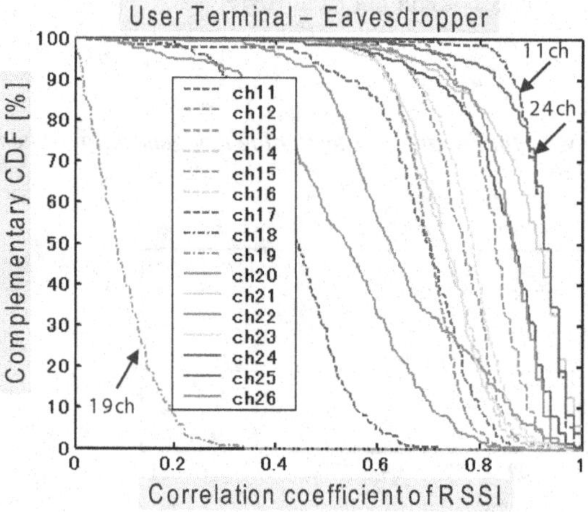


*Figure 37. RSSI correlation between User Terminal and Eavesdropper with the frequencies switching key generation scheme*

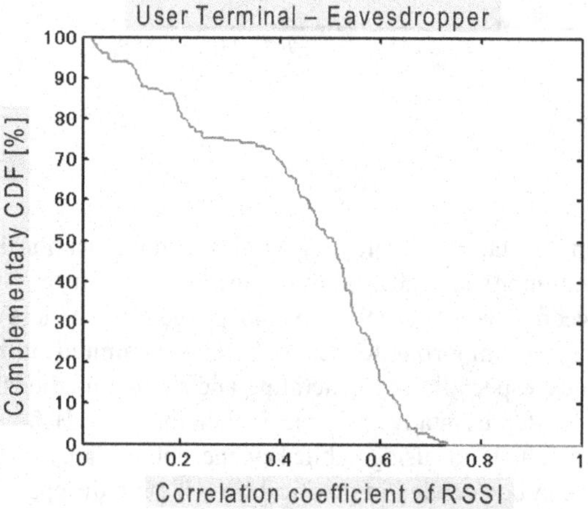

*Figure 38. Disagreement bit distribution with the frequencies switching key generation scheme*

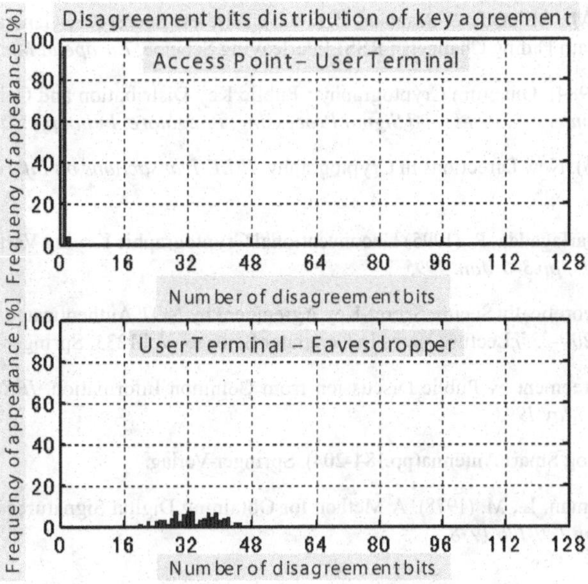


There are "unconditional (information-theoretic) secure methods," such as based on quantum cryptography (Bennett & Brassard, 1984), noisy communication channels (Maurer, 1997) (Maurer, 1993), fluctuation of radio wave channels (Hershey, Hassan & Yarlagadda, 1995). They attract attention recently. Even assuming that eavesdroppers have infinite computational power, these methods are secure since they are based on information-theoretic security.

Although the quantum cryptography for key distribution protocol such as BB84 became popular, it is unsuitable for wireless communication systems because it basically uses optical fibers.

A noisy communication channel model may be a more realistic solution, but it is not practical solution because the length of key depends on the noise environment between the terminals. If some noise environment is correlated with another environment, the key length is shorter than uncorrelated environment.

For wireless communication systems, the fluctuation of the radio wave channel response that was introduced in this chapter is the most practical solution.

The field of this research, especially establishment of other types of key generating and sharing system, evaluation of this type of system will be more important and growing up more and more for wireless system.

## CONCLUSION

In this chapter, the key generating and sharing system using smart antenna is introduced focusing on the experimental results. This system has a potential to achieve the "unconditional security."

Through the experimental results, the possibility of "unconditional secure" is shown at any environments. To improve the security of this key generating and sharing system additionally, optimization of key disagreement correction process and installation of privacy amplification process are considered.

## REFERENCES

Aono, T., Higuchi, K., Ohira, T., Komiyama, B., & Sasaoka, H. (2005). Wireless Secret Key Generation Exploiting Reactance-Domain Scalar Response of Multipath Fading Channels. *IEEE Transactions on Antenna and Propagation, Vol. 53, pp.3776-3784, No. 11, Nov. 2005*

Aono, T., Higuchi, K., Ohira, T., Komiyama, B., & Sasaoka, H. (2006). IEEE802.15.4 ESPARSKey (Encryption Scheme Parasite Array Radiator Secret Key), *Electronics and Communications in Japan, Part 1, Vol. 89, pp.31-44, No. 12, 2006*

Aono, T., Higuchi, K., Taromaru, M., Ohira, T., & Sasaoka, H. (2005). Wireless Secret Key Generation Exploiting the Reactance-domain Scalar Response of Multipath Fading Channels : RSSI Interleaving Scheme, *European Microwaves Conference*

Bennett, C., H., & Brassard, G. (1984). Quantum Cryptography : Public Key Distribution and Coin Tossing, *Proceeding of IEEE International Conference on Computers, Systems, and Signal Processing, Bangalore, India, pp.175*

Diffie, W. & Hellman, M., E. (1976). New Directions in Cryptography, *IEEE Transactions on Information Theory, Vol. 22, pp.644-654, No. 6, Nov 1976*

Hershey, J., E., Hassan, A., A., & Yarlagadda, R. (1995) Unconventional Cryptographic Keying Variable Management, *IEEE Transactions on Communications, Vol. 43, pp.3-6, Jan. 1995*

Maurer, U. (1997) Information-Theoretically Secure Secret-Key Agreement by NOT Authenticated Public Discussion, *Advances in Cryptology-EUROCRYPT'97 (pp.209-225),*Lecture Notes in Computer Science, Vol.1233, Springer-Verlag

Maurer, U. (1993) Secret Key Agreement by Public Discussion from Common Information, *IEEE Transactions on Information Theory, Vol. 39, pp.733-742, No. 3, May 1993*

Ohira, T. & Cheng, J. (2004). Analog Smart Antenna(pp.184-204). Springer-Verlag.

Rivest, R., L., Shamir, A., & Adleman, L., M. (1978). A Method for Obtaining Digital Signatures and Public-key Cryptosystems, *Communications of the ACM, 21, pp.120-126 1978*

# Chapter XXI
# Smart Antennas for Automatic Radio Frequency Identification Readers

**Nemai Chandra Karmakar**
*Monash University, Australia*

## ABSTRACT

*Various smart antennas developed for automatic radio frequency identification (RFID) readers are presented. The main smart antennas types of RFID readers are switched beam, phased array, adaptive beamfsorming and multiple input multiple output (MIMO) antennas. New development in the millimeter wave frequency band—60 GHz and above— exploits micro-electromechanical system (MEMS) devices and nano-components. Realizing the important of RFID applications in the 900 MHz frequency band, a 3×2-element planar phased array antenna has been designed in a compact package at Monash University. The antenna covers 860-960 GHz frequency band with more than 10 dB input return loss, 12 dBi broadside gain and up to 40° elevation beam scanning with a 4-bit reflection type phase shifter array. Once implemented in the mass market, RFID smart antennas will contribute tremendously in the areas of RFID tag reading rates, collision mitigation, location finding of items and capacity improvement of the RFID system.*

## INTRODUCTION

The Radio Frequency Identification (RFID) system is a new wireless data transmission and reception technique for automatic identification, asset tracking, security surveillance and many other emerging applications. An RFID system consists of three major components: a reader or integrator, which sends interrogation signals to an RFID **transmitter responder** (transponder) or tag, which is to be identified; an RFID tag, which contains the identification code; and middleware, which maintains the interface and the software protocol to encode and decode the identification data from the reader into a mainframe or a personal computer. Figure 1 below illustrates a generic block diagram of the RFID system. At the dawn of the new millennium, as barcodes and other means for identification and asset tracking are becoming inadequate for recent demands, RFID technology has been facilitating logistics, supply chain management, asset tracking, security access control, intelligent transportation and many other areas at an accelerated pace. A recent Google search of the terminology 'RFID' brought up thirty eight million hits. This large huge number of URLs represents the significant activities and applications of RFID in various sectors in either commercial domains or government agencies.

RFID technology is an off-shoot miniaturized version of the 'identification, friend or foe (IFF)' radar system developed by British defence during World War II. This radar technology used backscattered signals to identify and/or discriminate friendly targets from enemy targets and enabled decisions to attack appropriate targets. While low frequency RFID tags use strong magnetic coupling by being in proximity to the RFID reader's coil antennas, all ultra high frequency (UHF) and microwave RFID readers and tags are based on the radar principle of sending far-field electromagnetic (EM) interrogating signals from the readers and receiving the back-scattered modulated signals with the unique identification code of the tag. Thus identification of items, human beings and animals is possible in all weather conditions and off line-of-sight communication.

RFID was first proposed by H. Stockman (Stockman, 1948) who introduced the RFID system in his landmark paper "Communication by Means of Reflected Power". Stockman advocated that considerable research and development work was required to solve the basic problems of wireless identification by means of reflected power. A complementary article on the history of RFID can be found in Landt (2001).

Similar to radar technology, RFID is a multi-disciplinary technology which encompasses a variety of disciplines: (i) RF and microwave engineering, (ii) RF and digital integrated circuits, (iii) antenna design, and (iv) signal processing software and computer engineering. The latter encodes and decodes analog signals into meaningful codes for identification. According to Lai *et al* (2005), "The fact that RFID reading operation requires the combined interdisciplinary knowledge of RF circuits, antennas, propagation, scattering, system, middleware, server software, and business process engineering is so overwhelming that it is hard to find one single system integrator knowledgeable about them all. .... In view of the aforesaid situation, this present invention (RFID system) seeks to create and introduce novel technologies, namely redundant networked multimedia technology, auto-ranging technology, auto-planning technology, smart active antenna technology, plus novel RFID tag technology, to consolidate the knowledge of all these different disciplines into a comprehensive product family."

Due to the flexibility and numerous advantages of RFID systems compared to barcodes and other identification systems available so far, RFIDs are now becoming a major player in retail and government organisations. Patronization of the RFID technology by organisations such as Wal-Mart, K-Mart, the USA Department of Defense, Coles Myer in Australia and similar consortia in Europe and Asia has accelerated the progress of RFID technology significantly in the new millennium. As a result, significant momentum in the research and development of RFID technology has developed within a short period of time. The RFID market has surpassed the billion dollar mark recently (Das & Harrop, 2006), and this growth is exponential, with diverse emerging applications in sectors including medicine and health care, agriculture, livestock, logistics, postal deliveries, security and surveillance and retail chains. Today, RFID is being researched and investigated by both industry and academic scientists and engineers around the world. Recently, a consortium of the Canadian RFID industry has put a proposal to the Universities Commission on the education of fresh graduates with knowledge about RFID (GTA, 2007). The Massachusetts Institute of Technology (MIT) has founded the AUTO-ID centre to standardize RFID, thus enabling faster introduction of RFID into the mainstream of retail chain identification and asset management (McFarlane & Sheffi, 2003; Karkkainen, & Ala-Risku, 2003). The synergies of implementing and promoting RFID technology in all sectors of business and day to day life have overcome the boundaries of country, organisation and discipline.

As a wireless system, RFID has undergone close scrutiny for reliability and security (EPCglobal, Inc., 2006). With the advent of new anti-collision and security protocols, efficient antennas and RF and microwave systems, these problems are being delineated and solved. Smart antennas have been playing a significant role in capacity and signal quality enhancement for wireless mobile communications, mobile ad-hoc networks and mobile satellite communications systems. The advent of smart antennas has brought many benefits for the communications industries as many value added

*Figure 1. Generic RFID system*

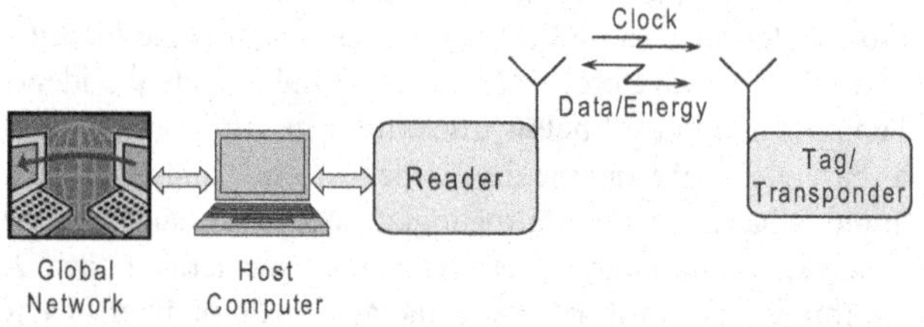

services can be accommodated in modern mobile communications. The spatial and polarization diversities exploited from smart antennas have added new and unique dimensions in wireless communications. The advantages of smart antennas have also been incorporated in RFID systems. Smart antennas are used in RFID readers where multiple antennas and associated signal processing units are easy to implement. Even multiple antennas are proposed in RFID tags to improve reading rate and accuracy (Ingram, 2003).

This chapter concerns smart antennas for RFID readers to enhance the performance of automatic identification systems, asset tracking in real time and inventory control in warehouses. It is important to note that smart antennas for RFID readers are currently in the developmental stage. To the best of the author's knowledge, such smart antenna systems for RFID have not yet emerged as the mainstream enabling technology. Therefore, smart antenna implementation systems for RFID readers are either still on the drawing boards of designers or under investigation, mainly by various commercial research groups as patents, be they either conceptual or physical developments.

The chapter on smart antennas for RFID readers has been organised as shown in the flow chart in Figure 2 below. An introduction to RFID is presented first. The limitations of barcodes as the currently available mainstream identification system are presented next, followed by RFID as the replacement and enabling technology for barcodes in the new millennium. The significance of smart antennas for RFID readers for the improvement of the three major application areas of asset tracking and management, anti-collision, and supply chain management is presented. Literature reviews of these application- specific smart antennas for RFID readers are analysed next. An in-house developed phased array antenna for 900 MHz band RFID readers is then presented, followed by conclusions.

## RFID AS BARCODE REPLACEMENT TECHNOLOGY

RFID systems potentially overcome the key disadvantages of bar codes: the necessity for line-of-sight scanning of individual items and the presence of human operator to direct the reading process. Because RFID tags are read remotely a human operator is not generally required to direct the reading process. This makes it practical to read RFID tags at many more points in a logistic or other chain, and to read them more often, thereby locating chain failures more rapidly. Several "anti-collision" algorithms have also been developed to allow the simultaneous reading of multiple RFID tags (EPCglobal, 2005; Law *et al*, 2000; Ward & Compton, 1992). RFID technology therefore additionally offers the possibility of automatically identifying multiple items at once, greatly speeding up the monitoring process. This multiple-reading capacity also has some disadvantages; it means that with RFID identification, every individual item must have its own, unique identification number. While all items of a particular type can be marked with the same barcode, this is not the case for RFID tags. RFID tags have much greater information content than barcodes and must be more easily customised, since every RFID tag must be unique.

The key barrier to barcode replacement by RFID is therefore an inability to produce suitable tags cheaply. In particular, the chip on such tags must be eliminated and replaced with a fully-printable technology – that is, a printable, multi-bit, chipless tag must be developed. This must be achieved whilst retaining (or improving upon) all of the other necessary properties of such tags, such as a reasonable reading-distance, orientation invariance in the tag's response to interrogation, and suitable collision avoidance measures. In effect, it is necessary to make the transponders as cheap as possible while building the required sophistication into the reader. This sophistication in the reader comes from smart but fast computational algorithms in signal processing chips and smart antennas for readers.

## RFID READERS

RFID readers should comprise *smart antenna systems*, dedicated digital signal processing units and embedded systems to efficiently read multiple tags with faster speed and highest accuracy. The system should also allow easy integration of RFID readers in data networks alongside middleware. Networking of these readers should comply with standardized data transfer protocols. However, the range and speed of data communications, and the operational frequency and coding techniques used by RFID readers vary greatly with application -specific requirements. The operating frequency for the application-specific RFID systems spreads over extremely low frequency (ELF) systems of a few hundred KHz up to microwave and millimetre wave frequencies such as 60 GHz and above (Tuovinen and Vaha-Heikkila, 2006). This wide range of applications and frequency bands creates a problem for ubiquitous RFID systems and acceptance of global standards. In Generation 2 tag systems a unified approach has been formulated to remove some of these barriers and

*Figure 2. Document map of smart antennas for RFID readers*

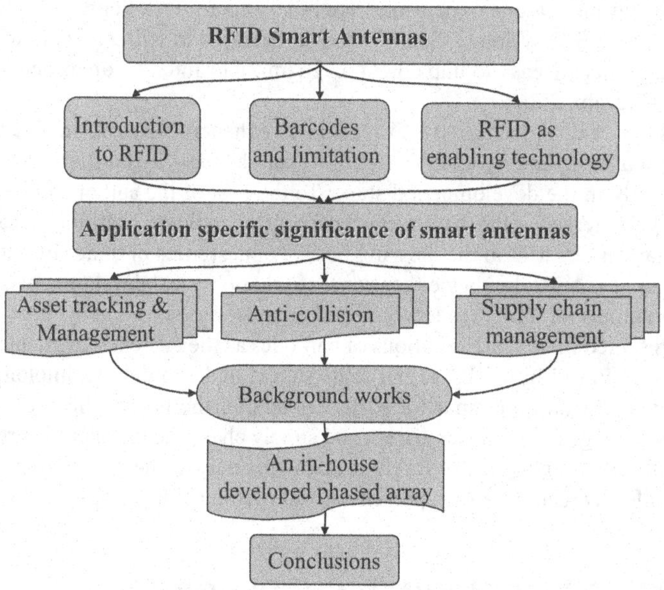


*Figure 3. Complete block diagram of RFID reader*

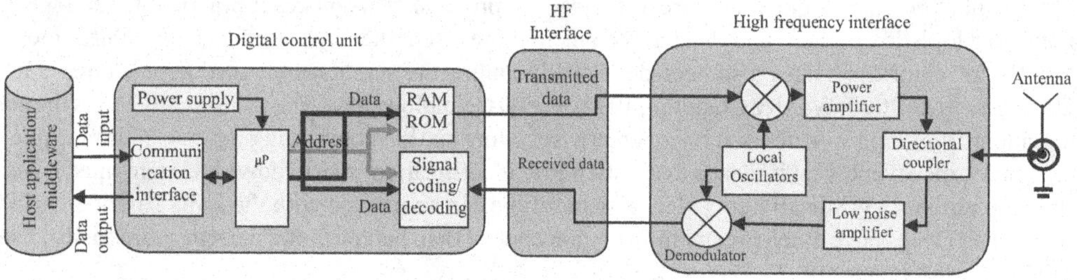


make the tag universal like the barcodes. To this end RFID readers play a significant role in reading tags with various formats and standards.

Figure 3 shows a generic RFID reader system. An RFID reader comprises three main parts shown. These main three functional blocks are: (i) control section; (ii) high frequency (HF) interface; and (iii) antenna. At the user end, the reader is connected to the host application such as enterprise software. The control section of the RFID reader performs digital signal processing and procedures over the received data from the RFID transponder. The control section also enables the reader to communicate with the transponders, wirelessly performing modulation, anti-collision procedures, and finally, decoding the received data from the transponders. These data are sent to the HF interface to interrogate tags (read) or to reprogram the tag (write). This section usually consists of a microprocessor, a memory block, analog-to-digital converters and a communication block for the software application. The HF interface of the reader transmits the interrogating signal to the tag and receives the returned coded echo from the tag. The HF interface comprises two signal paths corresponding to two directional data flows to and from the transponder. The local oscillator generates the RF carrier signal, a modulator modulates the signal, the modulated signal is amplified by the power amplifier, and the amplified signal is transmitted through the antenna. A directional coupler separates the system's transmitted interrogating signal and the received weak coded echoes from the tag (Pozar, 2005). An alternate choice is to use a duplexer switch to separate the transmit and receive paths when a single antenna is used in a monostatic radar configuration. The low noise amplifier (LNA) increases the received echo's amplitude before the signal is decoded in the demodulator. Another LNA can be used to amplify the signal after modulation of the signal. Different demodulation techniques are used to decode the received data from the transponder. Most RFID systems operate using binary phase shift keying (BPSK) (Kocer, Flynn, & Long-Range, 2005) and amplitude shift keying (ASK) (Karmakar *et al* 2006).

*RFID Reader Antennas.* The efficiency of the RFID reader's capability for interrogation and detection is highly dependant on the performance of the reader antenna. A comprehensive literature review (Preradovic & Karmakar, 2007) has suggested two types of RFID reader antennas: (i) fixed beam arrays and (ii) scanned arrays. A **fixed beam** antenna has a unique and fixed beam radiation pattern (Padhi et al, 2001). Most RFID readers are equipped with omni-directional or wide beamwidth antennas in order to cover as much area as possible as their interrogation zones. Several fixed beam antennas are also used, and can be commonly found in Alien Technology readers (Alien Technology Corporation, 2005). The antennas are easy to install and do not need any switching electronics and associated logic control to steer their beams. However, they pick up multipath signals when receiving transponders' backscattered signals. This situation usually leads to reading errors during interrogations.

RFID tags are subject to multipath interference—an inherent problem of electromagnetic signals. Multi-path interference makes an RFID tag unreadable even if it is within the reading range of the reader. To solve this problem, a new type of antenna technology is used that can electronically control the main beam emitted from the reader's antenna. This smart antenna technology incorporated in RFID readers reduces reflections from surroundings and thus minimizes degradation of the system performance due to multipath and other undesired effects.

**Smart antennas** are scanned beam antenna systems, which point beams to selective transponders within their main lobe radiation zones, thus reduce reading errors and collisions among tags. This technique exploits *spatial diversity,* and in many cases polarisation *diversity* of tags. The directed beam also reduces the effects of multipath fading (Ingram, 2003). This new approach to RFID antenna technology is being adopted by RFID manufacturers such as Omron Corporation, Japan (Nakamura & Seddon, 2006), RFID Inc. (Profibus, 2007) and RFSAW, USA (RF Analyst, 2003). In March 2006 Omron Corporation developed a new electronically controlled antenna technology in their UHF band RFID reader systems (Nakamura & Seddon, 2006). Recently, RFID Inc. (Profibus, 2007) has introduced a 32-element compact package smart antenna with associated switches at 125 KHz. This new product advances RFID applications in factory automation, process controls and original equipment manufacturer (OEM) markets. However, the technical details of these smart antennas for RFID readers are not available to readers due to commercial in confidence considerations. In this chapter every effort has been made to present technical information on smart antennas for RFID readers and this is provided in subsequent sections.

## APPLICATION SPECIFIC SMART ANTENNAS FOR RFID READERS

As stated above there are few smart antennas for RFID readers currently available on the market. This is due to the facts that (i) RFID technology is application specific and (ii) the operating frequency spans from extremely low frequency (ELF) of a few hundred KHz up to mm-wave frequencies of 60 GHz and above (Tuovinen and Vaha-Heikkila, 2006) depending on applications. A comprehensive literature survey by the author and a colleague (Karmakar and Bialkowski, 2000) suggests the following classifications of smart antennas for RFID readers: (i) switched beam array; (ii) phased array and (iii) adaptive antenna array. Unlike the low gain omni-directional antennas, directional steerable antennas are medium to high gain antennas and tend to have a gain value larger than 8 to 20 dBi. These antennas are made up of an array of small antenna elements, often for aesthetic reasons. The advantage of most directional antennas as opposed to omni-directional antennas is that the former tend to have better performance and suffer fewer signals fading (Ingram, 2003). The disadvantage of most directional antennas is that they require beam steering and switching electronics and associated beam-forming algorithms for adaptive antennas to point the focused beam of the antenna in the desired directions and nulls toward the interferers. Such antennas are therefore much more expensive that fixed beam antennas. In addition, higher gain directional antennas are usually much larger than omni-directional antennas. Nevertheless, while size and cost are disadvantages, the improved performance of the directional antennas outweighs that of fixed beam antennas.

Electronically steerable antennas are fixed with supporting platforms such as gantries. Electronic beam steering can be achieved by (i) switching on certain antenna elements and turning off the rest as in switched beam array antennas; (ii) changing the phase excitations of individual elements as in phased array antennas; and (iii) adaptively controlling the weight vectors of individual elements using array signal processing chips. Adaptive antennas can also be sub-classified as multiple input multiple output (MIMO) or multiple input single output (MISO). Finally, by combining switching and phase shifting the turned on subset of the antenna element, a switched beam phased array is formed. Figure 4 shows

Figure 4. Extended classification of smart antennas for RFID readers

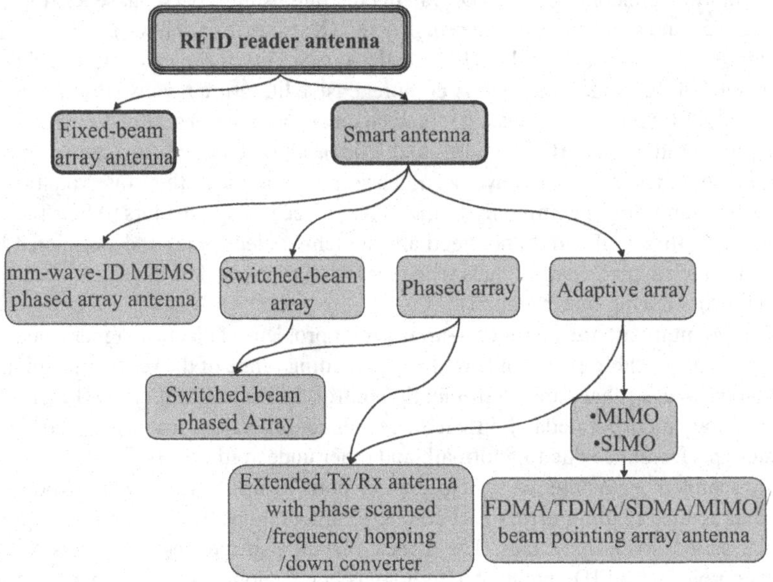


classifications of smart antennas for RFID readers. Following are detailed accounts of the available smart antennas for RFID applications.

## Switched Beam Array Antenna

A sequentially-fed switched beam antenna array (Hercht & Storch, 2007) is proposed to find the location of gaming chips with embedded RFID tags for gambling applications. The antenna array is capable of generating various subsets of beams in different directions. The RFID tagged gaming chips are located on a Blackjack, poker or other gaming table. A planar array of multiple elements is placed under the gaming table. A subset of antenna elements is sequentially turned on and off to form sectorized directional beams to illuminate the RFID tagged gaming chips, enabling the antenna to identify and locate a particular gaming chip. As an example of the excitation mechanism, the sub-set of antenna array **1** in a given zone is turned on to form a beam to illuminate the tags (called group **1**) that subsequently respond to that first antenna **1**. The data associated with antenna **1** are captured and recorded as antenna **1** activation. Next an adjacent antenna **2** is excited in the same reading zone and data are captured as antenna **2** activation. These two data sets are compared by an automatic programme and the results of the comparison reveal which tags are in group **1** but not in group **2**. A proximity advantage of the tag to the reader antennas is exploited to learn which tag is within a particular interrogation zone and which tag is leaving a particular zone. Thus the location of the gaming chip can be determined in real time. The complexity of the algorithm and the antenna array configuration depend on the applications used in gaming systems. For example, a Blackjack game has limited and more controlled chip placement areas on the table surface than those of other gambling systems. Therefore, Blackjack would need a less complicated switched beam antenna array and sequence excitations that a gaming table with more positions. Therefore, linear arrays of 3 elements up to planar arrays of 13 by 13 antenna elements are proposed as possible solutions for different gaming applications (Hercht & Storch, 2007).

A switched beam proximity magnetic reader has been proposed by Kowalski, Serra & Charrat (2005) to identify gas bottles and other metal containers with switchable inductive coils in the reader. The coils are sequentially turned on and off until the item is identified.

Lee (2005) proposes a switched beam coil antenna array configuration for RFID readers for smart shelves for retail shops. Figure 5 shows a schematic block diagram of the smart array reader. The reader has an N-number of coil antennas which can be placed under the platform of the racks. The items to be identified will be placed on the platform of the coil antenna. Antennas will be sequentially switched on one at a time and will go through frequent recalibrations for matching and tuning depending on the load changes on the coil antenna. This load changing happens when shoppers pick up items from the shelves and replace them after inspection. The sensing unit records the voltage levels received from the antenna and sends this information to the reader controller unit. The reader controller unit will send appropriate commands to the antenna matching unit to maximise the received power from the coil antenna. The directional coupler separates the transmit path from the receive path. The reader controller is connected to the I/O port and data communication bus to another server for the application software for identification and location. An anti-collision protocol with 'tag talk first' is implemented to discriminate multiple tags within the reader's zone of a coil antenna. The reader can be connected to many antennas and can retrieve the tag information and location of particular items in real time. The sensing unit is a half wave peak voltage detector and the matching circuits are made from switched capacitor banks, varactor diodes controlled by the reader controller, and switched inductors. This real time tracking and identification of items has great benefits for inventory control and supply chain management.

Chung and Lui (2004) propose some types of loop antenna arrays for detection of objects. The loop antennas are placed at different angles in different planes in a 3D rectangular volumetric space so that the diversity of the individual loops can be exploited to maximise the reading of RFID tags. These antennas are suitable for baggage tracking at check-points of ports of entry, inventory tracking in a warehouse scenario, factory or warehouse inventory control, security identification and access control. Chung and Lui (2004) also propose a curtain antenna with five loop antenna elements and associated matching and switching circuits similar to the configuration shown in Figure 5 above. In an extended version of the antenna configuration, an elongated antenna element with back-to-back loop arrangement connected to appropriate filters is also proposed to maximise the reading range. The antennas can operate over wide frequency bands of 125 KHz, 13.56 MHz, 915 MHz or 2.45 GHz depending on the tuning capability of the loop antenna and its configurations (physical dimensions). A reading distance of 1 meter with 2 antenna arrays with a power of 25 watts is claimed. In another embodiment of the switched loop antenna for RFID readers for metal containers, the loop antenna is made of cables with three undulations and the loop portions all in series and coupled to the tuner circuit. This mechanism well defines the detection region approximating the volume defined by the base and walls of the container. The coupling means of the antenna include a tuning circuit, a filter and a switch for selectively connecting a particular loop antenna to the external processor.

Lai *et al* (2005) propose a redundant networked multi-media RFID system which uses wireless local area network (LAN) and Ethernet connections simultaneously. The system is capable of providing output videos, graphics or image data so that after reading RFID tags, one or more pictures preloaded into the system databases can be downloaded instantly for visual verification by logistics handlers, security guards or customs officers in real time. The reader is equipped with an active smart antenna for transmission and reception. In the transmitting antenna a power amplifier is incorporated for cable loss compensation and to boost the transmitting power. In the receiver chain a low noise amplifier (LNA) is added to the antenna to improve the received signal's quality and power. The smart antenna for the reader has wider varieties of intelligence such as frequency hopping, time slotting, antenna positioning and beam scanning, subset antenna switching and polarisation diversity to exploit the maximum signal readability from multiple tags. A master synchronization controller controls all functionalities mentioned above for the antenna arrays. The antenna system can operate in multiple transmit and single receive mode or single transmit and multiple receive mode. The various transmitting antennas are configured in such a way that either spatial diversity collimation or temporal diversity collimation is achieved via sequential switching of the subsets of the antenna array. These spatial and temporal diversities will increase the reading probability of moving tags. As an extension of the above method, the author proposes multiple transmit multiple receive, in which the active transmit antennas operate in different frequency channels or hopping frequency channels. In this arrangement, the multiple transmitting antennas are not time slotted in operation, but all are actually working at the same time. This will enable true simultaneous multi-detection bistatic reader operation.

Mendolia *et al* (2005) in their US patent entitled "RF ID tag reader utilizing a scanning antenna system and method" claimed three inventions of the adaptive antenna systems for RFID readers: (i) a 9-element cylindrical switched beam phased array antenna (ii) a single band electronically steerable parasitic array radiator (ESPAR) and (iii) a dual-band ESPAR. In this section only the switched beam phased array antenna is discussed. The latter two antenna types will be discussed in Section 4.3. Figure 6 illustrates the claimed embodiment of the 9-element switched beam antenna in a simplified block diagram. The proposed antenna configuration has some advantages over the conventional phased array antenna, being cost efficient, power efficient, simple and compact in configuration, and finally, protocol efficient. Figure 6 is the simple configuration derived from Mendolia et al (2005). The array antenna has 9 dual polarised stack patch antenna elements, which are arranged in a circular grid. The antenna and associated antenna RF electronics are housed in a radome supported by a round baseplate. The top parasitic rectangular patch antenna is a 1.4 inch thick brass plate supported by a 1.5 inch dielectric slab having dielectric constant of 2.8±7% and loss tangent of 0.002. The lower exciting rectangular patch is made of copper plate, which is supported by another dielectric slab with similar dielectric

*Figure 5. A switched beam antenna reader for retail chain inventory and supply chain management*

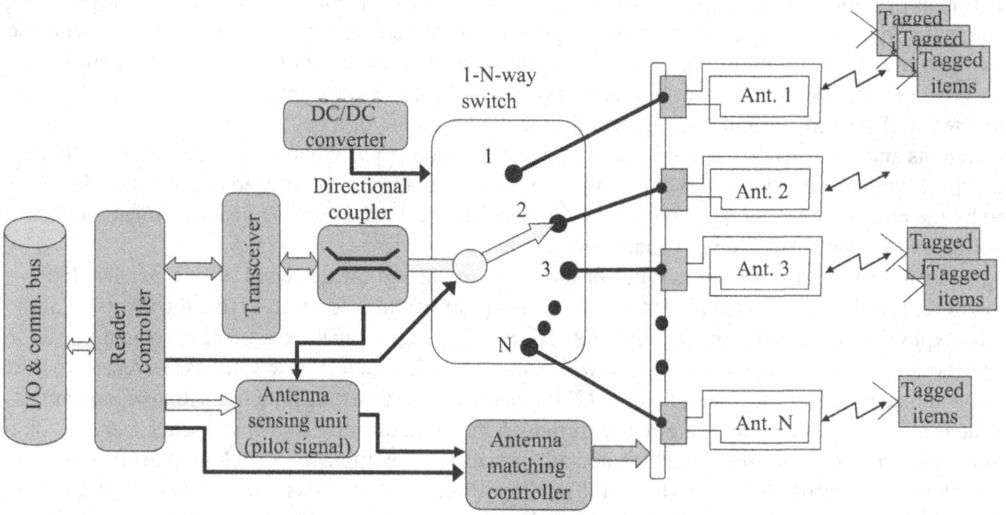


properties to the top one. Two orthogonal feed probes are inserted through the ground plane and soldered to the exciting patch. The 90° phase shifter is used to change the polarisation of the antenna from vertical to horizontal, to slanted polarisations to ±45°. The RF switches connected to the antennas permit any of the three consecutive antennas to be active and properly phased at any time. The switch is an MMIC switch and connected in consecutive 1, 4, 7 fashion so that at any time three consecutive elements are energised to generate a directional beam. The switching electronics receive the command from the micro-controller based on the feedback of the pilot signal from the RF transceiver. The digital electronics are connected to the host computer via a serial I/O interface such as RS232. The phase shifter is a tunable dielectric phase shifter made of two thin film SMD Parascan surface mounted varactor diodes.

Mendolia *et al* (2005) propose a polarizer to generate four different polarizations—horizontal, vertical, and ±45° slanted polarizations to maximise reading rates using polarization diversity. In this polarizer the dual feed lines of the antenna are connected to two single pole double throw (SPDT) switches followed by a quadrature hybrid coupler and another set of SPDT switches.

Karmakar and Bialkowski (1999) have built an 8-element switched beam phased array antenna for mobile satellite communications. For insights into the antenna design and the RF electronics of the switched beam phased array antenna, readers are referred to this article.

Marsh *et al* (1996) propose an embodiment of a switchable transmit and receive reader antenna system composed of two transmitting panel antennas and one receiving panel antenna. The panel antennas are a 4×2-element microstrip patch antenna array. The transmit interrogating signal is a narrowband signal at different bandwidth and the receiving identification signal from the tag is a broad bandwidth signal so that the tag is responsive to one or more interrogating signals. The patch antenna array is design at 800 – 1000 MHz band and covers the features of polarization diversity in transmission and reception. The transmitting interrogating antenna has facilities with different frequency switching in a narrowband manner. The frequency switching is carried out in an alternate manner so that the tag receives two distinct signals from the interrogating antennas. The interrogating signal is also more directive than that of the received signals. The beamwidth of the receiving reader antenna is at least five times that of the transmitting antenna. Polarization diversity and beam switching are implemented in the reader antenna so that there are no overlapping nulls of the E-field of the transmitting antenna, and as a consequence no tag is missed out of the reading process. Thus this proposed system increases the probability of identification. The use of different frequencies and different polarizations is very useful in a situation when many articles are thrown in a shopping trolley randomly. In such a situation, different frequency transmissions and different polarized antennas will pick up signals from all transponders.

Thus far switched beam antenna arrays for RFID readers have been discussed. The next embodiment of smart antennas for RFID readers is phase scanned array antennas.

## Digital Beamforming Phased Array Antennas

In relation to digital phase scanned array antennas, Salonen and Sydanheimo (2002) propose a non-uniform antenna array based on binomially distributed current excitations to achieve a -37 dB sidelobe level from a 5-element antenna array. The amplitude is controlled with variable amplifiers/attenuators connected to individual antenna elements in the array. Beam steering is done with an array of 4-bit digital phase shifters as shown in Figure 7.

The phase shifter is an array of four bit reflection type phase shifters comprised of quadrature hybrid couplers and reactive loadings at its coupling and isolation ports. The transmission phase of the hybrid coupler changes with switching RF pin diodes ON and OFF. The phase shifter bits are 22.5, 45, 90 and 180°, giving a resolution of ±11.25° phase shift. The attenuation and phase shift of each antenna element is controlled by a control bus which is eight bits wide. With this control bit, a theoretical fine beam steering angle of 3.75° is achieved. The detection of the received signal is based on the highest power levels from the tag for the reader antenna. The microprocessor controller of attenuations and phase shifts of each element makes the antenna adaptive. By controlling the current distributions and phase shifts to individual elements, any possible 3D steerable beam pattern can be adaptively generated. The adaptive beam can point the main beam toward the desired signal and nulls towards interferers. Detailed physical layer development of phase shifters, beamforming networks and antenna arrays are presented in Section 5.

## Adaptive Antenna Arrays

- Electronically steerable parasitic array radiator (ESPAR)

As mentioned above, Mendolia *et al* (2005) have developed three adaptive antenna systems for RFID readers. The 9-element cylindrical switched beam phased array antenna has been discussed in the section on switched beam array antennas above. The other two antenna types—a single band electronically steerable parasitic array radiator (ESPAR) and a dual-band ESPAR—are discussed here.

The next embodiments are single RF port electronically steerable switched parasitic monopole and dipole antennas. The salient feature of the ESPAR antenna is that, having only a single RF port, the adaptive antenna needs only one A/D converter. This is a significant cost saving measure in terms of hardware and software implementation. The detailed features of an ESPAR can be found in Sun, Hirata, Ohira & Karmakar (2004), where the authors provide details of the construction, algorithm development and prototype testing of an ESPAR antenna array for a mobile ad-hoc computing network. Figure 8a shows a single band ESPAR antenna with 7 monopoles. Figure 8b shows a dual-band ESPAR antenna with lower monopoles for the low frequency operation and small monopoles for the high frequency operation. The lower frequency band is at 1.6-1.7 GHz and the upper frequency band of operation is 2.4-2.5 GHz. Conventional adaptive antennas cannot operate in dual-band as proposed here. The central single port RF monopole antenna has the same length of the low frequency monopoles. An example of beam patterns generated by a 7-element ESPAR antenna derived from Sun and Karmakar (2004) is shown in Figure 9. The beam pattern can be adaptively shaped in any form pointing the maximum power to the desired tag and nulls to the interferers enabling anti-collision to be effectively implemented with an ESPAR antenna. According to the current embodiment of the ESPAR antenna for RFID readers, the radiated power can also be concentrated in the near field thus maximising the reading of the tag in the vicinity of the reader.

## Maximum Ratio Combining (MRC) Adaptive Antenna Array

While the ESPAR antenna is the simplest embodiment of an adaptive antenna array, Wang *et al* (2004) invented a very complex embodiment of adaptive antenna array. Figure 10 shows a detailed schematic diagram of an adaptive array antenna with multiple antenna elements, modulators, weight vector generator, RF receiver and automatic gain control (AGC) loop. The adaptive antenna combines signals received from multiple tags from the vicinity of the reader. The *smart antenna processing module* is connected to the antennas with transmit and receive (T/R) switches so that the same antenna can be used for transmitting signals to the tags and receiving signals from the tags. The main objective of the *smart antenna processing module* shown in **Figure 10** is to maximise the received signal-to-noise ratio (SNR) based weights determined by maximal ratio combination (MRC). The antenna uses a closed loop signal processing operation for antenna weight computation (IW and QW) and signal combination (Σ). Though the same antennae can be used to receive and transmit with a T/R switch (not shown here), Wang *et al* (2004) present the receive mode adaptive antenna array only.

*Figure 6. A 9-element switched beam phased circular array*

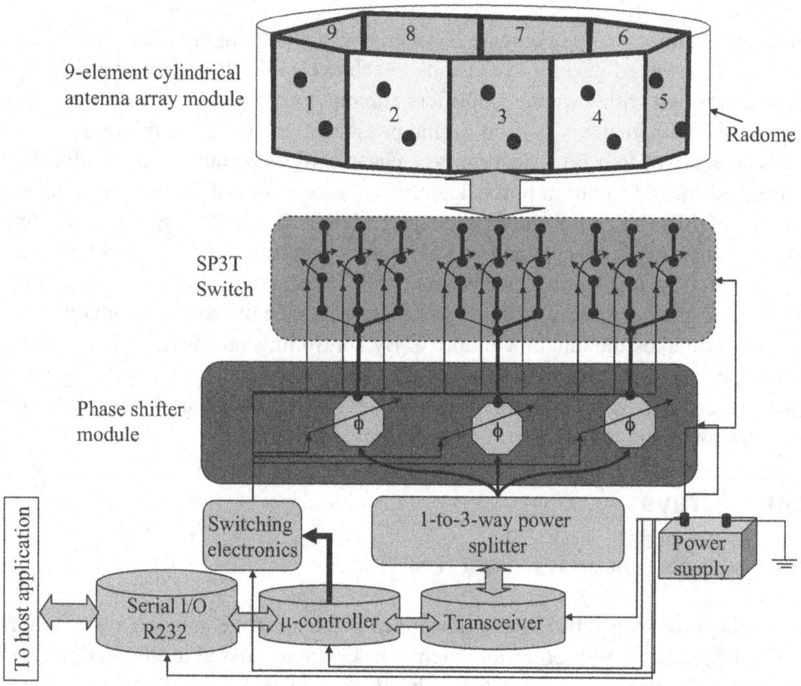


In the *receive path* of the closed loop feedback operation, the signals received from tags are amplified with low noise amplifiers (LNAs) and down-converted in the respective down converters. Each of the down-converters multiplies the output of the respective amplifiers by a local oscillator's in-phase (LO-I) and a quadrature phase signal (LO-Q). The resultant I and Q signals pass through the lowpass filters (LPFs) followed by the AGC loop which normalises the signal level before the MRC algorithm. The AGC provides consistent performance for the smart antenna processing unit at different input signal levels. At the output of the variable gain amplifiers, a power detector combines the power of I and Q signals and compares the signal power to a threshold value. The output of the comparison is fed to the AGC loop, which adjusts the gain of the variable amplifier accordingly for a consistent output. Thus the MRC algorithm-based beam forming processor is able to work at different input signal levels.

*MRC Beam Forming*: In MRC beamforming the processor adjusts the antenna weight adaptively by correlating each of the received signals with a combined signal as the received signal arrives. The correlation time determines the received signal's post-detection SNR of the weight computation. Increasing the correlation time allows optimal antenna weights to be achieved in very low SNR environments. The SNR improvement comes from both the antenna combining gain and the diversity gain as for example the polarisation diversity. Thus dual polarized antennas for transmission and reception is an added feature of the adaptive antenna. With MRC antenna weights, the received signals from different antennas can be coherently combined (i.e. in phase) while uncorrelated noise from different antennas is combined incoherently. As a result, the SNR increases after combining. Additionally, the signal received by some antennas could experience fading, by which the signal strength could be reduced significantly. Combining signals from all the antennas reduces the probability of the signal fading in the output signal and thereby achieves diversity gain. The authors claim significant SNR improvement in a fading channel. For example, with an array of 2-elements and in a Rayleigh fading environment, 8 to 9 dB of SNR gain is achieved for 802.11b WLAN signals. This SNR gain increases to 12-14 with a 4-element adaptive antenna within the same environment.

- Multiple Input Multiple Output (MIMO) Antenna System

*Figure 7. Adaptive digital beamforming network for RFID reader smart antenna*

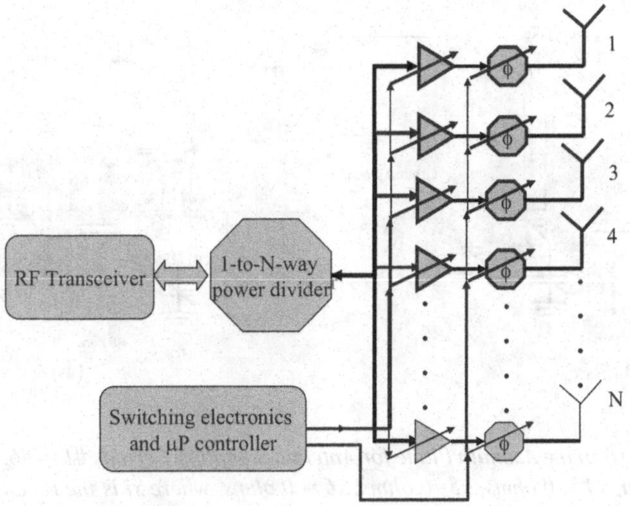


Ingram (2003) has proposed a multiple input-multiple output (MIMO) antenna system for RFID tag identification to maximise system capacity in terms of bit/s/Hz. Ingram is the first researcher who has proposed multiple antennas for RFID tags for capacity improvement and anti-collision. The physical layer development of a multiple antenna based tag is reported by Collins *et al* (2005). A set of multiple antennas is used for the transmitter and multiple antennas are used in the receiver. Figure 11 shows the actual MIMO RFID system where N transmitting antennas are transmitting signals to L number of tags (Griffin & Durgin, 2007). The backscattered signals from the L number of tags are received by N number of receiving antennas in the reader.

A tag with at least a signal antenna or an array of antennas is placed in between the transmitter and receiver. The algorithms to compute the weight from both transmit and receive antennas are represented by the channel antenna gain as follows:

$$H = \begin{vmatrix} H_{11} & H_{12} \\ H_{21} & H_{22} \end{vmatrix} \quad \text{and} \quad G = \begin{vmatrix} G_{11} & G_{12} \\ G_{21} & G_{22} \end{vmatrix} \tag{1}$$

where, $H_{ij}$ represents the channel gain of the transmit antenna and $G_{ij}$ represents the channel gains of the receive antenna. A computing device comprised of a microprocessor or a computer identifies the product channel matrix:

$$C_p = HG. \tag{2}$$

The identification of $C_p$ can be performed by sending a pilot signal from each transmitting antenna and measuring the responses at receiving antennas. Blind channel method can also be used to avoid the overhead with pilot signal. The computer device then performs a singular value decomposition or approximate identification of $C_p$, given by:

$$C_p = U\Sigma V^H \tag{3}$$

where, the columns of U are the left singular vectors, Σ is a diagonal matrix of the singular values of $C_p$, the columns of V are the right singular matrix and the superscript 'H' means conjugate transpose. The received output signal of the receiver is expressed as:

$$z(t) = XC_pW^H m(t)\cos(\omega_c + \omega_o t + \theta) + b\cos(\omega_c t + \theta') + n(t) \tag{4}$$

*Figure 8. ESPAR antenna (a) single band and (b) dual-band*

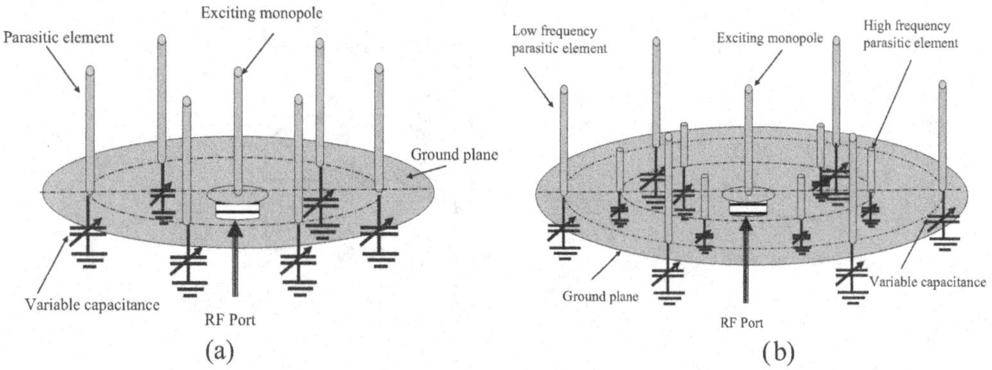


*Figure 9. Beam Patterns (a) to (f) in the Azimuth Plane for Antenna Sampling Periods #1 to #6. Load Reactance: x1 = -90 ohm, x2 = 0 ohms, x3 = 0 ohm, x4 = 0 ohms, x5=0 ohms, x6 = 0 ohms, where xi is the reactive loading of the varactor connected to the parasitic element i, and i = 1...6.*

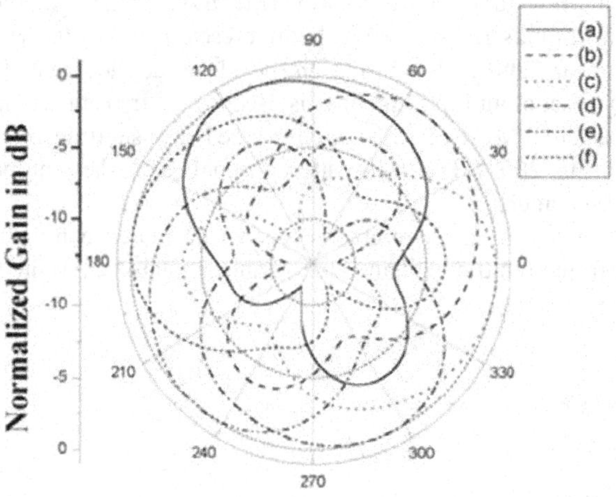


where, $X = [X_1, X_2]$ the weight vector of the transmitting antenna, $W = [W_1, W_2]$ is the weight vector of the receiving antenna, m(t) is the information waveform, $\omega_c$ is the RF carrier angular frequency, $\omega_0/2\pi$ is the pulse repetition rate of a periodic square wave, $\theta$ and $\theta'$ are sideband and carrier phases, respectively, $b \cos(\omega_c t+\theta')$ is the carrier component and n(t) is the additive thermal noise. The other sidebands, which are also proportional to $XC_P W^H$, are not shown.

Griffin & Durgin (2007) propose new analytic probability density function (PDF) expressions and several RFID tag design guidelines to exploit the reduction in small scale fading inherently available in the backscatter channel. According to Griffin and Durgin, in the RFID system uplink scenario, the backscatter channel is a pinhole channel and experiences small-scale fading, even under line-of-sight conditions. This is due to the fact that indoor operation is a cluttered reader environment and the tagged objects are inhomogeneous in nature. Ingram (2003) also states that even with a highly directional antenna array with beam steering capability this situation cannot be improved much. Griffin and Durgin (2007) propose the RFID channel as *M*×*L*×*N* dyadic backscatter channel consists of *M* transmitter (TX), *L* RF tags and *N* receiver (RX) antennas as illustrated in Figure 11. They assume the forward and backward links to be narrowband and present the baseband signal as:

$$\vec{y}(t)=\frac{1}{2}\widetilde{H}^{b}(t)\widetilde{S}(t)\widetilde{H}^{f}(t)\vec{x}(t)+\vec{n}(t) \tag{5}$$

*Figure 10. Schematic diagram of an adaptive array antenna for RFID reader*

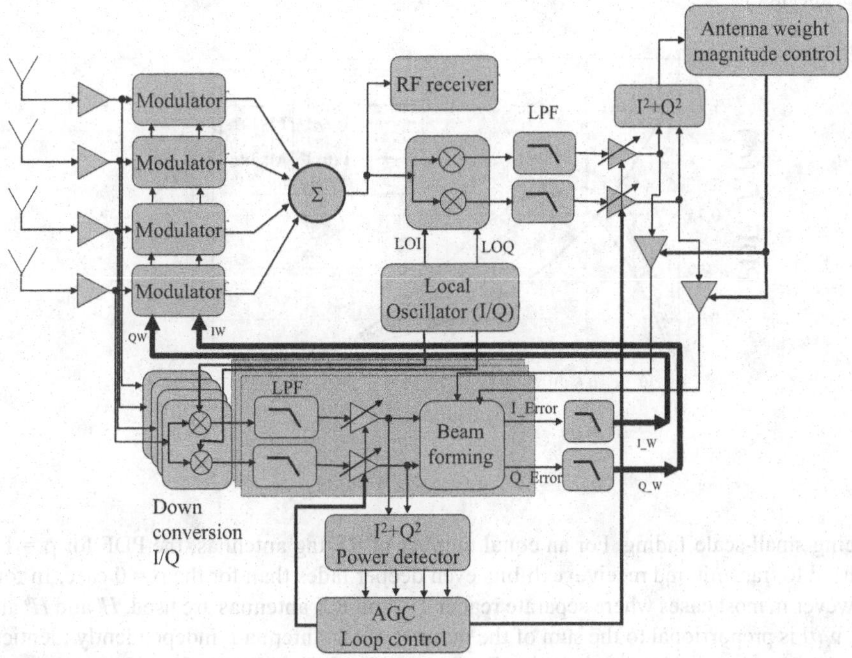


*Figure 11. (a) The general M×L×N dyadic backscatter channel with M transmitter antennas, L RF tag antennas, and N receiver antennas. (b) The signal received at the n-th receiving antenna is the sum of ML Gaussian products, L of which are statistically independent (Griffin & Durgin, 2007, Copyright IEEE, 2007).*

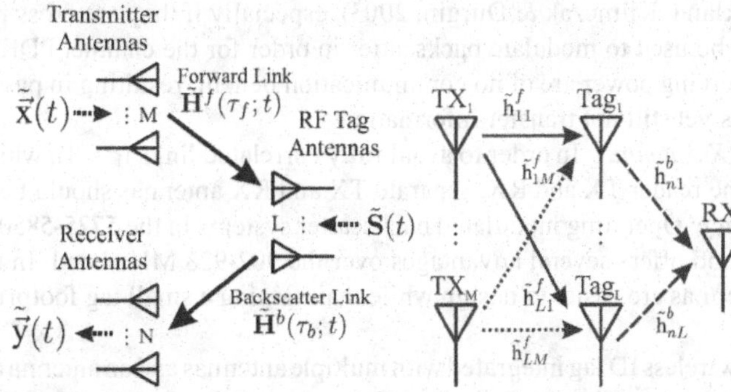


where y(t) is an $N \times 1$ vector of received complex baseband signals, $H^b(t)$ is the $N \times L$ complex baseband channel impulse response matrix of the backscatter link, and $H^f(t)$ is the $L \times M$ complex baseband channel impulse response matrix of the forward link [3]. S(t) is a narrowband $L \times L$ matrix that describes the time-varying modulation that an RF tag places on the radio signals absorbed and scattered by the L tag antennas. $x(t)$ is an $M \times 1$ vector of signals transmitted from the TX antennas, and $n(t)$ is an $N \times 1$ matrix of noise components.

Figure 12 shows the probability density function (PDF) $f_\alpha(\alpha)$ where $\alpha$ is the channel envelop. Here we assume both forward and reverse channel links are Rayleigh distributed for ρ representing the correlation coefficient between the elements $H^b(t)$ and $H^f(t)$. Figure12 shows that the PDF of the dyadic (two folds) backscatter channel has deeper, more frequent fades than a one-way (1x1x1) Rayleigh channel. However, the PDF improves as the number of RF tag antennas

*Figure 12. Plots of the envelop PDF at the nth receiving antenna with ρ {0, 1} (Griffin & Durgin, 2007, Copyright IEEE, 2007. Used with permission.)*

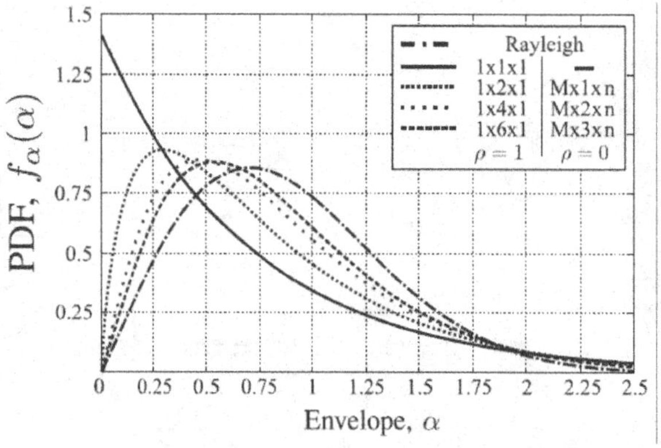


is increased, reducing small-scale fading. For an equal number of RF tag antennas, the PDF for ρ = 1 (when a single reader antenna is used to transmit and receive) exhibits even deeper fades than for the ρ = 0 case. In former case, $H^f$ is identical to $H^b$; however in most cases where separate reader TX and RX antennas are used, $H^f$ and $H^b$ are uncorrelated. In alternate terms, $y_n(t)$ is proportional to the sum of the number of tag antenna *L* independently identically distributed (i.i.d.) product terms that each represent a set of spatially separated and, therefore, independent propagation paths. As L increases, additional independent propagation paths reduce the probability that the envelope will fade. In other words, increasing the number of RF tag antennas increases the number of independent pinholes through which a signal may propagate. As L becomes very large, (3) approaches a Rayleigh distribution.

Based on the analyses and results presented above, Griffin and Durgin (2007) propose the following RF tag antenna design guidelines:

(i) *Multiple RF Tag Antennas:* at least two RF tag antennas should be used. Closely spaced antennas can have low envelope correlation (Auckland, Kilmczak & Durgin, 2003), especially if they are cross polarized. Furthermore, each RF tag antenna must be used to modulate backscatter in order for the channel PDF to improve. Additional antennas used only for receiving power are of no communication benefit, resulting in passive RFID tags that can power-up in some locations yet still not transfer information.

(ii) *Separate Reader TX and RX Antennas:* In order to avoid fully correlated links (ρ = 1), which can only occur if the same antenna is used for the reader TX and RX, separate TX and RX antennas should be used.

(iii) *Higher Operating Frequency:* Operating modulated backscatter systems in the 5725-5850 MHz industrial, scientific, and medical (ISM) band offers several advantages over the 902-928 MHz band. In this band, RF tags with multiple, uncorrelated antennas are easier to design while maintaining a small tag footprint.

Maeda *et al* (2007) propose a wireless ID tag integrated with multiple antennas and an antenna direction sensor in order to improve the accuracy in location measurement of wireless ID tags. A gravity sensor such as a gyroscope or magnetic sensor can be used as the direction sensor in addition to the data received at the base station and the polarization of the tag antenna. Along with the estimation algorithm, a three orthogonal antenna arrangement in the ID tag is exploited in the location measurement.

In the measurement set up four base stations with highly directional log periodic antennas focusing at the centre of the room were used. The reader/base station antennas were vertically polarised. Two types of tag antennas—a half wavelength dipole and a 17 mm diameter loop antenna to simulate the tag antenna were set at x-, y- and z-orientations for polarization diversity. The received power P(d), in each direction of X, Y (horizontal polarisation) and Z (vertical polarisation) was measured. The procedure was performed for each direction of the antenna. A cumulative probability distribution was derived from a distance error function. The 17 mm diameter loop antenna was predicted to be as good as the half wavelength dipole antenna, meaning that a tag with three dimensional orientation antennas can effectively be used for location measurement. Figure 13 shows the error distance of the three orientations and the average error of approximately 2.26 meters. Therefore, Ingram's (2003) proposition to use multiple antennas in RFID tags and Griffin

*Figure 13. Cumulative errors distance of tag with 3-orthogoanl antennas with 4 base station antenna located in 4 corners of a room with shelves: (a) half-wavelength dipole antenna and (b) 17 mm loop antenna. Measurement is at 300 MHz (Maeda, Matsumoto, & Yoshida, Copyright IEEE, 2007. Used with permission).*

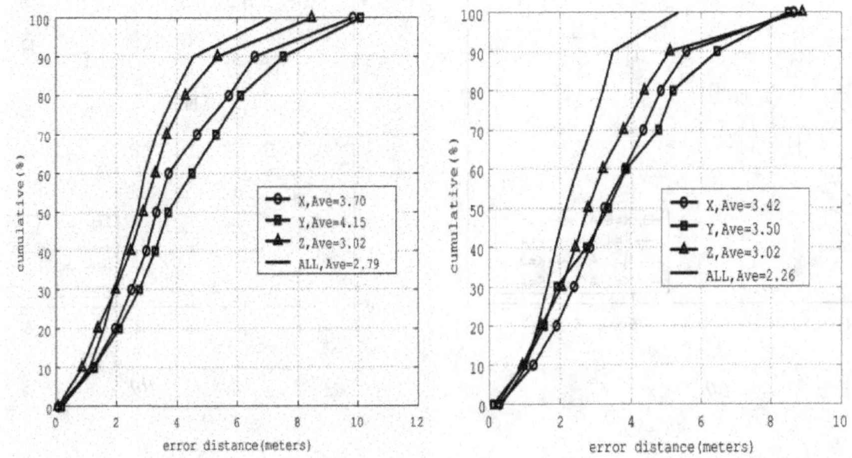


and Durgin's (2007) proposition to push the operating frequency to 5.8 GHz ISM band to miniaturize and fit a plurality of antennas in RFID tags are well justified by the experimental findings by Maeda *et al* (2007).

- RFID Anti-collision Algorithm with Multi-packet Reception

The objective of an efficient RFID reader is to read as many tags as possible within its reading range. However, the reader suffers from anti-collision problems when many tags are residing in close proximity. RFID tags as miniaturized devices have limitations in computational power, memory and communication bandwidth. To overcome the anti-collision problem while reading multiple tags in close proximity in terms of location and frequency of operation, two anti-collision algorithms are usually used—the slotted Aloha algorithm (EPCglobal, 2005) and the binary tree algorithm (Law, Lee & Siu, 2000). However, their tag reading rates are not high enough to simultaneously recognize a large volume of tags, especially when time is crucial in delay-sensitive applications such as a warehouse scenario where many items should be identified instantly.

Lee (2007) proposes multi-packet reception (MPR) capability in an RFID reader to enhance the RFID tag reading range. This is done with an antenna array which allows the reception of multiple responses transmitted by the tag simultaneously. Ward and Compton (1992) analyzed the throughput enhancement of S-Aloha packet radio networks with adaptive antenna array. In Lee's model (Lee, 2007) the reader is equipped with an array of multiple antenna elements as shown in Figure 13. There are N tags in the vicinity the reader. The reader operates in the downlink sending interrogation message to the tags. M responding tags are synchronized in packet transmission because they transmit in response to the request message broadcast by the reader. In Lee's MPR model, the MPR capability of a reader is characterized by a pair of positive integers *{F, K}* where $F \leq K$: *F* and *K* represent reception capability (*throughput or successful reading*) and collision-free capability (*number of antenna elements or independent channels*) respectively. Assuming *i* is the number of tags responding to a given slot, there are four possible cases: (i) idle slot where $i = 0$, (ii) $1 < i < F$, all *i* tags are correctly identified; (iii) $F < i < K$, only F tags among *i* tags are correctly identified; (iv) $i > K$, collision occurs and no signal can be decoded. All *i* tags are correctly identified. In general, a receiver MPR system cannot separate signals whose number is more than the number of antenna elements. Thus *K* is assumed to be equal to the number of antenna elements and *F* is bounded by *K*. The detailed of both algorithms can be found in (Lee, 2007). Figure 14 shows throughputs (successful reading) of RFID tags with the two algorithms—binary tree and S-Aloha protocols with various *F* and *K* combinations.

*Figure 14. Throughput of (a) binary tree (number of bits indicating the initial ID range of interests R = 12, tag density ρ = 0.5) and (b) S-Aloha (optimal response rate λ and total number of tags N = 130). "A" and "S" stand for "Analysis" and "Simulation" respectively. (Courtesy Lee, J. Seoul University, Korea. Used with permission).*

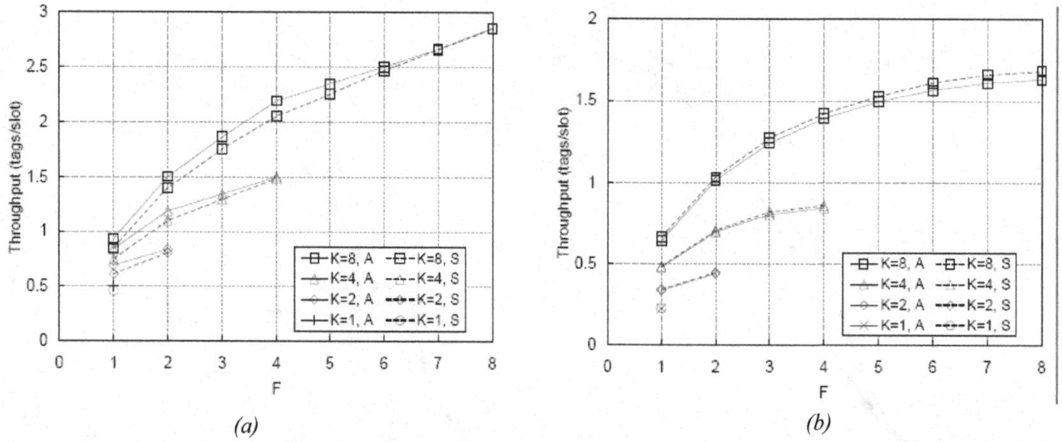


*Figure 15. Maximum throughput vs degree of freedom (no. of antenna elements) (Lee, Das & Kim, copyright IEEE 2004. Used with Permission)*

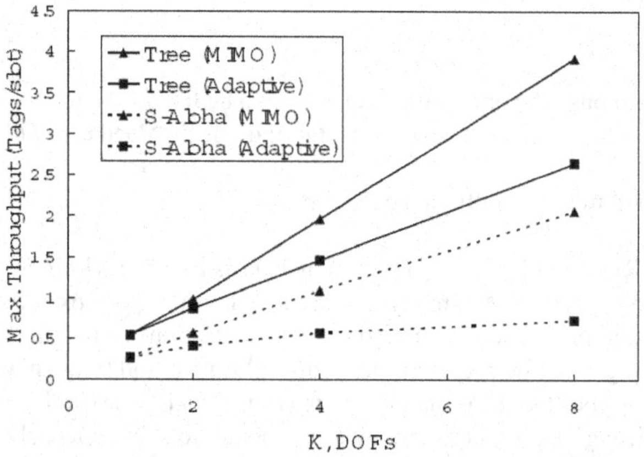


As can be seen with the degree of freedom (F), the throughputs have increased significantly. Here K = 1 represents the baseline case without MPR capability. Therefore, channel estimation and signal separation techniques (which determine *F*) play as important a role in improving system performance as the receiving antenna (*K*).

In their previous work Lee *et al* (2004) have analyzed RFID throughputs with two anti-collision algorithms—S-Aloha and Binary tree splitting and two smart antenna systems—adaptive array antennas and MIMO antennas. The results are produced in Figure 15. As can be seen in the figure, the throughput (tags/slot) has increased with the binary tree split and MIMO antenna combination. This is due to the fact that in MIMO antennas, *K* number of elements possess *K* number of degrees of freedom giving *K* number of independent channels. On the other hand, a K-element adaptive antenna array nullifies *K-1* interferers.

- Millimeter-wave Identification System

So far various smart antenna systems and MIMO antenna systems for improvements in the areas of RFID system capacity, location finding of tags, tag reading rate and anti-collision have been discussed. Research in these areas has

*Figure 16. MMID phased array antenna: (a) layout and (b) beam patterns (blue lines: co-polar and red lines: cross-polar patterns) (Tuovinen and Vaha-Heikkila 2006, copyright IEEE 2006. Used with permission).*

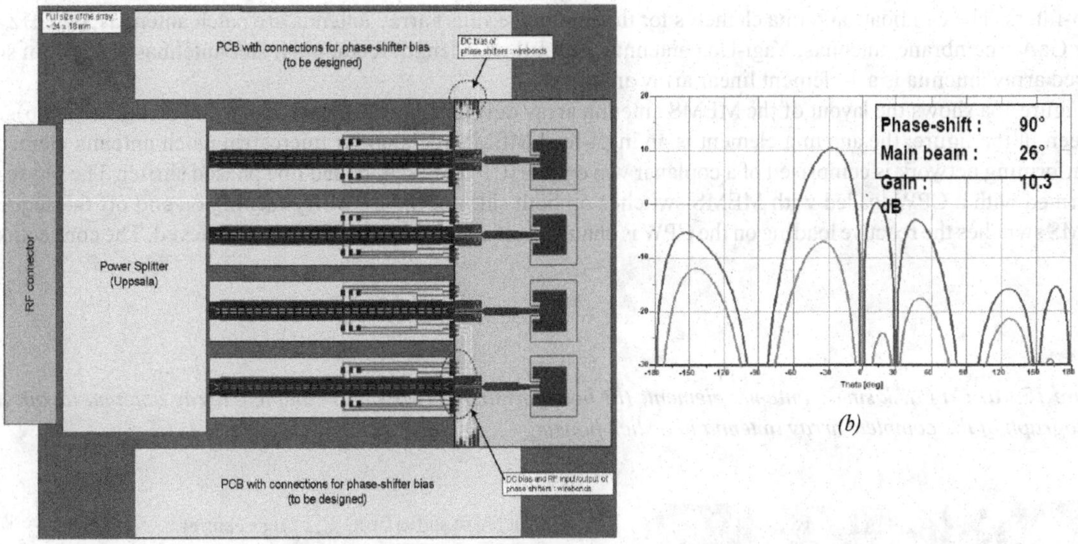


*(a)* *(b)*

*Table 1. Specification requirements of 900 MHz RFID reader phased scan array antenna*

| | |
|---|---|
| Frequency | 910 MHz |
| 1o dB RL Bandwidth | 100 MHz |
| Beam scanning | 3D azimuth and elevation beam scanning |
| Beamwidth | 42° |
| Peak gain | 12 dBi |
| Weight | Less tan 1 kg |
| Dimensions | 3 cm×50 cm × 50 cm |

been presented in detail. All aforementioned tag systems use frequency bands from a few hundred KHz up to 2.45 GHz. Griffin and Durgin (2007) propose a new class of 5.8 GHz frequency band RFID systems to accommodate multiple antennas in RFID tags for throughput improvement. While most RFID researchers and developers have been striving to develop low frequency and microwave RFID tag systems, an ambitious project has been undertaken by VTT – Millilab to develop a mm-wave identification system at 60 GHz frequency band and above. MMW-ID enjoys many advantages over conventional ELF and microwave frequency RFID systems. The MMW-ID advantages are as follows:

- Extremely miniaturized tags
- Novel remote sensor principle based on mm-wave frequency carriers
- Extremely large bandwidth
- Beam steering in mm-wave will open up many new initiatives and demands in micro-electro-mechanical system (MEMS) based antennas, actuators, oscillators and many other components.

This new technology will also open up new application areas of RFID such as wireless embedded health monitoring systems of patients, such as body temperature and heart beat recording in real time, in-room environment monitoring, ultra wide band short range data transmission, in-car and car-to-car traffic management, intelligent transportation and many more.

Tuovinen and Vaha-Heikkila (2006) propose a MMW-ID system based on MEMS devices including membrane antennas, switches, phase shifters, oscillators and amplifiers. Their research is on the whole gamut of physical layer and protocol developments at MM-wave frequencies. On-going activities (as at 2006) are in the areas of beam steering antenna development for RFID readers, mm-wave ID tags and nano-components such as mm-wave nano-oscillators and nano-filters. The candidate antenna elements for the mm-wave smart array antenna are patch antennas, 45 GHz and 60 GHz GaAs membrane antennas, Yagi-Uda antennas and full wavelength rectangular slot antennas. The beam steering phased array antenna is a 4-element linear array antenna.

Figure 16a shows the layout of the MEMS antenna array developed by Tuovinen and Vaha-Heikkila (2006). As can be seen in the figure, the antenna element is an inset-feed MEMS rectangular microstrip patch antenna element. The beamforming network is comprised of a coplanar waveguide (CPW) based loaded line phased shifter. The phase shifter is formed with a CPW loaded with MEMS switches on both sides of the CPW. By turning on and off the sequence of MEMS switches the reactive loading on the CPW is changed and required phase shifts are achieved. The connection from

*Figure 17. (a) Half disk single antenna element; (b) beamforming network; (c) complete array antenna layout and (d) photograph of the complete array antenna in a slick housing*

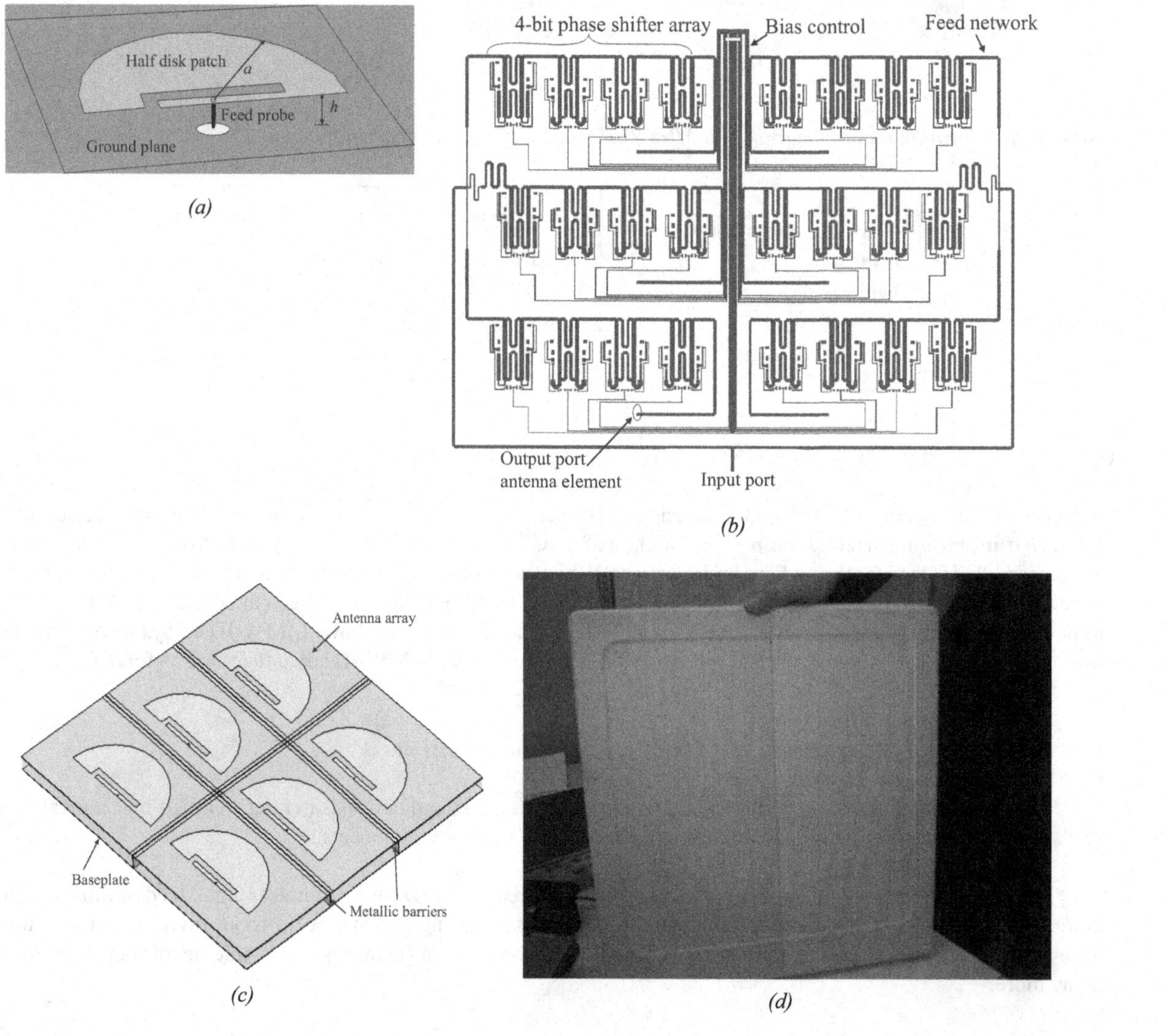

the microstrip line feed antenna element to the CPW based beamforming network is achieved via a transition between microstrip to CPW (Simons, 2001). The dc bias connections to the MEMS switches are via the DC bias control from both ends of the transition as shown in Figure 16a. The four phase shifters are connected to a 1-to-4-way power splitter developed by Uppsala University. The total area of the MEMS smart antenna array of 2.4 cm × 1.8 cm is extremely small compared to the conventional phased array antennas in mm-wave frequency bands, and this antenna has significant advantages for portable and embedded electronic gadgets. The resultant beam from the antenna is shown in Figure 16(b). Up to 42° beam squint is obtained with an appropriate phase excitation for the 4-element antenna array. The expected reading distance is 1~2 meters with a path loss of about -80~-100 dB. Once fully developed, this MMW-ID will open up a new horizon for identification systems.

## DEVELOPMENT OF PHASED ARRAY ANTENNA FOR 900 MHZ RFID READER

So far various smart antenna systems for RFID readers have been discussed, including the MM-ID system developed by VTT (www.vtt.fi). All published research has provided only superficial information in the commercial in confidence context, and the relevant information is available only in patent applications. Scholarly publications are seldom found

*Figure 18. (a) s-parameter vs frequency of 1-to-6-way power divider; (b) input return loss of complete antenna array; (c) CST Microwave Studio simulated broadside gain pattern of the array and (d) measured squinted radiation patterns of the array (Pink: +20° and blue:-30° beams).*

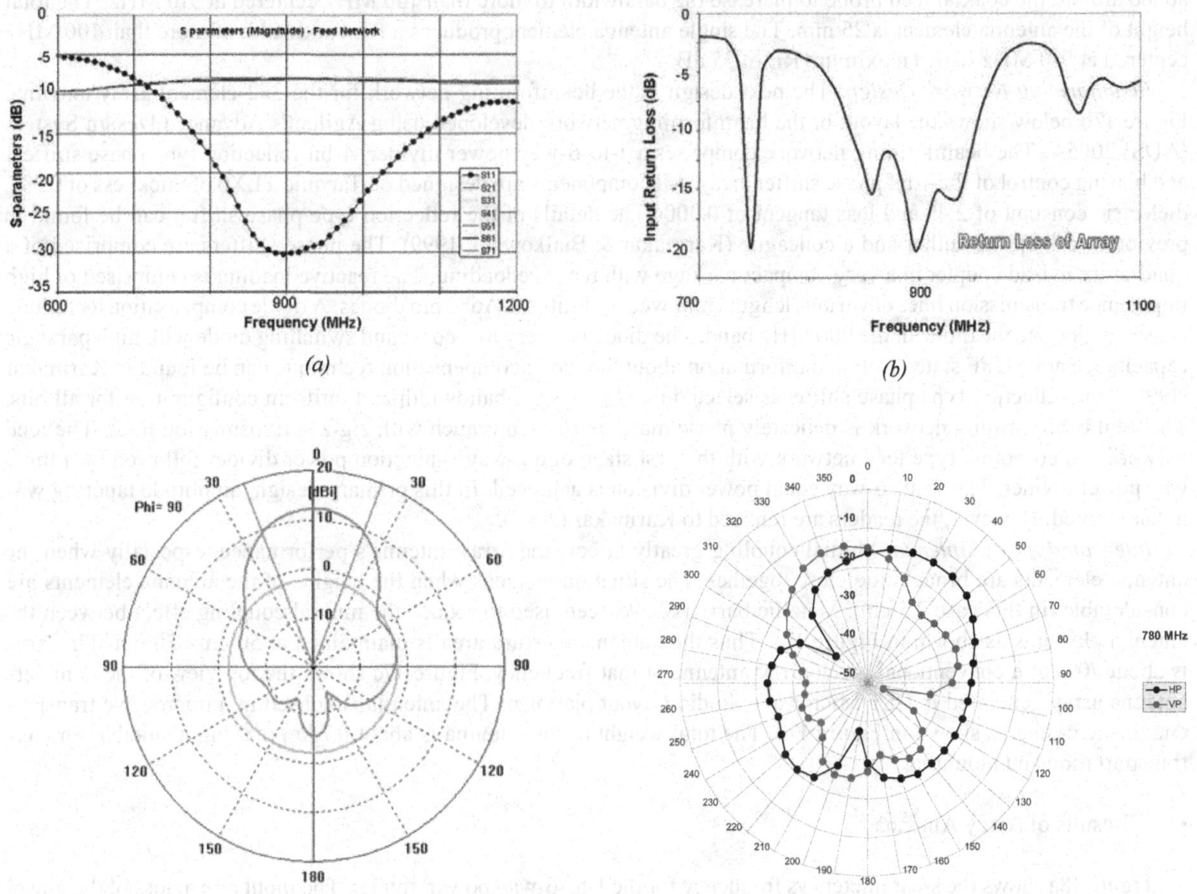

in this field of research. In this section, the technical development of a planar phase scanned antenna developed by the Monash RFID Research Group led by the author (Karmakar, 2007) is presented. The 900 MHz frequency band has many applications in RFID systems. The application areas are: 868-870 MHz for short range devices in CEPT countries including Australia and most European countries that recognize the U.S. participation in the CEPT Radio Amateur License (http://www.arrl.org/FandES/field/regulations/io/#cept, accessed 11 Sep. 2007); 902 – 928 MHz ISM band RFID for North America; 918-926 MHz RFID for Australia; and 950-965 MHz for Japan (Mendolia *et al*, 2005; Clampitt, 2006). Therefore, there is a demand for a smart reader antenna which can cover all these bands of operations. To fulfill this requirement the author has developed a 3×2-element planar phased array antenna operating at 900 MHz frequency band covering 100 MHz bandwidth. The objective is to make a low cost, low profile and light weight antenna for easy mounting and portable operation, aimed at mass markets worldwide. These research and development activities in RFID have also trained personnel for this potential RFID market.

- Array Antenna Design

To make the antenna compact and fully planar the most suitable candidate is the microstrip patch antenna, but the conventional microstrip patch antenna is a half-wavelength resonator. At 900 MHz, the antenna element is sizable compared to the wavelength and is not portable and easily mounted. Therefore, a crescent antenna (Deshmukh & Kumar, 2006) has been used as the most suitable candidate for portability and ease of mounting. The salient features of the developed array antenna are provided produced in Table 1 below.

*Antenna Element Design.* The antenna element is designed with full-wave electromagnetic solver CST Microwave Studio. Fig 17a shows the CST Microwave Studio generated half disk antenna with slot loading for broad bandwidth applications. The single antenna element is designed on FR4 substrate with a dielectric constant of 4.4 and loss tangent of 0.002. The thickness of the material is 0.787 mm. The radius of the half disk antenna is 8 cm. An L-shaped slot is added around the coaxial feed probe to increase the bandwidth to more than 100 MHz centered at 910 MHz. The total height of the antenna element is 25 mm. The single antenna element produces a RL bandwidth of more than 100 MHz centered at 910 MHz with a maximum RL of 35 dB.

*Beamforming Network Design.* The next design is the beamforming network for the 3×2-element array antenna. Figure 17b below shows the layout of the beamforming network developed using Agilent's Advanced Design System (ADS) 2005A. The beamforming network comprises a 1-to-6-way power divider, 4-bit reflection type phase shifters and biasing control of the 4-bit phase shifter array. All components are designed on Taconic TLX0 of thickness of 0.787, dielectric constant of 2.45 and loss tangent of 0.0006.The details of the reflection type phase shifter can be found in previous work by the author and a colleague (Karmakar & Bialkowski, 1999). The phase shifters are comprised of a quadrature hybrid coupler in a very compact package with reactive loading. The reactive loading is comprised of high impedance transmission lines of various lengths followed by Philips BA682 pin diodes. A diode compensation technique is used to operate the diode in the 900 MHz band. The diode is a very low cost band switching diode with high parasitic capacitance at its OFF state. Detailed information about the diode compensation technique can be found in Karmakar (1999). The reflection type phase shifter is selected due to its large bandwidth and uniform configuration for all bits. The total beamforming network is delicately phase matched to each branch with zig-zag transmission lines. The feed network is a corporate type feed network with the first stage of a 2-way T-junction power divider followed by a three way power divider. Thus 1-to-6-way equal power division is achieved. In this primary design amplitude tapering was not attempted. However, the readers are referred to Karmakar (2003).

*Integrated Array Antenna.* Mutual coupling greatly affects the array antenna's performance, especially when the antenna elements are brought too close together. The situation worsens when the heights of the antenna elements are considerable (in this case 2.5 cm). Metallic barriers have been used to reduce the mutual coupling effect between the antenna elements as shown in Figure 17c. Thus the antenna aperture area is maintained at 50 cm× 50 cm. This area is about 70% of a conventional patch array antenna at that frequency. Figure 17c shows the top view of the complete antenna array generated in CST Microwave Studio Layout platform. The antenna is housed in a microwave transparent plastic casing as shown in Figure 17d. The total weight of the antenna is about 0.7 kg, making it suitable for easy transportation and mounting.

- Results of Array Antenna

Figure 18a shows the s-parameters vs frequency for the 1-to-6-way power divider. The input return loss of the power divider is about 30 dB and the transmission loss for each output port is about -8 dB. The output ports are well phase matched at the center frequency of 910 MHz and fall linearly with the frequency. The maximum phase offset at the band edge is less than ±15°. The phase shifters have an average insertion loss of 3.5 dB in both on and off states and the maximum phase deviation of about ±18° for the four bit phase shifter array. This quantization error in phased array antennas is quite common. Fig 18b shows the input return loss of the assembled array antenna. The antenna is very well matched with more than 10 dB return loss in the prescribed bandwidth of 100 MHz centered at 900 MHz.

Figure 18c shows the radiation pattern of the array antenna in the broadside direction. As the figure shows, the antenna generates a beam of 42° beamwidth with a gain of 12 dBi. Figure 18d shows the beam patterns when they are scanned at +20° and -30° from the broadside direction. The results indicate that the antenna is capable of scanning main beams in 3D orientation. However, due to the compact design with only 6 antenna elements the degree of freedom is confined to about ±40° elevation scan angles from the broadside direction. With the number of antenna elements this scan range can be improved at the expense of more complex beamforming network, associated switching electronics and a larger antenna aperture.

## CONCLUSION

RFID is an enabling technology which is transforming the identification, asset tracking, security surveillance, medicine and many other industries at an accelerated pace. This chapter has presented a comprehensive review of smart antennas for modern RFID readers. This comprehensive review has covered new ground as follows:

(i) Analysis, synthesis and classification of various smart antennas for RFID readers developed recently, mainly by various commercial organizations. This classification is based on the unique features of the application specific RFID readers. They are: location finding for asset tracking and management in both retail chains and warehouse scenarios; anti-collision improvement using slotted-Aloha and binary tree algorithms and MIMO antenna systems; ESPAR and MRC adaptive antennas for maximising S/N ratios of receive signals from the tags and improving reading rate utilizing polarization and spatial diversity in smart antennas. To the best of the author's knowledge such a comprehensive classification based on the available literature from various sources has not been presented previously. Smart antennas for RFID readers have created a new field of research, similar to the smart antennas for mobile communications a decade ago.

(ii) Ingram (2003) has proposed multiple antennas in a tag to improve throughput. She has also recommended that MIMO antennas are more efficient in improving throughputs than a directional adaptive antenna. This finding has been reinforced by Griffin and Durgin (2007) in their throughput calculation, assuming a dyadic channel model to be the RFID propagation environment. They have suggested pushing the frequency of operation in the 5.8 GHz ISM band to exploit the miniaturized tag antennas at the higher frequency.

(iii) The scholarly literature available on smart antennas for RFID readers is lacking and it is very difficult to find technical information about recent developments. To overcome this shortcoming in smart antenna development in the educational arena, the author has investigated a 3×2-element compact planar phase scanned antenna array for RFID readers. A detailed design guideline for the component level physical layer development of the smart antenna has been presented. The antenna is low cost, compact and lightweight for portable applications and ease of mounting. The antenna is designed at the 900 MHz frequency band with 100 MHz bandwidth to cover a plethora of applications in this frequency band. The development certainly has a potential international market.

Finally, tremendous strides have been evolving around RFID technology. These developments will make the RFID technology more ubiquitous, accurate, high speed and user friendly. Smart antennas for RFID readers have shed light on the solution of existing problems for various applications such as collision avoidance, location determination of tags, and reading capacity improvement of RFID readers. Almost every type of smart antenna has been developed to fulfil the demands of RFID applications. Now it is time for antenna designers to reap the benefit of this new enabling technology with innovative and active participation in this field of research.

## ACKNOWLEDGMENT

The work is supported by Australian Research Council's Discovery Project Grant # DP665523: *Chipless RFID for Barcode Replacement.* The software supports from Agilent and CST are also acknowledged. Contract works by Muhammad Saqib Ikram and Sushim Mukul Roy are also acknowledged.

## REFERENCES

Alien Technology Corporation (2005), BAP ALR-2850 reader data sheet, 2005. Retrieved from http://www.alientechnology.com/products/rfid_readers.php in February 2006.

Auckland, D.T., Kilmczak, W. & Durgin, G.D. (2003, October). "Maximizing Throughput with Ultra-Compact Diversity Antennas", *Proc. of IEEE VTC 2003-Fall*, vol. 1. Orlando, FL, USA: IEEE, Oct. 2003, pp. 178–182.

BALOGH RFID (2003), HYPER X LMB-6012 RFID reader data sheet, Jan 2003. Retrieved from http://www.balogh.cc/HyperX/Support/PDF/LMB6012-6013.pdf in September 2007

Calmpitt, H.G. (2006). *RFID Certification Textbook, 2nd Ed.*, Arlington Heights, IL, American RFID Solution LLC.

Chung, K.K. T. & Liu, S. (2004), Antenna arrangement for RFID smart tags, *US Patent no. US 6,703,935 B1*, March 9, 2004

Collins, T.J., Gurney, D.P., Kuffner, S.L. & Rachwalski, R.S. (2005). Method and apparatus for multiple frequency RFID tag architecture, *US Patent no. US 2005/0052283 A1*, Mar. 10, 2005

Das, R. & Harrop, P. (2006) *RFID Forecasts, Players & Opportunities 2006 – 2016*, IDTechEx, London, UK,. Retrieved from http://www.idtechex.com/products/en/ view.asp?productcategoryid=93 on 11 September 2007.

Deshmukh, A. A. & Kumar, G. (2006). "Various slot loaded broadband and compact circular microstrip antennas" *Microwave and Optical Technology Letters*, 48(3), 435-439.

EPCglobal, (2005) EPCTM Radio-Frequency Identity Protocols Class-1 Generation-2 UHF RFID Protocol for Communications at 860 MHz - 960 MHz, Version 1.0.9, January 2005.

EPCglobal, Inc. (2006), The EPCglobal Network: Overview of Design, Benefits and Security, EPCglobal white paper. Retrieved on 11 April 2006 from http://www.epcglobalinc.org/

Griffin, J.D. & Durgin, G.D. (2007). "Reduced fading for RFID tags with multiple antennas", *IEEE Antenna Propagation Society International Symposium Digest* 2007, July Honolulu, USA.

GTA (2007). RFID Industry Group RFID Applications Training and RFID Deployment Lab Request Background Paper January, 2007

Hercht, K. & Storch, L. (2007). Sequenced antenna array for determining where gaming chips with embedded RFID tags are located on a Blackjack, poker or other gaming tables and for myriad other RFID application, *USA Patent no. US 2007/0035399 A1*, Feb. 15, 2007

Ingram, M.A. (2003). Smart reflection antenna system and method, US patent no. US 6,509,836, B1, Jan. 21, 2003

Karkkainen, M. & Ala-Risku, T. (2003) "Automatic Identification – Applications and Technologies", *Logistics Research Network 8th Annual Conference*, London UK, September 2003.

Karmakar, N.C., & Roy, S.M. (2007). "Development of a planar phased array antenna for RFID readers at 900 MHz", *Proc. 2nd International Conference on Sensing Technology (ICST 2007)* (pp. 202-204). Palmerston North, New Zealand.

Karmakar, N C & Bialkowski, M E (2000), "Electronically Steerable Array Antennas for Mobile Satellite Communications - A Review", *Proceedings of the IEEE International Conference on Phased Array Systems and Technology*, Dana Point, CA., USA, May 21-25, 2000, pp. 81-84.

Karmakar, N C and Bialkowski, M E (1999), "A Compact Switched-Beam Array Antenna for Mobile Satellite Communications", *Microwave and Optical Technology Letters*, 21(3), 186-191.

Karmakar, N C and Bialkowski, M. E. (1999). "A Compact Switched-Beam Array Antenna for Mobile Satellite Communications", *Microwave and Optical Technology Letters*, 21(3), 186-191.

Karmakar, N C and Bialkowski, M. E. (1999). "An L-Band 90° Hybrid-Coupled Phase Shifter Using UHF Band PIN Diodes", *Microwave and Optical Technology Letters*, 21(1), 51-54.

Karmakar, N. C., Roy, S. M., Preradovic, S., Vo, T. D., Jenvey, S. (2006) "Development of Low-Cost Active RFID Tag at 2.4Ghz", *36th European Microwave Conference*, pp: 1602-1605, 10-15 September, 2006.

Karmakar, N.C. (1999). *Antennas for Mobile Satellite Communications*, Unpublished doctoral dissertation, The University of Queensland, Australia.

Karmakar, N.C. (2003). "A Low Sidelobe Sub-Array of Portable VSATS" *Proc. 2003 IEEE International Antennas and Propagation Symposium and URSI North American Radio Science Meeting*, Columbus, Ohio, USA, June 22-27, 2003.

Kocer, F, Flynn, M.P. & Long-Range, A. (2005), "RFID IC with On-Chip ADC in 0.25 /spl mu/m CMOS", *Digest of Papers 2005 IEEE Radio Frequency Integrated Circuits (RFIC) Symposium*, pp: 361-364, 12-14 June, Long Beach, California, USA, 2005.

Kowalski, J. Serra D. & Charrat B. (2005), RFID-UHF Integrated Circuit, *US Patent no. US2005/0186904 A1*, Aug. 25, 2005

Lai, K.Y.A, Wang, O. Y.T., Wan, T.K.P., Wong, H.F.E., Tsang, N.M., Ma, P.M.J., Ko, P.M.J. & Cheung, C.C. D. (2005), Radio frequency identification (rfid) system, *European Patent Application EP 1 724 707 A2*, 22 July 2005.

Landt. L. (2001). Shrouds of Time. The History of RFID, An AIM Inc. publications, October 2001. Retrieved September 11, 2007 from http://www.aimglobal.org/technologies/rfid/resources/shrouds_of_time.pdf.

Law, C. Lee, K. & Siu, K.Y. (2000). "Efficient Memoryless Protocol for Tag Identification". *Proc. 4th ACM International Workshop on Discrete Algorithms and Methods for Mobile Computing and Communications,* August 2000.

Lee, D.V. (2005) RFID reader with multiple antenna selection and automated antenna matching, *US Patent No. 6,903,656 B1*, June 7 2005

Lee, J. (2007). Wireless *Networks Characterizations: Interference, Collision, and Localization*. Unpublished doctoral dissertation, Soul University, Korea.

Lee, J., Das, S.K. & Kim, K.A. (2004). "Analysis of RFID anti-collision algorithms using smart antennas" *Sensys '04*, Nov. 3-5, 2004, Baltimore, Maryland, USA, 265-266.

Maeda, T., Matsumoto, T. & Yoshida, H. (2007, July). "A study on the accuracy of location measurement of wireless ID tags integrated with multiple antennas and an antenna direction sensor", *IEEE Aps Digest 2007*, Honolulu, USA.

Marsh, M.J.C., Lenarcik, A., Van Zyl, C.A. Van Schalkwyk, A.C. & Oosthuizen, M.J.R. (1996). Detection of multiple articles, *US Patent no.* 5,519,381, may 21, 1996.

McFarlane, D. & Sheffi, Y. (2003). "The Impact of Automatic Identification on Supply Chain Operations", *International Journal of Logistics Management*, 14(1), 407 – 424.

Mendolia, G. Gupta, O. & Toit, C.F. (2005). RF ID tag reader utilizing a scanning antenna system and method, *US Patent No. US 2005/0113138 A1*, May 26, 2005

Nakamura, T. & Seddon, J. (2006) Omron Develops World's First Antenna Technology That Boosts UHF RFID Tag Read Performance, *Omron Corporation press releases*. Retrieved from http://www.omron.com/news/n_270306.htmlC in June 2007.

Padhi, S. K., Karmakar, N. C., Law, C. L., Aditya, S. (2001). "A Dual Polarized Aperture Coupled Microstrip Patch Antenna with High Isolation for RFID Applications", *2001. IEEE Antennas and Propagation Society International Symposium*, vol. 2, pp. 2-5, Boston, USA, July 2001

Pozar, D. M. (2005). *Microwave Engineering – 3rd Edition,* John Wiley & Sons, Inc., NY, USA, 2005.

Preradovic, S. & Karmakar, N.C. (2007). Modern RFID Readers – A Path to Universal Standard. *Microwave Journal, 13.* Retrieved from http://www.mwjournal.com/Journal/Issues.asp?Id=68)

Profibus (2007) Device Net-Ethernet IP-Modbus-Remote I/O, Retrieved from http://www.rfidinc.com in June 2007.

RFID Analyst (2003). Issue 26, March 2003. Retrieved from http://rfid.idtechex.com/ documents/en/sla.asp? documentid=67 in Sep 2007.

Salonen, P. & Sydanheimo, L., (2002). "A 2.45 GHz digital beam-forming antenna for RFID reader", *Proc. 55th Vehicular Technology Conference, 2002*, 4, 1766 - 1770

Simons, R (2001). *Coplanar Waveguide Circuits, Components, and Systems*, New York, Wiley.

Stockman, H. (1948). "Communication by Means of Reflected Power", *Proceedings of the IRE*, 1196-1204. October 1948.

Sun, C. & Karmakar, N. C. (2004). "Direction of Arrival Estimation with a Novel Single Port Smart Antenna", *EURASIP Journal on Applied Signal Processing*, 2004(9), 1364-1375.

Sun, C., Hirata, A, Ohira, T & Karmakar, N. C. (2004). "Fast Beamforming of Electronically Steerable Parasitic Array Radiator Antennas: Theory and Experiment", *IEEE Trans. Antena. Propagt.*, 52(7), 1819-1832.

Tuovinen, J. and Vaha-Heikkila, T (2006), "MMID-Millimetre-Wave Identification using RF-MEMS and Nanocomponents", *2006 European Microwave Conference Workshop Notes WS14 RF MEMS and RF Microsystem Application and Development in Europe, Manchester, UK, 2006.*

Wang, J.J.M., Yang, C.C. & Lin, W.Y. (2004). Method and apparatus for GPS signal receiving that employs a frequency-division-multiplexed phased array communication mechanism, *US Patent no. US 6,784*,831 B1, Aug. 31, 2004.

Ward, J & Compton, R.T. Jr (1992), "Improving the Performance of a Slotted ALOHA Packet Radio Network with an Adaptive Array". *IEEE Trans. Commun.*, 40(2), February 1992.

## Section IV

# Experiments and Implementations

# Chapter XXII
# Field Programmable Gate Array Based Testbed for Investigating Multiple Input Multiple Output Signal Transmission in Indoor Environments

**Konstanty Bialkowski**
*University of Queensland, Australia*

**Adam Postula**
*University of Queensland, Australia*

**Amin Abbosh**
*University of Queensland, Australia*

**Marek Bialkowski**
*University of Queensland, Australia*

## ABSTRACT

*This chapter introduces the concept of Multiple Input Multiple Output (MIMO) wireless communication system and the necessity to use a testbed to evaluate its performance. A comprehensive review of different types of testbeds available in the literature is presented. Next, the design and development of a 2×2 MIMO testbed which uses in-house built antennas, commercially available RF chips for an RF front end and a Field Programmable Gate Array (FPGA) for based signal processing is described. The operation of the developed testbed is verified using a Channel Emulator. The testing is done for the case of a simple Alamouti QPSK based encoding and decoding scheme of baseband signals.*

## EVOLUTION OF SMART ANTENNAS AND MIMO

In the past two decades, wireless communication systems have grown with an unprecedented speed from radio paging, cordless telephone, and cellular telephony to multimedia platforms offering voice and video streaming, interactive services, and even a global positioning information of the user. One undesired outcome of this expansion is a heavy utilization of the available frequency spectrum. A particular pressure comes from new multimedia applications, which require larger operational bandwidth for their implementations. Conventional coding and modulation techniques that are based on frequency, time or code division have difficulty to provide a suitable solution to this problem. For example, allocating new users or new applications leads to consuming of an additional frequency band in frequency division multiplexing systems. In time division and code division multiplexing systems this results in an increased level of a man-made noise, which in turn adversely affects the signal to noise (SNR) ratio and thus the quality of the received signal.

As the conventional coding and modulation techniques are unable to overcome the problem associated with the limited frequency spectrum, the designers turn to the space/angle domain to improve capacity and reliability of wireless systems. This approach is already in use in many currently available terrestrial and satellite communication systems. The best known example is the cellular telephony. By creating physically separated cells, the same frequency spectrum is re-used only at an expense of small interference between adjacent cells. A further extension of the cellular concept is via the introduction of sectors within cells. This task is realized by antennas at base stations having sectoral radiation patterns instead of omnidirectional ones. More advanced versions of this concept are antennas serving a variable width sector. This is required, to overcome a bottleneck caused by an excessive number of users in a particular time of the day in this sector (Rosol, 1995). Such an intelligent antenna system is an example of the simplest type of smart antenna (Liberti and Rappaport, 1999). A more advanced form, known as an adaptive antenna, can follow individual mobile users by forming narrow beams towards them. At the same time, they can produce nulls or low side lobes towards undesired users. In all of these cases, multiple element antennas (MEAs) are instrumental to divide space into angular slots and adapt their width in time so that the wireless system can serve an increased number of users. It is apparent that such intelligent systems are capable to fight co-channel interference and thus, they are able to improve the quality of a communication link.

All of these benefits are possible if an MEA systems are capable to distinguish between signals coming from desired and undesired directions. Such a task is relatively easy to accomplish under Line of Sight (LOS) signal propagation conditions. As a result, adaptive antennas rely on LOS conditions for their proper operation. In order to achieve this condition in practice, base station antennas are located on elevated platforms so their operation is unaffected by the presence of surrounding objects. Such conditions are usually difficult to fulfill for mobile units. This is because they operate at low heights and therefore their operation is affected by the presence of various scattering, refracting or reflecting objects that surround them.

Meeting LOS conditions becomes more challenging for indoor scenarios, as the signal's propagation is affected by walls, indoor furniture, equipment and humans. Under such conditions, signals of similar strength arrive from many directions on antennas. As a result, there is not much benefit from forming narrow beams. This is because there are too many directions at which these beams need to be formed.

It has been pointed out in (Foschini, 1996; Telatar, 1999; Gesbert et al., 2003; Paulraj et al., 2004) that this unfavorable situation for an adaptive antenna employing a beamforming strategy can be resolved in a positive manner when both the transmitter (TX) and the receiver (RX) employ multiple element antennas. The reason is that the rich-in-scattering Non-Line of Sight (NLOS) signal propagation conditions offer multiple virtual channels between the transmitter and receiver, which fade independently. If the propagation conditions change sufficiently slowly the receiver is capable to learn about the virtual channels from training sequences (the data known both to the transmitter and receiver) and utilize them in an advantageous manner. Telatar (1999) has shown that if such channels are independent and identically distributed (have identical statistical properties), the capacity is M-fold greater than of a conventional system equipped with single antennas at the TX and RX sides, where M is the minimum number of TX/RX antennas. Such a system is named a multiple input multiple output (MIMO) system. Following this new designation, smart antennas, depending on whether they are use at the transmitter or receiver, are designated as MISO or SIMO, while the conventional fixed beam antenna systems are named as SISO systems.

From the above it is apparent that multiple element antennas accompanied by suitable signal processing algorithms at one or two sides of a communication link can lead to substantial gains in capacity of wireless systems without additional frequency spectrum. These antenna systems are also capable of offering a better quality communication link via diversity when the signals in virtual channels are uncorrelated (Jakes, 1974; Rappaport, 2002). The benefits occur when the amplitude of a correlation coefficient is in the order of 0.7 or smaller. The signals appearing in virtual channels (branches) are of a similar strength.

It is apparent that MEA used at one or two sides of communication link can offer various modes of operation of wireless system. These include beamforming, multiplexing and diversity. The choice of a particular mode depends needs on the actual properties of a communication channel.

In order to test benefits of MEA in real world, a prototype system capable to implement various modulation, coding and transmission schemes is required. Such a system is called a MIMO testbed or MIMO demonstrator. This chapter reports on various MIMO testbeds which have been reported in the MIMO literature. Next, it describes the design, development and testing of an FPGA-based MIMO testbed for operation at 2.45GHz or 5GHz, which has been developed by the authors.

This presentation is outlined as follows. Firstly the MIMO system is introduced. Next, the motivation for developing MIMO testbeds is given. A review of MIMO testbeds, which have been developed by researchers in different parts of the world is presented. Next, details concerning the design and development of the FPGA-based 2×2 MIMO testbed, which was produced by the authors, are described. The results of testing of various sub-systems of this 2×2 MIMO testbed are shown.

## MIMO SYSTEM

### Configuration and Principles of Operation

Multiple Input Multiple Output (MIMO) system is a radio system formed by multiple antennas both at the transmitter and receiver sides with many signals transmitted between the two sides. As such, it represents a sophisticated generation of smart antenna. Such a system can operate over narrow or wide frequency bands. Here the considerations are restricted to the narrowband case. The wideband MIMO system can be obtained by applying Orthogonal Frequency Division Multiplexing (OFDM) to MIMO (Engels, 2002; Prasad, 2004). This MIMO-OFDM technique allows for treating a wideband MIMO system as a set of many narrowband MIMO sub-systems. Here considerations are focused on the narrowband case, as the expansion to MIMO-OFDM is apparent via the use of a MIMO-OFDM technique.

Fig. 1 shows the configuration of a narrowband $N_t \times N_r$ MIMO system that is constituted by $N_t$ transmitting (TX) antennas and $N_r$ receiving (RX) antennas.

At a chosen frequency **f**, the channel between the transmitter and the receiver is described by the complex channel matrix **H**. The relationship between the received and transmitted signals and noise entering the receiver is given as:

$$\mathbf{y} = H\,\mathbf{x} + \mathbf{n} \tag{1}$$

where **x** the transmitted signal vector, **y** the received signal vector, and **n** is the noise vector.

In (1), the coefficients of the **H** matrix can be interpreted as transmission coefficients between individual transmitting and receiving antennas.

The signals **x** , **y** and **n** vary randomly in time so are the channel properties. As a result, **H** , **y** and **n** regarded as random processes. At a given instant of time are random variables whose behavior is described by suitable probability density functions.

*Figure 1. Configuration of a narrowband MIMO system*

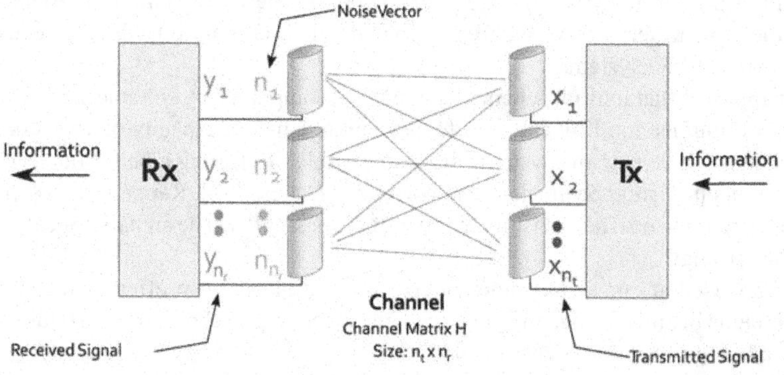

In (1),**x** represents raw data, which can be suitably encoded before being transmitted. Linear coding schemes are of particular value to MIMO systems because of their simplicity. For a linear coding scheme, the relationship between the vector of transmitted symbols, the vectors of received symbols and noise is given by the modified expression:

$$\mathbf{y} = HW\,\mathbf{x} + \mathbf{n} \tag{2}$$

where **x** is the $N_S\times1$ vector of transmitted symbols, and **y** and **n** are the $N_r\times1$ vectors of received symbols and noise. **W** is the $N_t \times N_S$ linear precoding matrix. The column dimension $N_s$ of **W** can be selected smaller than $N_t$. This is useful if, for example, the propagation environment can support $N_S$ virtual channels.

## MIMO Capacity

An instantaneous capacity of a communication system (given in bits/second/Hz) is defined as the amount of information in bits that can be passed between the transmitter and the receiver in one second using a bandwidth of one Hertz. In the case of the conventional SISO system, instantaneous capacity is given by the standard Shannon formula for capacity (Shannon, 1948; Fano, 1961).

$$C_I = \log_2(1 + \rho\,|h|^2) \tag{3}$$

where ρ is the average signal to noise ratio at the receiver and $|h|^2$ is the normalized channel power transfer characteristic. Note that as $|h|^2$ varies in time so does the instantaneous capacity.

In the case of MIMO, $|h|^2$ in expression (3) needs to be replaced by a factor that takes into account the channel matrix and the transmission scheme. This modification leads to the generalized Shannon capacity. In order to obtain the capacity with respect to some period of time, a suitable average over channel matrix realization over this time is formed. The resulting capacity is known as the ergodic capacity.

An expression for the ergodic capacity of MIMO system with $N_t$ antennas at the transmitter and $N_r$ antennas at the receiver, which uses an arbitrary signal transmission scheme was presented in (Telatar, 1999; Gesbert et al., 2003) and is rewritten here as (4):

$$C = \max_{tr\{\mathbf{Q}\}\le P_T;\, q_{nn}\ge 0} E_{\mathbf{H}}\left\{\log_2 \det(\mathbf{I}_{N_R} + \frac{1}{P_N}\mathbf{HQH}^\dagger)\right\} \tag{4}$$

where $E_H\{(\cdot)\}$ denotes the expectation value with respect to the channel matrix **H**, $\mathbf{Q} = E\{xx^\dagger\}$ is the covariance matrix of transmitted signals defining a transmission scheme, $E_H\{(\cdot)\}$ denotes the expectation value over time and value over time and $(\cdot)^\dagger$is the conjugate and transpose operation of the matrix or so-called Hermitian operation. $\mathbf{I}_{N_r}$ is the $N_r\times N_r$ identity matrix. The transmitted signals are constrained by the average total power tr{**Q**}$\le P_T$, where tr{**Q**} is the sum of the diagonal elements of matrix **Q.** $P_N$ is the noise power at receiving antennas.

As observed in (4), MIMO capacity depends both on the statistical properties of the **H** matrix coefficients as well as on the choice of the transmission scheme, which is defined by the covariance matrix **Q** of the transmitted data. Telatar (1999) has focused his considerations on so-called independent transmission scheme in which each branch of transmitted data $x_1, x_2, \ldots, x_{N_t}$ is independent and of equal power. In this case, the covariance matrix **Q** is equal to the $N_t \times N_t$ identity matrix scaled by the factor $P_T / N_t$. For this transmission scheme, the ergodic capacity of MIMO is given by (5):

$$C = E_{\mathbf{H}}\left\{\log_2 \det\left(\mathbf{I}_{N_r} + \frac{\rho}{N_t}\mathbf{HH}^\dagger\right)\right\} \tag{5}$$

where ñ represents SNR.

In order to compare the MIMO capacity for given channel conditions versus the one for a reference channel, the expression (5) requires a suitable normalization of the channel matrix **H** (Bialkowski et al., 2006a).

Assuming that the MIMO system operates under ideal Non-Line of Sight conditions, for which the entries of the channel matrix **H** are represented by independent identically distributed (i.i.d) Rayleigh variables, Telatar (1999) has shown that using an independent transmission scheme, the capacity of MIMO is $N_S$ times greater than the capacity of SISO system, $N_s = \min(N_t, N_r)$.This result is obtained by considering matrix:

$$\mathbf{W}=\begin{cases}\mathbf{HH}^{\dagger} \text{ for } N_r < N_t \\ \mathbf{H}^{\dagger}\mathbf{H} \text{ for } N_t < N_r\end{cases}$$

which is decomposed into $N$s eigenvalues $\lambda_1, \lambda_2, \ldots, \lambda_{Ns}$.

Using the determinant property that det(**I**+**AB**) =det(**I**+**BA**), (5) can by rewritten as shown in (6).

$$C = E_{\lambda}\left\{\sum_{n=1}^{N_s}\log_2\left(1+\frac{\rho}{N_t}\lambda_n\right)\right\} \tag{6}$$

## MIMO TESTBED

### Motivation for MIMO Testbed

The answer to how well a practical MIMO system can approximate the theoretical capacity limits given by expressions (3) and (5) can be provided by a MIMO testbed. Such a system can offer an assessment of the MIMO concept in real signal propagation environments including actual modulation and coding schemes. Its extra advantage against computer simulations is that it can point out complexity issues early in the design stage. This is important from a marketing perspective, as the MIMO potential can be tested with respect to entering mass production.

### MIMO Testbed Requirements

The design and development of MIMO testbed has to fulfill a number of requirements. These are addressed as follows.

#### High Performance Baseband Processor

Achieving an increased capacity of MIMO requires the knowledge of the channel matrix at least at the receiver side. The expressions for capacity (1) and (2) assume that the channel matrix entries are perfectly known at the receiver. In practice, this assumption is realized in an approximate manner by sending pilot symbols as part of each transmitted data packet and by applying a suitable channel estimation technique (Biguesh and Gershman, 2006).

This procedure needs to be accomplished in real-time. The reason is that realizing high capacities given by (1) and (2) relies on a quasi-static behavior of the channel during estimating the channel matrix and decoding the information data. Maintaining this assumption in practice can be accomplished by fast data processing. This requirement favours a Field Programmable Gate Array (FPGA) as a baseband signal processor because of its capability to perform parallel signal processing on multiple signals at the MIMO receiver. This component is the main limiting factor in the performance limit of the testbed.

#### Integrated RF Front End

In addition to the baseband processor, a MIMO system requires radio frequency (RF) front ends for each of the transmitting/receiving antennas. The RF front end is responsible for performing the signal up- and down-conversion process using either a *heterodyning* or *direct conversion* (homodyne) scheme. The differences between the two are as follows.

The *heterodyne* process involves mixing a baseband signal with an intermediate frequency (IF) first, and then with a second (RF-IF) to convert the signal to the required frequency (RF) band. After each stage of up-conversion (or down-conversion), filtering is required to prevent image frequencies from interfering within the desired band. Due to the difficulty of creating filters with sharp cut-off frequencies at the RF, a multi-stage conversion is used. The baseband signal is generated by a digital to analogue converter at the transmitter, and by an analogue to digital converter at the receiver.

*Direct conversion* involves mixing two baseband signals, an in-phase component (I) and a quadrature component (Q), with the carrier frequency. This approach removes the need for intermediate frequencies, as the image frequency cancels at DC when down-converting. This method does however require double the amount of converters (ADC/DAC).

If the system is to be portable, the RF hardware needs to be compact and integrated. This is because the approach involving the assembling of individual RF components usually leads to an increased size.

The demand for integrated MIMO transceivers is driven by the fact that many new wireless standards include a requirement or option of accommodating MIMO. Examples include the IEEE 802.16 (WiMAX) for fixed broadband wireless access (Erceg et al., 2001), IEEE 802.11n (Wireless LAN) for local area networking (Mannion, 2004; Abraham et al., 2005), and UMTS 3G long term evolution (LTE) (Dahlman et al., 2006). 3G LTE is being developed by the Third-Generation Partnership Project (3GPP) for mobile applications (Lucent et al., 2001).

## Compact Antenna Size

The works of Telatar (1999) and of other researchers working in the field of information theory have paid little attention to antennas and their operation in MIMO systems. Their theories assumed only reasonable antenna spacing to avoid signal correlation. The typical recommendation for narrowband operation is the antenna elements need to be separated by half-wavelength. Practical considerations demand small size antennas and their tight spacing if MIMO transceivers are to be portable and compact. However, this requirement leads to stronger mutual coupling of antenna elements and a possible increase in correlation of the transmitted/received signals. These in turn have an impact on MIMO capacity.

Other considerations concern the choice of number of antennas. The proper number of Tx and Rx antennas is crucial with respect to the estimation of the channel matrix. For a more efficient estimation process, a favorable choice is to have a larger number of receiving antennas. A MIMO testbed should provide a suitable provision to investigate these issues.

## Coding Scheme

The codes aimed for use in MIMO systems carry a special name of Space-Time Codes, which are abbreviated as STC. The choice of STC codes is governed by two approaches, the one that aims at data rate maximization, and the other that aims at diversity maximization. In practical cases the boundaries between the two may be less clear. When one aims for multiplexing and thus maximizing data rate, independent signals over individual virtual channels need to be transmitted. However, to protect them against errors, which are caused by channel fading (and noise plus interference), individual data streams may be joint-encoded. By doing this, the encoding scheme enters the second approach of diversity because the outage probability is minimized. Because the level of redundancy is increased, independence of data streams is reduced and so is the capacity. Ultimately joint coding may result in MIMO capacity equivalent to that of SISO. Therefore the coding scheme designer has to be aware of this situation.

Examples of STC coding schemes include BLAST (Foschini, 1996; Foshini and Gans, 1998) for spatial multiplexing and Alamouti (Alamouti, 1998) for spatial diversity. As the latter one will be applied in our 2×2 MIMO testbed, a short introduction of this scheme is made here. Firstly, it has to be noted that Alamouti's transmit diversity scheme is the first example of a space-time code which requires only linear processing at the receiver (as given by expression (2)). The motivation for its introduction has been the demand for an alternative scheme to the receive diversity which is commonly used at base stations of cellular systems. The Alamouti transmission diversity scheme is attractive because similarly as in the receive diversity schemes it uses two antennas at a base station and one antenna at a mobile. The diversity gain of the Alamouti scheme is equivalent to that of the mobile equipped with two receiving antennas and using the Maximal Ratio Combining receive diversity scheme. It has to be noted that Alamouti has extended his original scheme to the case of multiple element receiving antennas. In turn, Tarokh et al. (1999) extended this diversity scheme to MIMO systems having an arbitrary number of TX and RX antennas. Another extension of the Alamouti scheme is the differential Alamouti scheme (Tarokh and Jafarkhani, 2000). This scheme does not require channel estimation at the receiver, unlike the previous encoding algorithms. There is however a performance penalty of 3dB with respect to a coherent receiver. With respect to an Alamouti encoding scheme, both the transmitter and receiver are modified for this scheme.

## Review of Existing Testbeds

MIMO testbeds can be classified into two main categories, *online* and *offline*. An *online* system is able to fully process the input and output data streams in real-time, as they are produced. An *offline* system, on the other hand, does not process continuously incoming data in real-time and so it saves it for future processing. Most offline systems operate as multiple channel sounders. In the most basic case, they are used to record the channel matrix of the channel. Commercially available channel sounders, such as the MEDAV RUSK ATM channel sounder (Jungnickel et al., 2003; Channel Sounder, 2003), can be used to this purpose.

The benefits from using an *offline* are due to the fact that several alternative methods of decoding can be applied and compared using the same received signal. In order to use this to an advantage, some systems are configured to work in two ways, when the hardware buffers are connected to a high performance processing platform, then the system is *online*, but can also be stored and post processed in MATLAB (or other simulation software).

The historical development of testbeds has gone through a number of technological changes. Initially, testbeds made use of a digital capture card and a PC. A digital capture card contains a buffer which lets a PC process the data offline, thereby using a multi-channel capture card, realization of a simple channel sounder or testbed is possible. In order to accelerate the processing of the captured data the next developments concerned the use of high speed digital signal processors (DSP). This has been advanced by Application Specific Integrated Circuits (ASIC) or Field Programmable Gate Arrays (FPGA).

The shortfall of the DSP approach is that it requires a much higher clock rate than that of the signal bandwidth, as several operations are required to prepare data for transmission or reception. As a result, a 100MHz signal collected from multiple antennas may require a DSP with a clock speed of several GHz. In addition, scalability as more antennas are added, or additional processing, cannot be easily provided. Performance in a testbed for a DSP based solution would be limited by the clock speed of the DSP.

An ASIC based solution is capable of very high data throughput, however they are not able to be customized once produced, meaning all possible implementation types need to be thought of in advance, before the custom integrated circuit is produced. FPGAs, on the other hand, offer performance similar to the ASIC based designs. Their advantage is that they are reconfigurable on demand, for example, when a new design idea needs to be tested. Sometimes they are considered a reconfigurable or prototyping version of ASIC based designs. This is due to the synthesis process being very similar, so when the disadvantage of power or inadequate performance arrives, the additional steps to produce an ASIC based design can be performed.

In the hardware based design (ASIC/FPGA), operations occur simultaneously in parallel, therefore the clock rate required to process an incoming signal is usually the same, or less than that of the bandwidth of the incoming signal. For a wide band MIMO, hardware based designs using ASIC/FPGA technology are the only practical solution. Due to parallel nature, a 100MHz ASIC/FPGA–based hardware design can easily outperform a 3GHz based DSP design. The availability of technology with higher processing bandwidths has been the main engine for conversion of *offline* systems to *online* ones. Performance of FPGA and ASIC based testbeds would be limited by the number of operations that can be performed in parallel, therefore the size of the chip (in logic gates) is important.

## Examples of Testbeds

Current literature on MIMO shows quite an extensive number of MIMO demonstrator/testbeds. Here, an overview of selected examples, which give a good representation of the current status of MIMO testbeds, is provided.

The first narrowband MIMO prototype was reported by Bell labs in 1998 (Wolniansky et al., 1998). This demonstrator was built to obtain initial measurements to verify the predicted gains of a MIMO transmission. The system supported eight transmit and twelve receive antennas. Signal processing was performed offline on a frame-by-frame basis. Bell Labs called their system BLAST – Bell Labs Layered Space Time, and proposed a number of methods to encode the signals across space and time. The original V-BLAST testbed used a carrier of 1.9GHz and a symbol data rate of 30 ksymbol/sec, and processed data in real time.

The spin-off of Bell Labs, Lucent, was responsible for developing the first prototype of broadband MIMO (Wolniansky et al., 1998). In further developments, Lucent has shown the application of MIMO as an extension to the CDMA based systems (Adjoudani et al., 2003) (Lucent US) (Garrett et al., 2004) (Lucent Australia). Their design uses a combination of digital signal processors and field programmable gate arrays, which enabled operation in real time. In turn, IOSpan Wireless/Stamford (Sampath et al., 2002) has been the first to develop and demonstrate MIMO-OFDM system prototype.

*Table 1.*

| | Mode | # Tx/Rx | Platform Type | Freq. (GHz) | Sig. Rate | Year |
|---|---|---|---|---|---|---|
| Lucent (Wolniansky et al. 1998) | Offline | 8/12 | DSP | 1.9 | 30kHz | 1998 |
| BYU (Wallace et al. 2003) | Offline | 4/4 | DSP | 2.4 | 1.25MHz | 2001 |
| Bristol U (Horseman et al. 2003) | Offline | 4/6 | DSP | 5.2 | 12MHz | 2001 |
| Motorola (Batariere et al. 2001) | Online | 2/2 | FPGA/PC | 3.65 | 20MHz | 2001 |
| IOSpan/Stamford (Gesbert et al. 2002) | Online | 3/2 | ASIC | 2.5 | 2MHz | 2001 |
| Lucent/ETH (Adjoudani et al. 2003) | Online | 4/4 | FPGA | 2 | 2MHz | 2002 |
| Rice (Murphy et al. 2003) | Online | 4/4 | FPGA | 2.5/5 | 20MHz | 2003 |
| TU (van Zelst and Schenk 2004) | Online | 3/3 | PC | 5 | 20MHz | 2003 |
| UCLA (Rao et al. 2004) | Online | 4/4 | FPGA/PC | 5 | 20Mhz | 2004 |
| ETH (Haene et al. 2004) | Online | 4/4 | FPGA | 2.5/5 | 20MHz | 2004 |
| — (Perels et al. 2005) | Online | 4/4 | ASIC | 5.2 | 20MHz | 2005 |
| IMEC (Wouters et al. 2004) | Online | 2/2 | FPGA | 5.2 | 20MHz | 2004 |
| CSIRO (Suzuki et al. 2005) | Offline | 4/4 | FPGA/PC | 5 | 40MHz | 2005 |
| Alberta U (Goud Jr et al. 2006) | Online | 4/4 | FPGA/PC | 2.4 | 12.5MHz | 2006 |
| UQ Current / (Possible) | Online | 2/2 | FPGA | 2.4/5 | 12.5MHz (50MHz) | 2006 2007 |

Table 1 shows a number of MIMO testbed platforms, which were developed in different parts of the world.

Here we make brief comments on representatives of these systems. These include a DSP-based offline system at BYU (Wallace et al., 2003; Wallace et al., 2004), the FPGA-based quasi real-time testbed developed at CSIRO (Suzuki et al., 2005; 2006), the FPGA-based online systems developed at UCLA (Rao et al. 2004), Rice (Murphy et al., 2003; 2006) and Alberta University (Goud Jr et al., 2006), and wideband systems at Bristol (Piechocki et al., 2001; Horseman et al., 2003) and ETH-Zurich (Haene et al., 2004).

The BYU MIMO testbed (Wallace et al. 2004) is one of the early examples of MIMO demonstrators. Its purpose has been obtaining the channel matrix for indoor or outdoor environments. It can accommodate up to 16×16 antennas and a frequency of operation of 0.8 to 6GHz, however experiments are restricted to the 4×4 mode and the 2.45GHz band. The transmitter digitally generates N unique binary (±1) signal codes. These signals are up-converted and amplified in the RF chassis before being fed into N transmit antennas. The receiver uses an RF chassis to amplify and down-convert the signals from each of the M antennas. The resulting M intermediate frequency (IF) signals are low-pass filtered, amplified, and sampled using a 1.25 MSmp/s A/D card and stored on a PC. Synchronization between Tx and Rx is resolved through a 10MHz reference signal passed between them. The system is equipped with two different antenna arrays. The first design is a 4-element dual-polarization patch array with half-wavelength element spacing while the second array consists of a square metal plate with a two-dimensional grid of 33×33 holes spaced at roughly 1.5 cm intervals. Monopole antennas are placed in the holes to achieve a wide variety of array geometries.

The raw data collected using the measurement platform is processed to obtain estimates of the time-variant channel matrix. The technique consists of 3 basic steps: (1) code search, (2) carrier recovery, and (3) channel estimation.

Extensive amounts of data have been collected from indoor field trials, which were performed in one of the BYU computer labs. The obtained data served the purpose of the following investigations: channel stationarity, transfer matrix element statistics, and channel spatial correlation. Additionally, the effects of such factors as antenna element polarization and directivity on the capacity were studied. Other investigated issues concerned the relationship between multipath richness and path loss as well as their joint role in determining channel capacity.

The MIMO testbed produced by CSIRO, Australia (Suzuki et al., 2005; 2006) can be viewed as an advanced form of the BYU MIMO testbed. It is an 8×8 MIMO demonstrator, however its typical use is restricted to the 4×4 mode. By using a modular approach to its RF front ends, it can serve 2.4GHz, 5.2GHz and 40GHz band. Because the processing of data occurs on a PC in MATLAB, which is not capable of processing the packets in real time, it offers a "quasi" real-time mode of operation. The PC used is embedded into each MIMO transceiver, and is also used for displaying data. CSIRO has made its testbed available for research and teaching purposes to the Australian universities.

Researchers (Rao et al., 2004) at the University of California, Los Angeles have described three MEA systems (MIMO and SIMO) operating over various frequency bands as part of their research and education program. One of them is a DSP-based real-time 2×2 MIMO testbed operating in the 5GHz band. This testbed was developed to provide a platform that combines the flexibility of a software-defined baseband modem with an RF pass-band front-end. In this system, the data is modulated offline on a PC, and stored in a set of memory boards connected to the DSP. Upon command, the data is sent in real time to the DAC converters at a rate of 25 Msamples/s. The analog signal is upconverted to the carrier frequency of 5.25 GHz. The transmitter uses a two-step upconversion with an intermediate frequency of 1.75 GHz. The receiver uses a two-step downconversion with an IF of 2.45 GHz that feeds an integrated receiver chip (Maxim MAX2701) for final downconversion to baseband. The baseband analog signal is sampled at 25 or 50 Msamples/s (Analog Devices AD6644). The resulting sample streams are captured and stored in real time on memory boards and transferred offline to a PC for decoding. The memory boards allow for 140 ms of realtime transmission. The operation of the testbed is controlled by means of a graphical user interface (GUI). The software allows for configuring hardware parameters such as transmit power at individual antennas, and software-based parameters such as the modulation process that generates the data to be transmitted. The recent focus of the UCLA team is on MIMO-OFDM modulation, for which a software modem is written in C++. The radio hardware is reconfigurable to enable a wide variety of experimental scenarios. For instance, ideal synchronization can be accomplished using a common timing device. Onboard FPGAs at the receiver can be programmed to monitor the radio channel, identifying a start of transmission, providing gain control, and triggering data capture.

A MIMO testbed that has been developed at the Rice University in Houston, Texas is a 2×2 demonstrator operating at 2.4 GHz. This testbed is based on field programmable gate array (FPGA) development boards. Each FPGA board has two digital to analogue converters (DACs) and two analogue to digital converters (ADCs). Off-the-shelf RF up/downconverter boards from National Instruments are used. A novel feature of the Rice University testbed is its ability to incorporate commercial RF channel emulators. Each emulator can model fading channels such as Rayleigh, Ricean and Nakagami. The investigations described in (Murphy et al., 2003) mainly concern various challenges with hardware implementation of the MIMO concept. Initially, experiments have been limited to the implementation of the IEEE 802.11b standard and the 2×1 Alamouti coding scheme. In the most recent developments, the RF transceiver has been replaced with the MAX2829 IC for 2×2 MIMO OFDM demonstrations (Murphy et al., 2006).

The MIMO testbed developed at University of Alberta, Edmonton, Canada (Goud Jr et al., 2006) is a portable 4 × 4 MIMO testbed that is based on field programmable gate arrays (FPGAs). It operates in the 902 - 928 MHz ISM band and aims at real-time characterization of MIMO wireless channels (4×4 channel matrix) in a flat-fading environment. The system is fully mobile as the transmitter and receiver units are equipped with batteries and voltage regulation circuits. The transmitter is made up of a baseband board connected to an upconverter RF board that is wired to 4 Tx antennas. The baseband board generates 4 orthogonal m-sequence overlaid Walsh codes for each of the transmitting antennas using FPGAs. A spreading gain of 32 allows accurate estimation of the coefficients of the MIMO channel matrix. The codes are modulated using binary phase shift keying (BPSK) and are upconverted to the 900MHz ISM band by the RF board. The receiver has 4 antennas connected to a downconverter RF board which shifts the received signals to an intermediate frequency (IF). This IF signals are sampled by the baseband board and downconverted internally in the FPGA for further baseband processing. The FPGA recovers the chip timing using a suitable synchronization algorithm. Once synchronization is achieved, the correlations of all possible paths of the 4×4 MIMO channel are computed to obtain the coefficients of the channel matrix. 15625 matrices per second are determined. A computer is used for the real-time display of the MIMO channel matrices and statistics.

The MIMO testbed developed at the University of Bristol is the Turbo MIMO-OFDM system (Piechocki et al., 2001; Horseman et al., 2003). This system operates at 5 GHz and uses a DSP micro-processor development board for the baseband processing. The system uses a novel scheme for timing recovery and obtaining channel state information (channel matrix) at the receiver. These processes are accomplished through the use of training sequences (preambles) that start every frame of data. Each transmitter has a preamble that is orthogonal to all others and has an exclusive timing slot. At the receiver, the signal from each receiver antenna is processed by an autocorrelation routine. This routine determines the peak autocorrelation timing for each preamble and uses the information it obtains to calculate the channel state information.

The research team at ETH-Z (Haene et al., 2004) has developed MIMO-OFDM testbed. Its configuration is shown in Fig. 2.

*Figure 2. Configuration of a narrowband MIMO system*

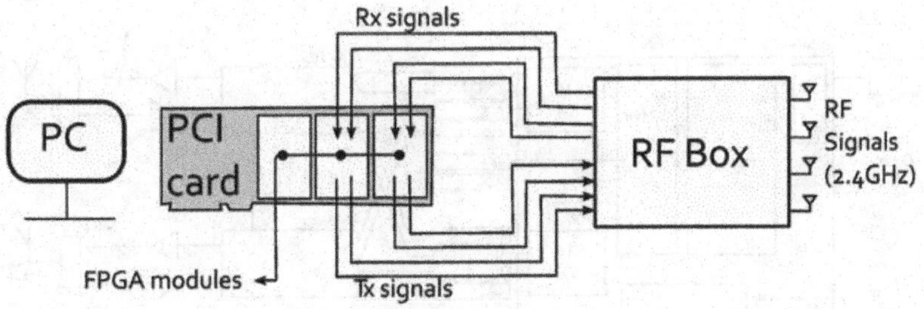


The setup supports a unidirectional point-to-point link under the control of a PHY-layer API which is implemented in MATLAB on a host-PC. This API allows to conduct PHY-layer measurements, but is not well suited as a basis for the implementation of a MAC layer, due to the absence of embedded processor resources appropriate for MAC layer processing. The employed PCI platform hosts three hardware modules: a large FPGA (Xilinx XC2V6000-6) module for baseband signal processing and two converter modules. On the converter modules an FPGA (Xilinx XC2V1000-4) produces an intermediate carrier frequency (IF) of 20 MHz. By producing this signal at twice the sampling frequency, the total number of converters is reduced by a factor of two compared with traditional baseband systems with separate I and Q components. Both the 14 bit DACs and the 12 bit ADCs run at a sampling rate of 80Msps.

The MIMO transceiver chain is comprised of four SISO super-heterodyne RF chains with an analogue IF of 475 MHz. The chains are built from discrete RF components and support 20MHz bandwidth channels with centre frequencies in the 2.4GHz ISM band. The received signal power at the ADCs is regulated by digitally controllable analogue attenuators, extending the dynamic range of the receive chains by 31 dB. In conjunction with the 12 bit resolution of the ADCs, the useful dynamic range amounts to approximately 70 dB. The testbed is controlled from a host-PC through multiple layers of software that allow communication with the hardware. The software stack includes a hardware driver for the PCI board, a transceiver-specific API that provides several functions (for the digital baseband configuration and the transmission or reception of OFDM frames), and a MATLAB interface (MEX) function to call these API functions. Demo applications, configuration tools, and measurement sequences are programmed as MATLAB scripts.

## UNIVERSITY OF QUEENSLAND 2×2 MIMO TESTBED

The MIMO testbed developed at the University of Queensland is a FPGA-based 2×2 MIMO demonstrator. With respect to processing of baseband signals, it is similar to testbeds produced by the Rice and Alberta universities as it employs an FPGA board. With respect to the design of RF front-ends, it shows similarities with the UCLA and Rice MIMO testbed, as it relies on the use of the commercially available transceiver chip MAX2829 from Maxim (2004). This integrated RF module allows the operation in 2.4GHz and 5GHz bands. The MAX2828/9 chip provides a common reference oscillator for use at different antenna pairs and thus facilitates common demodulation phase.

In order to verify its operation at the baseband level, it includes an option of using an in-house developed channel emulator. In difference to other testbeds, in addition to gathering channel state information (channel matrix properties) it implements specific encoding and decoding schemes. Also it employs both training-based and semi-blind channel (Bialkowski and Postula, 2007) estimation channel procedures. With respect to coding, it specifically uses the Alamouti coding scheme. Its extra advantage is it offers a real-time display of operation. This function is enabled through the use of a 100Mbit ethernet network connection, which interfaces the system to a PC. In order to enable real-world experiments in 2.4GHz and 5GHz bands it employs wideband planar monopole antennas of very compact size. The configuration of one of the UQ MIMO transceivers is shown in Fig. 3.

As seen in Fig. 3, the UQ MIMO testbed design is based on two main modules, the baseband signal processor and the RF front end. The FPGA signal processing module is designed around the Altera Stratix II S260 chip (Altera 2004). The RF front end module performing direct conversion between baseband and the 2.45GHz or 5GHz frequency band employs the MAX2829 IC (Maxim 2004) chip.

*Figure 3. Configuration of one of the transceivers of the 2×2 MIMO testbed developed*

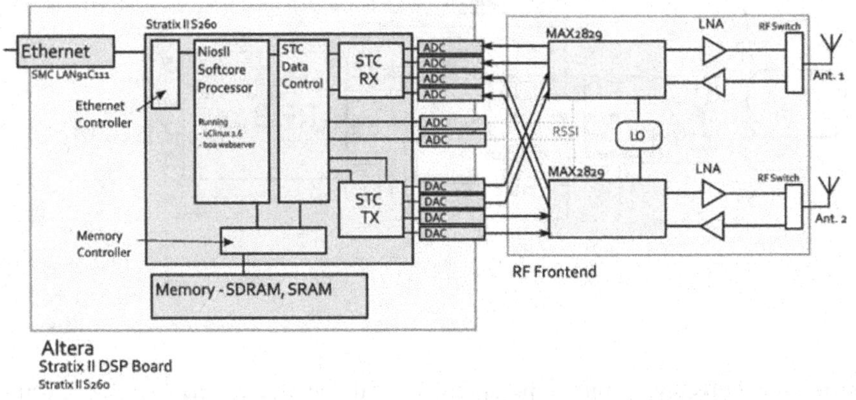


The middle part between FPGA and RF front end is a set of analogue to digital (ADC) and digital to analogue (DAC) converters. In the transmission mode they convert the digital signals processed in the FPGA to the analogue signals that are required by the RF transmitter. In the receive mode, analog signals are converted to digital for processing by the FPGA. Besides the MAX2829 IC chip, the RF frontend includes amplifiers, switches and antennas. These are present in the right hand side of the schematic in Fig. 3.

For post processing of results and visualization, special buffers and a soft-core processor running uClinux (2007) for network connectivity is included. In the current developmental stage, the testbed uses QPSK for signal modulation and the Alamouti scheme for space time coding. Also, the performance with respect to the differential Alamouti scheme is compared. Visualization of results on PC is accomplished using the embedded Ethernet controller, as shown in the left hand side of the schematic in Fig. 3.

Details of each major component of this system are described as follows.

## RF Front End

A photograph of the manufactured RF front end board is shown in Fig. 4.

The design of this board follows the MAX2829 design guidelines, as offered by Maxim (2004). The board includes additional IC chips that are added for conversion of the RSSI analogue signals to digital signals using a BB 75MHz ADC.

*Figure 4. Photograph of the RF front-end board*

Besides various chips, the photograph shows a number of coaxial ports for transmitted/received signals, an input for a reference oscillator, and digital connectors for monitoring and controlling the MAX2829 chip modes.

For direct conversion, the MAX2829 transceiver IC requires two analogue signals, an in-phase (I) and quadrature (Q) signal. As a result, four analog to digital converters (ADC) and four digital to analogue converters (DAC) are required to process baseband signals for a 2×2 MIMO system. In addition, another ADC is needed to read the RSSI signal from the MAX2829 IC for automatic gain control. A common reference oscillator is used for the two MAX2829 ICs. This arrangement is important for phase coherence of the received signals from different antennas. An internal phase locked loop (PLL) inside the MAX2829 produces the required frequency for conversion from IF to the required RF band (2.45GHz or 5GHz). The maximum bandwidth of the transmitted and received signals in the designed testbed is 100MHz.

During transmission, the RF board takes the I and Q signals and outputs them at 2.45GHz or 5GHz at the two antennas. For reception, the board takes the two RF signals (at selected frequency) and converts them to I and Q output signals to be processed in baseband.

## Antennas

For the testbed, two types of compact wideband antennas, which easily serve the two specified bands of 2.4GHz and 5GHz. The chosen varieties are shown in Fig. 5 and include: compact planar monopoles antennas, which feature omni-directional characteristics, and tapered slot antennas with directional properties. These antennas are capable to operate from 2.4 to more than 10GHz.

The structure shown in Fig. 5(a) is identified as a planar electric monopole. It is created by a planar conducting surface formed by the intersection of two ellipses. In this structure, the surface electric current flowing on an elliptical shaped conductor is regarded as the primary source of radiation. In turn, the structure shown in Fig. 5(b), which is complementary to that in Fig. 5(a), is named here as planar magnetic monopole. In this structure, the electric field (or its surface magnetic current equivalent) in an elliptical shaped slot is regarded as the primary source of radiated field.

These planar monopoles are designed and developed using Rogers RT6010LM substrate featuring a dielectric constant of 10.2 and a loss tangent of 0.0023, 0.64mm thickness plus 17μm thick conductive coating. The two antennas have dimensions 20mm×30mm and thus represent a very compact size. The measured gain of the two antennas is around 0.5dB at 2.4 GHz and 1.7dB at 5 GHz indicating a near omni-directional performance. Their radiation efficiency, is more than 90% for the two frequency bands, as proved by full-wave electromagnetic simulations.

The other the type of antenna, shown in Fig. 5(c), was designed for the testbed. It is needed to test performance of the MIMO systems when using directive antennas in a multipath environment. Similarly as for the planar monopoles the same substrate materials were used during its design. The designed antenna is of a compact size of 50mm×50mm. The measured gain of the antenna is 3dB and 7dB at 2.4 GHz and 5 GHz respectively, indicating directive properties. Radiation efficiency is more than 90% at the two frequency bands.

## FPGA Baseband

A photograph of the baseband hardware of the UQ MIMO testbed is shown in Fig. 6.

*Figure 5. Configurations of planar monopole antennas (a), (b) and planar tapered slot antennas (c)*

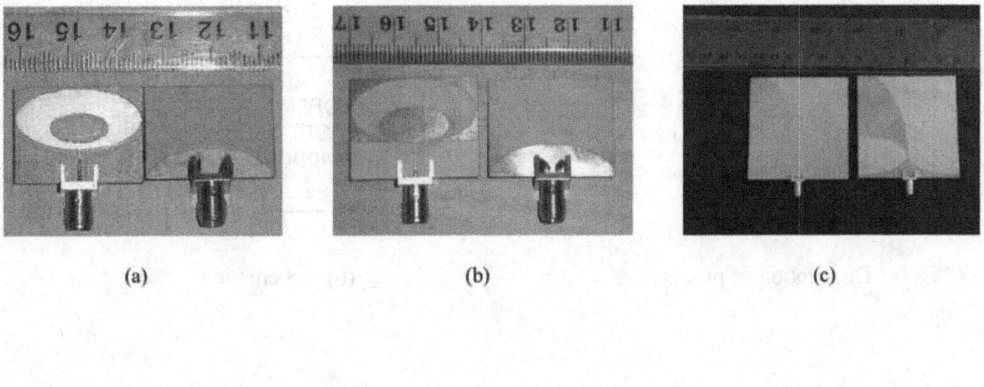

(a) (b) (c)

*Figure 6. Photograph of the FPGA baseband processor board*

Two boards with a total of two Stratix II 2S60 FPGAs, four high-speed Analogue to Digital converters (ADC), capable of 125MSmp/sec (12bits) and four Digital to Analogue converters (DAC), capable of 165MSmp/sec (14bits) are used. The DACs and ADCs run at 100MSmp/sec, as the chosen clock speed of 100MHz is assumed in the entire MIMO testbed design. Note that the purpose of the second FPGA is primarily for additional high speed ADC and DAC.

For high speed communication between boards, two wires using low voltage differential signals (LVDS) are used. This connection method allows for a signal rate of 1Gbit/sec over a pair of wires without experiencing adverse affects of mutual coupling over a parallel transmission bus. A 2×2 Alamouti single carrier QPSK modulated signal is used in the testbed. Only 5% of the available FPGA logic gates are utilized to perform all of the required tasks.

The baseband processing is accomplished using three main modules which include transmitter, receiver and packet controller. Their operation details are given as follows.

## Baseband Transmitter

When transmitting each packet, the data sequence is split between the two antennas, and then encoded and modulated. This is done by taking 4 bits (2 for 2× MIMO, and 2 for QPSK) and mapping them using the Alamouti space time code to two transmit symbols on 2 antennas, and modulating these symbols using QPSK modulation. This is done in a process as shown in Fig. 7(a), where a bit stream is first separated into symbols, which are then mapped into two output streams.

For the hardware implementation, the space time encoding is performed on the bit-stream directly, by taking a group of four bits $b_0, b_1, b_2$ and $b_3$ are encoded to 2×2 Alamouti as $Tx_1 = b_0, b_1, \bar{b}_2, \bar{b}_3$ and $Tx_2 = b_2, b_3, \bar{b}_0, \bar{b}_1$ the ($\bar{\bullet}$) operator implies a *not* operation on a single bit. These two bit streams are then modulated separately. This process is described schematically in Fig. 7(b). This schematic the z-1 a delay of a 1 bit period.

*Figure 7. Alamouti and QPSK encoding and modulation in the TX Module*

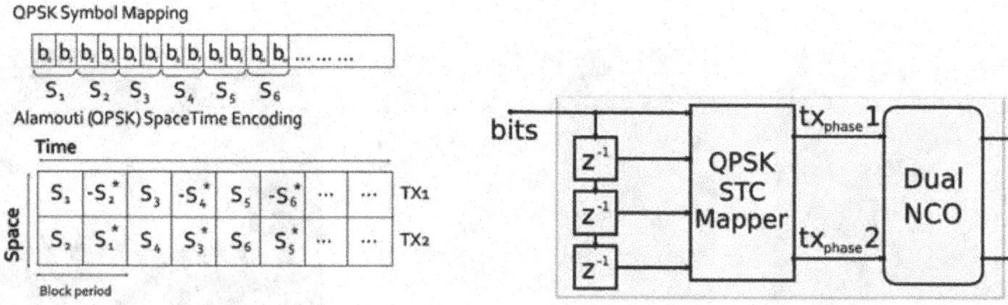


(a) Description of process (b) Schematic

*Figure 8. Schematic illustrating parallel channel estimation and data decoding*

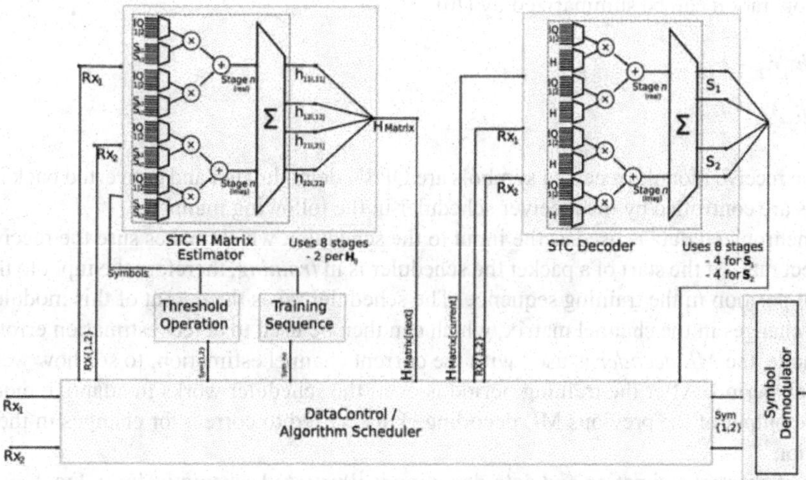


Differential Alamouti encoding is also considered for implementation in the MIMO testbed. Channel estimation not being necessary at the receiver for decoding the signal, making it a good candidate. For comparison purposes, the BPSK based differential Alamouti scheme is selected, due to it mapping pairs of blocks using a QPSK like constellation. The transmitter in the differential scheme encodes data using the mapping function in (7) and the function in (8).

$$M\ (00), (10), (01), (11) = (1\ 0), (0\ 1), (0\ -1), (-1\ 0) \qquad (7)$$

$$(s_3, s_4) = A_x(s_1, s_2) + B_x(-s_2^*, s_1^*) \qquad (8)$$

## Baseband Receiver

There are two modes of operation of the baseband receiver. These concern the channel matrix estimation and decoding of the data. The process is complicated by the requirement of synchronization. For channel matrix estimation both a training-based and a semi-blind channel estimation methods are used. The decoding of the MIMO signal is based on the maximum likelihood (ML) method. When implemented in FPGA the channel estimator and decoder operate in parallel, which is not possible in a DSP or software based implementation. In order to perform this, five sub-modules are used. These are ML Channel Estimator, ML Decoder, QPSK Demodulator, Training Sequence/Symbol Thresholder and Receiver Scheduler. Their operation is addressed as follows.

The *ML Channel Estimator* takes two symbols (either known when in training sequence mode, or thresholded versions of the received symbols) and the received signal data and produces a new channel estimation. In order to reduce the effects of noise, this new channel estimation is combined with a previous channel matrix estimation average using a weighting scheme.

The current symbol in the training sequence is provided by the training sequence module, and the thresholding operation is taken from the current ML decoding operation. The calculation of the new channel matrix estimation is shown in (9) below, and is performed over 8 clock cycles using a total of 32 multiplication operations. These 32 multiplication operations are formed from 8 complex multiplications.

$$\begin{aligned} h_{11} &= r_{11}s_1^* - r_{12}s_2 \\ h_{12} &= r_{11}s_2^* + r_{12}s_1 \\ h_{21} &= r_{21}s_1^* - r_{22}s_2 \\ h_{22} &= r_{21}s_2^* + r_{22}s_1 \end{aligned} \qquad (9)$$

The *ML decoder* takes the current estimated channel matrix and the current received data and produces the received symbol constellation. It does this over 8 clock cycles, and also performs a total of 32 multiplication operations (8 complex multiplications). This operation can be summarized by (10).

$$\begin{aligned} s_1 &= h_{11}^* r_{11} + h_{12} r_{21}^* + h_{21}^* r_{12}^* + h_{22} r_{22}^* \\ s_2 &= h_{12}^* r_{11} - h_{11} r_{21}^* + h_{22}^* r_{12} - h_{21} r_{22}^* \end{aligned} \quad (10)$$

On the output of the receive module, received symbols are QPSK demodulated and converted back into a bit stream. All of these operations are controlled by the receiver scheduler in the following manner.

Firstly, symbol synchronized data is used as the input to the scheduler, which makes sure the received symbol data is inputted at the correct rate. At the start of a packet the scheduler is in *training*, therefore the input to the *ML estimator* is based on the current position in the training sequence. The scheduler takes the output of this module and stores it in order to keep track of changes in the channel matrix, which can then be used to detect estimation errors.

During training mode, the *ML decoder* is used with the current channel estimation, to see how well the estimation works on the next block period. After the training period is over, the scheduler works in adaptive mode, therefore the estimator's input is the output of the previous ML decoding. This is used to correct for changes in the channel matrix during data transmission.

The entire process of channel estimation and data decoding is illustrated schematically in Fig. 8.

For the differential Alamouti scheme the decoding of the signal uses a different algorithm. This operation is shown in (11). A benefit of this scheme is that it can easily be extended to an arbitrary number of antennas.

$$\begin{aligned} s_1 &= \sum_{i=1}^{n_r} r_{(i,2t+1)} r_{(i,2t-1)}^* + r_{(i,2t+2)}^* r_{(i,2t)} \\ s_2 &= \sum_{i=1}^{n_r} r_{(i,2t+1)} r_{(i,2t)}^* + r_{(i,2t+2)}^* r_{(i,2t-1)} \end{aligned} \quad (11)$$

This operation, like the ML decoding and estimation method requires 8 complex multiplications for a 2×2 MIMO system, however it does not require any channel estimation. Compared to the semi-blind algorithm implemented above, this scheme will use half the amount of complex multipliers per block.

## Controlling Hardware/Visualization

The scheduling hardware (*Scheduler*) controls data packets for use by the transmitter, and oversees the reassembling of the received packets at the receiver module. At the transmitter, it is responsible for taking the input bit stream and prefixing the training sequence to every packet. At the receiver, the training sequence is recognized by the receiver module itself.

The scheduler module is accompanied by specialized hardware buffers and analysis modules. The specialized hardware buffers are created to store data for visualization. The signals resulting from the encoding and decoding processes are stored in these buffers, synchronized to each other. This approach allows for the visualizing device to have a synchronized display of a number of these signals, even when adequate bandwidth for real time display is not available. The synchronization is accomplished by buffers of a fixed size which can be reset (or cleared) synchronously. Once they are full, no further data is stored, until the next reset.

The post-processed data is ready for visualization on a PC. The connection to PC is via Ethernet and NIOS II soft-core processor. This processor uses the uClinux (2007) operating system. uClinux is selected due to its robust and high performance networking capability. The embedded HTTP server selected is the Boa webserver (Doolittle et al., 2005). Boa is a light weight HTTP server which allows for special programs (CGI) to be executed on certain requests. The CGI programs are connected via NIOS II general purpose IO (GPIO) to the controlling hardware and hardware buffers. The connection to the PC is made via a 100Mbit ethernet network connection. Using a web browser on the PC, a graphical interfacing using scalable vector graphics (SVG) is employed for interaction with the embedded web server to display and control what the hardware is sending and receiving.

Due to latency, only a quasi-real time display on PC is possible. This is because for 10ms of time domain data it takes 60ms between creation of data and its display. However this delay is only in the visualization, the FPGA hardware is capable of full real time continuous sending and receiving of data.

## Algorithm Description

Initially the operation of the 2×2 MIMO system is investigated at the baseband level. To this purpose, all of the above-described procedures for data encoding, modulation, demodulation and decoding are written in MATLAB. Once the developed algorithm passes successful testing, it is ported to C++ code using the ITPP framework (ITPP, 2007). The ITPP framework is a library for C++ which performs some of the functions existing in MATLAB. During conversion, some modules remain in vector form, and others accurately depict what would happen in hardware or a bit or number level.

Implementation in hardware incurs a number of challenges. The first concerns the complex number arithmetic to be converted to real number arithmetic. In the implemented approach, one complex multiplication is treated as 4 real number multiplications and 2 additions, as it is easier to schedule over multiple clock cycles than three multiplications and seven additions. Secondly, in hardware, a fixed point number system has to be implemented, instead of a floating point number system, which is typically used in simulation programs such as MATLAB. C++ offers a much easier use of fixed point data types than MATLAB.

There are considerable differences in how MATLAB algorithms function and FPGA hardware is used in a final implementation. In MATLAB, each operation (modulation, encoding, decoding, demodulation, channel estimation) occurs on the whole data set. In hardware each operation is performed on the data stream (as it arrives). This can be considered similar to the MATLAB Simulink model of simulation. C++ does not have as big a performance decrease when stream based processing is used, making it a good candidate for an intermediate implementation.

When converting from simulation to a hardware implementation, some additional requirements have to be taken into account. These concern synchronization, accurate channel estimation and packet detection. Without accurate channel estimation techniques, decoding a signal is impossible. Unlike a simulation platform, the receiver has no a priori information about the channel. In our case, this channel information is provided to the receiver by using training symbols prefixed to each packet of data. In general, a training sequence is a sequence of known data for the transmitter and receiver over time or frequency. In our case the sequence is over time. Ideally a training sequence should be inserted into the stream at a similar rate to the channel changes (or the channel coherence period, in block fading). The coherence period must be longer than the amount of time required to estimate the channel. If extra pilot symbols are not added then semi blind or blind estimation techniques are required. The algorithm used in our testbed is based on the Maximum Likelihood (ML) decoding and estimation technique.

The channel estimation procedure is implemented in two different ways. Initially a training sequence based approach is used. Next, a semi-blind approach, which continues to re-estimate the channel during data transmission.

A training sequence is prefixed to each block of data. To simplify the synchronization process, the first 4 symbols (2 blocks) of the training sequence are sent on only one antenna. This is like in a SISO system. After this, an additional 12 symbols/6 blocks (or 28 symbols/14 blocks) are sent using both antennas.

The semi-blind approach starts like the training sequence approach. After that the receiver enters an adaptive mode. In this mode, the received symbols are decoded using the previous valid channel matrix, and the result is quantized to the nearest correct symbol and used to estimate the new channel matrix. In order to reduce the effects of noise, the channel matrix used in next iterations, is a weighted average of the old and new channel matrix estimations. In our experiments, the ratio varies along the training sequence, and is between 1:8 and 1:32. During data transmission the choice of ratio depends on how big a change the previous and current estimations are from their previous values. The estimation which is less of a change is given a larger weight.

## Channel Emulator

Prior to launching real-world, real-time experiments, tests involving a MIMO channel emulator are performed. The reason for using an emulator is that real-time scenarios are inherently uncontrollable and unpredictable and therefore particularly unsuitable for the initial analysis of new algorithms. From a functional verification point of view, debugging calls for the ability to reproduce a problem and its history as accurately as possible to analyze the problem's origin. Emulators offer artificial but repeatable conditions and thus are useful to obtain an intuitive understanding for a prototype's behavior. In addition, emulators can avoid other undesirable adverse effects which are caused by noise and nonlinear characteristics of RF modules and frequency drifts causing phase synchronization problems. As overcoming these problems is only possible if all analogue components are eliminated, channel emulation is usually developed at the baseband level.

The basis for designing our channel emulator is the MIMO channel model, which generates a channel matrix. We use two approaches to provide the channel matrix data. One is based on the signal bounce scattering model (Bialkowski et al., 2006a; 2006b). The other one employs a set of channel matrices that are obtained from actual measurements performed with a vector network analyser (VNA) in an indoor environment. Initially, this channel emulator is produced in MATLAB. In the next step, this software emulator is extended to an "in-circuit" channel emulator, that is placed between the transmitter and receiver modules. It is implemented in time domain data and is unaware of the used modulation or coding schemes.

Next developments concern realizing data transmission, as given earlier by expression (1). In the present case, **x**(1) is assumed to be the instantaneous time domain signal transmitted. Due to the use of direct up conversion, the time domain signal has I and Q components, which together are treated as the complex time domain signal. This is then expanded as in (12):

$$\begin{bmatrix} y_1(t) & y_1(t+1) \\ y_2(t) & y_2(t+1) \end{bmatrix} = \begin{bmatrix} h_{11} & h_{12} \\ h_{21} & h_{22} \end{bmatrix} \begin{bmatrix} x_1(t) & x_1(t+1) \\ x_2(t) & x_2(t+1) \end{bmatrix} + \begin{bmatrix} n_1(t) & n_1(t+1) \\ n_2(t) & n_2(t+1) \end{bmatrix} \tag{12}$$

where **y** is the received signal vector, **x** is the transmitted signal vector, and **n** is the noise vector. Both **y** and **x** have two symbol periods for the same channel matrix **H** due to the application of Alamouti scheme. The hardware implementation of the channel emulator is illustrated in schematic form in Fig. 9.

*Figure 9. Schematic depiction of hardware implemented channel emulator*

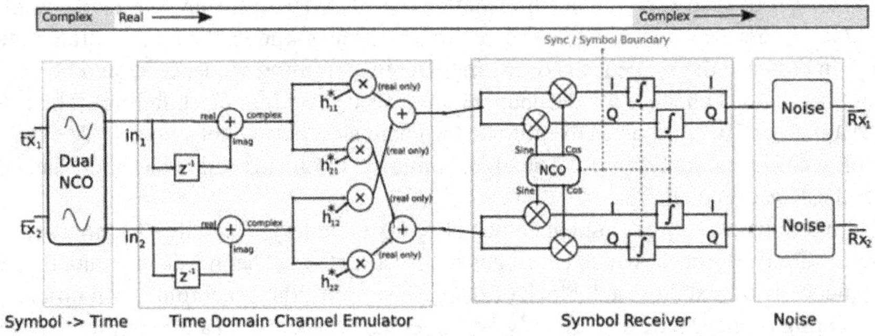


*Figure 10. Measurement setup for performing channel matrix measurements for the 2×2 MIMO in indoor environment*

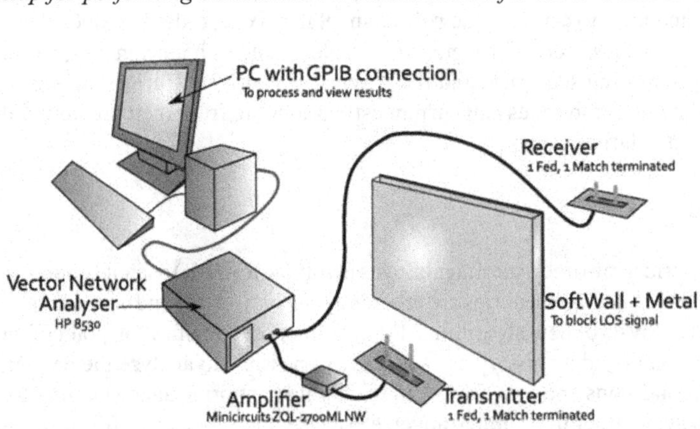

If only real signal data (no imaginary component is available), then the imaginary component Q can be synthesized by applying a 90ophase shift. In hardware this requires storing the previous values in a buffer and using the $d^{th}$ sample, where *d* corresponds to a $90^{0}$ phase shift of the carrier frequency. As the channel emulator operates on a stream, instead of using current and next, previous and current signals are used for channel emulation, as seen in (13).

$$y_1(t) = x_1(t)real(h_{11}) + x_1(t-d)imag(h_{11}) + x_2(t)real(h_{12}) + x_2(t-d)imag(h_{12})$$
$$y_2(t) = x_1(t)real(h_{21}) + x_1(t-d)imag(h_{21}) + x_2(t)real(h_{22}) + x_2(t-d)imag(h_{22}) \quad (13)$$

where *d* is the 90 degree phase delay of one symbol period.

In the *single bounce scattering model* (Uthansakul et al., 2006), the channel matrix **H** is obtained by considering a rectangular scattering environment of dimensions 200λ ×200λ (where ƒÉ denotes wavelength) with transmitter and receiver equipped in MEA located on opposite sides of the rectangle. 600 scatterers uniformly distributed within the rectangular region are assumed. For a frequency of 2.4GHz this is equivalent to a size of 24.4m × 24.4m. Each channel matrix realization is stored into a buffer of the FPGA, in order to perform emulation of the wireless channel. 1000 different channel matrices are saved in terms of real and imaginary parts.

An alternative is to use channel matrix information from actual measurements. The experimental setup used to measure the channels is shown in Fig. 10. In this setup, a vector network analyser (VNA) is used to record transmission coefficients between transmitting and receiving antennas.

The Tx/Rx antennas are two-element arrays of omni-directional monopoles that are separated by half a wavelength. A full set of measurements for every element of the channel matrix is performed at multiple locations several times each. ($Tx_1 \rightarrow Rx_1$, $Rx_1 \rightarrow Rx_2$, $Tx_2 \rightarrow Rx_1$, $Tx_2 \rightarrow Rx_2$). During these measurements, the VNA operates in S21 (transmission coefficient) measurement mode. In order to compensate for power losses, a Low Noise Amplifier (LNA) is used to increase the transmitted signal strength such that the received signal strength is adequate.

Using the setup shown in Fig. 10, four different receiving positions, each with a set of channel matrices for use in the channel emulator are collected. A total of 1024 matrices are generated.

The last issue concerns the generation of white noise **n**, as required in (1). In the undertaken approach, the noise data is pre-generated and stored in a table. The size of table used for the random numbers is 4096. This is due to the complexity in producing a controlled Gaussian based variable in logic gates.

## VERIFICATION

Verification of the 2×2 MIMO testbed operation is done separately with respect to baseband and RF modules including antennas. Most of the effort goes into the FPGA baseband module testing where the majority of signal processing tasks take place. When the channel emulation module is inserted between TX and RX baseband modules the situation is represented by the block diagram shown in Fig. 11.

### Baseband Module Verification

In the first step, the operation of the receiver is investigated when the transmitter sends a single QPSK signal over a single channel. A QPSK symbol modulated at 6.25MHz is implemented with two cycles per symbol, forming a 3.125Mbit data rate. Two cycles per symbol are used to improve the demodulation process. At this stage, the focus is on testing the synchronization procedure. This is of paramount importance in real implementations of hardware modules. A coaxial wire from the output of the DAC to the input of the DAC is introduced causing a delay between TX and RX. Beside the signal delay, the wire introduces about 2dB signal strength loss. This loss can easily be compensated with an automatic gain control module. However, this is not necessary, as this signal strength loss is quite small in comparison with the 14-bit capability of ADC at the receiver.

In the next step, the testing the receiver's operation is extended to the 2×2 MIMO mode, when the Alamouti coded QPSK modulated signal is transmitted. To test the initial operation, the channel estimation procedure is not included. Instead, in order for the receiver to decode the signal, the currently emulated channel matrix is made available to the receiver.

*Figure 11. Block Diagram of MIMO testbed when a channel emulator is placed between TX and RX*

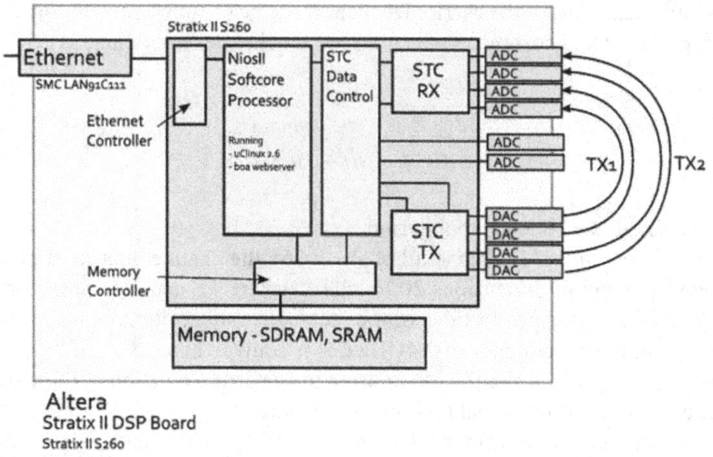


*Figure 12. Illustration of input and output signals over 2 antennas during testing 2×2 Alamouti scheme when perfect channel estimation is assumed*

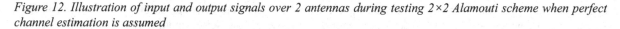

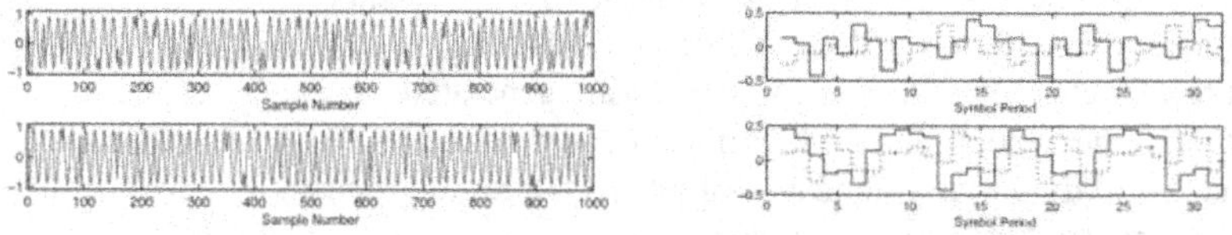

(a) Time Domain Received Signal

(b) Decoded I(solid) and Q(dashed) signals

Fig. 12 shows an example of typical input and output signals in this testing procedure. The upper and lower signals in this figure represent the received signal at antenna 1 and 2 respectively. Fig. 12(a) provides an example of a typical Alamouti encoded signal. In turn, Fig. 12(b) reveals the I and Q representation of the same signal.

When the channel estimation module is added to the receiver, a training sequence is added to the transmitted data. The training sequence initially is transmitted only by one transmitter, in order to reduce the complexity of channel estimation to that of a SISO system. That is, only the magnitude of the signal is reduced, and a phase shift occurs. As a result of the chosen synchronization on the $Tx_1$ to $Rx_1$, all terms of the estimated channel matrix **H** are made relative to $h_{11}$.

The next step concerns the situation in which the reception of data transmitted by two antennas is affected by the channel properties, as generated by the channel emulator. In this case, either the single bounce scattering model or the measured set of channel matrices acquired from indoor measurements are used. In these tests, quasi static fading is assumed. This means that only one channel matrix realization is employed for an entire packet transfer.

Fig. 13 illustrates the error of channel matrix estimation over the life time of a packet. The displayed graph shows only the first 2500 symbols for clarity of presentation. The data packet of 32kbits in size is sent through a channel with SNR of 8dB and a coherence period of 512 symbols (or 1024 bits). One of the current weaknesses of the algorithm is displayed in symbols 1024 to 1536, where the channel matrix estimation is wrong.

In the next tests the channel emulator is applied to verify performance for the case of dynamic channels. Two options of using fixed and floating point operations are investigated. For each set of simulation results, the following is evaluated:

*Figure 13. Illustration of the channel matrix estimation error in a block fading environment*

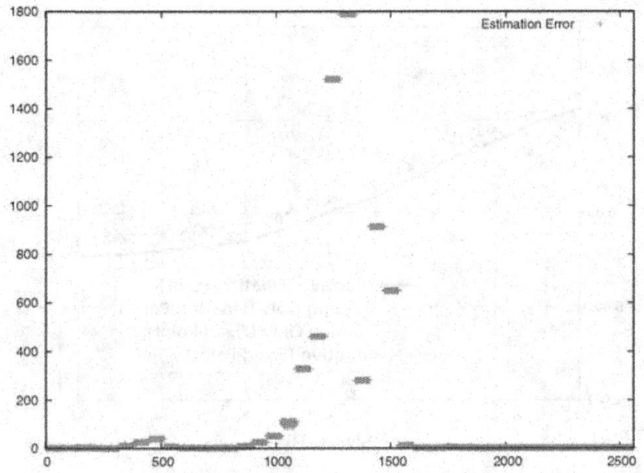


- Performance of the training-based MIMO channel estimation algorithm
- Performance of the semi-blind estimation algorithm
- The effect of quantization (due to fixed point operation in FPGA).
- Performance of semi-blind estimation algorithm with differential algorithm.

Identical received signals are used and decoded using the two separate receiver blocks. For the first two simulations, 64 bits of training sequence, 32 kbits of data and SNR values of 0dB to 12dB are assumed. To calculate Bit Error Rate (BER), 20k data packets are sent per simulation making a total of 640 kilobits of data.

Fig. 14 shows a comparison between the performance of the training sequence only (dashed line) and the adaptive algorithm (continuous line). Dashed line represents training sequence only, and continuous line represents the adaptive channel estimation technique. Also shown are quantization effects. The results are generated assuming the channel remains static during each packet.

The comparison between the two algorithms indicates a better performance of the adaptive algorithm, as represented by the lower values of BER.

With respect to the effects of quantization, the additional line being solid with 'X' represents quantized adaptive algorithm while the dashed with triangles line represents the quantized non adaptive algorithm. A quantization of 12 bits (real + imaginary) for both the received signals, as well as quantization of 24 bits for the channel matrix was used to investigate the quantization effect. The observed quantization errors are very small. The algorithms using quantized (12bit) data are only slightly worse than those using floating point arithmetic. In another experiment, the performance of the training based and semi-blind channel estimation algorithms under a block fading scenario is tested. In this case, a block fading period of 256 symbols is assumed. The results are shown in Fig. 15.

Similarly as for the data shown in Fig. 14, the dashed line represents the training sequence only algorithm (and dashed with triangles is the fixed point version) while the continuous line represents the adaptive channel estimation technique (and with crosses is the fixed point version).

As observed in Fig. 15, the BER of the non- adaptive algorithm is quite high and varies little with SNR. In turn, the BER for the adaptive algorithm is lower and improves with SNR. The results for the non-adaptive algorithm indicate that if channel is changing, then either the packet size must not exceed the block fading period, or pilot symbols must be inserted into data stream otherwise BER approaches 0.5. This problem is overcome by applying the proposed simple adaptive algorithm, which improves BER significantly.

A comparison between the BER performance of the differential method and the Alamouti based semi-blind scheme was also performed. This was done over the same two types of environments as shown previously in this section. The first channel is a quasi-static channel and the second is a block fading channel. Two major benefits of the differential scheme is, its complexity is less than half of training and semi-blind based receivers, and that errors propagation does not cascade. In the other approaches, if the channel is incorrectly estimated, then it is probable that until the next training

*Figure 14. BER performance of receiver in a static fading environment*

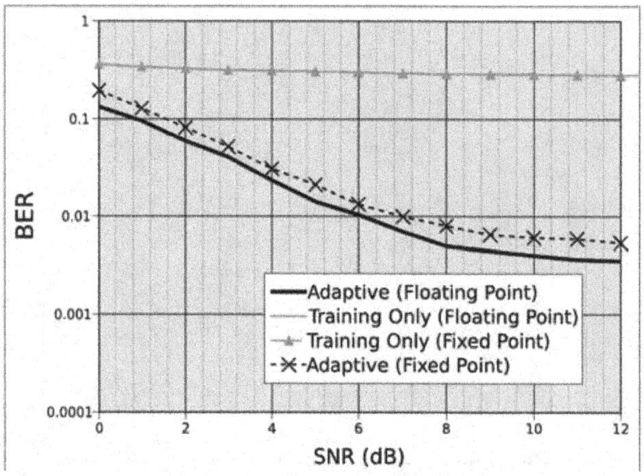


*Figure 15. BER performance of receiver in a static fading environment*

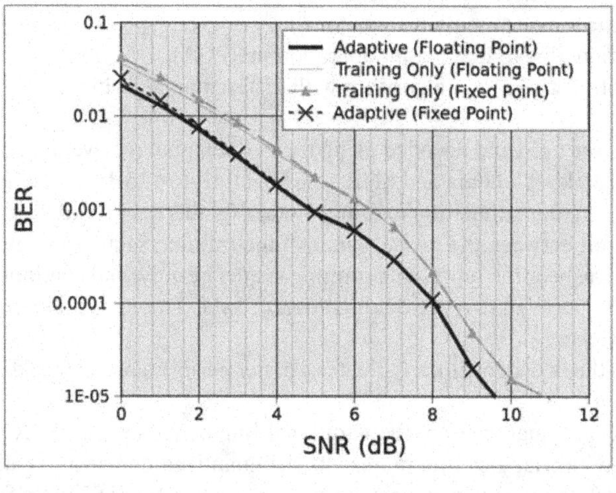


sequence the decoded data will be incorrect. In Fig. 16 it can be seen that the differential receiver has an error floor caused by block fading. The poor result in a block fading environment is well known in the literature, and due to this several extensions to the differential receiver are proposed (Tarasak et al., 2004; Liu et al., 2001; Schober and Lampe, 2002).

## Results Visualization

Visualization of various stages of signal transmission process is accomplished using interface between MIMO hardware and PC. The display concerns the time domain transmitted signal, the received signal constellation, the decoded signal constellation and the estimated channel matrices. Each new data can be displayed every 60-100ms allowing for almost real-time visualization. Fig. 17 illustrates the results when a packet of 32 thousand symbols was transmitted in a channel with 6dB signal to noise ratio (SNR). Fig. 17(a) shows the raw received symbols before MIMO decoding of the signal is performed. In turn, Fig. 17(b) presents results after decoding is accomplished. It is apparent that the signal is a QPSK signal. The presented results show that the decoder works correctly.

The visualization platform is also the method used to upload channel matrix data, and change the transmitted data. The displayed constellations shown in Fig. 17 and Fig. 18 show static channels for LOS and NLOS cases. In Fig. 18(a) the constellation is clipped due to the finite capture range of the ADC and fixed point number system.

*Figure 16. BER performance of differential receiver (compared to training only and adaptive algorithms) in a two fading environments environment*

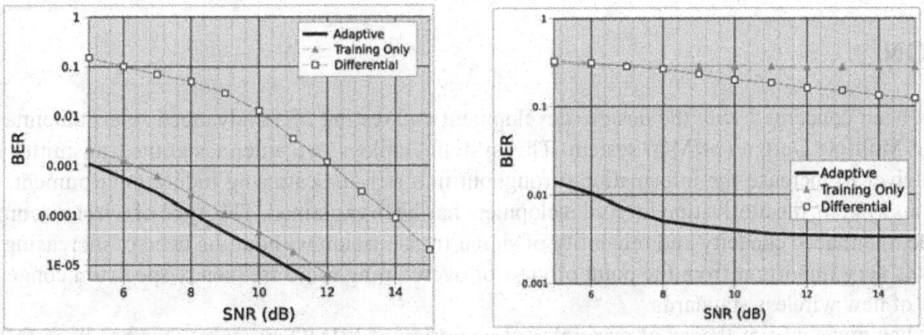


(a) Quasi Static Channel (b) Block Fading Channel

*Figure 17. Visualized data over the life of a packet - for a static LOS channel*

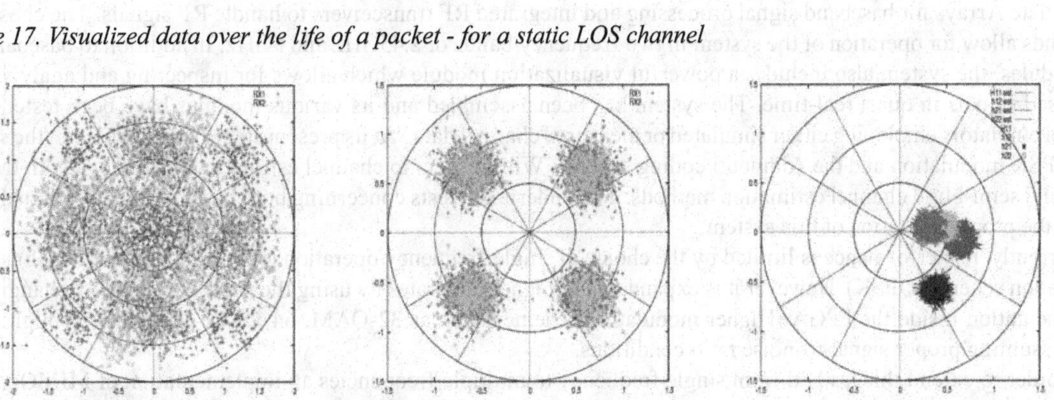

(a) Received signal (b) Decoded signal (c) Estimated Channel

*Figure 18. Visualized data over the life of a packet - for a static NLOS channel*

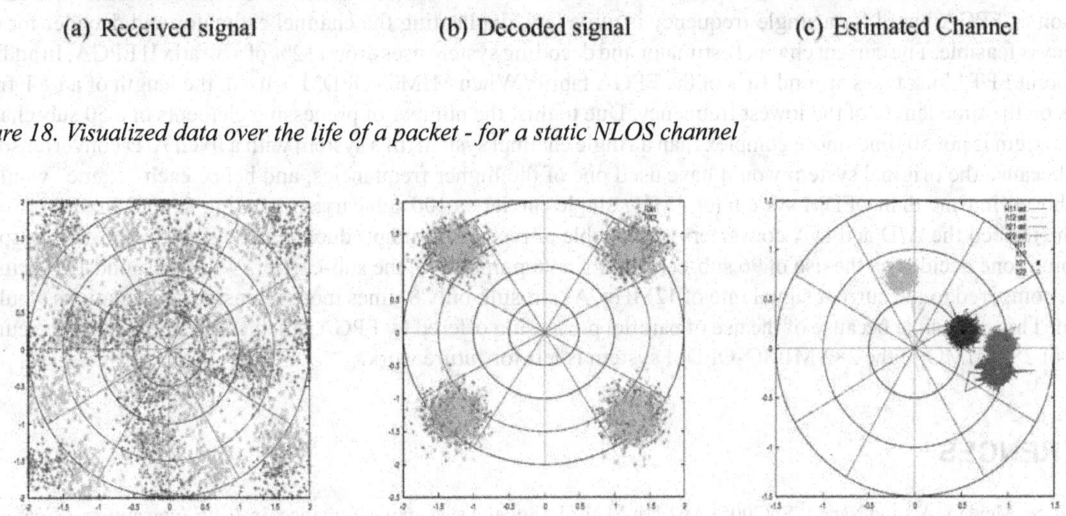

(a) Received signal (b) Decoded signal (c) Estimated channel

This result completes the verification process of the FPGA-based 2×2 MIMO testbed that has been developed at the University of Queensland.

## CONCLUSION

This chapter has been concerned with the design, development and testing of an advanced smart antenna in the form of Multiple Input Multiple Output (MIMO) system. This system employs two antennas at the transmitter and two antennas at the receiver to increase the information throughput in a rich-in-scattering indoor environment. Prior to the introduction of this system, the motivation for its development has been explained. This type of wireless communication system is capable to enhance capacity and reliability of signal transmission without the need of increasing operational bandwidth. This is very important from the point of view of overcoming radio frequency spectrum congestion caused by introductions of new wireless standards.

This chapter has given a brief theory of principles of operation of MIMO systems. This has been followed by an overview of several MIMO testbeds which were developed in different parts of the world. A particular emphasis has been given to various baseband signal processing and RF technologies used in these systems, which have evolved in recent years. The University of Queensland MIMO tesbed uses the most recent technologies which include Field Programmable Gate Arrays for baseband signal processing and integrated RF transceivers to handle RF signals. The chosen RF front ends allow for operation of the system in two frequency bands of 2.45GHz and 5GHz. In addition to baseband and RF modules, the system also includes a powerful visualization module which allows for inspecting and analyzing the processed signals in quasi real-time. The system has been assembled and its various modules have been tested using channel emulators employing either simulated or measured channel data. At its present stage of development, the system uses QPSK modulation and the Alamouti coding scheme. With respect to channel estimation, it employs both training based and semi-blind channel estimation methods. The undertaken tests concerning both baseband and RF signals have shown the proper operation of this system.

Currently, its performance is limited by the choice of single frequency operation (either 2.45GHz or 5GHz) and the modulation scheme (QPSK). However it is expandable to higher data rates by using the same hardware by changing the implementation inside the FPGA. Higher modulation schemes such as 32-QAM, 64-QAM can allow for higher data rates, assuming proper signal-to-noise ratio conditions.

In order to extend this testbed from single frequency to multiple frequencies an implementation of MIMO-OFDM technique is required. This can be accomplished via the use of an FFT block in the FPGA module. Since the current utilization of FPGA board for a single frequency is quite low, duplicating the channel estimator and decoder for each frequency is feasible. The current channel estimator and decoding system uses around 2% of a Stratix II FPGA. In addition a 1024 point FFT block takes around 10% of the FPGA fabric. When MIMO-OFDM is used, the length of a FFT frame depends on the time length of the lowest frequency. Due to this, the number of processing elements of a 50 sub-channel OFDM system is not 50 times more complex than a single channel system, in a system with a fixed A/D converter speed. This is because the original system would have used one of the higher frequencies, and hence each "frame" would be might shorter in time than OFDM system (eg 1MHz single carrier vs 100 sub carriers totaling to 100Mhz.)

In the testbed the A/D and D/A converters are capable of receiving and producing signals up to 100MHz in speed. Therefore if one decides on the use of 96 sub-carriers (of a 96 point FFT), the sub-carriers will be around 1MHz in signal rate, compared to the current signal rate of 12MHz. As a result, only 8 times more intensive computations would be required. This is feasible because of the use of parallel processing offered by FPGA. Further development of the current version of 2×2MIMO to the 2×2 MIMO-OFDM system is left for future work.

## REFERENCES

Abraham, S., Meylan, A., and Nanda, S. (2005). 802.11n MAC design and system performance. In *IEEE International Conference on Communications (ICC 2005)*, pages 2957-2961.

Adjoudani, A., Beck, E., Burg, A., Djuknic, G., Gvoth, T., Haessig, D., Manji, S., Milbrodt, M., Rupp, M., Samardzija, D., Siegel, A., Sizer, I., Tran, C., Walker, S., Wilkus, S., and Wolniansky, P. (2003). Prototype experience for MIMO BLAST over third-generation wireless system. *IEEE Selected Areas in Communications Journal*, 21(3):440-451.

Alamouti, S. (1998). A simple transmit diversity technique for wireless communications. *IEEE Selected Areas in Communications*

*Journal*, 16(8):1451-1458.

Altera (2004). Stratix II DSP Performance. White Paper version 1, Altera Corporation, San Jose, CA.

Batariere, M., Kepler, J., Krauss, T., Mukthavaram, S., Porter, J., and Vook, F. (2001). An experimental OFDM system for broadband mobile communications. *IEEE Vehicular Technology Conference*, 4:1947-1951.

Bialkowski, K. S. and Postula, A. (2007). Investigations into a semi-blind channel estimation technique for 2x2 MIMO system. In *Antennas and Propagation Symposium (APS)*, Hawai'i, USA.

Bialkowski, M., Uthansakul, P., and Bialkowski, K. (2006a). Investigations into the effect of LOS signal blocking on capacity of an indoor MIMO system. In *Proceeding of 16th International Conference on Microwaves, Radar and Wireless Communications*, pages 1-4. IEEE Press.

Bialkowski, M., Uthansakul, P., Bialkowski, K., and Durrani, S. (2006b). Investigating the performance of MIMO systems from an electromagnetic perspective. *Microwave and Optical Technology Letters*, 48(7):1233-1238.

Biguesh, M. and Gershman, A. (2006). Training-based MIMO channel estimation: a study of estimator tradeoffs and optimal training signals. *IEEE Transactions on Signal Processing*, 54(3):884-893.

Channel Sounder (2003). World of Channel Sounder. http://www.channelsounder.de/

Dahlman, E., Ekstrom, H., Furuskar, A., Jading, Y., Karlsson, J., Lundevall, M., and Parkvall, S. (2006). The 3G long-term evolution - radio interface concepts and performance evaluation. In *IEEE Vehicular Technology Conference*, volume 1, pages 137-141. IEEE.

Doolittle, L., Nelson, J., and Phillips, P. (2005). Boa Webserver. http://www.boa.org/.

Engels, M. (2002). *Wireless OFDM Systems*. Kluwer Academic Publishers, London.

Erceg, V., Hari, K., Smith, M., Baum, D., Sheikh, K., Tappenden, C., Costa, J., Bushue, C., Sarajedini, A., Schwartz, R., and Branlund, D. (2001). Channel models for fixed wireless applications. Technical report, IEEE 802.16 Work Group.

Fano, R. (1961). *Transmission of Information*, pages 168-178. John Wiley, New York.

Foschini, G. (1996). Layered Space-Time Architecture for Wireless Communication in a Fading Environment When Using Multiple Antennas. *Bell Laboratories Technical Journal*, 1(2):41-59.

Foshini, G. and Gans, M. (1998). On Limits of Wireless Communications in a Fading Environment when Using Multiple Antennas. *Wireless Personal Communications*, 6:311-335.

Garrett, D., Woodward, G., Davis, L., Knagge, G., and Nicol, C. (2004). A 28.8 mb/s 4x4 MIMO 3G high-speed downlink packet access receiver with normalized least mean square equalization. *IEEE International Conference on Solid-State Circuits*, 1:420-536.

Gesbert, D., Haumonte, L., Bolcskei, H., Krishnamoorthy, R., and Paulraj, A. (2002). Technologies and performance for non-line-of-sight broadband wireless access networks. *IEEE Communications Magazine*, 40(4):86-95.

Gesbert, D., Shafi, M., Shiu, D., Smith, P., and Naguib, A. (2003). From Theory to Practice: An Overview of MIMO. *IEEE Journal on Selected Areas in Communications*, 21(3):281-302.

Goud Jr, P., Hang, R., Truhachev, D., and Schlegel, C. (2006). A Portable MIMO Testbed and Selected Channel Measurements. *EURASIP Journal on Applied Signal Processing*, 1.

Haene, S., Perels, D., and Baum, D. (2004). Implementation aspects of a real-time multi-terminal MIMO-OFDM testbed. *IEEE Radio and Wireless Conference*, 1.

Horseman, T., Webber, J., Abdul-Aziz, M., Piechockim, R., Nix, A., Beach, M., and Fletcher, P. (2003). A testbed for evaluation of innovative turbo MIMO OFDM architectures. *Proc. 5th European Conference Personal Mobile Communications*, 1:453-457.

ITPP (2007). ITPP C++ library of mathematical, signal processing, speech processing and communication classes and functions. http://itpp.sourceforge.net/.

Jakes, W. C. (1974). *Microwave Mobile Communications*. John Wiley, New York.

Jungnickel, V., Pohl, V., and von Helmolt, C. (2003). Capacity of MIMO systems with closely spaced antennas. *IEEE Communications Letters*, 7(8):361-363.

Liberti, J. and Rappaport, T. (1999). *Smart Antennas for Wireless Communications: IS-95 and Third Generation CDMA Applications*. Prentice Hall PTR.

Liu, Z., Giannakis, G., and Hughes, B. (2001). Double differential space-time block coding for time-selective fading channels. *IEEE Trans. Communications*, 49(9):1529-1539.

Lucent, Nokia, Siemens, Ericsson, and Jeju, K. (2001). A standardized set of MIMO radio propagation channels. Technical report, 3GPP TSG-RAN WG1 23.

Mannion, P. (2004). IEEE pushes WLANs to 'nth' degree. *Electronic Engineering Times*, 1:8.

Maxim (2004). 802.11A and 802.11A/G RF Transceivers Support Pre-802.11N MIMO and Smart- Antenna Radio Systems. Technical report, Maxim Integrated Products, Sunnyvale, CA.

Murphy, P., Lou, F., Sabharwal, A., and Frantz, J. (2003). An FPGA based rapid prototyping platform for MIMO systems. *Proc. The 37th Asilomar Conference on Signals, Systems and Computers*, 1:900- 904.

Murphy, P., Sabharwal, A., and Aazhang, B. (2006). Design of warp: A flexible wireless open-access research platform. In *Proceedings of EUSIPCO.*

Paulraj, A., Gore, D., Nabar, R., and Bölcskei, H. (2004). An overview of MIMO communications-a key to gigabit wireless. *Proceedings of the IEEE*, 92(2):198-218.

Perels, D., Haene, S., Luethi, P., Burg, A., Felber, N., Fichtner, W., and Bolcskei, H. (2005). ASIC implementation of a MIMO-OFDM transceiver for 192 mbps WLANs.*European Solid-State Circuits Conference*, pages 215-218.

Piechocki, R., Fletcher, P., Nix, A., Canagarajah, C., and McGeehan, J. (2001). Performance evaluation of BLAST-OFDM enhanced Hiperlan/2 using simulated and measured channel data. *IEEE Electronics Letters*, 37(18):1137-1139.

Prasad, R. (2004). *OFDM for Wireless Communications Systems.* Artech House, London.

Rao, R., Zhu, W., Lang, S., Oberli, C., Browne, D., Bhatia, J., Frigon, J.-F., Wang, J., Gupta, P., Lee, H., Liu, D., Wong, S., Fitz, M., Daneshrad, B., and Takeshita, O. (2004). Multi-antenna testbeds for research and education in wireless communications. *IEEE Communications Magazine*, 42(12):72- 81.

Rappaport, T. S. (2002). *Wireless Communications: Principles and Practice.* Prentice Hall, 2nd edition.

Rosol, G. (1995). Base Station Antennas: Part 1, Part 2, Part 3. *Microwaves & RF*, 117-123.

Sampath, H., Talwar, S., Tellado, J., Erceg, V., and Paulraj, A. (2002). A fourth-generation MIMO- OFDM broadband wireless system: design, performance, and field trial results. *IEEE Communications Magazine*, 40(9):143-149.

Schober, R. and Lampe, L.-J. (2002). Noncoherent receivers for differential space-time modulation. *IEEE Trans. Communications*, 50(5):768-777.

Shannon, C. E. (1948). A Mathematical Theory of Communication. *Bell Systems Technical Journal*, 27.

Suzuki, H., Hedley, M., Daniels, G., and Yuan, J. (2006). Performance of MIMO-OFDM-BICM on measured indoor channels. *IEEE Vehicular Technology Conference*, 5:2073-2077.

Suzuki, H., Kendall, R., Hedley, M., Daniels, G., and Ryan, D. (2005). Demonstration of 4x4 MIMO data transmission on CSIRO ICT Center MIMO Testbed. In *Booklet of Abstracts for the 6th Australian Communications Theory Workshop*, page 35, Brisbane, Australia

Tarasak, P., Minn, H., and Bhargava, V. (2004). Improved approximate maximum-likelihood receiver for differential space-time block codes over rayleigh-fading channels.*IEEE Trans. Vehicular Technology*, 53(2):461-468.

Tarokh, V. and Jafarkhani, H. (2000). A differential detection scheme for transmit diversity. *IEEE Selected Areas in Communications Journal*, 18(7):1169-1174.

Tarokh, V., Jafarkhani, H., and Calderbank, A. (1999). Space-time block codes from orthogonal designs. *IEEE Trans. Information Theory*, 45(5):1456-1467.

Telatar, E. (1999). Capacity of Multiantenna Gaussian channels. *European Transactions on Telecommunication ETT*, 10(6):585-596.

uClinux (2007). Embedded Linux/Microcontroller Project. http://www.uclinux.org/

Uthansakul, P., Bialkowski, K., Bialkowski, M., and Postula, A. (2006). Assessing an FPGA implemented MIMO testbed with the use of channel emulator. *International Conference on Microwaves, Radar and Wireless Communications (MIKON)*, 1.

van Zelst, A. and Schenk, T. (2004). Implementation of a MIMO OFDM-based wireless LAN system. *IEEE Trans. Signal Processing*, 52(2):483-494.

Wallace, J., Jeffs, B., and Jensen, M. (2004). A real-time multiple antenna element testbed for MIMO algorithm development and assessment. *IEEE Antennas and Propagation Symposium*, 2:1716-1719 .

Wallace, J. W., Jensen, M. A., Swindlehurst, A. L., and Jeffs, B. D. (2003). Experimental Characterization of the MIMO Wireless Channel: Data Acquisition and Analysis. *IEEE Transactions on Wireless Communications*, 2(2):335-343.

Wolniansky, P., Foschini, G., Golden, G., and Valenzuela, R. (1998). V-BLAST: an architecture for realizing very high data rates over the rich-scattering wireless channel. In *IEEE Signals, Systems, and Electronics Symposium*, pages 295-300, Italy. IEEE.

Wouters, M., Bourdoux, A., Derore, S., Janssens, S., and Derudder, V. (2004). An Approach for real time prototyping of MIMO OFDM systems. In *Proc. 12th European Signal Processing Conference*, Austria.

# Chapter XXIII
# Ad Hoc Networks Testbed Using a Practice Smart Antenna with IEEE802.15.4 Wireless Modules

**Masahiro Watanabe**
*Mitsubishi Electric Corporation, Japan*

**Sadao Obana**
*ATR Adaptive Communications Research Laboratories, Japan*

**Takashi Watanabe**
*Shizuoka University, Japan*

## ABSTRACT

*Recent studies on directional media access protocols (MACs) using smart antennas for wireless ad hoc networks have shown that directional MACs outperform against traditional omini-directional MACs. Those studies evaluate the performance mainly on simulations, where antenna beam is assumed to be ideal, i.e., with neither side-lobes nor back-lobes. Propagation conditions are also assumed to be mathematical model without realistic fading. In this paper, we develop at first a testbed for directional MAC protocols which enables to investigate performance of MAC protocols in the real environment. It incorporates ESPAR as a practical smart antenna, IEEE802.15.4/ZigBee, GPS and gyro modules to allow easy installment of different MAC protocols. To our knowledge, it is the first compact testbed with a practical smart antenna for directional MACs. We implement a directional MAC protocol called SWAMP to evaluate it in the real environment. The empirical discussion based on the experimental results shows that the degradation of the protocol with ideal antennas, and that the protocol still achieves the SDMA effect of spatial reuse and the effect of communication range extension.*

## INTRODUCTION

Ad hoc networks (Jurdak, 2004) are the autonomous system of mobile nodes which share a single wireless channel to communicate with each other. The previous works on ad hoc networks assume the use of omni-directional antennas that transmit or receive signals equally well in all directions. Traditional MAC protocols, such as IEEE 802.11 DCF (Distributed Coordination Function) (ANSI/IEEE. Std, 1999), are designed for omni-directional antennas and cannot achieve high throughput in ad hoc networks because that waste a large portion of the network capacity.

On the other hand, smart antenna technology may have various potentials (Lehne, 1999). In particular, it can improve spatial reuse of the wireless channel, which allows nodes to communicate simultaneously without interference. Furthermore, the directional transmission concentrates signal power to the receiver, which enlarges the transmission range. Thus, it can potentially establish links between nodes far away from each other, and it prevents network partitions and the number of routing hops can be fewer than that of omni-directional antennas.

However these potentials smart antennas may have, a sophisticated MAC protocol is required to take advantage of these benefits in ad hoc networks. Recently, several directional MAC protocols, typically modifications of IEEE 802.11 DCF, have been proposed for ad hoc networks including our proposed MAC protocol called SWAMP (Takata, 2004). SWAMP provides both spatial reuse of the wireless channel and communication range extension by two types of access modes that utilize the directional beam effectively, and it contains a method of obtaining the location information of its neighbors.

For the real use to wireless devices, it should be evaluated the performance of proposed MAC protocol due to the effects of actual antenna pattern included both side-lodes and back-lobes, and actual propagation condition included realistic fading. So this time, we have developed a testbed for validation of the real environment, which is based on ESPAR antenna and IEEE802.15.4/ZigBee wireless module, and has the interface to GPS (Global Positioning System) for utilizing location information MAC protocol. And we embedded SWAMP into this testbed and evaluated in the actual environment

This paper summarizes SWAMP protocol and overview of testbed. In addition, we analyze the experimental result of SDMA and wider range communication due to SWAMP.

# BACKGROUND

## Directional MAC Protocols

Various MAC protocols using smart antennas or directional antennas, typically referred to as directional MAC protocols, have been proposed for ad hoc networks.

Ko et al. (2000) propose DMAC (Directional MAC) in which all frames are transmitted directionally except for the CTS (Clear To Send). Choudhury et al. (2002) propose MMAC (Multi-hop RTS MAC), which involves the multi-hop RTS (Request To Send) to take advantage of the higher gain obtained by directional antennas. These protocols, however, need various additional mechanisms to provide the location information and to forward the RTS.

In (Fahmy, 2002; Nasipuri, 2002) and (Takai, 2002), RTS is transmitted omni-directionally in order to find the receiver in case location information is not available. Each node estimates the direction of neighboring nodes for pointing the beam with AOA (Angle of Arrival) when it hears any signal. Because these protocols employ at least one omni-directional transmission, it limits the coverage area provided by directional transmissions and do not exploit one of the main benefits of directional antennas, i.e., the increase of the transmission range, either.

Ramanathan (2001) proposes circular directional transmission of periodic hello packets to obtain node information that is located farther away than the omni-directional transmission range. Korakis et al. (2003) proposes circular RTS, which scans all the area around the transmitter to find the addressed receiver and to tackle the deafness and the hidden-terminal problem arisen from directional transmissions. Bandyopadhyay et al. (2001) develops additional frames in order to determine the neighbor topology by recording the angle and signal strength under consideration of propagation conditions. Although these schemes attempt communication range extension, circular transmission increases the delay and incurs large control overhead.

In mobile ad hoc evaluation, Ramanathan (2005) shows a directional MAC testbed and simulation results, but not including actual experimental results. Although Nishida (2005) evaluated AODV (Ad hoc On demand Distance Vector routing) protocol based on IEEE802.11b DCF with omni-directional antennas in the experiment, not used directional antennas.

## SWAMP

SWAMP (Smart Antennas Based Wider-range Access MAC Protocol) is a MAC protocol for ad hoc networks using smart antennas based on IEEE 802.11 DCF, which utilizes the directional beam effectively to increase the spatial reuse and extend the transmission range.

## Neighbor Discovery

As described in Section I, a transmitter must know the location of the intended receiver to point the beam in the appropriate direction. SWAMP assumes that all nodes are equipped with a GPS (Global Positioning System) to determine their own locations.

A transmitter and a receiver exchange each other's location information by omni-directional RTS/CTS handshaking. Then, these nodes forward the location information that is obtained by the reception of the RTS or CTS to neighbors using an omni-directional beam. As a result, neighbors can obtain the location information of nodes located within an area at most two times farther away than that of the omni-directional antenna. We refer to this information that neighbors obtain as NHDI (Next Hop Direction Information).

Fig. 1 illustrates the propagation of locations with control frames. In the figure, node B is a transmitter and C is the intended receiver. After the RTS/CTS/SOF exchange, D and A can recognize not only locations of B and C, respectively, but also the initiation of communications between B and C.

Note that control frames should be modified in order to forward locations to neighbors of both the transmitter and receiver nodes. Moreover SWAMP requires the additional control frame SOF (Start Of Frame). Every node maintains an NHDI table with one entry for another node that can be obtained from NHDI in either CTS or SOF. Also note that the NHDI table of a node contains other nodes which the node cannot communicate directly with, which the node can communicate indirectly with by multi-hopping with an omni-directional beam, and which the node can communicate directly with a high gain directional beam to point their direction.

## Antenna Models

SWAMP provides four antenna beam forms. Each antenna beam is assumed to be able to point in any direction and can be formed during a SIFS (Short Inter Frame Space) interval as with in IEEE 802.11 DCF. Fig. 2 illustrates four ideal beam forms and each transmission range. Note that in the figure nodes can communicate when the transmitting beam and the receiving beam are at least tangential to each other. OB and DL are for the regular link communication in OC-mode, while DM and DH for the extended link communication in EC-mode.

## Access Modes

SWAMP consists of two access modes, OC-mode (Omni-directional transmission range Communication mode) and EC-mode (Extend omni-directional transmission range Communication mode). OC-mode is selected when the receiver is in the vicinity of the transmitter or when the transmitter does not know the location of the receiver. This mode mainly increases spatial reuse of the wireless channel. EC-mode is selected when a receiver is out of range of a transmitter's omni-directional beam. EC-mode extends the transmission range.

1) *OC-mode:* OC-mode is selected when the receiver node is located within the area of omni-directional transmission range or is not registered in the transmitter's NHDI table. Fig. 3 illustrates the OC-mode frame sequence with the corresponding beams. The RTS/CTS handshaking tries to reserve the wireless channel and to exchange the location information between the transmitter and the receiver. CTS/SOF forwards the NHDI in the neighborhood as shown in Fig. 1. These control frames are sent by an omni-directional beam, whereas DATA and ACK are sent by DLs that point beams towards each other.

Even if node A or D radiates to the omni-directional transmission range when node B is transmitting the data frame to C, A does not interfere with the on-going transmission between B and C because the directional beams of B and C are pointed towards each other. Therefore, A (and even D) does not have to wait for the ordinal NAV if its intended receiver is out of the B-C communication area. SWAMP assigns an omni-NAV shorter than the ordinary NAV to initiate the A's communication after the completion of SOF (Fig. 4). This can mitigate the exposed-terminal problem, and the hidden-terminal can communicate without interference to increase the spatial reuse of the wireless channel with ideal antenna models in Fig. 2.

*Figure 1. Control frame exchange*

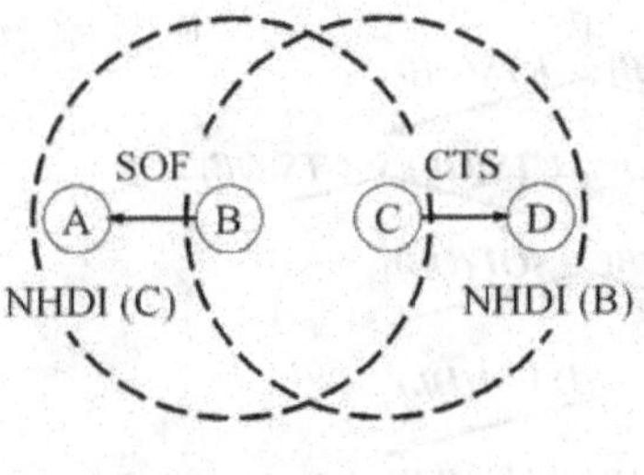


*Figure 2. Smart antenna beamform*

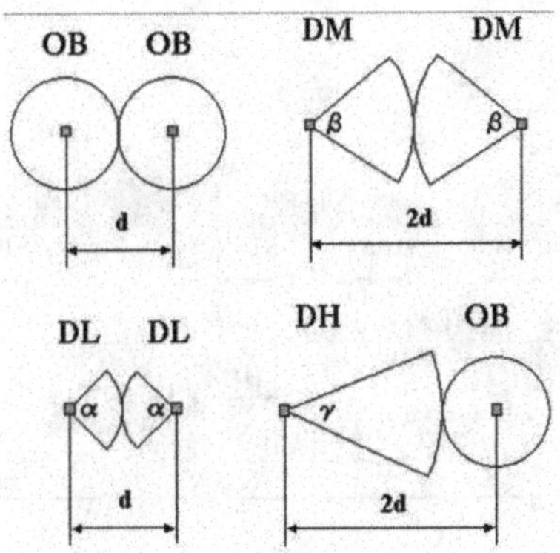


*Left side: Transmitting beamform, Right side: Receiving beamform*

2) *EC-mode:* EC-mode is selected when the receiver node has been already registered in the transmitter's NHDI table. Fig. 5 illustrates the EC-mode frame sequence with the corresponding beam. Because the transmitter has the prior knowledge of the direction of the intended receiver, the transmitter can determine the direction to point the beam toward the receiver. To perform communications between nodes at a distance of 2d, RTS is required to use the high gain beam form (DH) because the receiver node waits for signals with the omni-directional beam form (OB) in an idle state. After it sends RTS, the transmitter switches the beam form from DH to DM. After the receiver receives RTS, it also switches the form from OB to DM and points the beam toward the transmitter. When the transmitter fails EC-mode access over the EC-retry limit, the transmitter deletes the receiver information from its own NHDI table. In EC-mode, DNAV (Directional NAV) is used instead of NAV for virtual carrier sensing.

This mode exploits extension of the transmission range with the directional beam for all frames as discussed in (Korakis, 2003).

## MAIN THRUST OF THE CHAPTER

### Testbed

We have developed a compact wireless ad hoc network testbed for validation of the real environment, which is based on ESPAR antenna and IEEE802.15.4/ZigBee wireless module.

*Figure 3. OC-mode frame sequence (B to C)*

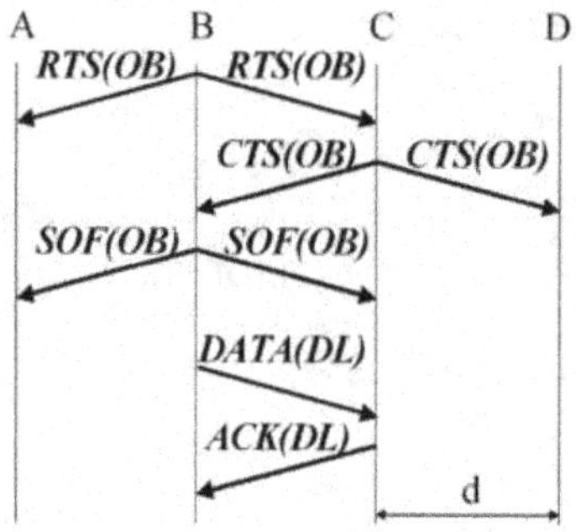


*Figure 4. SOF and omni-NAV*

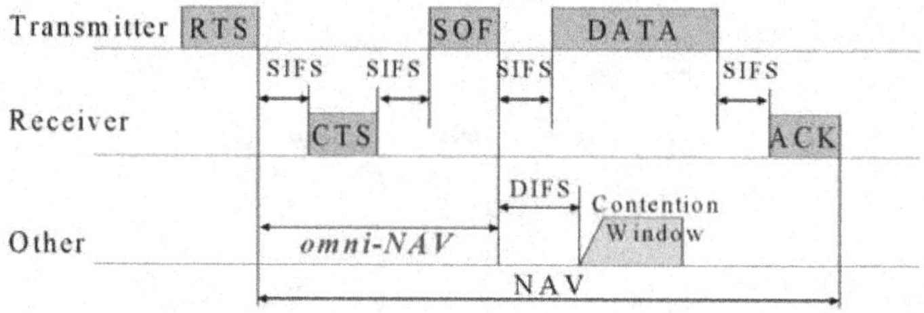


*Figure 5. EC-mode frame sequence (A to C)*

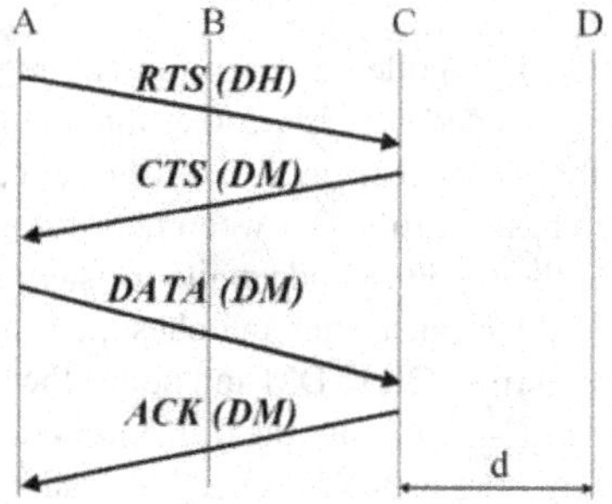


## ESPAR Antenna

ESPAR antenna is a smart antenna, which can form omni-directional beam pattern or directional beam pattern. An example of the specific structure is shown Fig.6. With a vertical monopole at the center, several parasitic elements are circularly arrayed in its vicinity. The parasitic elements may be called non-excited elements and or non-fed elements and are not connected to the feed circuit directly. They are electromagnetically coupled in space with the center element and other parasitic elements. At the bottom of parasitic element, a variable capacitance diode (varactor) is inserted in series and is grounded. A DC bias voltage is applied to each varactor via an RF choke or a large resistor. In contrast to the conventional phased array antenna in which the weight vector (amplitude and phase) for each element is direc-

*Figure 6. Structure of ESPAR antenna*

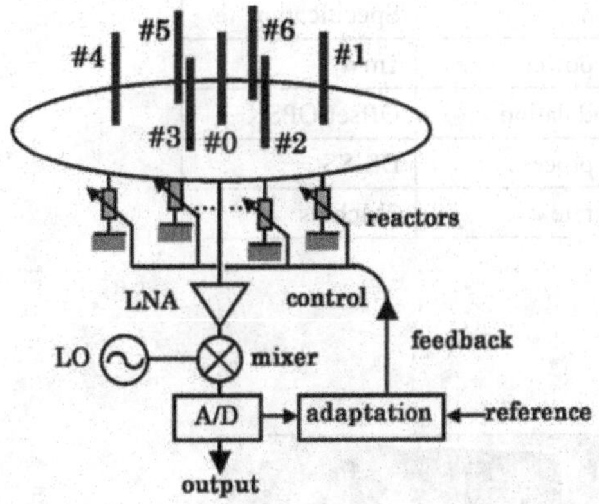


tory controlled, the ESPAR antenna does not have a weight circuit. Instead, the capacitance of the varactor loading the parasitic element is controlled. The ESPAR antenna differs fundamentally from the conventional phased array antenna in the following three aspects.

(1) one feed system
(2) Inter element coupling is the basis of beam formation, and
(3) Varactors are connected directly to the parasitic elements.

The RF signal is fed to the center element. Hence, only one transmitting and one receiving circuit are needed. By controlling the reactance value of the varactor by the applied DC bias voltage, beams with various radiation patterns are formed in the horizontal plane. Since these bias voltage are opposite to those of the variable capacitance diodes, no DC current flows under this voltage controlled operation. Therefore, the parasitic elements do not consume the DC and RF energies. In comparison to the conventional phased array antenna in which the radiating elements and the variable phase shifters are designed separately, the ESPAR antenna hardware configuration is substantially simplified. Hence, variable radiation pattern can be obtained at low cost and with low power consumption (Ohira, 2004).

## IEEE802.15.4/ZigBee Wireless Module

At the design of testbed for embedding proposed MAC protocol, it is important to choose the suitable type of wireless module. In wireless module, driver software for controlling its hardware should be open source for the modification due to the operation of proposed MAC protocol. In IEEE802.11 wireless LAN a/b/g, current driver software is closed source, so that we can't modify its driver software without special contract with chip maker. In IEEE802.15.4/ZigBee, driver software is open source, which specification for PHY and MAC layer has been authorized as international standardization and for network and application layer is currently offered as its standardization in ZigBee alliance. Although 250 Kbps data transmission rate of ZigBee is too small compared with that of WLAN, modification of ZigBee driver software is available and cheap cost and low DC current consumption, which make us choose ZigBee as wireless module for the testbed. The general specification of ZigBee chip CC2420 (manufactured by Chipcon company) is shown in Table 1.

In the design of software for embedding proposed MAC protocol, program library for controlling hardware of ZigBee chip includes carrier sense level adaptation, ESPAR antenna beam management, Tx power controlling, time-count and so on. Especially, the function of Tx power controlling (25dBmax dynamic range) offers the benefit for coordination of the EIRP (Effective Isotropic Radiation Power) between omni-directional beam and directional beam.

## System Configuration

Wireless ad hoc network testbed consists of notebook PC for command of data communication and data log collection of retry numbers, contention numbers and so on, ESPAR antenna as smart antenna and ZigBee wireless module, GPS and gyro for location information collection. Overview of the testbed is shown in Fig.7.

*Table 1. Zigbee: Chipcon CC2420*

| No. | Item | Specification |
|---|---|---|
| 1 | Tx power | 1mW |
| 2 | Modulation | Offset-QPSK |
| 3 | SS process | DS-SS |
| 4 | SS rate | 2Mchips |

*Figure 7. Overview of the testbed*

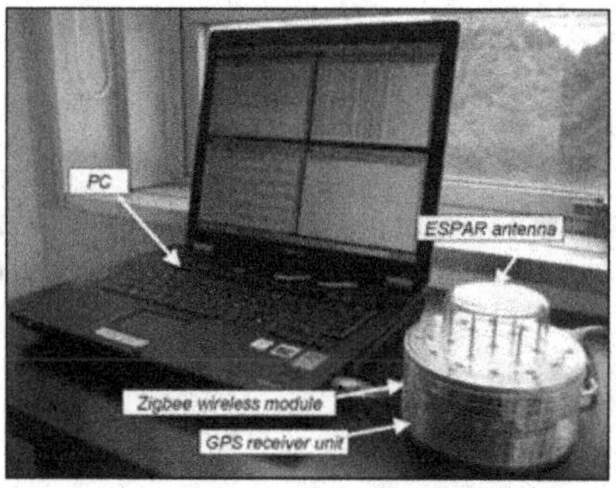


*Figure 8. Configulation of ZigBee data frame including SWAMP frame (control or data)*

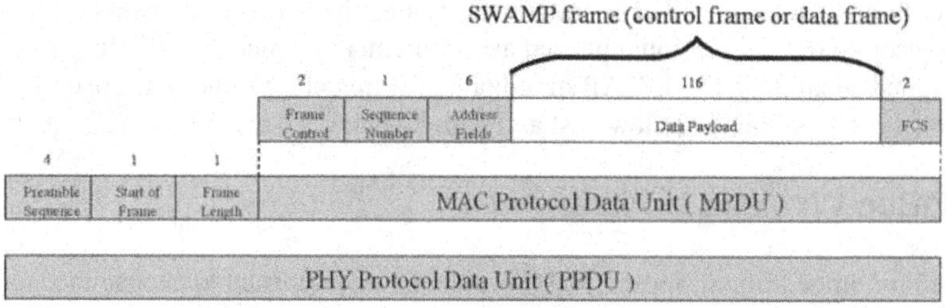


While we embed proposed MAC protocol: SWAMP on ZigBee wireless module, we adopted all the frames of SWAMP into ZigBee data frame without ZigBee control frame such as beacon. SWAMP frame including control frame (RTS/CTS/SOF) and data frame (DATA/ACK) should be taken as a capsule into the data payload of ZigBee data frame. When data size is over than maximum data payload size: 128 bytes, prural data frames will be transmitted continuously. Configulation of ZigBee data frame including SWAMP frame (control or data) is shown in Fig. 8.

Minimum received signal level: Smin has been measured to be -92dBm at 250Kbps fixed data rate with the description of 10% packets err rate at 500 times averaged transmitting data on coaxial wire line and variable attenuators without effect of fading. Measured packet err rate vs RSSI is shown in Fig. 9.

## SWAMP Experiment

### OC-Mode Evaluation

SWAMP OC-mode is expected for SDMA, which allows nodes to communicate simultaneously. Nevertheless, ESPAR antenna has side-lobes and back-lobes, of which gain is only about 10dB less than that of directional main beam, so that

*Figure 9. Measured packet err rate vs RSSI*

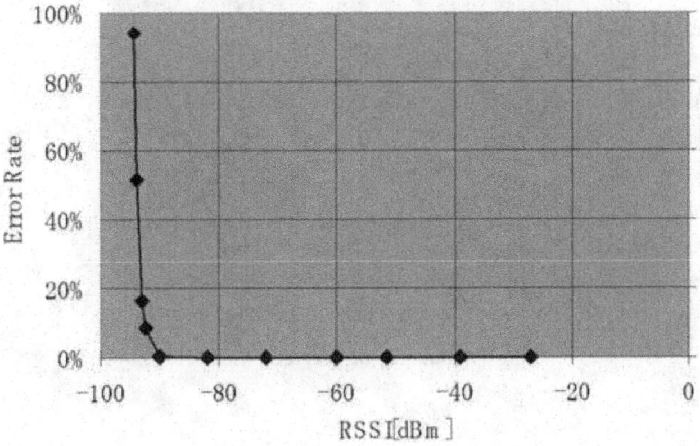


has more probability of interference compared with ideal directional antenna pattern. Actual ESPAR antenna pattern is shown in Fig. 10. Absolute antenna gains are 2dBi at omni-directional beam and 6dBi at directional main beam.

We prepare four nodes consists of two communication pairs situated on the line. The layout for 4 Nodes is shown in Fig. 11. Larger sector means directional main beam, smaller sector means back-lobe and circle line means omni-directional beam. In either case of communication pairs, the distance between two nodes (node1 to node2, and node 3 to node 4) is situated to be 6m due to the actual limitation in the park.

At SWAMP OC-mode, Tx power is controlled to be same EIRP between omni-directional beam and directional beam. In the actual field, we confirmed the same maximum communication distance 80m, when Tx power control levels are -14dB attenuator with each omni-directional beam and -25dB attenuator with each directional beam. The attenuator difference is 11dB, which includes a little bit correction 3dB, is different from calculated 8dB, which amounts twice (Tx and Rx) antenna gain difference between omni-directional beam 2dBi and directional beam 6dBi. RSSI fluctuation is often occurred due to the effect of fading and multipath in the actual field, which needs some correction between measured and calculated. Example data for RSSI vs communication distance is shown in Fig.12. Experiment conditions in the park are that antenna height is 55cm, Tx power is 0dBm without attenuation, on the grand including stones and grass and so on. RSSI under multipath is calculated by equation (1).

$$P_r = P_t G_t G_r \left[ D_d \left( \frac{\lambda}{4\pi r_d} \right) + D_r \left( \frac{\lambda}{4\pi r_r} \right) \Gamma v e^{-j\{k(r_d - r_r) + \phi\}} \right]^2 \qquad (1)$$

Where Pr is the received signal level, Pt is the transmitter power, Gt and Gr are the Tx and Rx boresight antenna gains, Dd and Dr are the direct wave and reflection wave directivities with respect to the Tx and Rx antennas, $r_d$ and $r_r$ are the optical path lengths of the direct wave and reflection wave from the grand, k is $2\pi/\lambda$, and $\Phi$ and $\Gamma v$ are the phase delay and reflection factor (electric field polarization) (Watanabe, 2004).

According to the result of experiment, RSSI fluctuated in a width including max. 7dB deviation from calculated.

Maximum communication distance including 3dB correction with several antenna pattern combinations each other is shown in Tab. 2.

In either case of antenna pattern combination each other, which are omni-directional beams or directional beams, we have measured each throughput in proportion to the distance (X) between 2 communication pairs. Experiment condition is that CBR125kbps, packet size is 512byte and other parameters are shown in Tab.2. The result of experiment is shown in Fig. 13 and Fig.14. In the combination of omni-directional beams each other, throughput degradation occurred at the range less than 80m due to interference between omni-directional beams each other, which agrees well in Tab. 2. On the other hand, in the combination of directional beams each other, throughput degradation occurred at the range less than 20m due to interference between directional main beam and back-lobe. This 20m plus 6m (two nodes distance in a communication pair) equals to 26m, so that is closely to 25m, which agrees well in Tab.2. While two nodes distance

*Figure 10. ESPAR antenna pattern*

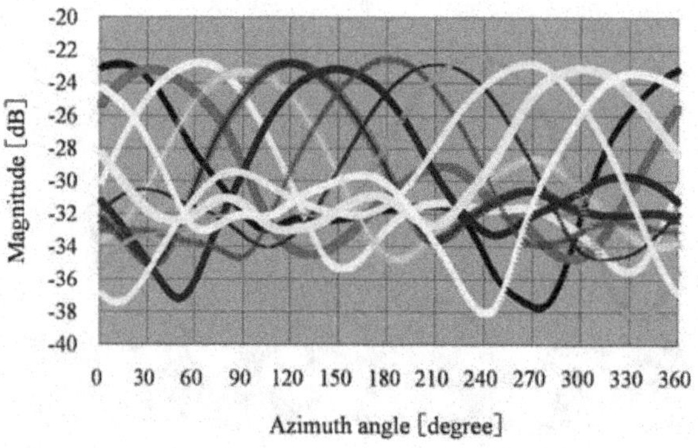


*Figure 11. Layout for 4 Nodes*

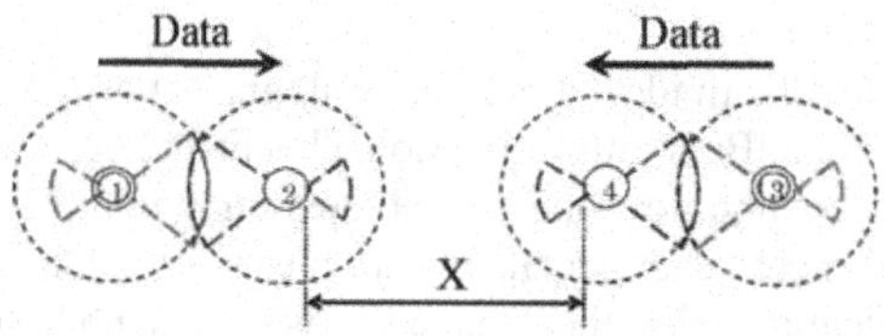


*Figure 12. RSSI vs communication distance*

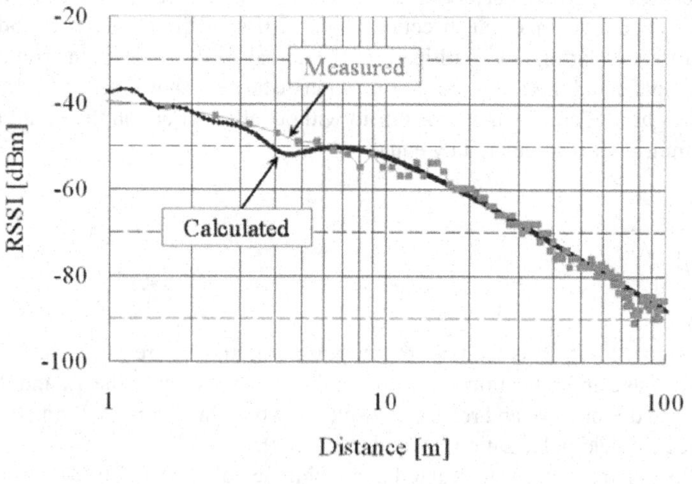


should be set 80m in a communication pair, interference distance (X) will be possible to close to be 8m, which is max. communication distance between back-lobe and back-lobe shown in Tab. 2.

Therefore, according to the result of experiment, we have verified the superiority of SDMA with directional beams rather than that with omni-directional beams.

## EC-Mode Evaluation

SWAMP EC-mode is expected for wider range communication, which owes Tx power concentration of directional beam towards to receiver node.

*Table 2. Maximum communication distance*

| No. | Item | Sector-Sector | | | Omni-Omni | - |
|---|---|---|---|---|---|---|
| | | Back-Back | Back-Main | Main-Main | | |
| f | Frequency | 2.405 | 2.405 | 2.405 | 2.405 | GHz |
| Pt | Tx power control | -25 | -25 | -25 | -14 | dBm |
| Lt | Tx loss | -2 | -2 | -2 | -2 | dB |
| Gt | Tx antenna gain | -4 | 6 | 6 | 2 | dBi |
| PtGt | EIRP | -31 | -21 | -21 | -14 | dBm |
| R | Max. comm. distance | 8 | 25 | 80 | 80 | m |
| – | Propagation loss | -58.1 | -68.0 | -78.1 | -78.1 | dB |
| Gr | Rx antenna gain | -4 | -4 | 6 | 2 | dBi |
| Lr | Rx loss | -2 | -2 | -2 | -2 | dB |
| Pr | Received signal level | -95 | -95 | -95 | -92 | dBm |
| Smin | Minimum received siganal level | -92 | -92 | -92 | -92 | dB |
| – | Correction (Pr-Smin) | -3 | -3 | -3 | 0 | dB |

*Figure 13. Throughput vs distance (X)*

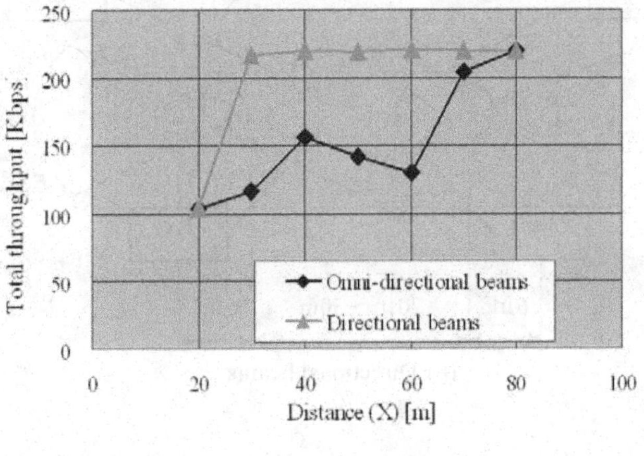


We prepare three nodes situated like a triangle, which formation is shown in Fig.15. Node B and node C are laid to be almost same distance from node A, so that the distance between node B and node C becomes to be longer than that distance (node A to node B, or node A to node C). In the assumption of experiment, node A is Tx in OC-mode with node B, node C is Tx in EC-mode with node B and node B is always Rx at both modes.

At the operation of SWAMP, OC-mode operation should be executed before EC-mode operation. In OC-mode, while node B is situated to be able to communicate with node A, although be unable to communicate with node C because of longer distance as mentioned above, that makes node C unaware of node B's location information. In EC-mode, after node A transmits SOF packet including node B's location information to node C due to OC-mode between node A and node B, that leads to be able to communicate from node C to node B in EC-mode.

We have measured the received packet numbers vs time at OC-mode and EC-mode simultaneously under the situation shown in Fig.16. Experiment condition is that CBR125kbps, packet size is 512byte, Tx power control levels in OC-mode are same in Tab.2, in EC-mode are without attenuator with directional beam. By the adjustment of node communication distance, measured RSSI in OC-mode is -89dBm, in EC-mode is -69dBm including about ±3dB fluctuation. The result of experiment is shown in Fig.15. Although, node C can't communicate with node B until OC-mode starts between node A and node B. Once, node C receives SOF packet from node A in OC-mode, node C can starts communication to node B in EC-mode at the time, which is about 44sec passed in Fig. 15.

*Figure 14. Node location related to communication distance (interference distance)*

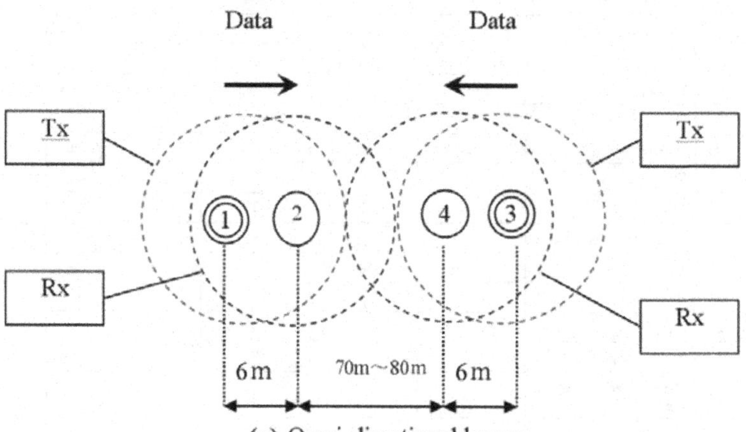


*(1) Omni-directional beams*

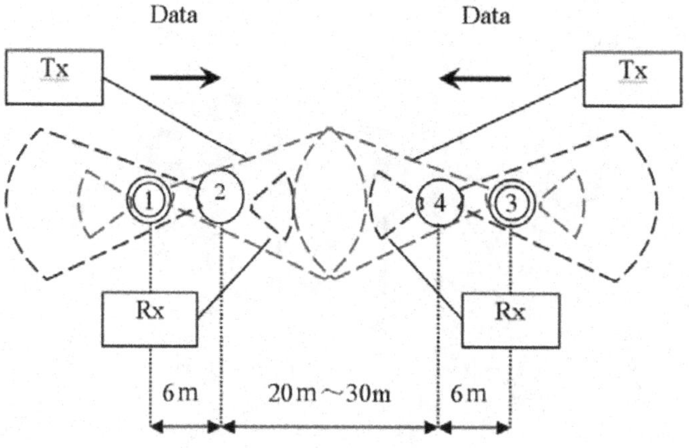


*(2) Directional beams*

Therefore, according to the result of experiment, we have confirmed the wider range communication with directional beams in EC-mode compared with OC-mode operation, due to transmission of SOF packet in OC-mode.

## CONCLUSION

We have developed a compact wireless ad hoc network testbed based on ESPAR antenna and IEEE802.15.4/ZigBee wireless module, which allow easy installment of different MAC protocols. We designed program library on driver software for controlling ZigBee hardware and embedded directional MAC protocol: SWAMP into the testbed for validation in the real environment.

According to the result of experiment, SWAMP OC-mode has showed the superiority of SDMA with directional beams rather than that with omni-directional beams, even if ESPAR antenna includes side-lobes and back-lobes. In either case of antenna pattern combination each other, which are omni-directional beams or directional beams, measured interference distances agree well with calculated distance. SWAMP EC-mode has showed extension of communication

*Figure 15. Situation for EC-mode evaluation*

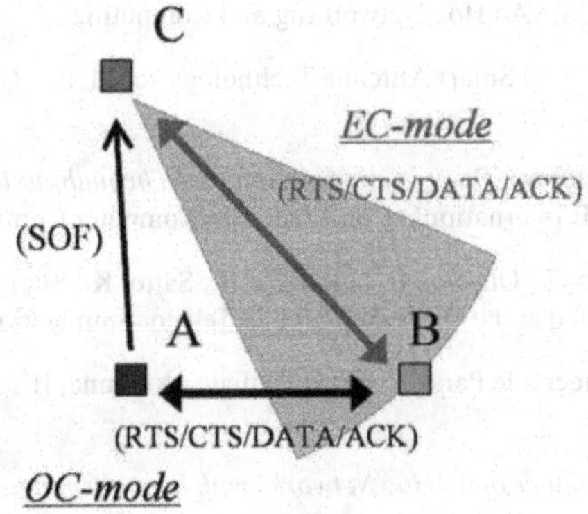


*Figure 16. Received packet numbers vs time*

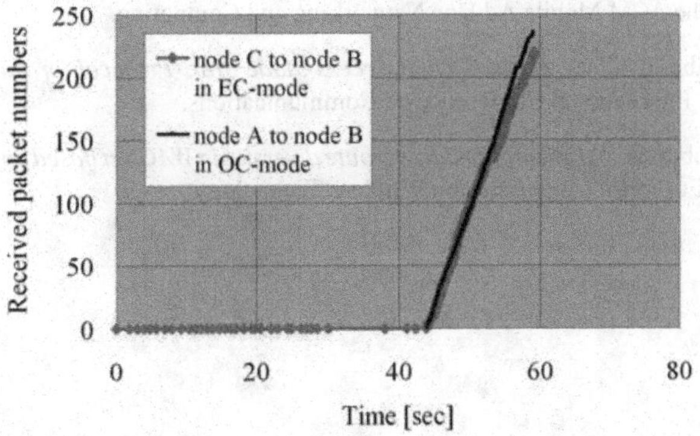


distance, even if nodes are unable to communicate in OC-mode because of packets unreachable longer distance. Once, transmission of SOF packet in OC-mode leads to be able to start communication in EC-mode, which made verify wider range communication.

## REFERENCES


ANSI/IEEE Std 802.11 (1999), *Wireless LAN Medium Access Control (MAC) and Physical Layer (PHY) specifications.*

Bandyopadhyay, S., Hasuike, K., Horisawa, S., & Tawara, S. (2001, November), *An Adaptive MAC Protocol for Wireless Ad Hoc Community Network (WACNet) Using Electronically Steerable Passive Array Radiator Antenna*, Paper presented at the IEEE Global Communications Conference.

Choudhury, R. R., Yang, X., Ramanathan, R. & Vaidya, N. H.(2002, September), *Using Directional Antennas for Medium Access Control in Ad Hoc Networks*, Paper presented at the ACM Mobile computing and Network.

Fahmy, N. S., Todd, T. D., & Kezys, V. (2002, April), *Ad Hoc Networks with Smart Antennas using IEEE 802.11-Based Protocols*, Paper presented at the IEEE International Conference on Communications.

Jurdak, R., Lopes, C. V., & Baldi, P. (2004), A Survey, Classification and Comparative Analysis of Medium Access Control Protocols for Ad Hoc Networks, *IEEE Communications Surveys and Tutorials*, 6(1)

Ko, Y. B., Shankarkumar, V., & Vaidya, N. H. (2000 March), *Medium Access Control Protocols Using Directional Antennas in Ad*

*Hoc Networks*, Paper presented at the EEE Conference on Computer Communications.

Korakis, T., Jakllari, G., & Tassiulas, L. (2003 June), *A MAC protocol for full exploitation of Directional Antennas in Ad-hoc Wireless Networks*, Paper presented at the ACM Mobile Ad Hoc Networking and Computing.

Lehne, P. H. & Pettersen, M. (1999), An Overview of Smart Antenna Technology for Mobile Communications Systems, *IEEE Communications Surveys and Tutorials*,12(4).

Nasipuri, A., Li, K., & Sappidi, U. R. (2002 October), *Power Consumption and Throughput in Mobile Ad Hoc Networks using Directional Antennas*, Paper presented at the IEEE International Conference on Computer Communications and Networking.

Nishida, T., Eguchi, K., Okamoto, Y., Warabino, T., Ohseki, T., Fukuhara, T., Saito, K., Sugiyama, K. (2005, June), *Multi-Hop Vehicle–to-Vehicle Communication*, Paper presented at the Workshop on ITS Telecommunications.

Ohira, T. &, Iigusa, K. (2004), Electronically Steerable Parastic Array Radiator Antenna, IEICE Transactions on Communications, J87-C(1), 12-31.

Ramanathan, R. (2001, October), *On the Performance of Ad Hoc Networks with Beamforming Antennas*, Paper presented at the ACM Mobile Ad Hoc Networking and Computing.

Ramanathan, R. (2005), Ad Hoc Networking With Directional Antennas: A Complete System Solution, IEEE Journal on Selected Areas in Communications, 23(3), 496-506.

Takai, M., Martin, J., Ren, A., & Bagrodia, R. (2002 June), *Directional Virtual Carrier Sensing for Directional Antennas in Mobile Ad Hoc Networks*, Paper presented at the ACM Mobile Ad Hoc Networking and Computing.

Takata, M., Nagashima, K., & Watanabe, T. (2004, June), *A Dual Access Mode MAC Protocol for Ad Hoc Networks Using Smart Antennas*, Paper presented at the IEEE International Conference on Communications.

Watanabe, M. & Tanaka, S. (2004, October), *Experimental results of route diversity in WACNet based on ESPAR antenna and 802.11b Ad hoc system*, Paper presented at the European Conference on Wireless Technology.

# Chapter XXIV
# Wideband Smart Antenna Avoiding Tapped-Delay Lines and Filters

**Monthippa Uthansakul**
*Suranaree University of Technology, Thailand*

**Marek E. Bialkowski**
*The University of Queensland, Australia*

## ABSTRACT

*This chapter introduces the alternative approach for wideband smart antenna in which the use of tapped-delay lines and frequency filters are avoidable, so called wideband spatial beamformer. Here, the principles of operation and performance of this type of beamformer is theoretically and experimentally examined. In addition, its future trends in education and commercial view points are identified at the end of this chapter. The authors hope that the purposed approach will not only benefit the smart antenna designers, but also inspire the researchers pursuing the uncomplicated beamformer operating in wide frequency band.*

## INTRODUCTION

In the past two decades, radio systems (also known as wireless systems) have grown with an unprecedented speed from early radio paging, cordless telephone, and cellular telephony to today's personal communication and mobile computing. This rapid expansion of radio systems has a profound impact on today's business world and people's daily lives. One undesired outcome is a heavy utilization of the available frequency spectrum. Because of this situation, a considerable interest has been shown in methods and techniques to overcome the limited frequency spectrum. One technique that is capable to increase the wireless system capacity without additional frequency spectrum is the smart antenna technique. Smart antennas are multiple element antennas accompanied by suitable signal processing algorithms either at the transmitter or receiver sides of a communication link. By pointing their beam towards a desired user and nulls or low side lobes towards interfering sources they are capable to considerably improve the quality of signal transmission in a multi-user environment. A significant value of smart antenna techniques in the efficient use of wireless spectrum has been addressed in (Kang, 2002; Jiang 1997). These multiple element antenna systems can also offer other advantages. These include the capability of minimizing the cost of establishing new wireless networks (Shao, 2003; Lee, 2005; Kawitkar, 2003), a better service quality (Hettak, 2000), and transparent operation across multi–technology wireless networks (Alexiou, 2004). It has to be noted that the benefits of smart antennas have been largely demonstrated for the case of narrowband communication systems. As the rapid growth of wireless technologies demands high bit rate data transmission, there is an interest in smart antennas which would operate over an increased frequency band. The design of such wideband intelligent antenna systems creates a challenge in terms of processing techniques and associated costs.

This book chapter gives a brief overview of wideband beamforming techniques. In particular, their advantages and disadvantages are discussed. As a result of these considerations the focus is on a wideband smart antenna system that relies on a fully spatial wideband beamforming technique. Full theoretical and experimental investigations into this wideband smart antenna system are presented.

The chapter is organized as follows. Firstly, shortfalls of narrowband smart antenna/beamforming techniques with respect to wideband signals are demonstrated via a suitable example. A number of wideband signal processing techniques are introduced and discussed to overcome this impairment. They are classified into three categories: space-time, space-frequency and fully spatial techniques. A brief comparison of these three wideband beamforming techniques is presented. As a result of this comparison, the fully spatial beamforming technique is selected for further considerations. Because of this choice, the main part of the chapter is devoted to a wideband spatial beamformer and its practical realization. The considerations commence with the introduction of configuration and the basic principles of operation of a wideband spatial beamformer that is created around a rectangular array of wideband antenna elements. The original beamforming algorithm, as reported in literature, is introduced and its shortfalls with respect to a small size array are pointed out. A suitable rectification of the original beamforming algorithm is proposed so it is valid for an arbitrary size array. The remaining considerations focus on small size arrays, which are easy to realize in practice. A 4×4 element wideband beamformer prototype is developed and tested over a specified frequency band. Various radiation patterns are realized by applying suitably devised signal weighting coefficients. The chapter is finalized with conclusions and remarks concerning future plans for wideband spatial beamformers.

## BACKGROUND

Most of considerations concerning smart antennas that can be found in the antenna or wireless communication literature are devoted to the narrowband variety. In this case, an array antenna is usually of linear type and formed by identical antenna elements spaced by half-wavelength at the centre frequency of a narrow operational band. Other configurations of arrays, such as rectangular, circular, or hexagonal, are also used. However, the operation is easier to explain for the linear array case. In many applications, this antenna array is placed in the horizontal plane and is aimed to transmit or receive signals in azimuth directions. In order to enhance reception of a desired signal and discriminate against undesired signals, the antenna beam is pointed towards a desired direction and nulls or low side lobes towards undesired directions. Usually these directions are known before the beam is formed. They are determined by suitable search methods that are based on Direction Of Arrival (DOA) estimation techniques (Liberti, 1999). For the narrowband case, the beam pattern is obtained using constant complex weighting coefficients that are applied to the received or transmitted signals at individual antenna elements. This beamforming algorithm works well for signals having a Fractional Bandwidth (FB) of a few percents.

This narrowband beamforming scheme becomes faulty when applied to wideband signals (Hefnawi, 2000; Uthansakul, 2004). The undesired effects include a main beam squinting, shifting nulls' locations and increasing sidelobe levels in the produced radiation pattern. These, in turn, adversely affect the quality of communication link.

The shortfalls of a narrowband beamformer when it is applied to a wideband signal are illustrated by simulation results presented in Figure 1. In the considered example, a signal has FB of 27.3% (from 1.9 to 2.5 GHz). The beamformer uses a LMS algorithm (Liberti, 1999; Kawitkar, 2005), to follow the direction of a desired signal. The array is formed by four elements spaced by half-wavelength at 2.2 GHz. The desired signal is coming from 30° off the array's boresight direction. Figure 1 shows that when the frequency changes from 1.9 to 2.5 GHz, the main beam and nulls' locations are shifted.

Figure 2 shows the results for the main beam direction deviations as a function of the operational bandwidth when the desired signals is assumed to arrive from 10°, 30° and 45° off the array boresight. It is apparent that the error in pointing the main beam to the desired direction becomes larger as the bandwidth of the signal increases. Furthermore, the error also depends on the direction of a desired signal. Larger errors occur for the signals coming from the far directions off the array's boresight.

The simulation results presented in Figures 1 and 2 confirm that the narrowband beamforming scheme is unsuitable when applied to wideband signals and thus alternative beamforming schemes are required to resolve this shortfall.

*Figure 1. Radiation pattern of narrowband beamformer employing 4×1 point-sources when applying a wideband signal having FB of 27.3%*

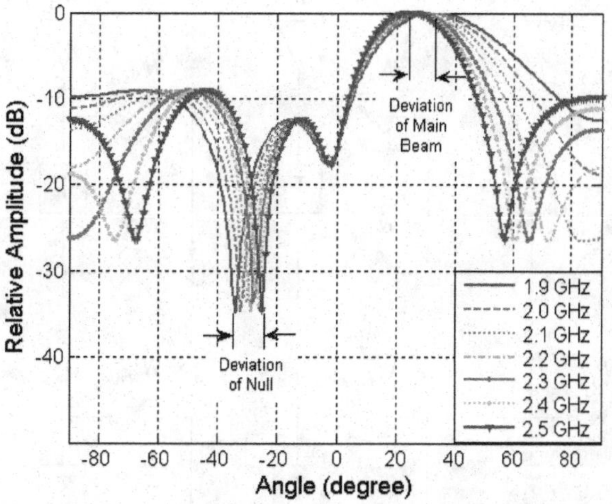


*Figure 2. Maximum deviation of the main beam direction for narrowband beamformer employing 4×1 point-sources when varying FB of the incoming signal from 0 to 30% and desired signal of 10°, 30° and 45°*

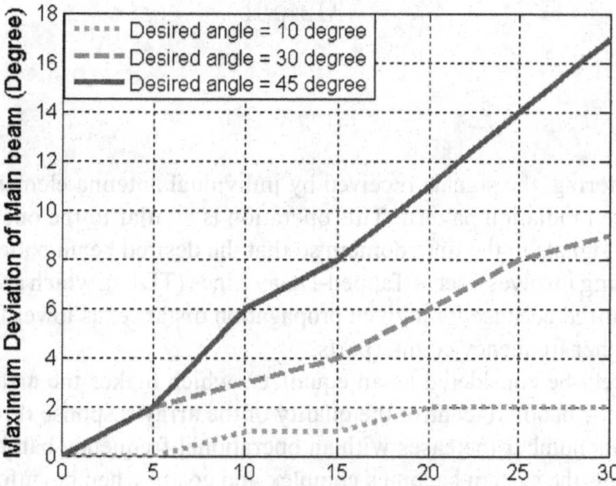


From literatures, the wideband beamformer techniques can be classified into three categories space-time, space-frequency and fully spatial beamforming. The brief description of each beamforming technique is presented as follows.

## A. Space-Time Beamformer

A space-time beamformer (Hong, 2004; Li, 2003; Godara, 1999), also called a spatio-temporal beamformer, is a two-dimensional (space/time) signal processor which combines spatial filtering with temporal filtering. This is illustrated in Figure 3.

*Figure 3. Space-time wideband beamformer constituted by a linear array with Nst antenna elements and K tapped-delay lines in each branch*

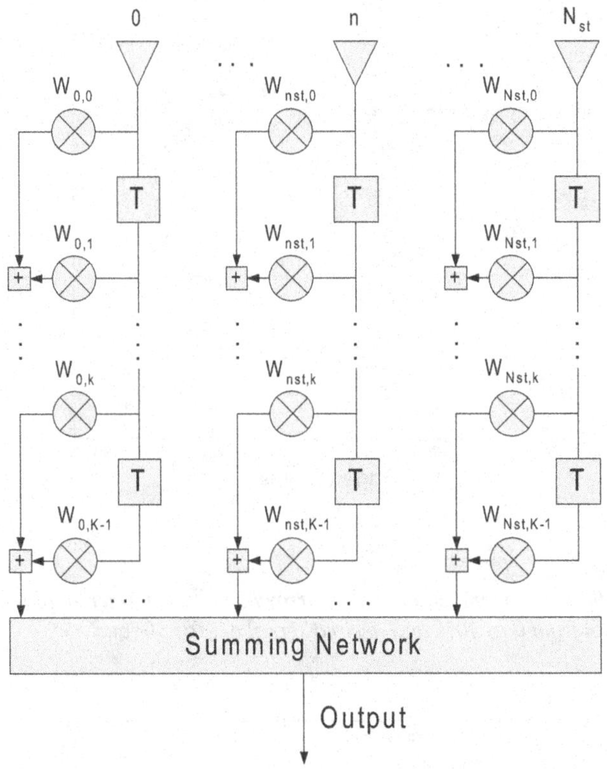


With regard to the spatial filtering, the signals received by individual antenna elements are multiplied by a set of complex weights to form a desired radiation pattern. This operation is similar to the one used in a narrowband smart antenna. However, to handle the signals in the time domain so that the desired beam pattern is maintained over a wide frequency band, the time processing involves a set of Tapped-Delay Lines (TDLs), which are used in each element branch of the array antenna. This is required because for a given propagation distance, as lower frequency signal components have less phase shift than the higher frequency counterparts.

This structure can alternatively be considered as an equalizer, which makes the array response having the same time delay across a wide frequency band. To control the quality of the array response throughout the band, the proper number of TDLs is required. This number increases with an operational frequency bandwidth (Vook, 1992; Mayhan, 1981; Materum, 2003). As a result, the system becomes complex and costly when beamforming is to be accomplished across an increased operational frequency band.

## B. Space-Frequency Beamformer

Wideband beamforming can be accomplished without TDL's by applying the space-frequency beamformer, which was described in (Hefnawi, 2001; Hefnawi, 1997). The configuration of this beamformer is shown in Figure 4.

This alternative beamformer consists of a linear array of antennas, which are spaced by half-wavelength at the center frequency of the band, and a set of filters. The signals received by each element in the array are decomposed into sets of non-overlapping (in frequency domain) narrowband signals using bandpass filters. Having decomposed a wideband signal into narrowband signals in each antenna element, the beamforming is performed for each sub-band using a narrowband weighting scheme. The only variation in comparison with the conventional narrowband beamformer is different array spacing in terms of the operational wavelength for each of the narrow sub-band. It has to be noted that in each of the sub-bands the array has to produce an approximately same radiation pattern.

*Figure 4. Space-frequency wideband beamformer constituted by a linear array with N antenna elements and J bandpass filters in each branch*

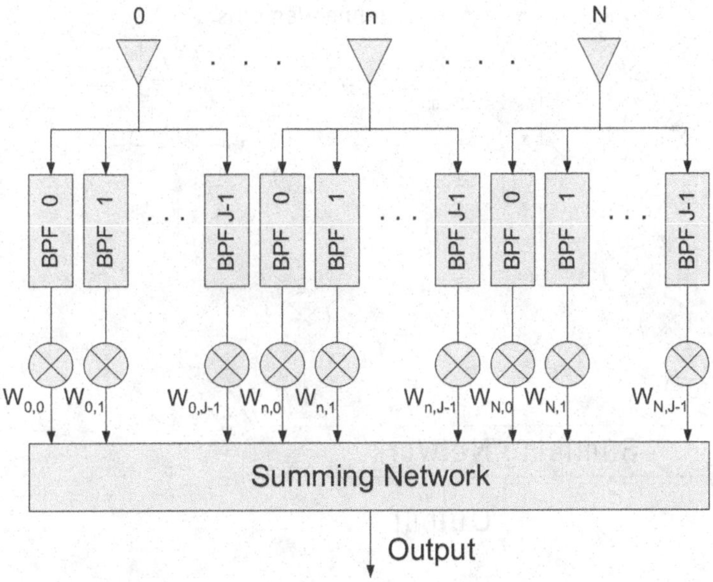


Because the narrowband beamforming algorithm works well for signals having only a few percent FB, the filters are required to have small pass bands. For a wideband signal, this requirement translates into a large number of narrowband filters that need to be included in each element of the array. These filters have to have sharp cut-off frequencies to avoid signal overlapping between the adjacent sub-bands.

## C. Fully Spatial Beamformer

A fully spatial beamformer is the third category of a beamforming technique for a wideband smart antenna. This beamforming technique was proposed in (Ghavami, 2002; Uthansakul, 2003; Uthansakul 2004). The configuration of this type of beamformer is illustrated in Figure 5.

In contrast to the two earlier introduced wideband processing schemes, this beamformer does not use delay networks or frequency filters. Instead, it employs a wideband two-dimensional (2–D) array antenna, in which the received or transmitted signals are multiplied by constant weighting coefficients, in a similar manner as it is done in a narrowband beamformer. As shown in Figure 5, a 2–D array antenna consists of identical wideband antenna elements arranged in a horizontal rectangular lattice with spacing equal to half-wavelength at the highest frequency of the designated band (Ghavami, 2000). The purpose of using a 2–D instead of 1–D array is explained as follows. As the signal is assumed to arrive from a direction parallel to the array's plane (the operation of this wideband beamformer for the case when the signal comes from other directions but not from the direction perpendicular to the array is also possible), the antenna elements receive the signal's replicas with different phases (delays). This results in a set of signals, which can be used for processing both in frequency and angular (space) domains.

In the original algorithm described by (Ghavami, 2000; Ghavami, 2002), the required weighting coefficients are calculated with the use of a 2-D Inverse Discrete Fourier Transform (IDFT) technique that is applied to the radiation pattern in frequency and angular domains. As the IDFT technique is applied to a symmetric function, the resulting coefficients for signals at individual antenna elements are given by *real numbers*. In practice, the real-valued weights can be realized using attenuators or amplifiers, which is of considerable practical advantage over the first two categories of beamformer employing complex weighting coefficients.

The validity of this signal weighting scheme was demonstrated for array antennas having a large number of elements (Ghavami, 2002). It has been pointed out by Uthansakul (2004) that this algorithm results in an erroneous radiation pat-

*Figure 5. Wideband spatial beamformer constituted by a $N_1 \times N_2$ rectangular array*

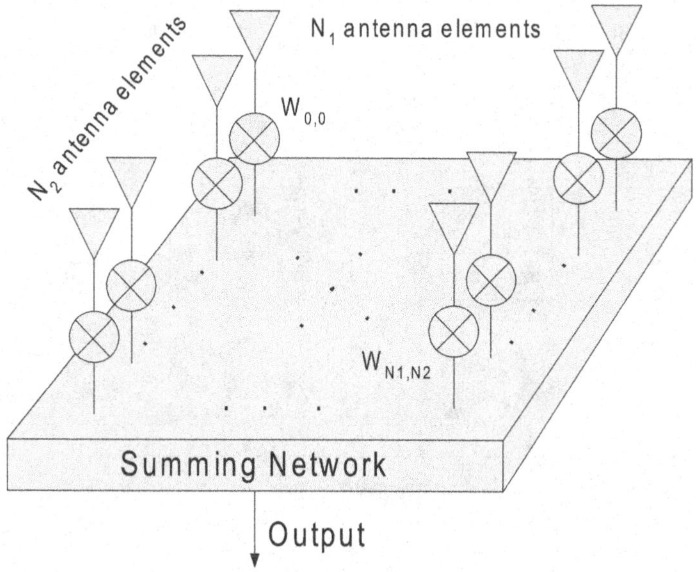


*Figure 6. Wideband spatial beamformer constituted by rectangular array followed by real valued weights and summing/dividing network*

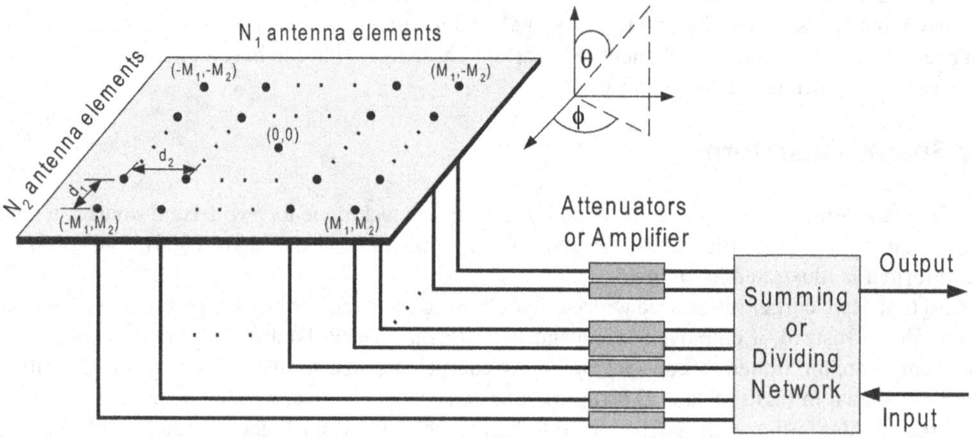


tern for small or moderate size arrays and thus it requires corrections. In addition, in the initial work (Ghavami, 2002), no attention was paid to the practical realization of wideband antenna elements forming the 2-D array. As a result, other practical aspects such as the effect of mutual coupling between array elements on the radiation pattern and thus the beamforming capabilities have not been considered.

The challenges associated with practical development of this fully spatial wideband beamformer are reported in this chapter. The presented material provides a detailed overview of research that has been carried out by the authors with respect to this fully spatial wideband beamforming system.

## PRINCIPLE OF OPERATION OF WIDEBAND SPATIAL BEAMFORMER

The general configuration of a wideband spatial beamformer is shown in Figure 6. It is formed by a 2-D array and a suitable signal processor.

The array's plane coincides with the azimuth plane (x–y plane, $\theta = 90°$) and its centre is located at the origin of the x-y coordinate system. As seen in Figure 6, the array includes $N_1 \times N_2$ antenna elements followed by attenuators or amplifiers that are responsible for producing weighting coefficients, and a summing/dividing network. The $N_1$ and $N_2$ are the numbers of antenna elements and, $d_1$ and $d_2$ represent inter-element spacing of the array in the two orthogonal directions. The array spacing is chosen as half–wavelength at the highest frequency ($f_h$) of the given band. Each antenna element is denoted by indices $m_1$ and $m_2$, where $-M_1 \le m_1 \le M_1$ and $-M_2 \le m_2 \le M_2$. The relation between $N$ and $M$ is $M_i = (N_i - 1)/2$.

### A. Beamforming Algorithm

Assuming that the signal arriving at the array comes from the ($\varphi$, $\theta$) direction (Figure 6), the phase of the signal at element ($m_1$, $m_2$) referring to the array centre is represented by

$$\psi(m_1, m_2) = \left(\frac{2\pi f}{c}\right)(d_1 m_1 \sin\phi + d_2 m_2 \cos\phi)\sin\theta \tag{1}$$

where $f$ and $c$ stand for frequency and the velocity of an electromagnetic wave in free space, respectively.

In the further analysis, it is assumed that the array elements are identical and the mutual coupling between them is neglected. The effect of mutual coupling effect on the beamforming performance is examined later in this chapter.

When signal is incident or transmitted in the plane defined by the elevation angle ($\theta$) of 90°, the array frequency-angle response can be expressed by

$$H(f,\phi) = G(f,\phi) \sum_{m_1=-M_1}^{M_1} \sum_{m_2=-M_2}^{M_2} w_{m_1,m_2} e^{j\left(\frac{2\pi f}{c}\right)(d_1 m_1 \sin\phi + d_2 m_2 \cos\phi)} \tag{2}$$

where $w_{m_1,m_2}$ are the weighting coefficients applied to signals at individual antenna elements and $G(f, \phi)$ represents the frequency-angle dependent gain of the array element.

In order to simplify (2), the two auxiliary variables ($u_1$ and $u_2$) are introduced as given by

$$u_1 = \frac{f d_1}{c} \sin\phi \tag{3a}$$

$$u_2 = \frac{f d_2}{c} \cos\phi \tag{3b}$$

By substituting (3), (2) can be rewritten as

$$H(u_1,u_2) = G(u_1,u_2) \sum_{m_1=-M_1}^{M_1} \sum_{m_2=-M_2}^{M_2} w_{m_1,m_2} e^{j2\pi u_1 m_1} e^{j2\pi u_2 m_2} \tag{4}$$

where $G(u_1,u_2)$ is obtained from $G(f,\varphi)$ using the auxiliary variables defined by (3).

The variables $u_1$ and $u_2$ are limited to the range of [–0.5, 0.5]. In turn, $H(u_1,u_2)$ can be considered as a desired radiation pattern with its main beam pointed towards a desired user over a given frequency band. Here, it is alternatively called an objective function. Full derivations of (4) can be found in Uthansakul (2006 b).

The close inspection of expression (4) indicates that $H(u_1,u_2)/G(u_1,u_2)$ represents a 2–D Discrete Fourier Transform (DFT). Therefore, the coefficients, $w_{m_1,m_2}$, can be determined by applying an IDFT to $H(u_1,u_2)/G(u_1,u_2)$. Because

$H(u_1,u_2)/G(u_1,u_2)$ can be represented as a symmetric function in the $u_1$-$u_2$ plane, an application of IDFT results in real-valued weighting coefficients.

The requirement for real-valued constant weighting coefficients can be met using either RF (analogue) or digital signal processing approaches. In the RF approach, attenuators or amplifiers can be used to realize these weighting coefficients. In the digital processing approach, the signals received or transmitted by individual antenna elements need to be multiplied by constant real numbers. It is apparent that the RF approach is very simple to realize, especially when attenuators are used as constant weights. However, one has to note that these weights (attenuators) have to be controlled using electronic means. The digital approach to creating weights usually requires an extra step which involves either a down-converting a signal from RF to baseband or up-converting from baseband to RF. To accomplish these functions, mixers and local oscillators are needed. The choice between RF and digital approaches depends on the availability of the particular technology. In our case, we used attenuators, as they were easy to produce using the available resources.

## Frequency-Angle Dependent Gain

As observed in (4), determining the coefficients, $w_{m_1,m_2}$, requires both the knowledge of the objective function $H(u_1,u_2)$ and the frequency-angle dependent gain of the antenna element, $G(u_1,u_2)$. With respect to the latter, $G(f, \phi)$, has to be firstly known in the angle-frequency domain. Because the beamforming algorithm neglects the effects of mutual coupling, $G(f, \phi)$ represents the frequency–angle dependent gain of an antenna element in isolation. For many antennas, the radiation pattern changes very little with frequency. This is opposite to the input impedance which usually is much more sensitive function of frequency. In such a case, $G(f, \phi)$ can be represented as a product of two functions as given by (5)

$$G(f,\phi) = G(f)\cdot G(\phi) \tag{5}$$

where $G(f)$ is the frequency dependent gain and $G(\varphi)$ is the angle dependent gain of the antenna. Note that the above expression does not include dependence on the elevation angle. Here, the elevation angle is fixed to 90°.

The radiation pattern representation of an array element, as given by (5), simplifies the task of programming of computer algorithms for predicting the performance of the introduced spatial beamformer. Also it is useful to simplify the measurement procedure of a real wideband antenna element. This procedure simplifies to the following steps.

1. Measure gain of a single antenna at boresight direction ($\varphi = 0°$) with respect to the frequency within the band of interest, thus yielding $G(f)$.
2. Measure gain of the same antenna by rotating it over the azimuth angles, f , for example from -90° to +90° when the operational frequency is given at the center frequency of the band, thus yielding $G(\varphi)$.

Note that $G(f)$ and $G(\varphi)$ obtained from simulations or measurements can be approximated by polynomial functions. An example of experimentally obtained frequency-dependent gain of a 40-mm square planar monopole located 3-mm above the 200–mm×200–mm ground plane that is represented by (5) and polynomial approximation is shown in (6)

$$\begin{aligned} G(f,\phi) &= G(f)\cdot G(\phi) \\ &= (28.4f^4 - 251f^3 + 829f^2 - 1211f + 659.8) \\ &\quad \cdot (5.1\times10^{-16}\phi^8 - 1.5\times10^{-16}\phi^7 - 1.1\times10^{-11}\phi^6 + 4.1\times10^{-13}\phi^5 \\ &\quad + 9.5\times10^{-8}\phi^4 + 10^{-8}\phi^3 - 4\times10^{-4}\phi^2 - 3.6\times10^{-5}\phi + 1) \end{aligned} \tag{6}$$

where frequency $f$ is given in GHz and angle $\varphi$ is given in degrees. This representation is valid for frequencies from 1.9 to 2.5 GHz and the azimuth angle from -90° to +90°. The choice of a planar monopole is dictated by the fact that it is able to operate over a wide frequency band (Ammann, 1999) and at the same time it is able to fulfill the requirement of tight spacing in the 2-D array, as required for the proper operation of wideband beamformer.

*Figure 7. Plot of objective function for 41×41–element beamformer designed for frequencies from 1.9 to 2.5 GHz and $d_1 = d_2 = 5$ cm, when the desired angle is 10° off boresight direction*

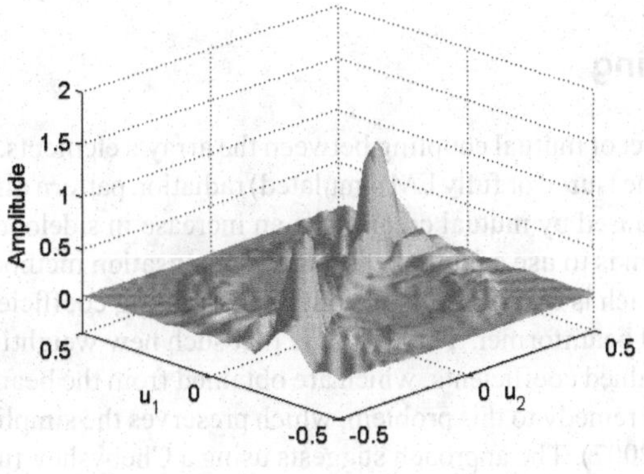


## C. Objective Function

As observed from (4), determining the weighting coefficients requires specifying the function $H(u_1,u_2)$ that represents a radiation pattern in the $u_1$-$u_2$ coordinate system. This objective function may carry information about the direction to which the main beam needs to be pointed. Also included can be the information about the side lobes level or specific directions at which nulls are to be pointed. This information is required to realize the weighting coefficients of the beamformer.

One example of $H(u_1,u_2)$ is shown in (7).

$$H(u_1,u_2)=\frac{\sin\left[\alpha\pi\left(\frac{u_1}{u_2}-\tan\phi_0\right)\right]}{\alpha\pi\left(\frac{u_1}{u_2}-\tan\phi_0\right)}\times\begin{cases}1 & \textit{desired region}\\ 1/\sqrt{10} & \textit{otherwise}\end{cases} \tag{7}$$

In this case, the objective function is given in the form of the Sinc function to obtain the maximum gain in DOA of the desired signal ($\phi_0$). The parameter α is related to the main beam first-null width. The method of calculating α for an arbitrary case can be found in Ghavami (2002).

Figure 7 shows the example of this objective function for the case of 41×41-beamformer aiming to maximize the gain at 10° off boresight direction for a signal spanning over the frequency band from 1.9 to 2.5 GHz.

## D. Determination of Weighting Coefficients

Having defined both $H(u_1,u_2)$ and $G(u_1,u_2)$, the next step concerns determining the weighting coefficients $w_{m_1,m_2}$. This task can be accomplished by the straight application of 2-D IDFT in the $u_1$-$u_2$ domain to (4). However, as shown in Uthansakul (2006 b), this procedure is invalid for a small size array. The remedy to this problem is an oversampling of $H(u_1,u_2)/G(u_1,u_2)$ while performing IDFT. Therefore the corrected algorithm for determining $w_{m_1,m_2}$ is given in (8):

$$w_{m_1'm_2'}=\left(1/N_{u_1}N_{u_2}\right)\sum_{u_1=-0.5}^{u_1=0.5}\sum_{u_2=-0.5}^{u_2=0.5}H(u_1,u_2)/G(u_1,u_2)e^{-j2\pi m_1' u_1}e^{-j2\pi m_2' u_2} \tag{8}$$

where $N_{u_1} > N_1$ and $N_{u_2} > N_{2.}$ Through detailed theoretical studies, it has been found that assuming $N_{u_1} = 2N_1$ and $N_{u_2} = 2N_2$ results in values for $w_{m_1,m_2}$, which properly represent the desired radiation pattern.

## E. Effect of Mutual Coupling

Formulas (4) and (8) neglect the effect of mutual coupling between the array's elements. When the weighting coefficients are calculated using (8), the actual (measured or fully EM simulated) radiation pattern differs from the one calculated from (2). One of undesired phenomena caused by mutual coupling is an increase in sidelobe levels in the radiation pattern.

A general remedy to this problem is to use a mutual coupling compensation method (e.g., Darwood, 1998; Steyskal, 1990). One problem with this approach is that it results in complex weighting coefficients, which work against the simplicity of the presented fully spatial beamformer. The reason is that such new weighting coefficients are more difficult to realize in practice than the real-valued coefficients, which are obtained from the beamforming algorithm that neglects mutual coupling effects. A practical remedy to this problem, which preserves the simplicity of the original beamforming scheme, was given in Uthansakul (2007). The approach suggests using a Chebyshev function instead of a Sinc function to represent the desired radiation pattern. The reason is that the Chebyshev function is capable of offering lower sidelobe levels. Therefore, when the mutual coupling is present (taken into account), the actual side lobe levels are lower that for the case when the Sinc functions is used as an objective function. This strategy preserves the real-valued weights.

## F. DOA Estimation of Wideband Signal Employing Fully Spatial Signal Processing

A smart antenna system, narrow or wideband, requires the information of DOA of the desired signal and the interfering signals. This information forms the basis for generating a radiation pattern, as given by the objective function $H(u_1,u_2)$.

Many existing DOA estimation methods (Wang, 1995; Xiao, 2004; Hong, 2003) are not suitable to be used for the spatial wideband beamformer. The reason is that they rely on the use of TDLs, which the presented wideband beamformer avoids. However, the DOA estimation method described by Friedlander (1993) looks attractive for the current application. This is because it concerns the direction finding for multiple wideband signals and does not rely on the use of TDLs. It applies an interpolation technique to generate a set of virtual arrays, each for a different frequency sub-band, having the same array manifold. The covariance matrices of these arrays are added up to produce a composite covariance

*Figure 8. Non-normalized radiation pattern of the 25×30–beamformer plotted for frequencies from 1.9 to 2.5 GHz when DOA of desired signal is -30° off boresight direction*

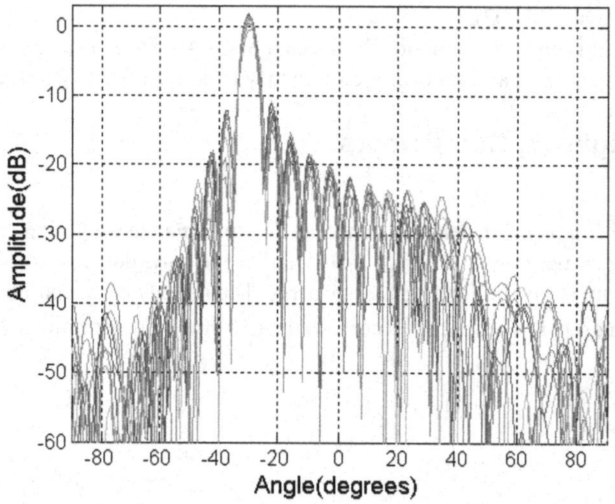

matrix. Then, DOA estimation is accomplished by eigen–decomposition of this composite covariance matrix using the conventional Multiple Signal Classification (MUSIC) algorithm. This DOA estimation technique is adapted for the fully spatial wideband beamformer, which is considered here. Full details of the extension of DOA algorithm from the 1–D to 2–D array case can be found in Uthansakul (2006 a).

## SIMULATED PERFORMANCES OF WIDEBAND SPATIAL BEAMFORMER

The computer simulated performances of wideband spatial beamformers employing 25×30 and 4×4 antenna elements are presented in Figures 8 and 9, respectively.

The presented results are not normalized. This means that the plotted radiation patterns include the array gain's variation with frequency, as observed by the varying peak values in the main beam. The frequency is varied from 1.9 to 2.5 GHz. A 40-mm square planar monopole located 3-mm above the ground plane is assumed to be a wideband antenna element. In the array, the elements are equally spaced by 5-cm which is half-wavelength at 3 GHz. The reason of assuming 3-GHz instead of 2.5-GHz as the upper frequency to calculate the array spacing is to avoid the edge effect when implementing IDFT technique in (7). The weighting coefficients used for simulations are calculated using (4), (6), (7) and (8).

The simulation results shown in Figures 8 and 9 indicate that the beamformer operates properly. Moreover, the beamforming algorithm, as given by (8) works well for an arbitrary (large and small) size array. The two arrays are able to direct their main beam in the desired direction (-30° for Figure 8 and -40° for Figure 9) over the given band, from 1.9 to 2.5 GHz. Also, nulls' locations and sidelobe levels do not vary considerably, especially within the angular range near the array's boresight in the vicinity of the main beam.

## PROTOTYPE OF WIDEBAND SPATIAL BEAMFORMER

The validity of the presented concept of wideband spatial beamformer is carried out by practical developments and experimental testing. The design is limited to the 4×4 element array case. The design frequency band is assumed to be 1.9 to 2.5 GHz, as used in earlier computer simulations.

*Figure 9. Non-normalized radiation pattern of the 4×4-beamformer plotted for frequencies from 1.9 to 2.5 GHz using modified beamforming algorithm when DOA is -40° off boresight direction*

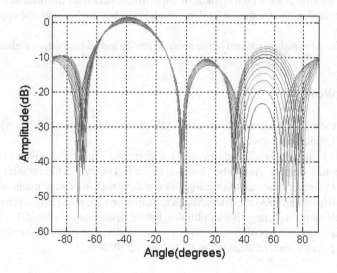

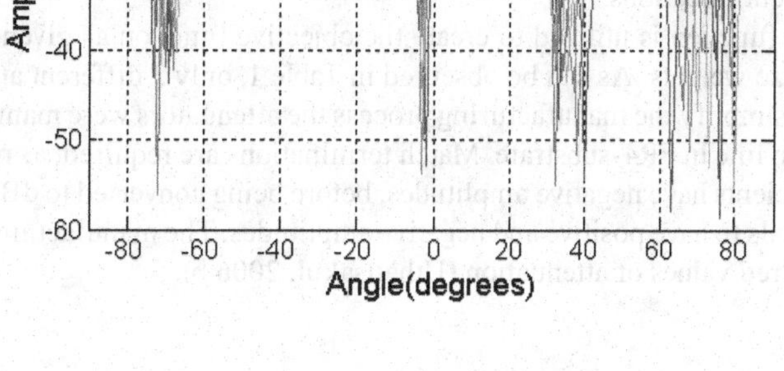

*Figure 10. Measured and simulated return loss for a 40–mm square planar monopole mounted 3–mm above a 200–mm square planar ground plane*

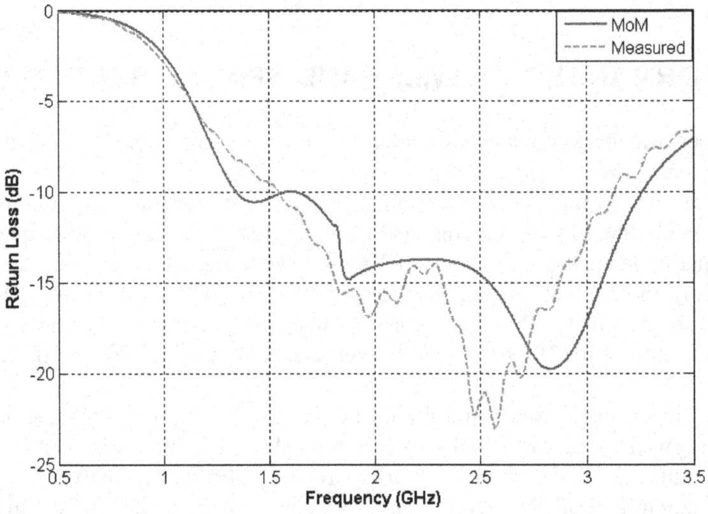


## A. Array Antenna

A square planar monopole of 40–mm×40–mm size, the same as used in the computer simulations, is chosen to be a radiating/receiving element of the array. It offers wideband performance and is easy to manufacture. Also it meets the requirement of tight array spacing which is defined as half-wavelength at the highest frequency of the operational band.

Figure 10 shows the measured and simulated return loss (dB) of the antenna. The result indicates that the antenna provides approximately 66.7% bandwidth for a return loss better than 10 dB. Its measured radiation patterns (Uthansakul, 2005) are nearly omnidirectional in H-plane. The behavior in the E-plane is typical for a monopole antenna. These patterns change only a little with frequency over the designated band of 1.9 - 2.5 GHz.

Having completed the design of a single antenna element, the next step concerns the design and construction of an array comprising of 4×4 elements of square planar monopoles. Figure 11 shows the photograph of the manufactured array. The antenna elements are made of 1.5-mm-thick brass sheet sized by 40-mm×40-mm. They are located 3-mm above the 200-mm×200-mm ground plane which is made of aluminium sheet with the thickness of 3-mm. The antenna elements are spaced by half-wavelength (50 mm) at 3 GHz, which is slightly higher than the upper frequency of the assumed band of 1.9 - 2.5 GHz.

Figure 12 shows the measured radiation pattern of the array when the individual antenna elements are fed with equal amplitude and phase.

## B. Beamforming Network

Table 1 shows an example of calculated weighting coefficients (dB for magnitude) for 4×4-beamformer with the purpose of steering main beam to different directions.

Note that for this case Sinc function is utilized to create the objective function as given in (7). For this prototype, attenuators are selected to realize weights. As can be observed in Table 1, only 3 different attenuation values (3, 5, and 10 dB) are required. In order to simplify the manufacturing process the attenuators were manufactured using sections of a stepped-impedance microstrip line in FR4-substrate. Match terminations are required, as represented by * in Table 1. Because some weighting coefficients have negative amplitudes, before being converted to dB values, the rat-race hybrid is employed to separate the signals to have positive and negative amplitudes. The manufactured components have shown small deviations from the required values of attenuation (Uthansakul, 2006 b).

*Figure 11. A photograph of 4×4-array employing square planar monopoles and SMA connectors at the back of ground plane*

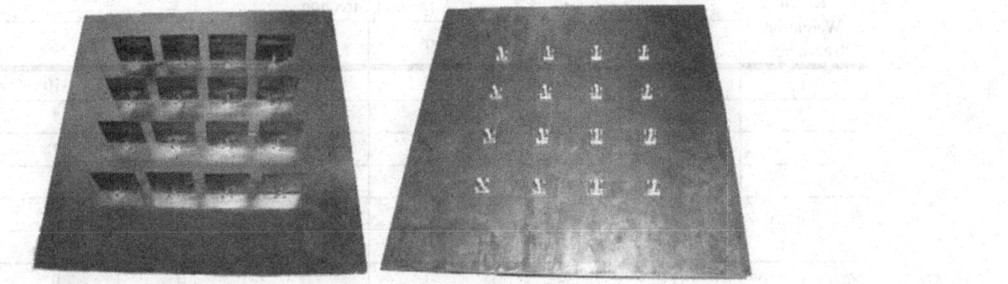

*Figure 12. Measured radiation pattern of 4x4–array employing square planar monopole plotted from 1.9 to 2.5 GHz, when the antenna elements are fed in-phase*

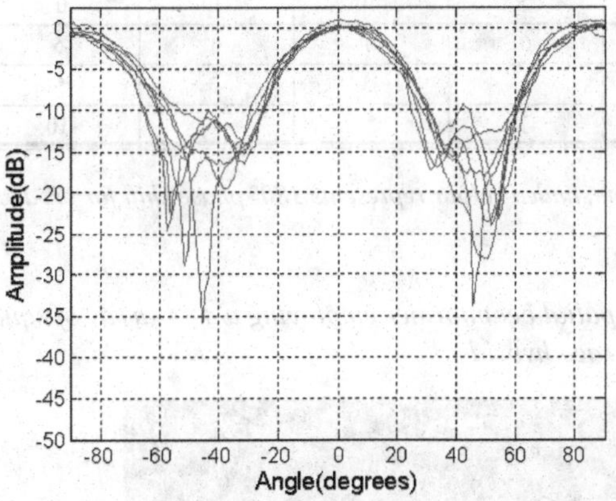


## C. Prototype Assembling

A photograph of the manufactured prototype is shown in Figure 13. The array antenna is supported by 4 pieces of 60-cm-high plastic poles above the base. At one side of the base, there are a number of plastic slots utilized to hold the microstrip attenuators. The rat-race hybrid and two 1:8-power splitters are mounted to the wooden base. The components are connected using a coaxial cable. Two 19–cm–long cables are employed to connect outputs of the hybrid to create ± amplitude of the signals fed to the splitters to form 16 signal channels. The 16 signals are fed to the microstrip attenuators accordingly using 16 coaxial cables with length of 14-cm. The attenuators outputs are fed to the array elements using 16 coaxial cables with length of 80-cm.

## D. Experimental Results

The assembled prototype is tested in an anechoic chamber. In the tests, the beamformer operates in the transmitting mode. A horn antenna located at a distance of 4–m from the beamformer is used to receive the transmitted signals. Non–normalized radiation patterns of the beamformer are used to observe performance at different frequencies.

The testing commences assuming the beamformer uses Sinc function to create the objective function, $H(u_1, u_2)$. Because the chosen 4×4–array has a two-fold symmetry, $N_1 = N_2$ and $d_1 = d_2$, the performance of the beamformer needs to be investigated only within the ±45° angular range. Outside this range, the beam formation can be judged with respect to the other side of the array. For example, the angle of 60° with respect to one side of the array can be viewed as -30°

*Table 1. Approximate weighting coefficients (dB) of 4×4–beamformer for various beam directions.*

| No. of Weighting Coefficient | Desired Direction | | | | | |
|---|---|---|---|---|---|---|
| | 0° | 10° | 20° | 30° | 40° | 45° |
| 1 | 0 | -3 | * | * | -10 | -10 |
| 2 | 0 | 0 | -3 | -3 | * | * |
| 3 | 0 | 0 | 0 | 0 | -5 | * |
| 4 | 0 | -5 | -10 | -10 | 0 | 0 |
| 5 | -5 | * | -5 | 0 | -10 | * |
| 6 | -3 | -10 | * | * | * | * |
| 7 | -3 | -3 | -3 | 0 | 0 | 0 |
| 8 | -5 | 0 | 0 | -3 | * | * |
| 9 | -5 | 0 | 0 | -3 | * | * |
| 10 | -3 | -3 | -3 | 0 | 0 | 0 |
| 11 | -3 | -10 | * | * | * | * |
| 12 | -5 | * | -5 | 0 | -10 | * |
| 13 | 0 | -5 | -10 | -10 | 0 | 0 |
| 14 | 0 | 0 | 0 | 0 | -5 | * |
| 15 | 0 | 0 | -3 | -3 | * | * |
| 16 | 0 | -3 | * | * | -10 | -10 |

*Note: * represents the matching termination; shaded colour represents 180° phase shift for producing negative amplitude*

*Figure 13. Photograph of a wideband spatial beamformer employing a 4×4-array of square planar monopoles, two 1:8-dividers, a set of attenuators and a rat-race hybrid*

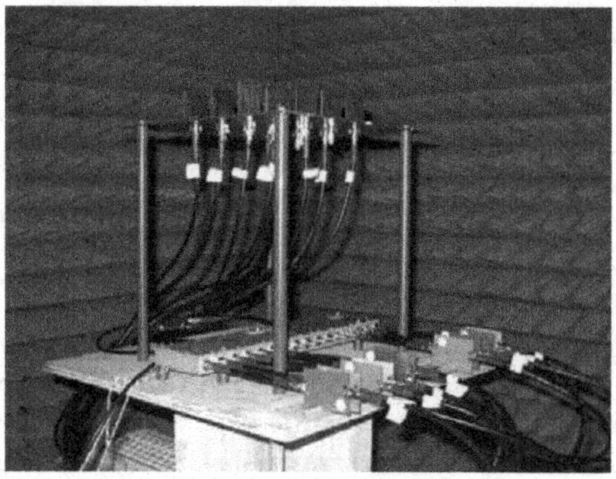

with respect to its other side. The DOA of desired signal is assumed to be known and given as -45°, -40°, -30°, -20°, -10°, 10°, 20°, 30°, 40° and 45°. The discrete frequencies chosen for measurements are 1.9, 2.0, 2.1, 2.2, 2.3, 2.4 and 2.5 GHz. The measured results for the radiation pattern are shown in Figure 14.

The results presented in Figure 14 indicate that the beamformer prototype is able to steering its main beam to the desired direction throughout the given frequency band. The observed differences concern sidelobe levels and nulls' locations. These are due to the non-ideal operation of the components (the 180 degree hybrid and the attenuators) in the beamforming network. This is confirmed by comparing the radiation pattern obtained for positive and negative angular directions off boresight directions, which in an ideal case would be symmetric.

The obtained results are good enough to state that the proposed wideband spatial beamformer functions properly and the beam formation can be accomplished without the use of TDLs, frequency filters, or phase shifters.

*Figure 14. Measured radiation pattern of the beamformer over the frequencies from 1.9 to 2.5 GHz when pointing the main beam from 10°(a), -10°(b), 20°(c), -20°(d), 30°(e), -30°(f), 40°(g), -40°(h), 45°(i) and -45°(j). The objective function was created using Sinc function.*

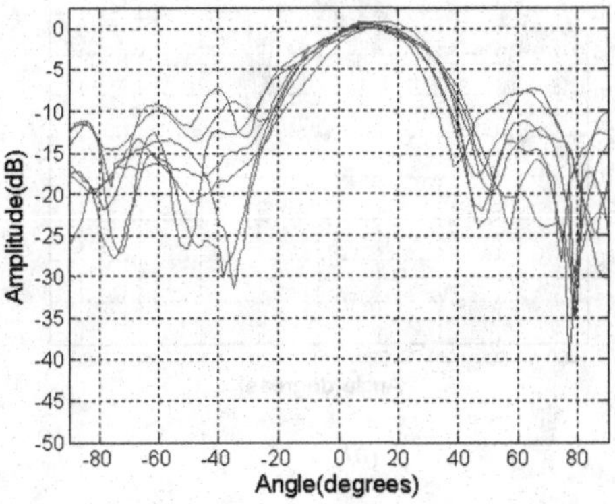


*(a)*

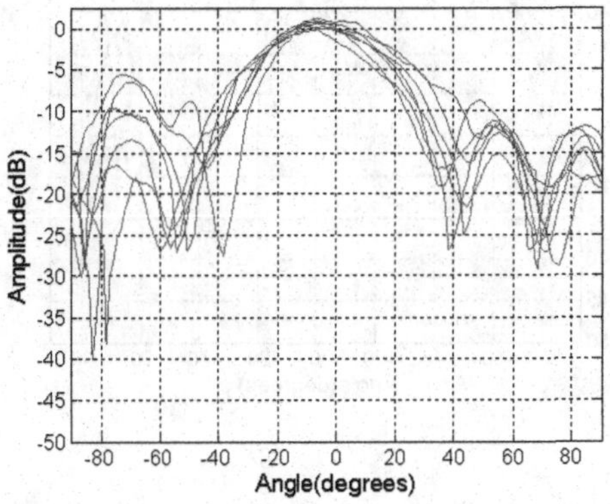


*(b)*

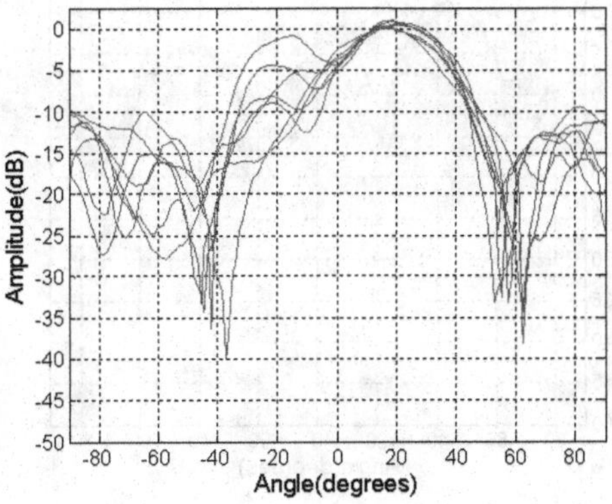


*(c)*

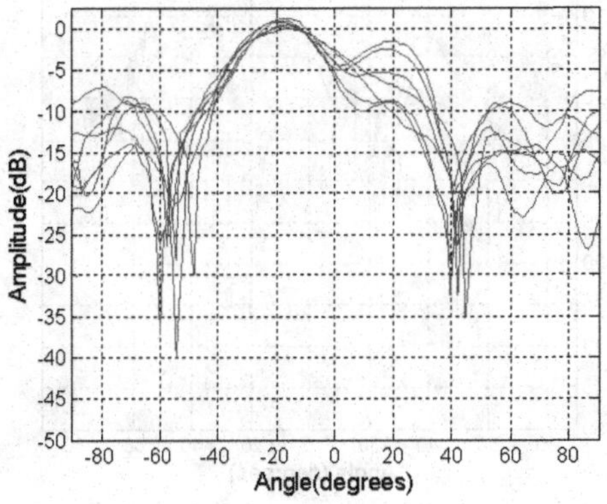


*(d)*

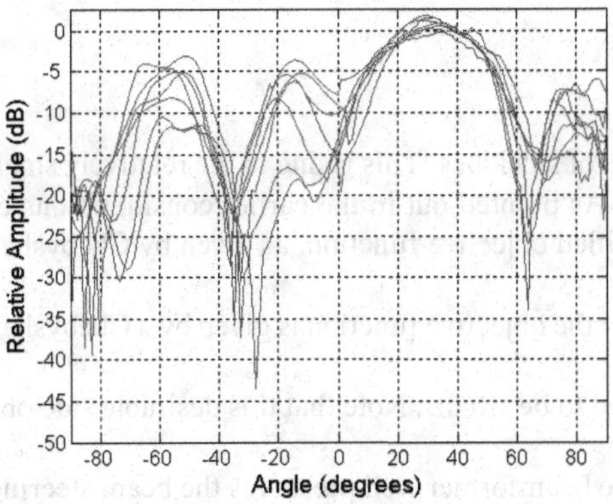


*(e)*

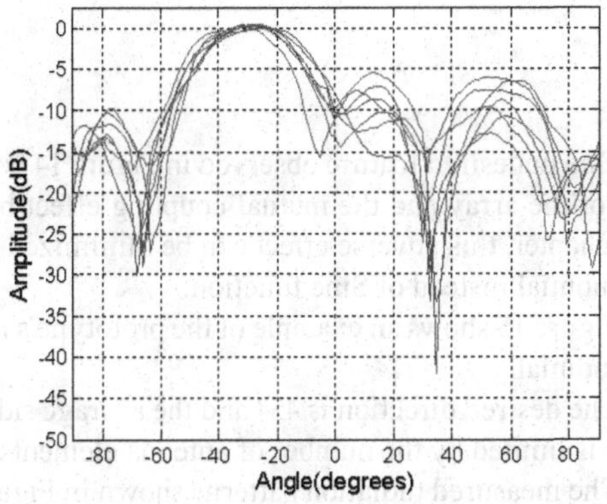


*(f)*

*continued on following page*

*Figure 14. continued*

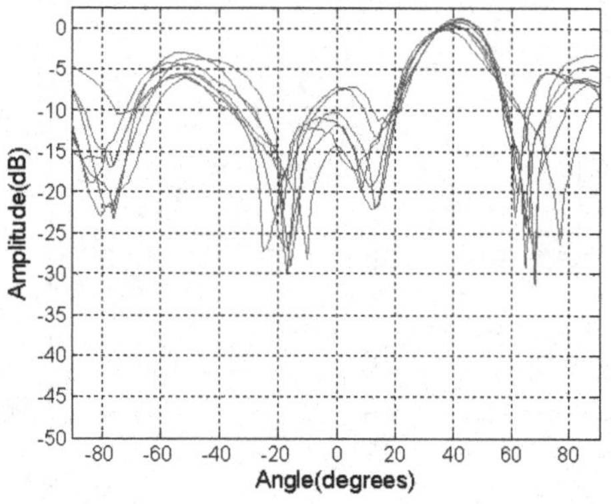


*(g)*

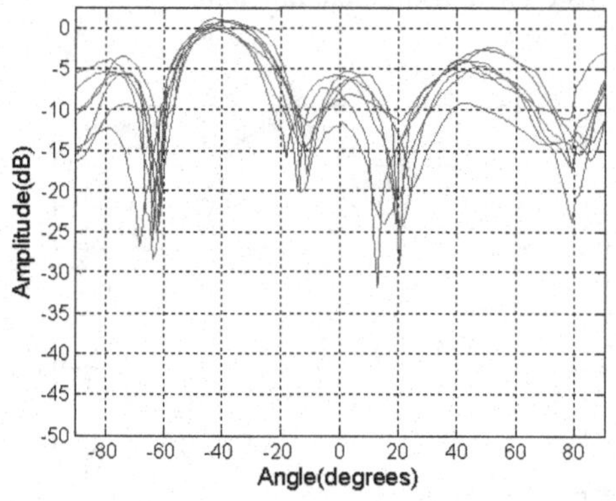


*(h)*

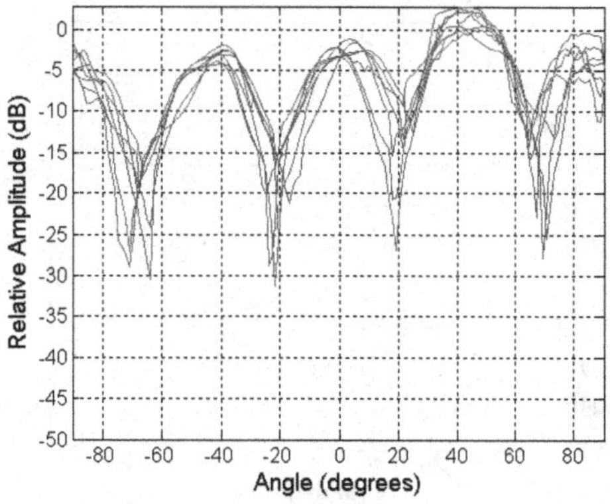


*(i)*

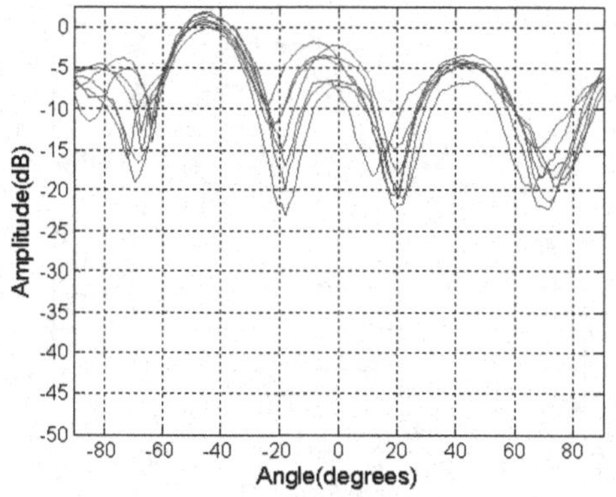


*(j)*

One undesired feature observed in Figure 14 is a relatively high level of sidelobes. This is due to the relatively small size of the array and the mutual coupling effect between its elements. As pointed out in the earlier considerations of this chapter, this adverse effect can be minimized by employing a modified objective function, as given by Chebyshev polynomial instead of Sinc function.

Figure 15 shows an example of the prototype's radiation pattern when the objective function is given by a Chebyshev polynomial.

The desired direction is 45° and the average sidelobe level is assumed to be -10 dB. Note that this desirable sidelobe level is limited by the number of antenna elements in the 4x4 array.

The measured radiation patterns shown in Figure 15 indicate that the beamformer well preserves the beam steering capability. Also its sidelobe levels are lower compared with those observed in the radiation pattern presented in Figure 14(i). This result confirms that the effect of mutual coupling can be reduced by suitably modifying the objective function of a wideband spatial beamformer. However, the replacement of the Sinc function by the Chebushev polynomial is at the expense of a broader main beam and a larger variation of values of the weighting coefficients.

*Figure 15. Measured radiation pattern of the beamformer when pointing the main beam to +45°. The objective function was created by Chebyshev polynomial.*

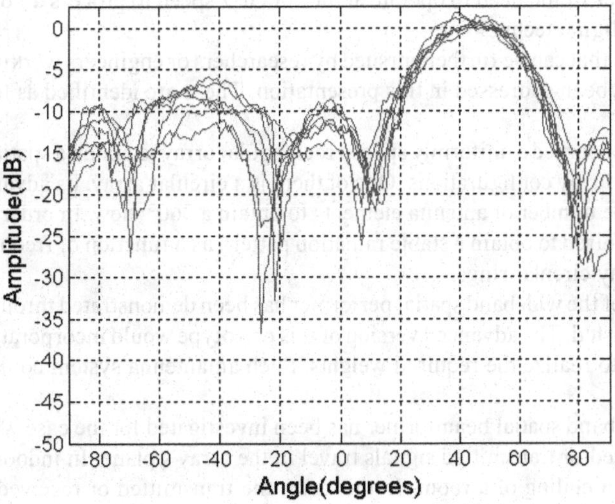


## CONCLUSION AND FUTURE TRENDS

Smart antennas are aimed for wireless communication systems to overcome problems associated with multiuser interference. There is a growing interest is in both narrow and wideband varieties of these intelligent antenna systems. Most of the works available in the literature report on narrowband smart antennas. Unfortunately, the narrowband beamforming techniques are unsuitable to apply to wideband signals.

The wideband beam formation can be accomplished by applying space-time, space-frequency or fully spatial signal beamforming techniques. The first two require a large number of TDLs or band pass filters. As a result, their practical realization becomes complicated and expensive. A fully spatial wideband beamformer is an attractive alternative because of a simple processing scheme and a low developmental cost.

This chapter has presented the principles of operation and performance of a wideband beamformer employing a fully spatial signal processing method. The beamformer employs a rectangular array of antennas and an IDFT technique for the signal processing. The required weighting coefficients are real-valued and can be realized in practice using attenuators or amplifiers, when the RF implementation of this beamformer is of concern. Initial operation of this wideband beamformer has been investigated through computer simulations. Next stages have concerned its experimental validation. A full prototype consisting of a 4x4–array of square planar monopoles followed by a feeding network for operation in the 1.9-2.5GHz band has been constructed and tested. The experimental results have shown that the beamformer is able to steer its main beam to the desired direction and the obtained radiation patterns are approximately the same throughout the assumed band. These results have fully confirmed that a wideband beamforming can be accomplished in a relatively simple manner without using filters of delay lines.

Because of the use of passive weights (attenuators) the developed beamformer can work both in receive and transmit modes. Alternatively, the constant weights can be realized with amplifiers (having flat gain characteristic). However, this choice restricts the operation of the beamformer to either transmit or receive modes because of the unilateral characteristics of the amplifiers. The extension to the two modes of operation would require duplexers or switches. A similar problem would occur with the digital implementation. Because of requirement for down or up-conversion between RF and baseband domains, the digital beamformer would work either in the transmission of reception modes.

When comparing the analogue (RF) and digital approaches, it is apparent that the digital beamforming is very simple. However, this comes at the expense of using of either down or up-conversion stages. In practice, these stages require local oscillators, amplifiers and mixers.

One can also think about another option which is the direct digitizing of the RF signals. This option is not easy to accomplish in practice because of a very high sampling rate which would be required in the case of the investigated beamformer. The considered prototype works in the GHz frequency band with signals coming from many antenna elements. As a result, the sampler would need to operate at the 100GHz speed to process all of these signals. This is not feasible at the current stage of digital technology.

There are a number of issues that can be further pursued by researchers or engineers working in the field of wideband smart antennas, which have not been addressed in this presentation. These are identified as follows.

1. The presented work has concerned a uniformly spaced rectangular array to obtain a spatial beamformer. Of interest there could be some other array configurations. One of them is a circular array. Its advantage over the rectangular array is that minimizes the number of antenna elements to obtain a 360° view. In order to obtain enough degrees of freedom, which are required to obtain a stable radiation pattern as a function of frequency, such an array would need to be formed by many circular rings.
2. The validity of operation of the wideband spatial performer has been demonstrated through the prototype, in which weights are manually adjusted. The advanced version of this prototype would incorporate electronically controlled attenuators or amplifiers to realize the required weights. Such an antenna system could be fully controlled by a microprocessor.
3. The operation of the wideband spatial beamformer has been investigated for the case when the array is in the azimuth plane and the received or transmitted signals travel in the array's plane. In indoor environments this smart antenna can be placed at a ceiling of a room. In this case, the transmitted or received signals are at the angles deviating from the array's plane. It would be worthwhile to investigate the design and operation of a wideband spatial beamformer operating in such a new environment.

## REFERENCES


Alexiou, A., & Haardt, M. (2004). Smart antenna technologies for future wireless systems: trends and challenges. *IEEE Communications Magazine, 42*, 90–97.

Ammann, M. J. (1999). Square planar monopole antenna. *IEE Conference on Antennas and Propagation.* (pp. 37-40).

Darwood, P., Fletcher, P. N., & Hilton, G. S. (1998). Mutual coupling compensation in small planar array antennas. *IEE Microwaves, Antennas and Propagation, 145(1)*, 1–6.

Friedlander, B., & Weiss, A. J. (1993). Direction finding for wide-band signal using an interpolated array. *IEEE Transactions on Signal Processing, 41(4)*, 1618-1634.

Ghavami, M. (2002). Wideband smart antenna theory using rectangular array structures. *IEEE Transactions on Signal Processing, 50(9)*, 2143-2151.

Godara, L. C., & Jahromi, M. R. S. (1999). Limitations and capabilities of frequency domain broadband constrained beamforming schemes. *IEEE Transactions on Signal Processing, 47(9)*, 2386-2395.

Hefnawi, M., & Delisle, G. Y. (1997). Adaptive array performance with focussing technique. *IEEE Antennas and Propagation Society International Symposium: Vol. 2*, (pp. 1016-1019).

Hefnawi, M., & Delisle, G. Y. (2000). Impact of wideband CDMA signals on smart antena systems. *IEEE International Conference on Personal Wireless Communications*, (pp. 5-8).

Hefnawi, M., & Delisle, G. Y. (2001). Performance analysis of wideband smart antenna systems using different frequency compensation techniques. *6th IEEE Symposium on Computers and Communications*, (pp. 237-242).

Hettak, K., & Delisle, G. Y. (2000). Smart antenna for capacity enhancement in indoor wireless communications at millimeter waves. *IEEE Conference on Vehicular Technology: vol. 3* (pp. 2152-2156).

Hong, T. D., & Russer, P. (2003). Wideband direction-of-arrival estimation and beamforming for smart antennas system. *Radio and Wireless Conference.* (pp. 127–130).

Hong, T. D., & Russer, P. (2004). Signal processing for wideband smart antenna array applications. *IEEE Microwave Magazine, 5*, 57-67.

Jiang, Y., & Bhargava, V. K. (1997). Application of smart antenna techniques in cellular mobile systems. *IEEE Pacific Conference on Communications, Computers and Signal Processing: vol. 1* (pp. 20–22).

Kang, M., Alouini, M. S. & Yang, L. (2002). Outage probability and spectrum efficiency of cellular mobile radio systems with smart antennas. *IEEE Transaction on Communications*: *50(12)*, 1871–1877.

Kawitkar, R. S., & Shevgaonkar, R. K. (2003). Design of smart antenna testbed prototype. *6th International Symposium on Antennas, Propagation, and EM Theory* (pp. 299–302).

Kawitkar, R. S., & Wakde, D. G. (2005). Smart antenna array analysis using LMS algorithm. *IEEE International Symposium on Microwaves, Antenna, Propagation and EMC Technologies for Wireless Communications: Vol. 1*, (pp. 370-374).

Lee, D., & Ng, W. T. (2005). Beamforming system for 3G and 4G wireless LAN applications. *European Circuit Theory and Design: Vol. 3* (pp. III/137–III/140).

Li, H. J., & Liu, T. Y. (2003). Comparison of beamforming techniques for W-CDMA communication systems. *IEEE Transactions on Vehicular Technology, 52*, 752-760.

Liberti, J. C., & Rappaport, T. S. (1999). *Smart antenna for wireless communications: IS-95 and third generation CDMA applications*. Prentice Hall PTR.

Materum, L. Y., & Marciano, J. S. (2003). Wideband nulling capability estimate of a tapped delay line beamformer. *IEEE Topical Conference on Wireless Communication Technology*, (pp. 386-387).

Mayhan, J. T., Simmons, A. J., & Cummings, W. C. (1981). Wide-band adaptive antenna nulling using tapped delay lines. *IEEE Transactions on Antenna and Propagation, 29(6)*, 923-936.

Shao, W., Xie, J., & Xang, G. (2003). Structure and implementation of smart antennas based on software radio. *IEEE International Conference on Systems, Man and Cybernetics: vol. 2* (pp. 1938–1943).

Steyskal, H., & Herd, J. S. (1990). Mutual coupling compensation in small array antennas. *IEEE Transactions on Antennas and Propagation, 38(12)*, 1971-1975.

Uthansakul, M., & Bialkowski, M. E. (2003). A smart antenna with non-uniform components for a wideband communication systems. *Asia-Pacific Microwave Conference: Vol. 3* (pp. 1542-1545).

Uthansakul. M., & Bialkowski, M. E. (2004). Impact of wideband signals on smart antenna system. *15th International Conference on Microwaves, Radar and Wireless Communications: Vol. 2*. (pp. 501-504).

Uthansakul, M., & Bialkowski, M. E. (2005). Investigations into a wideband spatial beamformer employing a rectangular array of planar monopoles. *IEEE Antennas and Propagation Magazine, 47(5)*, 91-99.

Uthansakul, M., & Bialkowski, M. E. (2006 a). DOA estimation of a wideband signal using a 2-D array antenna with spatial processing capability. *International Conference on Wireless Broadband and Ultra Wideband Communications*, (pp. 1–6). University of Technology Sydney Press.

Uthansakul, M., & Bialkowski, M. E. (2006 b). Fully spatial wide-band beamforming using a rectangular array of planar monopoles. *IEEE Transactions on Antennas and Propagation, 54(2)*, 527-533.

Uthansakul, M., & Bialkowski, M. E. (2007). A wideband smart antenna employing spatial signal processing. *Polish Journal of Telecommunications and Information Technology, 1*, 13-17.

Vook, F. W., & Compton, R. T. (1992). Bandwidth performance of linear adaptive arrays with tapped delay-line processing. *IEEE Transactions on Aerospace and Electronic System, 28(3)*, 901-908.

Wang, F., Lo, T., Litva, J., & Read, W. (1995). Performance of DF techniques with a VHF antenna array. *IEEE Transaction on Aerospace and Electronic Systems, 31*, 685–694.

Xiao, Y., Ma. L., & Khorasani, K., (2004). A novel wideband DOA estimation technique based on harmonic source model for a uniform linear array. *International Symposium on Circuits and Systems: Vol. 3* (pp. 23–26).

# Chapter XXV
# Omni-, Sector, and Adaptive Modes of Compact Array Antenna

**Jun Cheng**
*Doshisha University, Japan*

**Eddy Taillefer**
*Doshisha University, Japan*

**Takashi Ohira**
*Toyohashi University of Technology, Japan*

## ABSTRACT

*Three working modes, omni-, sector and adaptive modes, for a compact array antenna are introduced. The compact array antenna is an electronically steerable parasitic array radiator (Espar) antenna, which has only a single-port output, and carries out signal combination in space by electromagnetic mutual coupling among array elements. These features of the antenna significantly reduce its cost, size, complexity, and power consumption, and make it applicable to mobile user terminals. Signal processing algorithms are developed for the antenna. An omnipattern is given by an equal-voltage single-source power maximization algorithm. Six sector patterns are formed by a single-source power maximization algorithm. Adaptive patterns are obtained by a trained adaptive control algorithm and a blind adaptive control algorithm, respectively. The experiments verified the omnipattern, these six sector patterns and the adaptive patterns. It is hope that understanding of the antenna's working modes will help researcher for a better design and control of array antennas for mobile user terminals.*

## INTRODUCTION

A wireless ad hoc network is a collection of wireless mobile terminals that dynamically form a temporary network without the use of any existing network infrastructure. Such a network has distinctive features, such as being infrastructure-free, growing or reduction in size, fragment or dismantlement as desired. In the network, each terminal is conventionally equipped with omnidirectional antennas. However, if sector antennas and adaptive antennas were employed, the network could provide several additional features such as high scalability, high resource efficiency, and free join/disjoint to a community network.

Recently, a compact array antenna, e.g., the electronically steerable parasitic array radiator (Espar) antenna (Cheng, 2001; Ohira, 2004), has shown the potential for application to mobile user terminals, since the antenna only uses a single active radio receiver, which significantly reduces the antenna's cost, size, complexity, and power consumption. For the Espar antenna, only a single radiator is connected to the receiver. This active radiator is surrounded by parasitic radiators loaded with variable reactors. The radiation directivity of the antenna is controlled by changing the reactance values. In the Espar antenna, the signal combination is carried out in space by electromagnetic mutual coupling among array elements, not in circuits. This permits compact implementation of the antenna.

The concept of the Espar antenna dates back to the early work to Harrington's model (Harrington, 1978), where the "electric length" of the element is adjusted by changing the element's loaded reactance, thus causing a change in the radiation pattern. Dinger (1984, 1986) demonstrated a reactive approach that uses planar parasitic elements for anti-jamming. Another single-port approach related to the Espar antenna is the switched parasitic antenna (Preston, 1998; 1999; Scott, 1999; Svantesson, 2001; Thiel, 2001; Vaughan, 1999).

For the Espar antenna, however, the signal on the surrounding parasitic elements cannot be observed. Only the single-port output can be observed. This differs from the conventional array antenna, where the received signal on each element is observed. This characteristic prevents the conventional algorithms for array antennas from being applicable to the Espar antenna.

In this chapter, we describe the design of the Espar antenna and its three working modes, omni-, sector and adaptive modes. We give a) an omnipattern forming algorithm, b) a sector pattern forming algorithm, and c) a trained adaptive control algorithm and a blind adaptive control algorithm. The experiments show that the Espar antenna provides omnipattern and sector patterns, which are the basic function as a smart antenna. Furthermore, the adaptive patterns given by the trained and blind adaptive control algorithms, respectively, verify that the antenna can steer its beam towards desired signal and null to interference.

In this chapter, we first introduce the basic structure and design of the Espar antenna. Then we give the omnipattern forming and the sector beamforming of the antenna. Moreover, we describe a trained and a blind adaptive control algorithm to give adaptive beamforming.

## ESPAR ANTENNA DESIGN

This section describes the structure and the design of the Espar antenna.

A 2.484GHz Espar antenna is illustrated in Figs. 1 and 2. The seven-element monopole elements are arranged on a finite circular ground structure. The centre element is the feed element. The remaining $M$ (= 6) elements are parasitic and make a 0.25λ ring around the centre element, where λ=12.07 cm is the free space wavelength corresponding to the operating frequency of 2.484 GHz. The bottom of each parasitic element is loaded with a variable reactance. A bias voltage $V_m$ on it adjusts the value of the reactance. Variable beamforming is carried out by controlling the six bias voltages (control voltages) $V_m$, ($m$=1, 2,…, $M$), and thus the values of the reactances. A skirting is used on the ground plane to reduce the main lobe elevation. The radius of the circular ground plane is 0.5λ, and the skirting height is 0.25λ (Ojiro, 2001), which provides maximum gain of the antenna's radiation in its horizontal direction.

The reactive-loaded design shown in Fig. 2 gives a variable range of reactance. A pair of varactor diodes (1SV287) is positioned in parallel to terminate the parasitic element. To prevent the RF current on each varactor from leaking back to the baseband circuits, a series resistor $R_1$ and shunt capacitor $C_1$ are inserted with an $R_1C_1$ time constant that is sufficient for the effective decoupling of DC and RF. Thanks to zero DC current consumption on the reverse-biased diodes, one can employ a high resistance, say 10 kΩ, for $R_1$. This allows one to assign low capacitance, say 3 pF, for $C_1$ so as not to degrade the feedback control response. Therefore, the high $R_1C_1$ time constant isolates the RF electromagnetic analysis from the biasing circuits, so that only the varactor diodes need be taken into account in designing each parasitic element. According to varactor diode 1SV287 specifications, the capacitance range of each diode ranges from about 0.7 pF to 9.0 pF as the bias voltage is changed from 20 V to -0.5 V. Thus the reactive-loaded design provides a range of reactance from -45.8 Ω to -3.6 Ω for the frequency 2.484GHz.

The reactance loaded in each of the parasitic elements electronically adjusts its element length and makes the monopole element appear as a director or a reflector, as in the Yagi–Uda array antenna (Thiel, 2001), depending on the negative or positive value of the reactance. The element appears as an effectively 'shorter' monopole (director) if a negative reactance is loaded, while a positive reactance provides an effectively 'longer' element (reflector). This action of the loaded reactance causes a change in the radiation pattern.

In the design of the Espar antenna, the physical length of each parasitic element should be 0.2315λ if a wavelength-shortening coefficient is considered for a 0.25λ monopole element, where the radius of each parasitic element is 0.01λ. The 'electric length' is adjustable by changing the element's loaded reactance. The varactor diode's junction capacitance

*Figure 1. A fabricated 7-element Espar antenna*

essentially offers only negative reactance as described above. Since a range of reactance that covers both positive and negative parts is desirable in practice, we incorporated a 24.7 Ω shift of the reactive load in the design. This shift allows a range of reactance [-21.1 Ω, 21.1 Ω]. Stray inductance due to chip mounting should also be considered, and thus the shift value should be less than 24.7 Ω. A 16 Ω shift is adopted in this antenna design, and this shift is not implemented by connecting an inductor in series but, more practically, by adding an additional length of 0.0116 λ to the element. The additional part plays the role of an inductor of 16 Ω (IEICE, 1989, p. 681). As a result, the physical length of each parasitic element is finally designed to 0.2431λ.

The Espar antenna was evaluated in a radio anechoic room. The six bias voltages are fed to the Espar antenna using six digital-to-analogue converters (D/As). In each of the D/A converters, the bias voltage value over the range [-0.5 V, 20 V] was encoded using 12 bits, which corresponds to the digital range [-2048, 2047]. For convenience, we represent bias voltage by its digital value through this chapter.

Our objective is to find the digital (bias) voltage vector $\mathbf{V}=[V_1, V_2, \ldots, V_M]$ for the omni, sector and adaptive patterns.

## OMNIPATTERN FORMING

In this section, we describe how to find a bias voltage vector such that the Espar antenna behaves as an omnipattern with its gain as large as possible. Theoretically, the Espar antenna provides its omniradiation pattern when all of the bias voltages are set to the same value. This is due to the symmetric configuration of the antenna. Therefore, the bias voltage vector should be set to $V_0\mathbf{1}^M$ to form an omnipattern, where $\mathbf{1}^M$ is an $M$-dimensional column vector whose $M$ components are all one. Given a single-signal source, forming an omnipattern involves finding a digital voltage value $V_0$ over [-2048, 2047] such that the received power is maximum. This is called equal-voltage single-source power maximization for forming an omnipattern. Specifically, let $P$ denote the power of the received signal, which is a function of the bias voltage vector. The voltage vector $\mathbf{V}_{\text{omni}}=[V_{\text{omni}}, V_{\text{omni}}, \ldots, V_{\text{omni}}]^T$, where the superscript $^T$ is the transpose operation, for an omnipattern is given by

$$\mathbf{V}_{\text{omni}} = \underset{-2048 \le V_0 \le 2047}{\operatorname{argmax}} \; P(V_0\mathbf{1}^M).$$

Omnipattern forming experiments were conducted in a radio anechoic chamber. A continuous wave, generated by an RF network analyser, was transmitted at an RF frequency through a standard Horn antenna. In the receive site, only the centre element of the Espar antenna was connected to an input port of the network analyser, which measured the power of the received signal. We observed that the voltage $V_{\text{omni}}$=272 gives maximum power. Freezing the bias voltage vector $\mathbf{V}_{\text{omni}}$=[272, 272, 272, 272, 272, 272]$^T$, the antenna pattern was measured by rotating the platform on which the

*Figure 2. Espar cross section and its reactance circuit*

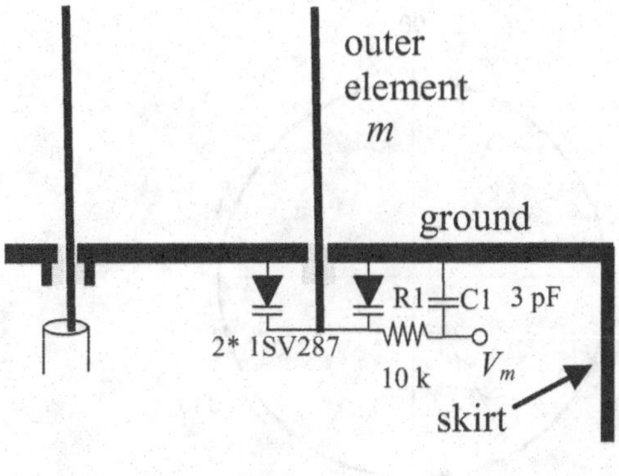


Espar antenna was mounted. It is an omnipattern with large gain (see Fig. 3). Over all angles [0°, 360°), the average gain of the pattern is -0.83 dBi and the standard deviation is 0.22 dB. The maximum and minimum gains are -0.40 dBi and -1.20dBi, respectively.

## SECTOR BEAMFORMING

In the previous section, we discussed the omnipattern of the Espar antenna. In this section, we shift the emphasis away from the omnimode to the sector-mode. We describe a single-source power maximization technique to form a sector pattern in a given direction.

### Single-Source Power Maximization

This section gives a single-source power maximization algorithm to form a sector pattern towards the direction of an impinging signal. Let V(*n*) denote the bias voltage vector impinging on the loaded reactances at time *n*. The vector **V**(*n*) is *M*-dimensional. According to the gradient-based method (Cheng, 2001), the update value of the voltage vector at time *n*+1 is computed using the simple recursive relation

$$\mathbf{V}(n+1) = \mathbf{V}(n) + \mu \nabla P(n) / P(n) \, , \quad n = 1,2,\ldots,N \, , \qquad (1)$$

where μ is a positive real-valued constant that controls the convergence speed, and $\nabla P(n)$ is a gradient vector of received signal power *P*(*n*). The power *P*(*n*) is a function of the control voltage vector and written as

$$P(n) = P(V_1(n)\, V_2(n)\, \ldots, V_M(n)) \, .$$

In (1), the gradient vector $\nabla P(n)$ is normalised by the power *P*(*n*), which eliminates the effect of the absolute power value on the convergence result 0.

Since the signals on passive elements cannot be observed, the gradient vector is estimated using finite difference approximations of derivatives (Cheng, 2001) as follows. The partial derivative $\partial P(n) / \partial V_m$ with respect to the control voltage $V_m$ is approximated to a change of the power, for *m* = 1, 2, ..., *M*,

$$\frac{\partial P(n)}{\partial V_m} = \frac{P\big(V_1(n), \ \ldots, \ V_m(n) + \Delta V_m, \ \ldots, \ V_M(n)\big) - P\big(V_1(n), \ \ldots, \ V_m(n), \ \ldots, \ V_M(n)\big)}{\Delta Vm} \qquad (2)$$

*Figure 3. Omnipattern corresponding to the bias voltage vector* **V**=[272,272,…,272]$^T$

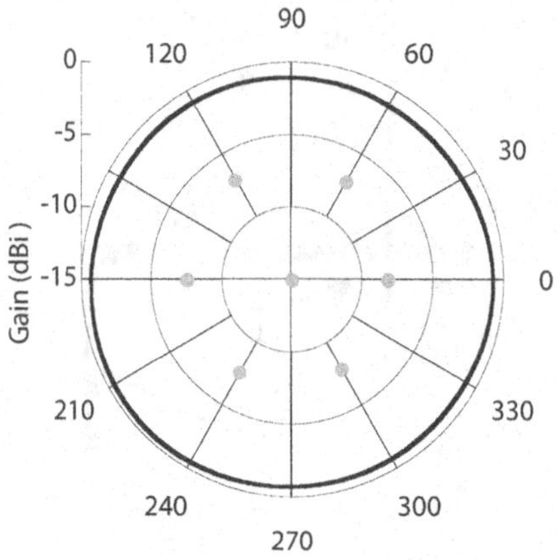


by incrementing $V_m(n)$ to $V_m(n)+\Delta V_m$, where $\Delta V_m$ is a perturbation size for bias voltage $V_m$. Here we assume that $\Delta V_1 = \Delta V_2 = \ldots = \Delta V_M \equiv \Delta V$ . Substituting this estimate of the gradient vector into (1), the control voltage vector **V**($n$) is calculated. Repeating these steps from $n$ = 1 to $N$, we obtain, for a sufficiently large $N$, a good voltage vector **V**($N$+1) in the sense that the power of the received signal is maximised. Experimental results below show that the voltage vector generates a sector pattern in the direction of the single-signal source.

## Experimental Results of Sector Beamforming

Sector beamforming experiments were conducted in a radio anechoic chamber with the same experimental setup as omnipattern forming above. We observed the sector beamforming procedure of (1) when a single source transmitted a signal in the direction 0°. The initial value of the digital voltage vector **V**(1) was set to zero, i.e. **V**(1)=[0, 0, …, 0]$^T$. After $N$=200 block iterations with $\mu = 600$ and $\Delta V$=40 (see (1) and (2)), the value of the digital voltage vector became **V**=[2047, 2047, -2048, -1506, -2048, 2047]$^T$. By freezing the voltage vector, we measured the antenna pattern and found that the main lobe of this pattern pointed in the desired source direction of 0° (see Fig. 4). As a result, this was the 0° sector pattern to be found. The gain in the beam direction 0° was 5.6 dBi, and the 3 dB beamwidth of the pattern was about 70.8°. It is interesting to observe the convergence voltage vector. The bias voltages observed on the loads of elements 1, 2, and 6 are all 2047, the maximum value in the range of digital voltage values. Those on elements 3, 4, and 5 are -2048, -1506,

*Figure 4. Sector pattern formed by single-source power maximization technique; the voltage vector is* **V**=*[2047, 2047, -2048, -1506, -2048, 2047]$^T$*

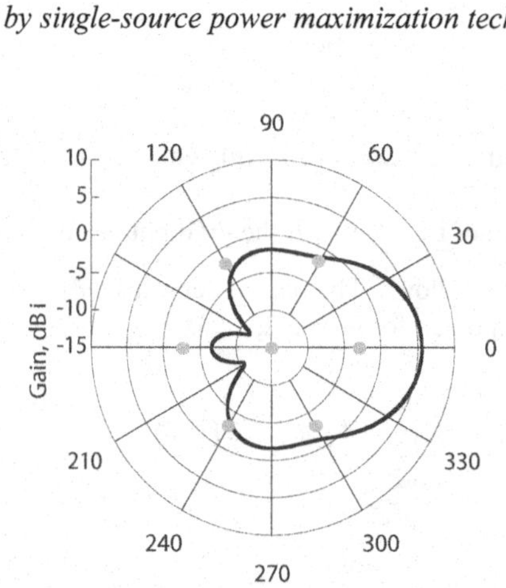

*Figure 5. Six sector patterns*

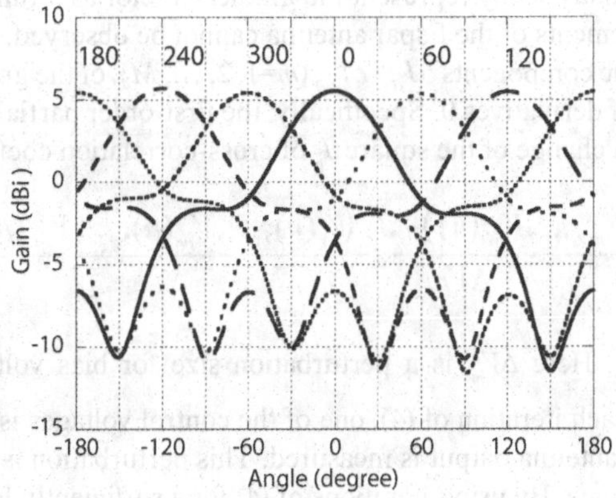


and -2048, equal to or approaching the minimum value. This implies that elements 1, 2, and 6 seem to play the role of director, while elements 3, 4, and 5 seem to play the role of reflector, as with the Yagi–Uda array antenna.

Similarly, by using the single-source power maximization technique, we obtained 6 digital voltage vectors and their corresponding sector patterns at every 60°. (see Fig. 5) The average gain at their main lobe was 5.5 dBi. It is seen that the patterns had almost the same shape, i.e. a 60° shift version of each other. This phenomenon arises from the symmetrical configuration of the antenna (see Figs. 1 and 2). The average value of the 3dB beamwidths for these patterns was 72.4°.

## ADAPTIVE BEAMFORMING

In this section, we propose two adaptive beamforming algorithms, a trained adaptive beamforming algorithm and a blind adaptive beamforming algorithm. Both of the algorithms make the Espar antenna adaptively steers its beam towards desired signal and its nulls to interference signals.

### Trained Adaptive Beamforming

#### Criterion and Trained Algorithm

The conventional criterion such as MMSE (Minimum Mean Square Error) is useless for control of Espar antenna, since the signals in the parasitic radiators can't be observed. In this section, we propose Maximum Square of Cross-correlation Coefficient (MCCC) as a trained criterion to optimize the bias voltages for adaptive beamforming of the Espar antenna. The criterion is defined by maximize the square of cross-correlation coefficient (Cheng, 2002)

$$J_\rho = \frac{|\mathbf{y}^H\mathbf{r}|^2}{\mathbf{y}^H\mathbf{y}\mathbf{r}^H\mathbf{r}}. \quad (3)$$

Here the superscript $^H$ is the transpose and conjugate operation, $\boldsymbol{r} = [r_1, r_2, \ldots, r_K]^T$ is a reference signal vector with $K$-dimension, and $\mathbf{y}(n) = [y_1(n), y_2(n), \ldots y_K(n)]^T$ is a received signal vector also with $K$-dimension.

The optimization of the bias voltages is based on the gradient-based method (Cheng, 2001). The optimization is utilized by using the recursion

$$\mathbf{V}(n+1) = \mathbf{V}(n) + \mu\nabla J_\rho\ , n = 1, 2, \ldots, N, \quad (4)$$

where $\nabla J_\rho$ is a gradient vector of $J_\rho$.

There may be some difficulty when we compute the gradient vector $\nabla J_{\rho}$. This arises from, as we have stated above, the fact that a) it may not be easy to analytically represent the gradient vector as a function of voltage vector **V**, and b) the signal vector impinging on the elements of the Espar antenna cannot be observed.

We here derive an estimation of the components $\partial J_{\rho} / \partial V_m$, ($m$=1, 2, …, $M$), of the gradient vector $\nabla J_{\rho}$ in (4) by the use of finite-difference approximations of derivatives 0. Specifically, the first-order partial derivative $\partial J_{\rho} / \partial V_m$ with respect to the voltage $V_m$ is approximated to a change of the square $J_{\rho}$ of cross-correlation coefficient as

$$\frac{\partial J_{\rho}(n)}{\partial V_m} = \frac{J_{\rho}(V_1(n),\ldots,V_m(n)+\Delta V_m,\ldots,V_M(n)) - J_{\rho}(V_1(n),\ldots,V_m(n),\ldots,V_M(n))}{\Delta Vm} \qquad (5)$$

by incrementing $V_m(n)$ to $V_m(n)+\Delta V_m$. Here $\Delta V_m$ is a perturbation size for bias voltage $V_m$, and it is assumed that $\Delta V_1 = \Delta V_2 = \ldots = \Delta V_M \equiv \Delta V$. In each iteration of (4), one of the control voltages is incremented by a small amount $\Delta V$, and the change of $J_{\rho}$ in the Espar antenna output is measured. This perturbation is done sequentially through all of the terminations of the parasitic elements. By using iterations of (4) for a sufficiently large $N$, we obtain a good voltage vector **V**($N$+1) in the sense that the square $J_{\rho}$ of cross-correlation coefficient of (3) is maximized.

As shown in (5), only one component of the gradient vector $\nabla J_{\rho}$ is calculated at a time from the output of the antenna. All the components of control voltage vector **V** are sequentially perturbed in order to get one gradient vector in a single iteration of (4). This sequential perturbation of the voltage requires $M$+1 times transmission of the reference vector **r**($n$) (with length $K$) for a single iteration. Thus the total $K(M+1)N$ symbols are required for $N$ (block) iterations. The flowchart of the proposed adaptive algorithm is illustrated in Fig. 6.

## Experimental Results of Trained Algorithm

An experimental setup is needed to verify the adaptive beamforming of the Espar antenna. The layout of the experiment, which takes place in a large radio anechoic chamber, is shown in Fig. 7. The transmission antennas and the test Espar

*Figure 6. Flowchart of the adaptive algorithm ($J = J_{\rho}$ or $J_m$)*

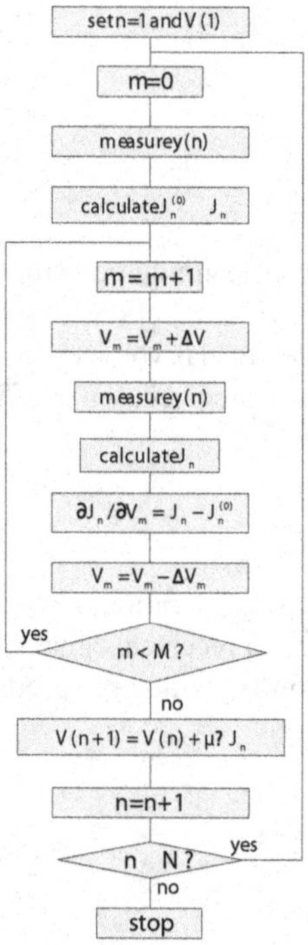

*Figure 7. Experimental setup of the trained and blind adaptive algorithm*

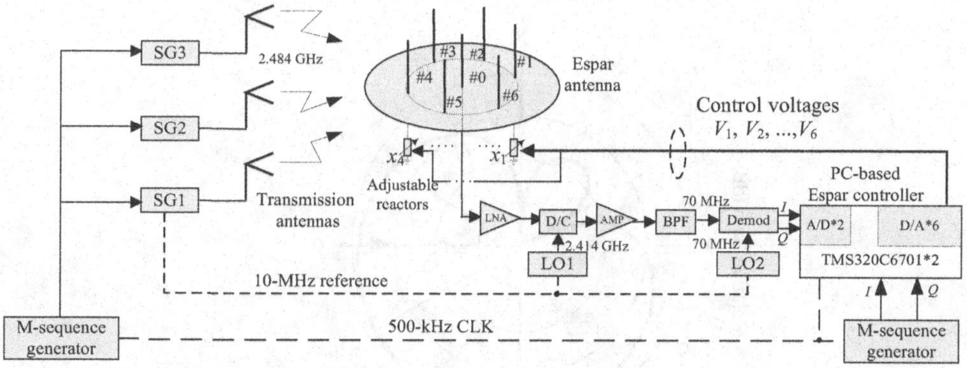


antenna are installed on the same horizontal plane. The data, which are 15-, 21- , and 23-order M-sequences, are modulated in BPSK using three signal generators (SGs), respectively. The modulated signals are transmitted at a frequency of 2.484 GHz from each transmission antenna.

In the receiver, only the center element of the Espar antenna is connected to a lower-noise amplifier (LNA) module. The 2.484-GHz RF signal is directly converted to a 70-MHz intermediate frequency (IF) by a down-converter (D/C) and local oscillators (LO) 1. The IF signal is routed into the band pass filter (BPF) and demodulated into orthogonal base-band signals in the in-phase (I) and quadrature-phase (Q) channels.

The base-band signals are then digitized individually with 12-bit resolution by analog-to-digital converters (A/Ds) at the rate of 500 kHz. A PC-based Espar controller is used to optimize the control voltages and then feed the voltages to the Espar antenna. A signal processing board is equipped in the PC-based Espar controller. The board contains two TMS320C6701 digital signal processors, and is used for all of the signal processing.

In our experiment, it is assumed that the desired signal is generated from SG1 (see Fig. 7). In the receiver, LO1 and LO2 were synchronized with the desired signal in their frequency by a 10-MHz reference signal. The M-sequences generated in send and receiver were synchronized by a 500-kHz reference signal. The power of the transmission signals was adjusted so that the signal-to-interference ratio (SIR) was 3 dB at the receiving site without declaration.

For the trained adaptive beamforming algorithm (with criterion of (3)), let us consider the case where there were two signals from the DoAs of 0° (desired signal) and 90° (interference). At the initial step, the value of each digital control voltage was set to zero. After $N = 50$ block iterations of (4), we obtained a digital control voltage vector $\mathbf{V} = [454, 2031, -49, -1994, -2048, -1913]^T$. Freezing the convergent voltage vector, we measured the radiation pattern. The measured pattern of Fig. 8 shows that the beam was steered to 0° at the desired direction, while a deep null was formed toward the interference signal at 90°. The power ratio of beam (at 0°) to null (90°) (BNR) was 38.6 dB. The result verified that the Espar antenna can adaptively form its beam to the desired signal, and its null toward the interference signal. Figures 9 and 10 display the variation of the square $J_\rho$ of cross-correlation coefficient and the control voltages during the adaptive control procedure.

In addition, the situation of three signals is considered in our experiment. Figure 11 is the measured radiation patterns after $N = 50$ block iterations with the BNRs 26.23 and 29.45 dB at the interferences 300° and 255°, respectively.

## Blind Adaptive Beamforming

### Criterion and Blind Algorithm

We propose a MoMent-based Criterion (MMC) for blind adaptive beamforming for the Espar antenna. The blind criterion is to maximize the objective function

*Figure 8. Adaptive antenna pattern (trained) after N=50 block iterations. The desired signal is from 0° and the interference is form 90°.*

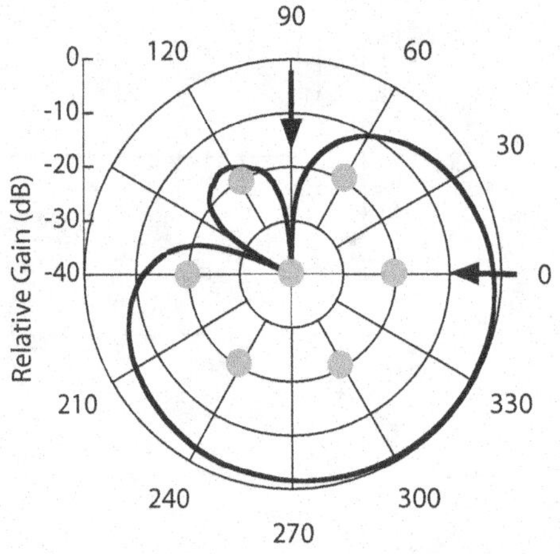


*Figure 9. Convergence curve associated with Figure 8*

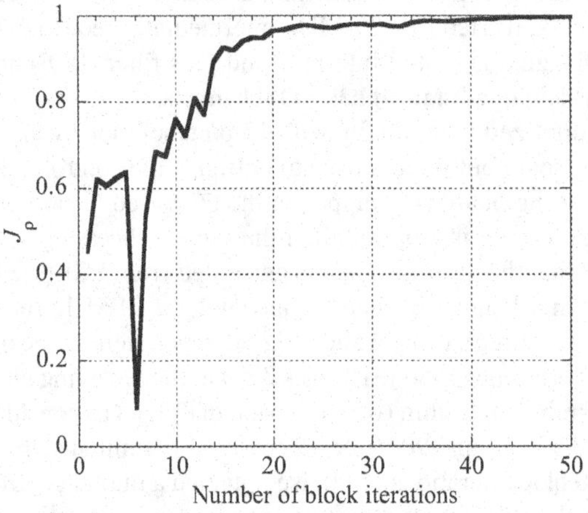


*Figure 10. Variation of bias voltages during iteration associated with Figure 8*

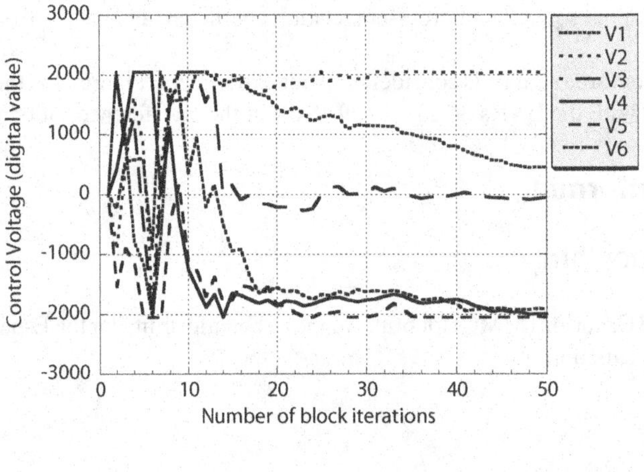

*Figure 11. Adaptive antenna pattern (trained) after N=50 block iterations; the desired signal is from 30° and the interferences are form 255° and 300°*

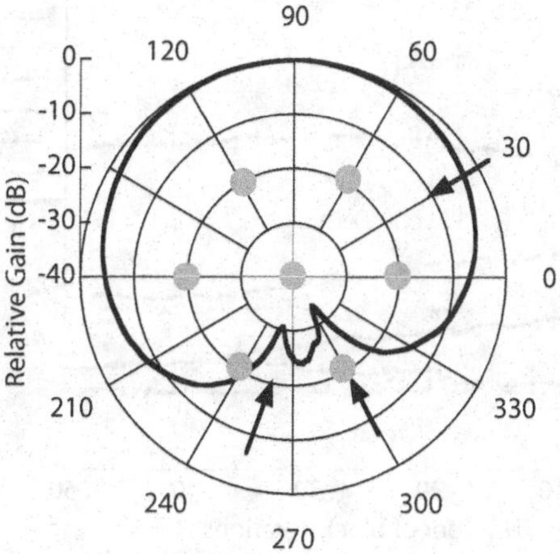


*Figure 12. Adaptive antenna pattern (blind) after N=50 block iterations; the desired signal is from 0° and the interference is from 225°*

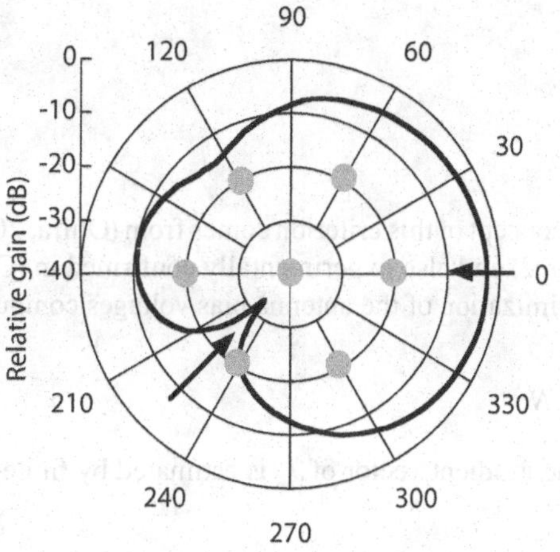


*Figure 13. Convergence curve associated with Figure 12*

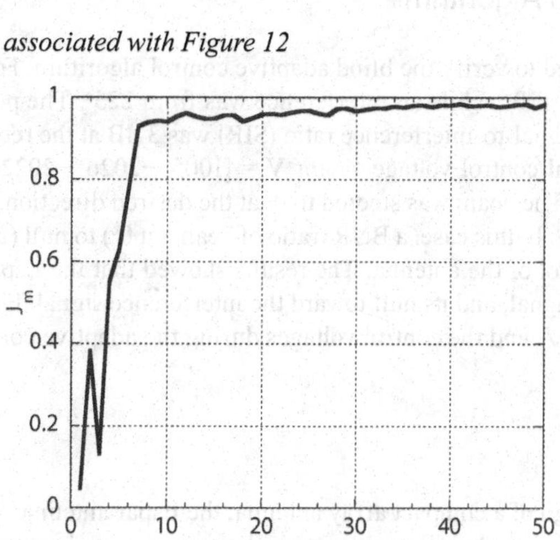

*Figure 14. Variation of bias voltage during iteration associated with Figure 12*

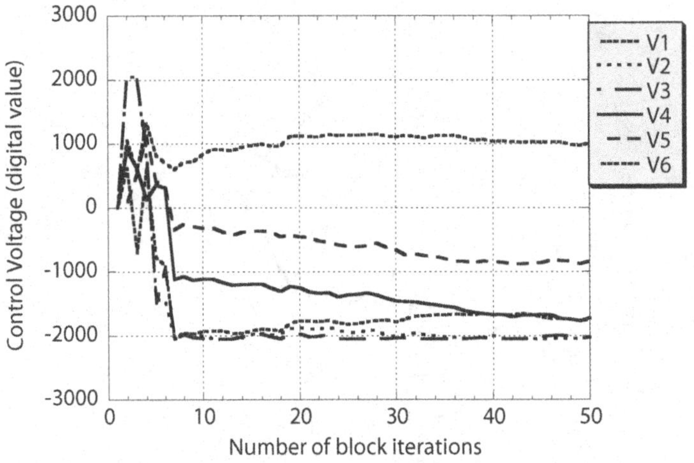


$$J_m = \frac{\left(\sum_{k=1}^{K} | y_k(n) |^2\right)^2}{K\sum_{k=1}^{K} | y_k(n) |^4} \tag{6}$$

without training sequence. The original concept of this criterion comes from (Ohira, 2004, Eq. (10)], whose mathematical proof was primarily shown in (Ohira, 2002) and also experimentally confirmed in (Cheng, 2002).

Similar to trained algorithm, the optimization of the antenna bias voltages computed by the recursive relation

$$\mathbf{V}(n+1) = \mathbf{V}(n) + \mu\nabla J_m \text{ , } n = 1, 2, ..., N \tag{7}$$

where $\nabla J_m$ is a gradient vector of $J_m$. The gradient vector of $J_m$ is estimated by finite-difference approximations of derivatives similar in (5).

## Experimental Results of Blind Algorithm

The same setup shown in Fig. 7 is used to verify the blind adaptive control algorithm. For the algorithm with criterion of (6), a signal was set from the DoA of 0°, while an interference was from 225°. The powers of the signal and the interference were adjusted so that the signal-to-interference ratio (SIR) was 3 dB at the received side. After $N = 50$ block iterations of (7), we obtained a digital control voltage vector $\mathbf{V} = [1005, -2026, -2022, -1723, -842, -1749]^T$, which gave a radiation pattern (see Fig. 12). The beam was steered to 0° at the desired direction, while a deep null was formed towards the interference signal at 225°. In this case, a BNR (ratio of beam (at 0°) to null (225°)) of 30.6 dB was obtained. This verified the blind adaptive control of the antenna. The results showed that the Espar antenna can adaptively and blindly form its beam to the desired signal, and its null toward the interference signal. Figures 13 and 14 illustrated the variation of the higher-order moment $J_m$ and the control voltages during the adaptive control procedure.

## CONCLUSION

We introduced the structure and design of a compact array antenna, the Espar antenna. We gave an omnipattern forming algorithm and a sector pattern forming algorithm, a trained adaptive control algorithm and a blind adaptive control algorithm for the antenna. The experiments in a radio anechoic chamber provided an omnipattern with a large gain, and provided six sector-patterns with maximum gains at beam directions. We also verified that the Espar antenna can adaptively form its beam to the desired signal, and its nulls toward the interference signals, by either trained beamforming or blind beamforming, respectively. The Espar antenna with those signal processing algorithms above is of great importance in a variety of applications in wireless communications.

The estimation of direction-of-arrival by the Espar antenna is also an interesting topic. Details on the estimation can be found in, e.g., (Plapous, 2004; Taillefer, 2008).

## REFERENCES

Cheng, J., Kamiya, Y., & Ohira, T. (2001). Adaptive beamforming of Espar antenna based on steepest gradient algorithm. *IEICE Trans. on Communications*, E84-B (7), 1790–1800.

Cheng, J., Hashiguchi, M., Iigusa, K., Furuhi, T., Hirata, A., & Ohira, T., (2002). Adaptive Espar antenna experiments using MCCC and MMC criteria. *Proc. IEICE General Conference* (p. 133).

Cheng, J., Iigusa, K., Hashiguchi, M., & Ohira, T. (2002). Blind aerial beamforming based on a higher-order maximum moment criterion (Part II: Experiments). *Proc. Asia-Pacific Microwave Conference*, (pp.185-188), Kyoto, Japan.

Cheng, J., Hashiguchi, M., Iigusa, K., & Ohira, T. (2003). Electronically steerable parasitic array radiator antenna for omni- and sector-pattern forming applications to wireless ad hoc networks. *IEE Proc.-Microwaves, Antennas & Propagation*, 150 (4), 203-208.

Dinger, R. J. (1984). Reactively steered adaptive array using microstrip patch elements at 4GHz. *IEEE Trans. on Antennas and Propagation*, AP-32 (8), 848–856.

Dinger, R. J. (1986). A planar version of a 4.0 GHz reactively steered adaptive array. *IEEE Trans. on Antennas and Propagation*, AP-34 (3), 427–431.

Harrington, R. F. (1978). Reactively controlled directive arrays. *IEEE Trans. on Antennas and Propagation*, AP-26 (3), 390–395.

Hudson, J.E. (1991) *Adaptive array principle*, Peter Peregrinus Ltd..

The IEICE, (Ed.). (1989). *Antenna engineering handbook* Ohmsha, Ltd Tokyo. (in Japanese)

Ohira, T. (2002). Blind aerial beamforming based on a higher-order maximum moment criterion (Part I: Theory), *Proc. Asia-Pacific Microwave Conf.* (pp. 802-811).

Ohira, T. & Iigusa, K. (2004). Electronically steerable parasitic array radiator antenna. *Electronics and Communications in Japan (Part 2: Electronics)*, 87(10), 25-45.

Ohira, T., & Cheng, J. (2004). Analog smart antennas, in S. Chandran (Ed.), *Adaptive Antenna Arrays: Trends and Applications* (pp.184-204). Berlin: Springer Verlag.

Ojiro, Y., Kawakami, H., Gyoda, K., and Ohira, T. (2001). Improvement of elevation directivity for ESPAR antennas with finite ground plane. *IEEE Int. Symp. on Antennas and Propagation*, (pp. 18–21).

Plapous, C., Cheng, J., Taillefer, E., Hirata, A., & Ohira, T. (2004). Reactance domain MUSIC algorithm for electronically steerable parasitic array radiator. I*EEE Trans. on Antennas Propagation*. 52(12), 3257–3264.

Preston, S. T., Thiel, D. V., Smith, T. A., O'Keefe, S. G., & Lu, J. W. (1998). Base-station tracking in mobile communications using a switch parasitic antenna array. *IEEE Trans. on Antennas and Propagation*, AP-46(6), 841–844.

Preston, S. T., Thiel, D. V., & Lu, J. W. (1999). A multibeam antenna using switched parasitic and switched active elements for space-division multiple access applications. *IEICE Trans. Electron.*, E82-C (7), 1202–1210.

Scott, N. L., Leonard-Taylor, M. O., & Vaughan, R. G. (1999). Diversity gain from a single-port adaptive antenna using switched parasitic elements illustrated with a wire and monopole prototype. *IEEE Trans. on Antennas and Propagation*, AP-47 (6), 1066–1070.

Svantesson, T., & Wennstrom, M. (2001). High-resolution direction finding using a switched parasitic antenna. *Proc. 11th IEEE Signal Processing Workshop on Statistical Signal Processing* (pp. 508–511).

Taillefer, E., Cheng, J., & Ohira, T., (2008). Direction of arrival estimation with compact array antennas: a reactance switching approach. In C. Sun (Ed.), *Handbook on advancements in Smart Antenna Technologies for Wireless Networks*, Idea Group, Inc.

Thiel, D. V., & Smith, S. (2001). *Switched Parasitic Antennas for Cellular Communications*, Artech House, Antennas and Propagation Library.

Vaughan, R. (1999). Switched parasitic elements for antenna diversity. *IEEE Trans. on Antennas and Propagation*, AP-47 (2), 399–405.

# About the Contributors

**Chen Sun** received the BEng from Northwestern Polytechnical University, Xi'an, China (2000) and a PhD from Nanyang Technological University, Singapore (2005), both in electrical engineering. From November 2002 to March 2003, he was with ATR Adaptive Communications Research Laboratories, Japan as a student intern. From August 2004 to May 2008 he was a researcher at ATR Wave Engineering Laboratories, Japan. In June 2008, he joined Ubiquitous Mobile Communications Group, National Institute of Information and Communications Technology (NICT) Japan as an expert researcher. His research interests include cognitive wireless system, smart antennas and communication theory.

**Jun Cheng** received the BS and MS degrees in telecommunications engineering from Xidian University, Xi'an, China, in 1984 and 1987, respectively, and the PhD degree in electrical engineering from Doshisha University, Kyoto, Japan, in 2000. From 1987 to 1994, he was an Assistant Professor and Lecturer in the Department of Information Engineering, Xidian University. From 1995 to 1996, he was an associate professor in the National Key Laboratory on Integrated Service Network, Xidian University. In April 2000, he joined ATR Adaptive Communications Research Laboratories, Kyoto, Japan, where he was a Visiting Researcher. From August 2002 to June 2003, he was working as a staff engineer at the R&D Center, Panasonic Mobile Communications Co., Ltd. (formerly Wireless Solution Labs., Matsushita Communication Industrial Co., Ltd.), Yokosuka, Japan. From July 2003 to March 2004, he was a staff engineer at Next-Generation Mobile Communications Development Center, Matsushita Electric Industrial Co., Ltd., Yokosuka, Japan. In April 2004, he joined Doshisha University, Kyoto, Japan. Currently, he is an associate professor in the Department of Intelligent Information Engineering and Sciences, Doshisha University, and a Visiting Researcher in ATR Wave Engineering Laboratories, Kyoto, Japan. His research interests are in the areas of communications theory, information theory, coding theory, array signal processing, and radio communication systems.

**Takashi Ohira** received BE and DE degrees in communication engineering from Osaka University, Osaka, Japan, in 1978 and 1983. In 1983, he joined NTT Electrical Communication Laboratories, Yokosuka, Japan, where he was engaged in research on monolithic integration of microwave semiconductor devices and circuits. He developed GaAs MMIC transponder modules and microwave beamforming networks aboard multi-beam communication satellites, Engineering Test Satellite VI (ETS-VI) and ETS-VIII, at NTT Wireless Systems Laboratories, Yokosuka, Japan. From 1999, he was engaged in research on wireless adhoc networks and microwave analog adaptive antennas at ATR Adaptive Communications Research Laboratories, Kyoto, Japan. Concurrently he was a consulting engineer for National Space Development Agency (NASDA) ETS-VIII Project in 1999, and an Invited Lecturer for Osaka University in 2000 to 2001. From 2005, he was director of ATR Wave Engineering Laboratories, Kyoto, Japan, and an invited professor of Tokyo Institute of Technology, Tokyo, Japan. Currently he is a professor of Toyohashi University of Technology, Toyohashi, Japan. He serves as URSI Commission CVice Chair, APMC2006 International Steering Committee Chair, EuMA Award Committee member, and EuMC2006 Technical Program Committee member. He is an IEEE fellow. He co-authored Monolithic Microwave Integrated Circuits (Tokyo: IEICE, 1997). Ohira was awarded 1986 IEICE Shinohara Prize, 1998 APMC Japan Microwave Prize, and 2004 IEICE Electronics Society Prize.

* * *

**Amin M. Abbosh** received the MSc degree in communications systems and the PhD in microwave engineering in 1991 and 1996 respectively both from Mosul University. He worked as a head of Computer & Information Engineering Department, Mosul University, until 2003. In 2004, he joined the Griffith University and then the University of Queensland, Australia, as a research fellow. His research interests include antennas, radio wave propagation, microwave devices, and ultra wideband wireless systems.

**Moeness Amin** received his PhD degree in 1984 from University of Colorado, Boulder. He has been on the Faculty of Villanova University since 1985, where he is now a Professor in the Department of Electrical and Computer Engineering and the Director of the Center for Advanced Communications. Amin is a fellow of the IEEE and International Society of Optical Engineers; Recipient of the IEEE Third Millennium Medal; Distinguished Lecturer of the IEEE Signal Processing Society for 2003; Member of the Franklin Institute Committee on Science and the Arts; Recipient of the 1997 Villanova University Outstanding Faculty Research Award; Recipient of the 1997 IEEE Philadelphia Section Service Award.

**Tomoyuki Aono** received BE degree from Kansai University, Osaka, Japan in 1991. In 1991 he joined Mitsubishi Electric Corp., Hyogo, Japan, where he was engaged in design and development of wireless communication systems and the direction finding systems to government offices. He had been engaged in research and development on wireless communication systems and adaptive array antennas at ATR Wave Engineering Laboratories in Kyoto, Japan. Now, he belongs to Mitsubishi Electric Corp. Aono is a member of IEICE of Japan.

**Biljana Badic** received the Dipl.-Ing. degree in electrical engineering from the Graz University of Technology, Austria. She obtained her Dr.techn. degree at Vienna University of Technology, Austria, where she worked as a research and teaching assistant. The focus of her recent work is wireless communications and MIMO technologies. In five years working experience, Biljana has been published a large number of conference and journal papers. She participated on COST 273 meetings (2003-2005) and some of her presented research was published in the book **"Mobile Broadband Multimedia Networks: Techniques, Models and Tools for 4G"** by Luis M. Correia.

**Gerhard Bauch** received the Dipl.-Ing. and Dr.-Ing. degree in Electrical Engineering from Munich University of Technology in 1995 and 2001, respectively, and the Diplom-Volkswirt degree from FernUniversitaet Hagen in 2001. In 2002 he joined DoCoMo Euro-Labs, Munich, Germany, where he is currently manager of the Advanced Radio Transmission Group. In 2007 he was also appointed Research Fellow at DoCoMo Euro-Labs. Since October 2003 he has also been an adjunct professor at Munich University of Technology. Furthermore, he lectured as guest professor at the University of Udine, Italy, and the Alpen-Adria University Klagenfurt, Austria.

**Konstanty S. Bialkowski** received his dual BEng/BSc degree with first class of Honors from the University of Queensland, Australia in 2003. At present, he is a PhD student in the School of Information Technology and Electrical Engineering at the University of Queensland. His research interests include wireless communication, multiple element antenna systems and digital signal processing. During his PhD studies he has been a recipient of the Australian Postgraduate Research Award, Morris Gunn Scholarship award, Richard Jago Fellowship award, and NICTA scholarship.

**Marek E. Bialkowski** received the M.EngSc degree (1974) in applied mathematics and the PhD degree (1979) in electrical engineering both from the Warsaw University of Technology and a higher doctorate (D.Sc. Eng.) in computer science and electrical engineering from the University of Queensland (2000). He held teaching and research appointments at universities in Poland, Ireland, Australia, UK, Canada, Singapore, Hong Kong and Switzerland. At present he is a tenured chair professor in the School of Information Technology and Electrical Engineering at the University of Queensland. His research interests include technologies and signal processing techniques for smart antennas and MIMO systems, antennas for mobile cellular and satellite communications, low profile antennas for reception of satellite broadcast TV programs, conventional and spatial power combining techniques, six-port vector network analyzers, and medical and industrial applications of microwaves. He has published 480 technical papers, several book chapters and one book. He is a fellow of IEEE.

**Steven D. Blostein** received his BS degree in electrical engineering from Cornell University, Ithaca, NY, in 1983, and the MS and PhD degrees in electrical and computer engineering from the University of Illinois, Urbana-Champaign, in 1985 and 1988, respectively. He has been on the Faculty at Queen's University since 1988 and currently holds the position of Professor and Head of the Department of Electrical and Computer Engineering. From 1999-2003, he was the leader of the Multirate Wireless Data Access Major Project sponsored by the Canadian Institute for Telecommunications Research. He has also been a consultant to industry and government in the areas of image compression, target tracking, radar imaging and wireless communications. His current interests lie in the application of signal processing to wireless communications systems, including smart antennas, MIMO systems, and space-time-frequency processing for MIMO-OFDM systems. He served as chair of IEEE Kingston Section in 1993-94, chair of the Biennial Symposium on Communications in 2000 and 2006, as associate editor for IEEE Transactions on Image Processing from 1996-2000, and currently serves as associate editor of IEEE Transactions on Wireless Communications. He is a registered professional engineer in Ontario.

**Santana Burintramart** was born in Bangkok, Thailand. He received the BS degree from the Chulachomklao Royal Military Academy, **Nakhon-Nayok,** Thailand, in 1998 and a MS degree from Syracuse University, Syracuse, New York, in 2004, where he is currently pursuing a PhD degree in the Department of Electrical Engineering. He has joined the Royal Thai Army (RTA) since 1998. His current research interest includes adaptive array signal processing related to smart antenna, radar, wireless communication and space-time adaptive processing.

**Huawei Chen** was born in Henan, China, in 1977. He received the BE degree from Henan Normal University, Xinxiang, China, in 1999 and MS and PhD degrees from Northwestern Polytechnical University, Xi'an, China, in 2002 and 2004, respectively. In 2004, he joined the Institute of Acoustics, Nanjing University, Nanjing, China, as a postdoctoral researcher. Since 2005, he has been with the Centre for Signal Processing, School of Electrical and Electronic Engineering, Nanyang Technological University, Singapore, as a research fellow. His research interests include statistical and adaptive signal processing, array signal processing.

**Sheng Chen** received his BEng degree in control engineering from Chinese Petroleum University, China, in 1982 and his PhD degree in control engineering from the City University, London, UK, in 1986. In 2005, he was awarded the Doctor of Sciences (DSc) degree by the University of Southampton, Southampton, UK. He joined the School of Electronics and Computer Science, University of Southampton, Southampton, UK, in September 1999. He previously held research and academic appointments at the University of Sheffield, Sheffield, the University of Edinburgh, Edinburgh, and the University of Portsmouth, Portsmouth, all in the UK. Professor Chen's research works include wireless communications, machine learning and neural networks, finite-precision digital controller design, and evolutionary computation methods. He has published over 300 research papers. In the database of the world's most highly cited researchers, compiled by Institute for Scientific Information (ISI) of the USA, Chen is on the list of the highly cited researchers in the engineering category.

**Woon Hau Chin** received the BEng and MEng degrees in electrical engineering from the National University of Singapore in 1999 and 2000, respectively, and the PhD degree in electrical engineering from Imperial College London, UK, in 2004. Since 2000, he has been with the Institute for Infocomm Research, Singapore (formerly the Centre for Wireless Communications), where he is currently involved in industrial projects on wireless LANs and Beyond 3G systems. He is concurrently an adjunct assistant professor at the National University of Singapore. His research interests are space–time signal processing, multicarrier systems, statistical signal processing, and cooperative systems.

**Jinho Choi** (Senior Member of IEEE) was born in Seoul, Korea. He received BE (magna cum laude) degree in electronics engineering in 1989 from Sogang University, Seoul, and the MSE and PhD degrees in electrical engineering from Korea Advanced Institute of Science and Technology (KAIST), Daejeon, in 1991 and 1994, respectively. He is now with Institute of Advanced Telecommunications (IAT), University of Wales Swansea, Swansea, United Kingdom, as a professor. His research interests include wireless communications and array/statistical signal processing. He authored a book entitled *Adaptive and Iterative Signal Processing in Communications*

(Cambridge University Press, 2006). Choi received the 1999 Best Paper Award for Signal Processing from EURASIP and a Senior Member of IEEE. Currently, he is an editor of *Journal of Communications and Networks (JCN)* and an associate editor of *IEEE Transactions on Vehicular Technology.*

**Salman Durrani** received the BSc (1st class honours) degree in electrical engineering from the University of Engineering & Technology, Lahore, Pakistan in 2000 and the PhD degree in electrical engineering from the University of Queensland, Brisbane, Australia in Nov. 2004. Since March 2005, he has been a lecturer in the Department of Engineering, The Australian National University, Canberra, Australia. His research interests are in channel modeling, wireless ad-hoc networks, MIMO and smart antenna systems and code division multiple access (CDMA) systems. He has 24 publications to date in refereed international journals and conferences. He is a member of IEEE and a graduate member of Institution of Engineers, Australia.

**Meng Hwa Er** received the BEng degree in electrical engineering with 1st class honors from the National University of Singapore in 1981, and the PhD degree in electrical and computer engineering from the University of Newcastle, Australia, in 1986. He joined the Nanyang Technological Institute/University in 1985 and was promoted to a full professor in 1996. He was founder of the Centre for Signal Processing and appointed its first director from 1995 to 1997. Since 1996, he has been dean, School of Electrical and Electronic Engineering. He was dean of College of Engineering from 2001 to 2004 and helped promote multi-disciplinary education and research among the engineering schools. He was also a deputy president of Nanyang Technological University. Currently, he is the associate provost of Nanyang Technological University. His research interests include array signal processing, satellite communications, computer vision and optimization techniques. He has published over 240 papers in international journals and conference proceedings. He served as an associate editor of the *IEEE Transactions on Signal Processing* from 1997 to 1998 and is a member of the editorial board, *IEEE Signal Processing Magazine* from 2005 to 2007. He was the general co-chair of the IEEE International Conference on Image Processing, 2004. He also serves as a member of the Board of Governors of Singapore Polytechnic and the Board of Directors of Singapore Technologies Electronics Limited, BITwave Limited, PowerGrid Pte Ltd, DSO National Laboratories and Centre for Strategic Infocomm Technologies. He is also a member of the Professional Engineers Board of Singapore.

**Pingyi Fan,** received the BS and MS degrees from the Department of Mathematics of Hebei University in 1985 and Nankai University in 1990, respectively, received his PhD degree from the Department of Electronic Engineering, Tsinghua University, Beijing, China in 1994. From August 1997 to March 1998, he visited Hong Kong University of Science and Technology as research associate. From May 1998 to October 1999, he visited University of Delaware, USA as research fellow. In March 2005, he visited NICT of Japan as visiting professor. From June 2005 to July 2005, August 2006 to Sept. 2006, he visited Hong Kong University of Science and Technology. He was promoted as full professor at Tsinghua University in 2002. Fan is a member of IEEE and an oversea member of IEICE. He has participated in organizing many international conferences including the TPC Chair of IEEE International Symposium of Multi-Dimensional Mobile Computing 2004 (MDMC'04), Conference Chair of IASTED CSNA2007 and TPC member of IEEE ICC05, VTC07, Globecom08 etc. He is currently serving as an editor of *IEEE Transactions on Wireless Communications, Wiley Wireless Communications and Mobile Computing, Interscience International Journal of Ad Hoc and Ubiquitous Computing.* He is also a reviewer of more than 14 international journals, including 10 IEEE Journals and three Europe journals. His main research interests include B3G technology in wireless communications such as MIMO, OFDM, Multicarrier CDMA, space time coding, and LDPC design etc., network coding, network information theory and cross layer design etc.

**Christian Guthy** was born in Munich, Germany in 1979. He received the BSc and Dipl.-Ing. degrees (the latter with honors) in electrical engineering from the Technische Universität München (TUM), Munich, Germany in 2004 and 2005, respectively. Since 2005 he has been with the Associate Institute for Signal Processing at TUM, where he is currently working towards his PhD degree. His main research interests include signal processing algorithms for next generation wireless communication systems with focus on multiantenna and multicarrier systems.

**Toru Hashimoto** received BE and ME degrees from Kobe University, Japan, in 1995 and 1997, respectively. In 1997, he joined Mitsubishi Electric Corporation, where he engaged in design and development of communication systems for government and municipal offices. From 2005, he engaged in research and development of ultra-high speed giga-bit rate wireless LAN systems at ATR Wave Engineering Laboratories in Kyoto, Japan.

**Akifumi Hirata** received BE and ME degrees in electrical engineering from Osaka Prefecture University, Osaka, Japan, in 1994 and 1996. In 1996, he joined SANYO Electric Co., Ltd., Gifu, Japan, and was engaged in research and development on telecommunication systems. From 2001 to 2004, he was engaged in research on microwave analog adaptive antennas for mobile user terminals at ATR Wave Engineering Laboratories, Kyoto, Japan. His current interests are broadband wireless systems for mobile user terminals with an array antenna. He is a recipient of the 2002 IEICE Young Engineer Award.

**Thomas Hunziker** received the Diploma and PhD degrees in electrical engineering from the Swiss Federal Institute of Technology (ETH) Zurich in 1992 and 2002, respectively. He was a research assistant at the Communication Technology Laboratory of ETH Zurich from 1997 until 2002. From 2002 to 2006, he was a researcher at the Advanced Telecommunications Research Institute International (ATR), Kyoto, Japan. He is now with the Communications Laboratory, University of Kassel, Germany.

**Jesús Ibáñez-Díaz** was born in Santander, Spain, in 1971. He received the Telecomm Engineer degree and the doctor degree from the Universidad de Cantabria, Spain, in 1995 and 2004, respectively. In 1995 he joined the Departamento de Ingeniería de Comunicaciones at the Universidad de Cantabria, Spain, where he is currently an associate professor. His research interest includes digital signal processing and digital communication applied to MIMO systems and sensor networks.

**Joakim Jaldén** was born in Gävle, Sweden on May 16, 1976. He received the PhD and MSc in electrical engineering from the Royal Institute of Technology (KTH), Stockholm, Sweden in 2006 and 2002 respectively. His PhD position at KTH was funded by a grant, "Honor Graduate Student Position," awarded by the dean office. Between September 2000 and May 2002 he studied at Stanford University, CA, USA, where he also conducted the research for his MSc thesis. For his work on MIMO communications, Jaldén has been awarded the IEEE Signal Processing (SP) Society's 2006 Young Author Best Paper Award and the first price in the Student Paper Contest at the 2007 International Conference on Acoustics, Speech and Signal Processing (ICASSP).

**Nemai Chandra Karmakar** (S'91–M'91–SM'99) obtained the MSc degree in (EE) from the University of Saskatchewan, Canada in 1991 and the PhD degree from the University of Queensland, Australia in 1999 and MHEd from Griffith University, Australia in 1997. He is a Senior Lecturer in the Department of Electrical and Computer Systems Engineering, Monash University, Australia. Karmakar's research interests cover areas such as RFID, smart antennas for RFID, mobile and satellite communications, EBG assisted RF devices, broadband microstrip antennas and arrays and beam-forming networks. He has published more than 160 referred journal and conference papers, and six book chapters. He is a member of the editorial board of *International Journal of RF and Microwave Computer Aided Engineering, Wiley Interscience*.

**Qinghua Li** received the BE degree in radio engineering from South China University of Technology, Guangzhou, China, the ME in signal processing from Tsinghua University, Beijing, China, and the PhD degree in electrical engineering from Texas A&M University, College Station, Texas, in 1992, 1995, and 2001 respectively. From 1995 to 1996, he was with Guangdong Telecommunication Academy of Science and Technology, China, where he was involved in the development of telephone exchange networks. Since February 2001, he has been with the wireless research group of Intel Corporation, Santa Clara, CA. His research interests are in the areas of MIMO, SDMA, relay network, ultra-wide band communications, wireless channel modeling, multiuser detection, interference mitigation, channel coding, and media access control protocols.

**Xin Li** received his PhD degree from University of Electronic Science and Technology of China in 2005. He is a post-doctor at the Center for Advanced Communications, Villanova University. From 1999 to 2001, he was a RF engineer at Beijing Datang Technology Co., where he participated in drafting and amending the TD-SCDMA standard, and in prototyping the demo system. In 2002, he was an engineer focusing on RF front-end module for dual-mode handsets at Zhejiang Holley Commun. Co. In 2005, he was with Alcatel Shanghai Bell R&I Center as a scientist for prototyping of MIMO-OFDM systems. Li's research interests include signal processing and applications, wireless network protocol development, wireless channel modeling, embedded system design, and circuit and antenna design.

**Xintian Eddie Lin** was born in Chengdu, China on July 1, 1969. He received the BS degree in physics from the University of Science and Technology of China in 1988. He then obtained the MS and the PhD degrees in physics from University of California, San Diego in 1989 and 1995 respectively. From April 1995 to September 2000, he was a research staff and postdoc fellow at Stanford Linear Accelerator Center. In October 2000, he joined Intel and became a wireless communication architect. His major interest is in the area of electromagnetic theory, photonic bandgap material and wireless communications, especially MIMO OFDM close loop feedback. He also worked extensively on antennas, platform noise and radio coexistence on notebook platform.

**Mohammad Madihian** received the PhD in electronic engineering from Shizuoka University, Japan, in 1983. He joined NEC Central Research Laboratories, Kawasaki, Japan, where he worked on research and development of Si and GaAs device-based digital as well as microwave and millimeter-wave monolithic IC's. In 1999, he moved to NEC Laboratories America, Inc., Princeton, New Jersey, and is presently the department head and chief patent officer. He conducts PHY/MAC layer signal processing activities for high-speed wireless networks and personal communications applications. He has authored or co-authored more than 130 scientific publications including 20 invited talks, and holds 35 Japan/US patents. Madihian has received the IEEE MTT-S Best Paper Microwave Prize in 1988, and the IEEE Fellow Award in 1998. He holds eight NEC Distinguished R&D Achievement Awards. He has served as guest editor to the *IEEE Journal of Solid-State Circuits, Japan IEICE Transactions on Electronics*, and *IEEE Transactions on Microwave Theory and Techniques*. He is presently serving on the IEEE Speaker's Bureau, IEEE Compound Semiconductor IC Symposium (CSICS) Executive Committee, IEEE Radio and Wireless Conference Steering Committee, IEEE International Microwave Symposium (IMS) Technical Program Committee, IEEE MTT-6 Subcommittee, IEEE MTT Editorial Board, and Technical Program Committee of International Conference on Solid State Devices and Materials (SSDM). Madihian is an adjunct professor at Electrical and Computer Engineering Department, Drexel University, Philadelphia, Pennsylvania.

**Patrick Mitran** received the bachelor's and master's degrees in electrical engineering, in 2001 and 2002, respectively, from McGill University, Montreal, Canada, and a PhD degree from the School of Engineering and Applied Sciences, Harvard University, Cambridge, MA in 2006. In 2005, he interned as a research scientist for Intel Corporation in the Radio Communications Lab. In 2007 he was a lecturer in the School of Engineering and Applied Sciences, Harvard University. Currently he is with the Department of Electrical and Computer Engineering at the University of Waterloo (Waterloo, Canada) at the rank of assistant professor. His current interests are in the study of cooperation and cognition in wireless networks both from an information theoretical viewpoint as well as coding theory and signal processing perspectives. This includes such aspects as capacity analysis, distributed source/channel coding and network coding.

**Sadao Obana** received his BE, ME, and PhD (Eng.) degrees from Keio Univ. Tokyo, Japan, in 1976, 1978 and 1993 respectively. After joining KDDI (former KDD) in 1978, he was engaged in R&D in the field of packet exchange systems, network architecture, OSI (Open Systems Interconnection) protocols, database, distributed processing, network management and ITS (Intelligent Transport Systems). In 2004, he joined Advanced Telecommunication Research Institute International (ATR) and is now a director of Adaptive Communications Research Labs in ATR. His current research areas are wireless mobile ah-hoc network, ubiquitous wireless sensor network and ITS (Intelligent Transport Systems). He received an Award of Minister of Education, Culture, Sports, Science and Technology in 2001. Obana is a fellow of Information Processing Society of Japan (IPSJ).

**Hideki Ochiai** received the BE degree in communication engineering from Osaka University, Osaka, Japan, in 1996 and the ME and PhD degrees in information and communication engineering from The University of Tokyo, Tokyo, Japan, in 1998 and 2001, respectively. From 2001 to 2003, he was with the Department of Information and Communication Engineering, The University of Electro-Communications, Tokyo. Since April 2003, he has been with the Department of Electrical and Computer Engineering, Yokohama National University, Yokohama, Japan, where he is currently an associate professor. From 2003 to 2004, he was a visiting scientist at the Division of Engineering and Applied Sciences, Harvard University, Cambridge, MA. Ochiai currently serves as an editor for *IEEE Transactions on Wireless Communications.*

**Björn Ottersten** was born in Stockholm, Sweden, 1961. He received the MS degree in electrical engineering and applied physics from Linköping University, Linköping, Sweden, in 1986. In 1989 he received the PhD degree in electrical engineering from Stanford University, Stanford, CA. Ottersten has held research positions at the Department of Electrical Engineering, Linköping University, the Information Systems Laboratory, Stanford University, and the Katholieke Universiteit Leuven, Leuven. During 96/97, Ottersten was director of research at ArrayComm Inc, San Jose, CA. He has co-authored papers that received an IEEE Signal Processing Society Best Paper Award in 1993, 2001, and 2006. In 1991 he was appointed professor of Signal Processing at the Royal Institute of Technology (KTH), Stockholm and he is currently dean of the School of Electrical Engineering at KTH. From 1992 to 2004 he was head of the department for Signals, Sensors, and Systems at KTH. Ottersten is also a visiting professor at the University of Luxembourg. Ottersten has served as associate editor for the *IEEE Transactions on Signal Processing* and on the editorial board of *IEEE Signal Processing Magazine.* He is currently editor-in-chief of *EURASIP Signal Processing Journal* and a member of the editorial board of *EURASIP Journal of Applied Signal Processing.* Ottersten is a fellow of the IEEE. His research interests include wireless communications, stochastic signal processing, sensor array processing, and time series analysis.

**H. Vincent Poor** is the Michael Henry Strater University professor of Electrical Engineering and dean of the School of Engineering and Applied Science at Princeton University. His primary research interests are in the area of wireless networks and related fields. Among his publications in these areas is the recent book, *MIMO Wireless Communications* (Cambridge, 2007), co-authored with Ezio Biglieri, et al. Poor is a member of the U.S. National Academy of Engineering, a fellow of the American Academy of Arts & Sciences, and a former Guggenheim fellow. He is also a fellow of the IEEE, the Institute of Mathematical Statistics, and other scientific and technical organizations. He is a former president of the IEEE Information Theory Society, and a former editor-in-chief of the *IEEE Transactions on Information Theory.* Recent recognition of his work includes the 2005 IEEE Education Medal and the 2007 IEEE Marconi Prize Paper Award.

**Adam Postula** received MS degree in electrical engineering from Warsaw University of Technology, Poland in 1974 and PhD degree from Poznan University of Technology in 1981. He worked from 1983 to 1992 as an electronic system designer in ABB Sweden and as a researcher in Royal Institute of Technology in Stockholm, Sweden. He also led development of EDA tools at Swedish Institute of Microelectronics. From 1995 he works as an academic at University of Queensland. His research interests include digital system design methodology, reconfigurable systems, and wireless embedded system design.

**Ignacio Santamaría** received his Telecommunication Engineer Degree and his PhD in electrical engineering from the Polytechnic University of Madrid, Spain in 1991 and 1995, respectively. In 1992 he joined the Departamento de Ingeniería de Comunicaciones, Universidad de Cantabria, Spain, where he is currently full professor. In 2000 and 2004, he spent visiting periods at the Computational NeuroEngineering Laboratory (CNEL), University of Florida. Santamaría has more than 100 publications in refereed journals and international conference papers. His current research interests include signal processing algorithms for wireless communication systems, MIMO systems, multivariate statistical techniques and machine learning theories. He has been involved in several national and international research projects on these topics.

**Magdalena Salazar-Palma** (M'89–SM'01) was born in Granada, Spain. She received the degree and PhD degree in ingeniero de telecomunicación from the Universidad Politécnica de Madrid, Madrid, Spain. She is currently a Catedrático (full professor) with the Departamento de Teoría de la Señal y Comunicaciones (Signal Theory and Communications), Escuela Politécnica Superior (College of Engineering), Universidad Carlos III de Madrid, Madrid, Spain. She has authored four books, 20 contributions for chapters and articles in books, 48 papers in international scientific journals, and 172 papers in international conferences, symposiums, and workshops, 13 contributions for academic books and notes, 56 papers in national conferences, and over 75 project reports, short course notes, etc. She has delivered numerous invited presentations, lectures, and seminars. She has lectured in several short courses, some of them in the frame of Programs of the European Community. She has participated at different levels (researcher or director) in a total of 62 projects and contracts, financed by international, European, and national institutions and companies. She has developed her research in the areas of electromagnetic-field theory; computational and numerical methods for microwave passive components and antenna analysis; network and filter theory and design; and design, simulation, optimization, implementation, and measurement of microwave circuits both in waveguide and integrated (hybrid and monolithic) technologies. She has been a member of the editorial board of two scientific journals. Salazar-Palma is member of the Technical Program Committee of several international and national symposiums and reviewer for different international scientific journals, symposiums, and editorial companies. She has assisted the Comisión Interministerial de Ciencia y Tecnología (Spain National Board of Research) in the evaluation of projects. She has also served in several evaluation panels of the Commission of the European Communities. She has been associate editor for the IEEE ANTENNAS AND WIRELESS PROPAGATION LETTERS. Since 1989, she has served the IEEE under different volunteer positions: vice chairperson and chairperson of the IEEE Spain Section Antennas and Propagation Society (AP-S)/Microwave Theory and Techniques Society (MTT-S) Joint Chapter, chairperson of the IEEE Spain Section, member of the IEEE Region 8 Committee, member of the IEEE Region 8 Nominations and Appointments Subcommittee, chairperson of the IEEE Region 8 Conference Coordination Subcommittee, chairperson of the IEEE Women in Engineering (WIE) Committee, liaison between the IEEE WIE Committee and the IEEE Regional Activities Board, and member of the IEEE Ethics and Member Conduct Committee. She is currently the membership development officer of the IEEE Spain Section, a member of the IEEE WIE Committee and a member of the IEEE AP-S Administrative Committee (AdCom). She was the recipient of two individual research awards.

**Aimin Sang** received a PhD from the University of Texas at Austin in 2001. His PhD dissertation is on the measurement-based traffic management for QoS guarantee in multi-service networks. From May 2000 to July 2002, he was a member of technical staff and software engineer at Santera System Inc., a startup company in designing and implementing the next-generation multi-service gateway. His duty was to design, implement, and test core traffic management algorithms on the switch fabric and control boards, integrating IP routing, ATM switching, and Class 4 and 5 telephony switching functionalities for multi-service Internet access at the Central Offices. From July 2002 to Nov. 2002, he was a post-doc at UT-Austin, researching on VPN provisioning and ad hoc sensor networks. He joined NEC Lab America in Nov. 2002. Sang is currently a research staff member in Broadband & Mobile Networking Department, NEC Lab. America, focusing on cross-layer design of 4G wireless systems, such as 4G Cellular base station, WiMax/WLAN systems, and their inter-networking architecture. His duty is to develop the core technologies including radio resource management and QoS schemes over an IP-optimized MC-CDMA or OFCDM/MIMO air interfaces. He is also interested in mobile Internet services and personal area wireless systems.

**Tapan K. Sarkar** received his B.Tech degree from the Indian Institute of Technology, Kharagpur, in 1969, the MSc.E. degree from the University of New Brunswick, Fredericton, NB, Canada, in 1971, and the MS and PhD degrees from Syracuse University, Syracuse, NY, in 1975. From 1975 to 1976, he was with the TACO Division of the General Instruments Corporation. He was with the Rochester Institute of Technology, Rochester, NY, from 1976 to 1985. He was a research fellow at the Gordon McKay Laboratory, Harvard University, Cambridge, MA, from 1977 to 1978. He is now a professor in the Department of Electrical and Computer Engineering, Syracuse University. His current research interests deal with numerical solutions of operator equations arising in electromagnetics and signal processing with application to system design. He obtained one of the "best solution" awards in May 1977 at the Rome Air Development Center (RADC) Spectral Estimation Workshop. He has authored or coauthored more than 280 journal articles and numerous conference papers and 32 chapters in books and fifteen books, including his most recent ones, *Iterative and Self Adaptive Finite-Elements in Electromagnetic Modeling* (Boston, MA: Artech House, 1998), *Wavelet Applications in Electromagnetics and Signal Processing* (Boston, MA: Artech House, 2002) and *Smart Antennas* (John Wiley & Sons, 2003). Sarkar is a registered professional engineer in New York. He received the Best Paper Award of the IEEE Transactions on Electromagnetic Compatibility in 1979 and in the 1997 National Radar Conference. He received the College of Engineering Research Award in 1996 and the Chancellor's Citation for Excellence in Research in 1998 at Syracuse University. He was an associate editor for feature articles of the IEEE Antennas and Propagation Society Newsletter (1986-1988). He was the Chairman of the Inter-commission Working Group of International URSI on Time Domain Metrology (1990–1996). He was a distinguished lecturer for the Antennas and Propagation Society from 2000-2003. He is currently a member of the IEEE Electromagnetics Award board and an associate editor for the *IEEE Transactions on Antennas and Propagation.* He is the vice president of the Applied Computational Electromagnetics Society (ACES) and the technical chair for the combined IEEE 2005 Wireless Conference along with ACES to be held in Hawaii. He is on the editorial board of *Journal of Electromagnetic Waves and Applications* and *Microwave and Optical Technology Letters*. He is a member of Sigma Xi and International Union of Radio Science Commissions A and B. He received Docteur Honoris Causa both from Universite Blaise Pascal, Clermont Ferrand, France in 1998 and from Politechnic University of Madrid, Madrid, Spain in 2004. He received the medal of the *friend of the city of Clermont Ferrand*, France, in 2000.

**Wee Ser** received his BSc (Hon) and PhD degrees, both in electrical and electronic engineering, from the Loughborough University, UK, in 1978 and 1982 respectively. He joined the Defence Science Organization (DSO), Singapore in 1982, and became the head of the Communications Research Division in 1993. In 1997, he joined NTU as an associate professor and was since appointed as the director of the Centre for Signal Processing at NTU. Ser Wee was a recipient of the Colombo Plan scholarship (undergraduate studies) and the PSC postgraduate scholarship. He was awarded the IEE Prize during his studies in UK. While in DSO, he was a recipient of the prestigious Defence Technology (Individual) Prize in 1991 and the DSO Excellent Award in 1992. He has served in several international and national advisory and technical committees and as reviewers to several international journals. He has published more than 80 papers in international journals and conferences. He holds five patents and has four pending patents. His research interests include microphone array and smart antenna array signal processing, signals classification techniques, EEG based brain signal analysis, emotion signal analysis, and channel estimation techniques.

**Constantin Siriteanu** was born in Sibiu, Romania, in 1972. He received his BS and MS degrees in Electrical Engineering, from "Gheorghe Asachi" Technical University, Iasi, Romania, in 1995 and 1996, respectively. In 2006 he received his PhD degree in Electrical Engineering from Queen's University, Kingston, Canada, for work on smart antenna array performance–complexity tradeoffs using eigencombining. Between September 2006 and February 2008, Dr. Siriteanu was a Post-Doctoral Research Fellow with Hanyang University, Seoul, Korea, where he also taught a graduate course on MIMO systems. Between March and June 2008, Dr. Siriteanu was a visiting professor at Kyungpook National University, Daegu, Korea. Since June 2008, Dr. Siriteanu has been on the faculty at the School of Computer Science and Engineering, Seoul National University, Korea. His research interests are in performance–complexity tradeoffs for multi-antenna adaptive wireless transceivers.

**Eddy Taillefer** was born in Les Abymes, Guadeloupe. He received a MSc in the field of electrical and computer sciences from École Nationale Supérieure de Sciences Appliquées et de Technologie (ENSSAT), France, in 2001. He worked with ATR Spoken Language Translation Research Laboratories in 2000 and with ATR Adaptive Communications Research Laboratories in 2001. Since 2002, he has been engaged in research on adaptive antenna signal processing at ATR Wave Engineering Laboratories, Kyoto, Japan. Since 2006, he is working towards the PhD degree in the Department of Intelligent Information Engineering and Sciences, Doshisha University, Kyoto. He is a member of IEEE and IEICE.

**Vahid Tarokh** received the PhD degree in electrical engineering from the University of Waterloo, Waterloo, ON, Canada, in 1995. He is currently a Perkins professor of Applied Mathematics and Hammond Vinton Hayes Senior Fellow of Electrical Engineering at Harvard University, Cambridge, MA, where he defines and supervises research in communications, networking, and signal processing. Tarokh has been one of the "Top 10 Most Cited Authors in Computer Science" since 2003, according to the ISI Web of Science. He is the recipient of a number of major awards and two honorary degrees.

**Makoto Taromaru** was born in Fukuoka, Japan in 1962. He received BE and ME degrees in electronics engineering from Tokyo Institute of Technology, Tokyo, Japan, in 1985 and 1987, respectively, and PhD degree from Kyushu Institute of Technology, Fukuoka, Japan, in 1997. In 1987, he joined Kyushu Matsushita Electric Corporation where he worked on the development of radio signal processing for Personal Handy-phone System (PHS) and other digital cordless telephone systems. Between 2001 and 2004, he was with the Faculty of Engineering, Kyushu Sangyo University, Fukuoka, Japan, where he was an associate professor of the department of Electronics. Since April 2004, he has been with ATR Wave Engineering Laboratories, Kyoto, Japan, and he is currently the head of the Department of Wireless Communication Systems. His research interests include radio communication systems, especially on diversity and adaptive antenna systems. Taromaru is a member of IEEE and IEICE.

**Pedro Tejera** received the telecommunication engineering degree from the Universidad Politécnica de Madrid (UPM) in 1999. In 1999 he worked as a network manager for the mobile communications provider Viag Interkom GmbH in Munich, Germany. From 2000 to 2002 he worked at the Einrichtung für Systeme der Kommunikationstechnik (Fraunhofer Gesellschaft) as a researcher in the field of digital subscriber line. In 2002 he joined the signal processing group at the Technische Universität München (TUM) where he is currently working towards his PhD.

**Monthippa Uthansakul** received the BEng degree (1997) in Telecommunication Engineering from Suranaree University of Technology, Thailand, MEng degree (1999) in electrical engineering from Chulalongkorn University, Thailand, and PhD degree (2007) in Information Technology and Electrical Engineering from The University of Queensland, Australia. She received 2nd prize Young Scientist Award from 16th International Conference on Microwaves, Radar and Wireless Communications, Poland, in 2006. At present, she is lecturer in School of Telecommunication Engineering, Suranaree University of Technology, Thailand. Her research interests include wideband/narrowband smart antennas, automatic switch beam antenna, DOA finder, microwave components, application of smart antenna on WLAN.

**Wolfgang Utschick** completed several industrial education programs before he received the diploma and doctoral degrees both with honors in electrical engineering from the TUM. In this period he held a scholarship of the Bavarian Ministry of Education for exceptional students. From 1998 - 2002 he co-directed the SP Group at the Institute of Circuit Theory and Signal Processing at the TUM. Since 2000 he is consulting in 3 GPP standardization in the field of multi-element antenna systems. In 2002 Utschick has been appointed professor at the TUM where he is head of the Associate Institute for Signal Processing. He gives courses on Signal Processing, Stochastic Processes and Optimization Theory in the field of digital communications. In 2006, Utschick has been awarded for his excellent teaching records and in 2007 he has received the ITG Award 2007 of the German Society for Information Technology (ITG). Wolfgang Utschick is SM of the VDE/ITG and SM of the IEEE where he is currently instrumental as associate editor for *T-CAS*.

**Javier Vía** received his telecommunication engineer degree and his PhD in electrical engineering from the University of Cantabria, Spain in 2002 and 2007, respectively. In 2002 he joined the Department of Communications Engineering, University of Cantabria, Spain, where he is currently assistant professor. In 2006 he spent a visiting period at the Smart Antennas Research Group (SARG), Stanford University. His current research interests include blind channel estimation and equalization in wireless communication systems, multivariate statistical analysis and kernel methods.

**Branka Vucetic** received the BSEE, MSEE, and PhD degrees in 1972, 1978, and 1982, respectively, in electrical engineering, from the University of Belgrade, Belgrade, Yugoslavia. During her career she has held various research and academic positions in Yugoslavia, Australia, and the UK. Since 1986, she has been with the University of Sydney, School of Electrical and Information Engineering in Sydney, Australia. She is currently the director of the Telecommunications Laboratory at the University of Sydney. Her research interests include wireless communications, digital communication theory, coding, and multiuser detection. In the past decade she has been working on a number of industry sponsored projects in wireless communications and mobile Internet. She has taught a wide range of undergraduate, postgraduate, and continuing education courses worldwide. Vucetic published four books and more than two hundred papers in telecommunications journals and conference proceedings.

**Genyuan Wang** received his BSc and MS degrees in mathematics from Shaanxi Normal University, Xi'an, China, in 1985 and 1988, respectively, and his PhD degree in electrical engineering from Xidian University, Xi'an China, in 1998. From July, 1988 to September 1994, he worked at Shaanxi Normal University as an assistant professor and then an associate professor. From September 1994 to May 1998, he worked at Xidian University as a research assistant. From June 1988 to December 2003, he was a post-doctoral fellow at Department of Electrical and Computer Engineering, University of Delaware. From January 2004 to April 2006, he was a research associate at the Center for Advanced Communications, Villanova University. Since May, 2006, he has been with Navini Networks as a senior system engineer. His research interests are radar imaging and radar signal processing, adaptive filter, OFDM system, channel equalization and space-time coding.

**Xiaodong Wang** received the PhD degree in electrical engineering from Princeton University. He is now on the faculty of the Department of Electrical Engineering, Columbia University. Wang's research interests fall in the general areas of computing, signal processing and communications, and has published extensively in these areas. Among his publications is a book entitled "Wireless Communication Systems: Advanced Techniques for Signal Reception," published by Prentice Hall in 2003. His current research interests include wireless communications, statistical signal processing, and genomic signal processing. Wang received the 1999 NSF CAREER Award, and the 2001 IEEE Communications Society and Information Theory Society Joint Paper Award. He has served as an associate editor for the *IEEE Transactions on Communications*, the *IEEE Transactions on Wireless Communications*, the *IEEE Transactions on Signal Processing*, and the *IEEE Transactions on Information Theory*.

**Masahiro Watanabe** was received BE degree in electrical engineering and Electronics from Kyoto Institute of Technology, in Kyoto, Japan, 1982. In 1982, he joined to Mitsubishi Electric Corporation, where he was engaged in development of Radar System and Inter Vehicle Communication System. In 2002, he joined to ATR Adaptive Communications Research Laboratories, Kyoto, Japan, where he was a senior researcher, and was engaged in research of Ad hoc Networks. From 2006, he is currently a visiting researcher of ATR Adaptive Communications Research Laboratories, Kyoto, Japan. Since 2006, He has been engaged in research of Sensing Information Processing and Ad hoc Networks as a senior researcher in Mitsubishi Electric Corporation Advanced Technology R&D Center, Hyogo, Japan. He is an IEICE member.

**Takashi Watanabe** received the ME and DE degrees from Osaka University, Japan in 1984, and 1987, respectively. In 1987, he joined the Faculty of Engineering, Tokushima University as an assistant professor. In 1990, he moved to the Faculty of Information, Shizuoka University. He was a visiting researcher at University of California, Irvine in 1995. He is currently a professor of Faculty of Informatics, Shizuoka University. His interests include computer networks and distributed system, especially MAC/Routing protocols for ad hoc and sensor networks. He is a member of Institute of Electronics, Information and Communication Engineers of Japan (IEICE), Information Processing Society of Japan (IPSJ), IEEE Communications/Computer Society and ACM SIGMOBILE. He is currently chair of the special interest groups of mobile computing and ubiquitous communications (MBL) of IPSJ.

**Jack H. Winters** received his PhD in electrical engineering from The Ohio State University in 1981. He was with AT&T in the research area for more than 20 years, where he was Division Manager of the Wireless Systems Research Division of AT&T Labs-Research, where he was primarily involved in research on smart antennas, and was then Chief Scientist at Motia, Inc., for four years, where he guided the research and development of their smart antenna products. He is a fellow of the IEEE, a former IEEE distinguished lecturer for both IEEE Communications and the Vehicular Technology Societies, area editor for Transmission Systems for the IEEE Transactions on Communications, and a New Jersey Inventor of the Year for 2001. He is currently a consultant and has over 30 issued patents and 50 journal publications.

**Xiang-Gen Xia** received his BS degree in mathematics from Nanjing Normal University, Nanjing, China, and his MS degree in mathematics from Nankai University, Tianjin, China, and his PhD degree in electrical engineering from the University of Southern California, Los Angeles, in 1983, 1986, and 1992, respectively. He was a senior/research staff member at Hughes Research Laboratories, Malibu, California, during 1995-1996. In September 1996, he joined the Department of Electrical and Computer Engineering, University of Delaware, Newark, Delaware, where he is the Charles Black Evans professor. He was a visiting professor at the Chinese University of Hong Kong during 2002-2003, where he is an adjunct professor. Before 1995, he held visiting positions in a few institutions. His current research interests include space-time coding, MIMO and OFDM systems, and SAR and ISAR imaging. Xia has over 160 refereed journal articles published and accepted, and seven U.S. patents awarded and is the author of the book Modulated Coding for Intersymbol Interference Channels (New York, Marcel Dekker, 2000). Xia received the National Science Foundation (NSF) Faculty Early Career Development (CAREER) Program Award in 1997, the Office of Naval Research (ONR) Young Investigator Award in 1998, and the Outstanding Overseas Young Investigator Award from the National Nature Science Foundation of China in 2001. He also received the Outstanding Junior Faculty Award of the Engineering School of the University of Delaware in 2001. He is currently an associate editor of the *IEEE Transactions on Wireless Communications*, *IEEE Transactions on Vehicular Technology*, the *IEEE Signal Processing Letters*, the *Journal of Communications and Networks (JCN)*, and *Journal of Communications (JCM)*. He was a guest editor of "Space-Time Coding and Its Applications" in the *EURASIP Journal of Applied Signal Processing* in 2002. He served as an associate editor of the *IEEE Transactions on Signal Processing* during 1996 to 2003, the *IEEE Transactions on Mobile Computing* during 2001 to 2004, and the *EURASIP Journal of Applied Signal Processing* during 2001 to 2004. He is also a member of the Sensor Array and Multichannel (SAM) Technical Committee in the IEEE Signal Processing Society. He is the general co-chair of ICASSP 2005 in Philadelphia.

**Nuri Yilmazer** received the BS degree from Cukurova University, Adana, Turkey in 1996, and the MS degree from University of Florida, Gainesville, Florida in 2000. He got his PhD degree from the Department of Electrical Engineering at Syracuse University, Syracuse, New York, in 2006. He worked as a post doctoral research associate in Computational Electromagnetics Lab at Syracuse University, Syracuse, NY from 2006 to August 2007. He is currently working as an assistant professor in Electrical Engineering and Computer Science department at Texas A&M-Kingsville, Kingsville, TX. His current research interests include adaptive array processing, signal processing in electromagnetic applications, and smart antennas.

**Zhu Liang Yu** received his BEng in 1995 and MEng in 1998, both in electronic engineering from the Nanjing University of Aeronautics and Astronautics China. He received his PhD in 2006 from Nanyang Technological University, Singapore. He worked in Shanghai BELL Co. Ltd as a software engineer from 1998 to 2000. He joined Center for Signal Processing, Nanyang Technological University from 2000 as a research engineer, then as a research fellow. His research interests include array signal processing, acoustic signal processing and adaptive signal processing.

**Guosen Yue** received the BS degree in physics and the MS degree in electrical engineering from Nanjing University, Nanjing, China in 1994 and 1997, and the PhD degree from Texas A&M University, College Station, TX, in 2004. Since August 2004, he has been a staff member with the Mobile and Signal Processing Department, NEC Laboratories American, Princeton, New Jersey, conducting research on broadband wireless systems and mobile networks. His research interests are in the general area of wireless communications and signal processing, specifically in channel coding, OFDM, MIMO, and iterative processing. He is a member of the IEEE. He served as the Technical Program Committee (TPC) subcommittee chair for IEEE RWS 2008, TPC member for IEEE RWS 2006, 2007, IEEE WCNC 2007, IEEE Sarnoff Symposium 2007, and IEEE ICC 2008.

**Chau Yuen** received the BEng and PhD degree from Nanyang Technological University, Singapore in 2000 and 2004 respectively. He was the recipient of Lee Kuan Yew Gold Medal, Institution of Electrical Engineers (IEE) Book Prize, and many other prizes. He was a post-doctoral fellow in Lucent Technologies Bell Labs, Murray Hill during 2005. Currently, he works in Institute for Infocomm Research, Singapore, where he has been developing the 802.11n Wireless LAN system, and actively participating in the standardization of the 3GPP Long Term Evolution (LTE). He also serves as an associate editor for IEEE Transactions of Vehicular Technology. His present research interests include multiple-input multiple-output (MIMO) coding, multi-user MIMO and wireless ad hoc network.

**Jianzhong (Charlie) Zhang** received the BS degrees in both electrical engineering and Applied Physics from Tsinghua University, Beijing, China in 1995, the MS degree in electrical engineering from Clemson University in 1998, and the PhD degree in electrical ngineering from University of Wisconsin at Madison in May 2003. He is currently a senior staff engineer with Samsung. Before he joined Samsung, he was a principal staff engineer with Motorola, and was the technical lead of the 3GPP HSPA standardization effort, focusing on areas such as MBMS, VoIP optimization, MIMO, etc. He also worked with Nokia from June 2001 to March 2006, where he was a senior research engineer and lead Nokia's phyical layer contribution to the IEEE 802.16e standard on topics such as LDPC codes, space-time-frequency coding and limited feedback based MIMO precoding.

**Yimin Zhang** received his PhD degree from the University of Tsukuba, Tsukuba, Japan, in 1988. He is currently a research associate professor at the Center for Advanced Communications, Villanova University, Villanova. His past positions include assistant professor at the Department of Radio Engineering, Southeast University, Nanjing, China; technical manager at the Communication Laboratory Japan, Kawasaki, Japan; visiting researcher at ATR Adaptive Communications Research Laboratories, Kyoto, Japan. Zhang's research interests lie in the area of statistical signal and array processing for communications and radar applications. He is a senior member of IEEE and serves as an associate editor for *IEEE Signal Processing Letters.*

**Zhendong Zhou** received his BE and ME degrees in information engineering from Zhejiang University, Hangzhou, China, in 2000 and 2003, respectively. Since 2003, he has been with the School of Electrical and Information Engineering, the University of Sydney, NSW, Australia, where he obtained his PhD degree in electrical engineering in 2007 and is currently working as a research fellow in the Telecommunication Laboratory. His research interests include adaptive modulation and coding in MIMO systems, cooperative communications and OFDMA synchronization and interference cancellation.

# Index

N



O



P



Q



R



S